THE HAMLYN PHOTOGRAPHIC GUIDE TO
BIRDS
OF THE
WORLD

THE HAMLYN PHOTOGRAPHIC GUIDE TO
BIRDS
OF THE
WORLD

Foreword by Dr Christopher Perrins

General Editor Dr Andrew Gosler

HAMLYN

Frontispiece

Hobby *(Falco subbuteo). A summer visitor to the Palaearctic from Afrotropical and Oriental regions.*

This edition first published in 1991 by
Paul Hamlyn Publishing Limited,
part of Reed International Books Limited,
Michelin House, 81 Fulham Road, London SW3 6RB

ISBN 0 600 57239 0

Produced by Mandarin Offset
Printed in Hong Kong

CONTENTS

Wood Duck *(Aix sponsa). A spectacular inhabitant of wooded wetland habitats in the Nearctic region.*

FOREWORD

Birds are the most popular group of animals; many more people go bird-watching than mammal-watching. This is perhaps surprising when one considers that we are mammals! Yet we are rather unusual mammals, not least because we rely, for our perception of the world, mainly on sight and sound; many mammals rely much less on sight and much more on their sense of smell, a sense in which we perform poorly in comparison with most other mammals. Further, the majority of mammals seem to be largely colour-blind while we have very good colour vision. Such senses, of course, match the 'needs' of the organism concerned, having been developed by natural selection. Many mammals live underground or are mainly active after dark – when vision, especially colour vision, is of little use but the sense of smell extremely important.

In contrast to most mammals – but not man – birds are largely diurnal. Since they fly, they need good early warning of objects which they are approaching. Hence they see well; good colour vision heightens this perception and aids them in their search for food. Hearing is a very useful sense for communication since, especially in thick woodland, it is not always easy to keep in visual contact with other birds. Birds have developed the sense of hearing – together with the matching transmitter, voice – to a high degree. Many species communicate with each other largely by voice and their songs can be very intricate. To our ears, some of these songs are very beautiful. Poets such as Shelley and Wordsworth have immortalized the songs of the Skylark and the Nightingale.

Most birds have less need for a well-developed sense of smell since this is less useful to a creature moving rapidly through the air. So, like us, most birds rely mainly on sight, with sound as their second sense and smell a poor third. It is not therefore surprising that, because we can perceive more of their world than we can of our fellow mammals', we find birds more enjoyable to watch than mammals.

They are not merely enjoyable to watch; because we can perceive more of their world, birds are in this sense easier to study than mammals. Much scientific knowledge, especially in the fields of behaviour and ecology, has come from observations of birds. The works of people such as Huxley, Lorenz and Tinbergen were based very largely on their studies of birds. We must, however, be careful not to fall into a trap. We must not assume that, because we can perceive some aspects of the birds' world, it follows that we necessarily 'see' such things in the same way that birds do. For example, the bird songs which we regard as beautiful are not necessarily beautiful to birds; to them, they function to identify others of their species and the song of one male may be perceived not as beautiful but as threatening by other males.

Before we can make useful observations on birds, we must know what species we are watching and something about them. This is what makes this book so useful; it makes it possible for us to get a much better idea of the birds of the world and so enables us to better enjoy and understand the birds we see. It describes a wide range of species from all over the world, emphasizing, in the main, those species which we are most likely to encounter. The text is written by experienced ornithologists and gives an authoritative account of these birds and the ways in which they live. Most of these are illustrated by colour photographs which show the birds in their natural habitats. Modern bird photographers have attained such a high standard that most of us might as well just throw our cameras away! The exception to this lies in many of the tropical parts of the world where the threatened flora and fauna is still poorly known. Here the editors have sometimes obtained the only existing photograph of a species.

Whether we make intensely detailed studies of their lives or merely enjoy their antics at the bird table, birds are an important part of the make-up of our planet and one that we can ill-afford to lose. Yet that is just what we are in danger of doing. We are not in imminent danger of losing all our birds, of course, but we are in danger of losing a very high proportion of them. The International Council for Bird Preservation list 1,029 species as threatened with extinction, some 11% of all the 9,000 or so bird species in the world! Many of these are birds that live in the great tropical rainforests which are, together with all their other fauna and flora, being destroyed at such a great rate at the present time. Indeed, even this 1,029 is not a complete list of those birds which we could lose; many others, although not in immediate danger, could become so if most of these forests are removed. Man's greed and over-population, leading to such extensive over-exploitation of our planet, is a threat, not only to the birds, but eventually to us also. So there are serious selfish reasons for trying to bring this plundering to a halt.

More subtle effects of man on birds have proved useful to man. It was the crash of peregrine populations and those of other birds of prey which prompted the studies which showed the hazards of the widespread use of pesticides and how they built up in animals until they reached dangerous levels; man is not immune from accumulating high levels of these dangerous compounds. These studies on birds led to restrictions on the use of certain chemicals and greater monitoring of the introduction of new ones. This lesson, important as it was, has not yet been fully heeded. Many developing countries have not yet listened to these warnings and many so-called developed ones still do not have the controls necessary to avoid all major pollution problems. Somehow many people still tend to think of pesticides as killing organisms that we do not want, but having no effect on those we do – or on ourselves. Would that life was as simple as that!

This book enables us to better understand the problems facing birds today, to recognize the species and it encourages us to learn more about what they are doing. Hopefully also it will lead to yet greater understanding and appreciation of our world and greater support for saving the environment. Birds and the planet need our concern.

Dr Christopher Perrins
Edward Grey Institute of Field Ornithology

INTRODUCTION

We live on a shrinking planet. The ease with which bird-watchers can travel in search of birds increases almost annually as more of our world becomes accessible to tourism. Often, however, the opening up of areas which previously could be reached only after a long journey on foot or by boat has been possible because of industrial or agricultural development such as timber extraction at the expense of the very habitat needed by the birds. While the majority of bird-watchers live in the developed world, the majority of bird species don't. They live in the primary habitats of the subtropics. Of 9,213 species recognized in this book, some 6,700 or about 70% occur wholly or partly in tropical regions and many others which breed at temperate northern latitudes migrate to the tropics to winter. Many of the resident tropical species are highly specialized ecologically with particular requirements for feeding and breeding. Indeed in some respects it is this which allows such great species diversity in these areas, and particularly in the rainforests where 300 species may occur together in 1 sq km ($\frac{1}{3}$ sq mile). But this also makes them vulnerable, since they are unable to survive when these complex and diverse habitats are changed or lost. Habitat loss is now the major threat to birds worldwide and because of the former inaccessibility of such areas, it is often the most endangered species that we know least about. Many species have become extinct in recent times – birds about whose ecology and behaviour we can only guess.

This book has three main aims. The first is to present the reader with an overview of the world of birds – a world of diverse sights and sounds which is often colourful and exciting, and which is always intriguing. The second is to provide in one volume a useful reference to birds by regions so that, wherever the reader finds himself or herself, at least many of the commoner species may be identified and the relationships and similarities between these and the birds of other regions may be recognized; in short, each may be placed in a global context. Having said this, it must be realized that no one book can hope to cover all of the world's birds in depth. Some species, particularly in the tropics, are very poorly known. For example, in some cases we do not know what they eat; in others the nest has never been found, or the immature plumage is unknown; and in many, the geographical distribution is known only vaguely. Finally, it is hoped that the book will stimulate both a greater interest in, and concern for, birds and their environment and so contribute to their conservation.

The introductory pages briefly describe the *evolution* of birds from their reptilian ancestors, and discuss how the great diversity of birds which we see today has developed over time through natural selection and *adaptation*. We shall also see how this process has resulted in patterns which we can use to unravel the evolutionary history of birds. These may be patterns of similarities between related species which indicate a common ancestry, or patterns in space in the form of a particular geographical distribution shown by a species or group of species. We shall then see how these patterns can be used to produce a *classification* of birds which reflects that evolutionary history. The next section introduces the 173 families of birds into which the species are grouped, and gives an overview of characteristics shared by the various members of each family. Throughout the book the families are generally arranged in the putative evolutionary order introduced in this section. The remainder, and bulk of the book, presents a representative selection of species by regions, followed by a complete checklist of the world's birds indicating where each species occurs and whether it is currently considered endangered by the International Council for Bird Preservation.

Ornithologists recognize six major zoogeographical regions of the world. Each is defined by the kinds of birds which are found there so that, although there is naturally some overlap in species between these areas, many will be distinct or *endemic* to that region. The regions are the Nearctic (North America), the Palaearctic (Eurasia), the Neotropical (South America), the Afrotropical (Africa south of the Atlas mountains), the Oriental (India and South-East Asia) and the Australasian (Australia, New Guinea, the Pacific islands and New Zealand and here also including Antarctica). In addition, the Caribbean is sometimes

King Penguin *(Aptenodytes patagonicus) breeding colony. The diversity of birds reflects variation on a theme. In penguins this variation is pushed to the limit.*

recognized as distinct from the Neotropical Region, and the Madagascan from the Afrotropical Region, because the bird communities or *avifaunas* of these areas contain many endemic species and indeed some endemic families. The Palaearctic shares about 12.5% of its breeding species (particularly those with circumpolar distributions) and about 35% of its genera (see p 14) with the Nearctic, so that they are sometimes referred to together as the Holarctic.

The birds of the six major regions are described by recognized authorities on the birds of each area. Each region is introduced by a short essay which describes the avifauna, its origins and relationships to the other regions; the principal habitats and threats to them; and organizations involved in the study and conservation of birds in the region. Each bird is identified by both its scientific name and an English or vernacular name. In some cases the vernacular name is not standard and more than one is given. A short description then appears facing its photograph, together with details of its distribution and status worldwide. Where a species occurs in more than one region this is mentioned in the text although it occurs only once in the book. Similar species, either in the same or a different region, whether they are illustrated in the book or not, are also listed, and where appropriate, a page reference is given. Wherever possible, the photographs have been selected to show not just the plumage of the bird, but also some aspect of its biology. Photographs show adult males unless otherwise stated.

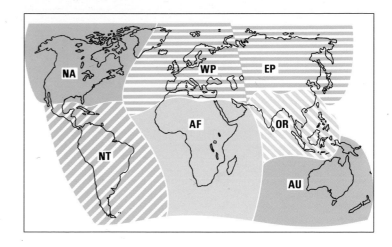

The six zoogeographical regions *used in this book. The Palaearctic is subdivided into Western and Eastern.*

WP Western Palaearctic including N Africa & Middle East
EP Eastern Palaearctic including Japan & N China
AF Afrotropical region including Red Sea
OR Oriental region including Himalayas, Philippines, to Wallace's Line
AU East of Wallace's Line to E New Guinea, Australia, New Zealand
NA Nearctic, south to Rio Grande
NT Neotropical, Latin and S America, West Indies

BIRD EVOLUTION

In a steamy world of cycads, tree ferns, giant dinosaurs and flying reptiles, and which will later become Bavaria, a bird about the size of a small crow dies and drops from its perch into the warm waters of a tropical lagoon. It is rapidly covered by fine, calcareous silt, so reducing the rate of decay, and it is fossilized. Here it lies for the next 150 million years. The sediments of the lagoon are moved hundreds of miles north by continental drift and raised hundreds of feet above sea-level by mountain-building forces which divide continents as if they were paper. Then, in 1861, the limestone matrix in which the bird rested for so many years is split by workmen quarrying stone for the lithographic printing industry.

The fossil, which is named *Archeopteryx lithographica*, is quickly recognized as one of the most significant in the history of palaeontology, for it is no bird with which we in modern times are familiar and it shows numerous characteristics which link birds unquestionably with their reptilian ancestors. The whole skeleton is essentially like that of a small dinosaur. The jaws are not modified into a beak as in modern birds, but are bony and carry numerous undifferentiated teeth. There is no fusion of vertebrae or reduction of the tail, which consists of a long series of vertebral elements tapering to a point. The breast bone (sternum) carries no keel for the attachment of the huge breast muscles necessary for powered flight. So why is it so obviously a bird? Because around the bones of the forelimbs and tail are the clear, unmistakable impressions of feathers which are structurally identical to those of any modern bird, and feathers are unique to birds. Indeed it is the possession of feathers which defines birds as a class, so that any organism bearing feathers is by definition a bird.

From the fossil record, which is generally poor for small, terrestrial vertebrates, we know that birds evolved rapidly from those early beginnings. The Cretaceous period (130-65 million years ago), which saw the peak of development and extinction of the dinosaurs and the rise of flowering plants, also saw the rise and extinction of many toothed birds. Because of the conditions needed for fossilization we know only of aquatic species such as the flightless, diver-like *Hesperornis* or the tern-like *Ichthyornis* (which might not have been toothed), but there must have been many terrestrial forms also. Both of these birds show much reduction of the tail and *Ichthyornis* also shows a well-developed sternal keel indicative of strong flight. However, the evolution of toothless, beaked birds must also have occurred rapidly during the Cretaceous as by the start of the Eocene period, 65 million years ago, many modern groups such as divers, grebes, cormorants, pelicans, flamingos, ibises, rails and sandpipers had already appeared. Again there is the bias in the fossil record towards aquatic forms. By the end of the Eocene, 36 million years ago, at least 20 modern orders (see p 14) had appeared. Fossils show that modern flightless birds, the ratites (Ostriches and so on), first appeared during the Eocene. However, they must in fact have arisen before the break-up of the southern supercontinent of Gondwanaland about 100 million years ago since the closely related Emu (p 282), Ostrich (p 186) and the rheas (p 96) must have evolved in isolation in Australia, Africa and South America respectively. This discrepancy between the fossil record and theory is almost certainly due to the fact that these are birds of arid habitats in which conditions are rarely suitable for fossilization.

Fossils also show that before the Jurassic there were no birds. Many species appeared and went extinct over the next 100 million years. There can be no doubt that evolution occurred. For example, the Cretaceous fossils show that conditions were suitable for the preservation of marine species at that time and yet important marine families such as penguins (Spheniscidae), gannets (Sulidae), shearwaters (Procellariidae) and petrels (Hydrobatidae) do not appear in the record until the Eocene or even later. But how does evolution occur?

To understand this we need to recognize that each species shows characteristics which fit it for a particular lifestyle. These are called *adaptations*. So, for example, all modern birds show numerous adaptations for flight, swimming birds have webbed feet and marine species have special salt-secreting glands. In the next section we shall look in more detail at the adaptations of birds, but for now it is sufficient to realize that this is the key to evolution. Given that common experience shows that every organism produces more offspring than can ultimately survive to breed, and given that there is variation among individuals which reflects variation in the genetic blueprint (that is, their characteristics are inherited from their parents), we should expect that some individuals will be more likely to survive by virtue of their inherited characteristics than will others. This simple expectation of differential survival is called *natural selection* to differentiate it from the related *artificial selection* carried out for centuries by human breeders of plants and animals.

If environmental conditions change or if there is little competition from other species, for example during a period of colonization, then different individuals in a population may survive better in slightly different habitats. Natural selection will then emphasize the differences between these subpopulations, resulting in new adaptations. Hybrids which are intermediate in form may be at a disadvantage because they are less well adapted to discretely differing habitats, so that there will be selection for distinct breeding groups. From this point they are genetically isolated and are effectively new species. This process of *speciation* can occur whenever populations are isolated genetically (that is, they are prevented from interbreeding). Of course selection can only act on existing variation and the origin of that variation is genetic mutation which occurs randomly in time and space through the population. Although the origin of variation is random, the selection of traits which adapt individuals better to their environment is not random. If it were, evolution would not occur. Nor is it goal directed. There is no 'purpose' to evolution, it is simply the inevitable outcome of genetic variation and nonrandom mortality.

Genetic isolation is considered the major cause of speciation in birds and is called *allopatric* speciation. If speciation occurs from within a population it is said to be *sympatric* but there is little evidence for this in birds. Examples of the great *adaptive radiation* which can occur when species are isolated and free to adapt to a variety of unfilled ecological niches can be seen today in the finches of the Galapagos and the Hawaiian honeycreepers (p 45). We shall consider adaptive radiation in more depth in the next section.

ADAPTATIONS OF BIRDS

In the last section we saw how natural selection has adapted birds to their environment. Some adaptations, such as the webbed feet of aquatic species, are so obvious as to be hardly worthy of mention, but birds show a great diversity of adaptations, many of which are superbly subtle and would leave the greatest aircraft designers dumbstruck with awe.

It is easy for the ornithologist to be overwhelmed by the apparent diversity of birds and yet, despite the diversity of colour and other features, the class shows remarkable consistency of structure. It is useful, for example, to compare birds with mammals, which range in size from the Pygmy Shrew, weighing 2.3 g (0.08 oz) to the Blue Whale at 136,200 kg (134 tons), some 59 million times heavier. In contrast, birds range in size from the Scintillant Hummingbird (p 118) weighing 2.25 g (0.08 oz) to the Ostrich (p 186) at about 150 kg (330 lb), only about 66,000 times heavier. Even the largest extinct bird does not bring the upper limit for the class to anywhere near that of mammals. Nor is there more variation in form than in mammals which, apart from those already mentioned, include the bats which can be every bit as agile on the wing as most birds. But birds, more than any other vertebrate group, are supremely adapted for flight and it is the need to reduce weight without a loss of structural strength which has been the overriding constraint in the evolution of birds.

The principle in bird structure of high strength to low weight is seen at its best in the skeleton itself. The bones are generally hollow and extremely light but strengthened internally by cross-bracing or by a honeycomb structure of bone. There has been considerable fusing of bones to reduce flexibility and increase strength. This has been most apparent in the spinal column, but the bones of the hand also are greatly fused and there are only three digits. Further cross-bracing occurs across the ribs, each of which shows a backward-pointing projection – the uncinate process, which overlies the next rib. In some diving birds such as auks (Alcidae), which endure great external pressures at depths of more than 150 m (490 ft), these processes are very long to prevent the rib-cage from collapsing. Movement of the wing is reduced at the shoulder to just that necessary for flight and this has allowed a reduction in the mass of some muscles on the back.

The skull has been greatly modified. The bony, toothed jaws of *Archeopteryx* have been replaced by a slender beak of bone over which sits a horny covering which in most birds is continuously growing and wearing away. The beak or bill can be enormously strong for its light weight. For example, in the Hawfinch (p 180), which cracks olive and cherry stones in its bill, point pressures of more than 50 kg (110 lb) must be generated. The muscle mass of the jaws is also reduced and concentrated near to the centre of gravity of the head. Much of the space in the head is taken up by the huge eyes, which are separated only by a thin sheet or *septum* of bone. The position of the eyes in the head varies between species depending on the relative importance of binocular vision for accurate distance judgement and monocular vision allowing a wide field of view. Hence, in predators, the eyes typically face forwards, giving a rather human appearance in owls (Strigidae) or downwards under the bill in herons (Ardeidae), while they point to the sides in more vulnerable species. The ultimate such development is probably seen in the Woodcock (p 160) which has a 360° field of view. Despite this, accurate distance judgement is essential for any fast-flying animal, especially one which inhabits dense cover, and in fact all birds have some binocular vision since in most species the eyes have some movement in their orbit. All diurnal species see the world in colour, and in all birds the optical resolution is at least as good as ours, and in many cases considerably better.

A fuller appreciation of the quality of eyesight in birds is gained by considering, for example, what a fishing Gannet (p 142) actually has to do to survive. From a height of up to 40 m (130 ft), travelling at 30 km/hr (20 mph) above a strong sea with a heavy swell and with a near gale-force wind blowing, it must spot fish less than 30 cm (12 in) in length, allow for refraction and dive headlong like a missile into the water among its prey. Even allowing for the fact that fish are probably caught on the bird's way back to the surface, there is little room for error. Faced with this prospect, it is probably not surprising that they prefer to fish shoals within 25 m (82 ft) of the surface.

Birds' sense of hearing is also acute, for while the total frequency range of sensitivity is often similar to our own, the resolution (the ability to differentiate sounds in time) can be ten times better than ours. This means in practice that some species can distinguish two sounds which are only a ten-thousandth of a second apart. In addition, many can distinguish between two sounds which differ in frequency by less than 1%. Generally, songbirds (see p 37) have a higher upper limit to their frequency range than do other birds. Although none is sensitive to ultrasonic vibration, some can hear extremely low-frequency sounds, well below our receptive ability. The auditory ability of birds, especially songbirds, means that they can receive a great deal of

Great Spotted Woodpecker (*Picoides major*). *All birds show adaptations to their specific ecology. In woodpeckers the bill, skull, tongue, feet and tail all show adaptation for wood-boring.*

information from samples songs. We still can only guess at the full information content of songs such as that produced by the Reed Warbler (Sylviidae, p 172). The eyesight and hearing of birds compensate them for their senses of taste and smell which in general are rather poor.

The musculature and internal organs also show numerous adaptations both to save weight and to concentrate the mass close to an optimum centre of gravity which lies beneath the wings when in flight. Hence, food is ground not by teeth, which would require a heavy jaw for support, but in the gizzard, a modification of the anterior part of the stomach containing grit, which the bird ingests for the purpose. This lies very close to the centre of gravity. The major parts of the wing and leg muscles lie close to the body, and rely on long tendons to transmit their power to the extremities of the limbs. The reproductive system can be regarded in terms of weight saving since not only does the female shed the weight of each egg shortly after its development, but the gonads shrink outside the breeding season to a fraction of their weight when breeding. There have also been extreme modifications of the respiratory and vascular systems to such a peak of efficiency that active flight is possible at altitudes where any mammal would quickly suffocate if it had not already died of the extreme cold. For example, geese (Anatidae) and waders (Scolopacidae) have been recorded migrating at altitudes of 7,500-9,000 m (25,000-30,000 ft).

Despite all these modifications, perhaps the most remarkable is still that uniquely avian invention – the feather. There has been much speculation as to the origins of this remarkable structure. This is mainly because it is difficult to see what purpose it might have served in an animal lacking the power of flight. Remember that the lack of a sternal keel in *Archeopteryx* suggested that it was not capable of prolonged powered flight, and yet it had a wing essentially the same as a modern bird. Indeed the asymmetry of the primary flight quills (see p 17) is only satisfactorily explained if the bird was capable of flapping flight. Of course, feathers serve the four roles of providing an aerofoil for flight, contouring the body into a stream-lined shape, providing a surface for the colours of birds and insulating the body against the cold by trapping a layer of air between them and the skin. As anyone who has ever slept beneath a down-filled cover will know, feathers are extremely efficient in this last role. The feather consists of a central shaft or *rachis* off which run numerous parallel barbs. Each barb carries numerous tiny hooks or *barbules* which latch the barb securely to those alongside it. The feather is a triumph of design. It is light, strong and flexible so that it tends to bend rather than break, and when damaged the barbs are simply zipped together by stroking it quickly through the bill.

So far only adaptations which are general to all birds have been considered, but there are many striking adaptations related to particular ways of life. Indeed, birds provide biologists with so many fine examples of adaptation that we often find ourselves looking (as if out of habit) for adaptive explanations for avian features when in fact there might not always be one. However, the rest of this section describes a few specific examples of adaptation whose validity is not questioned.

When a newly evolved group first occupies a new type of niche, there is often rapid and diverse radiation of adaptive forms to fill the many newly available niches. When this happens in birds, the features most closely related to feeding, and particularly the bill and legs, are those which evolve most rapidly. Some examples of this were briefly mentioned in the last section, but one of the best examples is that of the hummingbirds (Trochilidae,

Greater Flamingo *(Phoenicopterus ruber). The breeding biology of flamingos shows many adaptations, such as the production of a milk to feed the young until the highly modified bill has developed fully.*

12

pp 74, 118) of the Neotropics and Nearctic. Here, a novel form of locomotion has given access to a vast array of new and unexploited feeding opportunities in the form of nectar from flowers inaccessible to other birds. In fact this 'high octane' fuel is necessary to meet the considerable energy demands of hummingbird flight! This adaptation has resulted in a great array of bill lengths and shapes and a radiation into 332 species in 114 genera (see p 14 for these terms). A parallel development has occurred in the passerine family of sunbirds (Nectariniidae, p 230) of the Old World where adaptation to the same resource has produced similar or *convergent* characteristics from very different origins and a radiation into 120 species.

Variation in adaptation of the bill and feet, of course, tends to be greater between the orders of birds, which differ much more in ecology, than between closely related species. Examples of this are the strong, hooked bills and powerful talon-bearing feet of raptors (p 23), the spoon-like sifting bill and long wading legs of the Spoonbill (p 144) and the diversity of shape of bill and length of leg and bill in the waders or shorebirds (Scolopacidae and others). A remarkable suite of adaptations is seen in the flamingos (p 22). Here, the bill and tongue are greatly modified for filter-feeding and the birds specialize on microscopic plants and animals suspended in the water. In addition, the legs are extremely long, enabling them to feed in quite deep water. However, adaptation to this specialized niche has required the development of a further adaptation, namely, the production of a 'milk' from the parents' crop. This is because the chicks do not develop their unique bills until they are close to fledging and are therefore unable to feed for themselves. The Greater and Lesser Flamingo chicks may be fed for up to 75 days. Milk production is very rare in birds. Only one other group of birds – the pigeons (Columbidae, p 29) – has evolved the production of milk (though undoubtedly for different reasons than the flamingos) and this has allowed some members of that group to breed almost throughout the year even at temperate latitudes.

Another group with extensive modification for a specialized niche is the woodpeckers (Picidae, p 35) which feed on wood-boring invertebrates in timber. Here, radiation into 200 species has occurred because the resource (dead wood) is found world-wide, rather than because it offers diverse opportunities for the evolution of different species. Hence the major structural variation is in size – small species feed on smaller branches than large species. However, the adaptations for exploiting this niche, which are general to the family, are noteworthy. The bill is straight, is sharpened to a fine, chisel-like point and is reinforced along its length. The bone of the skull is reinforced at the base of the bill to withstand the immense forces incurred by the cranium while drilling into the timber of live hardwood trees. The tongue is extremely long, sticky and often barbed for extracting insect larvae from their chambers. The bony base of the tongue, called the *hyoid apparatus*, which is used to project the rest of the tongue, is so long that it curves back over the top of the cranium. The legs are short and strong and the *zygodactyl* feet (two toes facing forward, two back) are armed with strong, sharp, curved claws which allow the bird to grip the vertical surface of the tree. The tail feathers are pointed and stiffened so that, together with the feet, they act as a tripod which supports the bird when feeding. Woodpeckers are typical of the kind of ecological adaptation seen in birds, where modification occurs not to a single characteristic, but to a whole array or suite of characteristics.

This account has generally been restricted to adaptations of morphology, that is, form. In fact one sees adaptation in every area of bird biology in breeding systems, migration patterns, nest building and even in courtship behaviour, and the reader will find many more examples throughout this book.

BIRD CLASSIFICATION

Wanting to name the birds we see is our common experience. It is also essential before we can say anything more interesting about them, but to do so we need to recognize the limits of each species. How else, for example, would we know that the small, long-tailed, yellow, blue-headed bird in the field is a race of the Yellow Wagtail (p 170) rather than a separate species?

Living organisms do not come in *all possible* shapes and sizes. For example, consider the variation between hummingbirds and Ostriches. There are not, and never have been, all possible intermediate forms, but discrete groups which we know, to give some examples, as ducks, owls, thrushes and warblers. There are many reasons for this, including adaptive and genetic constraints. So we attempt to identify these different, discrete entities and to group them according to similarities, and separate them according to dissimilarities – in short, to classify them. This natural process is sometimes called 'box-in-box' classification because we arrange species together in groups or boxes which are similarly arranged into larger more inclusive groups to form a hierarchy. The science of classification is called taxonomy or systematics, and the different boxes in the hierarchy are called *taxa* (singular, *taxon*).

The *species* is the basic unit of taxonomy. Species are defined as groups of similar organisms which are, under natural circumstances, capable of interbreeding, and which would not normally interbreed with members of other species so defined. Hence species are ultimately defined by genetic similarity. In some cases, species might be interfertile, but because of behaviour or geography are effectively isolated genetically and would not normally form hybrids. Sometimes man has broken down these isolating mechanisms and produced hybrids which may be fertile so that it is not always easy to define the limits of a species. Despite this, it is important to realize that only the species level of taxonomy is defined biologically, that is, has real biological significance. All other levels are artificial constructions which may be more or less natural groupings. Within the species we might recognize geographical races or subspecies. Subspecies are completely interfertile and produce fertile offspring but are usually geographically isolated when breeding. Sometimes they may partially overlap geographically to form a hybrid zone such as in the Hooded and Carrion Crows (p 182).

Species are grouped into *genera* (singular, genus). Each species is recognized by a unique name within its genus and the two together form the scientific name of the species. The genus is always given first with the first letter a capital, the species second in lower-case letters. Scientific names are latinized constructions and are given in a distinct typeface, usually in italics. Because it is unique to the species, it is unambiguous, unlike the vernacular name. For example, the name 'Robin' means quite different things to an American, an Australian and a European, while *Turdus migratorius*, *Petroica phoenicea* and *Erithacus rubecula*, the 'Robins' of each of these areas, each have only one meaning. The scientific name is called a *binomial*. A third name is added to identify each subspecies. For example, the Yellow Wagtail is *Motacilla flava*, but there are several races; the blue-headed race *flava* may be recognized by the trinomial *Motacilla flava flava* to distinguish it from, say, the black-headed race *Motacilla flava feldegg* of the Balkans and Middle East. Genera are grouped into families (here 173); families are recognized by the ending '-idae'.

Wagtails are members of the family *Motacillidae* (Wagtails and Pipits). Families are grouped into orders (27 are recognized in this book), recognized by the ending '-iformes', in the wagtail case the *Passeriformes* (perching birds). Orders are grouped into classes. Birds form the class *Aves*. Other levels such as suborder, subfamily or superspecies may be inserted as the taxonomist sees fit to represent the relationships among groups better. A taxon may be defined which contains only one species, in which case it is said to be *monotypic*. The rules for naming birds are set out in the *International Code of Zoological Nomenclature*.

This all sounds very straightforward, but how do we go about grouping species? Given any group of objects, there are numerous ways in which we might group them. To take a trivial example, say we wanted to classify all the objects on the dinner table. They may all be related to food, but some are for cutting, others are containers or hold napkins. If we were to classify according to their composition, we should group the crockery together, distinct from cutlery. Where we place the napkin ring depends on whether it is metal, china or wood. Perhaps it would be better to classify according to how we use each object: bowls with spoons, plates with forks and so on; the napkin ring being perhaps a monotypic genus? Any such decisions would be quite arbitrary and we could manipulate the classification to emphasize anything we wanted. Returning to birds, we find that there are thousands of characteristics that we could use to classify – size, number of feathers, eye colour, toe-nail width, bill shape. We clearly need a way to make objective decisions about relationships.

Similarities between species exist either because they reflect a common ancestry, or because they show adaptations for a similar lifestyle (ecology). In the latter case such characteristics are said to be *convergent* and are of no use in classification, but if we classify so as to reflect evolutionary history – a *phyletic* or *phylogenetic* classification rather than a *phenetic* one – we should recognize natural relationships between species. Modern taxonomy aims to reflect the evolutionary relationships between taxa. To do this we should look ideally for characteristics which are not likely to change with time through adaptation to slight differences in environment (which might result in convergence). Hence while bill size and shape and leg length are poor characteristics on which to form groups (see, for example, Scolopacidae), subtle differences in the arrangement of bones in the palate, the type of syrinx (voice-box) or direct comparisons of the DNA (genetic blueprint) should be of greater use. A successful classification is one which groups together all descendants of a common ancestor and is said to be 'natural'. It can be recognized partly by its predictive power. That is, if new characteristics (say feather chemistry) are studied, perhaps because new techniques are available, a natural classification should give similar groupings based on the new characteristics while an unnatural classification would show poorer correspondence between groups derived according to different character sets. For this reason, as new information is obtained, classifications may change, and new groups, possibly requiring name changes, may be formed. Hence while scientific names may be more standard than the vernacular, they are still liable to change from time to time. This should not be resented as 'change for change's sake', but recognized and accepted as essential to the dynamic science of taxonomy.

Bird Classification

This diagram explains the classification of birds by tracing one species,
the Yellow Wagtail, through the classification system.

Orders

The class *Aves* is divided into 27 orders. Some examples are given below.

Ostriches
Struthioniformes   Penguins *Sphenisciformes*   Full-webbed Swimmers
Pelecaniformes   Diurnal Birds of Prey *Falconiformes*   Kingfishers and allies
Coraciiformes   Perching Birds *Passeriformes*

Families

The *Passeriformes* are the largest order with 74 families. Some examples are given below.

Manakins *Pipridae*   Lyrebirds *Menuridae*   Swallows and Martins
Hirundinidae   Wagtails and Pipits *Motacillidae*   Tits and Chickadees
Paridae   Hawaiian Honeycreepers
Parulidae

Genera

The *Motacillidae* family contains 6 genera. Some examples are given below.

Tree Pipit *Anthus trivialis*   Yellow Wagtail *Motacilla flava*

Species

The genus *Motacilla* contains 10 species. An example is given below.

Yellow Wagtail *Motacilla flava*

BIRD IDENTIFICATION

This section describes how to identify birds. Several schemes have been devised in the past to aid identification, especially in the use of memory joggers to remind the observer of which characteristics to look for. Although this is useful, the important point is to approach the problem systematically, noting the relative size, colour and so on of each part of the bird in turn. With practice one finds that it is not necessary to notice every feature to obtain a correct identification, since with some knowledge of bird classification and the range of possibilities at a particular site there will be a limited range of characteristics that need to be noted. For example, if we can quickly place the bird in its family – say auks (Alcidae), and we are in a breeding colony on the west coast of Britain, there are only three or four possibilities and we need only note the bill shape (see p 164). This might seem a trivial example of what can seem a nightmare of frustration in tropical rainforest, but the same principles apply to both. The first essential is that the observer is fully conversant with the terms used to describe the parts of a bird. These generally refer to the various feather tracts or *pterylae*, but also include the so-called soft parts of the bird – bill, eye and leg. These terms are used throughout this book and are presented here.

When describing a bird for the first time, it is useful to start with its size and shape, relating it to birds with which one is already familiar such as sparrow, thrush, pigeon, crow and so on. Then record length and shape of bill – short, medium, long, straight, decurved and so on, and legs – medium and so on. The upperparts of the head, the forehead, lores, crown, ear-coverts, nape, mantle, back, rump, uppertail-coverts, tail and wings are then described from the upper mandible to the tail. If all are the same, say simply 'upperparts olive-brown'. Note any conspicuous stripes on the head and wings, such as eye stripe, supercilium, wing-bars (and on which feathers, e.g. tips of median and greater coverts), outer tail feathers and so on, before moving on to the underparts. The underparts are similarly described from the anterior to posterior of the bird: lower mandible, chin, throat, breast, flanks, belly, undertail-coverts and tail and any underwing pattern if the bird was seen in flight. Bill, eye, eye-ring and leg colour may also be important. Again if the underparts are uniformly coloured simply say so, e.g. 'underparts white except for orange-red chin, throat and breast'.

This will seem a daunting prospect to the beginner but with practice it becomes second nature – as if instinctive – and only the key points need to be noted, as in many of the species accounts in this book. This relates back to the predictive power of classification mentioned in the last section. If we can quickly recognize our mystery bird as a woodpecker (Picidae) we needn't worry about leg colour (or the depth of water that it was wading in!) as this is not diagnostic in this group, and if, because of our location and because it is black and white, we can say that it is a *Picoides* species we will find that other characteristics can be ignored, such as the colour of outer tail feathers, whereas the distribution of black and white, its height in the tree and any calls or drumming may be important. The crucial or diagnostic characteristics to identify a particular species are called the 'field-marks'.

To learn the parts of a bird, it is useful to know something of bird structure, and especially that of the wing since contrasting patches or bars on the wings are often diagnostic. For present purposes, there are essentially two kinds of feather – the flight feathers, consisting of the primaries and secondaries of the wing (*remiges*) and tail feathers (*rectrices*), and the rest which are mostly contour feathers, giving an aerodynamic shape to the bird and aerofoil shape to the wing. The flight feathers are attached to the bones of the wing (primaries to the hand, secondaries to the forearm) and tail, while contour feathers are attached only to the skin. The primaries and secondaries slightly overlap each other like tiles on a roof so that the innermost secondary is uppermost and the outermost primary lowest and adjacent to the body when the wing is closed. The number of primaries varies little between taxa (usually 10), whilst the number of secondaries varies enormously (8-40+) depending on the length of the forearm. Overlying the primaries and secondaries are the primary and secondary coverts. The longest in each group, and those lying immediately over the remiges, are the greater coverts. Above these are the median coverts and above these are the lesser coverts. Whilst the primary coverts are known as greater, median and lesser primary coverts, those of the secondaries are usually known simply as the greater, median and lesser coverts. Similar groups of wing-lining coverts are found on the underwing. Finally, on the leading edge of the wing, and often partially overlying the primary coverts, are the three feathers of the alula or 'bastard wing'. These are attached directly to the bones of the thumb and increase the efficiency of the wing at low speeds by forming a slot in the leading edge of the wing.

The plumage is periodically replaced in the moult. Moults occur annually in most species but there is enormous variation between taxa and even populations. In many species, juveniles have a distinct plumage which is replaced during the first year, and in some there may be distinct seasonal plumages which may reflect the need for greater crypsis at some times of year. In addition, and especially during the breeding season, the sexes may differ dramatically in plumage. In general, males are brighter than females but the reverse also occurs (e.g. Phalaropidae), and is related to role reversal in breeding. Hence, while there may be 9,213 species, there might be several times as many plumages as this. The species accounts describe where juvenile and adult, male and female plumages differ.

Behaviour may also be crucial to identification and contributes to the 'jizz' of the bird. Jizz is the whole impression which a bird gives the observer and consists of the sight, sound and behaviour. Jizz enables an experienced bird-watcher to identify birds on only the most fleeting of glimpses when particular field-marks may be obscured. Learning to recognize birds by jizz comes naturally in time but only after many enjoyable hours of practice for which there is no substitute.

The species accounts should be used together with the full checklist to obtain a list of the species for a given region. However, note that the six major regions are very large, and few species occur throughout a specific region.

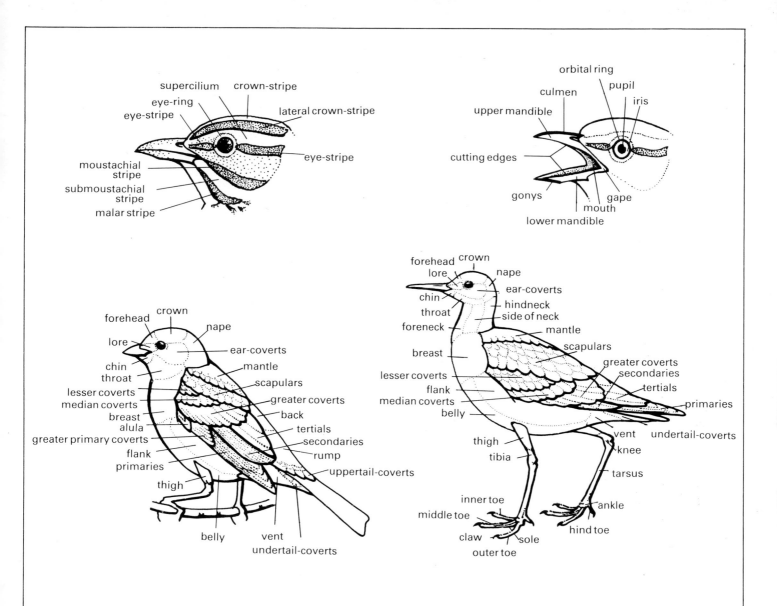

supercilium
crown-stripe
eye-ring
eye-stripe
lateral crown-stripe
moustachial stripe
submoustachial stripe
malar stripe
eye-stripe

orbital ring
culmen
pupil
iris
upper mandible
cutting edges
gonys
gape
mouth
lower mandible

forehead
crown
lore
nape
chin
ear-coverts
throat
lesser coverts
median coverts
breast
alula
greater primary coverts
flank
primaries
thigh
belly
vent
undertail-coverts
mantle
scapulars
greater coverts
back
tertials
secondaries
rump
uppertail-coverts

forehead
crown
lore
nape
chin
ear-coverts
throat
hindneck
foreneck
side of neck
breast
mantle
lesser coverts
scapulars
flank
greater coverts
median coverts
secondaries
belly
tertials
thigh
primaries
tibia
vent
undertail-coverts
knee
tarsus
inner toe
middle toe
ankle
claw
sole
hind toe
outer toe

CHART OF UPPERWING

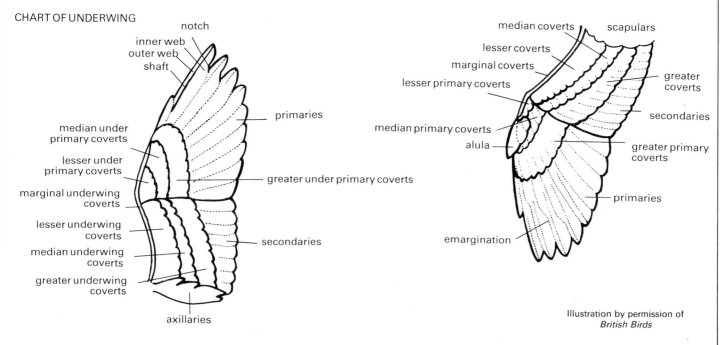

CHART OF UNDERWING

notch
inner web
outer web
shaft
primaries
median under primary coverts
lesser under primary coverts
marginal underwing coverts
greater under primary coverts
lesser underwing coverts
median underwing coverts
secondaries
greater underwing coverts
axillaries

median coverts
scapulars
lesser coverts
marginal coverts
greater coverts
lesser primary coverts
secondaries
median primary coverts
alula
greater primary coverts
primaries
emargination

Illustration by permission of
British Birds

1 Ostrich
2 Greater Rhea
3 Southern Cassowary
4 Emu
5 Brown Kiwi
6 Tawny-breasted Tinamou

BIRD FAMILIES OF THE WORLD

In the following descriptions of bird families, information about weights, incubation times, breeding habits and broods is omitted where no information is available. ● indicates threatened species.

For an explanation of terms used in this section see Glossary on p 372

OSTRICHES
Struthioniformes
Massive, flightless, unkeeled sternum, 2 toes, 1 family.

1 Ostriches Struthionidae
Struthio camelus (monotypic). See p 186.

Description: massive, flightless running birds (largest ratites). Long neck and legs, 2 toes, thighs bare.

RHEAS
Rheiformes
Massive, flightless, unkeeled sternum, 3 toes, 1 family.

2 Rheas Rheidae
2 species, 2 genera, e.g. Greater Rhea *Rhea americana*

Description: largest birds of New World, standing 90 cm (35 in) high, Wt: c. 10 kg (22 lb). Massive, flightless, powerful running birds (up to 50 km/hr, 30 mph), long neck and legs, 3 toes, thighs feathered, very short wings, tail absent. Plumage soft and filmy, grey or grey-brown. Monomorphic, females smaller. **Range and habitat:** NT. Dry plains and scrubland. **Migration:** non-migratory. **Breeding:** polygamous. Nest is a scrape surrounded by twigs. Eggs: 20-50 laid by 2-12 females. Incubation by male for 35-40 days. Broods: 1. Young are precocial, defended by male. **Food:** mostly vegetable, also some small animals.

CASSOWARIES AND EMUS
Casuariiformes
Massive, flightless, unkeeled sternum, 3 toes, 2 families.

3 Cassowaries Casuariidae
3 species, 1 genus, e.g. Southern Cassowary *Casuarius casuarius*

Description: massive flightless birds standing 110-80 cm (43-71 in), Wt: up to 54 kg (120 lb). Head carries bony casque arising from skull. Head and neck bare, brightly coloured and often wattled. Plumage largely black, coarse, heavy and drooping, wing feathers reduced to bare quills, tail more or less absent. Legs powerful, 3 toes – innermost armed with long sharp nail. Monomorphic, females larger. **Range and habitat:** AU. Dense tropical forest. **Migration:** non-migratory. **Breeding:** nest is a clearing on forest floor. Eggs: 3-6 pale to dark green, smooth or rough. Incubation by male, 30-50 days. Broods: 1. Young are precocial, attended by male or both, for c. 4 months. **Food:** uses casque as a shovel to find plant shoots and fruit, insects and small vertebrates on forest floor.

4 Emus Dromaiidae
Dromaius novaehollandiae (monotypic). See p 282.

Description: massive, flightless birds (2nd largest ratites). Long neck and legs, 3 toes, neck partly bare, coloured, closely related to cassowaries.

KIWIS
Apterygiformes
Medium, flightless, unkeeled sternum, 3 toes, 1 family.

5 Kiwis Apterygidae
3 species, 1 genus, e.g. Brown Kiwi *Apteryx australis*

Description: medium-sized, flightless strangely mammal-like birds. L: 45-84 cm (18-33 in), 30 cm (12 in) high, Wt: 1.3-4 kg (2 lb 14 oz-8 lb 13 oz). Plumage coarse, hairlike. Wings rudimentary, 4-5 cm ($1\frac{1}{2}$-2 in), tailless, legs short and stout, 3 toes. Bill long, slightly decurved and flexible with open nostrils at tip. Brown or greyish, monomorphic but female larger than male. **Range and habitat:** AU. Forest and scrub. **Migration:** non-migratory. **Breeding:** nest is a burrow dug largely by male. Eggs: 1-2 huge and yolky, 0.5 kg (1 lb 2 oz) (18-25% female wt). Incubation by male, 74-84 days. Broods: 1. Young are precocial. **Food:** uses well-developed sense of smell to find plant seeds, fruit and a wide range of soil invertebrates.

TINAMOUS
Tinamiformes
Primitive ground birds, keeled sternum, 3 or 4 toes, 1 family. Superficially like gamebirds (Galliformes).

6 Tinamous Tinamidae
46 species (1●), 9 genera, e.g. Tawny-breasted Tinamou *Nothocercus julius*

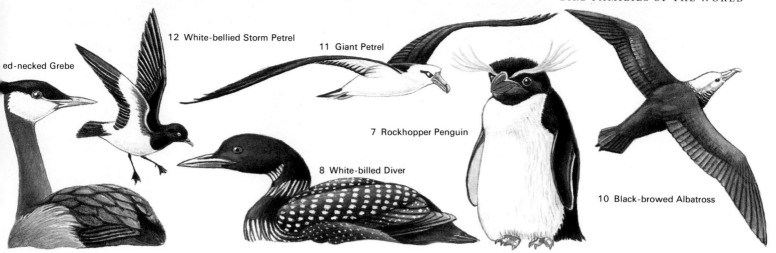

12 White-bellied Storm Petrel

11 Giant Petrel

ed-necked Grebe

7 Rockhopper Penguin

8 White-billed Diver

10 Black-browed Albatross

Description: small to medium, ground-dwelling, primitive palate. L:15-49 cm (6-19 in), Wt:450-2,300 g (1-5 lb). Legs short, poor and reluctant fliers, neck curved in flight. Bill thin, slightly curved. Tail short and soft, often covered by rump feathers. Plumage barred brownish, mottled. Monomorphic, females larger. **Range and habitat:** NT. Dense forest, scrub and grasslands below 4,000 m (13,000 ft). **Migration:** non-migratory. **Breeding:** nest is a scrape. Eggs: 1-12, large − 21-68 g ($\frac{3}{4}$-2$\frac{1}{2}$ oz) − glossy, unicoloured but variable in colour. Incubation by male for 19-20 days. Broods: 1. Young are precocial, protected by male. **Food:** fruit, seeds, insects and small vertebrates.

PENGUINS
Sphenisciformes
Highly modified flightless swimming seabirds, 1 family.

7 Penguins Spheniscidae
18 species, 6 genera, e.g. Rockhopper Penguin *Eudyptes chrysocome*

Description: medium to large flightless seabirds with paddle-like wings, used for propulsion under water, and heavy webbed feet set far back on body. L:40-115 cm (16-45 in), Wt:1-30 kg (2 lb 3 oz-66 lb). Strong swimmers (up to 20 knots). Bill medium-sized, stout. Plumage very uniform, generally blue-grey or blue-black above and white below, short, dense, lacking defined tracts. Most species more or less monomorphic. Some species suffer dramatic weight loss during incubation. **Range and habitat:** AF, AN, AU, NT. Marine, breeding on ice, rock and sandy shorelines. **Migration:** some migratory species, those of warmer waters are non-migratory. **Breeding:** varies greatly between species. Nest is on ground, in burrow or none. Eggs: 1-2, whitish. Incubation by male or both for 33-62 days. Broods: 1. Young are semi-altricial, creched. **Food:** fish, squid and crustacea.

DIVERS
Gaviformes
Swimming and diving birds, well-developed tail, neck stripes, 1 family.

8 Divers Gaviidae
5 species, 1 genus, e.g. White-billed Diver *Gavia adamsii*

Description: large swimming and diving birds, 3 front toes fully webbed, legs set far back, walk clumsily. L:53-69 cm (21-7 in), Wt: 1-2.4 kg (2 lb 3 oz-5 lb 5 oz). Strong swimmers. Tarsi laterally compressed. Bill medium-sized, sharply pointed. Body long, neck long and thin. Wings relatively short, pointed. Tail short but well developed. Plumage dense, soft and velvety on head and neck, hard and compact on body, black, white, greys above, white below. Breeding and winter plumages differ. Monomorphic, males larger. Young have 2 coats of nestling down. **Range and habitat:** Holarctic: WP, EP, NA. Breeds on inland freshwater habitats, winters on the sea. **Migration:** strongly migratory. **Breeding:** little courtship. Nest is of little or no material, by water, site chosen by male. Eggs: 2 (rarely 3), olive-green/brown.

Incubation by both male and female, 27-9 days. Broods: 1. Young are precocial, fledge at 50-65 days. **Food:** mainly fish, also frogs and aquatic invertebrates.

GREBES
Podicipediformes
Swimming and diving birds, tail reduced to a tuft, 1 family.

9 Grebes Podicipedidae
21 species (2●), 5 genera, e.g. Red-necked Grebe *Podiceps grisegena*

Description: small-medium to medium-large swimming and diving birds, 4 toes lobed, legs set far back. L:23-76 cm (9-30 in), Wt: 150-1,500 g (5$\frac{1}{4}$ oz-3 lb 5 oz). Tarsi laterally compressed. Bill short-medium to long, sharply pointed, nostril: a narrow slit. Body short to medium, neck long and thin. Wings short, pointed, flight often poor, some flightless. Tail vestigial. Plumage dense. Breeding and winter plumages usually differ. Breeding: greys, browns, reds, black and white above, tufts and crests on head used in courtship. Winter: usually grey and white. Monomorphic, males slightly larger. **Range and habitat:** worldwide except PA and AN. Open freshwater habitats, some winter on the sea. **Migration:** some strong migrants. **Breeding:** courtship highly ritualized. Nest is colonial or territorial, floating platform of weed. Eggs: 2-6, whitish but stain rapidly. Incubation by both, 21-8 days. Broods: 1-2 (rarely 3). Young are precocial, fledge at 44-80 days. **Food:** fish and aquatic invertebrates.

TUBE-NOSED SWIMMERS
Procellariiformes
Marine, tube-like nostril (tubinares), good sense of smell. Come to land only to breed, care of young shared. 4 families.

10 Albatrosses Diomedeidae
14 species, 2 genera, e.g. Black-browed Albatross *Diomedia melanophris*

Description: very large gliding seabirds. L:68-135 cm (27-53 in), Wt: 3-12 kg (6 lb 10 oz-26 lb 7 oz). Bill medium-long, strong, plated and hook-tipped. Nostril tubes open laterally. Wings very long, slender and pointed. Wingspan: 178-350 cm (70-138 in). Many secondary flight feathers. Legs short-medium, 3 forward toes fully webbed. Tail short and round to long and pointed. Monomorphic, some brown or grey, others white with dark upperparts. **Range and habitat:** EP, AF, AU, NT, PA, AN. Oceanic, breeding on oceanic islands. **Migration:** strongly migratory. **Breeding:** courtship ritualized. Nest is colonial, mound of soil or vegetation, or none. Eggs: 1 white. Incubation by both for 65-79 days. Broods: 1. Young are semi-altricial, fledge at 120-278 days. **Food:** fish and crustacea.

11 Petrels and Shearwaters Procellariidae
72 species (8●), 14 genera, e.g. Southern Giant Petrel *Macronectes giganteus*

13 Common Diving Petrel

14 White-tailed Tropicbird

15 (Eastern) White Pelican

17 Flightless Cormorant

18 American Darter

16 Cape Gannet

Description: small to large gliding seabirds with long, slender wings — up to 2 m (79 in) wingspan. L: 28-90 cm (11-35 in), Wt: 130 g-4 kg (4½ oz-8 lb 13 oz). Tail short. Bill short to medium-long, stout or slender, hook-tipped, nostril tubes dorsal. Legs short, 3 toes fully webbed. Plumage black, white, brown, grey or a combination with dark upperparts, light underparts. Monomorphic but males generally larger than females. **Range and habitat:** oceanic worldwide, breeding on oceanic islands. **Migration:** migratory, some strongly so. **Breeding:** more or less colonial. Nest on cliff ledges or in underground burrows. Eggs: 1, 25-237 g (1-8 oz). Incubation by both for 43-60 days. Broods: 1. Young are semi-altricial, fledge at 70-105 days. **Food:** fish, squid and crustacea.

12 Storm Petrels Hydrobatidae
20 species, 8 genera, e.g. White-bellied Storm Petrel *Fregetta grallaria*

Description: the smallest seabirds, with relatively long, pointed wings. L: 14-26 cm (5½-10 in), Wt: 25-68 g (1-2½ oz). Wingspan: 32-56 cm (13-22 in). Fluttering or bat-like flight especially when feeding. Legs medium-sized, slender. 3 toes fully webbed. Bill short to medium, slender, plated, nostril tubes mounted on top of bill. Tail short to medium, rounded, square or forked. Plumage dense, dark browns, greys and white. Monomorphic apart from size which varies with species. **Range and habitat:** oceans worldwide except Arctic seas, breeding on oceanic islands. **Migration:** most are migratory. **Breeding:** nest is more or less colonial, in burrows and rock crevices. Eggs: 1, white/whitish. Incubation by both for 40-50 days. Broods: 1. Young are semi-altricial, fledge at 56-73 days. **Food:** fish, squid and plankton taken from surface of water.

13 Diving Petrels Pelecanoididae
4 species, 1 genus, e.g. Common Diving Petrel *Pelecanoides urinatrix*

Description: in southern oceans ecologically analogous to northern Auks (Alcidae). Medium-sized diving seabirds with chunky bodies, short wings, tail and legs. L: 18-25 cm (7-10 in), Wt: 105-46 g (3¾-5 oz). Wingspan: 30-8 cm (12-15 in). 3 toes fully webbed. Bill short, stout, hooked. Nostril tubes open upwards. Plumage black or grey above, white below. Monomorphic. **Range and habitat:** AU, NT, AN. Marine, breeding on coasts and islands. **Migration:** non-migratory. **Breeding:** nest is colonial in burrows. Eggs: 1, whitish. Incubation by both parents for 45-53 days. Broods: 1. Young are semi-altricial. **Food:** small marine animals which they chase under water.

FULL-WEBBED SWIMMERS
Pelecaniformes
Seabirds with 4 toes webbed (totipalmate), more or less expandable throat (gular) pouch. Care of young shared. 6 families.

14 Tropicbirds Phaethontidae
3 species, 1 genus, e.g. White-tailed Tropicbird *Phaethon lepturus*

Description: large white seabirds with black on head and wings.

L: 80-110 cm (31-43 in) including tail streamers of 40-55 cm (16-22 in), Wt: 430-730 g (15 oz-1 lb 10 oz). Legs short. Bill red or yellow, long, pointed, strong and slightly decurved. Strong fliers with long, slender wings. Wingspan: 90-110 cm (35-43 in). Swim and walk poorly. Monomorphic, males larger. **Range and habitat:** tropical oceans and oceanic islands worldwide. **Migration:** partly migratory. **Breeding:** elaborate courtship display including undulation of tail streamers. Nest is colonial, on bare ground or in cliff or tree-holes. Eggs: 1, red-brown, blotched. Incubation by both for 40-6 days. Broods: 1. Young are altricial, fledge at 77-100 days. **Food:** fish (especially flying fish) and squid caught by plunge-diving.

15 Pelicans Pelecanidae
8 species, 1 genus, e.g. (Eastern) White Pelican *Pelecanus onocratalus*

Description: large waterbirds with long necks and flattened, hooked bills with huge, extensible gular pouch. L: 127-70 cm (50-67 in), Wt: 2.5-15 kg (5 lb 8 oz-33 lb). Legs short but feet large and strong. Tail short. Plumage of most species white with black wing tips, often with a wash of pink or orange on body, 1 species brown. Strong fliers, soarers, wings long and broad. Wingspan: 2-2.8 m (79-110 in). Monomorphic but males larger than females. **Range and habitat:** WP, EP, AF, OR, AU, NA, NT. Coasts and inland lakes. **Migration:** some migratory. **Breeding:** nest colonial, of sticks in tree or on ground. Male collects material, female builds nest. Eggs: 1-4, white. Incubation by both for 30-7 days. Broods: 1. Young are altricial, cared for by both for 50-70 days. **Food:** fish caught by plunge-diving or surface fishing, scooping prey into pouch. Sometimes birds fish cooperatively.

16 Gannets and Boobies Sulidae
9 species (1●), 3 genera, e.g. Cape Gannet *Morus capensis*

Description: large powerful seabirds, with long, strong bill. L: 66-100 cm (26-39 in), Wt: 900 g-3.6 kg (2-7 lb 15 oz). Long, pointed wings. Wingspan: 1.4-1.7 m (55-67 in); short legs, large feet. Generally adapted for plunge-diving. Small gular pouch. Bare skin on face and throat often highly coloured, as are feet. Plumage white and black or brown, juveniles brown and adult plumage acquired after several moults. Monomorphic, females usually larger. **Range and habitat:** WP, AF, OR, AU, NA, NT, PA. Oceanic and oceanic islands. **Migration:** migratory in temperate latitudes, in tropics more sedentary. **Breeding:** much ritual display at colonies. Most nests are in dense colonies, of weed on ground or sticks in trees depending on species. Eggs: 1-4 depending on species, whitish. Incubation by both for 42-55 days. Broods: 1. Young are altricial, fledge at c. 90 days. **Food:** fish and squid, often caught by plunge-diving from great height.

17 Cormorants Phalacrocoracidae
33 species, 2 genera, e.g. Flightless Cormorant *Nannopterum harrisi*

Description: large, often powerful, waterbirds with long tubular hooked bill, and long necks and tails. L: 50-100 cm (20-39 in),

22 Whale-headed Stork (Shoebill)
24 African Spoonbill
23 Jabiru
20 Goliath Heron
21 Hammerkop
19 Great Frigatebird

Wt: 900 g-4.9 kg (2-11 lb). Legs short, set far back, feet large. Strong swimmers, swimming low in water. Often crested. Strong fliers, wings medium-sized, broad. Wingspan: 80-160 cm (31-63 in). 1 flightless. Plumage black, white, greys and browns. Northern species tend to be darker. Wings held out to dry. Monomorphic. **Range and habitat:** WP, EP, AF, OR, AU, NA, NT, few at high latitudes. A wide variety of coastal and inland wetland habitats. **Migration:** northern species migratory. **Breeding:** nest is colonial, large mound of twigs or weed collected by male, built by female on cliff or in tree. Eggs: 1-6, chalky blue. Incubation on feet by both for 22-6 days. Broods: 1. Young are altricial: 50-70 days. **Food:** chases fish, amphibia and aquatic invertebrates under water.

18 Anhingas Anhingidae
4 species, 1 genus, e.g. American Darter *Anhinga anhinga*

Description: also known as snakebirds or darters, these large birds strongly resemble cormorants (Phalacrocoracidae) but neck very long and sinuous, body longer and bill long, straight and sharply pointed. L: 85-90 cm (33-5 in), Wt: 900 g-2.6 kg (2-5 lb 12 oz). Legs short and set far back, feet large. Strong swimmers, often swimming with only head and neck showing. Strong fliers and soarers. Wings long and broad. Wingspan: 120-7 cm (47-50 in) Tail very long. As in cormorants the wings are held out to dry after swimming as the bird lacks a preen gland. Dimorphic, males with black and white on neck and wings, females with light brown. **Range and habitat:** AF, OR, AU, southern NA, NT. A variety of freshwater habitats. **Migration:** migratory in Nearctic. **Breeding:** nest is colonial, bulky nest of sticks in tree near water. Eggs: 4 (rarely 2-6), elliptical, pale green, chalky and streaked red. Incubation by both for 26-30 days. Broods: 1. Young are altricial, fledge at 35-49 days. **Food:** chases fish which it spears with bill under water, amphibia and aquatic invertebrates.

19 Frigatebirds Fregatidae
5 species, 1 genus, e.g. Great Frigatebird *Fregata minor*

Description: large, extremely light, manoeuvrable seabirds with very long, slender wings. L: 79-104 cm (31-41 in), Wt: 750 g-1.63 kg (1 lb 10 oz-3 lb 9 oz). Wingspan: 176-230 cm (69-90 in). Legs very short. Feet small, webs reduced, especially to hind toe. Plumage has poor waterproofing and birds cannot swim. Bill long, hooked. Tail long, forked. Despite their expertise in the air they are relatively helpless on land. Dimorphic, males 25-30% lighter than females, generally brownish black with distensible, red throat (gular sac) inflated and vibrated in impressive courtship display. Female generally brown with variable white underparts and without gular sac. Gregarious. **Range and habitat:** tropical oceans and oceanic islands worldwide. **Migration:** local wandering but non-migratory. **Breeding:** nest is colonial, mostly of sticks accumulated by both during courtship, built in tree or sometimes on ground. Eggs: 1, large, white. Incubation by both for 55 days. Broods: 1. Young are altricial, dependent for 6-11 months. **Food:** fish and squid, also frequently kleptoparasites of pelicans and boobies.

HERONS AND ALLIES
Ciconiiformes
Long bill, neck, legs, feet unwebbed or front toes partially so. 6 families in 5 suborders.

Ardeae

20 Herons and Bitterns Ardeidae
60 species, 17 genera, e.g. Goliath Heron *Ardea goliath*

Description: small-medium to very large waterbirds with long, spear-shaped bill, long neck and legs. L: 30-140 cm (12-55 in), Wt: 100 g-c. 3 kg (3½ oz-6 lb 10 oz). Narrow-bodied. Some nocturnal. 4 long toes, unwebbed, middle claw has comb. Strong, manoeuvrable fliers, large, broad, rounded wings. Flies with neck bent in S shape. Tail short. Plumage loose, soft, powder down. Head may carry long plumes. Breeding plumage may include long display plumes. Patterned simply in browns (some highly cryptic), black, white (some entirely white) and greys. Monomorphic but males larger. **Range and habitat:** worldwide except AN. Fresh and salt wetland habitats. **Migration:** migratory in temperate latitudes. **Breeding:** nest is solitary or colonial, pile of sticks in tree or on ground, built by female. Eggs, depending on species: 2-7, whitish to blue, sometimes streaked. Incubation by both for 18-30 days. Broods: 1. Young are semi-altricial, fledge at 35-56 days. **Food:** fish, also small vertebrates and insects. Hunts by stalking, usually in water.

Scopi

21 Hammerkop Scopidae
Scopus umbretta (monotypic). See p 186.

Description: medium-sized, heron-like bird with anvil-shaped head.

Balaenicipites

22 Whale-headed Stork (Shoebill) Balaenicipitidae
Balaeniceps rex (monotypic). See p 188.

Description: large, stork-like bird with massive bill. Neck carried in heron-like S shape in flight.

Ciconiae

23 Storks Ciconiidae
19 species, 5 genera, e.g. Jabiru *Ephippiorhynchus mycteria*

Description: large to very large birds with long, pointed heavy bills varying in shape, colour and patterning. L: 70-150 cm (28-59 in), Wt: 2-9 kg (4 lb 7 oz-19 lb 13 oz). Long necks and legs. Front toes partly webbed. Tail short, rounded. Body heavy. Plumage variously black, white and greys, boldly patterned, face and neck bare in some. Strong fliers, wings large, and broad. Wingspan: 145-320 cm (57-126 cm). Neck level in flight. Monomorphic. **Range and habitat:** WP, EP, AF, OR, AU, NA, NT. A wide variety of habitats, usually open and often near water. **Migration:** northern species are strongly migratory. **Breeding:** nest is

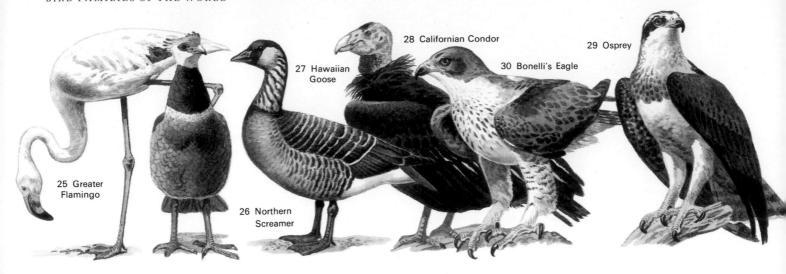

28 Californian Condor

27 Hawaiian Goose

29 Osprey

30 Bonelli's Eagle

25 Greater Flamingo

26 Northern Screamer

colonial or solitary, of sticks in tree or on cliff. Eggs: 3-5 (rarely 1), white becoming stained. Incubation by both for 30-50 days. Broods: 1. Young are altricial, fledge at 50-126 days. **Food:** fish, small vertebrates, insects and carrion depending on species.

24 Ibises and Spoonbills Threskiornithidae
31 species (6●), 18 genera, e.g. African Spoonbill *Platalea alba*

Description: medium to large long-legged birds with highly modified long bills – spoon-shaped (spoonbills), or slender and decurved (ibises). L:48-110 cm (19-43 in). Neck moderately long, level in flight. Flight strong and direct, wings broad and rounded. Tail short. Front toes webbed at base. May show bare skin on face or head, some crested. Plumage generally white, black, browns or reds. Monomorphic, males often larger. **Range and habitat:** WP, EP, AF, OR, AU, southern NA, NT. Plains, savannah grasslands, swamps and lakeside habitats. **Migration:** some species are strongly nomadic, others migratory. **Breeding:** most nests are colonial. Eggs: 2-5, white or blue, sometimes spotted. Incubation by both or mostly female for 21 days. Broods: 1. Young are altricial, fledge at 20-30 days. **Food:** aquatic animals, some seeds depending on species.

Phoenicopteri

25 Flamingos Phoenicopteridae
5 species, 3 genera, e.g. Greater Flamingo *Phoenicopterus ruber*

Description: large, long-lived wading birds with very long legs and necks and highly modified bills adapted for filter feeding. L:80-145 cm (31-57 in), Wt:2-3 kg (4 lb 7 oz-6 lb 10 oz). The feet are small, hind toe reduced, front toes fully webbed. Slender, graceful birds, fly rapidly with long neck outstretched, legs trailing behind. The wings are relatively short and pointed. Swim well when necessary. Plumage white to red depending on species and diet. Monomorphic. **Range and habitat:** WP, EP, AF, OR, NT. Fresh and salt water, shallow lagoons, lakes and estuaries. **Migration:** some migratory. **Breeding:** monogamous. Nest is colonial, of mud. Eggs: 1-2, white. Incubation by both for 28 days. Broods: 1. Young are precocial, cared for by both parents for 75 days and raised on 'milk' from parent's crop. **Food:** minute suspended aquatic plants and animals sieved by drawing water through filtering lamellae inside bill using muscular tongue.

WATERFOWL
Anseriformes
A well-defined order of semi-aquatic birds in 2 markedly different families and suborders. All share round, open nostrils, a feathered preen gland, well-developed down, 11 primaries and no incubation patch.

Anhimae

26 Screamers Anhimidae
3 species, 2 genera, e.g. Northern Screamer *Chauna chavaria*

Description: large, heavy-bodied birds with medium-length neck, crested or spiked head and small fowl-like bill. Medium-heavy legs with large, partially webbed feet to walk on floating vegetation. L:70-90 cm (28-35 in), Wt:2-5 kg (4 lb 7 oz-11 lb). Bones are highly pneumatized and they have many subcutaneous air sacs. Strong but slow fliers and soarers, the broad, rounded wings carry 2 large spurs at the bend of the wing. Plumage black to grey, brown. Monomorphic. Rather gregarious. **Range and habitat:** NT. Shallow ponds, marshes, wet grasslands. **Migration:** non-migratory. **Breeding:** nest is large platform on ground near water built by both parents. Eggs: 4-7, whitish with buff or green tinge. Incubation probably shared, takes 40-5 days. Young are precocial, fledge at 60-70 days. **Food:** largely aquatic plants.

Anseres

27 Ducks, Geese and Swans Anatidae
147 species (12●), 42 genera, e.g. Hawaiian Goose *Branta sandvicensis*

Description: a large and rather diverse family divided into 11 subfamilies. Small-medium to very large aquatic or semi-aquatic birds. L:30-150 cm (12-59 in), Wt:250 g-15 kg (9 oz-33 lb). Bill usually short, flat, broad, blunt with rounded serrations along edges. Neck medium to very long. Legs short, strong, 4 toes, front toes fully webbed, hind toe reduced, sometimes webbed. Good swimmers and fliers. Wings narrow and pointed. Tail short. Plumage either dimorphic: when female cryptic, male conspicuous; or monomorphic. Plumage highly variable. Most flightless during wing moult. Male may moult into cryptic temporary or 'eclipse' plumage at this time. **Range and habitat:** worldwide except AN. Wide variety of habitats, usually near water. **Migration:** many strongly migratory species, especially northern Holarctic breeders. Some have special moult migration. **Breeding:** ritualized display in many species, territorial. Nest is variable, platform of vegetation near water, rocky ledge, tree-holes or crown, often lined with down from female. Eggs: 4-14, white or pale. Incubation by female for 18-39 days. Broods: 1. Young are precocial, fledge at 21-110 days. **Food:** wide range of aquatic plants and animals (mostly invertebrate).

DIURNAL BIRDS OF PREY
Falconiformes
A large order of diurnal raptorial birds. 5 families.

28 New World Vultures Cathartidae
7 species (1●), 5 genera, e.g. Californian Condor *Gymnogyps californianus*

Description: large to very large scavengers. L:60-130 cm (24-51 in), Wt:900 g-14 kg (2-30 lb 13 oz). The bill is heavy, rounded and hooked, forest species have good sense of smell to find food. Others follow these to carcass. Head bare, often coloured, sometimes wattled. Condors have a ruff. Wings long and broad for soaring. Wingspan: 1.5-3.2 m (59-126 in). Legs and toes moderately strong, claws weak. Plumage usually black or brown, 1 species also white. Monomorphic except in condors in which the male is larger, also male Andean Condor has a caruncle on its head

31 Secretary Bird
36 Brown Mesite
35 Common Pheasant
34 Great Curassow
33 Mallee Fowl
32 Laggar Falcon

lacking in female. Gregarious at roost and at food, but nests solitarily. **Range and habitat:** NA, NT. A variety of open and in tropics also forest habitats. **Migration:** migratory in high latitudes. **Breeding:** nest is not built, but located in variety of sites such as rocky ledge, in caves or on bare ground. Eggs: 1-3. Incubation by both for 32-58 days. Broods: 1. Young are semi-altricial, fledge at 70-175 days. **Food:** carrion and animal waste.

29 Osprey Pandionidae
Pandion haliaetus (monotypic). See p 150.

Description: a large fish-eating hawk which catches fish in its talons. Underside of foot especially rough to aid grip.

30 Hawks and Eagles Accipitridae
225 species (11●), 56 genera, e.g. Bonelli's Eagle *Hieraaetus fasciatus*

Description: medium to large diurnal raptors and scavengers. L:25-115 cm (10-45 in), Wt:100 g-12.5 kg (3½oz-27 lb 9 oz). Nostrils in bare fleshy cere. Head and neck may be bare in scavengers allowing them to reach into carcass without soiling plumage. Eyesight often exceptional with good binocular vision. Neck usually short. Bill hooked and strong, claws hooked as talons and strong in most, comparatively weak in vultures. Legs medium-sized, tarsus bare. Powerful fliers, especially soaring, wings broad and rounded, often long. Tail medium to long. Mostly monomorphic but females generally larger. Colour varied, most browns, black, white and greys. **Range and habitat:** worldwide except AN. A wide variety of habitats. **Migration:** many strongly migratory species, especially Holarctic breeders. **Breeding:** solitary. Nest is a large mass of sticks in a tree or on a cliff ledge. Eggs: 1-7, white or marked with brown. Incubation depending on species: by both male and female (usually female does most of incubating) or female for 32-50 days. Broods: 1. Young are semi-altricial, fledge at 26-120 days. Allocation of parental duties between sexes differs among species. **Food:** a wide range of live animal prey killed with talons, or carrion depending on species.

31 Secretary Bird Sagittariidae
Sagittarius serpentarius (monotypic). See p 192.

Description: large, long-legged raptors, hunting on foot, weak toes and claws.

32 Falcons and Caracaras Falconidae
61 species (4●), 10 genera, e.g. Laggar Falcon *Falco jugger*

Description: small to large diurnal raptors. L:15-63 cm (6-25 in), Wt:110 g-2 kg (3¾oz-4 lb 7 oz). Bill hooked, strong and usually notched. Nostrils in bare fleshy cere. Eye-ring bare. Eyesight often exceptional with good binocular vision. Neck short. Legs medium to long, tarsus bare. Feet strong with sharp talons. Flight often very rapid and expert, wings medium-sized and pointed. Tail medium to long. Plumage varies but usually dull upperparts, paler barred underparts. Males usually brighter and smaller than females. **Range and habitat:** worldwide except AN. A wide variety of habitats. **Migration:** many strongly migratory species, especially Holarctic breeders. **Breeding:** solitary or colonial. Nest is of sticks in tree or on cliff, sometimes old crows' nest. Eggs: 2-6. Incubation by both but mostly by female for 25-49 days. Broods: 1. Young are semi-altricial. Male parent brings most food, female tends young for 25-49 days in nest, 14-60 days after fledging. **Food:** a wide range of live animal prey caught with talons (falcons) or carrion (caracaras).

FOWL-LIKE BIRDS
Galliformes
Chicken-like birds. 3 families.

33 Megapodes Megapodiae
19 species, 6 genera, e.g. Mallee Fowl *Leipoa ocellata*

Description: medium to large ground-dwelling birds with large body, medium to long tail, sometimes vaulted, legs and feet large and strong, 4 toes with strong nails. L:25-75 cm (10-29 in), Wt:900 g-8 kg (2 lb-17 lb 10 oz). Head small, bill short, strong and overlapping at tip. 1 crested species, others bare-headed, coloured skin, with wattles or casques. Wings short, rounded, rarely fly. Plumage sombre, brownish or black. 1 species has pink underparts. **Range and habitat:** OR, AU. Forests and dry scrublands. **Migration:** non-migratory. **Breeding:** 6-24 eggs are buried in sand, in volcanic ash or in great mound of rotting vegetation. Incubation is achieved by heat from sun or bacterial action. In some species the temperature in the mound is regulated by the male. Young are precocial, receiving no care after hatching. **Food:** wide variety of invertebrates, seeds, fruit and roots.

34 Curassows and Guans Cracidae
45 species (6●), 8 genera, e.g. Great Curassow *Crax rubra*

Description: medium to large, the most arboreal Galliformes, feed on ground until disturbed. L:50-100 cm (20-39 in), Wt:470 g-4.8 kg (1 lb 1 oz-10 lb 9 oz). Most crested with curved feathers. Some with other head ornaments. Lores bare, throat bare and wattled in some. Bill heavy and bent, sometimes with shiny cere, often with enlarged base to bill. The wings are short, rounded, flight fast but laboured. Legs long and strong and the toes heavy, long and with bent claws. Strong runners. Tail long. Most monomorphic, some differ in iris colour. Plumage sombre browns, black, white. **Range and habitat:** NT and southern NA. Mostly lowland forest. **Migration:** non-migratory. **Breeding:** varies between species but mostly nest is simple pile of sticks on tree branch or rarely on the ground. Eggs: 2-3 (rarely 4), large, white, rough or smooth-shelled. Incubation by female for 22-34 days. Young are precocial, flying after 3 or 4 days, fed mostly animal matter. **Food:** largely seeds and fruit, some insects.

35 Pheasants and Grouse Phasianidae
210 species (21●), 61 genera, e.g. Common Pheasant *Phasianus colchicus*

38 Plains Wanderer

37 Little Button-quail

39 Manchurian Crane

40 Limpkin

41 White-winged Trumpeter

42 Moorhen

Description: a large family with 5 well-defined subfamilies containing many gamebird and domesticated species and some breathtakingly beautiful forms. Small (quail) to large (turkeys) birds. L:12-110 cm (4½-43 in), Wt:43 g-18 kg (1½ oz-39 lb 11 oz). Feed and nest on ground, most roost in trees. Body bulky, head small with short neck. Bill short, strong, decurved. Head often wattled, sometimes bare. Wings short, round, broad. Strong manoeuvrable fliers. Tail short, rounded or square to very long, graduated. Some with exaggerated tail or uppertail-coverts (peacock). Legs medium-sized, strong, tarsus sometimes carries 1 or more spurs. 4 toes, strong, hind toe reduced and elevated, adapted for scratching soil and litter. Legs and feet bare or feathered (grouse). Plumage and size monomorphic or strikingly dimorphic with male larger than female. **Range and habitat:** WP, EP, AF, OR, AU, NA, NT. A wide variety of both open and forest habitats. **Migration:** some migratory species. **Breeding:** there is a great diversity of breeding systems including leks, polygyny and solitary nesting, and ritualized display. Nest is generally a scrape, and is rarely in trees. Eggs: 2-22, whitish, light brown or olive. Incubation by female (rarely both) for 16-28 days. Broods: 1. Young are precocial, tended generally by female (rarely both) for 30-100 days. **Food:** a variety of invertebrates, seeds, fruit and roots.

CRANES, RAILS AND ALLIES
Gruiformes

A diverse order containing 11 families in 8 suborders ranging from forms like Ciconiiformes and Galliformes to Charadriiformes, and probably polyphyletic. They share a number of anatomical characteristics.

Mesoenatides

36 Mesites Mesitornithidae
3 species (3●), 2 genera, e.g. Brown Mesite *Mesitornis unicolor*

Description: medium-sized, mostly terrestrial, gregarious birds which run well and take flight only when disturbed. L:25-8 cm (10-11 in). Reduced collar bones, fly poorly. Bill slender, straight or decurved. Medium-sized, strong legs and feet, 4 level toes, wings short and rounded, tail rounded, long and broad. Plumage mostly browns and greys with 5 pairs of powder down patches. Monomorphic, or dimorphic with female brighter. Complex social organization and breeding systems but little known. **Range and habitat:** Madagascar (AF). Forests and scrublands. **Migration:** non-migratory. **Breeding:** rarely observed. Nest is of sticks low in a bush. Eggs: 1-3. Incubation by female, male or both. Young are precocial. **Food:** seeds, insects and some fruit.

Turnices

37 Button-quails Turnicidae
15 species, 2 genera, e.g. Little Button-quail *Turnix sylvatica*

Description: small ground-dwellers characterized by absence of hind toe. L:11-18 cm (4½-7 in), Wt:c.50-80 g (1¼-2¾ oz). Superficially like quails (Galliformes). Like true quails, the bill is short and slender, the

wings short and rounded and the tail very short. No crop, preen gland is feathered, feathers have aftershaft. Plumage brown, grey and black. Monomorphic but females larger, brighter and more aggressive. **Range and habitat:** AF, OR, AU, southern WP and EP. Open forest and grasslands. **Migration:** mostly non-migratory. **Breeding:** reversed, courtship by female. Nest is usually a lined scrape, some more elaborate. Eggs: 4 (rarely 2). Incubation by male for 12-13 days. Young are precocial, tended by male, fledge at c.14 days. **Food:** seeds and insects.

38 Plains Wanderer Pedionomidae
Pedionomus torquatus (monotypic)

Description: like button-quails but foot with 4 toes though hind toe reduced and raised slightly. Also differences in internal anatomy which cause doubt of its correct taxonomic position – some grouping it with button-quails, others taking it out from the Gruiformes altogether. Female much larger and brighter. L: male: 150 cm (59 in), female: 170 cm (67 in). Plumage reds and browns, cryptic. Legs longer than button-quails, bill weaker and moderate length. Wings short and round, tail very short. Flies reluctantly with whirring flight. Often stands elevated on toes. **Range and habitat:** AU. Short grasslands. **Migration:** probably nomadic. **Breeding:** nest is scantily lined depression on ground. Eggs: 3-4, white and blotched. Incubated by male for 23 days. Young are precocial, raised by male. **Food:** invertebrates, seeds and other vegetable matter.

Grues

39 Cranes Gruidae
15 species (6●), 4 genera, e.g. Manchurian Crane *Grus japonensis*

Description: large, superficially like Ciconiiformes. L:76-152 cm (30-60 in), Ht:90-180 cm (35-71 in), Wt:2.7-10.5 kg (5 lb 15 oz-23 lb 2 oz). Neck carried level in flight, feet trailing behind. Long neck and legs, terrestrial (never perch in trees). Bill straight, medium to long. Some with bare coloured patches on head and neck. Flight powerful and easy. Wings long and broad. Wingspan: 1.5-2.7 m (59-106 in) with elaborate secondaries overhanging tail (which is short) used in display. Feet moderate with 4 toes, hind toe raised. Monomorphic, males larger. Plumage largely white, black, grey and browns. **Range and habitat:** WP, EP, AF, OR, AU, NA. Mostly shallow wetlands (breeding), short grasslands (winter). **Migration:** northern species migratory. **Breeding:** elaborate courtship displays. Nest is usually bulky of aquatic vegetation on marshy ground, rarely placed low in tree. Eggs: 2 (rarely 3-4), oval. Incubation by both for 30-40 days. Broods: 1. Young are precocial, usually only 1 fledged, cared for by both for 55-105 days. **Food:** a wide range of small animals, seeds, fruits, roots.

40 Limpkin Aramidae
Aramus guarauna (monotypic). See p 60.

Description: medium-sized wading bird with long bill and legs from marshlands of NT and southern NA.

48 African Jacana
45 Sun-bittern
46 Red-legged Seriema
47 Kori Bustard
43 African (Peters') Finfoot
44 Kagu

41 Trumpeters Psophiidae

3 species, 1 genus, e.g. White-winged Trumpeter *Psophia leucoptera*

Description: medium to large gregarious and noisy birds with small heads, short, stout, curved bills, medium-sized necks and long legs. L: 43-53 cm (17-21 in), Wt: 1-1.5 kg (2 lb 3 oz-3 lb 5 oz). Hind toe reduced and elevated. Fast runners. Tail stumpy. Feed on ground but nest and roost in trees. Wings broad. Laborious flight. Monomorphic. Plumage: chiefly black and velvety, especially on head and neck which have glossy purplish, green or bronze sheen. Outer webs of secondaries and tertials form hair-like filaments over back. **Range and habitat:** NT. Lowland rainforest. **Migration:** non-migratory. **Breeding:** little information available. Nest is in tree-hole or palm crown. Eggs: 6-10, whitish. Incubation by female. Young are precocial. **Food:** fallen fruit and insects on ground.

42 Rails and Coots Rallidae

123 species (8●), 18 genera, e.g. Moorhen *Gallinula chloropus*

Description: small to large, mostly secretive (many nocturnal), aquatic or semi-aquatic. L: 10-60 cm (4-24 in), Wt: 30 g-3.3 kg (1 oz-7 lb 4 oz). Most strong fliers, some flightless island endemics. Typically narrow-bodied, legs long and strong, 4 toes, sometimes lobe-webbed. Head may carry a frontal shield, usually coloured. Bill short to long, generally strong. Wings rounded, tail short and soft, sometimes almost absent. Flightless during wing moult. Plumage loose in texture. Monomorphic, males sometimes much larger. Colours vary from black and dark blue or purple to browns, reds and greys. Some cryptic spotted patterning. **Range and habitat:** worldwide except AN. A wide variety of wetland habitats, also damp forest and meadows. **Migration:** some strongly migratory species. **Breeding:** often strongly territorial, sometimes with nest-helpers. Nest is a pile of sticks or aquatic vegetation on ground or floating, usually hidden. Eggs: 1-14, variable in colour and pattern. Incubation by female or female and male, 20-30 days. Broods: 1-3. Young are precocial, raised by both (rarely female alone), c. 60 days. **Food:** small animals, seeds, fruit and aquatic vegetation depending on species.

Heliornithes

43 Sungrebes or Finfoots Heliornithidae

3 species, 3 genera, e.g. African (Peters') Finfoot *Podica senegalensis*

Description: distribution indicates antiquity. Medium-large aquatic, elongated body, long neck and pointed bill; short, strong legs. L: 30-60 cm (12-24 in). Toes broadly lobed and brightly coloured. Tail long, graduated and stiff. The wings bear a thorn-like claw on the first digit of the hand. Plumage rail-like, brownish above, lighter below. Male usually larger and brighter. Solitary, shy, secretive, reminiscent of Anhingidae. Well adapted for diving though usually seen swimming or walking, or perched on branch over water. Flight is laboured and close to water. **Range and habitat:** 1 species in each of AF, OR, NT. Tropical forest lakes and rivers. **Migration:** non-migratory or partially migratory (OR).

Breeding: nest is shallow twiggy platform lined with leaves on branch just above water. Eggs: 2-4 (rarely 5-6), cream or reddish brown. Incubated by both for 11 days (remarkably short for size of bird). Broods: 1. Young are altricial, cared for by both or male who carries them in skin pouches between flank feathers and underwing, even in flight. **Food:** fish, amphibia, aquatic invertebrates, leaves and seeds.

Rhynocheti

44 Kagu Rhynochetidae

Rhynochetos jubatus (monotypic)
Threatened (only a few hundred remain).

Description: superficially resembles herons (Ardeidae) but biochemical evidence places it closest to the Sun-bittern (Eurypigidae). L: 55 cm (22 in). Bill: L: 6.3 cm (2.5 in). Bill and legs long, red. The head is loosely and copiously crested and the ash-grey plumage is also generally loose and barred with darker grey and browns. Powerful and beautiful pre-dawn song. Monomorphic, males slightly larger. Wings broad, rounded but short, more or less flightless. Feet strong, hind toe reduced, fast runner. Move in loose flocks. **Range and habitat:** New Caledonia (PA). Dense montane forests, near still water and in rocky ravines. **Migration:** non-migratory. **Breeding:** nest is thin layer of dead leaves on ground. Eggs: 1, blotched. Probably incubated by both, for 35-40 days. Young are precocial but semi-dependent for nearly 100 days. **Food:** taps ground to find invertebrates, especially earthworms, which it digs up with its strong bill.

Eurypygae

45 Sun-bittern Eurypygidae

Eurypyga helias (monotypic). see p 108.

Description: a large, semi-arboreal wading bird from wet forests of Central and South America. Long legs, bill and neck. Remarkable dance and display.

Cariamae

46 Seriemas Cariamidae

2 species, 2 genera, e.g. Red-legged Seriema *Cariama cristata*

Description: large, ground-dwelling, long legs, small semi-palmate feet (reduced and elevated hind toe). L: 80-90 cm (31-5 in). Run fast with head down when disturbed, otherwise stalk prey slowly in small flocks. Bill: short, strong, decurved and broad. Bill and legs red or black. Head carries erectile crest (negligible in 1 species). Wings short, rounded, tail long. Flies reluctantly. Plumage is soft and hairy on neck, brownish grey and barred. Underparts paler. More or less monomorphic. **Range and habitat:** NT. Grasslands and brush forest. **Migration:** non-migratory. **Breeding:** nest is platform of sticks up to 3 m (10 ft) above ground. Eggs: 2, whitish and blotched. Incubation by both for 25-6 days. Broods: 1. Young are semi-altricial, tended by both in the nest till well grown. **Food:** diverse, but largely fruits and small animals.

50 Crab Plover

52 Ibisbill

53 Avocet

51 Palaearctic (European) Oystercatcher

54 Stone-curlew

49 Painted Snipe

Otides

47 Bustards Otididae

22 species (1●), 9 genera, e.g. Kori Bustard *Ardeotis kori*

Description: large, bulky, ground-dwelling, family includes the heaviest flying bird (Kori Bustard). L:40-120 cm (16-47 in), Wt:500 g-18 kg (1 lb 2 oz-39 lb 11 oz). Legs stout, hind toe absent. 3 toes broad, nails flattened. Bill usually short, blunt and flattened. Neck long, tail short and wings broad. Wingspan: 1-2.5 m (39-98 in). No preen gland, but produce dense powder down. Upperparts generally cryptic, browns, greys, buff, black, white, underparts generally white. Males generally larger, sometimes strikingly so. Males also often brighter. Often gregarious, shy in areas where persecuted for game, otherwise approachable. May take several years to reach maturity. **Range and habitat:** WP, EP, AF, OR, AU. Open plains, grasslands, scrublands. **Migration:** some migratory. **Breeding:** diverse mating systems. Nest is a scrape made by female. Eggs: 1-2 or 2-6 depending on species, olive to reddish. Wt:41-146 g (1½ oz-5 oz). Incubated by female for 20-5 days. Broods: 1. Young are cared for by female for 35-55 days. **Food:** omnivorous.

SHOREBIRDS, GULLS, AUKS
Charadriiformes

A diverse group of aquatic/semi-aquatic, often marine birds, often web-footed with reduced and elevated or absent hind toe. The order shows great adaptive diversity of bill form. 16 families in 3 suborders.

Charadrii (Waders or Shorebirds)

48 Jacanas Jacanidae

8 species, 6 genera, e.g. African Jacana *Actophilornis africanus*

Description: small to medium-large rail-like (Rallidae) birds with long legs and 4 very long toes. L:15-30 cm (6-12 in), 1 is 50 cm (20 in), Wt:40-230 g (1½-8 oz). Walks on floating vegetation such as lily pads. Runs, swims and dives well. Wings short, rounded, and carrying a carpal spur, flight laboured. Bill short, head sometimes has frontal shield or wattles. Plumage very varied, browns, black, white, greens, yellow, often darker below. Monomorphic but females larger. **Range and habitat:** AF, OR, AU, NT. Lake and riverside marshes. **Migration:** largely non-migratory, 1 strong migrant. **Breeding:** roles reversed, some polyandrous. Nest is simple platform built by male on floating vegetation. Eggs: 3-4, polished, streaked and lined. Incubated by male for 21-6 days. Broods: 1. Young are precocial, tended by male (carried under wings) for 85-110 days. **Food:** small animals and seeds of aquatic plants.

49 Painted Snipe Rostratulidae

2 species, 2 genera, e.g. Painted Snipe *Rostratula benghalensis*

Description: small-medium, long-legged, secretive waders. L:20-5 cm (8-10 in), Wt:76-165 g (2¾-5¾ oz). Bill medium-sized, L:36-54 mm (1½-2 in) slightly decurved, swollen tip. Wings short, rounded, flight weak. Plumage greens, browns, reds and white in exquisite pattern, round yellow spots on wing feathers shown in display. Females brighter and larger, otherwise similar. **Range and habitat:** EP, AF, OR, AU, NT. Reedy swamps and paddy-fields. **Migration:** locally or partially migratory. **Breeding:** depending on species, females may be polyandrous. Nest is pad of woven plant material built by male in dense cover. Eggs: 1-2 or 4 (rarely 5-6) depending on species, cream to buff, spotted black and brown. Incubation by male, possibly also female, for 19 days. Young are precocial, tended by male. **Food:** small swamp invertebrates and seeds.

50 Crab Plover Dromadidae

Dromus ardeola (monotypic). See p 202.

Description: unusual, medium-large, long-legged, black and white wader with massive, compressed, heron-like bill. Coasts of Indian ocean.

51 Oystercatchers Haematopodidae

11 species (1●), 1 genus, e.g. Palaearctic (European) Oystercatcher *Haematopus ostralegus*

Description: large black or pied waders with long, straight, powerful red bill which varies with diet. L:40-5 cm (16-18 in), Wt:400-700 g (14 oz-1 lb 8½ oz). Monomorphic but female usually larger with longer bill than male. **Range and habitat:** WP, EP, AF, OR, AU, NA, NT. Variety of coastal habitats, sometimes inland. **Migration:** some strongly migratory, most movements local or sedentary. **Breeding:** nest is a scrape. Eggs: 2-4, cryptic. Incubation shared for 24-7 days. Broods: 1. Young are precocial, fledge at 35-42 days. **Food:** invertebrates, chiefly bivalves.

52 Ibisbill Ibidorhynchidae

Ibidorhyncha struthersii (monotypic). See p 248.

Description: unusual medium-large wader with long, decurved red bill. Found in mountain riverbeds above 1,700 m (5,500 ft) in central southern Asia.

53 Avocets and Stilts Recurvirostridae

13 species (1●), 3 genera, e.g. Avocet *Recurvirostra avosetta*

Description: medium-sized waders with long, slender legs and bills, straight (stilts) or upcurved (avocets). L:30-50 cm (12-20 in), Wt:140-435 g (5-15 oz). Wings long and pointed, tail short, 3 toes partly webbed, hind toe vestigial. Plumage chiefly white, black, grey, brown. Monomorphic. **Range and habitat:** WP, EP, AF, OR, AU, NA, NT. Open marshes near fresh and brackish water. **Migration:** northern populations migratory. **Breeding:** colonial. Nest is a scrape, sometimes lined. Eggs: 4 (rarely 2-5), cryptic. Incubation by both for 22-8 days. Broods: 1. Young are precocial, fledge at 28-42 days. **Food:** aquatic invertebrates and small vertebrates. Feeds by sweeping water with bill (avocets) or probing (stilts).

54 Stone-curlews or Thick-knees Burhinidae

9 species, 2 genera, e.g. Stone-curlew *Burhinus oedicnemus*

59 Snowy Sheathbill
57 Dunlin
58 Rufous-bellied Seedsnipe
56 Ringed Plover
55 Black-winged Pratincole

Description: medium to large semi-aquatic waders with large head and very large yellow or orange eyes (partially nocturnal). L: 32-55 cm (13-22 in), Wt: c. 1 kg (2 lb 3 oz). Bill medium to large, L: 33-79 cm (13-31 in), straight or upcurved, stout to massive, according to ecology. Wings long, broad, rounded and boldly patterned, strong flight. Legs long and thick. 3 partially webbed toes. Ecologically similar to bustards (Otidae). Plumage generally brown, streaked and cryptic. Monomorphic. **Range and habitat:** WP, AF, OR, AU, NT. Open, sparsely vegetated sites both near water and in arid or semi-arid areas. **Migration:** northern populations migratory, nomadic in tropics. **Breeding:** nest is a shallow scrape. Eggs: 2, whitish to brown. Incubation shared, 25-7 days. Broods: 1, rarely 2. Young are precocial with large legs, attended by both parents for c. 40 days. **Food:** small vertebrates, terrestrial and coastal invertebrates, partly depending on species.

55 Coursers and Pratincoles Glareolidae
17 species, 5 genera, e.g. Black-winged Pratincole *Glareola nordmanni*

Description: extraordinary medium-sized waders in 2 distinct groups: coursers – long-legged, ground-dwellers, 3 toes, run fast, rarely fly, bill medium-sized, thin, decurved, tail short, square; pratincoles – short-legged, strong fliers, long, pointed wings, forked tail, swallow-like, bill short, stout, wide gape. L: 15-30 cm (6-12 in), Wt: 70-100 g (2½-3½ oz). Plumage browns, white, black and greys. Monomorphic. **Range and habitat:** WP, EP, AF, OR, AU. Arid or semi-arid open or sparsely scrubby areas, usually near water. **Migration:** generally nomadic outside breeding season. **Breeding:** nest is a scrape. Eggs: 2-3, cryptic. Incubation by female or shared, 17-31 days. Broods: 1. Young are precocial, fledge at 25-35 days. **Food:** insects and some other invertebrates, aerial insects in pratincoles.

56 Plovers Charadriidae
66 species (2●), 9 genera, e.g. Ringed Plover *Charadrius hiaticula*

Description: small to medium, semi-aquatic, compact, thick neck, large eyes. L: 14-41 cm (5½-16 in), Wt: 34-296 g (1¼-10½ oz). Bill short to medium, stout, straight. Wings long, pointed, sometimes spurred at carpal, flight strong and fast. Tail short, legs medium to long. Hind toe vestigial or absent. Fast runners. Plumage variable, contrasting, black, white, greys, browns, yellows, greens. Monomorphic, males often brighter (sometimes female). **Range and habitat:** worldwide except AN. Wide variety of coastal and inland open habitats. **Migration:** most species are strongly migratory. **Breeding:** varied mating systems: monogamy, polyandry, sex reversal, polygyny. Nest is a scrape. Eggs: 4 (rarely 2-5), cryptic. Incubation shared (or male or female) for 18-38 days. Broods: 1-3. Young are precocial, fledge at 21-42 days. **Food:** wide range of invertebrates, some plant matter.

57 Sandpipers and Snipe Scolopacidae
86 species (5●), 23 genera, e.g. Dunlin *Calidris alpina*

Description: small to large waders. L: 13-66 cm (5-26 in), Wt: 18-1,040 g (⅔ oz-2 lb 5 oz). Short-medium to very long, straight or curved bill and medium to very long legs. Bill and leg length related to ecology. Generally monomorphic but females usually larger with longer bill than male. Plumage browns, greys and reds and often with distinct breeding plumage. **Range and habitat:** worldwide except AN. Wide variety of coastal and inland open habitats, usually near water. **Migration:** many Arctic breeders and most species strongly migratory. **Breeding:** mostly monogamous, also leks and polygamy. Nest is a scrape, in grass tussock or rarely in tree-hole. Eggs: 4 (rarely 2-3), cryptic. Incubation by female or shared by male and female for 18-30 days. Broods: 1, rarely 2. Young are precocial, fledge at 16-50 days. **Food:** wide range of invertebrates, rarely some plant matter.

58 Seedsnipe Thinocoridae
4 species, 2 genera, e.g. Rufous-bellied Seedsnipe *Attagis gayi*

Description: small to medium-sized, ground-dwelling, gregarious. L: 15-27 cm (6-10½ in), Wt: 60-400 g (2-14 oz). Plump, short legs. Bill short, stout, conical, nostril protected by shield-like covers. Wings long, pointed. Flight strong, agile, fast. Tail medium-sized. Hind toe short, front toes long. Plumage generally browns, greys, cryptic. Monomorphic, males brighter. **Range and habitat:** NT. Sparsely vegetated highlands and fields. **Migration:** some species migratory. **Breeding:** nest is a lined scrape. Eggs: 4, cryptic. Incubation by female, c. 26 days. Eggs covered by female on departure. Young are precocial, fly at 49 days. **Food:** seeds, succulent vegetation and some insects.

59 Sheathbills Chionididae
2 species, 1 genus, e.g. Snowy Sheathbill *Chionis alba*

Description: medium-sized, pigeon-like shorebirds most closely related to Lari. L: 38-41 cm (15-16 in), Wt: 290-550 g (10 oz-1 lb 3½ oz). Named after saddle-shaped, horny, yellowish sheath covering the base of the short, stout bill. Generally found in association with penguin colonies. Generally walk on ground. Fly and swim strongly. Legs short, strong. Front toes slightly webbed, hind toe reduced, elevated. Plumage pure white, monomorphic. Gregarious, very social and tame. **Range and habitat:** AN. Rocky coasts. **Migration:** locally nomadic. **Breeding:** solitary, though still feeding socially. Nest is well hidden in rock crevices. Eggs: 2-4, brown, flecked. Incubation shared by male and female for 28-32 days. Broods: 1. Young are precocial, fledge at 50-60 days. Most pairs raise only 1 or 2 chicks. **Food:** omnivorous, especially scavenging.

Lari (Gulls, Terns and Skuas)

60 Skuas or Jaegers Stercorariidae
7 species, 2 genera, e.g. Arctic Skua *Stercorarius parasiticus*

Description: medium to large, slender to bulky. L: 40-60 cm (16-24 in), Wt: 250 g-1.9 kg (9 oz-4 lb 3 oz). Bill strong, hooked, nostrils in cere. Wings long, pointed. Flight strong, fast, stiffer-winged than gulls, light and agile to heavy. Tail medium to long with elongated central tail feathers – generally longer in the lighter species. Legs medium-sized.

62 Black Skimmer

60 Arctic Skua

61 Kittiwake

64 Black-bellied Sandgrouse

63 Brunnich's Guillemot

4 toes, 3 webbed, hind toe reduced and elevated. Lighter species white with dark upperparts (light and dark phases), heavy species uniform brown. White primary bases show in flight. Monomorphic. **Range and habitat:** oceans worldwide. Marine, breeds on tundra and barren coasts. **Migration:** strongly migratory or nomadic. **Breeding:** monogamous, pairing for life, or rarely polygamous, rarely colonial. Nest is a scrape. Eggs: 2 (rarely 1, 3. Clutches of 3 rarely hatch as skuas have 2 brood patches). Cryptic. Incubation shared for 30 days. Broods: 1. Young are semi-precocial, fledge at 45-55 days. **Food:** wide range of fish, crustacea, small mammals, eggs and young of other birds, also kleptoparasites of terns and other seabirds.

61 Gulls and Terns Laridae

91 species (2●), 17 genera, e.g. Kittiwake *Rissa tridactyla*

Description: generally white to grey, often with darker head and black wing tips. 3 webbed toes, hind toe reduced. Monomorphic, males larger. Wings long, pointed, flight strong. Distinct immature and subadult plumages. A large family with 2 well-defined subfamilies. Gulls (Larinae): medium to large, more bulky, good swimmers, tail generally medium-sized, squarish. L:25-78 cm (10-31 in), Wt:90 g-2 kg (3¼ oz-4 lb 7 oz). Bill heavy, hooked, medium-sized. Terns (Sterninae): small to large-sized, more slender, swim rarely. L:20-56 cm (8-22 in), Wt:50-700 g (1¾ oz-1 lb 8½ oz). Tail generally forked, medium to long, flight more buoyant, bill long, more slender, pointed. **Range and habitat:** coasts worldwide (fewer in tropics). **Migration:** many strongly migratory species (especially terns), the Arctic Tern, p 162, migrating further than any other species. **Breeding:** colonial. Nest is a lined scrape or grassy platform on cliff ledge. Eggs: 1-4, cryptic. Incubation shared for 18-35 days. Broods: 1. Young are precocial, fledge at 21-56 days. **Food:** omnivorous, including scavenging and kleptoparasitism, (gulls) or fish (terns – caught by plunge-diving).

62 Skimmers Rhynchopidae

3 species, 1 genus, e.g. Black Skimmer *Rhynchops niger*

Description: large, tern-like birds with specialized bill. L:40-50 cm (16-20 in), Wt:300-400 g (10½-14 oz). Bill straight, deep, blunt and laterally compressed, lower mandible up to 25% longer than upper (see Food). Unique among birds in that in bright light the pupil of the eye closes to a vertical slit. Wings long, pointed, broad-based. Flight powerful. Tail short to medium, forked. 3 toes webbed, small. Plumage dark grey to brown upperparts, white underparts. Monomorphic, males 25% larger. **Range and habitat:** AF, OR, NT. Sheltered sandy coasts, estuaries and large rivers. **Migration:** partly migratory. **Breeding:** colonial. Nest is a scrape on ground. Eggs: 3 or 4, cryptic. Incubation shared for 21 days. Young are precocial, 25-30 days. Young fly at about 35 days by which time the difference between mandibles is sufficient for them to feed. **Food:** chiefly small fish, also aquatic invertebrates. Fish caught by flying low over water with lower mandible cutting surface. This is snapped shut and up on contact with prey.

Alcae (Auks)

63 Auks Alcidae

22 species, 12 genera, e.g. Brunnich's Guillemot *Uria lomvia*

Description: small to medium-large seabirds with bulky, compact bodies, short necks, small, pointed wings. L:16-45 cm (6½-18 in), Wt:90 g-1 kg (3¼ oz-2 lb 3 oz). Powerful flight with rapid wingbeat due to heavy wing loading. Legs short, strong, set far back – especially in cliff-ledge nesters which shuffle on tarsi rather than walking. 3 webbed toes often with sharp claws, no hind toe. Strong swimmers and divers (using wings under water), some diving to great depths to fish. Bills diverse, always strong, pointed and shallow (e.g. guillemot) to blunt and very deep (e.g. puffins). Bill may vary seasonally with bright or contrasting bill sheath developed in breeding season, lost after breeding. Plumage generally dark grey or black upperparts, white underparts. Head sometimes crested. Monomorphic. **Range and habitat:** northern oceans and coasts outside tropics worldwide. Mostly off rocky coasts. **Migration:** partly migratory or nomadic. **Breeding:** most are colonial. Nest on cliff ledges, in rock crevices or burrows. Eggs: 1-2, white, ovoid (burrows) or cryptic, pear-shaped (ledge nesters), Wt:16-110 g (½-3¾ oz), incubated by female or shared for 29-42 days. Broods: 1. Young are precocial or semi-precocial, nestling period very variable depending on species' ecology: 2-50 days. **Food:** fish and crustacea.

PIGEONS AND SANDGROUSE
Columbiformes

Chunky land birds with long, pointed wings, short legs, small feet, small heads and bills. Limited distribution of down. Young fed by regurgitation. 2 families.

Pterocletes

64 Sandgrouse Pteroclididae

16 species, 2 genera, e.g. Black-bellied Sandgrouse *Pterocles orientalis*

Description: medium-sized, ground-dwelling, long, plump-bodied birds with powerful, straight flight. L:27-48 cm (10½-19 in), Wt:150-400 g (5¼-14 oz). Short, feathered legs. Toes short, sometimes partially webbed. Fast runners. Hind toe reduced or absent. Body and bill base insulated by dense down undercoat. Tail pointed, often with very long centre tail feathers. Plumage: dimorphic, males cryptic upperparts of barred browns, greys, orange; underparts banded black, chestnut or white. Females smaller and barred, more cryptic. Highly gregarious. **Range and habitat:** southern WP and EP, AF, OR. Sandy or rocky deserts and sparsely vegetated plains. **Migration:** some migrate irruptively. **Breeding:** nest is a scrape. Eggs: 3, cryptic. Incubation shared (male at night): 21-31 days. Broods: 1-3. Young are precocial, fly at 28-35 days. Males provide water to chicks by soaking modified belly feathers in water and carrying it to chicks which drink directly from these. **Food:** chiefly small seeds, also some other plant matter and small invertebrates. Drinks by sucking water into bill.

69 Green Turaco

68 Rose-ringed Parakeet

66 Black-capped Lory

67 Sulphur-crested Cockatoo

65 Rock Dove

Columbae

65 Doves and Pigeons Columbidae

301 species (16●), 41 genera, e.g. Rock Dove *Columba livia*

Description: small to large, arboreal and/or ground-dwelling, plump-bodied with powerful, straight flight. L:15-82 cm (6-32 in), Wt: 30 g-2.4 kg (1 lb-5 lb 5 oz). Nostrils in fleshy cere. Legs and feet short. 4 level, bare and unwebbed toes. Tail short, square to long, pointed. Plumage highly varied in colour and pattern, some crested. Most monomorphic. **Range and habitat:** worldwide except far northern latitudes and AN. Mostly forest and scrublands, some in more open habitats and cliffs. **Migration:** some (especially northern) species strongly migratory. **Breeding:** nest is twig platform, sometimes in tree-hole. Eggs: 2 (rarely 1), white. Incubation shared for 14-28 days. Broods: 2-3. Young are altricial or semi-altricial, in nest up to 35 days fed on pigeon 'milk', formed from sloughed cells of the crop lining. This allows many species to have protracted breeding seasons. **Food:** vegetable matter: fruit, leaves and seeds.

PARROTS AND RELATIVES
Psittaciformes

Parrots and allies. Tropical land birds defined by: powerful hooked bill with cere, upper mandible articulates with skull, short tarsus covered in granular scales, 4 toes − 2 front, 2 hind (zygodactyl), distinctive distribution of feathers and moult. Often long-lived and some good mimics. Eggs: white, incubated for 16-32 days. Young altricial in nest: 21-90 days. The 3 families are often treated within a single enlarged family, Psittacidae.

66 Lories Loridae

55 species (7●), 11 genera, e.g. Black-capped Lory *Lorius lory*

Description: small to medium parrots much modified for feeding on nectar and pollen on which they specialize. L:15-32 cm (6-13 in). The tongue is modified and brush-tipped allowing them to reach into flowers. In so doing, they are important pollinators for some plant species. The gut is also modified. Wings medium-sized, pointed, flight fast, direct. Tail short, square to long, pointed. Plumage bright greens, reds, blues in bold patches. Monomorphic. Noisy and social. **Range and habitat:** OR, AU, PA. A wide variety of forest and scrubland habitats including gardens. **Migration:** many species migratory or nomadic. **Breeding:** little known, solitary or colonial. Nest is generally a tree-hole. Eggs: 2-4. Incubation by female. **Food:** nectar, pollen, plant shoots and fruits.

67 Cockatoos Cacatuidae

19 species (4●), 6 genera, e.g. Sulphur-crested Cockatoo *Cacatua galerita*

Description: medium to large parrots, simply coloured, white to blackish, pinks and yellow with an erectile crest, sometimes very large. L:33-69 cm (13-27 in). Bill medium to massive with thick, fleshy tongue. Some with a brightly coloured bare facial patch. Wings long, pointed.

Tail medium-sized, square or rounded. Monomorphic or dimorphic, male larger. Some highly gregarious, noisy. **Range and habitat:** OR, AU, PA. A wide variety of forest, scrub and grassland habitats. **Migration:** generally non-migratory or nomadic, some partly migratory. **Breeding:** nest is a tree-hollow. Eggs: 1-3. Incubation shared for c. 30 days. Young in nest for up to 90 days. **Food:** a wide range of plant matter, fruits, buds, shoots, seeds, corms and some invertebrates.

68 Parrots Psittacidae

268 species (26●), 59 genera, e.g. Rose-ringed Parakeet *Psittacula krameri*

Description: a diverse family of small to large parrots including 1 flightless form (Kakapo). L:9-100 cm (3½-39 in). Wings medium-sized, rounded to long, pointed. Tail short, square to very long, pointed. Bill small to massive, with thick fleshy tongue. Lores often bare. Plumage varied and gaudy greens, reds, blues, yellows in bold patches. Generally monomorphic or male larger. Solitary to highly social, noisy. **Range and habitat:** AF, OR, AU, NT. A wide variety of forest, scrub and grassland habitats. **Migration:** most non-migratory. **Breeding:** nest is generally a tree-hollow. Eggs: 1-12. Incubation by female, 17-35 days. Young in nest for 21-70 days. **Food:** a wide range of plant matter, fruits, buds, shoots, seeds, corms and some invertebrates.

CUCKOOS AND ALLIES
Cuculiformes

Land birds defined by: 4 toes − 2 front, 2 hind (zygodactyl, but see Hoatzin below), 8-10 tail feathers, upper mandible unmovable, not hooked, no cere.

69 Turacos Musophagidae

20 species, 6 genera, e.g. Green Turaco *Tauraco persa*

Description: medium to large, elongated arboreal birds. L:35-75 cm (14-29 in), Wt:230-950 g (8 oz-2 lb 1 oz). Bill short, stout, pointed with diverse location and shape of nostril. Head small, often crested, neck medium-sized. Wings short and rounded, flight weak, tail long and broad. Legs medium-sized, walks with great agility through tree canopy. Plumage soft, variously: greens, blues, reds, white, brown and grey. Monomorphic. **Range and habitat:** AF. Dense forests and forest edge. **Migration:** non-migratory. **Breeding:** nest is a flat platform of twigs at 5-20 m (16-66 ft) height. Eggs: 2-3, glossy white or pale green/blue. Incubation shared for 21-4 days. Young are semi-altricial with well-developed carpal claw, leave nest after 10-12 days, fly after about 30 days. **Food:** chiefly fruit, also seeds, buds and invertebrates.

70 Cuckoos Cuculidae

139 species, 40 genera, e.g. Great Spotted Cuckoo *Clamator glandarius*

Description: a diverse family of medium to large arboreal or rarely ground-dwelling (roadrunner) birds, including many nest parasites, especially in Old World (1 in New World). L:17-65 cm (6½-26 in), Wt: 30-700 g (1 lb-1 lb 8½ oz). Slender birds with long tail. Legs short to

70 Great Spotted Cuckoo
72 Barn Owl
75 Tawny Frogmouth
71 Hoatzin
73 Great Grey Owl
74 Oilbird

medium. Bill stout, medium-sized, decurved, pointed, sometimes deep. Plumage generally streaked greys, browns, sometimes black, green or yellow. Most monomorphic. **Range and habitat:** a variety of woodland, scrub and marsh habitats. **Migration:** northern species strongly migratory. **Breeding:** nest is a simple platform of sticks in tree or on ground, or that of host. Eggs: 10-15, possibly more, in parasitic species, colour often mimics that of host eggs, very small relative to body-size in parasites. In non-parasites, 2-5, bluish or whitish. Incubation shared or by host for 11-16 days. Young are altricial, fledge at 16-24 days, may kill or evict host young or unhatched eggs in nest. **Food:** chiefly insects, especially prey considered toxic by other birds (e.g. hairy caterpillars). In some species also small vertebrates and plant matter.

71 Hoatzin Opisthocomidae
Opisthocomus hoazin (monotypic). See p 114.

Description: strangely reminiscent of Archeopteryx, the Hoatzin is an extraordinary, large inhabitant of riverine forests of NT. Combination of apparently primitive features such as claws on carpal of young birds, and specialized (derived) features adapted to feeding on tough leaves and fruits of a limited range of plants has led to great debate as to its true evolutionary and systematic relationships. It is certainly advanced rather than primitive. In the past it has variously been included within 8 different orders, including the Gruiformes, Galliformes, Columbiformes and Cuculiformes. Biochemical evidence suggests the last. It is sometimes treated as a monotypic order.

OWLS Strigiformes
Raptorial species, often nocturnal with many adaptations for silent hunting, e.g. exceptional hearing, soft plumage allowing silent flight. Large eyes face forwards, set in facial disc, cannot move within socket. Feet strong with sharp claws. 4 toes, the last reversible. Bill short, strongly hooked. Resemblance to raptors (Falconiformes) is superficial. 2 families.

72 Barn Owls Tytonidae
14 species, 2 genera, e.g. Barn Owl *Tyto alba*

Description: medium-sized owls with characteristic heart-shaped facial disc and relatively small eyes but mainly nocturnal. L:23-53 cm (9-21 in), Wt:180-1,280 g (6½oz-2lb 13oz). Head large, body slender but loose plumage gives deceptive impression of size. Legs long, tarsus feathered but feet bare, middle toe has comb claw. Wings very broad, rounded, hence light, buoyant flight. Tail medium-sized, shorter than legs, feathers emarginated, primary flight feathers not emarginated. Plumage brown to golden brown, grey above, whitish underparts. Monomorphic but females often larger. **Range and habitat:** WP, AF, OR, AU, NA, NT. A variety of forest and open (including cultivated) habitats. **Migration:** largely non-migratory. **Breeding:** nest is in tree-hole, building or ground, will use nestboxes. Eggs: 4-7, white. Incubation by female for 27-34 days. Broods: 1-2. Young are semi-altricial in nest, fledge at 49-56

days. **Food:** a wide range of small vertebrates caught with talons and generally swallowed whole, also large insects.

73 Owls Strigidae
160 species (3●), 22 genera, e.g. Great Grey Owl *Strix nebulosa*

Description: small to large owls with both nocturnal and diurnal forms. L:12-71 cm (4½-28 in), Wt:40g-4kg (1½oz-8lb 13oz). Eyes very large. Facial disc(s) rounded. Head large, sometimes carries ear-tufts. Legs medium to long but shorter than tail, tarsus and often feet feathered, except in fishing species which have roughened feet. Claws often very long. Arboreal and ground-dwelling species. Flight strong and silent on large, broad wings. Flight feathers fringed to reduce sound, especially in nocturnal species. Plumage soft, dense, gives false impression of size, generally cryptic mottled and barred browns, greys. Monomorphic, female often larger. **Range and habitat:** worldwide except AN. A wide range of forest, grassland, desert and tundra habitats. **Migration:** most species sedentary. **Breeding:** nest is in tree-hollows or old nests of other birds, e.g. crows (Corvidae). Eggs: 1-7, white. Incubation by female or shared for 15-35 days. Broods: 1. Young are semi-altricial, in nest for 24-52 days but may leave nest before they can fly. **Food:** a wide range of small vertebrates including fish, also earthworms and large insects. Prey caught with talons, generally swallowed whole.

NIGHTJARS AND ALLIES Caprimulgiformes
A well-defined (natural) order of nocturnal or crepuscular, mostly insectivorous, cryptic birds with very wide gape. Flight light and bat-like. Feet small and weak. Intermediate in form (and systematic order) between owls and swifts. 5 families.

74 Oilbird Steatornithidae
Steatornis caripensis (monotypic). See p 116.

Description: a largely frugivorous species, with a longer and more decurved bill than others of the order, but still retaining the rictal bristles indicative of its insectivorous ancestry. The Oilbird is the only nocturnal frugivorous bird. It breeds in large, noisy colonies in caves on Trinidad and NT.

75 Frogmouths Podargidae
13 species, 2 genera, e.g. Tawny Frogmouth *Podargus strigoides*

Description: medium to large with broad, rounded wings and tail, the latter longish. L:23-54 cm (9-21 in). Owl-like patterning, superb cryptic coloration of browns and greys. Eyes very large and coloured in some. Flight strong for short periods. The bill is boat-shaped with a huge gape and rictal bristles which led to the earlier belief that they hunted aerial insects on the wing. However, this is not the case: most prey are caught on the ground or from branches. Despite apparent evidence of their ancestry from nightjars, there are numerous internal anatomical differences such as the number of cervical vertebrae, toe bones, type of syrinx and palate. Feet small with long middle toe (3 forward, 1 hind).

76 Common Potoo

77 Mountain Owlet-nightjar

81 Sword-billed Hummingbird

80 Crested Tree Swift

78 African Dusky Nightjar

79 Alpine Swift

Monomorphic. At roost during the day some adopt a posture making best use of their crypsis and mimicking a broken branch. **Range and habitat:** OR, AU. Mostly tropical forest. **Migration:** partly migratory. **Breeding:** nest is a platform of sticks and feathers on a horizontal bough of a tree. Eggs: 1-3, white. Incubation by female or shared for 30 days. Young are semi-altricial, in nest for 30 days. **Food:** chiefly insects but according to species; also small vertebrates.

76 Potoos Nyctibiidae
5 species, 1 genus, e.g. Common Potoo *Nyctibius griseus*

Description: medium to large, long, slender birds. L: 23-50 cm (9-20 in). Solitary, nocturnal. Small bill is terminally decurved. Has a tooth on the edge of the upper mandible. Lacks rictal bristles. The mouth is very large. Eyes very large, iris sometimes yellow or dark brown but reflecting orange at night. Wings long and rounded, tail long and pointed. Legs very short. Feet small and weak, toes long and slender, lacking comb-claw. Highly cryptic plumage of browns and greys. Monomorphic. Flight strong and direct. **Range and habitat:** NT and southern NA. Forest. **Migration:** non-migratory. **Breeding:** little known. Nest is a small depression in tree-hollow. Eggs: 1 (rarely 2), white. Incubated in rather upright posture for at least 33 days. Young are semi-altricial, leave nest at c. 25 days but fly at 47-51 days. **Food:** aerial insects.

77 Owlet-nightjars Aegothelidae
8 species, 1 genus, e.g. Mountain Owlet-nightjar *Aegotheles albertisi*

Description: small to medium, solitary, nocturnal birds closely related to frogmouths (Podargidae). L: 20-30 cm (8-12 in), Wt: 45-115 g (1½-4 oz). The bill is similar to frogmouths but smaller and weaker and is largely hidden by forehead feathering. Includes rictal bristles. Plumage cryptic, generally like frogmouths but includes many stiff, partly erectile filoplumes on forehead and lores, and some on chin. Wings long, rounded. Flight strong and direct. Tail long and pointed. Feet small and weak, toes and claws long, slender, lacking comb-claw. Foraging seems intermediate between frogmouths and nightjars, and includes both hawking and (probably a majority) ground feeding. Monomorphic. **Range and habitat:** AU. Tropical forest, also scrublands in Australia. **Migration:** non-migratory. **Breeding:** little known. Nest is in tree-hollow. Eggs: 3-4, white. Young are semi-altricial. **Food:** aerial insects and ground invertebrates.

78 Nightjars Caprimulgidae
82 species, 18 genera, e.g. African Dusky Nightjar *Caprimulgus pectoralis*

Description: medium-sized, mostly solitary, nocturnal or crepuscular. L: 19-29 cm (7½-11½ in), Wt: 40-120 g (1½-4 oz). Head flattened. Bill short, weak with very wide gape and usually also rictal bristles. Eyes large. Legs short, sometimes feathered. Wings long, pointed. Flight light, agile, bat-like. Tail medium to long, square or forked. 2 species with long pennant feathers on wing or tail. Feet small, weak with comb-claw on middle toe. Plumage cryptic, barred browns and greys. Often dimorphic, males having white patches shown in display flight. 1 species hibernates.

Range and habitat: worldwide except far northern Holarctic and AN. A wide variety of forest, scrub and semi-arid habitats. **Migration:** some species strongly migratory, migrating in loose flocks. **Breeding:** nest is on ground, no material. Eggs: 1-2, cryptic, white to pinkish, marbled or blotched black, brown or violet. Incubation shared for 16-19 days. Broods: 1-2. Young are semi-altricial, in nest for 16-20 days. **Food:** aerial insects.

SWIFTS AND HUMMINGBIRDS
Apodiformes
An uneasy grouping of species in 2 natural groups or suborders – i.e. sharing many well-defined or unique characteristics. All are highly agile fliers with small feet but there are many internal differences between the suborders. 3 families in 2 suborders.

Apodi
79 Swifts Apodidae
92 species, 21 genera, e.g. Alpine Swift *Tachymarptis melba*

Description: small to medium, fast-flying aerial birds with long, scythe-shaped wings, and short to medium-sized forked tails, sometimes spiny. L: 10-30 cm (4-12 in), Wt: 9-150 g (¼-5¼ oz). Feet tiny with very sharp, curved claws. Many have reversible hind toe and perch with difficulty. Bill very small but gape wide. Body slender and short-necked. Plumage dense, dark browns, black and sometimes white throat, flanks or rump. Monomorphic. Often gregarious and noisy with screaming display flights. Probably the most aerial of birds, even copulation taking place on the wing. **Range and habitat:** worldwide except high latitudes, PA and AN. Aerial feeders coming to rest only to nest. **Migration:** many long-distance migrants. **Breeding:** solitary or colonial, some with nest-helpers. Nest is aerial flotsam cemented together with saliva. The nest of 1 species considered edible. Eggs: 1-6, white. Incubation shared for 17-28 days. Broods: 1-2. Young are altricial, in nest for 34-72 days, becoming torpid when conditions are unfavourable for the parents to hunt. **Food:** aerial insects.

80 Tree Swifts Hemiprocnidae
3 species, 1 genus, e.g. Crested Tree Swift *Hemiprocne longipennis*

Description: small to medium swifts, less highly adapted to an aerial lifestyle than the Apodidae, and superficially resembling swallows (Hirundinidae) because of their outer tail streamers, more manoeuvrable flight and their perching abilities. L: 17-33 cm (6½-13 in), Wt: 70-120 g (2½-4 oz). General body form is like true swifts but the head carries a long erectile crest, the eyes are large and the hind toe of the tiny foot is not reversible. Perches and roosts high in trees. Plumage dense, soft, glossy blue-greys to browns. Dimorphic. **Range and habitat:** OR, AU. Open woodland and forest edges. **Migration:** non-migratory. **Breeding:** habits poorly known. Eggs: 1, white, cemented into a tiny cup-shaped nest, itself cemented to the side of a lateral tree branch. Incubation shared. **Food:** aerial insects.

31

82 Speckled Mousebird

83 Blue-crowned Trogon

86 Blue-crowned Motmot

85 Cuban Tody

84 (Eurasian) Kingfisher

87 Swallow-tailed Bee-eater

Trochili

81 Hummingbirds Trochilidae

332 species (6●), 114 genera, e.g. Sword-billed Hummingbird *Ensifera ensifera*

Description: a huge but natural group showing diverse adaptive radiation into an otherwise unexploited ecological niche. Tiny to small, the family includes the smallest birds known. L: 5.8-21.7 cm (2¼-8½ in), Wt: 2-20 g (¹⁄₁₆-¾ oz). Hummingbirds show a number of adaptations giving them access to flowers from which they suck nectar (in so doing they are important pollinators for some plant species). In particular, a modified arrangement of the wing bones means birds make more than 70 wing-beats/sec allowing precision hovering in any direction including backwards. They are also capable of very fast flight. In addition, a long, thin bill and protrusible tongue allows access to the nectaries at the base of the flower's corolla tube. The family shows great diversity of bill size and shape, which are adapted to feeding from different flower types. Plumage dimorphic, often brilliantly coloured in metallic greens and reds, sometimes crested in males, females more drab. Tail short, square to long, forked. Feet tiny, weak, allow perching but not walking. **Range and habitat:** NA, NT. Forest, savannah scrublands and desert. **Migration:** some strongly migratory species. **Breeding:** polygynous, highly territorial, or lek. Nest is either a deep cup of plant fibres and spiders' webs astride a stem or twig, or a pendant nest attached by a cobweb to the underside of a leaf and counter-weighted by debris at the base. Eggs: 2, white. Incubation and care by female (except 1 recorded case) for 14-23 days. Young are altricial, in nest for 18-38 days. **Food:** nectar and small insects.

COLIES OR MOUSEBIRDS
Coliiformes

A monotypic order showing no close affinity to any other.

82 Mousebirds Coliidae

6 species, 2 genera, e.g. Speckled Mousebird *Colius striatus*

Description: small birds with long, pointed, graduated tails. L: 30-5 cm (12-14 in) including tail which constitutes 66% of length, Wt: 45-55 g (1½-2 oz). Wings short, rounded, flight fast and level for short distances. Legs short with stout feet. 4 toes usually directed forwards but outer toe is reversible. Bill short, stout, decurved and hooked. Usually perch hanging vertically but can also perch normally. Hop and run fast on ground also. Plumage soft, crested, variously marked greys and browns. Diverse moult patterns. Monomorphic. Highly gregarious. **Range and habitat:** AF. Scrublands and forest edges, also gardens and cultivation. **Migration:** non-migratory. **Breeding:** solitary or loosely colonial. Nest is open cup, usually in thorn bush, sometimes bulky and untidy. Eggs: 3 (rarely 1-8), whitish streaked blackish or brownish. Incubation shared, sitting tightly for 11-14 days. Young are altricial, in nest for 15-20 days, fed by regurgitation. Care by female or shared. **Food:** a variety of plant matter including seeds, fruit, shoots and leaves.

TROGONS
Trogoniformes

Among the most beautiful of birds, no clear relationships with other orders.

83 Trogons Trogonidae

39 species, 7 genera, e.g. Blue-crowned Trogon *Trogon curucui*

Description: medium-sized solitary, arboreal birds showing remarkably little variation considering their widespread distribution and movements. L: 23-38 cm (9-15 in). The legs and feet are small and weak, the latter unique in that, although zygodactyl, it is the 1st and 2nd toes which point backwards rather than the 1st and 4th. Bill short, stout, decurved (serrated in NT species) and very broad-based with rictal bristles. Wings short, roundish. Tail long, truncated. Plumage bright and soft, metallic except in OR, greens, blues, reds, yellows, black and white. Dimorphic, females duller. **Range and habitat:** pantropical: AF, OR, NT. Tropical forest up to 3,000 m (9,850 ft). **Migration:** non-migratory. **Breeding:** nest is a hole in tree, stump or termite mound. Eggs: 2-3 (rarely 4), unmarked glossy white, cream, buff, brown, pale blue or green. Incubation shared for 17-19 days. Young are altricial in nest for 17-18 days. **Food:** chiefly insects, in NT also fruit.

KINGFISHERS AND ALLIES
Coraciiformes

A large diverse order of carnivorous land birds, usually brightly coloured birds with large bills and usually with 2 front toes (of 4) joined for all or part of their length (syndactyl). 10 families in 4 suborders. Little or no nest sanitation is practised.

Alcedines

84 Kingfishers Alcedinidae

92 species, 14 genera, e.g. (Eurasian) Kingfisher *Alcedo atthis*

Description: small to large, compact birds with large head, long, stout and pointed bill. L: 10-45 cm (4-18 in), Wt: 8-500 g (¼ oz-1 lb 2 oz). Neck short. Wings short, rounded. Flight fast, direct. Tail short to medium, square to rounded. Legs and feet short and weak. The 3rd and 4th toes are joined for most of length, the 2nd and 3rd partially joined. The 2nd toe is vestigial or absent in some species. Plumage generally bright and conspicuous, sometimes metallic, blues, greens, reds, orange, sometimes cryptic. Most more or less monomorphic. Generally solitary. Piscivores, hunt by diving into water to take prey. **Range and habitat:** worldwide except high latitudes and AN. A wide variety of forest, scrub and savannah habitats, often near water. **Migration:** mostly non-migratory. **Breeding:** some with nest-helpers. Nest is a chamber at end of long tunnel excavated in river-bank, or tree-hole. Eggs: 2-4 in tropics, 6-10 in temperate latitudes, white. Incubation shared for 18-22 days. Broods: 1-2. Young are altricial, in nest for 20-30 days. **Food:** fish, small vertebrates or insects depending on species.

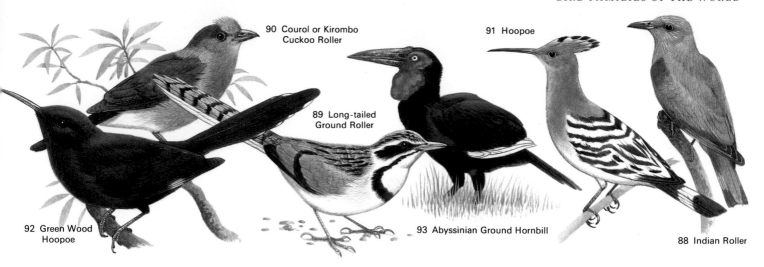

90 Courol or Kirombo Cuckoo Roller

91 Hoopoe

89 Long-tailed Ground Roller

92 Green Wood Hoopoe

93 Abyssinian Ground Hornbill

88 Indian Roller

85 Todies Todidae

5 species, 1 genus, e.g. Cuban Tody *Todus multicolor*

Description: exquisite tiny, compact, short-necked birds with long, flattened, broad, red bill. L:9-12 cm ($3\frac{1}{2}$-$4\frac{1}{2}$in), Wt:5.4 g ($\frac{1}{4}$oz). Wings short, rounded. Tail short, square. Plumage varies little, green upperparts, white underparts with yellow and pink patches, and red throat patch. Some with small blue patch below ear. Fossil evidence indicates that present distribution is a remnant of a formerly widespread range. Monomorphic. Tame, arboreal, usually paired. **Range and habitat:** endemic to Caribbean (NT). Forests, often near water. **Migration:** non-migratory. **Breeding:** highly territorial, may have nest-helpers. Nest is a chamber in bank at end of 30 cm (12 in) tunnel. Eggs: 2-5, white, 25% of female weight. Incubation shared for rather long period: 21-2 days. Young are altricial in nest for 19-20 days, emerge sooner if helpers present. Fed extremely frequently. **Food:** insects, very high feeding rate.

86 Motmots Momotidae

9 species, 6 genera, e.g. Blue-crowned Motmot *Momotus momota*

Description: small to medium birds most closely related to Todidae. L:17-45 cm ($6\frac{1}{2}$-18 in) including long tail. Unlike todies, the bill is strong, decurved, long and pointed. Wings short, rounded. Tail long, stiff, sometimes with bare shaft for part of its length, giving a racket-tip. Legs and feet short, toes as kingfishers (Alcedinidae). Plumage loose textured, bright greens, blues, browns, often with a black breast-spot and black eye stripe. More or less monomorphic. Solitary or paired, perch motionless for long periods. **Range and habitat:** NT. Tropical forest, often near streams. **Migration:** non-migratory. **Breeding:** nest is a chamber at end of a very long tunnel. Eggs: 3-4, white. Incubation shared for 15-22 days. Young are altricial, in nest for 24-38 days. **Food:** insects caught by flying out from a perch (sallying), also some reptiles and fruit.

Meropes

87 Bee-eaters Meropidae

23 species, 3 genera, e.g. Swallow-tailed Bee-eater *Merops hirundineus*

Description: small to medium but long, slender, gregarious birds with long, pointed, decurved bills. L:17-35 cm ($6\frac{1}{2}$-14 in), Wt:15-85 g ($\frac{1}{2}$-3 oz). Wings large, pointed, flight strong and agile. Tail long, usually with elongated centre tail feathers. Legs short, feet small and weak. Feet syndactyl. Perch on branches and telegraph wires from which to hunt. Plumage brilliant greens, blues, yellows and reds, in bold patches with a black mask and often a contrasting throat patch. Monomorphic, males brighter. **Range and habitat:** southern WP and EP, AF, OR, AU. Open country including open forest and scrub, 6 species in rainforest. **Migration:** temperate breeders are migratory. **Breeding:** colonial or solitary (forest species), some with nest-helpers. Nest is a chamber at end of tunnel up to 3 m (10ft) long in sandbank. Eggs: 2-7 (2-4 in tropics), white. Incubation shared for 18-23 days. Broods: 1. Young are altricial, in nest for 27-32 days. **Food:** aerial insects, especially venomous worker bees which are rubbed on a branch to remove the sting.

Coraci

88 Rollers Coraciidae

12 species, 2 genera, e.g. Indian Roller *Coracias benghalensis*

Description: probably most closely related to Bee-eaters. Medium to large, thickset birds with powerful, medium to long or short, broad, hooked bills. L:25-45 cm (10-18 in), Wt: c.100-200 g (c.$3\frac{1}{2}$-7 oz). Head large. Wings long, pointed. Flight strong, agile. Tail usually short to medium, square, sometimes with long outer feathers. Legs short with small, weak feet. Inner toes syndactyl but outer toe free. Plumage gaudy blues, greens and greys. More or less monomorphic. Generally solitary, arboreal. Noisy. **Range and habitat:** WP, EP, AF, OR, AU. Forests, scrublands and grasslands, including cultivation. **Migration:** some migratory species. **Breeding:** rolling courtship display flight. Nest is in tree-hollows, burrows, old walls. Eggs: 4, glossy white, Wt:10-17 g ($\frac{1}{4}$-$\frac{1}{2}$oz). Incubation shared for 17-19 days. Young are altricial, in nest for 25-30 days. **Food:** insects by sallying, also small vertebrates.

89 Ground Rollers Brachypteraciidae

5 species (4●), 3 genera, e.g. Long-tailed Ground Roller *Uratelornis chimaera*

Description: medium-sized. L:24-45 cm ($9\frac{1}{2}$-18 in), 4 species less than 30 cm (12 in), solitary or in small groups, crepuscular ground-dwelling birds closely related to (and often treated with) rollers (Coraciidae). However, the legs are much longer and the wings shorter and rounder than in that family. Tail very long and truncated. Bill long, decurved, stout and hooked as rollers. Flight is strong and rolling in display, but birds tend to run rather than fly when disturbed. Plumage more cryptic than rollers with streaked browns, yellow, greens, but also patches of blue. **Range and habitat:** Madagascar (AF). Dense forest and scrub. **Migration:** non-migratory. **Breeding:** little known. Nest is a chamber at end of sloping tunnel in bank. Eggs: 3-4, white. Incubation probably shared. **Food:** insects and small vertebrates on ground.

90 Courol or Kirombo Cuckoo Roller Leptosomatidae

Leptosomus discolor (monotypic).

Description: large, noisy, rather gregarious, arboreal roller-like bird. L:42 cm (17 in). It is unusual in the presence of powder down patches on either side of the rump, and the 4th toe which is rotated to the side. Feathers of the lores face forwards and upwards, so concealing the base of the bill. Dimorphic: male iridescent green, head grey, underparts white. Female greenish with brown head and underparts spotted. **Range and habitat:** Madagascar and the Comoro Islands (AF). Evergreen forests. **Migration:** non-migratory. **Breeding:** poorly known. Nest is in tree-cavity. **Eggs:** 2 cream-buff. **Food:** insects and small vertebrates.

91 Hoopoe Upupidae

Upupa epops (monotypic). See p 166.

Description: common bird of open habitats in central and southern WP, and AF except Sahara. Feeds on ground on invertebrates and small

95 Spotted Puffbird

98 Red-billed Toucan

97 Black-throated Honeyguide

99 Grey-faced Woodpecker

96 Crimson-breasted Barbet

94 Rufous-tailed Jacamar

vertebrates. Long, decurved bill used to probe ground for food. Feet syndactyl. Its systematic relationships are somewhat unclear as it shares some characteristics with Passeriformes rather than Coraciiformes.

92 Wood Hoopoes Phoeniculidae
8 species, 2 genera, e.g. Green Wood Hoopoe *Phoeniculus purpureus*

Description: medium to large, arboreal, solitary or in small groups. L: 23-46 cm (9-18 in). Bill long, thin, slightly to strongly decurved. Legs short, thick, sometimes feathered. Toes with long, curved claws, toes 3 and 4 fused at base. Bill and legs vary: red, orange, yellow or black depending on species, population and age. Tail very long and graduated. Wings rounded. Hops on ground unlike Hoopoe (Upupidae) which walks but generally flies strongly from tree to tree. Plumage generally black or dark green/purple with metallic sheen. **Range and habitat:** AF. Forest, wooded savannah. **Migration:** non-migratory. **Breeding:** solitary to cooperative with nest-helpers. Nest is in tree-hollows. Eggs: 2-4 (rarely 5), blue, spotted. Incubation by female for 17-18 days. Young are altricial, in nest for 28-30 days. **Food:** insects and small fruits.

Bucerotes
93 Hornbills Bucerotidae
48 species, 2 genera, e.g. Abyssinian Ground Hornbill *Bucorvus abyssinicus*

Description: a distinct group with many unique features. Medium to very large, arboreal or ground-dwelling birds with massive bills, often carrying a horny casque which may be as large as the bill itself. L: 38-160 cm (15-63 in), Wt: 85 g-4 kg (3 oz-8 lb 13 oz). The casque is light and hollow but supported internally by bone. Bill generally long, pointed, powerful and gently decurved. Bill and casque usually coloured red or yellow. Colours vary with age. Fused axis and atlas vertebrae help to support the weight of the bill. Hornbills have eyelashes and sometimes coloured (according to sex) bare facial skin and wattles. Neck medium to long. Wings long, broad. Flight strong but laboured, especially in larger species. Tail long, broad, rounded. Legs short to medium (shorter in arboreal species). Plumage coarse, loose-webbed, generally browns, greys, black, boldly marked. Monomorphic or female duller, males up to 10% larger with 15-20% larger bill. Females generally moult during incubation. **Range and habitat:** AF, OR, New Guinea (AU). Forests and wooded savannah. **Migration:** non-migratory. **Breeding:** may have nest-helpers. Nest is usually a hollow tree (except ground hornbills) into which the female is sealed so that only a vertical slit remains. She is fed through this by the male during incubation and until the young are half-grown. Eggs: 1-7 (fewer in larger species), white. Incubation is for 25-40 days. Young are altricial, in nest for 45-86 days. **Food:** chiefly insects, fruit and small vertebrates.

WOODPECKERS AND ALLIES
Piciformes
An order generally defined by zygodactyl foot and internal anatomy. 6 families in 2 suborders.

Galbulae
94 Jacamars Galbulidae
17 species, 5 genera, e.g. Rufous-tailed Jacamar *Galbula ruficauda*

Description: small to medium, slender, graceful, highly active exciting birds with long, thin, sword-like bills. L: 15-31 cm (6-12 in) but largest has long tail, Wt: 20-70 g ($\frac{3}{4}$-2$\frac{1}{2}$ oz). Legs short. Feet small, 1 species has 3 toes only (having lost the inner hind toe). Tail long, truncated or pointed. Wings short, rounded. Plumage loose-webbed, usually metallic green, bronze or purple above, rufous orange (most) or black below, often with white throat. Slight dimorphism, males brighter. **Range and habitat:** NT. Forest and wooded savannah. **Migration:** non-migratory. **Breeding:** known for few species. Solitary. Nest is a nest-burrow in bank or rootplate of fallen tree. Eggs: 3-4, glossy white. Incubation shared during day, female at night for 20-30 days. Young are semi-altricial (unlike others of the order), in nest for 19-26 days. **Food:** insects caught by sallying.

95 Puffbirds Bucconidae
32 species, 10 genera, e.g. Spotted Puffbird *Bucco tamatia*

Description: small to medium, stolid, rather inactive, generally arboreal birds with large head and short neck. L: 14-29 cm (5$\frac{1}{2}$-11$\frac{1}{2}$ in), Wt: 30-90 g (1-3$\frac{3}{4}$ oz). Bill medium-sized, strong, decurved or hooked, flat, with prominent rictal bristles. Legs short, feet zygodactyl. Wings short, rounded. Tail medium to long, square. Plumage soft and often puffed out when perched, dull browns, greys and white, sometimes streaked or spotted and often with a breast band. More or less monomorphic. **Range and habitat:** NT. Forest and wooded savannah. **Migration:** non-migratory. **Breeding:** poorly known. No nest sanitation practised. Nest is a tunnel dug by both parents in a bank, termite mound or ground, or, more rarely, a tree-hole. Eggs: 2-3 (rarely 4), white. Incubation shared for unknown period. Young are altricial, in nest for 20-1 days. **Food:** chiefly insects caught on the wing by sallying, or on ground.

96 Barbets Capitonidae
81 species, 15 genera, e.g. Crimson-breasted Barbet *Megalaima haemacephala*

Description: small to medium, generally solitary or paired, arboreal and sometimes rather inactive birds with large heads, short necks. L: 9-33 cm (3$\frac{1}{2}$-13 in). Bill large, stout, decurved, pointed and often sharply toothed along the edge. Rictal bristles prominent, forming 'beard' in some. Legs generally short. Feet strong and zygodactyl. Hops clumsily on ground. Wings short and rounded, flight rather weak and for short distances. Tail short and square. Plumage brightly coloured in contrasting patches of greens, reds, blues and yellows. More or less monomorphic. **Range and habitat:** AF, OR, NT. A variety of forest, savannah and semi-arid habitats. **Migration:** non-migratory. **Breeding:** sometimes with helpers at the nest. Nest is a hole in a bank, termite mound or rotten tree-stump dug by both parents, or old woodpecker hole. Eggs: 2-5, white. Incubation shared for 12-19 days. Young are altricial, in nest for 20-35 days. In

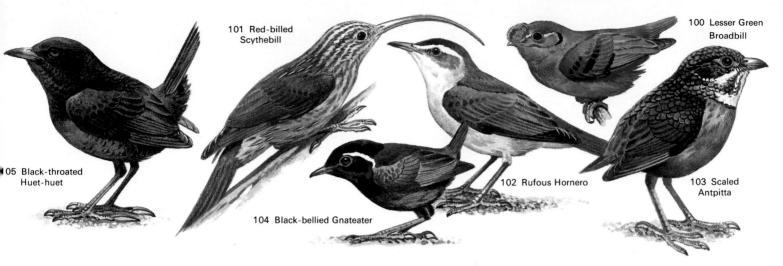

101 Red-billed Scythebill

100 Lesser Green Broadbill

05 Black-throated Huet-huet

102 Rufous Hornero

103 Scaled Antpitta

104 Black-bellied Gnateater

smaller species, young fed by regurgitation. **Food:** fruit, flowers, buds, insects. Larger species also take small vertebrates.

97 Honeyguides Indicatoridae

17 species, 4 genera, e.g. Black-throated Honeyguide *Indicator indicator*

Description: small to medium, solitary arboreal birds, some showing a unique relationship with man (see Food). L: 10-20 cm (4-8 in), Wt: 10-55 g ($\frac{1}{4}$-2 oz). Bill short, stout or thin, pointed. Legs short with strong, zygodactyl feet. Wings long, pointed. Flight strong, undulating. Tail graduated, rather long. Plumage generally drab greys and browns, paler underparts, some with yellow or orange patches or white outer tail feathers. Most more or less monomorphic. **Range and habitat:** AF, OR. Forest and wooded savannahs. **Migration:** seasonal short-distance movements. **Breeding:** nest parasites of hole-nesters such as barbets and woodpeckers, and of some flycatchers. Eggs: 1 per nest, white or blue (in 1 species to mimic host). Incubation 12-13 days. Young are altricial, in nest for 38-40 days, some have tooth on bill to kill host chicks if present. **Food:** insects and wax, especially beeswax. The Black-, or Greater, and the Scaly-throated Honeyguides instinctively lead man and other large mammals to the nests of honey-bees. The 'helper' then opens the nest to obtain honey. Some human tribes leave beeswax for the bird. It is not clear to what extent the birds depend on wax obtained in this way since attempts to study them are bedevilled by the fact that the birds switch to 'guiding' behaviour on the approach of any human observer.

98 Toucans Rhamphastidae

35 species, 6 genera, e.g. Red-billed Toucan *Rhamphastos tucanus*

Description: medium to large, generally solitary, arboreal birds with massive, brightly coloured bill, superficially resembling hornbills of the Old World but these are quite unrelated. L: 33-66 cm (13-26 in). Bill generally long, deep and with a markedly decurved culmen (ridge of upper mandible) with nostrils at base. The cutting edges of the bill (tomia) are sometimes saw-toothed. Although sometimes as long as the bird's body, the bill is extremely light and strengthened internally by numerous bony filaments. The tongue is also much modified. Bare skin round eye is brightly coloured. Legs medium to long, strong with stout zygodactyl feet. Wings short, rounded. Flight rather ungraceful. Tail long, graduated or rounded. Plumage coarse, gaudily patterned in contrasting patches of black, yellow, greens, blues and red. Most are monomorphic but male bill larger. **Range and habitat:** NT. Rainforest and more open wooded foothills. **Migration:** some altitudinal migrants. **Breeding:** nest is in tree-hollow. Eggs: 2-4, white. Incubation shared for 15-16 days. Young are altricial with thick heel pads to protect feet (which bear chicks' weight) against rough nest interior. In nest for 43-51 days, develop very slowly, e.g. eyes do not open until day 14. **Food:** chiefly fruit and insects, also other small animals.

Pici

99 Woodpeckers Picidae

200 species (5●), 27 genera, e.g. Grey-faced Woodpecker *Picus canus*

Description: a well-defined group with many shared characteristics which adapt them to wood-boring. Small to medium-large, generally solitary, mostly arboreal birds. L: 8-55 cm (3-22 in), Wt: 8-563 g ($\frac{1}{4}$ oz-1 lb 4 oz). Bill chisel-like, straight, pointed and strong. Skull strengthened. Tongue very long, protrusible and often brush-tipped. Tail stiffened as a support while hopping up tree-trunk except in piculets. Wings rather long, broad, pointed. Flight undulating. Head sometimes crested. Plumage generally rather coarse, boldly marked black, white, greens, browns, greys, yellow and red. Often barred. Generally dimorphic in size and with small colour differences between sexes. Wrynecks (Jynx species) superbly cryptic with fine barring. **Range and habitat:** WP, EP, AF, OR, AU, NA, NT. Wide range of forest, scrub and grasslands including cultivation. **Migration:** most non-migratory. **Breeding:** usually solitary, sometimes with nest-helpers. Nest is in excavated tree-hole. Eggs: 3-11 (rarely 2-14), glossy white. Incubation shared for 9-20 days. Broods: 1, rarely 2. Young are altricial, in nest for 18-25 days. **Food:** chiefly insects extracted from beneath tree bark or ground. Also fruit, seeds, sap and honey depending on species.

PERCHING BIRDS
Passeriformes

Perching land birds with 4 unwebbed, level toes − 3 forward, 1 hind. Toes generally separate. Most fly strongly. Characteristic internal anatomy and spermatozoa. Eggs usually incubated by female but young usually fed by both. Young altricial. Often dimorphic, with males larger and brighter. Many fine songsters. The largest order, with 74 families in 4 suborders.

Eurylaimi

100 Broadbills Eurylaimidae

14 species, 7 genera, e.g. Lesser Green Broadbill *Calyptomena viridis*

Description: small to medium, chunky, rounded wings, wide, coloured bill, wide head, large eyes, strong feet, short or graduated tail. L: 13-28 cm (5-11 in), Wt: 20.5-160 g ($\frac{3}{4}$-5$\frac{1}{2}$ oz). Strikingly coloured − reds, pinks, greens, blues, 1 wattled. Most dimorphic. **Range and habitat:** AF, OR. Tropical forests. **Migration:** non-migratory. **Breeding:** nest is pear-shaped, hanging bag with porched entrance. Eggs: 2-8, colour varied. Little known. **Food:** fruit, insects and small vertebrates.

Tyranni

101 Woodcreepers Dendrocolaptidae

52 species, 13 genera, e.g. Red-billed Scythebill *Campylorhamphus trochilirostris*

Description: small to medium, arboreal. L: 13.5-35 cm (5$\frac{1}{2}$-14 in), Wt: 12-120 g ($\frac{1}{2}$-4 oz). Stiff tail, strong claws, short legs, round wings. Bill short, straight to long, decurved. Plumage dull browns, cryptic, monomorphic. **Range and habitat:** NT. In forest and forest edge. **Migration:** non-migratory. **Breeding:** nest is in tree-holes. Eggs: 2-3,

111 Hooded Pitta

108 Andean Cock of the Rock

107 White-bearded Manakin

109 Sharpbill

106 Eastern Kingbird

110 Chilean Plantcutter

white. Incubation shared for 15-20 days. Young are in nest for c. 20 days. **Food:** invertebrates and small vertebrates taken largely from bark.

102 Ovenbirds Furnariidae
226 species, 37 genera, e.g. Rufous Hornero *Furnarius rufus*

Description: a diverse family. Front toes joined at base. Small to medium, wings short to long, round to pointed. L:10-26 cm (4-10 in), Wt:9-46 g ($\frac{1}{4}$-1$\frac{1}{2}$oz). Bill short to long, slender, pointed. Browns, greys, white, very varied pattern. Monomorphic. **Range and habitat:** NT. A wide variety of forest, scrub and grassland habitats. **Migration:** most species non-migratory. **Breeding:** nest is diverse including domed mud 'ovens', burrows and large twiggy nests. Eggs: 2-3, whitish. Incubation shared for 15-20 days. Young are in nest for 13-18 days. **Food:** invertebrates on ground or in trees.

103 Antbirds Formicariidae
238 species (6●), 52 genera, e.g. Scaled Antpitta *Grallaria guatimalensis*

Description: a diverse family. Front toes joined at base. Small to medium, arboreal or ground foragers. L:8-35 cm (3-14 in), Wt:9-75 g ($\frac{1}{4}$-2$\frac{3}{4}$oz). Wings short, round. Tail short to long. Legs short to long. Bill medium-sized, stout, hook-tipped. Eyes red to yellow. Soft plumage generally drab but diverse patterns, many have white patches exposed in disputes. Dimorphic. **Range and habitat:** NT. A wide variety of forest and scrubland habitats. **Migration:** non-migratory. **Breeding:** nest is an open cup in bush. Eggs: 2 (rarely 3), white, spotted. Incubation shared for 14-20 days. Young are in nest for 9-18 days. **Food:** fruit and small animals.

104 Gnateaters Conopophagidae
8 species, 1 genus, e.g. Black-bellied Gnateater *Conopophaga melanogaster*

Description: small, compact; short, round wings. L:11-14 cm (4$\frac{1}{2}$-5$\frac{1}{2}$in), Wt:20-3 g ($\frac{3}{4}$-1 oz). Long toes, front 3 joined at base, legs long and thin. Tail very short. Bill short. Plumage generally browns, drab, soft. Dimorphic. **Range and habitat:** NT. Forest undergrowth. **Migration:** non-migratory. **Breeding:** little known. Nest is a cup. Eggs: 2 (rarely 3), yellowish, spotted. Incubation shared. **Food:** insects.

105 Tapaculos Rhinocryptidae
30 species, 12 genera, e.g. Black-throated Huet-huet *Pteroptochos tarnii*

Description: family defined by nostril covered by movable flap. Small to medium, ground-dwelling, walk fast, legs and feet long, strong. L:11-25 cm (4$\frac{1}{2}$-10 in), Wt:c. 20 g (c. $\frac{3}{4}$oz). Fly rarely. Wings short, rounded; tail short to long, often held erect. Bill short, stout. Plumage soft, loose, drab red-browns. Monomorphic. **Range and habitat:** NT. Dense forest and scrubland undergrowth. **Migration:** non-migratory. **Breeding:** little known. Nest is on ground, domed cup or in tree-hole. Eggs: 2-4, white. Incubation shared. **Food:** chiefly invertebrates, some plant matter.

106 Tyrant Flycatchers Tyrannidae
396 species, 91 genera, e.g. Eastern Kingbird *Tyrannus tyrannus*

Description: ecologically diverse. Small to medium-sized, L:6-50 cm (2$\frac{1}{4}$-20 in), Wt:4.5-80 g ($\frac{1}{4}$-2$\frac{3}{4}$oz). Bill broad, flat, hook-tipped, rictal bristles. Large head, sometimes crested. Wings short to long, pointed. Tail medium-sized, sometimes forked. Legs and feet generally short, weak. Usually greens, browns and greys. Most monomorphic. **Range and habitat:** NA, NT. A wide range of terrestrial habitats. **Migration:** temperate species are migratory. **Breeding:** nest is diverse, includes cups (sometimes domed), purses in diverse sites. Eggs: 3-4 (rarely 2-8), whitish. All care by female for 14-20 days. Young are in nest 14-23 days. **Food:** aerial insects by sallying, tropical fruits and small vertebrates.

107 Manakins Pipridae
51 species, 17 genera, e.g. White-bearded Manakin *Manacus manacus*

Description: small, L:9-19 cm (3$\frac{1}{2}$-7$\frac{1}{2}$in), Wt:10-25 g ($\frac{1}{4}$-1 oz). Front toes partly joined at base. Wings short, tail short to long, bill short, broad, hook-tipped. Males bright black, white, reds, blues and yellows; females greenish. **Range and habitat:** NT. Tropical forest. **Migration:** non-migratory. **Breeding:** polygynous, sometimes leks. Nest is a frail cup. Eggs: 2. All care by female for 17-21 days. Young are in nest 13-15 days. **Food:** insects and fruit.

108 Cotingas Cotingidae
61 species (4●), 25 genera, e.g. Andean Cock of the Rock *Rupicola peruviana*

Description: small to medium, most arboreal. L:9-46 cm (3$\frac{1}{2}$-18 in), Wt:6-400 g ($\frac{1}{4}$-14 oz). Bill often flattened, short to long. Legs short, feet large, 2 front toes joined at base, wings and tail short to long. Some crested, wattled, very diverse patterns and colours, monomorphic or dimorphic, females duller. Some with powder downs. In the Passeriformes only these and the Artamidae have such feathers. **Range and habitat:** NT. Forest at all levels. **Migration:** non-migratory. **Breeding:** little known, many polygynous, ritualized courtship. Nest is varied, usually cupped. Eggs: 1-3, buff or olive, spotted. Incubation by female for 19-28 days. Young in nest for 21-44 days. **Food:** insects and fruit.

109 Sharpbill Oxyruncidae
Oxyruncus cristatus (monotypic). See p 132.

Description: small, close to Tyrannidae but bill conical, sharp. L:16.5-18 cm (6$\frac{1}{2}$-7 in). Front toes joined at base. Nostrils elongated, covered by flap. Rictal bristles replaced by fine bristly feathers. Monomorphic. Ecologically rather like tits. **Range and habitat:** NT. Rainforest. **Migration:** probably non-migratory. **Breeding:** little known. Nest first described in 1980: cup on branch. Eggs: Probably 2. Incubation probably by female for 14-24 days. Young are in nest for 25-30 days. **Food:** small fruits and invertebrates.

110 Plantcutters Phytotomidae
3 species, 1 genus, e.g. Chilean Plantcutter *Phytotoma rara*

Description: small, bill short, strong, conical, serrated. L:17-20 cm (6$\frac{1}{2}$-8 in). Legs short, large feet, front toes partly joined. Wings short,

112 Rifleman

113 Velvet Asity

116 Shore Lark

114 Superb Lyrebird

115 Rufous Scrubbird

117 Swallow

pointed, tail longish. Plumage grey or brown, reddish below. Eyes bright yellow or amber. Dimorphic. **Range and habitat:** western and southern NT. Scrub, low woodland and cultivation. **Migration:** partly migratory. **Breeding:** nest is a flat, untidy platform of root fibres. Eggs: 2-4, blue-green, spotted brown. Incubation by female. **Food:** a range of plant matter including young leaves, buds, seeds and fruit.

111 Pittas Pittidae
29 species, 1 genus, e.g. Hooded Pitta *Pitta sordida*

Description: small, compact, short necked, ground-dwelling but roost in trees. L:15-28 cm (6-11 in), Wt:42-218 g ($1\frac{1}{2}$-$7\frac{1}{2}$ oz). Fly strongly. Wings short, rounded. Tail very short. Long legs, large feet. Stout, slightly decurved, hook-tipped bill. Plumage loose, soft and brightly coloured. Colours vary but are often contrasting in bold patches and brighter on underparts. Dimorphic. **Range and habitat:** AF, OR, AU. Chiefly tropical forest and scrublands. **Migration:** some migratory species. **Breeding:** nest is a huge, untidy, domed cup of twigs, lined. Eggs: 3-5 (rarely 2-7), whitish, spotted. Incubation shared for 15-17 days. Young in nest for 14-21 days. **Food:** small animals on ground.

112 New Zealand Wrens Xenicidae
3 species, 2 genera, e.g. Rifleman *Acanthisitta chloris*

Description: very small, compact. L:8-10 cm (3-4 in), Wt:6.3-9 g ($\frac{1}{4}$ oz). Fly weakly but run around logs, tree roots, rocks. Wings short, rounded. Tail very short. Long legs with large feet. Bill medium-sized, very thin, pointed. Plumage soft, brownish or green, yellow, white. Monomorphic, males brighter; or dimorphic when male smaller (unusual in Passeriformes). **Range and habitat:** New Zealand (AU). Forest. **Migration:** non-migratory. **Breeding:** nest is a dome of loosely woven, abundant material, side entrance. Eggs: 2-5 (20% female wt), white, laid at 2-day intervals. Incubation shared for 19-21 days. Young are in nest for 24 days, slow growth rate. **Food:** invertebrates and some fruit.

113 Asities Philepittidae
4 species (1●), 2 genera, e.g. Velvet Asity *Philepitta castanea*

Description: small, compact, arboreal. L:9-16.5 cm ($3\frac{1}{2}$-$6\frac{1}{2}$ in). Wings and tail short, rounded. Strong flight. Legs and feet short and weak or long and strong. Bill various, generally pointed, stout, broad-based, medium-sized and slightly decurved. Plumage soft. Dimorphic: males blue and yellow or black with yellow on wing. Orbital ring area bare or with extraordinary wattle in breeding season. Females greenish. **Range and habitat:** Madagascar (AF). Forests. **Migration:** non-migratory. **Breeding:** largely unknown. The nest of the Velvet Asity is pear-shaped, suspended. Eggs: 3, white. **Food:** (all species) fruit and some insects.

Menurae

114 Lyrebirds Menuridae
2 species, 1 genus, e.g. Superb Lyrebird *Menura novaehollandiae*

Description: large, ground-dwelling, fast runners, poor flight but roost in trees. L: male: 89-100 cm (35-9 in), Wt:1.15 kg (2 lb 9 oz); female:

76-86 cm (30-4 in), 950 g (2 lb 1 oz). Solitary. Tail long and train-like, of 3 types of highly modified feathers in male; simpler in female. Legs long, feet large with long claws. Bill pointed, sharp and conical. Plumage dark brown above with bluish orbital ring, underparts paler. Highly dimorphic. Good mimics. **Range and habitat:** AU. Montane forest. **Migration:** non-migratory. **Breeding:** males polygynous, complex courtship and mating system. Nest is massive lined chamber of twigs up to 22 m (72 ft), Wt:c. 14 kg (30 lb 13 oz). Eggs: 1, grey, mottled, Wt:49-72 g ($1\frac{3}{4}$-$2\frac{1}{2}$ oz). Incubated by female for 50 days. Young are in nest for 47 days, all care by female but independent after 8 months. **Food:** a wide range of invertebrates from soil and rotting wood.

115 Scrubbirds Atrichornithidae
2 species, 1 genus, e.g. Rufous Scrubbird *Atrichornis rufescens*

Description: small to medium, ground-dwelling, fast runners with tail erect, but flight very poor and rarely seen. L:16-21 cm (6-8 in), Wt: 30-50 g (1-$1\frac{3}{4}$ oz). None-the-less it roosts in trees. Bill medium-sized, strong, pointed, curved culmen. Wings very short, rounded. Tail long, graduated. Legs and feet large and strong. Plumage rufous above, paler below, finely barred. Monomorphic, male larger. Good mimics. **Range and habitat:** AU. Forest edge and scrubland. **Migration:** non-migratory. **Breeding:** monogamous (may be polygamous in captivity). Nest is a loosely woven domed cup on or near ground. Eggs: 1, pinkish. All care by female. Incubation for 36-8 days. Young are in nest for 25 days. Stay with female till following year. **Food:** chiefly invertebrates, some small vertebrates and seeds.

Oscines (Songbirds)

116 Larks Alaudidae
79 species, 18 genera, e.g. Shore Lark *Eremophila alpestris*

Description: small, ground-dwelling. Legs and toes longish, claws long and straight. Back of tarsus rounded and with scales. L:12-24 cm ($4\frac{1}{2}$-$9\frac{1}{2}$ in), Wt:15-75 g ($\frac{1}{2}$-$2\frac{3}{4}$ oz). Wings pointed, innermost secondaries very long (sometimes as long as primaries). Tail medium-sized. Bill medium-sized, stout, pointed. Plumage generally variously streaked browns and greys, cryptic, white outer tail feathers. Underparts generally paler. Head often with small crest. Monomorphic, males larger. Excellent songsters, many singing on the wing in towering song-flights. **Range and habitat:** WP, EP, AF, OR, AU, NA and NT (1 species). A variety of open habitats including shorelines. **Migration:** many strong migrants. **Breeding:** nest is a cup on ground, some domed. Eggs: 2-6, speckled. Incubation by female for 11-16 days. Broods: 1-3. Young leave nest before they can fly at about 10 days. **Food:** seeds and invertebrates.

117 Swallows and Martins Hirundinidae
81 species (1●), 16 genera, e.g. Swallow *Hirundo rustica*

Description: small, slender birds superbly adapted to an aerial lifestyle, hence superficially resemble swifts (Apodidae). L:11.5-21.5 cm ($4\frac{1}{2}$-8 in), Wt:10-55 g ($\frac{1}{3}$-2 oz). Wings long, pointed. Tail short to long, forked, often with outer tail streamers. Legs and feet very small, reduced, but still used

121 Blue-backed Fairy Bluebird
120 Garden (Yellow-vented) Bulbul
118 Yellow Wagtail
123 Blue Vanga
122 Great Grey Shrike
119 Ground Cuckoo Shrike

for perching. Bill short, flattened, pointed, broad-based with rictal bristles. Plumage usually darker above. Most more or less monomorphic, males larger with longer tail streamers when present. Flight rapid and agile, much time spent on the wing. Gregarious with twittering or buzzing songs and contact calls. **Range and habitat:** worldwide except high latitudes. Diverse. A wide range of open habitats and over forests and streams. **Migration:** many strong migrants, particularly in temperate latitudes. Traditional herald of spring in northern temperate regions. Diurnal migrants. **Breeding:** often colonial. Nest is a variety of mud hive-like structures cemented to a rock overhang, with or without entrance passage, or mud and straw cup. Several species nest on buildings or in tree-holes or burrows in sand cliff. Eggs: 3-8 (rarely 1-2), white or white with red spots. Incubation by female or shared for 13-16 days. Broods: 1-3. Young are in nest for 16-24 (rarely 28) days. **Food:** aerial insects caught on the wing.

118 Wagtails and Pipits Motacillidae

58 species, 6 genera, e.g. Yellow Wagtail *Motacilla flava*

Description: small, slender-bodied, mostly ground-dwelling birds with long tails and longish legs. L: 12.5-22 cm (5-8½ in), Wt: 12-50 g (½-1¾ oz). Toes medium-sized, hind toe often long, nails short to long. Walk, never hop. Bill medium-sized, thin, pointed. Tail usually with paler outer feathers. Wings medium-sized, pointed. Flight strong and undulating. Plumage contrasting, boldly marked: black, white, yellow, blue, greys in wagtails, dimorphic; in pipits monomorphic, variously streaked browns. Many are gregarious and some form large communal roosts. **Range and habitat:** worldwide except PA and AN. A variety of open habitats including shorelines. **Migration:** many strong or partial migrants. **Breeding:** nest is an open cup on ground, in tree or rock cavity. Eggs: 2-7, white, grey or brown, speckled. Incubation by female (pipits) or shared (wagtails) for 12-20 days. Broods: 1-2. Young are in nest for 12-20 days. **Food:** chiefly insects, some small snails, seeds.

119 Cuckoo Shrikes Campephagidae

76 species (2●), 9 genera, e.g. Ground Cuckoo Shrike *Pteropodocys maxima*

Description: small to medium, solitary or gregarious, chiefly arboreal birds. L: 14-36 cm (5½-14 in), Wt where known: 20-111 g (¾-4 oz). Bill medium-sized, stout, slightly decurved, notched and hook-tipped, nostrils partly hidden by bristles. Legs short with weak feet. Wings medium-sized, pointed. Flight strong but unsustained. Tail long, graduated or rounded. Plumage soft. In males chiefly light and dark grey, black, white or black and red. Females paler or browner. Feathers of back and rump thickly matted, spiny and readily lost, possibly in defence. 1 species has seasonally varying plumages achieved by 2 moults each year. **Range and habitat:** EP, AF, OR, AU. Dense forest and some forest edge habitats. **Migration:** most non-migratory. **Breeding:** known for few species. Nest is a shallow cup of fine plant fibres, high in a fork or on a branch of a tree. Eggs: 2-5, pale green, marked with brown, grey or purple. Incubation by female or shared for 14-23 days. Young are in nest for 12-25 days depending on species. **Food:** insects, fruit and small vertebrates depending on species.

120 Bulbuls Pycnonotidae

119 species, 15 genera, e.g. Garden (or Yellow-vented) Bulbul *Pycnonotus barbatus*

Description: small to medium, generally gregarious, arboreal birds, either secretive and skulking (forests) or noisy and conspicuous (gardens). L: 15-28 cm (6-11 in), Wt: 20-65 g (¾-2¼ oz). Bill medium-sized, slender, pointed and slightly decurved with well-developed rictal bristles. Legs medium-sized. Head sometimes crested and with brightly coloured eyes. Wings short, concave. Tail medium to long. Plumage soft, long and fluffy especially on lower back. Nape feathers hair-like. Usually browns, greys, yellow or greens, sometimes with white, yellow or red patches on head and undertail-coverts. Most monomorphic, males sometimes brighter. **Range and habitat:** AF, OR. Forest and scrub, also some species have occupied cultivated and suburban habitats. **Migration:** northern species are migratory. **Breeding:** males may lek, may have helpers at the nest. Nest is shallow cup in tree fork, often lined. Eggs: 2-5 (rarely 1), pink or white, marked purple, brown or red. Incubation usually shared for 11-14 days. Young are in nest for 14-18 days. **Food:** fruit and buds, some insects.

121 Leafbirds and Ioras Irenidae

14 species, 3 genera, e.g. Blue-backed Fairy Bluebird *Irena puella*

Description: small to medium, arboreal, gregarious birds resembling bulbuls (Pycnonotidae) in fluffy texture of plumage, tendency to shed feathers on handling and presence of hair-like nape feathers. L: 14-27 cm (5½-10½ in, Wt: 10-90 g (¼-3¼ oz). Bill medium to long, pointed, decurved. Legs short, strong or slender. Wings short, rounded, but flight fast. Tail medium-sized, square or rounded, sometimes with long uppertail-coverts. Plumage generally brighter than bulbuls, blues, greens, yellows with contrasting black patches on throat or wings. Dimorphic, males brighter. **Range and habitat:** OR. A range of forest and scrubland habitats. **Migration:** non-migratory. **Breeding:** little known, some with elaborate courtship. Nest is an open cup of plant fibres or rough platform of sticks in tree. Eggs: 2-3, pinkish or pale green marked with red and purple. Incubation by female or shared. **Food:** fruit, buds, seeds, nectar and some insects.

122 Shrikes Laniidae

83 species (2●), 13 genera, e.g. Great Grey Shrike *Lanius excubitor*

Description: a large family containing a number of distinct groups. Small to medium, solitary (gregarious in helmet shrikes), arboreal birds with large heads. L: 15-38 cm (6-15 in), Wt: 10-87 g (¼-3 oz). Sharp, stout, notched and hooked bills with partly covered nostrils and well-developed rictal bristles; some have weaker bills. Aggressive predators analogous to raptors. Legs medium-sized, strong claws sharp. Wings rather rounded, tail medium to long. Flight strong. Plumage varied, often soft, greys and browns with black and white markings on wings and

126 Dipper

124 Waxwing

129 Alpine Accentor

127 Wren

125 Palm Chat

128 Northern Mockingbird

head, or bright reds, greens and yellows. Pronounced dimorphism to monomorphic. **Range and habitat:** WP, EP, AF, OR, NA. A wide range of forest, savannah and savannah woodland, grassland and cultivated habitats. **Migration:** some strongly migratory species, especially in northern latitudes. **Breeding:** nest is a varied, open cup of twigs and finer material, often lined. Eggs: 2-7, white, spotted. Incubation by female (sometimes shared) for 12-18 days. Broods: 1-2. Young are in nest for 16-20 days. **Food:** insects and other invertebrates, also small vertebrates depending on species.

123 Vanga Shrikes Vangidae
13 species (3●), 10 genera, e.g. Blue Vanga *Leptopterus madagascarinus*

Description: a poorly known group of small to medium, rather gregarious, arboreal shrike-like birds. L: 13-32 cm (5-13 in). Varied bill-shapes, bill of most is relatively massive, notched and hooked. In 1 it is thin and curved for probing, in several it is high, laterally compressed or inflated. In bill form they are a good example of adaptive radiation in isolation. Otherwise like shrikes but wings and tail rather long. Monomorphic or dimorphic. **Range and habitat:** Madagascar (AF). Forest, scrublands and mangrove swamps. **Migration:** non-migratory. **Breeding:** little known, nest-helpers may be present. Nest is a cup of twigs in a tree. Eggs: 2-3 (rarely 4), white or pale green, heavily spotted. Incubation probably shared. **Food:** chiefly insects.

124 Waxwings and allies Bombycillidae
8 species, 5 genera, e.g. Waxwing *Bombycilla garrulus*

Description: small to medium, highly gregarious, arboreal birds, most showing adaptations for fruit-eating. L: 8-24 cm (3-9½ in), Wt: 40-70 g (1½-2½ oz). Bill short, stout, slightly hooked with very wide gape. Legs and feet short, stout. Head often crested. Wings short to long, rounded to pointed. Tail short, square to long. Plumage soft, silky, vinaceous to greyish browns, yellow, black, red or greys. More or less monomorphic or highly dimorphic (Silky Flycatcher). **Range and habitat:** largely temperate WP, EP, NA, OR. Forest and scrubland habitats. **Migration:** some species irruptive. **Breeding:** solitary or loosely colonial. Nest is a cup of twigs and moss, finely lined, usually in conifer or thorn bush. Male may help in construction. Eggs: 5 (rarely 3-7), grey-blue finely spotted with black. Incubation shared for 12-18 days. Broods: 1. Young are in nest for 19-25 days, fed by regurgitation. **Food:** (waxwings) the most frugivorous northern temperate birds. Fruit, buds and some insects or (silky flycatchers) insects caught by spectacular sallying flights.

125 Palm Chat Dulidae
Dulus dominicus (monotypic).

Description: medium-sized, noisy, gregarious, arboreal species closely related to waxwings (Bombycillidae). L: 20 cm (8 in). Upperparts olive, underparts buff, heavily spotted. **Range and habitat:** Hispaniola and Gonave Islands in West Indies (NT). Open woodlands and cultivation. **Migration:** non-migratory. **Breeding:** the nest is a large, communal,

compartmented structure of twigs, lined with grass, in palm or pine trees. Eggs: 2-4, white, spotted grey. **Food:** fruits and flowers.

126 Dippers Cinclidae
5 species, 1 genus, e.g. Dipper *Cinclus cinclus*

Description: the only fully aquatic Passerine family, showing remarkably few obvious adaptations to this lifestyle. Small, solitary (but may roost communally), compact, short-necked birds. L: 15-17.5 cm (6-7 in), Wt: 60-80 g (2-2¾ oz). Bill slender, straight, compressed with slit-like nostrils. Legs medium length, with strong feet and claws. Often bobs while perched on rocks by water. Wings short, pointed, used in swimming underwater. Flight strong, direct. Tail short, square. Plumage soft and dense with thick layer of down. Uniform brown, grey or black, often with large white patch on throat, breast, head or back. Monomorphic. **Range and habitat:** WP, EP, OR, western NA and NT. Fast-flowing streams, usually in mountains. **Migration:** non-migratory. **Breeding:** nest is a large domed structure with side entrance of moss generally under ledge over running water. Eggs: 5 (rarely 4-6), white. Incubation by female for 16-17 days. Broods: 2. Young are in nest for 19-25 days. **Food:** chiefly aquatic insects for which it dives. Also other invertebrates, some small fish and tadpoles.

127 Wrens Troglodytidae
69 species, 17 genera, e.g. Wren *Troglodytes troglodytes*

Description: small, highly territorial, aggressive, compact birds. Most L: 7.5-12.5 cm (3-5 in), Wt: 8-15 g (¼-½ oz); 1 larger L: 16-22 cm (6½-8½ in). Bill long, thin, pointed. Legs rather long, feet strong. Wings and tail short, rounded. Tail often cocked. Flight strong, direct. Plumage soft, generally browns and greys, barred and spotted. Monomorphic, males larger. Polygynous species with very loud, complex songs. Monogamous species with simple buzzing songs. **Range and habitat:** WP, EP, NA, NT. Generally dense, low vegetation in a wide variety of habitats. **Migration:** some northern species migratory. **Breeding:** monogamy and polygyny. Nest is a domed structure with side entrance of twigs and grass suspended in vegetation, or in a cavity, may be bulky, usually constructed by male who may make several nests from which the female chooses. Eggs: 2-11 (fewer in tropics). Incubation by female for 12-20 days. Broods: 1-4 (in polygynous species, male may raise more broods than females). Young are in nest for 12-18 days. **Food:** small invertebrates.

128 Mockingbirds and Thrashers Mimidae
30 species, 12 genera, e.g. Northern Mockingbird *Mimus polyglottos*

Description: medium-sized, slender, arboreal and ground-dwelling, solitary. L: 20-30 cm (8-12 in). Bill medium to long, slender, pointed, straight to decurved. Legs long. Wings short to medium, rounded. Flight mostly rather weak. Tail long, often graduated. Plumage browns, greys, black and white. Paler underparts, often streaked, or uniform, some with black and white wings and tail. Good songsters, many mimic. **Range and habitat:** NA, NT. Scrublands and forest edge. **Migration:** northern

135 Blue-grey Gnatcatcher

132 Chestnut-backed Scimitar Babbler

131 Spotted Quail Thrush

133 Bearded Reedling

130 Swainson's Thrush

134 White-necked Bald Crow

species migratory. **Breeding:** monogamous and aggressively territorial, helpers may be present at the nest. Nest is a lined cup of twigs in dense bush or tree. Eggs: 2-5, varied pale and spotted or streaked. Incubation by female (rarely shared) for 12-13 days. Young are in nest for 12-13 (rarely longer) days. **Food:** insects, fruit and seeds on or near ground.

129 Accentors Prunellidae
12 species, 1 genus, e.g. Alpine Accentor *Prunella collaris*

Description: small, solitary to gregarious, inconspicuous, drably coloured, ground-dwelling birds. L:13-18 cm (5-7 in), Wt:18-26 g ($\frac{3}{4}$-1 oz). Mostly known from Palaearctic species. Biochemical evidence suggests that they may be closest to the weavers (Ploceidae). In general appearance they are rather like sparrows in that family though the bill is longer and more slender. Legs medium-sized, feet weak. Wings medium-sized, rounded to pointed, flight generally rapid. Tail medium-sized, square. Plumage greys, browns, streaked. Monomorphic, males larger. **Range and habitat:** WP, EP, northern AF, OR. Scrub, forest edges, cultivation. Most species at high altitudes. **Migration:** many altitudinal migrants. **Breeding:** diverse and flexible mating systems. Nest is an open cup of fine plant material low in vegetation or rock crevice. Eggs: 3-6, bluish. Incubation by female or shared for 11-15 days. Broods: 2-3. Young are in nest for 12-14 days. Some species parasitized by cuckoos (Cuculidae). **Food:** chiefly insects, also seeds and fruit in winter.

Note: families 130-43 are often treated as subfamilies (Turdinae to Pachycephalinae) of an enlarged and rather unwieldy family Muscicapidae.

130 Thrushes and Chats Turdidae
328 species (5●), 52 genera, e.g. Swainson's Thrush *Catharus ustulatus*

Description: small to medium, arboreal or ground-feeding. L:11-33 cm ($4\frac{1}{2}$-13 in), Wt:8-220 g ($\frac{1}{4}$-8 oz). Solitary when breeding, often gregarious in winter or on migration. Structurally similar. Bill medium to slender, pointed, straight, usually with well-developed rictal bristles. Legs medium to long. Feet strong to weak. Wings rounded to pointed. Flight strong, direct. Tail medium to long, square. Plumage very varied, browns, greys, black, white, reds and yellows. Males larger. Monomorphic or dimorphic in which females generally plain. Many good songsters. **Range and habitat:** worldwide except AN. A wide range of terrestrial habitats from tropical forest to desert. **Migration:** many migratory species, especially in northern latitudes, many nocturnal migrants. **Breeding:** monogamous. Nest is an open cup of grasses and twigs in vegetation, sometimes with mud lining, sometimes a tree-hole or rock cavity, built by female. Eggs: 2-6 (rarely 7), whitish, blue to green or buff, often spotted or streaked. Incubation by female for 12-15 days. Broods: 1-3. Young are in nest for 11-18 days, usually fledge with spotted first plumage. **Food:** invertebrates and fruit, especially in winter.

131 Logrunners Orthonychidae
19 species, 9 genera, e.g. Spotted Quail Thrush *Cinclosoma punctatum*

Description: small to medium, dumpy birds. L:10-30 cm (4-12 in), Wt:20-80 g ($\frac{3}{4}$-2$\frac{3}{4}$ oz). Poorly known because of their secretive ground-dwelling behaviour. Bill medium to stout, straight, pointed. Legs and feet robust. Wings short and rounded, tail medium-sized, some with spiny bare quill at feather tips. Plumage mostly brown, black or white or blue and white. Monomorphic or dimorphic. Noisy in breeding season. **Range and habitat:** OR, AU. Dense forest and scrubland. **Migration:** non-migratory or locally nomadic. **Breeding:** incompletely known. Nest is usually a cup or dome of plant fibres, placed low in vegetation, or a lined scrape. Eggs: 1-3, white or pale blue. Incubation: 17-21 days. Young in nest for 12-14 days. **Food:** invertebrates and seeds.

132 Babblers Timaliidae
255 species, 51 genera, e.g. Chestnut-backed Scimitar Babbler *Pomatorhinus montanus*

Description: a large and diverse assemblage with uncertain relationships. Small to medium, arboreal or ground-dwelling, solitary or highly gregarious. The latter with complex social organization. L:10-35 cm (4-14 in), Wt:5-150 g ($\frac{1}{4}$-5$\frac{1}{4}$ oz). Bill slender but strong, often decurved and with rictal bristles sometimes prominent. Legs medium to long, strong. Feet stout. Wings short, rounded. Flight rather weak. Tail medium-sized, square to long, graduated. Plumage soft, usually drab browns, also white, and a few boldly patterned. Most monomorphic. **Range and habitat:** AF, OR, AU, western NA (1 species). A very wide range of terrestrial habitats. **Migration:** most species non-migratory. **Breeding:** solitary or in groups. In the latter, helpers often present at the nest. Nest is an open cup or domed with side entrance in diverse sites. Eggs: 2-6, usually white or blue. Incubation by female or shared for 14-15 days. Young are in nest for 13-16 days. **Food:** invertebrates.

133 Parrotbills Panuridae
19 species, 3 genera, e.g. Bearded Reedling *Panurus biarmicus*

Description: small to medium, lively, gregarious birds especially outside breeding season. L:9-29 cm ($3\frac{1}{2}$-11$\frac{1}{2}$ in), Wt:5-36 g ($\frac{1}{4}$-1$\frac{1}{4}$ oz). Bill short, stout, decurved and laterally compressed, nostrils covered by bristles. Legs short, feet strong, 1 species unique among Passerines in having 3 toes only (outer toe is a stump syndactyl with the middle toe). Wings short, rounded. Flight strong, direct but some species fly reluctantly. Tail long, graduated. Plumage varied, usually shades of grey-brown with paler or white underparts, some with black on head, patterns sometimes complex. Dimorphic or monomorphic. **Range and habitat:** WP, EP, OR. Swamps, grass and bamboo thickets. **Migration:** non-migratory or only local movements. **Breeding:** little information. Nest is a deep cup of grass and hair bound with cobwebs, placed low in vegetation or reeds. Eggs: 2-7, pale, spotted red-brown. Incubation shared for 12-13 days. Broods: 2. Young in nest for 9-12 days. **Food:** insects, small seeds and fruits.

134 Bald Crows Picathartidae
2 species, 1 genus, e.g. White-necked Bald Crow *Picathartes gymnocephalus*

140 Little (Yellow) Thornbill

136 Rüppell's Warbler

137 Spotted Flycatcher

138 Brown-throated Wattle-eye

141 Pied Flycatcher

139 Variegated Wren

Description: little is known about these extraordinary birds. Our understanding of their natural history owes much to the descriptions of native Africans as few ornithologists have witnessed them at first hand. Medium-sized, L: 39-50 cm (15-20 in). Bill is long, strong, and crow-like. Legs long. Head bare and coloured red, black and blue or black and yellow. Upperparts black or grey with black wings. Underparts white or grey and white. Wings rounded, tail long, graduated. Monomorphic. Chiefly ground-dwelling, weak flight, rather quiet though gregarious. **Range and habitat:** western AF. Rainforests up to 2,100 m (6,900 ft), especially in caves. **Migration:** non-migratory. **Breeding:** little information available, colonial, possibly with nest-helpers. Nest is a massive lined cup of mud cemented to rock faces 2-4 m (6½-13 ft) high in caves, or in rock crevice. Eggs: 2, profusely spotted and blotched in brown and grey. Incubation shared by male and female. **Food:** chiefly invertebrates, also frogs near streams.

135 Gnatcatchers Polioptilidae
13 species, 3 genera, sometimes treated within the Sylviidae, e.g. Blue-grey Gnatcatcher *Polioptila caerulea*

Description: tiny, restless, arboreal and solitary birds. L: 10-12 cm (4-4½ in), Wt: less than 7 g (¼ oz). Long tails and slender, pointed bills. Legs long, thin. Wings pointed. Plumage mostly blue-grey or brown upperparts, white underparts. Some with black on head, many with white outer tail feathers. More or less monomorphic. **Range and habitat:** NA, NT. Forest understorey, open forest and semi-arid habitats. **Migration:** 1 migratory species, most non-migratory. **Breeding:** nest is an open cup of plant fibres bound with spiders' web, attached at base and side. Eggs: 3-5. Incubation shared for 13-15 days. Broods: 2 (may overlap in time). Young are in nest for 11-14 days. **Food:** chiefly insects gleaned from leaves.

136 Old World Warblers Sylviidae
362 species (4●), 60 genera, e.g. Rüppell's Warbler *Sylvia rueppelli*

Description: very small to medium. L: 9-16 cm (3½-6½ in), Wt: 5-20 g (¼-¾ oz) mostly. Bill medium-sized, pointed. Legs medium-sized, with strong or weak feet. Wings short, rounded to long, pointed. Tail medium to long, square to rounded. Plumage browns, greys, green or yellow, often streaked darker, some with black on head. Males usually larger. Generally monomorphic, some dimorphic. Many good songsters. **Range and habitat:** WP, EP, AF, OR, AU. A variety of forest, scrub and marshland habitats depending on species. **Migration:** many strongly migratory species, especially WP to AF and EP to OR. Nocturnal migrants. Some non-migratory. **Breeding:** monogamous or more rarely polygynous. Nest is elaborate, cupped or domed, usually placed low in dense vegetation. Eggs: 2-7, semi-cryptic. Incubation by female or shared for 12-14 days. Broods: 1-2. Young are in nest for 11-15 days. Some species heavily parasitized by cuckoos (Cuculidae). **Food:** chiefly insects, some fruit in autumn/winter.

137 Old World Flycatchers Muscicapidae
108 species (2●), 9 genera, e.g. Spotted Flycatcher *Muscicapa striata*

Description: small, solitary, arboreal birds analogous to the tyrant flycatchers (Tyrannidae) of the New World. L: 10-21 cm (4-8 in), Wt: 6-20 g (¼-¾ oz). Head and eyes rather large. Bill flattened, slightly hook-tipped, broad-based with well-developed rictal bristles. Legs short. Feet small and weak. Wings short to long and pointed. Flight strong, agile. Tail short to medium, square. Plumage varied, uniform browns to pied, bright blue, yellow or red. Males often larger. Monomorphic to strongly dimorphic. **Range and habitat:** WP, EP, AF, OR, AU, PA. Forest and scrublands. **Migration:** many strong migrants, nocturnal migrants. **Breeding:** monogamy and polygyny. Nest is an open cup on branch or in tree-hole or crevice. Eggs: 2-6 (fewer in tropics), whitish, greenish or buff, spotted. Incubation by female or shared for 10-15 days. Broods: 1-2. Young are in nest for 10-15 days. **Food:** insects taken by sallying (especially larger species) and gleaning.

138 Wattle-eyes and Puff-back Flycatchers Platysteiridae
31 species, 4 genera, often treated within the Muscicapinae, e.g. Brown-throated Wattle-eye *Platysteira cyanea*

Description: tiny to small, arboreal, solitary birds. L: 8-16 cm (3-6½ in), Wt: c. 5-20 g (¼-¾ oz). They differ in several respects from the Muscicapidae. The bill is similar but sometimes stouter. Bill slightly hook-tipped, and with prominent rictal bristles. The back feathers are often long and fluffy and raised in alarm. Several species show blue or red wattles around the eyes. Females tend to differ more between species than males and are believed in some species to take a more dominant role in courtship. Legs, feet and wings as Muscicapidae, tail very short to medium, square. Plumage generally contrasting blues, black or violet upperparts, white underparts, often with red or orange breast band. Some with white wing patches. Generally dimorphic, immature plumage sometimes distinct. **Range and habitat:** AF. A wide variety of forest, forest edge and savannah habitats. **Migration:** non-migratory. **Breeding:** little known. Nest is a small cup of plant fibres and cobwebs on a branch. Eggs: 2-3, pale green to blue, spotted or blotched grey, brown, violet or red. Incubation shared for 12-14 days. Young are in nest for 12-14 days. Some species are heavily parasitized by honeyguides (Indicatoridae) and cuckoos (Cuculidae). **Food:** chiefly insects.

139 Australasian Wrens Maluridae
24 species (3●), 4 genera, e.g. Variegated Wren *Malurus lamberti*

Description: small, active, gregarious birds with long tails carried cocked over back. L: 14-22 cm (5½-8½ in), Wt: 7-37 g (¼-1¼ oz). Bill generally thin, pointed, slightly decurved, weak. Legs medium-sized, thin. Wings short, rounded. Tail feathers may be sparsely barbed or 'decomposed'. Highly dimorphic in breeding season. Male plumage duller outside this time. Plumage basically browns, rufous and white. Breeding males often bright metallic blues, purples, red, chestnut and white. Females may also show blue. **Range and habitat:** AU. Forest and scrublands. **Migration:** non-migratory. **Breeding:** may be helped by young from earlier broods, and by unpaired males. Nest is usually domed with porched side entrance, of plant fibres, sometimes bulky; or a cup.

41

143 Golden Whistler

144 Long-tailed Tit

146 Crested Tit

142 Lemon-breasted Flycatcher

145 Penduline Tit

Eggs: 2-5, white speckled red-brown. Incubation by female for 12-15 days. Broods: 2, may overlap. Young are in nest for 10-12 days. Some species are heavily parasitized by cuckoos (Cuculidae). **Food:** small insects taken by gleaning.

140 Thornbills and Flyeaters Acanthizidae
71 species, 17 genera, e.g. Little (Yellow) Thornbill *Acanthiza nana*

Description: tiny to small, restless, arboreal to semi-terrestrial, solitary or gregarious birds often living in extended family groups or clans. L:8-20 cm (3-8 in), Wt:7-40 g ($\frac{1}{4}$-1$\frac{1}{2}$oz). Bill short to medium, stout to thin. Legs medium-sized, thin. Wings short, round to pointed. Tail medium-sized, broad and rounded. Plumage variously patterned, mostly browns, greys, greens, yellows, often with white in the tail. Monomorphic. **Range and habitat:** AU, PA. Wide variety of forest and arid scrubland habitats. **Migration:** most non-migratory, a few nomadic. **Breeding:** little known. Monogamous, some helped at the nest by failed breeders. Nest is a neat dome with side entrance, often porched, of plant fibres. Eggs: 2-3 (rarely 4), white, sometimes spotted. Incubation is for 15-20 days. Young are in nest for 15-20 days. **Food:** insects and seeds.

141 Monarchs and Fantails Monarchidae
133 species (3●), 20 genera, e.g. Pied Flycatcher *Arses kaupi*

Description: small, striking birds. L:12-30 cm (4$\frac{1}{2}$-12 in), Wt:15-40 g ($\frac{1}{2}$-1$\frac{1}{2}$oz). Solitary or paired, arboreal. Bill medium-sized, often flattened, notched and with a wide gape and small hook at tip. Rictal bristles prominent. Legs and feet small, weak. Wings medium to long, pointed. Tail square or graduated, medium to very long. Plumage often metallic black or grey, or chestnut with white underparts. Sometimes crested, frilled or wattled. Most are strongly dimorphic. **Range and habitat:** AF, OR, AU. A variety of forest habitats. **Migration:** non-migratory. **Breeding:** little known. Some with elaborate courtship displays. Nest is a small cup of plant fibres bound by cobwebs on branch or in fork, built by female or both. Eggs: 2-4, white, spotted brown. Incubation by female or shared for 14-17 days. Young are in nest for 12-18 days. **Food:** insects flushed from vegetation or by sallying, also some fruit.

142 Australasian Robins Eopsaltridae
39 species, 11 genera, often treated within Muscicapinae, e.g. Lemon-breasted Flycatcher *Microeca flavigaster*

Description: small, usually solitary, arboreal birds, probably so named as they reminded early settlers of the European robin: they are plump little birds often with red underparts and take their food from the ground. L:10-16.6 cm (4-6$\frac{1}{2}$ in). Characteristic upright posture with occasional flicks of drooping tail. Bill medium-sized, broad-based, flattened, slightly hook-tipped and with rictal bristles. Legs medium-sized, weak. Wings medium to long, pointed. Flight agile. Tail medium to long, square. Plumage black, brown or greenish upperparts, red, pink, white, yellow or grey underparts. Varying amounts of white on wings and tail. More or less monomorphic, females more drab and less variable. **Range and habitat:** AU, PA. A variety of scrubland and forest habitats.

Migration: many partial migrants. **Breeding:** nest is a tiny shallow, lined cup of moss bound with cobwebs on a branch up to 20 m (65 ft) and sometimes decorated. Eggs: 2-4, pale blue or green, spotted with red-brown to violet. Incubation by female for 12-14 days. **Food:** chiefly insects and their larvae caught on ground by dropping from a perch (robins) or by sallying (flycatchers).

143 Whistlers Pachycephalidae
52 species, 12 genera, e.g. Golden Whistler *Pachycephala pectoralis*

Description: small to medium, solitary or gregarious, arboreal, often conspicuous. L:12-25 cm (4$\frac{1}{2}$-10 in), Wt:15-100 g ($\frac{1}{2}$-3$\frac{1}{2}$oz). Head rounded, heavy with chunky body. A few are wattled. Bill medium-sized, stout, straight to decurved, some hook-tipped and with or without rictal bristles. Legs medium-sized, strong. Wings medium to long, pointed. Flight strong, undulating. Tail medium-sized, square. Plumage generally dull rufous browns and greys, with some green, white and black. Some with bright yellow underparts. Monomorphic or dimorphic, in most the male is brighter. Juvenile plumage sometimes distinct. Many with explosive songs. **Range and habitat:** AU, PA. Forest, scrublands, savannah and mangroves. **Migration:** some migratory species. **Breeding:** little known. Nest is a cup of sticks, plant fibres and cobwebs, in a tree-fork or cavity. Eggs: 2-4, glossy white to pinkish, spotted brown and grey. Incubation shared for c.17 days. Broods: 1-3. Young are in nest for 13-16 days. **Food:** insects and fruit.

144 Long-tailed Tits Aegithalidae
8 species, 3 genera, e.g. Long-tailed Tit *Aegithalos caudatus*

Description: tiny birds, 66% of whose length is tail. L:9-14 cm (3$\frac{1}{2}$-5$\frac{1}{2}$ in), Wt:5-7 g ($\frac{1}{4}$oz). Recent evidence suggests that they are most closely related to babblers (Timaliidae). Gregarious with complex social organization, often in large flocks. Arboreal, moving with agility through vegetation. Bill short, stout, conical and laterally compressed. Legs and feet small. Wings medium-sized. Flight undulating, light. Tail long, graduated, especially in juveniles. Plumage soft, black, greys, pinks, white and browns. Upperparts darker than underparts. Monomorphic, males slightly larger. **Range and habitat:** WP, EP, OR, NA. Forest, forest edge and scrub habitats. **Migration:** non-migratory. **Breeding:** defend clan territories, and have helpers at the nest. Nest is a purse-shaped globe of moss, lichens and cobwebs, feather-lined with side entrance in vegetation. Eggs: 6-10, white, spotted red. Incubation by female for 13-14 days. Broods: 1. Young in nest for 16-17 days. **Food:** chiefly insects.

145 Penduline Tits Remizidae
10 species, 4 genera, e.g. Penduline Tit *Remiz pendulinus*

Description: small, compact, arboreal birds behaviourally similar to the true tits (Paridae), moving acrobatically with great agility through vegetation. L:10-11 cm (4-4$\frac{1}{2}$ in), Wt:8-11 g ($\frac{1}{4}$-$\frac{1}{2}$oz). Solitary or gregarious. Bill small, conical with straight culmen, pointed. Legs medium-sized, feet strong. Wings medium-sized. Flight strong. Tail medium-sized, square. Plumage pale buff, chestnut, black, brown, grey and

148 Common Treecreeper

151 Thick-billed Flowerpecker

149 Stripe-headed Creeper

147 Nuthatch

150 White-throated Treecreeper

yellow depending on species. More or less monomorphic. **Range and habitat:** WP, EP, AF, NA. Sparsely wooded habitats, from semi-arid to marshland depending on species. **Migration:** non-migratory. **Breeding:** breeding groups, may have helpers at the nest. Nest is a pendulous purse of flower pappus, some with a false entrance. Built by male and female. Family named after their nests. Eggs: 5-10, white. Incubated by female for 12-14 days. Broods: 1. Young are in nest for 16-18 days. **Food:** insects and small seeds.

146 Tits and Chickadees Paridae
50 species, 3 genera, e.g. Crested Tit *Parus cristatus*

Description: among the best studied of birds. Small, compact, acrobatic and arboreal, inquisitive and gregarious birds forming mixed-species flocks. L:11.5-14 cm (4½-5½ in), Wt:6-20 g (¼-¾ oz). Bill short to medium, thin to stout, generally strong. Nostrils covered by bristles. Legs medium-sized. Feet strong. Wings medium-rounded to pointed. Flight strong. Tail medium-sized. Plumage varied. Generally, brown, white, grey, black and white, some with yellow, green and blue. Underparts generally paler. Most monomorphic, males larger and brighter. **Range and habitat:** WP, EP, AF, OR, NA. A variety of forest, forest edge and scrub habitats including cultivation and suburbia. **Migration:** some partial migrants. **Breeding:** monogamous, territorial. Nest is a cup of moss and hair in tree-hole, excavated by some species. Some readily accept nestboxes. Eggs: 4-12 (rarely 18), white, spotted red-brown. Incubation by female: 13-14 days. Broods: 1, rarely 2. Young are in nest for 17-20 days. **Food:** a wide variety of invertebrates, also buds and seeds, especially of trees, feeding from ground to canopy.

147 Nuthatches Sittidae
25 species, 4 genera, e.g. Nuthatch *Sitta europaea*

Description: small, compact, arboreal, agile and acrobatic, though less so than tits (Paridae), they can run swiftly up and down tree-trunks. L:9.5-20 cm (3¾-8 in), Wt:10-60 g (¼-2 oz). Many are highly territorial – some establishing territories within a few weeks of fledging. Bill long, straight or decurved and pointed. Legs short, stout. Feet strong with long, curved, strong nails. Wings longish, pointed. Flight strong. Tail short, square, not used in support. Plumage chiefly grey to blue above, white, brownish below. Some also with red. Head often striped. More or less monomorphic, males slightly larger. **Range and habitat:** WP, EP, OR, AU, NA. Natural and semi-natural forest habitats, also rocky habitats. **Migration:** most species non-migratory, some partial migrants. **Breeding:** monogamous, highly territorial. Nest is usually of dry leaves in a tree-hole or rock crevice. Several species reduce the entrance size by plastering it with mud. Eggs: 4-10, white with reddish spots. Incubation by female for 15-18 days. Broods: 1. Young are in nest for 23-5 days. **Food:** chiefly invertebrates gleaned from bark and leaves, also seeds.

148 Treecreepers Certhiidae
7 species, 2 genera, e.g. Common Treecreeper *Certhia familiaris*

Description: small, slender, arboreal and solitary (though sometimes roost communally and often join mixed-species flocks). L:12-15 cm (4½-6 in), Wt:7-16 g (¼-½ oz). Bill long, thin, pointed and decurved. Legs short. Toes with long, decurved claws. Wings short. Flight fast. Tail long, of stiff, pointed feathers used in support when foraging on bark. Plumage: upperparts brown, streaked with buff; underparts white. Monomorphic, males slightly larger. **Range and habitat:** WP, EP, AF, OR, NA. Forests. **Migration:** generally short-distance partial migrants. **Breeding:** nest is a lined cup of twigs and plant fibres in a crevice, e.g. behind tree bark, or on branch. Eggs: 5-6 (rarely 2-9), white, reddish spots. Incubation chiefly by female; 14-15 days. Broods: 1-2. Young in nest for 15-16 days. **Food:** small invertebrates gleaned from bark.

149 Philippine Creepers Rhabdornithidae
2 species, 1 genus, e.g. Stripe-headed Creeper *Rhabdornis mystacalis*

Description: among the least known of birds. Many geographical races are recognized. Usually in small flocks, sometimes with other species. Small birds with brown, streaked upperparts, underparts white with dark streaking on flanks. L:13-15 cm (5-6 in). Males greyer. 1 species with a black mask and white streaking on the head. Bill straight and slightly hook-tipped or decurved, stouter than in Certhiidae. Tongue brush-tipped. Legs rather long, elongated hind toe. Wings medium or long, pointed. Tail medium-sized, square with broad feathers. **Range and habitat:** OR. A variety of forest habitats. **Migration:** non-migratory. **Breeding:** unknown. Nest is a tree-hole. **Food:** insects from ground or gleaned from bark; birds also feed from flowers in the open.

150 Australian Creepers Climacteridae
7 species, 1 genus, e.g. White-throated Treecreeper *Climacteris leucophaea*

Description: small, solitary or in small groups, arboreal and ground-foraging birds. L:12-19 cm (4½-7½ in). Wt:20-40 g (¾-1½ oz). Bill long, decurved, pointed. Legs and feet strong, with long, curved claws. Wings pointed. Tail short, square, soft and not used in support. Plumage: brown to black upperparts with broad, pale wing-bar, underparts rufous or streaked. Most monomorphic. **Range and habitat:** AU. Forests. **Migration:** non-migratory. **Breeding:** solitary or communal with helpers at the nest. Nest is a cup of plant fibres and hair in tree-hole. Eggs: 2-3, white or pinkish with reddish spots. Incubation by female for 16-23 days. Broods: 1-2. Young are in nest for 25-6 days. **Food:** insects (especially ants), flowers, pollen and nectar.

151 Flowerpeckers Dicaeidae
58 species, 7 genera, e.g. Thick-billed Flowerpecker *Dicaeum agile*

Description: small, compact, with short neck, active and arboreal. L:8-15 cm (3-6 in), Wt:5-20 g (¼-¾ oz); 1 species 21 cm (8 in), 42 g (1½ oz). Bill short, stout and straight or long, thin and decurved. Distal 3rd of bill notched. Tongue tip tubular for nectar-feeding. Legs short. Wings medium-sized, pointed. Outermost primary sometimes vestigial. Flight fast, darting. Tail short, square. Plumage monomorphic, dull grey-green; or dimorphic with males glossy bright blues above yellows below, with

157 Parula Warbler

156 Bananaquit

153 Oriental White-eye

154 Fuscus Honeyeater

155 Rock Bunting

152 Southern White-bellied Sunbird

patches of red and yellow, females duller. **Range and habitat:** OR, AU. Treetops in forest and scrubland habitats. **Migration:** non-migratory. **Breeding:** nest is a lined cup in a tree-hole or dense vegetation, or pendant with a side entrance, all of plant fibres and cobwebs built by female; or a tunnel in a bank dug by both. Eggs: 1-3 (rarely 1-5), white, sometimes blotched. Incubation by female for c. 12 days. Young are in nest for c. 15 days. **Food:** small fruits (especially mistletoe berries, whose seeds the birds disperse), nectar and invertebrates.

152 Sunbirds Nectariniidae

120 species, 5 genera, e.g. Southern White-bellied Sunbird *Nectarinia talatala*

Description: small birds analogous to the hummingbirds (Trochilidae) of the New World. L: 9-30 cm (3½-12 in) including long tail, Wt: 5-20 g (¼-¾ oz). Solitary or paired, arboreal. Bill long, slender, decurved and pointed, serrated near tip. Nostrils covered by operculum, rictal bristles absent. Bills vary greatly between species which may specialize on different flowers. Tongue protrusible, tubular and divided at tip (not brush-tipped) for sucking nectar. Legs short to medium, strong. Wings short, rounded to pointed. Flight fast, dashing, sometimes hovering. Tail often sexually dimorphic: short, square (females) to very long, pointed or graduated (males). Plumage generally dimorphic. Males dark metallic purples, greens with red, yellow or orange underparts. Females duller olive, grey-green. **Range and habitat:** WP (1 species), AF, OR, AU. Forest, scrubland, cultivation and gardens. **Migration:** most non-migratory. **Breeding:** strictly monogamous unlike hummingbirds, highly territorial. Nest is an elaborate hanging purse with porched side entrance of plant materials and cobwebs (cup-shaped in spiderhunters). Eggs: 2 (rarely 3), white or bluish, heavily spotted. Incubation chiefly by female (shared in spiderhunters) for 13-15 days. Broods: 1-5. Young are in nest for 14-19 days. Some species are parasitized by cuckoos (Cuculidae). **Food:** nectar and small invertebrates, sometimes fruit. Perch while feeding.

153 White-eyes Zosteropidae

85 species (4●), 11 genera, e.g. Oriental White-eye *Zosterops palpebrosa*

Description: small, arboreal, restless and gregarious in flocks. L: 10-14 cm (4-5½ in), Wt: 8-31 g (¼-1 oz). Little variation across whole family. Bill slender, decurved and pointed. Tongue brush-tipped. Legs medium-sized, strong. Wings short, rounded, with outermost primary absent. Flight dashing. Tail moderate. Plumage green above, white/yellow below. Most with white eye-ring of tiny feathers, varying in size between species. Always monomorphic, but males often larger. **Range and habitat:** AF, OR, AU, PA. Forest, scrublands and gardens. **Migration:** mostly non-migratory. **Breeding:** monogamous, paired for life, communal species may have nest-helpers. Can have very rapid breeding cycle – 3 weeks from nest construction to fledging. Nest is a deep cup of plant material, suspended in a fork near the canopy. Eggs: 2-4, whitish. Incubation shared for 10-12 days. Broods: 2-4. Young are in nest for 10-12 days. **Food:** insects, fruit and nectar.

154 Honeyeaters Meliphagidae

172 species (1●), 39 genera, e.g. Fuscus Honeyeater *Meliphaga fusca*

Description: a diverse group showing evolution parallel to many other groups. Small to medium-large, arboreal, often gregarious and pugnacious birds. L: 8-45 cm (3-18 in), Wt: 6.5-150 g (¼-5¼ oz). Bill slender, pointed, medium to long and decurved. Tongue elaborate, protrusible and brush-tipped. Head sometimes with areas of bare skin and wattles. Legs short to medium. Wings long, pointed. Flight direct and dashing in small species to slow and rather clumsy in larger species. Tail medium to long. Plumage varied, of drab greens, browns, greys, yellows (monomorphic) or bold reds, black and white (dimorphic). Many have a conspicuous coloured patch behind the ears. **Range and habitat:** southern AF, AU. Forest and scrublands. **Migration:** some migratory species. **Breeding:** many colonial breeders with nest-helpers. Nest is cup-shaped or pendulous with a side entrance, of plant materials and cobwebs, often high in tree. Eggs: 2 (rarely 1-4), white to pale buff, spotted reddish. Incubation by female or shared for 13-17 days. Broods: rarely more than 2. Young in nest for 10-16 days. **Food:** nectar, insects and some fruit, both generalist and specialist species.

155 Buntings, Cardinals and Tanagers Emberizidae

583 species (6●), 147 genera, e.g. Rock Bunting *Emberiza cia*

Description: a large family with 5 well-defined subfamilies:
a. Buntings Emberizinae 286 species, 71 genera
b. Plush-capped Finch Catamblyrhynchinae monotypic
c. Cardinals and Grosbeaks Cardinalinae 47 species, 16 genera
d. Tanagers Thraupinae 248 species, 58 genera
e. Swallow Tanager Tersininae monotypic
Generally small, arboreal to terrestrial, solitary to gregarious. L: 9-28 cm (3½-11 in), Wt: 8.5-40 g (¼-1½ oz). Bills heavy with or without rictal bristles, short to medium, conical, pointed or hook-tipped (tanagers). Legs short to medium, feet often weak. Wings long, pointed, outer primary absent. Flight strong, direct, but not dashing. Tail medium to longish. Plumage varies from cryptic streaked browns (buntings) to stunning bright reds, blues, yellows and black (cardinals and tanagers). Monomorphic or dimorphic, males often larger. **Range and habitat:** worldwide except AU, PA, AN. Chiefly open habitats, desert, grassland, tundra, alpine meadows and cultivation, also open forest and forest edge. **Migration:** many strong migrants, especially northern species. **Breeding:** monogamous and polygamous. Nest is usually a woven cup of plant material in vegetation, often low down. Rarely a domed structure with a side entrance or in tunnel. Eggs: 4-5 (rarely 2-6), pure white to coloured and spotted. Incubation by female (or, less usually, shared) for 10-18 days. Broods: 1-3. Young are in nest for 9-24 days. **Food:** chiefly seeds and some fruits, nestlings fed largely on arthropods (especially insects).

156 Bananaquit Coerebidae

Coereba flaveola (monotypic). See p 138.

Description: abundant arboreal nectar and insect feeder of Caribbean

162 Zebra Finch

161 American Goldfinch

159 Red-eyed Vireo

163 Paradise Whydah

158 Akiapolaau

160 Common Grackle

and NT, found in a wide variety of wooded and garden habitats. Solitary, highly active, small birds sometimes treated within the Emberizidae.

157 New World Warblers Parulidae

125 species (3●), 27 genera, e.g. Parula Warbler *Parula americana*

Description: an important family of active, small, arboreal insectivores, analogous to the Sylviid warblers of the Old World but more closely related to the Emberizidae. L: 10-18 cm (4-7 in), Wt: 7-25 g (¼-1 oz). Generally solitary, flocking at certain times of the year. Bill thin, pointed, usually with rictal bristles. Legs medium-sized. Wings medium-sized, pointed, outermost primary absent. Tail medium-sized, rounded or square. Plumage varied, often bright blues, greens, yellows, orange, black and white. Temperate species most dimorphic and with seasonal variation. Some monomorphic. **Range and habitat:** NA, NT. Forests and scrublands. **Migration:** northern species are strong, nocturnal migrants. **Breeding:** nest is a cup in tree or bush (most); domed with side entrance and on ground or in tree-hole. Eggs: 4-5 (rarely 2-8), white to green, spotted brown. Incubation by female for 10-14 days. Broods: 1. Young are in nest for 8-12 days. Predation rates for nestlings may be high. **Food:** chiefly insects and other invertebrates, also some fruit and seeds.

158 Hawaiian Honeycreepers Drepanididae

20 species (another 16 recently extinct, 17●), 12 genera, e.g. Akiapolaau *Hemignathus munroi*

Description: small, mostly arboreal birds, solitary or in loose flocks. L: 10-20 cm (4-8 in), Wt: 10-45 g (¼-1½ oz). The family gives one of the best examples of adaptive radiation (comparable with Darwin's finches, Emberizidae, of Galapagos Islands) since their bills and ecology vary from heavy, hooked, seed-cracking bills to long, thin, decurved and pointed nectar-feeding bills (these also with tubular tongues). Bill never notched or serrated. Believed to have evolved from cardueline finches (Fringillidae). The family is threatened by human habitat destruction. Legs medium-sized. Wings pointed, outermost primary vestigial. Flight strong. Tail medium-sized. Plumage usually dimorphic, males brighter but nonmetallic. Colours include black, white, red, green, brown, yellow and dazzling orange. **Range and habitat:** Hawaiian Islands (PA). Native forest and scrublands. **Migration:** non-migratory. **Breeding:** little known for many. Nest is an open cup of twigs lined with plant fibres in a tree, shrub or grass clump, rarely in cavities. Eggs: 2-4, whitish, streaked red-brown. Incubation by female for 13-14 days. Young are in nest for 15-22 days. **Food:** chiefly insects and nectar, also fruit and seeds depending on species. Young fed seeds or seeds and insects.

159 Vireos Vireonidae

47 species, 4 genera, e.g. Red-eyed Vireo *Vireo olivaceus*

Description: small, arboreal and largely solitary birds, sometimes with mixed-species flocks with warblers (Parulidae) from which they are distinguished by chunkier appearance, less active behaviour and the bill form. L: 10.2-16.5 cm (4-6½ in), Wt: 9-39 g (¼-1¼ oz). Tropical species flock throughout the year. Bill adapted to insect-eating but rather heavy,

notched and hook-tipped, unlike warblers. Rictal bristles present. Species vary in bill size and shape. Legs short and strong. Wings rounded to pointed, outermost primary vestigial. Flight strong. Tail medium-sized, squarish. Plumage essentially greens, yellows, greys and white, often with wing-bars, eye-rings and eye stripes. Monomorphic. **Range and habitat:** NA, NT. Forests and scrublands. **Migration:** temperate species are migratory. Nocturnal migrants. **Breeding:** nest is an open cup suspended from fork in a branch, often low down. Eggs: 4-5 (2 in tropics), whitish, spotted brownish. Incubation shared (or, less usually, by female) for 11-13 days. Young are in nest for 11-13 days. Some species heavily parasitized by cowbirds (Icteridae). **Food:** insects and fruit.

160 New World Blackbirds Icteridae

96 species (1●), 23 genera, e.g. Common Grackle *Quiscalus quiscula*

Description: small to large, arboreal or ground-dwelling, usually gregarious and noisy birds with complex social organization. L: 15-53 cm (6-21 in), Wt: 20-454 g (¾ oz-1 lb). Bill unnotched, conical, long and pointed or short and stout, sometimes casqued, lacking rictal bristles. Legs medium to long, strong. Wings long, pointed, outermost primary absent. Flight strong. Tail medium to long, square or graduated and vaulted. Plumage generally dark glossy purples, black or browns; or boldly coloured in black, yellow, orange, red. Usually dimorphic, males larger. May have distinct immature plumage. **Range and habitat:** NA, NT. Forests, scrublands, grasslands and marshes. **Migration:** temperate species mostly nocturnal migrants. **Breeding:** varied mating systems: monogamous, polygynous, some species colonial, some with nest-helpers, some parasitic, e.g. on vireos (Vireonidae). Nests are extremely varied, from an open cup to long pendant purses, in reeds, shrubs, trees or on the ground. Eggs: 2-6, variable in colours and spotting. Incubation by female (rarely shared) or by host for 12-15 days. Broods: 1, rarely 2. Young are in nest for 9-35 days. **Food:** wide range of fruit and small animals.

161 Finches Fringillidae

124 species (3●), 19 genera, e.g. American Goldfinch *Carduelis tristis*

Description: small, arboreal and ground-feeding, solitary or gregarious birds adapted to seed-eating. L: 11-19 cm (4½-7½ in), Wt: 9-100 g (¼-3½ oz). There are 2 well-defined subfamilies: Fringillinae (3 species) and Carduelinae (121 species) Bill always strong, short to medium, conical, often pointed, sometimes crossed depending on diet. Rictal bristles may be reduced. Legs and feet short to medium, often slender, generally weak. Wings medium to long, rounded to pointed. Flight strong, undulating. Tail medium-sized, often notched. Plumage varies from cryptic streaked browns and greens to bright yellows, pink, reds. Monomorphic or dimorphic, males usually larger and brighter. **Range and habitat:** WP, EP, AF, OR, NA, NT, introduced to New Zealand (AU). Wide variety of terrestrial habitats, especially open forest, scrub and cultivation. **Migration:** many migratory species. **Breeding:** courtship and territoriality differ between subfamilies, some loosely colonial. Nest is an open cup of plant fibres in tree or shrub built by female. Eggs: 3-5,

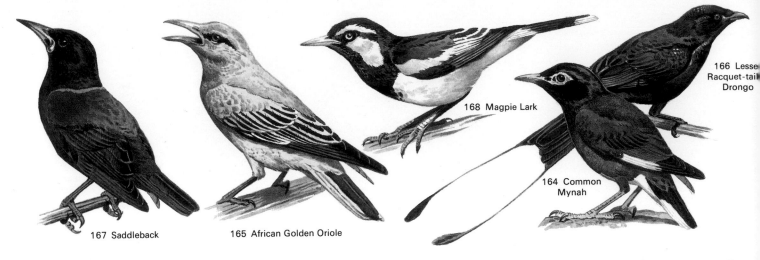

167 Saddleback   165 African Golden Oriole   168 Magpie Lark   164 Common Mynah   166 Lesser Racquet-tailed Drongo

whitish to dark green, spotted. Incubation by female for 12-14 days. Broods: 1-2, rarely 3. Young are in nest for 11-17 days. **Food:** chiefly seeds, nestlings may be fed on insects from parents' bill (Fringillinae) or seeds by regurgitation with or without insects (Carduelinae).

162 Waxbills Estrildidae
133 species, 29 genera, e.g. Zebra Finch *Poephila guttata*

Description: tiny to small, highly gregarious, active, largely ground-feeding birds with complex social behaviour. L: 9-14 cm (3½-5½ in) Wt: 6-13 g (¼-½ oz). Bill short, stout, pointed. Legs short to medium. Wings short, rounded to medium, pointed with reduced outermost primary. Flight fast, direct. Tail short to medium, square to rounded, some with extended, pointed central tail feathers. Plumage very varied in bright colours, usually dimorphic. **Range and habitat:** AF, OR, AU, introduced to Caribbean (NT). Chiefly grasslands, reedy marshes, tropical forests and forest edges. **Migration:** non-migratory. **Breeding:** many are colonial, have complex courtship and pair for life. Nest is a large, untidy, dome of plant material with a side entrance, built by both. Rarely, a tree-hole. Eggs: 4-8, white. Incubation usually shared for 10-14 days. Broods: opportunistic. Young are in nest for 14-21 days. Several species parasitized by whydahs (Ploceidae). **Food:** chiefly small seeds taken on ground, some fruit and insects (especially young).

163 Weavers and Sparrows Ploceidae
161 species (4●), 22 genera, e.g. Paradise Whydah *Vidua paradisaea*

Description: small to medium birds, the largest having long tails. L: 10-65 cm (4-26 in), Wt: c. 11-64 g (½-2¼ oz). Largely ground feeding, generally highly gregarious with complex social behaviour. Bill stout, conical, slightly decurved and without rictal bristles. Legs medium-sized, feet weak. Wings medium-rounded to pointed. Flight strong, rapid. Tail medium-square to very long and elaborate, used in display (whydahs). Plumage varied, streaked browns, greys to bold black, white, yellow, orange and red. Monomorphic to highly dimorphic. Males larger. **Range and habitat:** WP, EP, AF, OR, introduced to NA and AU. Wide range of habitats from desert to forest. Several species associate with man. **Migration:** non-migratory. **Breeding:** many colonial, polygamous, some parasitic and with elaborate courtship. Nest varies, generally a large, untidy structure (though sometimes intricately woven) of plant stems in hole or tree. Sometimes complex colonial nests. Eggs: 2-8, white to heavily spotted and varying greatly in colour. Incubation by female or shared for 10-17 days. Broods: 1-3. Young are in nest for 15-24 days. **Food:** chiefly seeds, also fruit, buds and insects.

164 Starlings Sturnidae
108 species (1●), 24 genera, e.g. Common Mynah *Acridotheres tristis*

Description: small to medium, arboreal or ground-feeding, highly gregarious birds with complex social behaviour. L: 16-45 cm (6½-18 in), Wt: 45-170 g (1½-6 oz). Bill long, strong and pointed, slightly arched. Stout and strong legs and feet. Walk on ground. Wings pointed. Flight strong and direct. Tail short to medium, square. Plumage diverse from grey and pied to purples, greens, yellows, often with glossy, metallic sheen. Juvenile plumages sometimes distinct. Eyes sometimes bright yellow. Some fine mimics. **Range and habitat:** WP, EP, AF, OR, PA, introduced into NA and AU. Forest, scrublands, grasslands, cultivation and suburbia. **Migration:** many migrants or partial migrants. **Breeding:** many colonial or loosely so. Nest is chiefly a cup of plant material in tree-hole, some species make domed, open cup, or pendulous nests. Eggs: 3-7, blue to white, sometimes spotted. Incubation by female or shared for 11-18 days. Broods: 1-2. Young are in nest for 18-30 days. **Food:** a wide range of plant and animal foods.

165 Orioles Oriolidae
26 species, 2 genera, e.g. African Golden Oriole *Oriolus auratus*

Description: medium-sized, arboreal, with strong, pointed bill, legs and feet, bright plumage – yellows, greens and black. L: 20-30 cm (8-12 in), Wt: c. 70 g (2½ oz). Wings pointed, tail medium to long. Solitary. Dimorphic, males larger. **Range and habitat:** WP (1 species), EP, AF, OR, AU. Forests and dense scrublands. **Migration:** some migratory species. **Breeding:** nest is a deep, grassy cup high in tree. Eggs: 2-5, green with reddish markings. Incubation sometimes shared for 14-15 days. Broods: 1. Young are in nest for 14-15 days. **Food:** insects and fruit.

166 Drongos Dicruridae
22 species, 2 genera, e.g. Lesser Racquet-tailed Drongo *Dicrurus remifer*

Description: medium-sized, solitary, active and arboreal birds with elaborate tails. L: 18-64 cm (7-25 in) including up to 30 cm (12 in) tail, Wt: c. 40-50 g (1½-1¾ oz). Often crested. Bill stout, arched and slightly hooked with prominent rictal bristles. Nostrils sometimes concealed by dense feathering. Legs and feet strong. Wings long, pointed. Flight strong but unsustained. Tail medium to very long, outer feathers splayed outwards or developed into racquet shape. Plumage generally black with green to purple iridescence. **Range and habitat:** AF, OR, AU. Open wooded habitats, savannah, forest edges and cultivation. **Migration:** some migratory species. **Breeding:** little known for many. Nest is a shallow cup in fork of a branch. Eggs: 2-4, often white speckled brown. Incubation by female or shared for 16 days. Broods: 2-3. **Food:** follow large mammals for disturbed insects. Also nectar.

167 Wattlebirds Callaeidae
2 species (1●), 2 genera, e.g. Saddleback *Creadion carunculatus*

Description: medium-sized, largely ground-dwelling birds. L: 22-50 cm (8½-20 in), Wt: 77-240 g (2¾-8½ oz). Exceptional for the sexual dimorphism of their bills. The extinct Huia of this family was the most extreme in this regard, the female's long, decurved bill being twice the length of the male's stout, conical bill. The gape carries orange or blue wattles of varying size. In the extant species: bill short, stout and hooked to long and decurved. Legs long, strong. Feet strong. Hops on ground. Wings short. Flight weak. Tail long, slightly vaulted. Plumage black, blue-grey or brown. **Range and habitat:** New Zealand (AU). Dense native forest. **Migration:** non-migratory. **Breeding:** elaborate courtship. Nest is a

169 Dusky Wood Swallow

173 Common Raven

171 Satin Bowerbird

172 Superb Bird of Paradise

170 Grey Butcher Bird

shallow, lined cup of sticks, on or near ground. Eggs: 2-3, grey, spotted and blotched purple and brown. Incubation by female for 18-25 days. Broods: 1. Young in nest for 27-8 days. **Food:** fruit, nectar and insects.

168 Magpie Larks Grallinidae
4 species, 3 genera, e.g. Magpie Lark *Grallina cyanoleuca*

Description: medium-sized, highly gregarious, rather graceful birds. L: 20-50 cm (8-20 in), Wt: 40-350 g (1½-12½ oz). Bill medium-sized, pointed, straight with uncovered nostrils. Legs long. Wings pointed. Flight poor. Tail long, square. Plumage grey or pied. Dimorphic in head patterns. Juvenile plumage distinct. **Range and habitat:** AU. Open forest, marshes and cultivation, often near water. **Migration:** non-migratory or nomadic migrants. **Breeding:** strongly territorial. Nests are deep bowls of mud strengthened with hair and plant fibres on a high tree branch, Wt: up to 1 kg (2 lb 3 oz). Eggs: 3-5 (rarely 8), white to pink, blotched red, brown or grey. Incubation shared for 17-18 days. Broods: 1-3. Young are semi-altricial, in nest for 19-23 days. **Food:** invertebrates, some seeds.

169 Wood Swallows Artamidae
10 species, 1 genus, e.g. Dusky Wood Swallow *Artamus cyanopterus*

Description: medium-sized, chunky, and highly gregarious birds. L: 12-23 cm (4½-9 in), Wt: 14-73 g (½-2½ oz). Bill stout, slightly decurved with a wide gape and without rictal bristles. Legs and feet short and strong. Wings very long and pointed. Flight agile and graceful. Sometimes soars. Tail short and square. Plumage solid colours of browns, greys and black. Paler below. Monomorphic. Has powder down plumes which break up to dress feathers. In the Passeriformes, only these and Cotingidae have such feathers. **Range and habitat:** OR, AU. Wooded open country, often near water. **Migration:** some species migratory. **Breeding:** colonial. May have nest-helpers. Nest is a fragile cup high in a bush or cavity in tree or rock. Eggs: 2-4, white, spotted red-brown. Incubation shared for 12-16 days. Young are semi-altricial, in nest for 16-20 days. **Food:** aerial insects.

170 Butcher Birds Cracticidae
10 species, 3 genera, e.g. Grey Butcher Bird *Cracticus torquatus*

Description: medium to large, solitary or gregarious and arboreal but often feed on ground. L: 25-50 cm (10-20 in), Wt: 80-140 g (2¾-5 oz). Bill large, straight, more or less hooked, with bare nostrils and prominent rictal bristles. Legs medium to long, strong. Wings long, pointed. Flight strong. Tail long, square. Plumage grey or pied. Monomorphic or dimorphic. **Range and habitat:** AU. Open forests, scrublands and grasslands. **Migration:** non-migratory or partial altitudinal migration. **Breeding:** rather poorly known. Strongly territorial. Nest is a large, lined, open cup of sticks high in a tree. Eggs: 3-5, green or blue, blotched and streaked brown. Incubated chiefly by female for 20 days. Broods: 1. Young are in nest for 28 days. **Food:** omnivorous but chiefly insects.

171 Bowerbirds Ptilonorhynchidae
18 species, 8 genera, e.g. Satin Bowerbird *Ptilonorhynchus violaceus*

Description: medium-sized, solitary, ground-dwelling birds. L: 21-38 cm (8-15 in), Wt: 70-230 g (2½-8 oz). Bill stout, strong, sometimes slightly decurved and hooked. Nostrils covered. Wings, tail and legs tend to be short. Legs strong, hind toe short. Plumage usually dramatically dimorphic. Females drab, cryptic, greens and browns. Males metallic, beautiful blues, yellow, orange or red. Some with crest on the nape. **Range and habitat:** AU. Forest and semi-forest. **Migration:** non-migratory. **Breeding:** highly polygynous, a few monogamous. In polygynous species, the male builds an elaborate stage or 'bower' in which he displays to the female. This is often decorated with coloured artefacts. The bower is totally independent of the nest in which the male takes no interest. Nest is a bulky cup of twigs and plant fibres in fork or crevice. Eggs: 1-2 (rarely 3), whitish to buff, plain or blotched at end. Incubation by female: 19-24 days. Broods: females, 1; males, several. Young are in nest for 18-21 days. Care by female alone in all polygynous species. **Food:** fruit, invertebrates (especially insects), small vertebrates.

172 Birds of Paradise Paradisaeidae
42 species, 20 genera, e.g. Superb Bird of Paradise *Lophorina superba*

Description: medium to large, stunningly beautiful, solitary, arboreal birds with diverse and spectacular plumage. L: 15-110 cm (6-43 in). Bill medium to long, straight to decurved, stout. Legs medium-sized, stout. Wings medium-sized, rounded. Flight weak. Tail short, square to very long, highly modified. Plumage highly dimorphic. Females generally dull browns and greys. Males enormously varied, often metallic with elaborated flank, back, tail or tail-covert feathers depending on species. Colours from red, yellow, orange to blues and greens. Some with coloured wattles or bare skin on head. **Range and habitat:** AU. Tropical forests, wooded savannahs. **Migration:** non-migratory. **Breeding:** little known for several species. Most polygynous, a few monogamous. Fantastic species-specific courtship displays, especially in polygynous species. Wild hybrids occur indicating genetic similarity. Nest is a bulky, shallow cup of plant material on a foundation of twigs, in a tree. Eggs: 1-2 (rarely 3), pale and streaked. Incubation by female for 17-21 days. Young in nest for 17-30 days. All care by female in polygynous species. **Food:** fruit, insects depending on species, also leaves, buds and small vertebrates.

173 Crows and Jays Corvidae
113 species (2●), 26 genera, e.g. Common Raven *Corvus corax*

Description: medium to large, family includes the largest Passeriformes. L: 15-71 cm (6-28 in), Wt: 80-1,500 g (2¾ oz-3 lb 5 oz). Strong black bill. Legs and feet medium-sized, strong. Crows: black/grey and white. Tail short to medium, long, broad wings. Jays: smaller, tail medium to long, graduated, brightly coloured black and white, greys, browns, blues. Nostrils covered by bristles. Monomorphic, males larger. **Range and habitat:** worldwide except high Arctic and AN. Very diverse. **Migration:** most non-migratory. **Breeding:** some colonial, some with nest-helpers. Nest is a platform of sticks in tree, sometimes domed, sometimes a tree-hole. Eggs: 3-10, pale. Incubation by female or shared for 16-22 days. Broods: 1. Young in nest for 20-45 days. **Food:** most omnivorous.

THE NEARCTIC REGION

The Nearctic Region extends from Alaska and the Aleutian Islands in the west to Greenland in the east, and from Ellesmere Island in the high Arctic to the northern edge of the tropical rainforest in central Mexico. It is bordered by the Arctic ocean to the north, the Bering sea and Pacific ocean to the west, the Atlantic to the east, and the Caribbean sea and Mexican rainforest to the south. The biological similarities of the Nearctic and Palaearctic are often emphasized by referring to the combined regions as the Holarctic. Similarities to the Neotropics are occasionally stressed by referring to the 'New World Region'. The Nearctic boundaries may be imprecise but its zoogeography is best understood in terms of a Nearctic avifauna interacting with a Palaearctic avifauna to the north and west and a Neotropical avifauna to the south.

Prior to the Mesozoic era North America was located at the equator of Pangaea, a single land mass containing all the modern continents. About 200 million years ago Pangaea broke apart, with North America and Eurasia forming the separate continent of Laurasia. North America and Eurasia separated early in the Tertiary period of the Cenozoic era (about 50 million years ago), but a European connection through Greenland persisted during the early Tertiary and an Asian connection has been renewed repeatedly across the Bering Strait. To the south three large islands with four substantial water gaps separated North America from South America throughout most of the Tertiary period, until about three million years ago. During this separation the southern half of North America was subtropical or tropical; even late in the Tertiary, as the climate cooled, southernmost North America remained tropical. Thus throughout most of the Tertiary two tropical avifaunas evolved in parallel, one in the Nearctic Region and one in the Neotropical. Elimination of the water gaps between the Nearctic and Neotropics coincided with climatic cooling, which eliminated all tropical habitat from the Nearctic. The change in habitat and formation of a land bridge led to an extensive interchange of tropical species between the Nearctic and Neotropics and explosive radiation of several families and subfamilies, such as wrens, wood warblers, mimic thrushes and vireos from the tropical Nearctic, and tyrant flycatchers, blackbirds, tanagers and hummingbirds from the Neotropics.

The major geographic features of the Nearctic are oriented longitudinally. The western mountains extend from Alaska to central Mexico and include the Rockies and lesser ranges. The central plains extend from Canada to the Gulf of Mexico. Several more or less isolated, minor mountain ranges in the north-east coalesce into the Appalachians, which extend south from Pennsylvania to Georgia and Alabama. Florida and Mexico project south toward the Neotropics and the Mississippi river flows south from Minnesota to the Gulf of Mexico.

The Nearctic is home to about 650 bird species in 62 families, three of which are introduced and none of which is endemic. These families can be divided into six groups: 1 waterbirds, which include the oceanic skuas, partially oceanic pelicans and cormorants, and the freshwater grebes; herons and bitterns; storks; ibises; flamingos; ducks, geese and swans; and rails, coots

American Wood Stork (or Wood Ibis) *colony. Their massive nests are placed at various heights above water from a few feet to the tree tops.*

48

and gallinules; **2** cosmopolitan families, such as the albatrosses, shearwaters, fulmars and petrels; gulls and terns; hawks, eagles and falcons; swifts; and swallows; **3** an indigenous element that evolved during the Tertiary isolation: wrens, mimic thrushes, vireos, wood warblers and American sparrows, also several small families and subfamilies: the New World vultures, quails and turkeys, dippers, gnatcatchers, waxwings, silky flycatchers, motmots, todies and the palmchat; **4** an old Holarctic element that evolved and dispersed during the Eocene when North America and Europe were connected through Greenland: cranes, pigeons, cuckoos, certain owls and caprimulgids, kingfishers, the blue jay group, cardueline finches and certain thrushes belong to this group; **5** a recent Holarctic element that has crossed the Bering Strait land bridge from Asia and includes the Barn Owl, larks, tits, nuthatches, creepers, kinglets, the Wrentit, pipits, some swallows and some corvids; **6** immigrants from the Neotropics such as the hummingbirds, tyrant flycatchers, tanagers and blackbirds.

Within the Nearctic are nine major biotic communities characterized by their dominant vegetation.

Tundra is the northernmost community, also found above 3,000 m (10,000 ft) in the western mountains. It is characterized by lichens, mosses, grasses and dwarf willow. Winters are extremely cold and precipitation is moderate to heavy. Birds of the tundra include ptarmigans, Snowy Owl, Shore Lark, Water Pipit and Lapland, Smith's and Snow Buntings.

Northern coniferous (boreal) forest is dominated by pines, spruces, hemlocks, firs and cedars. The forest may be dense with little or no understorey, or more open with an understorey of shrubs and herbaceous plants. Winters are cold, summers cool, and rainfall moderate to heavy. Characteristic birds include Goshawk, Olive-sided Flycatcher, Red-breasted Nuthatch, Brown Creeper, Wren, Golden-crowned Kinglet, Hermit Thrush, Yellow-rumped Warbler, Dark-eyed Junco, Crossbill and Pine Siskin.

The eastern deciduous forest extends from the southern edge of the coniferous forest, with which it mixes, south into the south-eastern Nearctic. It is dominated by beech, maple, basswood, oak, hickory and walnut. The understorey includes small shrubs and herbaceous plants. Winters are cool, summers warm, and rainfall moderate to heavy. Birds of the deciduous forest include Red-shouldered and Broad-winged Hawks, Barred Owl, Downy Woodpecker, Eastern Wood-pewee, Carolina Chickadee, White-breasted Nuthatch, Wood Thrush, Yellow-throated Vireo and Cerulean Warbler.

Grasslands occur in the interior and are characterized by abrupt seasonal changes with cold to very cold winters, warm to hot summers, and moderate to low rainfall. Swainson's and Ferruginous Hawks, Greater Prairie Chicken, Sharp-tailed Grouse, Long-billed Curlew, Burrowing Owl, Lark Bunting, Grasshopper Sparrow and Western Meadowlark are found throughout the grasslands; Sprague's Pipit, Baird's Sparrow, McCown's Longspur and Chestnut-collared Longspur are found only on the northern prairies.

In the western mountains above the grasslands and below the coniferous forest is the Pinyon–Juniper Woodland, a park-like woodland of small trees, pinyon pines, several species of juniper and yuccas. The woodland has warm summers, cool winters and little rainfall. Its birds include the Grey Flycatcher, Pinyon Jay, Plain Titmouse, Bushtit and Bewick's Wren.

Oaks (of the genus *Quercus*) characterize the south-western oak woodland. Trees are widely spaced, with grass and shrubs covering the ground. Summers are warm, winters cool, with moderate to low rainfall. Characteristic birds include Nuttall's Woodpecker, Bridled Titmouse, Hutton's Vireo and Virginia's and Black-throated Grey Warblers.

South-western California with its mild climate, dry summers and wet winters is characterized by large areas of dense brush and stunted trees. This is the chaparral. Occasional rock outcroppings and grassy areas provide relief from the brush, which is inhabited by the Wrentit, California Thrasher, Grey Vireo, Orange-crowned Warbler and Black-chinned, Sage and White-crowned Sparrows.

Further north in the Great Basin Plateau between the Rocky mountains and the Sierra Nevada–Cascade mountains is another brushy habitat dominated by sagebrush. Summers are very hot and dry. Winters are cool and dry. Sage Grouse, Sage Thrasher and Brewer's and Sage Sparrows are characteristic of sagebrush.

Lowlands throughout the south-western Nearctic are very dry and hot with widely spaced plants, 1-2 m ($3\frac{1}{4}$-$6\frac{1}{2}$ ft) high. The creosote bush is the dominant plant, but mesquite, paloverde, catclaw, ironwood, ocotillo, agaves, cacti, yuccas and grasses are also present. A surprising variety of birds including Gambel's Quail, Greater Roadrunner, Elf Owl, Lesser Nighthawk, Costa's Hummingbird, Gila Woodpecker, Vermilion Flycatcher, Verdin, Cactus Wren, Black-tailed Gnatcatcher, Bendire's, Crissal and LeConte's Thrashers and Phainopepla occur here.

Parts of these areas have not reached the characteristic community structure. Because of poor soil and frequent fires the south-eastern Nearctic remains coniferous rather than deciduous. Characteristic here are the Red-cockaded Woodpecker, Brown-headed Nuthatch and Bachman's Sparrow. Large areas of south-western Texas and nearby New Mexico are dominated by mesquite rather than creosote bush, and Golden-fronted Woodpecker and Black-crested Tufted Titmouse are characteristic of these. Along the coast of the Pacific north-west the coniferous forest is unusually wet and Chestnut-backed Chickadee and Varied Thrush are confined to this temperate rainforest.

Throughout these relatively stable communities are ephemeral communities that will develop toward the characteristic vegetation of the region. Such communities are frequent enough to have names and characteristic species. Beaches are inhabited by plovers, gulls and terns. Lakes are used by divers, swans, ducks and geese. Pied-billed Grebes and Black Terns feed on the open water of ponds, whereas bitterns, moorhens, rails, Sedge and Marsh Wrens and Red-winged Blackbirds nest in the cat-tails and sedges bordering the water. Shrublands are inhabited by Common Yellowthroats and Field Sparrows. Riparian woodlands, characteristic of streams and bottomlands in the western plains and desert south-west, are often inhabited by species characteristic of deciduous forest.

Throughout the Nearctic many natural biotic communities have been and continue to be disrupted by human interference, such as agriculture, forestry, flooding caused by dams, and deserts caused by overgrazing. However, these habitats may contain a rich variety of species that includes American Kestrel, Mourning Dove, kingbirds, American Crow, American Robin, bluebirds, Song Sparrow and blackbirds.

Several societies focus on Nearctic birds. The National Audubon Society (Membership Data Center, PO Box 52529, Boulder, CO 80322, USA) emphasizes birds but addresses broad conservation issues. The American Birding Association (PO Box 6599, Colorado Springs, CO 80934, USA) is devoted to birds and bird-watching. Enquiries about other societies should be directed to the Ornithological Societies of North America, PO Box 1897, Lawrence, KS 66044-8897, USA.

DIVERS Gaviformes

Great Northern Diver *Gavia immer* (Gaviidae) 71-89 cm (28-35 in). Black head, white necklace barred with black; black above with white checks on the back and white spots on the sides, white below; bill black, long, dagger-like. Sexes alike, female smaller. Winter and immature plumages grey above, shading to white below; bill light, with bluish cast. Sits low in water. Feet far back on body; walking posture upright, shuffling and awkward. Flight rapid with head and feet outstretched, body slightly hunched. Voice: loud yodelling; echoing quality. **Range and habitat:** Holarctic. Breeds on freshwater lakes, winters in coastal bays, estuaries, open lakes. **Similar species:** other divers, especially White-billed Diver (see p 142).

GREBES Podicipediformes

Western Grebe *Aechmophorus occidentalis* (Podicipedidae) 51-61 cm (20-4 in). Largest of Nearctic grebes. Distinguished by its long thin neck, long straight, slender bill and red eyes. Black crown, back of neck, and back; chin, throat, foreneck and breast white. Polymorphic with Mexican Grebe having yellow-orange bill; white face surrounding the eye and speckled white flanks, whereas 'Western' Grebe has dark greenish yellow bill, dark face and dark flanks. Courtship behaviour includes spectacular aquatic displays. Call, a shrill 'creet-creet' in 'Western' Grebe (single 'creet' in Mexican Grebe). **Range and habitat:** reeds and rushes along edges of freshwater lakes, west-central NA and PA coast. **Similar species:** Mexican Grebe, winter plumage Slavonian Grebe, Red-necked Grebe.

Slavonian Grebe *Podiceps auritus* (Podicipedidae) 30-8 cm (12-15 in). In breeding plumage both sexes have chrome-yellow 'horns' that extend from behind the eye to beyond the crown; head and back dark green; neck, breast and flanks rich chestnut; belly white. In winter plumage 'horns' absent, dark above with white cheek, chin, throat, breast and belly. Slender, straight bill, dark in summer, paler in winter. Red eye. Sexes alike. Voice: series of loud, weird croaks and chattering, followed by several prolonged shrieks. In flight shows white speculum. **Range and habitat:** Holarctic. Breeds on northern marshes, ponds, quiet rivers. Winters along coasts and open lakes. **Similar species:** Black-necked Grebe; in winter plumage, Red-necked and Western Grebes.

Pied-billed Grebe *Podilymbus podiceps* (Podicipedidae) 30-8 cm (12-15 in). Most widespread NA grebe. Small, stocky, short-necked brown grebe with a fluffy white rear, named for its cream-coloured, chicken-like bill encircled with a black band; narrow white eye-ring, black throat. In winter bill unmarked and throat brown. Sexes alike. Juvenile is conspicuously striped on head, less so on body. Often swims with only head above water. Voice: frequent, lengthy series of whinnying sounds, slowing to a mellow 'cow-cow-cow-cowp-cowp'. **Range and habitat:** freshwater with emergent vegetation from north, central and southern NA, winters on saltwater bays, southern NA into NT. **Similar species:** all other grebes have slender bills (see pp 96, 142, 282).

TUBE-NOSED SWIMMERS Procellariiformes

Black-footed Albatross *Diomedea nigripes* (Diomedeidae) 68-78 cm (27-31 in), wingspan 193-213 cm (76-84 in). White around base of bill, under and behind eye, and white-shafted primary feathers, otherwise sooty brown; bill greenish brown, feet black. Individuals become whiter with age, white spreads on to forehead and over crown, belly becomes pale, undertail-coverts and rump may become white; bill fades to yellowish or pinkish and feet become dusky yellow. Sexes alike. Voice: groans, squeals and shrieks. **Range and habitat:** PA, pelagic except when breeding on Hawaiian islands and Torishima off Japan. Seen along NA coast all year, most commonly in summer and autumn. **Similar species:** Short-tailed Albatross.

Greater Shearwater *Puffinus gravis* (Procellariidae) 48 cm (19 in). Dark, grey-brown cap and back with pale feather-edgings giving a scaly appearance; flight feathers black; white throat, collar and underparts, with diagnostic dark smudge on belly; underwings white with dark outline; narrow white rump patch; tail short, rounded with dark undertail-coverts; bill dark, long, slender and hooked. In flight, wings held straight and stiff, long glides punctuated by bouts of slow flapping. Sexes alike. Generally silent, occasional groans and squeals. **Range and habitat:** pelagic, except when nesting in burrows on Tristan da Cunha, migrates north in spring and summer to western Atlantic. **Similar species:** Cory's Shearwater, Black-capped Petrel.

Sooty Shearwater *Puffinus griseus* (Procellariidae) 40-5 cm (16-18 in). Uniformly sooty brown with white wing-linings; slender, black bill; black feet. More widespread than Greater Shearwater, but less apt to occur in flocks. In flight wings appear crescent-shaped, several quick flaps followed by curving glide. Dives more frequently than other shear-waters, swims under water with its feet and folded wings. Sexes alike. Voice: nasal, bleating call when competing for fish at sea. **Range and habitat:** nests in burrows on islands in southern Southern Hemisphere, migrates north along Atlantic and PA coasts. Pelagic on continental shelf. **Similar species:** Northern Fulmar, Pale-footed and Short-tailed Shearwaters, dark jaegers (see pp 68, 142, 284).

Wilson's Storm Petrel *Oceanites oceanicus* (Hydrobatidae) 18 cm (7 in). Head and body black with large, square white rump patch that extends to the undertail-coverts; long legs with yellow webbing between toes. In flight feet extend beyond tail, which is square or slightly rounded. Sexes alike. Glides only occasionally. Wings rounded and wing-beat swallow-like. Feeds with wings raised in a V and feet paddling on the surface of the ocean. **Range and habitat:** nests in burrows or rock crevices on AN and nearby islands. Migrates north into Atlantic, Indian and PA oceans. Pelagic. **Similar species:** other all dark or white-rumped storm petrels (see p 142).

FULL-WEBBED SWIMMERS Pelecaniformes

Brown Pelican *Pelecanus occidentalis* (Pelecanidae) 106-37 cm (42-54 in). Long grey bill and large black pouch suspended from bill and throat. Bare skin around eyes and base of bill is grey to black. Adults have white head, yellowish forehead, chestnut neck and grey-brown body. Nonbreeding birds have less yellow and no chestnut. Sexes alike, male slightly larger. Legs and feet black. Catches fish by plunging head-first from the air or scooping with the bill while sitting on the surface. Immatures entirely grey-brown with white underparts. Usually silent, occasional croaks. Nestlings squeal. **Range and habitat:** coastal, southern NA to northern and western NT. **Similar species:** white pelicans (see pp 142, 186, 238, 286).

American White Pelican *Pelecanus erythrorhynchos* (Pelecanidae) 137-78 cm (54-70 in). The long, salmon-coloured bill with pouch suspended below mark this as a pelican. Plumage entirely white, except for black primary coverts, primaries and outer secondaries; bill brighter during breeding season; legs and feet orange-red. Develops vertical horny plate projecting from midway along upper mandible during breeding season, also pale yellow patch on chest and pale yellow crest on nape that turns to grey by the end of the season. On water wings often held folded, but elevated. In flight head rests on chest. Sexes alike. Immatures dusky with dark, streaked crown. Voice: in colony low groan, nestlings whining grunts. **Range and habitat:** nests in widely scattered colonies on isolated lakes in central NA. Winters along coast of south-eastern and south-western NA. **Similar species:** Brown Pelican, other white pelicans (see pp 142, 186, 238, 286).

Great Northern Diver

Western Grebe in courtship display

Slavonian Grebe

Pied-billed Grebe

Greater Shearwater

Black-footed Albatross

Wilson's Storm Petrel

Sooty Shearwater

Brown Pelican

American White Pelican

American Darter *Anhinga anhinga* (Anhingidae) 81-91 cm (32-6 in). Small head, long neck and tail; long, stiletto bill used to spear fish; unique curve in neck resulting from hinge mechanism of vertebrae; black plumage. Male uniformly black with white plumes on back and wings, yellow bill. Female has buff-coloured head, neck and breast with black body and white plumes. Immature browner than female. In flight neck outstretched, alternately flaps and glides, may soar to great height. Often swims with only head and neck above water. **Range and habitat:** cypress swamps, wooded ponds, rivers of south-eastern NA, eastern NT. **Similar species:** cormorants, darters (see pp 186, 238, 286).

Double-crested Cormorant *Phalacrocorax auritus* (Phalacrocoracidae) 66-81 cm (26-32 in). Small head; long, slender, often crooked neck; long tail; dark plumage, iridescent green and purple on back; yellow-orange throat pouch; bill long, narrow, black. During spring feather tufts project from crown. When perched, often spreads wings. Sexes alike. Immatures flat brown above, pale below. In flight neck outstretched, body horizontal. Often swims with bill and head tilted up. Voice: pig-like grunts at nest, otherwise silent. **Range and habitat:** coasts, large fresh-water lakes, rivers of northern and central NA. Winters along central and southern NA coasts. **Similar species:** other cormorants, darters (see pp 186, 238).

Magnificent Frigatebird *Fregata magnificens* (Fregatidae) 94-114 cm (37-45 in). Long pointed wings, angled at wrist; tail scissor-like, often folded in a point. Glides effortlessly. Wingbeats infrequent, deep and deliberate, may soar to great heights, dives at high speed. Male black with red, inflatable throat pouch during breeding season, feathers of back long, purple. Female black with white breast. Immature black with white head and breast. Bill long, slender, hooked. Hoarse cackles; with pouch inflated, whistling whinny. **Range and habitat:** oceanic coasts and islands of south-eastern NA, tropical Atlantic, eastern PA. **Similar species:** other frigatebirds (see pp 238, 286).

HERONS AND ALLIES Ciconiiformes

American Bittern *Botaurus lentiginosus* (Ardeidae) 58-68 cm (23-7 in). Medium-sized solitary, secretive, heron most often seen flying low over marsh, where its uniformly brown back and dark brown wing tips are diagnostic. Occasionally seen standing at water's edge with bill pointing straight up, its light brown underparts streaked with dark brown blending with its cat-tail background. Sexes alike. Immatures lighter brown. Usually silent, but during spring a resonant 'pump-per-lunk' repeated from 3 to 5 times. When flushed 'kok-kok-kok'. **Range and habitat:** fresh-water or brackish marshes of northern and central NA. Winters central and southern NA into Central America. **Similar species:** immature night herons, Eurasian Bittern (see p 144).

Green-backed Heron *Butorides striatus* (Ardeidae) 38-56 cm (15-22 in). Adults with rich chestnut head and sides, white streaks down throat, white underparts; black crown, dark bill, white eye-line and moustache; green-black wings and back; yellow eye; conspicuous orange-yellow legs. Sexes alike. Immature browner, heavily streaked below with buff edges to wing-coverts. Voice: loud 'skeeowp'. **Range and habitat:** wet woodlands, streams, lakeshores, marshes and swamps from east-central to south-central NA, also west-central NA south along Pacific coast. In winter southern NA to northern NT, AF, OR. **Similar species:** American Bittern, Least Bittern, Little Blue Heron.

Great Egret *Casmerodius albus* (Ardeidae) 94-104 cm (37-41 in). Large bird, entirely white plumage, yellow bill and black legs and feet. In breeding plumage up to 54 long lacy plumes extend down back and beyond tail. In nonbreeding plumage, no plumes, bill less yellow. Sexes similar. Hoarse croak, also 'cuk-cuk-cuk'. **Range and habitat:** fresh and saltwater marshes of east-central and south-eastern NA. Winters southern NA to northern NT also AF, EP, OR, AU.

Similar species: other white egrets, white morph of Great Blue Heron, immature Little Blue Heron.

Snowy Egret *Egretta thula* (Ardeidae) 56-66 cm (22-6 in). Small white egret; bill black, facial skin yellow; legs black with brilliant yellow toes. When breeding has lacy, recurved plumes on head, back and breast. Contrast between feet and toes less in immature, may have yellow stripe up rear of legs. Forages by rushing about, stabbing into water or stirring water with 1 foot then stabbing into the turbulence. **Range and habitat:** wetlands, east-central NA south to southern NT, local in south-central and south-western NA. Winters in southern part of range. **Similar species:** other white egrets especially Litte Egret, immature Little Blue Heron (see p 144).

Black-crowned Night Heron *Nycticorax nycticorax* (Ardeidae) 58-66 cm (23-6 in). Short-billed, short-necked, short-legged nocturnal heron; black crown and back; narrow white plumes extend backward from crown; striking red eye; face and underparts white, wings grey. Sexes similar. Immature grey-brown with light spots above and light streaks below. Hunched profile when perched. In flight deep wingbeats, head retracted forming straight line with back, toes extend beyond tail. Voice: flat 'kwawk'. **Range and habitat:** wetlands, central and east-central NA to southern NT, southern WP, EP into AF, OR. **Similar species:** American Bittern, immature Yellow-crowned Night Heron.

Yellow-crowned Night Heron *Nycticorax violaceus* (Ardeidae) 56-71 cm (22-8 in). Less common than Black-crowned Night Heron, more diurnal. Adult pale grey with black head, white patch behind eye; yellowish forehead fades to white over the crown. Stout, thick, pinkish bill. Sexes similar. Immature is slate-brown with scattered small, light spots dorsally and fine light streaks ventrally; yellowish legs. Loud high-pitched 'quak'. **Range and habitat:** cypress and mangrove swamps, lush marshes, east-central NA to northern NT. Winters in southern part of range. **Similar species:** American Bittern, immature Black-crowned Night Heron.

Great Blue Heron *Ardea herodias* (Ardeidae) 127-37 cm (50-4 in). Pointed yellow bill, white forehead and crown, bordered with black; short dark plumes project backward from black nape; white face, neck cinnamon-grey with white ventral stripe spotted with dark; body and wings blue-grey; legs very long, dark. Sexes similar, male larger. Immature pale; black crown without plumes. Wingbeat deep and slow. All-white morph with yellow bill and legs in Florida. Harsh, guttural croaks. **Range and habitat:** wetlands of central and southern NA. Winters from southern NA to north-western NT. **Similar species:** other dark herons, Great Egret, Sandhill Crane (see p 60).

Little Blue Heron *Egretta caerulea* (Ardeidae) 63-73 cm (25-9 in). Slate-blue, with maroon head and neck; bluish bill has dark tip; blue-grey, unfeathered lores; eyes yellow; legs bluish green. Slow, deliberate foraging behaviour. Sexes similar. Immature has dark-tipped, bluish bill, blue-grey lores, plumage entirely white and legs pale. Moulting immatures pied blue and white. Occasional low clucks, screams when fighting. **Range and habitat:** coastal and freshwater marshes and swamps, south-eastern NA to southern NT. **Similar species:** immature resembles white egrets. Adult resembles Reddish Egret, Tricoloured Heron.

Tricoloured (Louisiana) Heron *Egretta tricolor* (Ardeidae) 61-71 cm (24-8 in). Long bill and neck, slim profile. Slate-blue body, white belly, flanks, underwings, rump; head and neck dark, mottled white and red-brown stripe, from chin to breast, short white plumes project from nape. In breeding plumage cinnamon to buff plumes extend over rump. Sexes similar. Immature lacks plumes. In shallow water runs after small fish, also stalks prey slowly, often in deep water. Noisy, harsh croaks, groans. **Range and habitat:** coastal, south-eastern NA to northern NT. **Similar species:** Little Blue Heron, Reddish Egret.

American Darter drying wings

Double-crested Cormorant juvenile

Magnificent Frigatebird male (left) and female

American Bittern

Green-backed Heron

Great Egret

Snowy Egret

Black-crowned Night Heron

Yellow-crowned Night Heron

Great Blue Heron

Little Blue Heron

Tricoloured Heron

Wood Stork *Mycteria americana* (Ciconiidae) 89-114 cm (35-45 in). Large, long-legged wading bird with white plumage and black primaries, secondaries and tail; naked skin of head and neck black and scaly; bill black and decurved toward tip; legs black, feet pink. Sexes similar. Flies with neck extended, legs trailing, often soars in large flocks. Immature has yellow bill and greyish brown, feathered head and neck. Nesting adults hiss, young noisy. Threatened by habitat destruction. **Range and habitat:** fresh and saltwater coastal marshes from south-eastern NA to southern NT. Nests in cypress. **Similar species:** White Ibis, white pelicans, Whooping Crane (see pp 50, 60).

White Ibis *Eudocimus albus* (Threskiornithidae) 56-71 cm (22-8 in). Adult white with black-tipped primaries; facial skin and decurved bill red; legs red during breeding season, grey otherwise. Sexes similar. Immature brown above, finely marked with white on head and neck, white belly and rump; bill brown to red. Flies with head and neck extended, feet trailing behind, rapid wingbeats alternate with gliding, often in large flocks that assume a V or linear formation. **Range and habitat:** coastal wetlands, also grassy and ploughed fields from south-eastern NA to northern NT. **Similar species:** Wood Stork, immature resembles Tricoloured Heron, immature Glossy Ibis.

WATERFOWL Anseriformes

Trumpeter Swan *Cygnus buccinator* (Anatidae) 150-83 cm (59-72 in). Largest North American swan. White except black bill; long, flat bill and forehead profile (Tundra Swan has rounded profile); edges of mandibles salmon-coloured. When alert neck held erect, otherwise slightly curved and drooped with kink at base. Autumn immature is dusky with black bill that is salmon on mid-dorsal surface. Once almost extinct, now recovering. Deep, sonorous, rasping 'ko-ko', trumpet-like. **Range and habitat:** nests locally in north-western and north-central NA. Winters in southern part of range. **Similar species:** other swans (see p 144).

Canada Goose *Branta canadensis* (Anatidae). Eleven sub-species range in size from the small 'Cackling' Canada Goose, 56-68.5 cm (22-7 in), to the 'Giant' Canada Goose, 81-122 cm (32-48 in). All have black head and neck with a white chin-strap. Grey-brown body, darker on wings and back; belly, flanks and tail-coverts white; rump and tail black. Sexes similar, males larger. Fly in V formation. Deep musical honking in larger subspecies, high-pitched yelping in smaller subspecies. **Range and habitat:** breeds throughout northern and central NA, locally to the south. Winters as far north as open water persists. **Similar species:** Brent Goose, Barnacle Goose (see p 146).

Snow Goose *Anser caerulescens* (Anatidae). Greater Snow Goose, 71-84 cm (28-33 in), is white. Lesser Snow Goose, 68-78 cm (27-31 in) has 2 colour morphs, a white form and a grey-bodied form with a white head and upper neck. Immature white morphs sooty above, white below. Immature dark morphs brownish grey, lighter below, with white chin spot. All have black primaries and red legs, feet and bills. Sexes similar, males larger. Noisy, shrill notes, also a soft honk. **Range and habitat:** Lesser breeds from EP to north-central NA. Winters from west-central NA to south-eastern NA. Greater breeds in north-eastern NA. Winters along east-central coast. **Similar species:** Ross's Goose, immature White-fronted Goose (see p 146).

Brent Goose *Branta bernicla* (Anatidae) 56-66 cm (22-6 in). Head, neck, breast and back black with a whitish patch on both sides of neck below throat. In the 'Black' subspecies found on Pacific coast, white patch may encircle neck, sides and underparts much darker than in Atlantic subspecies. Extreme lower belly, flanks and undertail-coverts white in both subspecies; legs and short bill black. Sexes similar, males larger. Drawn-out, rolling, throaty 'r-r-ronk, rr-ronk'. **Range and habitat:** breeds along Holarctic coasts, winters on coasts further south. Abundant in bays in winter.

Similar species: Canada Goose, Barnacle Goose (see p 146).

North American Black Duck *Anas rubripes* (Anatidae) 51-63 cm (20-5 in). Similar in size and shape to Mallard, with which it often hybridizes. Body dark sooty brown, lighter brown neck and head; in flight, white wing-linings contrast with dark body. Purplish blue speculum bordered with black. Feet and legs orange, brighter in male than female. His bill bright yellow, hers mottled with black, immature's olive-green. Voice: quacks. **Range and habitat:** favours brackish marshes and estuaries of north-eastern and north-central NA. Winters in south-central and south-eastern NA. **Similar species:** female Mallard (see p 146).

Gadwall *Anas strepera* (Anatidae) 45-57 cm (18-22 in). Non-descript grey-bodied duck with rusty scapulars, black tail-coverts, light brown head and neck. Female uniformly mottled brown. In flight both sexes show white belly, wing-linings outlined in dark feathers, and distinctive white speculum. Usually in small flocks, dives more than other dabbling ducks. In WP often with coots. Voice: male a low 'bek-bek', also whistles. Female quacks. **Range and habitat:** freshwater wetlands of north-western, central and west-central NA, eastern and northern WP. Winters in central and southern NA, AF, southern EP. **Similar species:** male and female resemble many other female ducks (see pp 146, 240, 288).

American Wigeon *Anas americana* (Anatidae) 45-58 cm (18-23 in). Prominent white forehead and crown, green patch from eye around nape; lower face and neck brown; breast and flanks pinkish buff, contrasting with white rear flank feathers and black tail-coverts. Male in eclipse and female brownish grey. Blue bill tipped black. Rapid, erratic flight shows elliptical white belly, white wing-linings and large white wing patches. Voice: male, 3 whistles, higher middle note; female, low quack. **Range and habitat:** ponds and marshes of northern and central NA. In winter coastal and southern NA and Central America. **Similar species:** Green-winged Teal, Eurasian Wigeon, Gadwall.

Blue-winged Teal *Anas discors* (Anatidae) 35-42 cm (14-17 in). The small size, fast and erratic flight and china-blue wing patches are distinctive in both sexes. Male's white crescent-shaped face patch contrasts with blue-grey head. Body tan with brown spots. White flank patch contrasts with black undertail-coverts. Female is small, mottled brown. Male in eclipse resembles female, but regains breeding plumage in mid-November. Voice: male a peeping whistle, female a faint quack. **Range and habitat:** nests in grass near prairie potholes and shallow marshes throughout central NA. Winters from southern NA to central NT. **Similar species:** other teal (see pp 146, 240).

Cinnamon Teal *Anas cyanoptera* (Anatidae) 37-43 cm (15-17 in). In breeding plumage, head and body of male are cinnamon-red. Feathers of back brown with buff highlights; red eyes. Females mottled brown. Both sexes have large chalky blue patches on upper surface of wings, paler than in Blue-winged Teal. Eclipse male resembles female. Flies in smaller flocks than other teal. Male gives low 'chuk-chuk-chuk', female weak quack. **Range and habitat:** marshes from west-central and central NA south, also central and southern NT. Winters south-western NA to northern NT. **Similar species:** other teal, Ruddy Duck (see pp 146, 190, 240).

Wood Stork

White Ibis

Trumpeter Swan

Canada Goose

Snow Goose

Brent Goose

North American Black Duck

Gadwall

American Wigeon

Blue-winged Teal

Cinnamon Teal

Ruddy Duck *Oxyura jamaicensis* (Anatidae) 38-40 cm (15-16 in). Short body and neck, large head and long, stiff tail are distinctive. Male chestnut with black neck and head, white cheek patch and bright, pale blue bill; grey in autumn and winter, but the light cheek patch contrasts with the grey head. Female grey with less strongly contrasting cheek crossed by a dark line. More apt to dive than fly. Voice: nasal sounds during breeding. **Range and habitat:** breeds on marshy lakes and ponds throughout plains, also west-central NA. Winters coast to coast from central NA to northern NT. Resident populations in West Indies and northern NT. **Similar species:** Masked Duck (see p 102).

Wood Duck *Aix sponsa* (Anatidae) 38-53 cm (15-21 in). Head green with iridescent green crest edged in white; bill and eye red-orange; white from chin and throat extends on to cheek and around neck; neck and breast maroon, flecked with white; belly white, flanks golden; back and wings dark, secondaries with white trailing edge. Female brownish green with light spots, unmarked grey crest, white around eye and on belly. Dark bill. Both sexes have long tails. Voice: male gives high-pitched, rising 'jeeeee'; female a squealing 'ww-e-e-ek'. **Range and habitat:** wooded swamps, rivers, ponds from east-central NA south, also west-central NA south along Pacific coast.

Canvasback *Aythya valisineria* (Anatidae) 48-56 cm (19-22 in). Male has white body framed by black tail, tail-coverts and breast; head and neck copper-red; eye red. Bill black; its sloping profile is diagnostic. Female and juvenile have brown heads and necks, but the light body and sloping profile are distinctive. Canvasbacks fly faster than most other ducks; the flicker created by the rapidly beating wings crossing the white body is distinctive. Male coos, growls; female quacks, purrs. **Range and habitat:** large inland lakes, coastal freshwater bays, north-western to central NA. Winters in central and southern NA. **Similar species:** Redhead, male Greater and Lesser Scaups (see p 148).

Redhead *Aythya americana* (Anatidae) 45-56 cm (18-22 in). Male's copper-red head has a rounded profile; bill blue with black tip; grey body with black breast, tail and tail-coverts. Female is uniformly brown, lighter than female scaup and Ring-necked Duck, with a white belly and grey bill tipped in black. Often occur singly, in pairs or in small flocks on migration. Flight rapid, low, straight. Lays eggs in nests of other ducks. Male, cat-like 'whee-oogh', purr; female, harsh 'squak'. **Range and habitat:** freshwater marshes of central and west-central NA. Winters from east-central and south-western NA south. **Similar species:** Canvasback, female scaup and Ring-necked Duck. **Not illustrated.**

Lesser Scaup *Aythya affinis* (Anatidae) 38-45 cm (15-18 in). Endemic to the western hemisphere. Dark head, neck, breast; grey back; white flanks, dark tail and tail-coverts; head with purple gloss; pale blue bill. Back darker with darker vermiculations on flanks than in Greater Scaup; white stripe extends only across secondaries. Female dark brown with white belly. Grey bill with white around base. Voice: soft 'whee-ooo' by courting male, otherwise loud whistles, scolding. **Range and habitat:** north-west to central NA. Winters along coasts and rivers, central NA to northern NT. More frequent in central NA and protected coastal areas than Greater Scaup. **Similar species:** Greater Scaup, Ring-necked Duck, Tufted Duck.

Ring-necked Duck *Aythya collaris* (Anatidae) 35-45 cm (14-18 in). Male's head, neck, back and tail black; bill grey, tipped black, encircled with white ring; flanks grey; white wedge separates chest from flank. In female bill is less marked; head and body brown, darker on crown and back. In both the head peaks toward back of crown. Forms large flocks on migration. Dabbles and dives. Male whistles; female growls. **Range and habitat:** small freshwater ponds, marshes, bogs northern and central NA. Winters from central NA south on both coasts into Central America, also interior south-east. **Similar species:** Scaup, Tufted Duck.

King Eider *Somateria spectabilis* (Anatidae) 53.5-61 cm (21-4 in) Stocky. White foreparts, black rear parts with white patches in wings. Modified scapulars project like sails. Pearl-grey crown, green cheeks, red bill and orange bill shield are unmistakable. Convex bill shield and rounded crown offer distinctive profile. Female warm brown, heavily barred. Immature male dusky with white foreparts. Gregarious, but in smaller flocks than other eiders. Dives up to 60 m (180 ft), only Oldsquaw dives more deeply. Voice: courting male dove-like cooing, female a low 'kuck'. **Range and habitat:** Coasts of north-western and north-central NA. Winters on the ocean and along north-western and north-eastern coasts. **Similar species:** Eider (p 148).

Barrow's Goldeneye *Bucephala islandica* (Anatidae) 40-51 cm (16-20 in). Black head with purple sheen and white crescent between black bill and golden eye; long nape feathers form partial crest; neck, breast, forward flanks white; back, tail, rear flanks black with white markings. Female and juvenile have brown heads with pale grey to white neck, breast and belly, darker back, flanks and tail; bill yellow to pinkish orange. White on wings, which whistle in flight. Mews and grunts during courtship. **Range and habitat:** breeds western mountains and north-eastern NA, western Iceland. Winters western, north-eastern and east-central coasts. **Similar species:** Common Goldeneye.

Harlequin Duck *Histrionicus histrionicus* (Anatidae) 35-48 cm (14-19 in). Short-billed duck. Breeding male blue-grey, bold white and chestnut pattern. Male in eclipse grey with white crescent at base of bill, traces of white breast band and white in wings. Female and autumn immatures grey-brown with white cheek and ear patches. Flight fast, erratic, low. Appear small, long-tailed, lack obvious white on wing. Squeaks and whistles. **Range and habitat:** north-eastern, north-western and west-central NA in turbulent streams. Winters along rocky coasts. **Similar species:** female Bufflehead, female scaup, female Surf Scoter.

Bufflehead *Bucephala albeola* (Anatidae) 33-9 cm (13-15 in). Small diving duck with large iridescent black head, large white semicircle wrapping over the nape from 1 eye to the other; back and rump black; breast, flanks, belly white. Flight reveals large white patches in the wings. Female has dark head with long white patch below eye, dark back, grey flanks, white chest and belly. White wing patches small. Male whistles; female croaks. **Range and habitat:** nests in old woodpecker holes from north-western to west-central and central NA. Winters in small flocks in protected coastal areas and ice-free lakes. **Similar species:** Hooded Merganser, Ruddy Duck, female Harlequin Duck.

Hooded Merganser *Lophodytes cucullatus* (Anatidae) 40-8 cm (16-19 in). Thin bill and crest give distinctive profile. Male has black head, white patch from behind eye to black-bordered crest; white breast separated from rusty flanks by 2 vertical black bars. In flight, male shows large white wing patches, female smaller patches, both appear long-tailed. Brownish grey female has rust-orange crest. Eclipse male like female, juveniles with small crest. Frog-like croak when courting. **Range and habitat:** nests in tree-cavities near fresh water in north-western, west-central NA, also north-eastern east-central NA. Winters on coasts, ice-free lakes. **Similar species:** Bufflehead.

Surf Scoter *Melanitta perspicillata* (Anatidae) 43-53 cm (17-21 in). Male has large black head, white patches on forehead and nape; deep orange, red, white and black bill extends on to face; body and wings black. Female brown with large bill and 2 light patches below eye. Juveniles 2 light facial patches, pale breast. First-winter males blacker than females, some bill colour, white nape. When diving uses partially folded wings in underwater 'flight'. Low whistles and grunts during courtship. **Range and habitat:** nests on ground near freshwater marsh or pond in northern NA. Winters along coasts and on Great Lakes. **Similar species:** other scoters, female Harlequin Duck (see p 148).

Ruddy Duck

Wood Duck drake behind

Canvasback

Lesser Scaup

Ring-necked Duck

King Eider

Barrow's Goldeneye

Harlequin Duck drake (left) and duck

Bufflehead

Hooded Merganser

Surf Scoter

DIURNAL BIRDS OF PREY Falconiformes

Turkey Vulture *Cathartes aura* (Cathartidae) 66-81 cm (26-32 in). Large, soars with wings in shallow V, outer primaries separated, often wobbles in turbulent air, rarely flaps; silvery grey flight feathers but blackish brown plumage on wings and body; long tail; legs dull orange. Head, small and naked, red in adults, grey in immatures. Lighter wing loading than Black Vulture so can live in cooler areas where thermals less strong. Grunts and hisses when disturbed. **Range and habitat:** central NA to southern NT. Winter in subtropical parts of range. **Similar species:** Black Vulture, eagles.

American Black Vulture *Coragyps atratus* (Cathartidae) 58-68 cm (23-7 in). Large, all-black with short, broad wings and a short broad tail. Holds wings horizontal, soars in tight circles quite often, often makes 4 or 5 laboured wingbeats between glides. White patches at base of primaries are conspicuous in flight. Head grey, feet pale. Sexes alike. Occasional grunts, hisses. **Range and habitat:** southeastern NA to southern NT. Less common than Turkey Vulture where both species present. **Similar species:** Turkey Vulture.

American Bald Eagle *Haliaeetus leucocephalus* (Accipitridae) 76-109 cm (30-43 in). Large, fish-eating and scavenging eagle with long broad wings; stout, hooked, yellow bill, eyes and feet also yellow; white head and tail with blackish brown body. Immature brown mottled with white, bill brownish, eyes pale yellow, feet yellow. Head and tail turn white in 4th or 5th year. Sexes alike. Hunts from perches overlooking water or occasionally on wing. Voice: harsh, high cackle. **Range and habitat:** breeds and winters on large bodies of water from north-eastern EP throughout NA. Population severely reduced by pesticides, now recovering. **Similar species:** African Fish Eagle (see p 150).

Sharp-shinned Hawk *Accipiter striatus* (Accipitridae) Male 25.5-29 cm (10-11½ in); female: 30-4 cm (12-13 in). Preys on small to medium-sized birds. Long legs, short rounded wings, long notched or squared tail. Adults blue-grey above, white below barred with rufous. Eyes yellow-orange, redder with age. Immatures brown above with white spots on mantle, cream below with brown streaks on breast, barring on flanks. Eyes pale yellow. Voice: shrill 'kik-kik-kik'. **Range and habitat:** nests throughout NA, winters central and southern NA, Central America. Western subspecies smaller. **Similar species:** Cooper's Hawk.

Cooper's Hawk *Accipiter cooperii* (Accipitridae) Male 38-44 cm (15-17 in); female 43-9 cm (17-19 in). Attacks medium-sized birds and mammals from ambush. Short rounded wings; long rounded tail; blue-grey above, white barred with rufous below; yellow-orange eyes, redder with advancing age. Immature brown, white spots on mantle, white underparts, streaked with brown; eyes pale yellow. Proportionately shorter wings, larger head than Sharp-shinned Hawk. Immature less heavily streaked than Sharp-shinned Hawk. Voice: flicker-like 'kek-kek-kek'. **Range and habitat:** breeds in central and southern NA. Winters in southern part of range to Central America. Western subspecies smaller. **Similar species:** Sharp-shinned Hawk.

Red-tailed Hawk *Buteo jamaicensis* (Accipitridae) 48-63 cm (19-25 in). Large, stocky, broad-winged, broad-tailed hawk with 2 colour phases, several subspecies. Light-phase adults dark head, throat and dorsum; brick-red tail; white breast, heavily barred belly. White wing-linings with dark wrists and patagial stripes. Dark phase: dark above and below with reddish tail and breast. Immature darker than adult with longer wings and tail, but tail reddish. Voice: asthmatic, descending 'keeerrrrr'. **Range and habitat:** NA and West Indies. Winter from central NA south. **Similar species:** Red-shouldered Hawk, Swainson's Hawk, dark-phase Ferruginous and Rough-legged Hawks.

Red-shouldered Hawk *Buteo lineatus* (Accipitridae) 43-61 cm (17-24 in). Slim with narrow wings, long tail. Dark above, head tinged rufous, blacker on back with white chequering; breast and shoulder rufous, belly and flanks barred rufous; tail with wide dark bands, narrow white bands. Sexes similar. Immature brown above mottled with white, light below streaked with brown. In flight shows white patch at base of outer primaries. Voice: loud 'kee-yar', pitch drops. **Range and habitat:** moist woodlands of east-central and south-central NA. Winters south NA. **Similar species:** Broad-winged, Red-tailed, Cooper's Hawks.

Broad-winged Hawk *Buteo platypterus* (Accipitridae) 35-48 cm (14-19 in). Both sexes stocky, dark brown above, whitish below with rufous barring. Tail has black and white stripes of equal width and a narrow white terminal band. Wing-lining white with dark border. Immatures dark above, often with white spotting, light below with brown streaks, tail crossed by 5 or 6 narrow dark bands. Often seen hunting from perch at edge of clearing. Voice: a high, whistled 'pweeeee', constant pitch, diminuendo. **Range and habitat:** deciduous forests from central and eastern NA. Winters Central America south to Brazil. **Similar species:** Red-shouldered Hawk.

Swainson's Hawk *Buteo swainsoni* (Accipitridae) 48-56 cm (19-22 in). When perched, folded wings extend beyond long tail. Light phase: brown above; tail grey-brown with from 5 to 7 narrow dark bands; throat white, breast brown; flanks, belly and legs cream; white wing-linings, dark flight feathers. Dark phase: dark with buff undertail-coverts. Immature dark brown above, white to buff below streaked and spotted with brown. Soars with wings in shallow V. Voice: shrill, plaintive 'kreeeeer'. **Range and habitat:** plains and sparse forest, central and west-central NA south. Winters in NT. Numbers declining. **Similar species:** Red-tailed Hawk, dark-phase Rough-legged and Ferruginous Hawks.

Ferruginous Hawk *Buteo regalis* (Accipitridae) 55-68 cm (22-7 in). 2 colour phases. Light phase; reddish brown above, pale head, white below with rufous barring on flanks; wing-linings white with dark crescent and narrow chestnut patagial stripe; legs feathered to toes with heavily barred chestnut feathers; tail white, reddish near tip. Melanistic phase (2-5%): clove-brown body, light tail. Immature dark above, mottled white with cinnamon on breast, white belly and tarsi. Tail white proximally, brownish distally. Voice: 'kree-a-ah'. **Range and habitat:** arid, open prairies central, west-central NA south. Winters in south-western NA. **Similar species:** Red-tailed, Swainson's and Rough-legged Hawks.

American Kestrel *Falco sparverius* (Falconidae) 19-20 cm (7½-8 in). Male rufous above, black barred rump, blue-grey wings, black primaries spotted with white; tail tipped with black and white bands; forehead and eyebrows blue-grey, throat and face white crossed by 2 vertical black bands; 2 black spots on nape; cinnamon below with black spotting, heavier on flanks and mid-section. Female has rufous wings, black barring on back, tail and below. Voice: high-pitched 'klee-klee-klee'. **Range and habitat:** common in NA along roadsides. In winter leaves edges of range. **Similar species:** Sharp-shinned Hawk, Merlin (see p 152).

Prairie Falcon *Falco mexicanus* (Falconidae) 35-45 cm (14-18 in). In flight dark wing-linings are distinctive. Adults sandy brown above with buff feather-edgings. Underparts buff to cream with brown spotting on breast and belly, barring on thighs. Sexes similar. Juveniles darker above and below with heavier streaking. Nests on cliff ledges. Hunts from perches or from low rapid flight. Captures prey on or near ground. Call a sharp 'kik, kik, kik', wails and whines during spectacular aerial courtship. **Range and habitat:** arid interior, west-central and central NA south. **Similar species:** Peregrine Falcon (see p 152).

■□□ **Turkey Vulture**
□■□ **American Black Vulture**
□□□ **American Bald Eagle**

■□□ **Sharp-shinned Hawk**
□■□ **Cooper's Hawk**
□□■ **Red-tailed Hawk** eating rattlesnake

■□□ **Red-shouldered Hawk**
□■□ **Broad-winged Hawk**
□□□ **Swainson's Hawk** (pale phase)

■□□ **Ferruginous Hawk**
□■□ **American Kestrel** (female)
□□■ **Prairie Falcon** with ground squirrel as prey

FOWL-LIKE BIRDS Galliformes

Northern Bobwhite *Colinus virginianus* (Phasianidae) 25 cm (10 in). Brown above streaked with chestnut; underparts white mottled with black. Male has black crown bordered by white stripe that extends from bill over the eye to side of black neck; a contrasting black stripe extends from bill under the eye; throat is white. Female's head pattern similar, but brown replaces black and buff replaces white. When frightened usually runs to heavy cover, freezes in group; if threat persists, flushes with whirr of short rounded wings. In cool weather aggregate in coveys of 8-25 birds. Voice: clear, whistled 'bob-white', rising at end. **Range and habitat:** open woodlands and brushy fields, east-central and central NA. **Similar species:** Grey Partridge (see p 154).

Wild Turkey *Meleagris gallopavo* (Phasianidae) 122-7 cm (48-50 in). Large, shy, ground-dwelling bird. Flies weakly. Body plumage iridescent bronze; primaries and secondaries dark brown and white barred; tail rust-coloured to brown, finely barred with black subterminal band and pale rust tip. Head unfeathered; in male, bluish with red wattles on throat and foreneck. Male also has a 'beard' of feathers projecting from the breast. Female smaller, slimmer, with less iridescence, smaller head and no 'beard'. Male gobbles and struts with wings drooped and tail fanned. Female clucks. **Range and habitat:** open woods and clearings, east-central and southern NA.

Sage Grouse *Centrocercus urophasianus* (Phasianidae) 68-86 cm (27-34 in). For most of year larger male and smaller female brown with pointed tails, black bellies and partially white undertail-coverts. Breeding male has large buff air sacs on upper breast, a black bib lies above and white collar surrounds the air sacs. Males display on communal leks where several hundred may gather, strutting with air sacs inflated and white collars expanded to contrast with drooping brown wings and black head. The tail forms a spiked backdrop. Male during courtship gives a 'plop-plop', female 'quak, quak'. **Range and habitat:** wherever big sagebrush occurs in western NA. **Similar species:** other grouse, all of which are smaller (see p 152).

Greater Prairie Chicken *Tympanuchus cupido* (Phasianidae) 43-5 cm (17-18 in). Both sexes brown and white barred with short brown tail. In spring male has elongated feathers which are erected to form ear-like structures above bright orange-red air sacs, which can be inflated. Orange combs above the eyes are erected and the male leans forward, stamps his feet, bobs and utters a booming, cooing sound, similar to blowing across the top of an empty bottle. Displays occur on traditional lek usually in early spring, from shortly before dawn to mid-morning, occasionally at dusk. **Range and habitat:** prairies of central and south-central NA. **Similar species:** Lesser Prairie Chicken, Sharp-tailed Grouse.

Spruce Grouse *Dendragapus canadensis* (Phasianidae) 33-40 cm (13-16 in). Small, unusually tame. Male has black breast and throat, black tail tipped with rust, white-tipped undertail-coverts, finely barred grey and black on back and rump; scarlet comb over each eye; sides, lower chest and belly finely barred dark brown and buff with white mottling. Male Franklin's, a subspecies of Rocky mountains, lacks rust tail tips, but has white-tipped uppertail-coverts. Female is brown with black barring, short fan-shaped tail with terminal buff band. Courtship includes strutting with spread tail oscillating side to side and short flights on whirring wings. **Range and habitat:** boreal forest of NA. **Similar species:** Ruffed Grouse, female Blue Grouse.

CRANES, RAILS AND ALLIES Gruiformes

Sandhill Crane *Grus canadensis* (Gruidae) Greater: 102-27 cm (40-50 in); Lesser and Cuban: 51-76 cm (20-30 in). Long neck and legs. Both sexes pale grey; nonmigratory, smaller southern subspecies darker than migratory larger northern subspecies. Dark green bill, yellow eye, unfeathered red crown; black legs. In spring adults smear iron-containing mud on feathers, giving plumage a rusty cast. Juveniles reddish brown, crown covered with brown feathers. Flies with neck and feet outstretched. Voice: loud, rolling, bell-like rattle that carries well. **Range and habitat:** shallow open wetlands and grassy uplands from eastern EP to north-central NA. South in scattered populations through central and western NA. Winters in south-western NA. Resident subspecies in south-eastern NA. **Similar species:** Whooping Crane.

Whooping Crane *Grus americana* (Gruidae) 127-52 cm (50-60 in). The tallest North American flying bird has white plumage except for black primaries; naked crown and cheeks are red; eyes yellow, bill greenish, legs black. Juveniles are cinnamon-brown with white belly and secondaries, black primaries and legs, pale bill, feathered crown. Population has increased from a low of 15 in 1937. Voice: exceptionally loud, higher-pitched and more continuous than that of Sandhill Crane. **Range and habitat:** breeds in Wood Buffalo National Park near Arctic circle in Northwest Territories, Canada and winters in Aransas National Wildlife Refuge in Texas. **Similar species:** Sandhill Crane.

Limpkin *Aramus guarauna* (Aramidae) 63-8 cm (25-7 in). Dull brown, heavily streaked and spotted with white; bill pale at base, dark at tip, long, slender, slightly decurved; neck long, legs long and dark. Flies with neck and feet outstretched. Wingbeat slow, with crane-like upward flick. Slinks quietly through marsh like rail, often pumps tail when walking. Voice: a loud, wailing scream 'kur-r-ee-ow', like cries of lost child, given most often at twilight, at night and on cloudy days. **Range and habitat:** locally common in freshwater marshes with open water channels, along marshy river-banks from south-eastern NA, West Indies, to southern NT. **Similar species:** immature night herons and ibises (see p 52).

American Coot *Fulica americana* (Rallidae) 33-40 cm (13-16 in). The white bill and frontal shield against the black head and slate-grey body are distinctive. Sides of undertail-coverts are white, forming a distinctive pattern from the rear. Legs greenish and lobed around each toe-bone. Head oscillates when swimming. Feeds on vegetation by dipping or diving. Runs across water when taking off. Chick has orange head and breast. Immature like adult, but paler with dusky bill. Voice: loud cackles, whistles, croaks, grunts, also loud 'kuk-kuk-kuk'. **Range and habitat:** freshwater marshes from central NA in Andes to north-western NT. Winters, often in large flocks, on ice-free fresh and salt water. **Similar species:** other coots, Moorhen, Purple Gallinule (see pp 154, 196).

American Purple Gallinule *Gallinula martinica* (Rallidae) 30-5 cm (12-14 in). Both sexes have deep purple head, neck and underparts blending through blue to bronze-green on back and wings; undertail-coverts white; bill red with yellow tip and pale blue frontal shield; legs yellow with long toes which spread weight as it walks across floating vegetation. Immature has buff head, dark brown back and wings, copper-coloured underparts, white undertail-coverts; legs yellow, bill dark. In flight legs and feet dangle. Associates with coots and moorhens, but more apt to stay in cover. **Range and habitat:** uncommon resident of freshwater marshes from south-eastern NA, West Indies, to central NT. Winters in tropical and subtropical parts of range. **Similar species:** Moorhen, immature American Coot (see p 154).

Northern Bobwhite

Wild Turkey

Sage Grouse male in courtship display

Greater Prairie Chicken male in courtship display

Spruce Grouse (female)

Sandhill Crane

Whooping Crane

Limpkin

American Coot

American Purple Gallinule

Virginia Rail *Rallus limicola* (Rallidae) 23 cm (9 in). Only small North American rail with long bill. Crown and back olive, feathers of back edged in buff, grey face, rust-coloured throat and breast; flanks strongly barred black and white; undertail-coverts white; long orange and black bill; legs brown with long toes. Immature has olive crown and back, blackish below, bill and legs dark. Voice: distinctive 'kidick-kidick', also descending laugh, various grunts. **Range and habitat:** local breeder in freshwater marshes from central NA to south-eastern NT. Winters in coastal saltmarshes. **Similar species:** Clapper Rail.

Clapper Rail *Rallus longirostris* (Rallidae) 35-40 cm (14-16 in). Olive or grey-brown with dark streaks above, grey cheeks, flanks with vertical grey and white bars, white undertail-coverts. Bird of Atlantic coast grey-brown throat and breast. Birds along Gulf and Pacific coast have rusty throat and breast, generally richer colour than Atlantic form. Bill orange and dark, long, slightly drooped. Voice: distinctive 'kek kek kek' repeated 20-5 times. Noisiest at dusk when emerges to feed on mudflats. **Range and habitat:** Atlantic and Pacific saltmarshes, central NA to northern NT, West Indies. **Similar species:** Virginia Rail.

King Rail *Rallus elegans* (Rallidae) 38-48 cm (15-19 in). Cinnamon replaces grey-brown of Clapper Rail. Rust-coloured cheeks; shoulders, breast bright rust, flanks barred black and white; dark streaks on back and tail more contrasting than in Clapper Rail; white undertail-coverts. Like Clapper Rail often cocks tail. Toes not webbed, but swims well and often. Flies rarely and with legs dangling. Voice: low, grunting 'bup-bup'. **Range and habitat:** freshwater marshes throughout east-central and south-eastern NA. Winters along coast in saltmarshes south to Cuba. **Similar species:** Clapper and Virginia Rails.

Sora (Sora Crake) *Porzana carolina* (Rallidae) 20.5-25.5 cm (8-10 in). Often feeding at edge of marsh openings. Olive-brown crown, back and wings with black spots; face black; cheeks, foreneck, upper breast grey, belly white, flanks barred white and brown; undertail-coverts white; bill short, yellow. When walking, flicks short tail up. Immature brown above, buff below, with barred flanks. Voice plaintive rising 'ker-wee', sharp 'keek', descending whinny. Noisiest at twilight. **Range and habitat:** marshes and wet meadows, north-central, central and south-western NA. Winters from southern NA to northern NT, also West Indies. **Similar species:** Virginia and Yellow Rails.

SHOREBIRDS, GULLS, AUKS Charadriiformes

Killdeer *Charadrius vociferus* (Charadriidae) 23-8 cm (9-11 in). Only North American plover with 2 black breast bands. White forehead bordered by black; crown, back and wings brown; white eyebrow and black moustache; white collar and underparts; uppertail-coverts rufous, tail rufous edged in black and white. Sexes alike. Young and immatures have single black breast band. Ringing, repeated 'kill-dee' or cries of 'deer-deer-deer'. **Range and habitat:** meadows, pastures and dry uplands of NA; also west-central NT. Winters from central NA into north-western NT. **Similar species:** other belted plovers (see pp 110, 156, 200, 294).

Semi-palmated Plover *Charadrius semipalmatus* (Charadriidae) 16.5-19 cm (6½-7½ in). Single black breast band and stubby orange and black bill. Forehead white surrounded with black; nape, back, wings colour of wet sand; white collar encircles neck, underparts white; legs orange. In flight white stripe extends along wings. Winter birds have brown breast bands, dark bills, dull legs. Sexes similar. Immature has black bill and brown breast band. Voice: plaintive, rising 'too-li'; on breeding grounds a liquid rattle. **Range and habitat:** tundra of NA. Winters on coasts from southern NA to southern NT. **Similar species:** all belted plovers (see pp 110, 156, 200, 294).

American (Lesser) Golden Plover *Pluvialis dominica* (Charadriidae) 24-8 cm (9½-11 in). Upperparts spotted black, yellow and gold; face, foreneck and underparts black; forehead, eyebrow, sides of neck and breast white. In winter speckled brown above, pale below. Juveniles heavily spotted with golden brown dorsally, underparts streaked and barred, eyebrow white, rump and tail dark. Never a wing-bar. Voice: plaintive 'queedle', dropping at end. **Range and habitat:** smaller, more brightly coloured western subspecies breeds from eastern EP to western NA, migrates Hawaii and South Pacific islands; larger subspecies breeds across northern NA, winters in southern NT. **Similar species:** European Golden Plover (see p 156).

Upland Sandpiper *Bartramia longicauda* (Scolopacidae) 28-32 cm (11-13 in). Small head with large eyes on long, thin neck; large body, long tail projecting well beyond the folded wings; long yellow legs; straight yellow bill. Brown above, feathers edged with buff; light below. In flight wings long, pointed, dark. Flaps below horizontal. After landing often extends wings vertically. Runs in spurts through grass, often perching on rocks or hillocks. Voice: mellow, whistled 'kip-ip-ip', song collection of reedy whistles and trills. **Range and habitat:** grassy prairies, open meadow north-western to central and east-central NA. Winters in Patagonia. **Similar species:** yellowlegs, Solitary Sandpiper.

Greater Yellowlegs *Tringa melanoleuca* (Scolopacidae) 32-8 cm (13-15 in). Long chrome-yellow legs. Both sexes black and grey above mottled with white; white below, heavily to lightly barred; bill slightly upturned; wings dark, rump white. In winter brownish grey above, throat and breast finely streaked with grey, white below. Juvenile blackish brown above with dense white spotting, white below with dark bars. Voice: 3 whistled syllables, in spring a yodelling 'whee-oodle'. **Range and habitat:** muskeg of northern NA. Winters along coasts central NA to southern NT. **Similar species:** Lesser Yellowlegs, Solitary Sandpiper, Common Greenshank, Common Redshank (see p 158).

Lesser Yellowlegs *Tringa flavipes* (Scolopacidae) 24-8 cm (9½-11 in). Often feeds in large loose flocks, unlike more solitary Greater Yellowlegs. Most distinctive feature is thin, straight black bill, about same length as head or slightly longer. Legs proportionately longer, giving a distinctly long-legged appearance. Belly white. Juveniles have fine grey streaks on breast and are browner above than adults. Voice: high, flat soft 'tew' or 'tew tew'. **Range and habitat:** marshes and bogs of north-west and north-central NA. Winters on coast from southern NA to southern NT. **Similar species:** Greater Yellowlegs, Solitary Sandpiper, Common Redshank, Common Greenshank (see p 158).

Spotted Sandpiper *Actitis macularia* (Scolopacidae) 18-20.5 cm (7-8 in). Breeding adults grey-brown above, flecked with dark; whitish eyebrow, dark eye stripe; white below with conspicuous dark spots; bill stout, pale orange, dark tip, droops slightly. In teetering and stiff-winged flight, shows white wing stripe, rounded tail with outer feathers barred white and black. Winter and juvenile plumages white below without spots, grey-brown above. Voice: whistled 'peet weet', 2nd note often repeated. **Range and habitat:** pebbly freshwater shores throughout NA, except southern extreme. Winters from southernmost NA to southern NT, has bred in WP. **Similar species:** Common Sandpiper, Solitary Sandpiper (see p 158).

Solitary Sandpiper *Tringa solitaria* (Scolopacidae) 19-23 cm (7½-9 in). Conspicuous white eye-ring, dark olive above flecked with white; dark tail, outer feathers barred with white; neck and breast heavily streaked with black, throat and belly white; short, straight, olive bill. Juveniles and winter adults, washed with brown and faintly streaked below, greyer above. Dark rump distinguishes it from yellowlegs. Swallow-like flight. Voice: whistled 'peet!' or 'peet-weet-weet!', higher than Spotted Sandpiper. **Range and habitat:** streamsides, wooded swamps, ponds and marshes throughout boreal NA. Winters from southern NA to south-eastern NT. **Similar species:** yellowlegs.

Virginia Rail

Clapper Rail

King Rail

Sora

Killdeer

Semi-palmated Plover

American (Lesser) Golden Plover

Upland Sandpiper

Greater Yellowlegs

Lesser Yellowlegs

Spotted Sandpiper

Solitary Sandpiper

Semi-palmated Sandpiper *Calidris pusilla* (Scolopacidae) 14-18 cm (5½-7 in). Most common of 'peeps' – small sandpipers. Greyish brown and black above with rufous-edged feathers, white below with dark brown streaks across upper breast and along flanks. In winter grey-brown above, whitish eyebrow, dusky face and sides of breast, white below. Sexes similar. Juveniles scaly buff and black above, scapulars rufous-edged, wing-coverts light-edged. Upperparts white with buff band across upper breast. Legs, feet and bill black. Voice: harsh 'chit'. **Range and habitat:** arctic tundra of NA. Winters along southern coasts of NT. **Similar species:** other 'peeps' (see pp 158, 250).

Western Sandpiper *Calidris mauri* (Scolopacidae) 15-18 cm (6-7 in). In breeding plumage, separated from Semi-palmated Sandpiper by rusty-margined scapulars; slightly longer bill, drooped at tip; rusty tinge to crown, ear-coverts and back; V-shaped spots across breast and along flanks. Feeds in deeper water than Semi-palmated. In winter (after late September) distinguished from Semi-palmated only by call, a high-pitched 'cheep'. Juvenile has bright rusty-edged scapulars, but moults rapidly. **Range and habitat:** arctic tundra of north-western NA. Winters along both coasts from central NA to NT. **Similar species:** other 'peeps' (see pp 158, 250).

Least Sandpiper *Calidris minutilla* (Scolopacidae) 13-15 cm (5-6 in). Dull yellow or greenish legs and feet separate it from other 'peeps'. Sooty brown above with narrow chestnut and buff feather-edges, which wear away leaving sooty brown appearance. In flight, wing-bar faint. Foreneck and upper breast pale buff with coarse brown streaking. Winter plumage dusky grey. Sexes similar. Juveniles have rufous edges to feathers of crown and back, incomplete white V on back. Voice: thin, drawn-out 'kreeeet'. **Range and habitat:** wet muddy or grassy areas in northern NA. Winters from southern coasts of NA to northern NT. **Similar species:** other 'peeps' (see pp 158, 250).

White-rumped Sandpiper *Calidris fuscicollis* (Scolopacidae) 18-20.5 cm (7-8 in). Folded wings extend beyond tail. In flight shows white wing-bar and white rump. Crown and ear-coverts edged chestnut, dorsal feathers edged buff and chestnut; white below with dark brown streaks on breast and flanks. Winter birds grey-brown with white eyebrow, rump and underparts. Sexes similar. Juveniles like breeding birds but with incomplete white V on the sides of the back. Feet black. Voice: high-pitched, thin, mouse-like 'jeet'. **Range and habitat:** tundra of north-western and north-central NA. Winters in southern NT. **Similar species:** other 'peeps' (see pp 158, 250).

Baird's Sandpiper *Calidris bairdii* (Scolopacidae) 18-19 cm (7-7½ in). Folded wings extend beyond tail. Face, foreneck, upper breast buff thinly streaked with brown; back blotchy grey-brown; underparts white; legs black; bill black, straight and slightly tapered. Winter plumage like White-rumped Sandpiper, but dark rump. Juvenile as adult, but more scaly than blotchy. Voice: low, soft 'krrrit', also a rolling trill. **Range and habitat:** tundra from north-eastern EP to north-eastern NA. Frequents muddy, sandy or grassy areas on migration, usually in small flocks. Winters in southern NT. **Similar species:** other 'peeps', Buff-breasted and Pectoral Sandpipers, Sanderling (see p 158).

Stilt Sandpiper *Micropalama himantopus* (Scolopacidae) 19-23.5 cm (7½-9 in). In flight white rump and tail and no wing-bar are distinctive. Black mottled with white above; ear-coverts and nape tinged chestnut, underparts white, heavily barred with black; legs greyish green to yellow; bill black, long, outer 3rd drooped. Winter plumage brownish grey above, whitish eyebrow and underparts. Sexes alike. Juvenile blackish brown above with buff feather-edges, foreneck and breast cream, faintly streaked with brown. Voice: monosyllabic 'querp'. **Range and habitat:** tundra of northern NA. Winters from extreme south NA to south NT. **Similar species:** Lesser Yellowlegs (see p 62).

Pectoral Sandpiper *Calidris melanotos* (Scolopacidae) 23-4 cm (9-9½ in). Dorsal feathers sooty brown with buff edges; dense dark brown streaking on foreneck and breast ends abruptly at lower breast, contrasts with white belly, flanks and undertail-coverts. Winter plumage greyer. Legs and feet yellow, occasionally green or brown; bill black. Sexes similar. Juveniles brighter than adults, incomplete buff V on sides of back, pectoral streaking suffused with buff. Voice: reedy 'trrrip trrrip'. **Range and habitat:** tundra from EP to north-central NA. On migration favours grassy wetlands. Winters in southern NT. **Similar species:** Least and Sharp-tailed Sandpipers.

Buff-breasted Sandpiper *Tryngites subruficollis* (Scolopacidae) 19-21.5 cm (7½-8½ in). Large eye set off by white eye-ring in buffy face, dorsal feathers dark brown with buff edging provide scaly appearance, underparts buff; crown speckled; bill short, tapered, pale at base, feathering extends far down bill; legs yellow. In flight buff body contrasts with white wing-linings, no wing-bar. Juveniles, dorsal scaly appearance enhanced by white edges to feathers. Voice: low trill 'prr-r-r-rret', sharp 'tik'. **Range and habitat:** tundra of northernmost NA. Favours grassy areas on migration. Winters in southern NT. **Similar species:** Upland Sandpiper, juvenile Baird's Sandpiper.

Short-billed Dowitcher *Limnodromus griseus* (Scolopacidae) 27-30.5 cm (10½-12 in). Characteristic chunky body, very long straight bill, white triangle on back. Subspecies include *hendersoni* (reddish brown underparts, some white on belly, lightly spotted with black from throat to undertail-coverts), *griseus* and *caurinus* (reddish underparts with dense spotting on breast, often including scalloped bars, white belly). All forms have barred flanks and barred tail feathers. Long-billed Dowitchers differ by salmon-red bellies and black tail feathers. Voice of Short-billed staccato 'tu-tu-tu'. **Range and habitat:** north-western (*caurinus*), north-central (*hendersoni*) aand north-eastern (*griseus*) NA. Winter along southern coasts to central NT. **Similar species:** Long-billed Dowitcher, Common Snipe (see p 160).

Willet *Catoptrophorus semipalmatus* (Scolopacidae) 36-41 cm (14-16 in). Black and white wings and white rump. Breeding plumage: eastern subspecies grey-brown finely barred with black; western larger, paler, greyer, may have pinkish wash on breast. Bill long, straight, heavy, pale pink or blue-grey with black tip. Sexes similar. Voice: musical 'pill-will-willet', loud 'kay-ee', repeated 'kip-kip-kip'. **Range and habitat:** western subspecies breeds in central NA, eastern along east-central and south-eastern coasts. Winters from central coasts south. **Similar species:** Greater Yellowlegs, winter plumage Hudsonian Godwit (see p 66).

Long-billed Curlew *Numenius americanus* (Scolopacidae) 51-66 cm (20-6 in). Exceptionally long, sickle-shaped bill. Warm brown mottled with black on upper parts, crown and head unstriped, underparts buff. Cinnamon wing-linings conspicuous in flight. Female much larger. Juvenile has shorter bill. Call a loud, clear, rising 'cur-lee', repeated. **Range and habitat:** breeds in grasslands of central and west-central NA, frequents marshes, mudflats, shorelines on migration. Winters southern NA to north-western NT. **Similar species:** Marbled Godwit, Whimbrel (has crown stripes), Curlew (see p 160).

Marbled Godwit *Limosa fedoa* (Scolopacidae) 40-51 cm (16-20 in). Slightly smaller than Long-billed Curlew, bill upturned, very long, salmon-coloured with dark tip; dark back, heavily mottled with buff; wing-linings, secondaries and inner primaries cinnamon. Outer primaries, primary coverts blackish; underparts cinnamon-buff and barred; tail finely barred black and cinnamon to buff. Juveniles and winter adults, underparts unmarked, may look paler or greyer. Voice: accented 'godwit', also 'raddica, raddica'. **Range and habitat:** breeds in wet meadows of central NA. Winters on southern coasts to west-central NT. **Similar species:** other godwits, Long-billed Curlew (see p 160).

Semi-palmated Sandpiper

Western Sandpiper

Least Sandpiper

White-rumped Sandpiper

Baird's Sandpiper

Stilt Sandpiper

Pectoral Sandpiper

Buff-breasted Sandpiper

Short-billed Dowitcher

Willet (eastern subspecies)

Long-billed Curlew

Marbled Godwit

Hudsonian Godwit *Limosa haemastica* (Scolopacidae) 35-40 cm (14-16 in). Very long, slightly upturned bill. Recognized in flight by black axillaries and wing-linings, narrow white stripe along wing, white tail with terminal black band. Mottled brownish black above, chestnut below barred with black and white. In winter brownish grey with white lower breast and belly. Juvenile brownish black feathers edged with cinnamon, buff below. Bill salmon tipped with black, paler in juveniles. Voice: loud high-pitched 'godwit!'. **Range and habitat:** breeds in north-western and north-central NA. Winters in southern NT. **Similar species:** other godwits, Willet (see pp 64, 160).

American Woodcock *Scolopax minor* (Scolopacidae) 25-30 cm (10-12 in). Plump shorebird of wet thickets, alder swales. Rusty grey forehead, face; black crown crossed by 4 rusty bars; back mottled rust, black, brown and grey; underparts rusty buff, unmarked; legs short, pale; bill long, pale. Eyes set high on head, providing binocular vision forward and backward. Performs courtship dance and flight in meadows during spring. Outer primaries narrowed to produce a twittering sound in flight. Voice: nasal 'peent', liquid song during courtship flight. **Range and habitat:** east-central and south-eastern NA. Winters in south-eastern NA. **Similar species:** Snipe, other woodcocks (see p 160).

Black-necked Stilt *Himantopus mexicanus* (Recurvirostridae) 34-9 cm (13-15 in). Black above; white forehead, patch behind eye, lower face, foreneck and underparts; needle-like black bill, very long pinkish red legs. In flight wings show black above and below, axillaries and tail white, neck and legs extended. Female and juvenile brownish black where male is black. Strides along shores or wades up to its belly picking insects off surface of water. Voice: sharp yipping. **Range and habitat:** freshwater and brackish ponds, open marshes, wet meadows along coasts from central NA to southern NT; locally in south-western NA. In winter withdraws from extremes of range. **Similar species:** Black-winged Stilt, Australian Stilt (see pp 156, 294).

American Avocet *Recurvirostra americana* (Recurvirostridae) 43-7 cm (17-18 in). White with large black patches on wings and back; head and neck cinnamon; black bill tapers to fine point and turns up, more so in females; legs long, grey-blue. In flight wing-linings white, outer half of wing, greater and median coverts black, secondaries white. Juvenile and winter plumages, head and neck greyish white. When feeding, rapidly sweeps bill back and forth over muddy water. **Range and habitat:** freshwater marshes and saline lakes and marshes, west-central and central NA south. Winters southern NA. **Similar species:** Avocet (see p 156).

Surfbird *Aphriza virgata* (Scolopacidae) 25.5 cm (10 in). In breeding plumage, black and white streaked and spotted above with chestnut and black scapular feathers; blackish spots and bars on white underparts, belly white; short, dark plover-like bill, with yellow base on lower mandible. Winter adults and juveniles brownish grey head, back, breast; dark spotting along flanks, white belly. All plumages show prominent white wing stripe and white tail in flight. Voice: 'kee-wee' or 'ke-wee-ek'. **Range and habitat:** breeds in north-western NA. Winters on Pacific coast from north-western NA to southern NT. **Similar species:** Black Turnstone, Rock Sandpiper.

Black Turnstone *Arenaria melanocephala* (Scolopacidae) 23 cm (9 in). Black head, breast and back; white spot between eye and bill; white flecks across forehead, crown, sides of head, neck and breast; belly white. Winter adults and juveniles lack white flecks; head, breast, flanks somewhat brownish. In flight shows striking pied pattern with white shoulders, white stripe across secondaries and inner primaries, white lower back and base of tail. Short dark bill, slightly upturned. Legs short, dark. Voice: grating rattle. **Range and habitat:** frequents rocky shores, also rotting kelp, north-western NA. Winters along western coasts of NA. **Similar species:** Ruddy Turnstorne, Surfbird (see p 156).

Wilson's Phalarope *Phalaropus tricolor* (Scolopacidae) 21.5-4 cm (8½-9½ in). Slim phalarope with white rump and tail, but no wing stripe. Breeding female, grey crown, nape; dark stripe through eye, down neck, white below; chestnut wash across foreneck and breast; back lead grey with chestnut stripes. Bill black, needle-like, longer than head. Male similar, paler. Juveniles brown mottled with buff above, buff wash on sides of breast. In winter, white forehead, dark eye-line, pale grey above, white below with yellow legs. Pronounced sex-role reversal. **Range and habitat:** shallow lakes and marshes in central, west-central, and south-western NA. Winters in southern NT. **Similar species:** other phalaropes.

Laughing Gull *Larus atricilla* (Laridae) 40-3 cm (16-17 in). Slim, long-winged gull with lead-grey mantle, black wing tips, white body, black head, white crescents above and below dark eye; bill red, legs dark. In winter head white, mottled with grey, bill black. Subadult grey, darker above, brown in mantle. Juvenile brown, back scaly. Subadult, juvenile have white tail with black band. Voice: 'hah-ha-ha-ha-hah-haah-haaaaa', drawn out into wail. **Range and habitat:** saltmarshes, coastal bays, beaches, east-central and southern NA, visits south-western NA. Winters southern NA to central NT. **Similar species:** other black-headed gulls (see pp 162, 250).

Franklin's Gull *Larus pipixcan* (Laridae) 33-8 cm (13-15 in). Small, compact. White stripe separates grey mantle from black outer primaries, which have white tips. White body, black head, white crescents above and below eye; bill scarlet, legs dark red. In winter half-hood covers head, eye crescents visible. Subadults have half-hood, eye crescents, grey mantle mottled with brown, greyish tail with black band, incomplete in 1st-year birds. Voice: shrill 'kuk-kuk-kuk', mewing, laughing cry. **Range and habitat:** prairie marshes of central NA. Winters along coasts of southern NA to central NT. **Similar species:** other black-headed gulls (see pp 162, 250).

Ring-billed Gull *Larus delawarensis* (Laridae) 45-51 cm (18-20 in). Black band encircles yellow bill. Yellow-green legs; head, body, tail white; mantle pale grey; wing tips black with white spots; eyes yellow. In winter neck streaked. Immature white with dusky spots, grey mantle mottled brown, incomplete subterminal tail band. Juvenile lighter than Herring Gull, mantle grey with brown carpal bar, dark secondary bar. Tail white with mottled base and incomplete subterminal band. Voice: 'kree-kree-kree' when squabbling. **Range and habitat:** central and south-eastern NA. Winters on Great Lakes, along Mississippi, and south-eastern coast. **Similar species:** other gulls.

Bonaparte's Gull *Larus philadelphia* (Laridae) 30-5 cm (12-14 in). Pointed wings; square tail; black head; narrow white eye crescents; white body; grey mantle with black tips to primaries; bill dark, legs red. Conspicuous white wedge in outer primaries in all plumages. In winter head white, black ear patch, legs pink. First winter: blackish red bill, wing has black trailing edge with narrow white border, dark secondary bar, narrow black tail band. Voice: nasal 'cheer', conversational notes. **Range and habitat:** nests in conifer trees, north-western and north-central NA. Winters along central and southern coasts and rivers. **Similar species:** other black-headed gulls (see pp 162, 250).

Glaucous Gull *Larus hyperboreus* (Laridae) 53-66 cm (21-6 in). Large, heavy-bodied gull with heavy yellow bill, yellow eye and yellow legs. Head, body, tail white; mantle pale grey; primaries white. 1st-winter birds white mottled with pale grey-buff, bill pink with dark tip, legs pink. 2nd-year birds some grey in mantle, by 3rd year mantle grey, some buff mottling in body plumage, bill dark-tipped. Voice: hollow 'cowp', also deep 'err-ul'. **Range and habitat:** nests colonially around sea cliffs along Holarctic coasts. Winters from arctic south along coast to mid-latitudes. **Similar species:** other white-winged gulls (see p 162).

Hudsonian Godwit

American Woodcock

Black-necked Stilt

American Avocet

Surfbird

Black Turnstone

Wilson's Phalarope

Laughing Gull

Franklin's Gull

Ring-billed Gull

Bonaparte's Gull

Glaucous Gull

Western Gull *Larus occidentalis* (Laridae) 51-8 cm (20-3 in). White-bodied gull with yellow bill, dark eye, pink legs, slate-grey (southern California) to medium grey (central and northern California) mantle with white trailing edge, and black wing tips with white spots. Underside of wing shows pronounced dark shadow along secondaries. In winter dusky on head. First-winter birds mottled grey-brown. Second winter, mottled brown, whiter below, some grey in mantle. Third winter, white with dusky smudges, mantle grey-brown. Voice: gutteral 'kuk-kuk-kuk', also 'whee whee whee' and 'ki-aa'. **Range and habitat:** west-central and south-western Pacific coast. Winters in southern part of range. **Similar species:** Slatey-backed and Lesser Black-backed Gulls (see p 162).

California Gull *Larus californicus* (Laridae) 45-51 cm (18-20 in). Grey mantle, darker than that of Herring Gull. Black wing tips with white spots; head and body white; yellow bill has black and red spots; legs yellow-green; eyes dark. Sexes similar. First winter: mottled grey-brown above, lighter below; dark primaries; pink legs; pink bill, dark-tipped. Second winter: white mottled with grey-brown, grey mantle mottled brown, primaries dark, legs grey-green, bill grey-pink with black tip. Third winter: more grey in mantle, less grey-brown mottling in white. Voice: squealing 'kiarr'. **Range and habitat:** lakes, rivers, farmland, central to south-western NA. Winters on Pacific coast from west-central to south-central NA. **Similar species:** other grey-backed gulls (see p 162).

Pomarine Skua *Stercorarius pomarinus* (Laridae) 51 cm (20 in). Flight slow, even, direct, with shallow wingbeats. Large head and bill, broad wings give bulky appearance. Light phase has light bill, dark cap, yellow wash to nape and cheeks, white below, mottled brown breast band extending along flanks, dark back and wings; blunt-tipped, twisted central tail feathers project beyond rest of tail; in flight base of primaries shows white. Dark phase best recognized by tail feathers, flight pattern, bulky appearance. Sexes similar. Juveniles brown, heavily barred below, lack distinctive rectrices; so do subadults, which are otherwise similar to adults. **Range and habitat:** Holarctic tundra. Winters at sea. **Similar species:** other skuas (see pp 160, 296).

Forster's Tern *Sterna forsteri* (Laridae) 35-42 cm (14-17 in). Silvery white upper surface of primaries, paler than light grey of mantle; tail grey, darkest around inner edge of fork, with outer web of outermost feathers white; black cap, white body, orange bill with black tip, orange legs. Sexes similar. In winter crown is white, black covers eye and ear to nape, which varies from pale grey to blackish. Juvenile has same mask with brownish upperparts. Bill brownish black, legs yellow. Voice: 'zreep', raspy buzz like that of a nighthawk, shrill 'pip-pip, pip, pip'. **Range and habitat:** breeds on interior marshes of central and west-central NA, also south-eastern coast. Winters along southern coasts. **Similar species:** Common Tern (see p 162).

Royal Tern *Thalasseus maximus* (Laridae) 45-53 cm (18-21 in). Large, common white-bodied tern with grey mantle and black cap, spotted with white on forehead, and black crest projecting from nape. In nonbreeding and immature plumage forehead is white. In western subspecies, white of forehead extends behind eye. Bill large, orange to yellow-orange. In flight underside of primaries shows light. Plunge-dives for fish. Voice: 'dhirrup', 'keer' and 'kaak'. **Range and habitat:** saltwater habitats of east-central and south-eastern NA, west AF. Nests on sandy beaches with other colonial waterbirds. Parents recognize their own eggs. Winters along southern NA coasts, coastal NT, also west AF. **Similar species:** Caspian and Elegant Terns.

Black Skimmer *Rynchops niger* (Laridae) 40-51 cm (16-20 in). Lower mandible longer than upper and laterally flattened to form blade; upper mandible shorter, wider; both mandibles red at base, black toward tip. Feeds by flying swiftly and evenly just above surface, with lower mandible slicing through water ahead of the bird, flipping small fish and crustaceans into mouth. Black above with white forehead, foreneck, wing-lining and underparts; tail short, slightly notched. Sexes similar. Juvenile similar to adult, but with mottled brown upperparts. Voice: nasal barking notes, soft 'keow, keow'. **Range and habitat:** nests with other seabirds on sand spits, barrier beaches and dredge-spoil islands from east-central NA to central NT. Winters from southern NA to central NT. **Similar species:** African Skimmer (see p 202).

Pigeon Guillemot *Cepphus columba* (Alcidae) 30-5 cm (12-14 in). Thin neck, rounded head, long pointed bill contribute to distinctive guillemot profile. In all plumages, large white wing patches with 1, sometimes 2, black wedges; bill dark red to black, mouth and feet brilliant orange-red. Breeding birds black. In winter, mottled grey and white with dark wing-linings. Sexes similar. Immatures dusky grey above, wing patches smaller than in adult, crossed by indistinct dark bar. Voice: whistled 'peeee'. **Range and habitat:** nests in natural cavities above intertidal zone from north-eastern EP and north-western NA to east-central EP and west-central NA. Winters at sea south of arctic. **Similar species:** Black Guillemot, winter murrelets (see p 164).

Tufted Puffin *Lunda cirrhata* (Alcidae) 38-40 cm (15-16 in). Large, laterally compressed, orange and green bill; white face with red eye-ring; long yellow tufts behind eyes; black crown and body; orange legs. Sexes similar. In winter, tufts reduced to short grey feathers, face dark, bill narrowed at base, orange only at tip. Subadults have mottled face, yellowish bill. Juveniles have dark bill, underparts vary from dark grey to white. Voice: growling 'errrr'. **Range and habitat:** nests on islands in burrows atop cliffs or on steep sea-facing slopes, north-eastern and east-central EP, north-western and west-central NA. Winters at sea. **Similar species:** Rhinoceros Auklet.

Rhinoceros Auklet *Cerorhinca monocerata* (Alcidae) 38 cm (15 in). Large alcid, sooty brown above, greyish brown throat, foreneck, breast and flanks, belly white. White plumes run down from above eye past ear, 2nd line of white plumes form moustache. Compared to other puffins, reddish orange bill is longer, narrower, shallower. Pale knob projecting from base of bill is absent in winter, as are facial plumes. Sexes similar. Immature like winter-plumage adult but darker, bill smaller and darker. Voice: low growl. **Range and habitat:** nests on sea-facing slopes in burrows, coming out only at night, north-eastern and east-central EP, north-western and west-central NA. Winters on inshore and offshore waters in breeding range south to south-western NA. **Similar species:** immature Tufted Puffin.

Brunnich's Guillemot *Uria lomvia* (Alcidae) 43-8 cm (17-19 in). Possibly the most abundant bird of the Northern Hemisphere. Slightly larger than the Guillemot, bill shorter and thicker with white line along base of upper mandible. Dark above, white underparts rise to acute angle on foreneck. In winter lower cheek and foreneck white, white line at base of upper mandible faint or lacking. Sexes similar. Uses wings under water in pursuit of fish prey. Voice: low, purring. **Range and habitat:** nests on rocky cliffs in enormous colonies along coasts, north-eastern EP to east-central EP, Also northern NA. Winters on open water near breeding range and somewhat further south. **Similar species:** Guillemot, Razorbill (see p 164).

Western Gull

California Gull

Pomarine Skua
(pale phase)

Forster's Tern
(first winter plumage)

Royal Tern with Sandwich Terns in background

Black Skimmer

Pigeon Guillemot

Tufted Puffin

Rhinoceros Auklet

Brunnich's Guillemot

PIGEONS AND SANDGROUSE Columbiformes

Common Ground Dove *Columbina passerina* (Columbidae) 16.5-18 cm (6½-7 in). Size of sparrow. Male light grey-brown with black spots on upperwing-coverts. Primaries flash rufous in flight. Short rounded tail black and brown with narrow white corners; legs pinkish orange, bill similar with black tip. Female and young paler, greyer, less strongly spotted. When flushed, flies erratic course with rapid wing-beat, drops to ground or alights suddenly after short distance. Voice: soft, repeated 'woo-oo'. **Range and habitat:** frequents open dry areas, but occurs in broad range of habitat from forest edges to gardens, extreme southern NA and Bermuda south through Lesser Antilles to central NT. **Similar species:** Inca Dove (see p 112).

Band-tailed Pigeon *Columba fasciata* (Columbidae) 35-8 cm (14-15 in). Similar to Rock Dove but tail longer, black at base with broad pale grey terminal band; body grey to bluish grey, with iridescent nape and white crescent on back of neck; head and upper breast grey-brown in females, pink or purple in males; yellow bill with black tip, yellow feet with black nails. Immature has grey to yellow bill and feet, plumage grey with buff to brown edging on wing-coverts and primaries, lacks iridescence. Voice: deep, owl-like 'whoo-whooooo', repeated. **Range and habitat:** north-western to south-western and south-central NA into north-western NT. **Similar species:** Red-billed Pigeon, also Mourning, Rock and White-winged Doves (see p 164).

Mourning Dove *Zenaida macroura* (Columbidae) 28-33 cm (11-13 in). Very common bird in North America, and increasing. Grey-brown above with some dark spots, pinkish buff on head, foreneck and underparts; long diamond-shaped tail with white feather tips; slim neck, small head with blue eye-ring, dark spot below ear, purplish iridescence on sides of neck (barely noticeable in female); legs red. Wings whistle on take-off. Immature browner than adults, scaly upperparts. Mournful cooing, often begins before dawn. **Range and habitat:** towns, roadsides, farms, open woods, scrub from north-western NA throughout central and southern NA to north-western NT. **Similar species:** other doves.

CUCKOOS AND ALLIES Cuculiformes

Yellow-billed Cuckoo *Coccyzus americanus* (Cuculidae) 28-33 cm (11-13 in). Long tail, slender profile. Brown above with rufous primaries visible along edge of folded wing and outer half of wing in flight; cheek, throat, underparts white; upper mandible dark, lower mandible yellow; inconspicuous yellow eye-ring; tail feathers graduated, white terminal spots producing series of large round spots along underside of folded tail. Voice: rapid, guttural 'ka-ka-ka-ka-ka-ka-ka-kow-kow-kowlp-kowlp-kowlp-kowlp', slowing at end. **Range and habitat:** woodlands, forest edges, thickets, orchards and farmlands central and southern NA, more common east of Rockies. Winters in north-eastern NT. **Similar species:** Black-billed and Mangrove Cuckoos.

Black-billed Cuckoo *Coccyzus erythropthalmus* (Cuculidae) 30 cm (12 in). Slender, with long tail. Grey-brown above, no rufous in wings; white cheeks, throat and underparts, black bill slightly decurved; red eye-ring surrounds dark eye; tail feathers with narrow white tips, graduated to produce series of narrow spots along underside of folded tail. Feet grey; 2 toes forward, 2 backward. Eye-ring of immature yellowish. Raises own young. Voice: fast rhythmic 'cucucu, cucucu, cucucu'. **Range and habitat:** deciduous woodlands, forest edges, orchards and thickets central and east-central NA. Winters in north-western NT. **Similar species:** Mangrove and Yellow-billed Cuckoos.

Greater Roadrunner (Roadrunner) *Geococcyx californianus* (Cuculidae) 51-61 cm (20-4 in). Large ground-dwelling cuckoo; can run at 24-32 kph (15-20 mph) to escape predation and to pursue lizards, snakes, birds and inverte-brates. Dark brown with narrow white streaks; belly and undertail-coverts cream; bushy crest; very long tail; bright blue and red patches behind eyes; large, long pointed bill; legs long, sturdy and blue-grey. Sexes similar. Immature has lighter, wider streaks and lighter underparts. Voice: hoarse, throaty 'coo, coo, coo, ooh, ooh, ooh'. **Range and habitat:** deserts, chaparral, grasslands, open woodlands and agricultural areas of south-central and south-western NA. **Similar species:** Lesser Roadrunner.

Smooth-billed Ani *Crotophaga ani* (Cuculidae) 35 cm (14 in). Large black bill with raised curved culmen, also long tail, distinctive. Black with bronze-brown edges to feathers of head and neck; iridescent green crescents in scapulars, lesser wing-coverts and feathers of upper back; flight feathers of wings and tail have purple gloss. Immature browner, especially on wings and tail. From 1 to 3 pairs build communal nest; females share incubation. Young fed by all parents. Whining whistle 'quee-ick, quee-ick', also chuckling notes. **Range and habitat:** dense brush near fields or marshes from south-eastern NA to east-central NT. Resident. **Similar species:** Groove-billed Ani (see p 114).

OWLS Strigiformes

Short-eared Owl *Asio flammeus* (Strigidae) 33-43 cm (13-17 in). Active at twilight. Flies low, coursing back and forth on long wings with buoyant, moth-like wingbeat. Rounded head; small facial discs; neck and upper breast heavily streaked with brown; buff belly with narrow, vertical brown streaks. Brown above, mottled with buff. In flight shows buff patch on upper surface of wings, black wrist mark on wing-lining. Feeds on small mammals. Voice: high, rasping 'wak, wak, wak', also 'toot-toot-' repeated 15-20 times. **Range and habitat:** grasslands, marshes, dunes, moors and tundra of northern NA, WP and EP. Winters in central and southern NA, southern part of Eurasian range. **Similar species:** Barn Owl, Long-eared Owl (see p 166).

Snowy Owl *Nyctea scandiaca* (Strigidae) 51-68 cm (20-7 in). Arctic predator. Adults white with yellow eyes, no ear-tufts; feet feathered to talons; feathers project on to bill. Most show some dusky bars on white plumage, females more than males. Vagrants to south tend to be lightly to heavily barred; often at airports. Feeds on mammals and birds up to size of Arctic Hare and Common Eider. Croaks, growls, whistles and hoots. **Range and habitat:** circumpolar Holarctic, throughout Arctic. Southward movement in winter erratic, depending on food supply. **Similar species:** white phase of Gyrfalcon, Arctic subspecies of Great Horned Owl.

Great Horned Owl *Bubo virginianus* (Strigidae) 45-63 cm (18-25 in). Very large, widely spaced prominent ear-tufts, large yellow eyes, white throat contrasts with dark brown mottled upperparts and white to buff underparts crossed by narrow dark brown bars; long, broad wings dark above, lined with buff. Overall colour varies from nearly white in Arctic to very dark forms of Pacific north-west. Diet of skunks, opossums, ducks, hawks and Great Blue Herons. Nests very early, often in old hawk nests. Voice: deep, low hoots 'hoo-hoo-hoooo hoo-hoo'. **Range and habitat:** resident in forests, open country and city parks throughout NA and NT. **Similar species:** Barred, Great Grey, Long-eared, Spotted and Snowy Owls (see p 166).

Great Grey Owl *Strix nebulosa* (Strigidae) 60-84 cm (24-33 in). Brownish grey, 'earless', relatively small yellow eyes set in prominent grey facial discs marked with concentric dark brown rings; body grey with heavy dark grey-brown streaking. White 'bow-tie' on the upper breast readily visible. Long tail is distinctive. Sexes similar. Low-pitched hoots, slowing toward end of series, infrequent and audible only at close range. **Range and habitat:** resident in northern coniferous forests of western and northern NA. Occasionally invades eastern Canada and northern United States in large numbers. **Similar species:** Barred, Great Horned and Spotted Owls (see p 72).

Common Ground Dove
Band-tailed Pigeon
Mourning Dove

Yellow-billed Cuckoo
Black-billed Cuckoo
Greater Roadrunner

Smooth-billed Ani
Short-eared Owl
Snowy Owl female on nest
Great Horned Owl
Great Grey Owl

Barred Owl *Strix varia* (Strigidae) 43-60 cm (17-24 in). A large owl distinguished by its large round head, lack of ear-tufts and 9-note hoot 'Who cooks for you, who cooks for you-all!', pitch dropping on 'all'. Grey-brown plumage crossed by transverse bars on the neck and upper breast and by vertical streaks on the lower breast and belly. Brown eyes are distinctive. Wings broad and short. Sexes similar. **Range and habitat:** wooded swamps and deep forest, where nests in tree-cavities and old hawk nests, in west-central, central, east-central and south eastern NA, recently south into south-western mountains. **Similar species:** Great Grey Owl, Spotted Owl (see p 70).

Northern Saw-whet Owl *Aegolius acadicus* (Strigidae) 18-21.5 cm (7-8½ in). Large round head. Large yellow eyes set in facial discs that lack the black edge found in most owls. Dark above flecked with white, buff below streaked with brown; short tail; dark bill. Sexes similar. Juveniles chocolate brown above, cinnamon-buff belly, white triangle on forehead. Hard to find, but uses same roost for days. Voice: 'too-too-too-too-too'. Can be whistled into view by imitation of its song. **Range and habitat:** common resident of boreal forests of central NA and montane forests of mountains in central and southern NA. Winters in and just south of breeding range. **Similar species:** Tengmalm's, Elf and Flammulated Owls, Ferruginous and Northern Pygmy-owls.

Burrowing Owl *Athene cunicularia* (Strigidae) 20.5-3 cm (8-9 in). Sandy-coloured head, back, wings. White throat and upper breast, latter crossed by single dark line; barred lower breast, belly; no ear-tufts on broad head; long legs, unfeathered on lower portion, short tail. Sexes similar. Occupies abandoned rodent burrows, which it modifies by digging with wings, beak and feet. When disturbed, bobs, bows and chatters. Active during day. Voice: high, mournful 'coo-coo-coo', or 'co-hoo'. **Range and habitat:** deserts and open grasslands of west-central and central NA south into NT, also resident in Florida. Winters from south-western NA south.

Eastern Screech Owl *Otus asio* (Strigidae) 18-25.5 cm (7-10 in). Only small owl of eastern North America with ear-tufts. Red morph is cinnamon rufous above with rufous facial discs, white below streaked and barred with rufous. Grey morph is grey-brown above streaked with brown, white eyebrows, lores and underparts, heavily streaked with dark brown. Both morphs have yellow eyes surrounded by facial discs bordered by a broad black stripe that continues across throat. Sexes similar. Juveniles barred brown and white all over, red morph having rufous edges to bars. Voice: tremulous whinny on same pitch or descending. **Range and habitat:** woodlands and groves from central and south-eastern NA south. Nests in cavities. **Similar species:** Western Screech Owl.

NIGHTJARS AND ALLIES Caprimulgiformes

Common Nighthawk *Chordeiles minor* (Caprimulgidae) 21.5-5.5 cm (8½-10 in). Long pointed wings with conspicuous white patches just beyond bend of wing identify this widespread species. Mottled upperparts, barred underparts; white throat patch and contrasting white subterminal band on tail. Males grey and black, females brownish, with throat and wing patches buff-white and no white in tail. Juveniles may lack throat patch. Feed in evening and at night when nasal, buzzy call only evidence of their presence. During day perch lengthwise on tree limbs, fence-posts, roofs. Voice: nasal 'peent'. **Range and habitat:** open country, open woods throughout NA, except northernmost areas. Winters in northern and central NT. **Similar species:** Antillean Nighthawk, Lesser Nighthawk and other nightjars (see pp 116, 166, 212, 254, 304).

Whip-poor-will *Caprimulgus vociferus* (Caprimulgidae) 23-5.5 cm (9-10 in). Known primarily for its ringing 'whip-poor-will', accented on the first and last syllables and repeated over and over. Widespread species of deciduous woodlands, feeds on flying insects, caught on wing at night, roosts on ground or lengthwise on a limb by day. Mottled grey-brown plumage; black chin; narrow white throat patch; rounded wings; large white patches on corners of tail. **Range and habitat:** central and east-central NA south, also south-western NA. Winters in south-central, south-eastern NA into north-western NT. **Similar species:** Chuck-will's-widow, Buff-collared Nightjar and other nightjars (see pp 116, 166, 212, 254, 304).

Common Poorwill *Phalaenoptilus nuttallii* (Caprimulgidae) 18-21.5 cm (7-8½ in). Smallest North American nightjar. Brown upperparts mottled with silver-grey and black; underparts black and grey with faint bars; throat white, contrasting with black face and upper breast; wings rounded; short tail has white corners. Sexes similar. Juveniles have less distinct throat and tail patches and a buff wash. Migratory status unknown, but can go into torpor to survive periods of adverse weather. Voice: 'pooooor-will!', accent on 2nd syllable. **Range and habitat:** upland arid areas of north-western and north-central NA south. Winters from south-western NA south, but wintering range poorly known. **Similar species:** Lesser Nighthawk, Whip-poor-will, other nightjars (see pp 116, 166, 212, 254, 304).

SWIFTS AND HUMMINGBIRDS Apodiformes

Chimney Swift *Chaetura pelagica* (Apodidae) 14 cm (5½ in). Small, brown, darkest above and on wings, lighter below, lightest on throat and upper breast. Wings long, narrow, mostly hand and primaries; tail short, stiff, slightly rounded when fanned, central shaft of each tail feather project beyond webbing as a spine; small dark bill, but large mouth. Nest small, of sticks glued together with saliva on inside of chimney or hollow tree. Rapid flight, bouts of shallow wingbeats interspersed with gliding, appears to beat wings alternately. Often flies in small, vociferous groups. Voice: rapid series of staccato chips. **Range and habitat:** open sky from central and south-central NA eastwards. Winters in upper Amazon basin. **Similar species:** other swifts (see pp 116, 166, 216, 254, 304).

Vaux's Swift *Chaetura vauxi* (Apodidae) 11.5 cm (4½ in). Small, brown, dark above and on wings, lighter below, lightest on throat and upper breast, tail with spines. Sexes similar. Western equivalent of Chimney Swift, but smaller, much lighter below. Commonly seen over woods in which it nests in hollow redwood stumps. Gradually it is adopting chimneys as forests are destroyed. Voice: similar to Chimney Swift, but somewhat higher-pitched and less vocal. **Range and habitat:** north-western and north-central NA. Winters from south-central NA to north-central NT. **Similar species:** other swifts (see pp 116, 166, 216, 254, 304).

White-throated Swift *Aeronautes saxatalis* (Apodidae) 15-18 cm (6-7 in). Large swift with bold black and white pattern. Throat, upper breast white, extending down midline to lower belly; 2 white patches on flanks just behind wings; prominent broad white band across trailing edge of secondaries; rest of head, body and wings black; tail slightly forked and without projecting spines. Voice: drawn out, descending 'skeeeeee'. **Range and habitat:** rocky cliffs of western mountains and desert canyons from north-central to south-central NA and Central America. Winters from south-western into south-central NA. **Similar species:** other swifts, Violet-green Swallow (see pp 116, 166, 216, 254, 304).

■□□
Barred Owl
□■□
Northern Saw-whet Owl
□□■
Burrowing Owl

■□□
Eastern Screech Owl (red phase)
□■■
Common Nighthawk

■■□
Whip-poor-will
□□■
Common Poorwill

■□□
Chimney Swift
□■□
Vaux's Swift
□□■
White-throated Swift

Ruby-throated Hummingbird *Archilochus colubris* (Trochilidae) 8-9 cm (3-3½ in). Only hummingbird in eastern North America. Iridescent green above, white below. Male has ruby-red gorget, can appear black; olive-buff sides and flanks; forked tail. Female white throat, sides, flanks; tail rounded with white feather tips. Bill long and dark. Male has looping, pendular courtship flight. Voice: high squeaks. **Range and habitat:** woodlands, parks and gardens from central and east-central NA south. Winters from south-central NA to north-central NT. **Similar species:** Rufous Hummingbird.

Calliope Hummingbird *Stellula calliope* (Trochilidae) 7-7.5 cm (2¾-3 in). Short bill; short, broad, unforked tail. Male has gorget of elongated rose-purple rays on white background; olive-buff sides, flanks. Female lacks gorget, throat covered with dusky spots. Buff wash on sides and breast, white corners on tail. Display flight, shallow U-shaped arc. Voice: brief 'pfft' at bottom of display, high 'tsip' when foraging. **Range and habitat:** meadows, thickets, brush in coniferous forests of west-central and south-western mountains of NA. Winters in Mexico. **Similar species:** female and juvenile Rufous and Allen's Hummingbirds.

Black-chinned Hummingbird *Archilochus alexandri* (Trochilidae) 7.5-9 cm (3-3½ in). In western USA replaces Ruby-throated Hummingbird, which male resembles, but with black gorget bordered below by iridescent purple-violet. Both sexes have white spot behind eye. Display flight is broad U-shaped arc. High-pitched warble; high, buzzy notes; wings with distinctive dry buzz in flight. **Range and habitat:** semi-arid country near water, west-central through south-western to south-central NA. Winters in south-western NA. **Similar species:** females like female Costa's or Ruby-throated Hummingbirds.

Broad-tailed Hummingbird *Selasphorus platycercus* (Trochilidae) 9-10 cm (3½-4 in). Male is only western hummingbird with green crown and rose gorget. Female has white throat spotted with brown, bright buff sides, rufous near base of outer rectrices. Tail is broad in both sexes. Deep U-shaped display flight. Voice: sharp 'tew'; wings of male produce distinctive loud, rattling trill, except during moult. **Range and habitat:** meadows, patches of flowers within conifer and aspen forests of south-western and south-central NA. Winters from south-central NA into Central America. **Similar species;** female Calliope, Rufous and Allen's Hummingbirds, also Ruby-throated Hummingbird.

Anna's Hummingbird *Calypt anna* (Trochilidae) 9-10 cm (3½-4 in). Unique, rose red gorget elongated at sides with rose red crown. Females have fine red spots on the throat, are green above and grey below; tail white at corners. Juveniles show white, unspotted throat. Long steep display dive ending with loud explosive popping in front of female. Voice: squeaky phrases; soft 'chick' when foraging, rapid 'chee-chee-chee'. **Range and habitat:** gardens, canyons, foothills and riparian woodlands, west-central and south-western NA. **Similar species:** female resembles female Black-chinned and Costa's Hummingbirds.

Rufous Hummingbird *Selasphorus rufus* (Trochilidae) 8-9 cm (3-3½ in). Adult males orange-rufous above, iridescent orange-red gorget, white upper breast, orange rufous wash on sides. Female green above, rufous sides and tail. White below, small red to golden green spots on throat. White tips on broad outer tail feathers. Young male some red on throat, green back. Vertical circular display with trilling buzz made by male's wings. Soft 'tchup'; excited, buzzy squeal 'zeeee-chuppity-chup'. **Range and habitat:** north-western and north-central NA. Winters in south-central NA. **Similar species:** Allen's Hummingbird, female Calliope and Broad-tailed Hummingbirds.

KINGFISHERS AND ALLIES Coraciiformes

Belted Kingfisher *Megaceryle alcyon* (Alcedinidae) 30-5 cm (12-14 in). Long, stout bill; large head; bushy crest.

Male blue-grey above, white throat, distinctive white collar, white underparts with blue-grey breast band. Female similar, with rufous on flanks and chest below blue-grey band. In flight white patch visible on upper surface of blackish primaries. Fishes by diving from a branch or from hovering flight. Distinctive rattle, often given in flight. **Range and habitat:** deep, irregular streams, lakes, sea coasts of NA. Nests in burrow. Winters along coasts, Mississippi valley, southern NA, West Indies and Bermuda. **Similar species:** other kingfishers (see pp 120, 208, 256, 304).

WOODPECKERS AND ALLIES Piciformes

Common Flicker *Colaptes auratus* (Picidae) 28-35 cm (11-14 in). Barred black and brown above, white rump, buff-white below with black spots and black pectoral band. Eastern subspecies: crown and nape grey, red nape; face, foreneck buff, black moustaches (male); wings and tail yellow below. Western: crown and nape brown, face grey, red moustaches (male); wings and tail salmon below. Forages on ground, especially for ants. Voice: loud territorial call 'wicka, wicka, wicka, wicka', also sharp 'klee-yer'. **Range and habitat:** throughout NA south of arctic, also Cuba. Winters throughout most of range. **Similar species:** Gila Woodpecker, female Red-bellied Woodpecker.

Downy Woodpecker *Picoides pubescens* (Picidae) 15-16.5 cm (6-6½ in). Black and white barred wings; white back; black crown, eye stripe and moustache; white eyebrow and lower face to nape; white below; black tail, white outer tail feather with 2 or more black bars. Male has red on nape. Short bill. Forages on branches and smaller trees. Voice: weak 'pik'; rapid rattle, speeding up toward end. **Range and habitat:** forests, woodlots, shade trees throughout NA, except extreme south-western and south-central NA. Leaves northernmost parts of range in winter. **Similar species:** other spotted woodpeckers (see pp 74, 168, 258).

Yellow-bellied Sapsucker *Sphyrapicus varius* (Picidae) 18-20.5 cm (7-8 in). Black and white barred above with white rump and distinctive white patches in wings; upper breast and belly yellowish; red crown; throat red in males, white in females. In Rockies has red nape. Immature, distinctive white rump and wing patches. Drills pits in trees, returns to collect sap and insects. Voice: nasal mewing note, downward 'cherrrr'. Drumming syncopated. **Range and habitat:** aspen woodlands, western, central and east-central NA. Winters along Pacific coast and throughout southern NA and Central America. **Similar species:** immature Red-breasted and Williamson's Sapsuckers, other black and white woodpeckers (see pp 74, 168, 258).

Three-toed Woodpecker *Picoides tridactylus* (Picidae) 20.5-23 cm (8-9 in). Back, wings and flanks brownish black and white barred, eastern form blacker. Head and face black with yellow crown, more extensive in male: eyebrow, whisker, throat and underparts white; tail black, outer feathers barred with white. Males heavier, longer-billed. Immature browner. Voice: 'pik', lower than Downy Woodpecker; 'pik-ik-ik-ik-ik' rattle. **Range and habitat:** resident in dense spruce-fir forest or swamp forests with dead trees. Circumpolar to southern edge of coniferous forest. Migrates south irruptively. **Similar species:** other spotted woodpeckers (see pp 74, 168, 258).

Pileated Woodpecker *Dryocopus pileatus* (Picidae) 42-8 cm (17-19 in). Large, black, with scarlet crest and white face. Black eye stripe extends to nape; black moustache turns down and connects to breast. Male has red on forehead and at base of moustache. Female has dark forehead. White wing-linings conspicuous in undulating flight. Immature greyish, streaked throat, orange crest, sexual distinctions as in adults. Voice: hollow, slow 'wuk, wuk, wuk'. **Range and habitat:** resident locally in mixed conifer-hardwood forests from north-western to south-western, central and eastern NA. **Similar species:** Ivory-billed Woodpecker.

■□□ **Ruby-throated Hummingbird**
■■□ **Calliope Hummingbird**
□□□ **Black-chinned Hummingbird**

■□□ **Broad-tailed Hummingbird** female
□■□ **Anna's Hummingbird**
□□■ **Rufous Hummingbird**

■□□ **Belted Kingfisher** female
□■□ **Common Flicker** female
□□■ **Downy Woodpecker**

■□□ **Yellow-bellied Sapsucker**
□■□ **Three-toed Woodpecker**
□□■ **Pileated Woodpecker** female

PERCHING BIRDS Passeriformes

Shore Lark *Eremophila alpestris* (Alaudidae) 16-17 cm (6-6½ in). Rather uniformly coloured lark with black and yellow face in summer. In breeding plumage, face pale yellow with black forehead and two upswept 'horns', black moustache and throat patch. Rest of head and back pinky-brown, lightly streaked black. Underparts whitish, tinged pinky-brown on sides of breast. Tail pinky-brown in centre with black sides. Head pattern mostly lost in winter. Gregarious in winter, running and walking on ground. Short warbling song mainly from ground, also 'tseep' call. **Range and habitat:** breeds circumpolar Holarctic tundra, open plains in NA, and open habitats at high altitude in WP and EP. Northern populations migrate to south for winter. **Similar species:** other larks (see p 168).

Eastern Kingbird *Tyrannus tyrannus* (Tyrannidae) 21.5-23 cm (8½-9 in). Head black, upper parts dark bluish-black, 2 indistinct whitish wing-bars, white below with grey wash on flanks and upper breast; tail dark with terminal white band; narrow red crown patch (seldom visible); bill broad, black. Sexes similar. Inhabits open environments with emergent perches from which it sallies forth after insects. Flies with shallow wingbeats. Nest in top of tree. Parents very aggressive. Voice: 'dzee-dzee-dzee' and 'dzeet', given frequently. **Range and habitat:** wood edges, fencerows, roadsides throughout most of NA except far north and desert south-west. Winters in NT. **Similar species:** Grey Kingbird.

Western Kingbird *Tyrannus verticalis* (Tyrannidae) 20.5-4 cm (8-9½ in). Elongate profile with heavy black bill. Head and nape ash-grey, throat and breast pale grey, lores and ear-coverts darker creating narrow mask; back and wings grey-green, underparts yellow, tail black with white outer webs in outermost tail feathers; orange-red crown patch, usually concealed. Sexes similar. Voice: sharp 'whit' or 'whit-ker-whit'. Flight song 'pkit-pkit-deedle-ot', with emphasis on 'deedle' high note. **Range and habitat:** open country with scattered trees, often on wire fences or telephone lines from central and west-central NA southwards. Winters south-central NA into Central America. **Similar species:** Cassin's, Couch's and Tropical Kingbirds (see p 130).

Great Crested Flycatcher *Myiarchus crinitus* (Tyrannidae) 18-20.5 cm (7-8 in). Common in eastern deciduous forests, more often heard than seen. Olive-brown above with grey throat and breast, bright yellow belly; wings dark brown with 2 white wing-bars and cinnamon rufous in primaries; tail brown with cinnamon rufous on inner web of tail feathers, visible when spread in flight. Sexes similar. When searching for insect prey or encountering other birds erects head feathers, leans forward with outstretched neck and bobs head slowly. Voice: loud, harsh, ascending 'wheeep'. **Range and habitat:** eastern and south-eastern NA west to prairies. Winters from extreme south-eastern NA to north-western NT. **Similar species:** Ash-throated, Brown-crested and Olivaceous Flycatchers.

Eastern Phoebe *Sayornis phoebe* (Tyrannidae) 16.5-18 cm (6½-7 in). 1st flycatcher to return to the streams, bridges and farms that are its summer home. Upright posture and wagging tail. Greyish black above; head, wings and tail slightly darker than back; whitish below, washed in pale yellow; bill black; no eye-ring, no wing-bars. Sexes similar. Immature has rufous wing-bars, yellow wash below. Voice: hoarse, clear 'fee-bee, fee-bee' with 2nd syllable alternately rising and falling, also clear, soft chip' note. **Range and habitat:** eastern, central and north-western NA. Winters in south-eastern and south-central NA. **Similar species:** *Empidonax* flycatchers, wood-pewees.

Willow (Traill's) Flycatcher *Empidonax traillii* (Tyrannidae) 13-16.5 cm (5-6½ in). Separated from Alder Flycatcher in 1973. Olive-brown above with brown wings and 2 prominent wing-bars; white below with olive wash on flanks; tail dark. Eye-ring narrow in eastern birds, usually lacking in western birds. Voice: an explosive 'fitz-bew' or 'fritz-be-yew', vaguely 3-syllabled, but emphasis always on 1st syllable, 2nd syllable lower. May be shortened to dry, burry 'rrrrrip'. also a thick 'whit'. **Range and habitat:** brushy fields, willow thickets, prairie woodlots, woodland edges in central, south-western and south-central NA. Uses drier, smaller, more open areas than Alder Flycatcher. Winters south to central NT. **Similar species:** other *Empidonax* flycatchers, wood-pewees.

Alder Flycatcher *Empidonax alnorum* (Tyrannidae) 16.5 cm (6½ in). Olive-green above, more green than Willow Flycatcher, 2 wing-bars on brown wing; olive wash on flanks, white throat and underparts, brown tail; eye-ring indistinct; bill long, broad, dark above, orange-yellow below. Sexes similar. Voice: burry, 2-syllable 'rrree-beea', 2nd syllable higher and stressed, 'a' ending faint, often inaudible. Occasionally shortened to ascending 'rrreeee', also flat 'peep'. **Range and habitat:** alder swamps, aspen thickets from northern and east-central NA. Wetter habitat than Willow Flycatcher. Winters in NT. **Similar species:** other *Empidonax* flycatchers, wood-pewees. **Not illustrated.**

Least Flycatcher *Empidonax minimus* (Tyrannidae) 12.5-14.5 cm (5-6 in). Grey-green back, greyest of the eastern *Empidonax*. Whitish underparts, occasionally a yellow wash on flanks and lower belly; 2 light wing-bars, prominent eye-ring. Sexes similar. 1st *Empidonax* to come north in the spring and last to migrate south. Voice: sharp, often repeated 'ch-bek', 2nd syllable stressed. Call a flat 'wit'. Often sings during spring migration. **Range and habitat:** open woods, aspen groves, orchards, shade trees across northern and central NA, in mountains to south-eastern NA. Winters from south-central NA to north-central NT. **Similar species:** other *Empidonax* flycatchers, wood-pewees.

Scissor-tailed Flycatcher *Tyrannus forficatus* (Tyrannidae) 30-8 cm (12-15 in). Pearl-grey flycatcher with very long pairs of outer tail feathers that are strikingly patterned in white and black; central tail feathers shorter and black; wings and uppertail-coverts black; flanks, belly and under-tail-coverts washed in salmon pink, wing-linings bright salmon pink. Female has shorter tail, is less colourful. Immature has short tail, brownish back. Voice: 'keck', also 'ka-leep', harsh kingbird-like sounds. **Range and habitat:** semi-open country of south-central NA. Often perched on wire fences or dead exposed branches. Winters from south-central, south-eastern NA to north-central NT. **Similar species:** Fork-tailed Flycatcher, immature similar to Western Kingbird (see p 130).

Olive-sided Flycatcher *Contopus borealis* (Tyrannidae) 18-20.5 cm (7-8 in). Perches on high, conspicuous dead branches where its olive flanks with white ventral stripe give it a distinctive vested appearance. White, cottony tufts behind wing also diagnostic. General coloration is dark greyish-olive with no eye-ring and no wing-bars. Larger, stouter than pewee with larger head. Voice: vigorous, whistled 'quick three beers', 1st note abrupt, inaudible at a distance. Call, low 'pip, pip, pip'. **Range and habitat:** conifers of northern and western NA. Winters in NT. **Similar species:** Smoke-coloured Pewee, wood-pewees.

Western Wood-pewee *Contopus sordidulus* (Tyrannidae) 12.5-15 cm (5-6 in). Replaces Eastern Wood-pewee in west. Greyish olive above and on flanks, wings dark with 2 prominent white wing-bars (buff in 1st-winter birds); lacks eye-ring; underparts whitish, tail dark; upper mandible dark, lower paler. Sexes similar. Voice: sequence of emphatic 'pheer-reet' alternating with less harsh, somewhat rolling 'phee-rr-reet'. Call burry, slightly descending 'pheeer' or more emphatic 'pheer-reet', repeated every 5-10 sec. **Range and habitat:** open deciduous or mixed forest from western into central NA. Winters in northern and central NT. **Similar species:** Eastern Wood-pewee (distinguishable only by voice), Smoke-coloured Pewee, Olive-sided Flycatcher.

Shore Lark

Eastern Kingbird

Western Kingbird

Great Crested Flycatcher

Eastern Phoebe

Willow Flycatcher

Least Flycatcher

Scissor-tailed Flycatcher

Olive-sided Flycatcher

Western Wood-pewee

Purple Martin *Progne subis* (Hirundinidae) 18-22 cm (7-8½ in). Largest North American swallow. Proportionately shorter, broader wings than other swallows. Male blue-black head and body, black wings and tail. Female and immatures grey below, lighter belly. Female with greyish forehead and collar, blue-black crown and back, black wings and tail. Tail in both sexes long, broad, forked. Glides in circles, alternating quick flaps and glides. Voice: complex, low gurgling. Call 'swee-swuh', 2nd note lower, in flight a rich, descending 'tyu'. **Range and habitat:** open or semi-open country, often near water from central and southern NA except in western mountains. Winters in NT. **Similar species:** in flight resembles European Starling (see p 182).

Northern Rough-winged Swallow *Stelgidopteryx serripennis* (Hirundinidae) 12.5-14.5 cm (5-6 in). Wide-spread, common, but small, solitary and inconspicuous. Brown above with dusky throat, white lower breast and belly; medium tail, slightly notched. Deep wingbeats with wing tips pulled back at end of each stroke, little gliding. Sexes similar. Voice: rough, rasping 'brrrt'. **Range and habitat:** near running water where it nests in drainpipes, cracks in bridges, old burrows of central and southern NA. Winters from south-central NA to northern NT. **Similar species:** Swallow (see p 168), immature Tree Swallow, Southern Rough-winged Swallow.

American Cliff Swallow *Hirundo pyrrhonota* (Hirundinidae) 12.5-15 cm (5-6 in). Chunky swallow, square tail. Cream forehead, blue-black crown and back, narrow white stripes on back, rust-coloured rump; throat chestnut shading to black at breast, underparts dull white to greyish along sides and at collar. Sexes similar. Juvenile has darker forehead, white feathers mixed into throat. May soar on flattened wings. Voice: low 'chur', harsh 'keer', and series of grating and squeaking notes given in flight. **Range and habitat:** NA except south-east and extreme north. Winters in NT. Nest jug-like, fixed to wall or cliff under overhang. **Similar species:** Cave Swallow.

Tree Swallow *Tachycineta bicolor* (Hirundinidae) 12.5-15 cm (5-6 in). Male metallic blue-green above, white below. First-year female brown, some metallic green above, white below. Second-year female metallic green above, some brown, especially on forehead. Juvenile grey-brown above, white below. Only NA songbird with delayed plumage maturity in female. Females breed in 1st year. First swallow to return in spring, last to leave in autumn. **Range and habitat:** forage and nest near water from treeline to central NA. Winters in southern NA. **Similar species:** Violet-green Swallow; Northern Rough-winged Swallow resembles immature Tree Swallow (see p 168).

Loggerhead Shrike *Lanius ludovicianus* (Laniidae) 20.5-5.5 cm (8-10 in). Predatory songbird. Large head; black, almost conical, hooked beak; grey above; wings and tail black, white at wrists and corners of tail; greyish white below; black mask extends over bill. Sexes similar. Juvenile greyish, finely barred on underparts. Tends to drop low after take-off and fly low to ground, then swoop up to next perch. Caches prey, often by impaling it. Voice: harsh, deliberate, repetitive notes, also single note 'shack'. **Range and habitat:** semi-open country, central and southern NA. Winters in southern NA. **Similar species:** Northern Mockingbird, Great Grey Shrike (see p 170).

Cedar Waxwing *Bombycilla cedrorum* (Bombycillidae) 16.5-20.5 cm (6½-8 in). Sculpted appearance. Crested, cinnamon to grey-brown, black face edged in white; chin glossy black in male, dull black in female; conical black bill. Yellowish flanks and belly; undertail-coverts white; tail grey with yellow terminal band; secondaries tipped with scarlet. Juvenile greyer than adult, diffusely streaked, especially on underparts; black mask and throat incomplete or lacking. Feeds on berries, but will flycatch. Gregarious. Voice: lisping 'hiss, sse, ssee'. **Range and habitat:** open woodlands, orchards from north-western and central NA.

Winters in central and southern NA to NT and West Indies. **Similar species:** Waxwing (see p 170).

North American Dipper *Cinclus mexicanus* (Cinclidae) 18-21.5 cm (7-8½ in). Stout, wren-like bird, thrush-size with short tail. Slate-grey with brownish wash on head; thin white crescents above and below eye; black bill. Sexes similar. Winter adults and juveniles have pale-tipped feathers on wings and underparts. Juvenile tinged rust, pale greyish throat, mottled below, bill yellow. Voice: rich, musical runs and trills with much repetition of phrases, also loud 'bzeet'. **Range and habitat:** resident western NA to north-western NT. Inhabits cold, rushing streams. Forages along edges and under water on bottom of stream for insect larvae and small fish. When flushed flies low along stream.

Sedge Wren *Cistothorus platensis* (Troglodytidae) 10-11.5 cm (4-4½ in). Dark brown crown faintly streaked with tan. Lacks white eye stripe of Marsh Wren and white dorsal stripes less prominent. Flanks orange-buff fading to buff on undertail-coverts, otherwise white below. Sexes similar. Voice: 2-3 deliberate notes followed by staccato chatter that descends at the end. Call, sharp 'slap'. **Range and habitat:** low shrubs of wet meadows and marsh edges, central and east central NA. Winters in south-eastern, south-central NA into NT. **Similar species:** Marsh Wren.

House Wren *Troglodytes aedon* (Troglodytidae) 11.5-14 cm (4½-5½ in). Small, aggressive, energetic wren, grey-brown above with dusky bars on wings and tail; underparts greyish white, lightest on throat, faint bars on flanks; tail cocked over back. Sexes similar. Male usually builds several nests from which female chooses one, lines it and lays eggs. Voice: musical gurgling; pitch rises, then falls in bubbling glissando. Scolding note a rapid grating sound. When singing, male perches with tail straight down. **Range and habitat:** common in bird houses, thickets, woodland openings and gardens of central and southern NA, throughout NT. Winters in southern NA. **Similar species:** Wren (see p 178).

Carolina Wren *Thryothorus ludovicianus* (Troglodytidae) 14-15 cm (5½-6 in). Large wren with rufous crown and upperparts, dark barring in wings and tail. White throat, buff-rust underparts; prominent white eyebrow; tail long, often cocked or switched back and forth; bill dark, long, slightly decurved. Sexes similar. Seldom still. Voice: loud, clear, usually 3-syllabled 'tea-kettle, tea-kettle, tea-kettle, tea'; also loud descending trill, 'tiirrrr'; scolding and twitters. Vocal in all kinds of weather throughout the year. **Range and habitat:** resident in yards and undergrowth near water and yards of east-central and south-eastern NA. **Similar species:** Bewick's and Marsh Wrens.

Cactus Wren *Campylorhynchus brunneicapillus* (Troglodytidae) 18-22 cm (7-8½ in). Largest North American wren, similar to small thrasher in size and shape. Rust-brown crown, broad white stripe over each eye; body brown, darker above, barred and spotted with black and white. Sexes similar. Feeds on ground, flies slowly with rapid wingbeats. Sings from top of cactus. Tail not usually cocked. Voice: throaty 'choo-choo-choo-choo' or 'chug-chug-chug-chug', much like a stalled car. **Range and habitat:** resident in south-western deserts, especially those with large cholla cactus, mesquite and palo verde. Also common in desert residential areas. **Similar species:** desert thrashers (see p 80).

Purple Martin

Northern Rough-winged Swallow

American Cliff Swallow

Tree Swallow at nest-hole

Loggerhead Shrike

Cedar Waxwing

North American Dipper

Sedge Wren

House Wren

Carolina Wren

Cactus Wren

Grey Catbird *Dumetella carolinensis* (Mimidae) 23 cm (9 in). Slim, long-tailed profile characteristic of the mimic thrushes; slate-grey above and below, black crown, chestnut undertail-coverts. Tends to hold tail above horizontal. Sexes similar. Juvenile brownish grey, with pale rufous undertail-coverts. Voice: disjointed phrases, without repetitions characteristic of thrashers and Northern Mockingbird. Cat-like mew, often flicks tail and leans forward when calling, also grating 'tchek-tchek'. **Range and habitat:** undergrowth, brush and gardens, always near dense cover, central into south-eastern NA. Winters in south-eastern and south-central NA into Central America. **Similar species:** female Brewer's and Rusty Blackbirds (see p 91).

Northern Mockingbird *Mimus polyglottos* (Mimidae) 28 cm (11 in). Grey above, white to light grey below; 2 white wing-bars contrast with black flight feathers, also large white patch across base of primaries visible in flight. Wingbeats slow, buoyant. Tail long, black, with white outer feathers. Sexes similar. Juvenile has faint spotting on breast. Voice: varied medley of notes and phrases, many of them imitations of birds and local noises, each repeated 3 or more times. Sings from high exposed perch, often at night. Call a loud 'tchak'. **Range and habitat:** thickets, woodland edges, towns, roadsides, gardens of central and southern NA. Northernmost birds may withdraw south in winter. **Similar species:** Townsend's Solitaire, shrikes (see p 170).

Sage Thrasher *Oreoscoptes montanus* (Mimidae) 20.5-3 cm (8-9 in). Smallest thrasher. Short, straight, slender bill. Grey-brown above, darker on wings and tail; 2 white wing-bars; white corners in tail. Eye yellow, with pale eyebrow above and moustache below; cream underparts blend to buff on lower belly and undertail-coverts; small wedges of brown on throat and breast grow to streaks on flanks. Sexes similar. Immatures browner, with more diffuse streaking and dark eye. Voice: medley of sweet, warbling notes, also a throaty 'chuck'. Sings from top of sagebrush and in flight. **Range and habitat:** open sage and scrub country of western NA. Winters in deserts of south-western NA. **Similar species:** Bendire's and Curve-billed Thrashers.

Curve-billed Thrasher *Toxostoma curvirostre* (Mimidae) 24-9 cm (9½-11½ in). Pale grey-brown above, grey-buff below with dark grey spots arranged in longitudinal streaks. Conspicuous white corners on long tail and white wing-bars may be lacking in some individuals, particularly those of southern and western Arizona. Long, decurved bill. Bright orange-red eye. Runs, hops on ground; digs with bill; low, swift flight between bushes. Voice: loud, carolling from top of cholla cactus; low trills; wren-like chattering; call loud, liquid 'whit-wheet'. **Range and habitat:** common in sparse brush, cacti of arid south-western and south-central NA, withdraws from northern part of range in winter. **Similar species:** Bendire's and Crissal thrashers.

Brown Thrasher *Toxostoma rufum* (Mimidae) 29 cm (11.5 in). Rufous above with 2 cream wing-bars. Face grey-brown; underparts white, washed with buff and heavily streaked with dark brown; eye yellow to orange; bill dark, with pale base on the lower mandible and slightly decurved. Forages for insects in leaf litter by tossing leaves over its back with bill. Voice: succession of musical couplets, rich and melodious. Usually sings from prominent, exposed perch. Call a harsh 'chack', also a 3-note whistle. **Range and habitat:** dry thickets, brushy pastures, second-growth woods, sometimes in garden shrubbery, central and east-central NA south. Winters in south-eastern NA. **Similar species:** brown-backed thrushes, Long-billed Thrasher.

Hermit Thrush *Catharus guttatus* (Turdidae) 19 cm (7½ in). Olive-brown above with rufous tail; face grey-brown with thin whitish eye-ring; whitish buff below, with large dark brown spots on breast forming narrow streaks on sides of throat and flanks; belly white. Sexes similar. Juveniles buff below. When perched often jerks tail up and drops it slowly. Frequently flicks wings. Voice: clear, flute-like, echoing; 3

or 4 phrases at different pitches, separated by pauses. Calls 'tuk, tuk, tuk', soft 'chuck', harsh 'pay'. **Range and habitat:** coniferous and mixed forests of northern NA south in western and eastern mountains. Winters in west-central and southern NA. **Similar species:** Fox Sparrow, other thrushes, Veery (see pp 88, 134, 176).

Eastern Bluebird *Sialia sialis* (Turdidae) 15-16.5 cm (6-6½ in). Bright blue crown, face, back, wings and tail; white chin; chestnut throat, breast, flanks; white belly and undertail-coverts. Female has brownish back, blue wings and tail; indistinct buff eye-ring; pale rust throat, breast, flanks; white belly and undertail-coverts. Juveniles heavily speckled grey, blue in wings and tail. Perches in hunched, vertical posture on exposed posts or branches. Makes short foraging flights to ground and back to perch. Voice: rich 3- or 4-note warble. Call a musical 'tru-ly'. **Range and habitat:** open woodlands from central and east-central NA south, also south-eastern Arizona. Winters in southern part of breeding range into Central America. Population increasing. **Similar species:** Mountain and Western Bluebirds.

Wood Thrush *Hylocichla mustelina* (Turdidae) 19-21.5 cm (7½-8½ in). Cinnamon rufous crown and nape, fading to olive-brown on uppertail-coverts and tail; narrow white eye-ring; ear-coverts finely streaked with white and brown; white underparts with extensive dark brown round spots on throat, breast, flanks, belly. Sexes similar. Juveniles similar, but with tawny olive or buff spots on head and upperparts; on wing-coverts these may form 1 or 2 indistinct wing-bars. Voice: rising, flute-like 'eee-lee-o-lay'. Calls, occasional guttural notes, liquid 'pip, pip, pip, pip', low 'quirt'. **Range and habitat:** deciduous woods, central and east-central NA south. Winters from south-central NA to north-west NT. **Similar species:** other thrushes, Veery (see p 134).

Swainson's Thrush *Catharus ustulatus* (Turdidae) 16-19 cm (6½-7½ in). Olive-brown (east) to russet (west) above; face, throat and breast creamy buff, appears spectacled; breast densely patterned with black round spots, becoming wedge-shaped on sides of throat; belly white, flanks washed olive-grey. Sexes similar. Juveniles darker above, with buff spots on back. Voice: rising, breezy series of flute-like phrases. Call a liquid 'whoit'. Night migrants give a distinctive, sharp 'queep'. **Range and habitat:** spruce forests of northern NA, south in western and eastern mountains. Winters from south-central NA to southern NT. **Similar species:** other thrushes, Veery (see pp 134, 176).

Varied Thrush *Ixoreus naevius* (Turdidae) 25.5 cm (10 in). Black crown, orange eyebrow above a black mask; upperparts black with 2 conspicuous orange wing-bars and orange and black flight feathers; brick-red below with black breast band. Female duller, grey breast band may be incomplete or lacking. Immature has indistinct breast band mottled with orange, eyebrow and wing-bars rusty, underparts have dusky spots. Voice: prolonged, melancholy, quavering whistle, a pause, and another note on a lower or higher pitch fading away at the end. Call, 'chuck'. **Range and habitat:** thick, wet coniferous forest of north-western and west-central NA. Winters from west-central to south-western NA. **Similar species:** American Robin.

American Robin *Turdus migratorius* (Turdidae) 23-8 cm (9-11 in). Grey back, rusty red breast, white chin streaked with black, white lower belly and undertail-coverts. Male has black head with white crescents above and below dark eye, black tail with white corners in eastern subspecies. Female paler. Juvenile has pale rust breast with large black spots, white throat, grey back with faint white stripes and pale wing-bars. Voice: loud, short, carolling phrases alternately rising and falling. More than 10 distinctly different calls, 'tut, tut, tut' most common. **Range and habitat:** woodland openings, but has adapted well to suburban lawns throughout NA, local in south. Winters in central and southern NA. **Similar species:** Eyebrowed, Rufous-backed and Varied Thrushes.

Grey Catbird

Northern Mockingbird

Sage Thrasher

Curve-billed Thrasher

Brown Thrasher

Hermit Thrush

Eastern Bluebird

Wood Thrush

Swainson's Thrush

Varied Thrush

American Robin

Blue-grey Gnatcatcher *Polioptila caerulea* (Polioptilidae) 11.5-12.5 cm (4½-5 in). Tiny, slender bird; thin black bill; bluish grey above, dark wings edged in white; pale grey below; long black tail with white outermost feathers and underside often cocked and flicked about. Narrow white eye-ring in both sexes. Male has black forehead and eyebrow. Voice: thin, banjo-like twang 'zpee'; song insect-like notes followed by warbled ditty. **Range and habitat:** treetops of moist forests, oak woods, chaparral, open pinyon and juniper woods of central and southern NA. Winters from southern NA south. **Similar species:** Black-tailed and Black-capped Gnatcatchers.

Golden-crowned Kinglet *Regulus satrapa* (Sylviidae) 9-10 cm (3½-4 in). Crown orange in males, yellow in females; in both sexes bordered with black, also white eyebrow. Body olive-grey, lighter below, brown wings with 2 white wing-bars; short tail. Forages quickly, may hover briefly. Nervous wing-flickering characteristic. Juveniles have dark crown. Voice: high-pitched 'tsee-tsee-tsee'. Song, series of high notes rising in pitch then descending into chatter. **Range and habitat:** breeds in cool coniferous forests of eastern and western NA. Winters in forests of central and southern NA. **Similar species:** Ruby-crowned Kinglet, Firecrest.

Ruby-crowned Kinglet *Regulus calendula* (Sylviidae) 11.5 cm (4½ in). Tiny. Olive-green, darker above, with prominent white eye-ring, 2 white wing-bars (buffy in juvenile), lower wing-bar bordered with black. Short tail. Male has scarlet crown patch, usually hidden. Forages rapidly, often hovers. Frequently flicks wings. Voice: musical; 3 or 4 high notes, several low notes, then a rollicking chant, 'tee, tee, tee, tew, tew, tew, tew, ti-dadee, ti-dadee, ti-dadee'; wren-like 'cack', husky 'zhi-dit'. **Range and habitat:** coniferous forests of western and northern NA. Winters along east- and west-central coasts and throughout southern NA to Central America. **Similar species:** Golden-crowned Kinglet, Goldcrest.

Tufted Titmouse *Parus bicolor* (Paridae) 11.5-14 cm (4½-5½ in). Small, crested; mouse-grey above, pale grey to white below with peach-coloured sides; large dark eye, conical black bill. North and east of central Texas, grey crest, black forehead. From southern and central Texas southwards, black crest, pale forehead. Sexes similar. Juveniles lack peach colour. Forages actively and acrobatically, often in mixed flocks. Visits feeders. Voice: clear, whistled 'peter, peter, peter. . .', repeated 6-10 times. Calls nasal, wheezy, complaining. **Range and habitat:** resident in coniferous and deciduous forests of east-central, south-central and south-eastern NA. Nests in cavities, including bird houses. **Similar species:** Plain Titmouse.

Black-capped Chickadee *Parus atricapillus* (Paridae) 14.5 cm (6 in). Black cap, black bib with ragged lower edge, white cheeks. Grey above, white below, buff flanks; wing-coverts edged white. Often forages in mixed flocks. Visits feeders. Voice: pure, whistled 'fe-bee', 2nd note lower. Call a clear 'chick-adee-dee-dee', other gargles and hisses. **Range and habitat:** mixed and deciduous woods, willow thickets, groves and shade trees. Nests in cavities including bird houses. Resident of north-western and central NA. **Similar species:** Boreal, Carolina and Mountain Chickadees, Marsh and Willow Tits (see p 178).

Mountain Chickadee *Parus gambeli* (Paridae) 15 cm (6 in). Black cap and throat. White forehead, eyebrow and cheeks. Grey above, white below. Birds of Rockies browner back, cinnamon sides; those from Great Basin more buff, further west greyer. Cavity nester, will use nestbox. Visits feeders. Voice: sweet, whistled song 'fee-bee-bay' like 'Three blind mice'. Call 'chick-a-dee-a-dee-a-dee'. **Range and habitat:** open coniferous forests of west-central and south-western mountains. Winters in valleys in oaks, cottonwoods and willows where it forages with migrating wood-warblers and vireos. **Similar species:** Black-capped Chickadee.

Verdin *Auriparus flaviceps* (Remizidae) 10-11.5 cm (4-4½ in). Very small; grey, paler below with bright yellow head in male, dull yellow in female, rufous at bend of wing (brighter in male than female); sharply pointed black bill. Immature brownish grey, paler below, lacks yellow head and rufous wing patch. Forages among terminal twigs, underside of leaves, flits about. Voice: loud 'tsee, seesee', call emphatic 'see-lip'. **Range and habitat:** resident in south-western deserts with mesquite, palo verde and thorny scrub, also riparian woodlands. **Similar species:** Common Bushtit, penduline tits (see pp 176, 272). **Not illustrated.**

White-breasted Nuthatch *Sitta carolinensis* (Sittidae) 15 cm (6 in). Crown and nape glossy black in male contrasting with bluish grey upperparts; female dull black with dull grey upperparts. Tail stubby with white corners visible in flight and display. Face and underparts white. Undertail-coverts rusty. Forages on tree-trunks, often creeping head downward. Visits feeders. Voice: series, nasal notes 'to-what, what, what,. . .'; also 'yank, yank,' soft 'kit, kit' while foraging. **Range and habitat:** resident in forests, woodlots, shade trees in central and southern NA; not central plains. **Similar species:** other nuthatches, Black-capped, Carolina and Mountain Chickadees (see pp 176, 272).

Red-breasted Nuthatch *Sitta canadensis* (Sittidae) 11.5-12 cm (4½-5 in). Crown and nape glossy black in male, lead-coloured or dull black in female. Prominent white eyebrow, black eye stripe; upperparts blue-grey; throat, lower face white, underparts rust-red (paler in females), white on undertail-coverts. Small size, stubby tail, large head, jerky flight between trees; may visit feeders. Voice: slow, nasal 'enk, enk, enk'. **Range and habitat:** resident in coniferous forests of central, south-western and north-eastern NA. Irregular irruptions south of breeding range. Nests in cavities of dead conifers, smears hole with pitch. **Similar species:** other nuthatches (see pp 176, 272).

Solitary Vireo *Vireo solitarius* (Vireonidae) 12.5-15 cm (5-6 in). Grey head, olive-green back. White eye stripe and eye-ring; 2 white wing-bars. White below, yellow and olive-grey wash on flanks. Subspecies of Rockies greyer; Pacific coast subspecies greener above, yellower below. Sexes similar. Voice: like Red-eyed Vireo, but more deliberate; 2- to-6-note phrases higher, sweeter; may burst into warble of 15-20 notes; also nasal 'see-a' or 'see-weep'. **Range and habitat:** mixed northern hardwoods of central NA, south in eastern and western mountains. Winters in south-eastern and south-central NA and Central America. **Similar species:** Black-capped, Grey and White-eyed Vireos.

Red-eyed Vireo *Vireo olivaceus* (Vireonidae) 14-16.5 cm (5½-6½ in). Grey crown bordered with black, olive nape, back, wings and tail; red eye, white eyebrow, black eye stripe; underparts white with yellowish olive wash on flanks. Sexes similar. Immatures have brown eyes. Central American subspecies yellower. Voice: short, robin-like phrases separated by deliberate pauses: 'cherries? (pause) Have some. (pause) Take a few.' Up to 40 phrases per minute. Call harsh 'chway'. Sings through heat of day well into mid-summer. **Range and habitat:** mixed and deciduous woods of north-western, central, south-central and south-eastern NA, Central America. Winters in Amazon basin. **Similar species:** Black-whiskered, Philadelphia and Warbling Vireos.

Warbling Vireo *Vireo gilvus* (Vireonidae) 12.5-15 cm (5-6 in). Light grey tinged with olive-green above, crown and nape paler; white below, usually with greenish yellow wash on flanks, particularly in immatures. Sexes similar. Juveniles often more buffy with faint buff wing-bar. Voice: husky, languid warble, pitch rises, tempo increases: 'one, two, three, four, fivesixseveneight', also wheezy, cat-like 'twee'. **Range and habitat:** deciduous woods, roadsides, aspen groves, riparian and alder thickets of north-western, central and southern NA. Winters from south-central NA to north-western NT. **Similar species:** Philadelphia and Red-eyed Vireos, Tennessee Warbler.

Blue-grey Gnatcatcher (female)

Golden-crowned Kinglet female

Ruby-crowned Kinglet

Tufted Titmouse

Black-capped Chickadee

Mountain Chickadee

White-breasted Nuthatch

Red-breasted Nuthatch

Solitary Vireo

Red-eyed Vireo

Warbling Vireo

Black and White Warbler *Mniotilta varia* (Parulidae) 11.5-14 cm (4½-5½ in). Boldly striped black and white above and below. Male has black facial striping, black throat, black streaks on sides, black spots on undertail-coverts. Female dull black, with brownish wash; face, throat, underparts mostly white; flanks thinly streaked with black. Immature like female. Creeps over branches probing cracks. Voice: high, lisping 'wee-wee', repeated 6 to 8 times, also sharp 'pit' and thin 'seet'. **Range and habitat:** deciduous and mixed forests from north-central and north-eastern NA south almost to Gulf coast. Winters south-eastern NA to northern NT. **Similar species:** Blackpoll and Black-throated Grey Warblers (see p 86).

Yellow-rumped Warbler *Dendroica coronata* (Parulidae) 12.5-15 cm (5-6 in). 'Myrtle' subspecies blue-grey above, streaked with black, yellow on crown, rump and shoulders, white in outer tail feathers; white eyebrow and eye-ring in black face; throat white; black upper breast, black streaked flanks. Female and winter male similar, but greyish brown. Winter female and immature browner, may lack yellow. 'Audubon' similar, but slate-grey, yellow throat, black breast and upper flanks, white wing patch; female, winter male, immature browner. Loose musical trill, loud 'check'. **Range and habitat:** conifer and mixed forests of northern and north-central NA, south in western mountains to south-central NA. Winters from east- and west-central NA through southern NA into Central America. **Similar species:** Cape May, Magnolia and Palm Warblers.

Magnolia Warbler *Dendroica magnolia* (Parulidae) 11.5-13.5 cm (4½-5½ in). Grey crown, black back, yellow rump, black tail with rectangular white patches; white eyebrow, black face, white crescent below eye; 2 wide white wing-bars. Throat and underparts yellow streaked black, under-tail-coverts white. Female similar, duller. Immature, autumn adult as female, head greyish with narrow white eye-ring, back olive-green, streaking faint on underparts, greyish breast band. Voice: short, rising 'weeta, weeta, weetsee', also 'tlep'. **Range and habitat:** small conifers of northern and east-central NA. Winters south-central NA to north-western NT, also West Indies. **Similar species:** Yellow-rumped Warbler, immature Prairie Warbler.

Northern Parula *Parula americana* (Parulidae) 11.5 cm (4½ in). Blue-grey above with greenish yellow patch on back, 2 white wing-bars and white patches on outer tail feathers; broken white eye-ring, yellow throat and breast with black and orange band across upper breast; belly and undertail-coverts white. Female duller, lacks breast band. Immature dull. Voice: climbing, buzzy trill that falls 'zzzzzzzzzurp'. **Range and habitat:** humid woods with hanging mosses for nest, east-central and south-eastern NA. Winters south-central and extreme south-eastern NA, Central America, West Indies. **Similar species:** Yellow-throated Warbler, immature Nashville Warbler, Tropical Parula.

Chestnut-sided Warbler *Dendroica pensylvanica* (Parulidae) 11.5-13 cm (4½-5 in). Lemon-yellow crown, yellow-green back streaked with black, white in outer tail feathers; dark wings, 2 yellow wing-bars; face black and white, underparts white with narrow chestnut stripes along flanks. Female paler. Immature yellow-green above, greyish white below with yellow wing-bars, grey face, prominent white eye-ring. Autumn adult yellow-green above without black face, chestnut flanks. Voice: loud, 'pleased, pleased, pleased to meet cha', accent on 'meet'; also 'chip'. **Range and habitat:** brushy pastures, slashings central and east-central NA, south in mountains. Winters in Central America. **Similar species:** immature Bay-breasted Warbler; immature resembles Ruby-crowned Kinglet (see p 82).

Palm Warbler *Dendroica palmarum* (Parulidae) 11.5-14.5 cm (4½-5½ in). Frequent tail-wagging. Chestnut crown, olive-brown face and back with dusky streaks and greenish yellow rump. Faint wing-bars. Dark tail has white corners. Eyebrow, throat, upper breast, undertail-coverts yellow.

Subspecies *palmarum* has lower breast, belly greyish white; *hypochrysea* bright yellow below. Both streaked with chestnut below. Sexes similar. Immature brown above, yellow undertail-coverts. Voice: weak, buzzy trill, also 'check'. **Range and habitat:** wooded borders of muskeg, north-central and north-eastern NA. Winters in south-eastern, south-central NA and Central America. Forages near ground in low bushes, weedy fields on migration. **Similar species:** Prairie and Yellow-rumped Warblers.

Black-throated Green Warbler *Dendroica virens* (Parulidae) 11.5-14 cm (4½-5½ in). Yellow face contrasts with yellow-green crown and back, black throat, breast and flanks. Greenish eye stripe and ear-coverts. White wing-bars, white in outer tail feathers. White below. Female similar, less black; chin and throat mostly yellow. Adult and immature in autumn have yellow face, dusky throat, breast, flanks. Voice: lazy 'zee, zee, zee, zoo, zee'; also high 'tsip'. **Range and habitat:** mixed and coniferous forest (birch and aspen in north, cypress in south), north-central and eastern NA. Winters in south-central and extreme south-eastern NA into Central America, Greater Antilles. **Similar species:** Golden-cheeked and Townsend's Warblers.

Yellow Warbler *Dendroica petechia* (Parulidae) 11.5-13 cm (4½-5 in). Yellow below, greenish-yellow above with 2 yellow wing-bars and yellow patches in outermost tail feathers; breast and flanks with narrow chestnut streaks; dark eye in yellow face. Female and immature male yellow below, yellowish green above; immature females drab olive-green. Subspecies vary in density of chestnut streaks and amount of chestnut on crown. Voice: lively 'sweet, sweet, oh so sweet'; also soft, clear 'chip'. **Range and habitat:** willow and alder thickets bordering wetlands, and in cleared woodland, brushy fields, suburban gardens, northern NA to northern NT. Winters from south-central NA to central NT. **Similar species:** Wilson's Warbler (see p 86).

Orange-crowned Warbler *Vermivora celata* (Parulidae) 11.5-14 cm (4½-5½ in). Olive-green above, olive washed in yellow below with yellow undertail-coverts; greenish yellow eyebrow, thin broken yellow eye-ring; dusky olive streaking on breast; orange crown patch, rarely visible. Western subspecies more yellow than eastern. Staccato trill, pitch often rises then falls. **Range and habitat:** brushy clearings, northern and central NA, south in western mountains, more common in west. Winters along both coasts from central to south-central NA and Central America. **Similar species:** Tennessee and Arctic Warblers, female Yellow and Wilson's Warblers (see p 86).

Nashville Warbler *Vermivora ruficapilla* (Parulidae) 11.5-13 cm (4½-5 in). Crown, nape, face grey with concealed chestnut crown patch and conspicuous white eye-ring; olive-green back, wings and tail; yellow throat and under-parts, except for whitish belly. Female, autumn male, immature pale grey or brownish grey head, pale yellow underparts, buff eye-ring. Voice: 2-parted, 1st part higher, slower, 'sebit, seebit, seebit, seebit, titititititi'; also sharp, metallic 'clink'. **Range and habitat:** cool, open, mixed woods, forest edges, bogs central NA south in western mountains. Winters in south-central and extreme south-eastern NA. **Similar species:** Connecticut, MacGillivray's, Mourning, Virginia's and Wilson's Warblers (see p 86).

Black-throated Blue Warbler *Dendroica caerulescens* (Parulidae) 12-14 cm (4½-5½ in). Male grey-blue above; black face, throat, upper breast and flanks; lower breast and belly white; wing and tail feathers dark, white patch in wing, white in outer tail feathers. Female olive-grey above, traces of blue on crown and uppertail-coverts; olive-yellow below; white eyebrow and white wing patch. Immatures resemble adults. Voice: slow, buzzy 'I am so lazeeee', last note rising, junco-like 'smack'. **Range and habitat:** undergrowth of deciduous and mixed woodlands of east-central NA, south in mountains. Winters in West Indies. **Similar species:** female Cerulean and Tennessee Warbler, Philadelphia Vireo.

Black and White Warbler

Yellow-rumped Warbler

Magnolia Warbler

Northern Parula
female

Chestnut-sided Warbler

Palm Warbler

Black-throated Green Warbler
(autumn plumage)

Yellow Warbler

Orange-crowned Warbler

Nashville Warbler

Black-throated Blue Warbler

Blackpoll Warbler *Dendroica striata* (Parulidae) 12.5-15 cm (5-6 in). Black cap contrasts with white cheek. Back olive-grey streaked with black; 2 white wing-bars; white patches in outermost tail feathers; white below, streaked with black on sides of neck and flank. Female greenish cap, pale eyebrow, olive-grey above with dusky streaks, white below, washed in yellow. Autumn adults and immatures resemble female, but greener above, greenish yellow below with faint streaks. Legs pinkish or straw-coloured. Voice: extremely high, deliberate 'zi-zi-zi-zi-zi-zi-zi-zi', crescendo followed by decrescendo, also loud 'smack'. **Range and habitat:** northern spruce-fir forest. Winters in central NT. **Similar species:** Black and White and Black-throated Grey Warblers, immature Pine and Bay-breasted Warblers.

Bay-breasted Warbler *Dendroica castanea* (Parulidae) 12.5-15 cm (5-6 in). Chestnut crown, black face bordered on neck with buff patches, chestnut throat, upper breast, flanks; greenish grey above, streaked with black, 2 white wing-bars, white patches in outer tail feathers; lower breast, belly, undertail-coverts buff. Female similar, chestnut reduced to wash, buff neck patch inconspicuous. Immature and autumn adult grey above with dusky streaks, buff below without streaks, trace of chestnut on flanks, dark legs (unlike pinkish legs of Blackpoll Warbler). Voice: similar to Black-and-white Warbler but shorter, thinner, single pitch; also mellow 'chip'. **Range and habitat:** follows outbreaks of spruce budworm across northern coniferous forests. Winters in northern NT. **Similar species:** Chestnut-sided Warbler, immature Blackpoll Warbler (see p 84).

Wilson's Warbler *Wilsonia pusilla* (Parulidae) 11-12.5 cm (4-5 in). Black cap bordered by yellow forehead and eyebrow; olive-green above, yellow below; dark eye. Female and immature olive-green above with yellow forehead and eyebrow, but with black cap absent or obscured by green feather-edging, yellow below. Rarely forages above 3 m (10 ft). Often twitches longish tail in circular motion, flicks wings like kinglet. Voice: thin, rapid chatter 'chi, chi, chi, chi, chet, chet', dropping in pitch at end, also flat 'chuck'. **Range and habitat:** moist tangles, thickets along wooded streams from treeline across northern NA, south in western mountains. Winters south-central and extreme south-western NA to north-western NT. **Similar species:** Yellow Warbler, female Hooded Warbler (see p 84).

Canada Warbler *Wilsonia canadensis* (Parulidae) 12.5-15 cm (5-6 in). Bluish grey above, yellow below. Face, sides of neck black with yellow 'spectacles'; necklace of black spots across breast; white undertail-coverts. Female similar, face grey, necklace faint. Immature like female, necklace further reduced, back often washed with olive-brown. Forages actively, darting after insects, rarely higher than 3 m (10 ft). Voice: explosive burst of staccato notes, irregularly arranged, but with 'ditchety' phrase and ending with single 'chip', also a loud 'chick'. **Range and habitat:** moist undergrowth of mature northern and eastern woodlands of central, north-eastern and east-central NA, south in eastern mountains. Winters in NT. **Similar species:** Kentucky Warbler.

Common Yellowthroat *Geothlypis trichas* (Parulidae) 11.5-14 cm (4½-5½ in). Broad black mask across forehead, face, on to sides of neck, bordered above by pale grey; brownish olive above; bright yellow chin, throat, upper breast; flanks buff-brown, belly off-white; undertail-coverts yellow. Female similar, lacks mask, but has faint eye-ring, ill-defined eye stripe, greyer above. Immature male browner than adult male, mask incomplete. Immature female browner than adult female, only hint of yellow throat. Voice: rollicking 'witchity-witchity-whitch', noticeable geographic variation. Flight song a jumble of notes ending with 'witchity-whitch'. Note, husky 'tchek'. **Range and habitat:** marshes, wet thickets, brushy fields throughout NA, except extreme north. Winters along east-central and southern coasts to northern NT, also West Indies. **Similar species:** Kentucky

Warbler, female Nashville Warbler, other species of yellowthroat found throughout Caribbean (see p 84).

Northern Waterthrush *Seiurus noveboracensis* (Parulidae) 12.5-15 cm (5-6 in). Yellowish buff eyebrow diagnostic. Olive-brown above, yellowish buff below with small dark brown spots on throat, larger dark brown spots and streaks on breast, flanks, belly. Forages on ground where it walks with teetering gait. Voice: vigorous, clear 'sweet, sweet, sweet, swee-wee-wee, chew, chew, chew', accelerating staccato ending. Call a sharp 'chip'. **Range and habitat:** cool, dark wooded swamps, brush bogs and wooded lake and streamsides from treeline to central NA. Winters from south-central NA to northern NT, also West Indies. **Similar species:** Louisiana Waterthrush, Ovenbird, Spotted Sandpiper, thrushes (see pp 62, 80).

Ovenbird *Seiurus aurocapillus* (Parulidae) 14-16.5 cm (5½-6½ in). Orange crown bordered with dark brown, olive-brown above, white below, densely marked with elongate brown spots on breast and flanks. Prominent white eye-ring. Legs pink. Female similar, slightly paler. Juveniles have 2 buff wing-bars. Walks. Voice: emphatic crescendo of 'teacher, teacher, teacher, teacher' always emphasizes the 1st syllable. Evening flight song a rapid jumble of bubbling, warbling notes ending with 'teacher, teacher'. Note, sharp 'chock'. **Range and habitat:** forages on ground in deciduous forest of west-central, central, north-eastern and east-central NA. Winters from south-eastern and south-central NA to northern NT. **Similar species:** thrushes, water thrushes (see p 80).

Mourning Warbler *Oporornis philadelphia* (Parulidae) 12.5-14 cm (5-5½ in). Grey hood covers head, nape, throat, becoming black on upper breast; some black by the eyes; olive above, yellow below. In autumn black veiled in grey. Female has dull grey-brown hood with broken eye-ring, olive above, yellow below. Immature has hood brownish olive, yellow tinge on throat, broken eye-ring, olive above, pale yellow below. Voice: loud, liquid chant 'chirry, chirry, chirry, chorry, chorry'; also a loud, low-pitched 'tcheck'. **Range and habitat:** shrubby undergrowth in woods, slashings, old fields from west-central to north-eastern NA, south in the eastern mountains. Winters from extreme south-central NA to northern NT. **Similar species:** Connecticut, MacGillivray's and Orange-crowned Warblers, female Common Yellowthroat.

Yellow-breasted Chat *Icteria virens* (Parulidae) 19 cm (7½ in). Black lores contrast with white 'spectacles' and white whisker; bill large, heavy, black; olive-green above with long tail; bright yellow chin, throat, breast; white belly and undertail-coverts. Sexes similar. Juvenile brownish or greyish olive above, no yellow below, throat and breast spotted. Elusive, foraging and singing from within brush. Voice: clear whistles, alternating with mews, grunts, rattles, cackles, squeaks, suggests mimic thrush but longer pauses between phrases. Often sings at night. Call 'whoit' or 'kook'. **Range and habitat:** brushy tangles, thickets and briars in low wet areas of central and southern NA. Winters from south-central NA to north-western NT.

American Redstart *Setophaga ruticilla* (Parulidae) 11-13.5 cm (4½-5½ in). Black above, also on throat and breast, lower belly white. Broad orange patch in wings and across all but central tail feathers; orange flanks. Female grey-brown head, olive-grey, dingy white below with yellow flanks; wing and tail patterns like male, but yellow. Male in 1st spring like female, but black patches and orange tinge in yellow markings. Voice: strident, sibilant 'zee, zee, zee, zee, zwee' (last note rising), alternating with 'zee, zee, zee, zee, zawaah' (last note slurring downward), also a sharp 'chick'. **Range and habitat:** 2nd-growth deciduous woodlands across northern, west-central and most of eastern NA. Winters from southern NA to central NT.

Blackpoll Warbler

Bay-breasted Warbler

Wilson's Warbler

Canada Warbler

Ovenbird

Common Yellowthroat

Northern Waterthrush

Mourning Warbler

Yellow-breasted Chat

American Redstart 2nd year male

Rose-breasted Grosbeak *Pheucticus ludovicianus* (Emberizidae) 18-21.5 cm (7-8½ in). Black above; rose red triangle on breast, enlarges with age; wings and tail black with white patches; white rump, flanks, belly, undertail-coverts; wing-linings rose. Female brown above, prominent light crown and eyebrow stripes; white wing patches, saffron wing-linings; grey-white below, streaked with brown. Winter male like female, but pink chest, black and white wings, rose wing-linings. Immature male like winter male, less pink, brown in wings. Voice: like American Robin, but more mellow, more rapid; squeaky 'ink'. **Range and habitat:** deciduous woods, thickets, north-central, central, north-eastern and east-central NA. Winters south-central to northern NT. **Similar species:** female Black-headed Grosbeak, female Purple Finch (see p 90).

Northern Cardinal *Cardinalis cardinalis* (Emberizidae) 19-23 cm (7½-9 in). Brilliant red, crested; red bill, surrounded by black face and throat. Female warm brown, shading to red on wings and tail; red bill, less black on face and throat. Immature grey-brown plumage, dark bill, little or no red in wings and tail. Voice: loud, throaty variable whistle. Female sings softer and less often than male. Call, metallic 'tsink'. **Range and habitat:** brushy undergrowth, east-central and southern NA. Expanding range north. Visits feeders. **Similar species:** Pyrrhuloxia, Summer and Scarlet Tanagers (see p 90).

Indigo Bunting *Passerina cyanea* (Emberizidae) 14 cm (5½ in). Brilliant blue, almost violet on head. Female brown, lighter below with faint streaking on breast. 1st-year male often mottled blue and olive-brown with buff-brown wing-bars. Immature birds as females, but streaked on breast. Voice: high, strident, with phrases sung in couplets, each on different pitch; also sharp, thin 'spit'. **Range and habitat:** brushy pastures, woodland edges east-central and southern NA. Winters south-eastern and south-central NA to northern NT. **Similar species:** Blue Grosbeak, females of other buntings (see pp 178, 274).

Rufous-sided Towhee *Pipilo erythrophthalmus* (Emberizidae) 18-22 cm (7-8½ in). Black back, throat, breast; rufous flanks, white belly; white in black wings and tail. Female similar, but brown where male black. Juvenile brown above, light below, heavily streaked. Western males have white spots on back and wings. Eastern and western subspecies have red eyes, southern white eyes. Voice: musical 'Drink your tea', last syllable higher, trilled. Western birds 'chup, chup, zeeee'. Call 'chwink' in east, 'chweee' in west. **Range and habitat:** undergrowth of central and southern NA, except short grass prairies. Winters southern NA.

Savannah Sparrow *Passerculus sandwichensis* (Emberizidae) 12.5-16.5 cm (5-6½ in). Grey-brown above, cream below; dark brown streaks over all, usually forming central spot on breast; white stripe across mid-crown, yellowish eyebrow; tail short, notched; legs pink. Western subspecies darker than eastern, yellow eyebrow diagnostic. Voice: lisping notes rising then descending trill, also high 'tsip'. **Range and habitat:** meadows, prairies, dunes and wet grasslands, northern, central, and south-west NA. Winters along east- and west-central coasts southern NA and Central America. **Similar species:** Lincoln's, Song and Vesper Sparrows.

Song Sparrow *Melospiza melodia* (Emberizidae) 14-18 cm (5½-7 in). Brown above, streaked black and grey; white below, with dark brown streaks on breast and flanks, forming spot on breast; crown brown with central grey stripe, broad greyish eyebrow, grey-brown ear-coverts. Tail long, rounded; pumped in flight. Legs pale. Immatures often lack breast spot. Voice: variable, musical or buzzy, usually starts with 3 or 4 clear notes. Call, nasal 'tchep'. **Range and habitat:** widespread in brush, wooded edges north-western, central, south-western and south-eastern NA. Winters central and southern NA. **Similar species:** Lincoln's, Savannah, Swamp Sparrows.

Chipping Sparrow *Spizella passerina* (Emberizidae) 12.5-14 cm (5-5½ in). Rufous cap, white eyebrow, black eye stripe, brown back streaked with black; grey cheeks, underparts, rump. Winter adult has black streaks in cap, often with light median stripe; eyebrow, eye stripe less contrasting; brown greyer. Juvenile browner, finely streaked on breast and flanks; retains this plumage well into autumn. Voice: mechanical trill on 1 pitch, also high, sweet 'tsip'. **Range and habitat:** open woods; conifers; farms; towns throughout NA, except Alaska and extreme southern United States. Winters in south-central NA and Central America. **Similar species:** other rufous-capped sparrows.

Vesper Sparrow *Pooecetes gramineus* (Emberizidae) 14-16.5 cm (5½-6½ in). White outer tail feathers in dark tail are diagnostic. Pale grey-brown above, with dark brown streaks; grey-white below, with buff wash on breast and flanks; fine dark streaks on throat, breast, flanks; whitish eye-ring and chestnut at bend of wing. Immature finely streaked. Voice: sweet, throaty, usually begins with 2 minor notes, rises to 2 higher-pitched notes followed by descending series of trills. Call, sharp 'chirp'. **Range and habitat:** prairies, meadows, roadsides of central and south-western NA. Winters in east-central and southern NA. **Similar species:** pipits, Savannah and Song Sparrows.

Fox Sparrow *Passerella iliaca* (Emberizidae) 16-19 cm (6½-7½ in). Eastern subspecies has rufous tail, rust on grey head and back, and on wings. Intermountain subspecies grey head and back, rufous tinged with grey; west coast subspecies dusky brown. All subspecies have heavily streaked breasts, often with central spot. Voice: melodious, usually introductory whistle, then series of sliding whistles and clear notes. Call, 'click' or 'chip'. **Range and habitat:** undergrowth, brush across northern NA, south in western mountains. Winters along west-central and south-western coast, east-central and south-eastern NA. **Similar species:** thrushes, especially Hermit Thrush (see p 80).

Dark-eyed Junco *Junco hyemalis* (Emberizidae) 12.5-16.5 cm (5-6½ in). All subspecies have pink bill, white outer tail feathers, white belly, dark eyes. 'Slate-coloured' (3 subspecies): slate-grey head, breast, flanks, back, wings, tail. 'White-winged' (1) similar, with 2 white wing-bars, more white in tail. 'Oregon' (2): black hood, brown back, pink sides. 'Pink-sided' (1): pale grey head, breast; dark lores; brownish back; pink flanks. 'Grey-headed' (2): light grey head, breast, flanks; dark lores; rusty back. Females paler or browner. Juveniles have streaked breasts. Voice: loose, musical trill, and 'smack'. **Range and habitat:** conifer and mixed woods, northern, western, east-central NA, south-eastern mountains. Winters central and southern NA. **Similar species:** Mexican Junco, Vesper and Lark Sparrows.

White-throated Sparrow *Zonotrichia albicollis* (Emberizidae) 16.5-19 cm (6½-7½ in). Polymorphic. 1 morph: black and white head stripes, grey face, white throat; other: tan and dark brown head stripes, grey-brown face, white throat. Both chestnut-brown above, wings streaked with black, yellow lores, grey breast, faintly streaked olive-brown flanks, white belly and wing-bars. Pure whistle, 2 notes followed by 3 3-syllable notes: 'Old Sam Peabody, Peabody, Peabody'; loud 'chink'. **Range and habitat:** coniferous forests, north-central, north-eastern and east-central NA. Winters in east-central and southern NA. **Similar species:** White-crowned Sparrow, immature Swamp Sparrow.

White-crowned Sparrow *Zonotrichia leucophrys* (Emberizidae) 14-18 cm (5½-7 in). Large, black and white striped crown with pearl-grey cheeks, neck, breast; back and wings brown with white and black streaks; long tail. Juveniles similar, but with cinnamon-brown and buff head striping. Voice: clear whistles in minor key, followed by husky, trilled whistles; numerous dialects; call, high 'seeet'. **Range and habitat:** shrubby areas, northern, west-central and south-western NA. Winters along Pacific coast and southern NA. **Similar species:** Golden-crowned Sparrow.

Rose-breasted Grosbeak

Northern Cardinal

Indigo Bunting
male singing

Rufous-sided Towhee

Savannah Sparrow

Song Sparrow

Chipping Sparrow

Vesper Sparrow

Fox Sparrow

Dark-eyed Junco

White-throated Sparrow

White-crowned Sparrow

Summer Tanager *Piranga rubra* (Emberizidae) 18-19.5 cm (7-8 in). Male rose red. Female and immature pale yellow below, yellowish olive above, more olive on wings, tail and back. Bill large, pale in both sexes, darker during nonbreeding season. First-spring males similar to females, washed or blotched with red. Second-spring males similar to adult males, but paler, more yellowish. Forages high in foliage; has particular fondness for bees. Voice: series of sweet, clear phrases similar to American Robin, but more deliberate, also a staccato 'pi-tuk'. **Range and habitat:** deciduous forest, especially oak groves of southern and east-central NA. Winters from south-central NA to central NT. **Similar species:** Cardinal, Scarlet Tanager; female resembles female Northern Oriole.

Scarlet Tanager *Piranga olivacea* (Emberizidae) 16.5-19 cm (6½-7½ in). Male glowing scarlet with jet black wings and tail. First-spring males duller orange-red, wings and tail brownish black. Female and immature green above, yellow below, grey-brown to blackish brown wings and tail. Autumn males resemble females, but green is brighter, some yellow on crown, yellow underparts have faint orange wash. Juveniles resemble female, but with yellow wing-bars. Forages high in foliage. Courting males display from low perches. Voice: hoarse carolling of 4 or 5 phrases, like an American Robin with a sore throat, also hoarse 'chip-burr'. **Range and habitat:** deciduous forests, especially oak groves of east-central and south-eastern NA. Winters from northern to central NT. **Similar species:** other tanagers (see p 136).

Bobolink *Dolichonyx oryzivorus* (Icteridae) 15-20.5 cm (6-8 in). Black head and body with golden nape, white scapulars, lower back and rump. Female buff-olive above with dark brown streaks, dark brown wings and tail with buff edges on wing-coverts. Buff-white below, finely streaked with dark brown on flanks and undertail-coverts; head greyish white with brown crown stripes and brown stripe behind dark eye; breast warm buff. Immatures similar, but yellowish below. Winter male like female, but darker above with yellow face. Voice: reedy introduction followed by bubbly, twanging jumble of notes, often given in flight. Call, sharp 'pink'. **Range and habitat:** open grassy areas of central NA. Winters in central NT. **Similar species:** female Red-winged Blackbird, Savannah Sparrow (see p 88).

Western Meadowlark *Sturnella neglecta* (Icteridae) 20.5-25.5 cm (8-10 in). Brown, buff, black streaked and mottled above and on flanks, head with prominent stripes; throat, breast and belly yellow, with black V across the breast; yellow in eyebrow; long conical bill. In flight several rapid wingbeats alternate with short glides, white outer tail feathers prominent. Voice: double-noted, gurgling and flute-like, quite unlike 2 slurred whistles of Eastern Meadowlark. Sings from atop tall weeds, fence-post, tree or in flight. Call, 'chupp'. **Range and habitat:** prairies, meadows, drier habitat than Eastern Meadowlark from central and south-central NA west. Withdraws southward in winter. **Similar species:** Eastern Meadowlark.

Red-winged Blackbird *Agelaius phoeniceus* (Icteridae) 19-24 cm (7½-9½ in). Male black with scarlet epaulettes bordered with buff; occasionally rufous in area of epaulette. Female smaller, dark brown above streaked with buff, light below heavily streaked with dark brown. Buff eyebrow and throat. First-spring male darker than female, epaulette orange flecked with black, buff eyebrow, light cheek and throat. Immature similar to female, more buff. Voice: liquid 'konk-a-ree'. Call, low 'chuck'. Males establish territories in marshes, wet meadows and thickets. Females choose territories; many males are polygamous. **Range and habitat:** marshes, wet meadows, NA except northernmost areas. Winters in central and southern NA. **Similar species:** Tricoloured Blackbird.

Brown-headed Cowbird *Molothrus ater* (Icteridae) 15-20 cm (6-8 in). Male has coffee-brown head, glossy black body. Female uniform grey-brown. Juvenile brown, lighter below, often streaked dark brown. Bill short and conical. Known to parasitize over 100 North American species; has reduced populations of vireos, wood-warblers, tanagers, orioles. Eggs not mimetic. Chicks hatch earlier than host, grow faster and larger, crowding out host chicks. Courtship includes a sleeked posture with bill pointed up and a bowing posture with wings spread and plumage fluffed. Voice: high-pitched, bubbling, with glissando in middle. Female has a loud harsh chatter. **Range and habitat:** north-central, central and southern NA. Winters in southern NA. **Similar species:** Bronzed Cowbird, other blackbirds (see p 136).

Brewer's Blackbird *Euphagus cyanocephalus* (Icteridae) 20.5-25.5 cm (8-10 in). Black with purple iridescence on head, green iridescence on body; bill conical, sharply pointed; eyes creamy white to pale yellow. Female grey-brown, lighter on head and neck; eyes dark brown. Juveniles resemble females. First-winter males may be blackish with grey eyebrows, some grey barring on back and chest. Often in large monospecific flocks, or with Red-winged Blackbirds. Voice: a creaking 'ksheeik', also metallic 'cheek'. **Range and habitat:** wherever trees and open space available together from central and south-central NA west. Winters in west-central and southern NA. **Similar species:** other blackbirds, Common Grackle, European Starling (see pp 136, 182).

Common Grackle *Quiscalus quiscula* (Icteridae) 28-34 cm (11-13 in). Large, with long profile; long, conical, slightly decurved bill; long, wedge-shaped, keeled tail, larger in males than females. Adults all-black; nearly uniform purple gloss in one subspecies, blue sheen on head contrasting with a bronze sheen on back in the other. Eyes of both subspecies, both sexes, yellow. Juveniles sooty, dark eyes. Highly gregarious. Male lowers wings, puffs plumage and bows before female. Voice: scraping, ascending 'kogubaleek'. Call, loud 'tchack'. **Range and habitat:** common in towns, croplands, groves, streamsides in central and eastern NA. Winters entirely in eastern NA. **Similar species:** blackbirds, cowbirds, grackles, European Starling (see pp 136, 182).

Boat-tailed Grackle *Quiscalus major* (Icteridae) 33-42 cm (13-16 in). Very large; long, wedge-shaped, keeled tail; all-black with iridescent blue, blue-green on back and breast. Eye colour varies from yellow on Atlantic coast to dark brown in peninsular Florida and the north-western Gulf coast; medium brown outlined with yellow along central Gulf coast. Females much smaller than males, dark brown with buff throat and breast, dark eye. Polygynous. Courting male spreads wings and tail; bobs and bows in front of female. Voice: harsh whistles and clucks. **Range and habitat:** resident in coastal areas near salt water, saltmarshes east-central and south-eastern NA. **Similar species:** Common and Great-tailed Grackles, Fish Crow.

Northern Oriole *Icterus galbula* (Icteridae) 18-20.5 cm (7-8 in). Formerly 2 species, 'Baltimore' and 'Bullock's' orioles. Both subspecies flame-orange and black with white wing patches that are much larger in 'Bullock's' subspecies. 'Baltimore' has black hood extending on to breast. 'Bullock's' has black crown, nape, eye-stripe and bib with orange eyebrow, cheeks and sides of neck. Female and young 'Baltimore' olive-brown above, orange-yellow below with 2 white wing-bars. Female and young 'Bullock's' grey above with 2 white wing-bars; yellow-orange face, throat, upper breast; grey-white belly. Voice: rich, piping whistles, somewhat disjointed. Call, low, whistled 'hew-li' ('Baltimore'); clear sweet 'kleek' ('Bullock's'). **Range and habitat:** open deciduous woodland of central and southern NA. Winters from south-central NA to northern NT. **Similar species:** other orioles (see p 136).

Bobolink

Western Meadowlark

Red-winged Blackbird

Brown-headed Cowbird

Brewer's Blackbird

Common Grackle

Northern Oriole ('Bullock's subspecies)

Scarlet Tanager

Boat-tailed Grackle

Summer Tanager

Pine Siskin *Carduelis pinus* (Fringillidae) 11.5-13.5 cm (4½-5½ in). Brown above, white below, finely streaked all over with dark brown; wings and tail dark brown with yellow at base of flight feathers and sides and base of tail feathers; tail deeply notched; bill thin, sharply pointed cone. Voice: long, rapid jumble of notes including a rising, buzzy 'zzzhrrreee' that is diagnostic. Call 'sweeeet'. **Range and habitat:** coniferous forest of northern and central NA, in western mountains to south-western NA. Winters in central and southern NA. Visits feeders. **Similar species:** female House Finch and Goldfinch, Spruce Siskin (see p 180).

American Goldfinch *Carduelis tristis* (Fringillidae) 11.5-14 cm (4½-5½ in). Breeding male brilliant yellow with black cap, black wings and tail, white in wings and tail, and white tail-coverts; bill, legs pink. Female, similar pattern in yellow-olive, black and white. Autumn and winter male resembles female, but whiter below; male has more yellow about head, female less. Juvenile like female, cinnamon-buff above and on wing-bars, wings and tail brownish black. Voice: sweet jumbled twittering. Flight call 'potato chip, potato chip' given between wingbeats. **Range and habitat:** weedy fields, streamsides, willow thickets across central and south-eastern NA, along south-western coast. Winters along west-central coast, across southern and east-central NA, largely leaves interior. **Similar species:** immature Indigo Bunting, Lawrence's and Dark-backed Goldfinches, Pine Siskin (see p 188).

House Finch *Carpodacus mexicanus* (Fringillidae) 12-14 cm (4½-5½ in). Male has red to orange forehead, crown, eyebrow, throat and rump; face and upper body grey-brown streaked with brown; off-white below with brown streaks. Female lacks reddish colour, head unpatterned brown. Tail brown with square end in both sexes. Voice: melodious, jumble of notes lasting more than 3 sec and ending with ascending note. Call, sweet 'cheet'. **Range and habitat:** abundant native, resident; west-central, south-western and south-central NA. Introduced into New York City in 1940, spreading north, south and west explosively. **Similar species:** Cassin's and Purple Finches.

Purple Finch *Carpodacus purpureus* (Fringillidae) 14-15 cm (5½-6 in). Male raspberry red head, throat, breast, back with underlying dark cheek, whisker and dorsal streaks; belly white, dark streaks, highlighted with red. Female streaked grey-brown above, prominent white eyebrow and cheek patch, white below with brown streaks. Both sexes have notched tail. Immature like female, but buff. Voice: rich, bubbling outburst of melodious notes; usually in couplets or triplets. Call, soft 'pick'. **Range and habitat:** north-western, central and north-eastern coniferous forests, south in far west in coastal forests. Winters from central and east-central NA south, also far west-central and south-western NA. **Similar species:** Cassin's and House Finches.

White-winged Crossbill *Loxia leucoptera* (Fringillidae) 15-17.5 cm (6-7 in). Male bright pink, lightest on belly; wings and tail black, white wing-bars. Female and juvenile grey with olive streaks, more heavily streaked above, white wing-bars; yellowish wash on breast and rump. Mandibles cross in both sexes, upper mandible to left. Voice: succession of long, loud trills at different pitches, often given in flight. Call, 'wink-wink-wink' with upward, questioning inflection. **Range and habitat:** northern coniferous forests of Holarctic. Wanders south in winter to central NA, where may visit feeders. **Similar species:** Pine Grosbeak, other crossbills.

Pine Grosbeak *Pinicola enucleator* (Fringillidae) 20.5-25.5 cm (8-10 in). Male pink with black wings and tail, white wing-bars, grey flanks and belly. Female grey with dull yellow-orange head and rump, black wings and tail, white wing-bars. Immature male similar to female, but head and rump reddish. All have stubby bill. Flight undulating. Voice: varied warble similar to Purple Finch's song. Call, whistled 'tee-tew-tew'. **Range and habitat:** coniferous

forests of Northern Hemisphere; south in North America in western mountains. May wander south in winter. Visits feeders. **Similar species:** White-winged Crossbill.

Evening Grosbeak *Coccothraustes vespertinus* (Fringillidae) 18-21.5 cm (7-8½ in). Yellow forehead and eyebrow, black crown; throat, face, neck and nape rich yellow-brown, becoming yellow on breast, flanks, belly, back and rump; wings and tail black, large white wing patches. Large bill apple-green in spring, chalky white in winter. Female grey, darker above, yellow wash below; white in tail feathers. Voice: wandering warble. Call, ringing 'cleer' or 'clee-ip'. **Range and habitat:** coniferous forests, central and western NA. Winters in deciduous and fruiting trees, central and southern NA. **Similar species:** American Goldfinch, Hawfinch.

Steller's Jay *Cyanocitta stelleri* (Corvidae) 33 cm (13 in). Crested jay, black on head, back, throat, breast, with deep blue wings, rump, tail and belly. Inland form has conspicuous white eyebrow, absent in form that inhabits Pacific coast. Bill and legs black. Voice: loud 'waah, waah, shack, shack, shack'; variety of calls and notes; soft, sweet song similar to 'whisper song' of American Robin; imitates calls of Red-shouldered Hawk and Golden Eagle. **Range and habitat:** resident in conifer and pine-oak forests of western NA. **Similar species:** Blue, Grey, Pinyon and Scrub Jays.

Scrub Jay *Aphelocoma coerulescens* (Corvidae) 28-33 cm (11-13 in). Crestless, long, slim jay, rich blue above with grey-brown on back; grey ear-coverts and underparts; white throat and upper breast with fine blue streaks and blue-grey necklace extending on to or crossing breast, depending on subspecies; undertail-coverts whitish to bluish. Florida subspecies nests in family groups. Voice: harsh, repeated 'ike, ike, ike', slight upward inflection; longer, rough, slightly metallic, sharply rising 'iennk'; rough, rapidly repeated 'quick-quick-quick'. **Range and habitat:** brushy areas, south-western and south-central NA. **Similar species:** Grey-breasted and Pinyon Jays.

Blue Jay *Cyanocitta cristata* (Corvidae) 28-32 cm (11-13 in). Blue crest, back, wings and tail, black bars across wing and tail, white wing-bars, white edging on tail feathers; black band across breast around neck and nape; face, throat, underparts white, washed with grey; black eye and eye stripe. Voice: harsh, slurring 'jaaay'; musical, bell-like 'queedle, queedle'; guttural rattling 'trrrrr'; mimics calls of Red-shouldered and Red-tailed Hawks. **Range and habitat:** common in oak-pine forests, suburbs, city parks and bird feeders from plains to east coast. Withdraws from northern areas in winter. **Similar species:** Steller's Jay.

Clark's Nutcracker *Nucifraga columbiana* (Corvidae) 30-3 cm (12-13 in). Profile resembles Common Flicker. Ash-grey body, black wings with white in secondaries, tail white with black central feathers; forehead, throat and undertail-coverts white, bill black, long, heavy, pointed. Young birds browner, especially on wing-coverts. Noisy, bold, often in small groups, walking and feeding on human scraps like crows. Voice: harsh, grating 'khaaa'. **Range and habitat:** resident coniferous forests near treeline in west-central and south-western NA; occasional invasions of lower elevations. **Similar species:** Grey Jay, Northern Mockingbird, shrikes (see pp 78, 80, 170).

American Crow *Corvus brachyrhynchos* (Corvidae) 43-53 cm (17-21 in). Black, slight purplish gloss; square tail; bill large, stout, black; legs black. Juvenile brownish. Often seen in small groups, typically with a sentinel on a high perch; wary of humans. Voice: distinctive 'caw' or 'cah', singly or in series; young birds more nasal than adults; mimics other birds, mammals and noises. **Range and habitat:** woodland, agricultural fields, river groves, recently suburbs and towns throughout NA except extreme north and extreme south-central. Winters in central and southern NA. **Similar species:** Chihuahuan and Common Ravens, other crows (see pp 182, 228, 278, 322).

Pine Siskin

American Goldfinch

House Finch

Purple Finch

White-winged Crossbill

Pine Grosbeak

Evening Grosbeak

Steller's Jay

Scrub Jay

Blue Jay

Clark's Nutcracker

American Crow

THE NEOTROPICAL REGION

South of the cool Mexican plateau lies tropical rainforest, marking the northern boundary of the zoogeographical zone known as the Neotropical Region, which extends from 20°N to Cape Horn at 57°S. There are three main faunistic subdivisions: South America, Central America (Mexico to Panama) and the West Indies. The latter two are much influenced by recent geological and biogeographical events. South America holds the key to the unequalled species richness of the region.

The South American continent lies mainly within the tropics and subtropics. However, the Andes, the world's longest mountain chain, spanning 10,000 km (6,200 miles) from the Sierra Nevada of Santa Marta in the north to the continent's southern-most tip, introduces climatic diversity at all latitudes. Thus a chain of vegetation zones runs the length of the continent, providing a pathway for temperate species into equatorial latitudes. The Andes are geologically part of the mountain chain running through western North America and Central America, resulting from uplift as an oceanic plate pushes below a continental plate. This activity is still taking place: the Santa Marta massif is only a million years old and the Andes hosts much volcanic and earthquake activity. The Andean chain is complex with several *cordilleras* with inter-Andine valleys which isolate highland areas. Thus habitat types such as Alpine *paramo* are discontinuous and this has provided scope for speciation. Some bird species are only found on certain 'highland islands' surrounded by valleys with tropical vegetation which presumably they cannot cross. Rivers drain the Andes eastwards and westwards, but there is also a huge area of inner drainage, once a vast lake, now the *altiplano* or *puna* in Peru and Bolivia with large lakes such as Titicaca − 8,300 sq km (3,200 sq mile) − and salt-plains which host endemic grebes, coots and flamingos.

As well as providing a north−south pathway, the Andes act as an east−west barrier for many tropical and subtropical species, resulting in significant differences between the avifaunas in the foothills and lowlands to either side.

The narrow Pacific slope and lowlands boast great extremes of rainfall. In western Colombia and northern Ecuador, the annual rainfall can exceed 10 m (33 ft). This region of wet forest, extending from the tropical lowlands into montane cloud forest, has very high biodiversity and is notable for the large number of tanager species, forming colourful mixed-species flocks: one of the most spectacular of bird-watching experiences. Further south, the cold Humboldt current causes the precipitation to fall offshore and seasonal sea mists provide the only moisture to the deserts which run southwards to northern Chile. This arid zone extends inland, crossing the Andes into central Argentina. The Humboldt current causes upwellings of nutrient-rich water, making the inshore waters highly productive and able to support huge numbers of seabirds.

Further south still are the southern beech forests of Chile which have a very distinctive but depauperate avifauna. The birds are, however, of Neotropical provenance, such as tapaculos and furnariids; unlike the plants and insects, which show Austral-asian affinities.

East of the Andes are two mountain blocks: the Tepui or Guianan-Venezuelan highlands and the Serra do Mar of eastern Brazil. Both support distinctive endemic avifaunas. Vegetation types in the eastern lowlands are also determined by rainfall. High annual rainfall with no long dry season gives rise to humid forest. The greatest of all of these is the vast Amazon rainforest which originally covered about 6 million sq km (nearly 2½ million sq miles), approximately the land area of the United States of America. Far from being one vast uniform forest, rainfall variation (wetter in the west), soils and relief produce very different vegetation types, each with a distinctive bird community. This patchwork of different habitats is one reason why Amazonia supports the world's richest avifauna. Other factors are natural changes which enhance this mosaic, from small-scale tree falls opening up niches to specialists such as jacamars, to changes in the courses of rivers which leave 'shadows' of their former routes in the form of a continuum of successional stages in the forest. In the longer term, climatic changes affect forest cover and cause fragmentation and conglomeration of habitat, increasing opportunities for speciation.

The richest area of all is western Amazonia, where over 500 species of birds can be found in a single locality, including over 20 species of parrot, 14 woodpeckers, over 45 antbirds and 15 cotingas. Despite this wealth, finding these birds can be a frustrating task for the visiting bird-watcher. In the semi-darkness of the great forest, periods of stillness alternate with encounters with mixed-species flocks when a bewildering variety of birds may present themselves for an all-too-short moment.

More seasonal areas may have open grassland, flooded for parts of the year. Both the *llanos* of Venezuela and Colombia and the *pantanal* in south-west Brazil and northern Argentina are examples of these wet-dry tropics. Large concentrations of waterbirds, such as storks and ibises, may be found there. Strips of gallery forest and stands of *Mauritia* palm provide important refuges for parrots. Where rainfall is less, deciduous woodland, scrub or open grassland predominates. These habitats are particularly important in the south. The natural vegetation of much of central Brazil is savannah (including *campos* and *cerrado*), home of the Greater Rhea and Red-legged Seriema. Pampas takes over in eastern Argentina, whilst to the south lies the bleak steppe of Patagonia.

More than 3,100 species of birds (including migrants from the Nearctic) occur in South America and there are new species described each year. In all, 88 families of birds breed in the Neotropics, of which a staggering 25 are endemic to the region. This high endemicity is partly because South America remained isolated from North America until just three million years ago, when the land bridge between Panama and Colombia formed, and many Neotropical families have not spread beyond the region. A major feature of the Neotropical avifauna is the predominance of the Deutero-oscines. Fewer than 50 species in this group occur in the Old World tropics, whereas over 500 are found in South America, including furnariids, woodcreepers, antbirds, cotingas and tyrant flycatchers.

Central America has a mixed avifauna. The northern half had, until recent geological time, been a tropical peninsula of North America with an island arc extending southwards. These islands were colonized by mobile North American families (such as thrushes) moving south and hummingbirds and tyrant fly-catchers moving north. Once established, the isthmus provided a

Montezuma Oropendala. *Related to the New World blackbirds and orioles, breeding colonies of oropendalas are a spectacular sight in the Neotropics.*

route for the movement of many other forest species northward and the radiation of northern families like wrens and wood warblers southward, facilitating the spread of such families into South America as well. The eastern and western coastal lowlands of Central America are isolated from each other by mountains. This barrier has led to the development of distinctive Pacific slope and Caribbean slope avifaunas. Central America also attracts many Nearctic migrants. For example, about a quarter of Costa Rica's birds are migrants, mainly from North America.

The West Indies are oceanic islands, never connected to continental land masses, and have an avifauna based on mobile families derived from the Nearctic, Central American and South American elements. These are typified by thrushes, vireos, wood warblers, tanagers, finches, icterids, tyrant flycatchers, hummingbirds, parrots and pigeons. It is noteworthy that not one species of antbird or furnariid is represented. Successful colonists island-hopped and distinctive island forms evolved. There is a very high level of endemicity, including one endemic family, the Todidae. Many of the endemic species are rare and extinctions have occurred, such as the Cuban Macaw. A turnover of species is an ecological phenomenon on small islands, but by far and away the most important threat now comes from the action of man,

directly through hunting or trapping or indirectly through habitat change. Although some island species are adaptable, their small population sizes increase their vulnerability greatly.

Elsewhere in the Neotropics, there are awesome threats to many species. In Brazil alone, the International Council for Bird Preservation have identified 121 species at threat, over half of which only occur in Brazil. In the lowlands, the rate of forest clearance for logging and ranching is causing great alarm. The formerly large tracts of grasslands in southern South America, home of species like the Black and White Monjita, have been degraded by heavy grazing and agricultural conversion. Even more precarious is the future of the forests of the densely populated Andean slopes and Central America, where the demand for land is increasing. Discontinuities in forest cover in these areas can fragment bird populations already restricted ecologically into narrow altitudinal belts and disrupt vertical migration, which may be important to the annual cycles of more Andean birds than we at present imagine.

The Neotropics remain the poorest known of all the avifaunal regions. The arrival in recent years of several excellent bird books for Costa Rica, Panama, Colombia, Venezuela and the Andean chain and the first volume of a continental handbook will do much to facilitate study. These texts also point out the enormous gaps in our knowledge and will certainly stimulate the next generation of field ornithologists.

RHEAS Rheiformes

Greater Rhea *Rhea americana* (Rheidae) 125 cm (49 in). Very large, flightless bird. Blackish crown, nape and upper back. Upperparts otherwise grey, whitish below. Neck sometimes has white patches. Females are smaller and darker. Normally in small groups, readily run off when alarmed. Several females usually lay eggs in same scrape on ground. **Range and habitat:** NT, eastern Brazil south to central Argentina. Lowland to mid-elevation grassy plains. **Similar species:** Lesser Rhea.

TINAMOUS Tinamiformes

Solitary Tinamou *Tinamus solitarius* (Tinamidae) 46 cm (18 in). Large tinamou. Brownish above, darker on crown, with wings, back and rump heavily barred black. Ochrous head and neck, speckled on side of head, whiter on throat and buff line on side of neck. Foreneck and breast grey, whiter on belly with vermiculated flanks. Usually solitary, very wary. **Range and habitat:** NT, eastern Brazil. Lowland forest. **Similar species:** White-throated Tinamou.

Undulated Tinamou *Crypturellus undulatus* (Tinamidae) 28 cm (11 in). Medium-sized tinamou with greyish bill and legs. Rather uniform greyish brown plumage with whitish throat and dark barring on flanks. At close quarters, fine black vermiculations on upperparts visible. Although wary, this ground-dwelling bird can be quite easy to see because it often 'freezes' when alarmed. Calls frequently, particularly at dawn and dusk, a melancholic 3-4-note whistle. **Range and habitat:** NT, Guiana to western Argentina. Common in seasonally flooded forest. **Similar species:** Little Tinamou.

Brown Tinamou *Crypturellus obsoletus* (Tinamidae) 28 cm (11 in). Small. Unmarked dark chestnut-brown above, crown blackish, sides of head brownish grey with greyer throat. Underparts rufous brown barred black on flanks. Legs yellowish grey. Shy, ground-dwelling bird which flies reluctantly, usually seen running away. Call, usually heard at dawn and dusk, a crescendo of tremulous whistles. **Range and habitat:** NT, Andean chain from Venezuela to Argentina, and Brazil. Occurs from tropical lowlands to edge of temperate zone in forest edge and secondary vegetation. **Similar species:** Little and Highland Tinamous.

PENGUINS Sphenisciformes

Humboldt Penguin *Spheniscus humboldti* (Spheniscidae) 68 cm (27 in). Medium-sized with stout blackish bill, fleshy at base. Head mainly black with narrow white line forming loop from base of bill, around sides of crown to throat. Single black horseshoe across top of breast and extending down flanks to thighs, otherwise underparts mainly white. Upperparts blackish grey. Juvenile has brownish head, grey cheeks, lacks white head stripe and underparts entirely whitish. **Range and habitat:** NT, Pacific coast of Peru and Chile to 33° S. **Similar species:** Magellanic Penguin.

Magellanic Penguin *Spheniscus magellanicus* (Spheniscidae) 71 cm (28 in). Medium-sized with blackish bill. Head mainly black with white line forming loop from base of bill, around sides of crown to throat. Upperparts blackish grey, broad black band across upper breast, with narrower black horseshoe below, with sides extending down flanks to thighs; rest of underparts mainly white. Juvenile has pale grey cheeks and a single broad, but diffuse, breast band. **Range and habitat:** NT, on Pacific and Atlantic coasts. Mainly to south of Humboldt Penguin, but overlaps in winter when range extends northwards. **Similar species:** Humboldt Penguin.

Galapagos Penguin *Spheniscus mendiculus* (Spheniscidae) 50 cm (20 in). Small. Bill blackish upper mandible, yellow lower. Head blackish with narrow white line extending from eye around side of cheeks to throat. Upperparts blackish grey. Underparts mainly white with broad blackish breast band and narrower lower band forming horseshoe which extends down sides of flanks to thighs. Juvenile duller, with greyish cheeks, lacks striped pattern. Call is donkey-like. **Range and habitat:** NT, restricted to Galapagos.

GREBES Podicipediformes

Short-winged Grebe *Rollandia micropterum* (Podicipedidae) 40 cm (16 in). Flightless. Adult in breeding plumage has blackish forehead and crown, giving crested appearance. Bill yellow with reddish tip. Ear-coverts tufted black. Hindneck chestnut. Rest of upperparts dark brown with blackish streaks. Rump rather chestnut. Underparts white with brownish flecks on flanks. Nonbreeding birds paler. Often in family groups. Runs across water if alarmed, sometimes exposing wing. Calls include squeals and whines, also a repeated 'gaeck'. **Range and habitat:** NT, series of lakes on border of Peru and Bolivia, including Titicaca. High Andean lakes with access to reed-beds or floating vegetation. **Not illustrated.**

Great Grebe *Podiceps major* (Podicipedidae) 61 cm (24 in). Large with very long blackish bill. In breeding plumage, head blackish, greyer on ear-coverts, lores and throat. Hindneck blackish, rest of neck chestnut. Upperparts dark brown, glossed blackish green. Underparts white, tinged chestnut-brown on both flanks and breast. Upperwings as upperparts, with whitish secondaries and bases of inner primaries. Underwing whitish. Nonbreeding birds rather paler, with whitish cheeks. **Range and habitat:** NT, Peruvian coast and southern Chile, southern Brazil southwards to Tierra del Fuego. Breeds on lowland lakes, moves to coastal areas outside breeding season.

Least Grebe *Podiceps dominicus* (Podicipedidae) 23 cm (9 in). Very small, dumpy grebe with black bill and yellow eye. Breeding plumage shows black crown and throat. Rest of head and neck dark grey. Upperparts grey-brown, paler below. Speculum white. Nonbreeders have entire head dark grey. Normally solitary or in pairs. Usually silent. Calls include short, nasal hoot and high-pitched chattering. **Range and habitat:** NA and NT, southern USA to Argentina. Freshwater habitats with aquatic vegetation. **Similar species:** Pied-billed Grebe (see p 50).

Puna Grebe *Podiceps taczanowski* (Podicipedidae) 35 cm (14 in). Flightless. Medium-sized, long-necked grebe with slender grey bill, slightly uptilted. Breeding birds show crest. Crown grey-brown. Nape blackish. Rest of upperparts brownish grey, paler on rump. Underparts whitish, particularly on flanks. Nonbreeders lack crest and black on nape. Usually in small groups. Call a whistling note. **Range and habitat:** NT, restricted to Lake Junin in Peru where population declining sharply and severely threatened. Remains in open water, entering edges of reed marsh to breed. **Similar species:** Silvery Grebe.

Greater Rhea

Solitary Tinamou

Undulated Tinamou

Brown Tinamou

Humboldt Penguin

Magellanic Penguin with juvenile

Galapagos Penguin

Great Grebe

Least Grebe

Puna Grebe

TUBE-NOSED SWIMMERS Procellariiformes

Waved Albatross *Diomedea irrorata* (Diomedeidae) 89 cm (35 in). Large. Uniform yellow bill and bluish legs. Head and neck white, with buffish tinged crown and hindneck. Upperparts chestnut-brown with fine paler bars, most prominent on rump and uppertail coverts. Uppersurface of wings and tail brown. Underparts as upperparts, with white of neck extending down to upper breast. Underwings pale, bordered by brown. Flaps more and soars less than other albatrosses. **Range and habitat:** NT, PA. Breeding restricted to Galapagos and Las Platas, Ecuador. Outside breeding season off Peruvian and Ecuador coasts, with nonbreeders present all year. Only truly tropical albatross. **Similar species:** Wandering Albatross, Giant Petrel (see p 284).

Southern Fulmar *Fulmarus glacialoides* (Procellariidae) 46-50 cm (18-20 in). Rather long-winged, stocky-bodied seabird. Tube-nose bill is stout, mainly pink with dark tip. Body plumage white, with greyer tone on nape, breast and flanks. Lores darker. Mantle, rump and tail bluish grey as are upperwings. Black trailing edge to inner wing. Outer wing shows variable amounts of black. Underwing mainly white. Flies in characteristic stiff-winged gliding manner. Often in groups. **Range and habitat:** southern oceans. Regularly occurs during austral winter off western coast of South America to 10° S. **Similar species:** Fulmar (see p 142).

Dark-rumped (Hawaiian) Petrel *Pterodroma phaeopygia* (Procellariidae) 43 cm (17 in). Large, long-winged petrel. Upperparts entirely dark brown apart from white forehead and (sometimes) white sides of rump. Dark cap extends downwards behind eye. Underside white apart from dark patch on axillaries and dark margin to wing, with diagonal bar crossing from carpel joint towards axillaries. Bill dark, legs pink. Has distinctive bounding flight of downward glide followed by short series of wing-flaps. **Range and habitat:** Pacific and NT. Breeds on Hawaii and Galapagos. Population decreasing. Outside breeding season, dispersal to open ocean within tropical zone, Galapagos birds to offshore regions between Mexico and northern Peru. **Similar species:** White-necked Petrel.

Audubon's Shearwater *Puffinus lherminieri* (Procellariidae) 30 cm (12 in). Small, stocky shearwater with broad wings. Bill dark, legs pink. Upperparts, tail and vent brown. Rest of underparts white with brown on nape extending downwards on to sides of breast. Underwing broadly margined brown with white centre, sometimes with very little white. Can occur in large flocks. Flight rapid followed by short glides. **Range and habitat:** tropical oceans. Breeds on Pacific islands including Galapagos, West Indies, Indian ocean islands including Seychelles and Maldives. Movements off eastern seaboard of USA to 40° S, Gulf of Mexico and off coast of Ecuador. **Similar species:** Little and Manx Shearwaters (see p 142).

Georgian Diving-petrel *Pelecanoides georgicus* (Pelecanoididae) 18-21 cm (7-8 in). Very small, dumpy bird with short wings. Bill black, legs blue with black webs. Dark head with whitish throat. Upperparts black with distinctive greyish white scapulars. Black upperwings, browner tone on primaries. Trailing edge of secondaries white. Underparts whitish, some grey on sides of breast and flanks. Underwing linings whitish, axillaries and flight feathers darker. Fly low over waves, wings whirring. Visit breeding sites at night. **Range and habitat:** subantarctic. Breeds on South Georgia and other subantarctic islands. Probably sedentary. **Similar species:** other diving-petrels.

FULL-WEBBED SWIMMERS Pelecaniformes

Blue-footed Booby *Sula nebouxii* (Sulidae) 76-84 cm (30-3 in). Medium-sized with bright blue feet. Pale brown head, white patch at top of back, scaly brown and white back, pale band across rump and brown lower rump and tail. Upperwings uniform dark brown. Underparts whitish with

brown tail and extensive white patch at base of underwing, extending as 2 white bars towards carpel joint. Rest of underwing brown. Immature has dull grey feet, darker head and breast but shows characteristic underwing pattern. Flight direct and powerful. Feeds in small groups close to shore by plunge-diving typical of gannets and boobies. Males give whistling call, female has hoarser call. **Range and habitat:** NT, Pacific coast from Mexico to Peru. Breeds on offshore islands. Normally stays close to land. Wanders beyond normal range when food supplies short. **Similar species:** Brown, Blue-faced and Peruvian Boobies.

Peruvian Booby *Sula variegata* (Sulidae) 74 cm (29 in). Small booby with white head and underparts. Black bare skin at base of bill. Upperparts brown with pale feather edgings and whitish uppertail-coverts and centre tail feathers. Underwing dark with white bar extending across wing from axillaries. Immature has pale brown body plumage. Sometimes in very large flocks with other seabirds. **Range and habitat:** NT, in Humboldt current region off Peruvian and Chilean coast, breeding on islands. Remains close to shore throughout year; periodic changes in sea currents change food supplies and cause dispersal beyond normal range. Intensive commercial fishing is causing a population decline. **Similar species:** Blue-footed Booby.

Red-footed Booby *Sula sula* (Sulidae) 66-77 cm (26-30 in). Variable species with several plumage forms. All have red legs and greyish bills with pink bases. White form is entirely white apart from black flight feathers and black patch on undersurface of wing at carpel joint. On the Galapagos, this form also has dark tail. Brown form is completely greyish brown. Also forms with white heads and/or white tails. Almost always in groups, frequently follows ships. **Range and habitat:** NT, PA, OR, AF and AU. Breeds on islands, but wanders far on feeding trips. **Similar species:** Blue-faced Booby.

Neotropic (Olivaceous) Cormorant *Phalacrocorax olivaceus* (Phalacrocoracidae) 64-73 cm (25-9 in). Large cormorant appearing all-black at distance. Long hooked bill is dark brown. Dull yellow naked skin on face bordered behind by narrow white margin. Breeding plumage shows white tufts on side of head. Immature greyish brown, paler on underparts, with breast whitish on younger birds. Rather long-tailed. Often in small flocks. **Range and habitat:** NT and southern NA. Marine and freshwater habitats, estuaries, lakes and rivers. Mainly lowland. May show some seasonal movements. Immatures wander. **Similar species:** Double-crested Cormorant (see p 52), Magellan Cormorant.

Guanay Cormorant *Phalacrocorax bougainvillii* (Phalacrocoracidae) 76 cm (30 in). Large cormorant with blue-black upperparts and neck, white chin and underparts. Nonbreeding birds are browner. Immatures are less pure white below. Breeds in huge colonies, sometimes of millions of birds. Large feeding flocks assemble, with other species such as boobies, at great height to search for shoals of fish. **Range and habitat:** NT. Occurs along the cold, food-rich Humboldt current off the Peruvian and Chilean coasts. Breeds on islands and some mainland sites. As with other seabirds endemic to this region, will disperse if food supplies fail. **Similar species:** Magellan Cormorant.

Imperial Shag (Blue-eyed Cormorant) *Phalacrocorax atriceps* (Phalacrocoracidae) 72 cm (28 in). Striking-looking cormorant with crest and dark blue-black upperparts contrasting with underparts entirely white from chin to undertail-coverts apart from dark thighs. Amount of white on face variable. White tips of scapulars in breeding plumage and white patch on centre of back, which the Falklands form lacks. Breeds in colonies and often forms large flocks outside breeding season. **Range and habitat:** NT and southern oceans. Northernmost populations breed southern tip of South America. Breeding sites may be used as roost sites during austral winter, although some dispersal to sea takes place. **Similar species:** Magellan Cormorant.

Waved Albatross

Southern Fulmar

Dark-rumped Petrel

Audubon's Shearwater

Blue-footed Booby

Peruvian Booby

Georgian Diving-Petrel

Neotropic (Olivaceous) Cormorant

Red-footed Booby white phase adult and immature

Guanay Cormorant juveniles at colony

Imperial Shag (Blue-eyed Cormorant)

HERONS AND ALLIES Ciconiiformes

Stripe-backed Bittern *Ixobrychus involucris* (Ardeidae) 30-3 cm (12-13 in). Very small heron. Yellowish green bill and legs. Plumage generally straw-coloured with back boldly streaked with black. Crown black. White throat with some white streaking on underparts. In flight, primaries show rufous with dark patch at base. Shy bird which clambers through marsh vegetation and tends to 'freeze' to spot to avoid detection. Flies at low height with rather slow wingbeats. **Range and habitat:** NT, Venezuela to southern Brazil. Lowland bird of both freshwater and brackish marshes. **Similar species:** Least Bittern, Zigzag Heron.

Rufescent Tiger Heron *Tigrisoma lineatum* (Ardeidae) 66-76 cm (26-30 in). Stocky heron with long, thickset neck. Yellowish powerful bill, rather short greenish legs. Adult plumage, attained after about 5 years, with rich chestnut head, neck and chest. Rest of upperparts finely barred black and buff. White and buff stripes extending from throat to belly, which is dull buff. Immature plumage strikingly different, boldly barred dark brown and golden buff above with underparts paler and less strongly marked. Calls include a series of hoots. Hunts alone. When flushed usually flies upwards to a tree perch. **Range and habitat:** NT, in lowland forests, freshwater lakes and open country in proximity of tree cover. **Similar species:** Fasciated Tiger and Bare-throated Tiger Herons.

Capped Heron *Pilherodius pileatus* (Ardeidae) 58 cm (23 in). Small stocky heron. Bluish bill and bright blue bare facial skin on lores and around eye contrast with white plumage, apart from black crown. Breeding birds show creamy tone on neck and breast. Immature plumage very similar, with crown streaked black and white. Mainly silent. Shy and usually solitary, often perches in trees. **Range and habitat:** NT, from Panama to southern Brazil and Paraguay. Occurs along river-banks and edges of small lakes. **Similar species:** Black-crowned Night Heron (see p 52), Cattle Egret (see p 186).

Cocoi Heron *Ardea cocoi* (Ardeidae) 102-30 cm (40-51 in). Large, slender, long-necked heron. Crown black, rest of head, neck and upper breast white with some fine black streaking on front of neck. Upperparts whitish with blackish flight feathers. Belly black with white thighs. Immature browner above with underparts largely white. Usually solitary. Stands motionless beside shallow water, but wary. Often seen perched in trees. Flight slow and heavy. Call a croak. **Range and habitat:** NT, from Panama southwards, mainly in lowlands in marshes, lakes and rivers, also coastal sites. **Similar species:** Great Blue Heron (see p 52).

Boat-billed Heron *Cochlearius cochlearius* (Ardeidae) 46-51 cm (18-20 in). Small but thickset heron with enormously broad, shovel-like bill. Black on crown and crest, extending to upper back, but forehead and cheeks white. Rest of upperparts pale grey (Central American form is darker grey). Breast whitish (buffish in Central America) and belly rufous with blackish flanks. Immature has no crest and is rufous brown above and buffish white below. Deep croaking call. Feeds at night, leaving tree cover at dusk. Roosts in groups but probably feeds mainly alone. **Range and habitat:** NT, locally from Mexico to northern Argentina. In lowland freshwater marshes, lagoons and mangrove swamps. **Similar species:** Black-crowned Night Heron (see p 52).

Maguari Stork *Ciconia maguari* (Ciconiidae) 97-102 cm (38-40 in). Large stork with straight grey bill tipped dark red. Long orange legs. Unfeathered loral region orange-red. Rest of plumage white apart from black lower back, tail, greater coverts and flight feathers. Pale iris. Juveniles black. Usually solitary or in small parties, sometimes with other species. Soaring flight. Largely silent apart from a double light whistle. **Range and habitat:** NT, mainly east of the Andes from Guianas to southern Argentina and Chile. Lowlands in open country. **Similar species:** Wood Stork (see p 54), King Vulture (see p 102).

Jabiru *Ephippiorhynchus mycteria* (Ciconiidae) 122-40 cm (48-55 in). Very large stork with heavy, slightly upturned black bill and long black legs. Head and neck bare black skin apart from tuft of white feathers on crown and reddish region of skin on lower neck. Rest of plumage white. Immature mainly greyish. Individuals wander widely, but large congregations also occur. Prey is captured by stalking. Silent apart from bill-clappering at nest. **Range and habitat:** NT, southern Mexico to northern Argentina. Local. Mainly occurs in open lowland wetlands. Individuals wander to coastal estuaries and large rivers.

Sharp-tailed Ibis *Cercibis oxycerca* (Threskiornithidae) 71-81 cm (28-32 in). Large stocky ibis with long, decurved dusky bill. Bare skin of lores and around eye is red. Orange bare skin on throat. Forehead and malar region pale grey. Rest of plumage blackish, glossy green. Thick-necked appearance from tuft of feathers running down hindneck. Proportionately long tail. Short reddish legs. Usually solitary or in small groups. Tends to fly just short distances. Call a loud double-noted nasal 'tur-dee'. **Range and habitat:** NT, lowland areas east of Andes from Guyana south to south-west Brazil. Open wet areas, including ricefields. **Similar species:** Green Ibis. **Not illustrated.**

Green Ibis *Mesembrinibis cayennensis* (Threskiornithidae) 56 cm (22 in). All-dark glossy green ibis, with greenish bill and short legs. Dark grey bare facial skin. Rather thick-necked appearance from feather tufts on hindneck. Immature is much duller. Usually solitary or in pairs. Wary, feeds on muddy patches on forest floor or edge of open water. Frequently perches in trees. Flight is stiff and jerky. Calls mainly at dusk and before dawn, often in flight, with loud, rolling, fluty 'clu-clu-clu' repeated several times. **Range and habitat:** NT, Honduras to northern Argentina. Occurs in forested areas close to rivers, gallery forest and swamps. Infrequent in open marshes. **Similar species:** Whispering and Glossy Ibises.

Scarlet Ibis *Eudocimus ruber* (Threskiornithidae) 56-61 cm (22-4 in). Adult has bare parts reddish (or bill black when breeding) and plumage entirely scarlet apart from black wing tips. Immature greyish brown upperparts with white rump and underparts, tinged pink with age, and whitish brown head and neck. Often in large, spectacular flocks, sometimes mixed with other species. Call a high-pitched, plaintive 'tior-tior', given mainly at nest or roost sites. **Range and habitat:** NT, lowlands of Colombia east to Guianas, including Trinidad, south to eastern Brazil. Wanders into Central America and West Indies. Marshes, ricefields and mangroves. **Similar species:** immature very similar to White Ibis.

White-faced Ibis *Plegadis chihi* (Threskiornithidae) 56 cm (22 in). Adult has reddish bill and brownish legs. Bare facial skin usually reddish, bordered by white feathers during breeding season. Rest of plumage chestnut with glossy greenish tone on upperparts when breeding. Outside breeding season, much duller with fine whitish streaking on head and neck. Red iris of adult is diagnostic. Immature has brownish eye, greyish brown bill and legs, and dull dark grey-brown plumage with greenish tone on wing. Usually solitary or in small groups. Flight is strong with periods of gliding between flaps. Normally silent. **Range and habitat:** NA and NT, USA to central Mexico, wintering southwards to El Salvador. Locally in South America south to central Argentina. Mainly freshwater marshes and flooded areas. **Similar species:** Glossy Ibis.

Stripe-backed Bittern

Rufescent Tiger Heron

Capped Heron

Cocoi Heron

Boat-billed Heron

Maguari Stork

Jabiru

Scarlet Ibis

Green Ibis

White-faced Ibis

WATERFOWL Anseriformes

Horned Screamer *Anhima cornuta* (Anhimidae) 86-94 cm (34-7 in). Huge, turkey-like bird. When perched, appears largely glossy black with white belly and white scaling on head and neck. At close quarters, horny quill visible on forehead. In flight, note large white patch on upperwing-coverts. Feeds on marsh vegetation, but spends a lot of time perched on top of bushes or even tall trees. Flies powerfully and adeptly, sometimes gliding at great height. Call an extraordinary loud deep braying sound. **Range and habitat:** NT, Venezuela south to central Brazil. Typical bird of swampy and lakeside vegetation of Amazon basin, locally elsewhere. **Similar species:** 2 other screamers.

Red-billed (Black-bellied) Whistling Duck *Dendrocygna autumnalis* (Anatidae) 48-53 cm (19-21 in). Typical long-necked and long-legged shape of pantropical whistling ducks. Bright red bill and red legs. Rather pale grey cheeks, chestnut-brown body apart from boldly black belly. Wings very distinctively patterned with brown and grey forewing, bold central white band across wing and black flight feathers. Tail black. Immatures are much duller. Southern populations show some grey on breast. Usually in flocks, resting during day and feeding during night. Call a loud, whistling 'chee-chi-chee'. Often perched high on dead branches of trees. **Range and habitat:** NT and NA, southern USA to northern Argentina. Both freshwater marshes and brackish areas, including mangroves. **Not illustrated.**

Flying Steamer Duck *Tachyeres patachonicus* (Anatidae) 66-71 cm (26-8 in). Large stocky duck. Plumage variable but generally mottled black and grey body with white belly and vent. Head of male is usually whitish in breeding season, variable but often mainly brownish afterwards and greyish brown with narrow white 'spectacle' during rest of year. Bill is orange during breeding season, but otherwise greyish. Female also shows great variation, but head generally shows some rufous on cheeks and white 'spectacle'. Broad white speculum on wing. Unlike other steamer ducks, can fly although some individuals and populations may do so only rarely, if at all. **Range and habitat:** NT, southern tip of South America and Falklands. Occurs both along coast and on lakes and rivers. Tends to leave inland sites during winter. **Similar species:** 3 other steamer ducks.

Muscovy Duck *Cairina moschata* (Anatidae) 66-84 cm (26-33 in). Large stocky duck, ancestor to familiar domestic bird. All-black plumage, rather glossy on upperparts, apart from white upper and lower wing-coverts which are striking in flight views. Female is smaller version of male. Immature is duller and lacks white wing markings. Shy bird, normally solitary or in small groups, which feeds in shallow water or at water's edge. Frequently perches on trees and usually nests in tree-holes. Generally silent, apart from quiet hisses or quacks. **Range and habitat:** NT, Mexico to northern Argentina. Lowland freshwater marshes, rivers and lakes, in forested areas. Local and much persecuted.

Torrent Duck *Merganetta armata* (Anatidae) 46 cm (18 in). Small, stream-lined duck with long tail. Male has striking black and white striped head pattern. Upperparts dark, finely barred. Elongate black-centred feathers on scapular and mantle. Speculum green with narrow white leading and trailing edge. Underparts variable according to geographical race. Female bluish grey above, cinnamon below. Juveniles greyish, paler below with barring on flanks. Usually in pairs or family groups. Agile swimmer and diver in fast-flowing water. Call a loud, shrill 'wheet'. **Range and habitat:** NT, fast-flowing rivers along entire Andean chain. Some altitudinal migration, although generally resident.

Chiloe Wigeon *Anas sibilatrix* (Anatidae) 43-54 cm (17-21 in). Medium-sized grazing and dabbling duck. Plumage of both sexes very similar. Dark, glossy greenish black head with extensive white area on face and smaller spot behind and below eye. Scaly breast, rufous flanks and pale fringed dark scapulars. In flight, note white greater coverts and glossy greenish black speculum. Rump white and tail black. Pale underwing and white belly. Immature is much duller. Usually in small flocks and very vocal, producing whistling notes. **Range and habitat:** NT, Argentina, southern Chile and Falklands. Common in freshwater lakes and marshes. Moult migration may occur and populations migrate northwards in austral winter.

Silver (Versicolor) Teal *Anas versicolor* (Anatidae) 38-43 cm (15-17 in). Small dabbling duck. Both sexes have similar plumage. Bill rather large and bluish with yellow at base, crown dark and creamy sides of head. Rest of body feathers dark with pale fringes, giving spotted effect on breast and barring on flanks. Vermiculated rump. Greenish speculum bordered by white. Andean race is larger and paler. Usually in small groups. Generally quiet, calls include whistles and quacks. **Range and habitat:** NT; breeds in southern South America, moving northward into Brazil in winter. Fairly common in freshwater marshes and lakes with fringing vegetation. Andean race occurs in high altitude lakes in Peru southwards.

Masked Duck *Oxyura dominica* (Anatidae) 30-35 cm (12-14 in). Small, stiff-tailed duck with bluish bill, dark at tip. Male has black face and crown which contrasts with rufous chestnut nape and breast. Rest of body mottled black and chestnut. Female has distinctive head pattern with dark crown and whitish cheeks crossed by 2 dark lines. Male eclipse plumage similar to female. In flight, white panel on wing of both sexes is diagnostic. Usually in pairs or small groups. **Range and habitat:** NT, thinly distributed from southern Texas to central Argentina. Mainly east of Andes. Occurs in marshes and small pools with floating vegetation. **Similar species:** Ruddy Duck (see p 56).

DIURNAL BIRDS OF PREY Falconiformes

Lesser Yellow-headed Vulture *Cathartes burrovianus* (Cathartidae) 58-66 cm (23-6 in). Medium-sized vulture with bare yellow head and blackish plumage. Flight feathers are slightly paler than rest of wing, giving 2-toned appearance. Quills of outer primaries are whitish, producing pale panel on upperwing surface. Flies in buoyant manner, normally at low height over open country, rarely in groups. **Range and habitat:** NT, Mexico to northern Argentina. Common in lowland savannah and other open terrain. May show seasonal movements. **Similar species:** Greater Yellow-headed and Turkey Vultures (see p 58).

King Vulture *Sarcoramphus papa* (Cathartidae) 71-81 cm (28-32 in). Large with 2 m (6½ ft) wingspan. Adult is white with black flight feathers, tail and rump. Unfeathered head and neck brightly coloured. Ruff greyish. Immatures dark brown, increasingly white with age on wing-linings and body. Broad wings and very short, square-ended tail. Soars steadily, often at great height and rarely with other species. **Range and habitat:** NT, southern Mexico to northern Argentina. Widely distributed in tropical forest zones. **Similar species:** Wood and Maguari Storks (see pp 54, 100).

Andean Condor *Vultur gryphus* (Cathartidae) 102-30 cm (40-51 in). Very large with 3 m (10 ft) wingspan. Black plumage with most of upperwing surface bright white. White ruff and reddish unfeathered head. Immature dark dull brown. Soaring flight with wings flattened, tips of primaries spread and upward-pointing, from cliff sides to great heights. Normally silent. Groups may congregate at carcasses. **Range and habitat:** NT, from Venezuela to Tierra del Fuego along Andes. In north of range, very localized and mainly above 3,000 m (9,850 ft), occurs at lower elevations in south.

Horned Screamer

Flying Steamer Duck

Muscovy Duck domestic bird with white breast

Chiloe Wigeon

Silver (Versicolor) Teal

Torrent Duck

Masked Duck

King Vulture

Lesser Yellow-headed Vulture

Andean Condor

Hook-billed Kite *Chondrohierax uncinatus* (Accipitridae) 41-6 cm (16-18 in). Confiding raptor usually seen perched. Hooked bill, greenish cere, bare skin in front of whitish eye. Males usually grey above, barred below, 2 broad pale bands on tail. Female usually brown above with rufous collar and barred rufous below. There is a melanistic form. Rather long-winged and long-tailed in flight, barred flight feathers visible from below. Generally perches below canopy. Call a musical whistle. Diet mainly snails. **Range and habitat:** NT, Mexico to northern Argentina. Lowland humid forest and wooded marshy areas. **Similar species:** several, including Grey Hawk.

Swallow-tailed Kite *Elanoides forficatus* (Accipitridae) 56-66 cm (22-6 in). Graceful, elegant raptor with deeply forked tail and pointed wings. White apart from black flight feathers, tail, uppersurface of wings and back. Usually in small groups, feeding on flying insects or small animals from canopy. Call a high-pitched 'klee-klee-klee', but normally silent. **Range and habitat:** NT and NA, from southern USA to northern Argentina. Northern populations migratory. Mainly lowlands and foothills, usually over forest.

Plumbeous Kite *Ictinia plumbea* (Accipitridae) 36 cm (14 in). Small long-winged raptor. Almost wholly dark grey with rufous patch on base of primaries and 2 or 3 narrow white bands at base of tail, which are difficult to see. Immature paler, with streaks below. Perches on bare branches at treetops, wings extending beyond tip of tail. Often in groups, sometimes in large flocks during migration. Catches insects and other small animals in air or from canopy. **Range and habitat:** NT, Mexico to northern Argentina. Central American populations mainly migratory. Common in tropical and subtropical zones in wide range of habitats from forest to open, wooded areas. **Similar species:** Mississippi Kite.

White Hawk *Leucopternis albicollis* (Accipitridae) 46-56 cm (18-22 in). Distinctive raptor with underparts entirely white, black wing tips and band across tail. Amount of black on upperwing and back variable according to race. Wings and tail rather broad. Usually solitary, soars low over canopy but most frequently encountered perched below canopy. Diet consists of snakes and other reptiles. Call a hoarse 'ker-wee'. **Range and habitat:** NT, southern Mexico to central Brazil. Widespread but generally uncommon in lowland forest. **Similar species:** Black-faced Hawk.

Savanna Hawk *Buteogallus meridionalis* (Accipitridae) 46-61 cm (18-24 in). Medium-sized raptor with very long wings and short tail. At rest appears almost uniformly rufous with rather faint dusky bars. In flight, greyish toned back contrasts with rich brown head and wings, latter with a bold black trailing edge. Black tail has white tip and median band. Immature is browner with buff mottling. Conspicuous bird which hunts from a perch or whilst soaring. Groups gather at grass fires or follow agricultural activities. **Range and habitat:** NT, Panama to northern Argentina. Common in open country, ranchland, sometimes along seashore. **Similar species:** Black-collared Hawk.

Roadside Hawk *Buteo magnirostris* (Accipitridae) 33-8 cm (13-15 in). Common raptor. Prominent yellow cere and eyes. Grey head, upperparts and chest. Rest of underparts barred brown and white. In flight, rufous base to primaries contrasts with rest of grey upperwing. Immatures brown above, buffy below with some streaking on breast. Rather short-winged, most often seen perched on exposed branches or posts. Flies sluggishly, with glides interspersed with flaps. Call a loud, drawn-out mew. **Range and habitat:** NT, Mexico to central Argentina. Widespread in forest edge and open habitats in lowlands and mid-elevations. **Similar species:** Broad-winged Hawk (see p 58), Grey Hawk.

Guiana Crested Eagle *Morphnus guianensis* (Accipitridae) 79-89 cm (31-5 in). Large, slim eagle with small crest. 2 plumage phases. Pale phase has pale grey head, neck and breast, dark lores, white belly, dark brown above and long tail with 3 pale bands. Underwing whitish with flight feathers barred. Dark phase has head, neck and breast dark grey, rest of underparts and wing-linings barred black and white. Like most large raptors, spends most of day perched. Call a high-pitched whistle, sometimes disyllabic. **Range and habitat:** NT, central America to northern Argentina. Rare and poorly known lowland forest species. **Similar species:** Harpy Eagle.

Harpy Eagle *Harpia harpyja* (Accipitridae) 89-102 cm (35-40 in). Huge with black, forked crest, grey head and powerful bill. Black chest band and white underparts with barred thighs. Upperparts blackish, slightly scaled with grey. 3 grey bands on black tail. Underwing barred. Immatures lack dark chest band. Rarely soars above canopy, normally hunts within cover, twisting and turning between trees to take monkeys and other large prey by surprise. Call a drawn-out wail. **Range and habitat:** NT, southern Mexico to northern Argentina. Rare forest species, mainly restricted to lowland area, requiring large tracts of suitable habitat. **Similar species:** Guiana Crested Eagle.

Ornate Hawk Eagle *Spizaetus ornatus* (Accipitridae) 58-64 cm (23-5 in). Hawk eagles have rather oval-shaped wings and longish tails. Adult Ornate has black crown and crest, chestnut sides of head and breast, bordered anteriorly by narrow black margin to white throat. Upperparts blackish, underparts barred. Tail has 3 grey bands. Underwings show barred flight feathers and spotted wing-linings. Immature is white on head and underparts. Does not usually soar at great height, tending to capture prey on short flights from subcanopy perch. Call a loud series of whistles. **Range and habitat:** NT, Mexico to northern Argentina. Uncommon in humid forest, mainly in lowlands. **Similar species:** Grey-bellied Goshawk.

Black (Yellow-throated) Caracara *Daptrius ater* (Falconidae) 43-8 cm (17-19 in). Slim, long-winged all-black raptor with white base of tail and conspicuous extensive bare orange skin on face. Yellow legs. Immature is duller with barred underparts and tail. Often in small groups, perched on exposed branches or on ground beside rivers in forested areas. Flies rapidly and directly. Call an eerie, rather hoarse, scream. **Range and habitat:** NT, Venezuela to central Brazil. Lowland forest, particularly along watercourses. **Similar species:** Red-throated Caracara.

Common Caracara *Polyborus plancus* (Falconidae) 51-61 cm (20-4 in). Large caracara with powerful bill. Red cere and bare facial skin. Black crown and bushy crest. Rest of head and neck white. Breast barred with rest of underparts black. Upperparts dark with barring on back and pale patches on outer primaries. Pale rump and lightly barred tail with subterminal band. Often in pairs or family groups walking on ground, hunting for carrion or small prey. Flight is direct, rarely soars. Call a harsh cackle. **Range and habitat:** NT and NA, southern USA to Tierra del Fuego. Open country in lowlands and mid-elevations. **Similar species:** Yellow-headed Caracara.

Yellow-headed Caracara *Milvago chimachima* (Falconidae) 41-6 cm (16-18 in). Slender-looking caracara with small pale bill. Creamy body plumage apart from brownish back and narrow line behind eye. Wings brown with pale patch on outer primaries. Pale rump and lightly barred tail with subterminal band. Immature shows some streaking on head and underparts. Usually solitary or in pairs, walking on ground or perched on exposed situations. Quarters ground with slow, deliberate flight and short glides. Call a long, harsh hiss. **Range and habitat:** NT, Central America to northern Argentina. Common in lowland and mid-elevation open country and beside large rivers.

Hook-billed Kite

Swallow-tailed Kite

Plumbeous Kite (immature)

White Hawk

Savanna Hawk

Roadside Hawk

Black (Yellow-throated) Caracara

Harpy Eagle

Ornate Hawk Eagle

Guiana Crested Eagle

Common Caracara

Yellow-headed Caracara

Laughing Falcon *Herpetotheres cachinnans* (Falconidae) 46-56 cm (18-22 in). Medium-sized raptor with large head and long tail. Very distinctive appearance with creamy head and underparts contrasting with bold dark 'bandit' mask and upperparts. Tail banded. Usually seen perched in exposed branches, often at top of isolated tree. Flies with series of wingbeats and glides, directly but slowly. Typical call a loud 'gua-co' repeated endlessly, often in duet, and especially after sunset. **Range and habitat:** NT, Mexico to northern Argentina. Mainly a lowland species in wooded areas and forest edges.

Lined Forest Falcon *Micrastur gilvicollis* (Falconidae) 30-6 cm (12-14 in). Forest falcons are typically short-winged, long-legged raptors which are agile hunters of small birds within forests. Lined Forest Falcon has orange cere, bare facial skin, yellow legs and whitish eyes. Upperparts are dark grey, underparts white with fine black barring. Black tail has 2 narrow white bands and white tip. Immatures are brown above. Call is loud 2-note bark. **Range and habitat:** NT, Amazonia. Uncommon in lowland tropical forest. **Similar species:** Barred and Slaty-backed Forest Falcons.

Bat Falcon *Falco rufigularis* (Falconidae) 23-30 cm (9-12 in). Small falcon. Appearing very dark with blackish hood and upperparts. Creamy throat extends along base of cheeks. Underparts black with fine white bars, rusty thighs and vent. Often in pairs, seen perched on exposed branches on tops of trees. Dashing flight with rapid wingbeats, dives and occasional soars, catching insects, small birds and bats on wing. Somewhat crepuscular. Call a high-pitched 'ki-ki-ki-ki-ki'. **Range and habitat:** NT, Mexico to northern Argentina. Mainly a lowland species in forest borders and wooded areas. **Similar species:** Orange-breasted and Aplomado Falcons.

Aplomado Falcon *Falco femoralis* (Falconidae) 36-43 cm (14-17 in). Rather slim-built, long-tailed falcon. General plumage tone pale, with grey upperparts, creamy ring around head, dark moustachial stripe. Creamy cheeks and breast, finely barred blackish underparts and pale rusty thighs. Tail is barred. Immature is browner and streaked below. Usually solitary. Tends to use low perches, sometimes hovers but normally catches prey in fast sallies, often attracted to bush fires. **Range and habitat:** NT and NA, southern USA to Tierra del Fuego. Mainly in dry open country at low elevations. **Similar species:** Bat and Orange-breasted Falcons.

FOWL-LIKE BIRDS Galliformes

Speckled (Variable) Chachalaca *Ortalis motmot* (Cracidae) 53 cm (21 in). Rather small, slim and long-tailed cracid. A variable species, generally brownish above, paler below with a speckled breast. Tail dark brown with chestnut outer tail feathers. Head and neck brown or greyish depending on race. Reddish bare malar region and grey bare skin around eye. Noisy bird, usually encountered in small groups in canopy beside small clearings and in secondary vegetation. **Range and habitat:** NT, Venezuela to central Brazil. Mainly a lowland species of forest edges and woodland. **Similar species:** Andean Guan.

Andean Guan *Penelope montagnii* (Cracidae) 61 cm (24 in). Rather short-tailed guan with bare red skin on chin inconspicuous. Plumage generally warm brown, although head and breast greyer with fine whitish markings. Usually in small groups. Timid bird normally keeping to tops of trees but may move into more open situations to find fruit. Produces a honking call and a mechanical rattle with its wings during display flights. **Range and habitat:** NT, along Andean chain from Venezuela to Argentina, in highland forest. **Similar species:** Band-tailed Guan, Speckled (Variable) Chachalaca.

Spix's Guan *Penelope jacquacu* (Cracidae) 89 cm (35 in). Large guan with ample flap of red bare skin on chin. Overall has dark plumage tone, glossed green on wings and tail, brown edged with white on head and breast, warmer brown on belly. Usually in pairs. Shy bird which normally stays close to canopy, occasionally coming to lower levels, and making a loud deep trumpeting when disturbed. Like other guans, produces mechanical wing rattle during displays. **Range and habitat:** NT, basins of Amazon and Orinoco rivers. Lowland forest, gallery forest and wooded areas. **Similar species:** Crested Guan.

Common (Blue-throated) Piping Guan *Aburria pipile* (Cracidae) 69 cm (27 in). Very distinctive cracid with bluish bare skin on chin and normally paler skin around face, bushy white crest and bold white patches on wing. Otherwise plumage wholly blackish. Usually in small groups, moving through treetops, often perched in open. Produces series of whistling notes and wing rattles during displays. **Range and habitat:** NT, Venezuela to northern Paraguay. Lowland humid forest.

Black Guan *Chamaepetes unicolor* (Cracidae) 64 cm (25 in). All-glossy black guan, duller below, with red legs and blue bare facial skin. Usually in pairs or family groups, normally in trees but sometimes coming to ground to feed on fallen fruit. Has harsh alarm call and also produces whistling notes and wing rattles during display. Like other guans, flies with series of flaps followed by glide. **Range and habitat:** NT, Central America. Occurs in high-elevation cloud forest and in patchy forest where not hunted.

Razor-billed Curassow *Crax mitu* (Cracidae) 89 cm (35 in). Large, stocky bird with long tail. Plumage mainly glossy black with belly and vent chestnut and tail broadly tipped white. Black crest and heavy red bill with greatly enlarged upper mandible, forming upward keel. Sexes alike. Normally recorded in small groups, on ground, but flying to trees if disturbed. Produces a humming song from a perch. **Range and habitat:** NT, Amazon basin in lowland forest. **Similar species:** Wattled, Salvin's and Lesser Razor-billed Curassows.

Great Curassow *Crax rubra* (Cracidae) 91 cm (36 in). Both sexes have curly crest, but male is black apart from yellow swollen base to bill and white belly and vent, whereas female is generally chestnut-brown with barred wings and tail and closely barred black and white head. However, female plumage is variable and can be much darker. Both sexes have greyish legs. Usually alone or in pairs and seen on ground. Normally runs off rather than flies if disturbed. Alarm calls include a high-pitched whistle. Males produce a deep humming song. **Range and habitat:** NT, Mexico to Ecuador. Humid lowland forest. **Similar species:** Blue-billed Curassow.

Red-billed Curassow *Crax blumenbachii* (Cracidae) 89 cm (35 in). Another crested curassow. Males have red base to bill, swollen to produce small knob and 2 wattles. Plumage glossy black, white belly and vent in males, chestnut in females. Females have white barring on crest and rufous mottling on wings. Usually in pairs, feeding in fruiting trees or shrubs or on fallen fruit on ground, fly to mid-level branches if disturbed. Male produces deep, far-carrying booming call. **Range and habitat:** NT, southeastern Brazil. Extremely rare in lowland primary forest, threatened by hunting and forest destruction. **Similar species:** Wattled Curassow.

Laughing Falcon

Lined Forest Falcon

Bat Falcon

Aplomado Falcon (male and female)

Speckled (Variable) Chachalaca

Andean Guan

Spix's Guan

Common (Blue-throated) Piping Guan

Black Guan

Razor-billed Curassow

Great Curassow (female)

Red-billed Curassow

Crested Bobwhite *Colinus cristatus* (Phasianidae) 20-3 cm (8-9 in). Common small terrestrial bird, both sexes with crest. Male has pale buff crest and head with darker crown and line behind eye. Upperparts vermiculated dark brown and black. Underparts heavily spotted white. Female duller. Usually in pairs or small groups, tending to run if alarmed or fly a short distance into cover. Often seen on roadsides. Call a rapid 'bob-bob-wheet', frequently heard. **Range and habitat:** NT, Central America to northern Brazil. Quite common, except at high elevations, in dry open country. **Similar species:** Northern Bobwhite (see p 60). **Not illustrated.**

Starred Wood Quail *Odontophorus stellatus* (Phasianidae) 24 cm (9½ in). Dumpy, terrestrial bird with conspicuous crest. Crown and crest bright chestnut. Upperparts greyish with lower back more olive-brown. Dark brown wings, spotted buff on wing-coverts. Throat grey, rest of underparts rufous with black barring on vent. Bare red skin around eye. Female has blackish brown crest. Groups walk in single file, scattering or 'freezing' if alarmed, only reluctantly flying. Call a far-carrying, rhythmic, rollicking chorus. **Range and habitat:** NT, western Amazonia. Lowland humid forest. **Similar species:** Marbled Wood Quail. **Not illustrated.**

CRANES, RAILS AND ALLIES Gruiformes

Grey-winged (Common) Trumpeter *Psophia crepitans* (Psophiidae) 48-56 cm (19-22 in). Shy, ground-dwelling bird with hunched appearance. Plumage blackish with long pale grey feathers covering back. Normally in groups, extremely wary, giving soft humming notes or harsh calls in alarm. Kept by forest Indians as 'guard dogs'. **Range and habitat:** NT, lowland Amazonian forest. **Similar species:** White-winged Trumpeter.

Grey-necked Wood Rail *Eulabeornis cajaneus* (Rallidae) 36-8 cm (14-15 in). Large rail, frequently heard. Rather powerful yellow bill, red eye and long red legs. Head and neck grey. Upperparts olive-brown, rump, tail and vent black. Breast and flanks rufous. Rufous primaries visible in flight. Generally solitary but forms noisy duets at dusk with loud cackling calls. Usually on ground but sometimes climbs into shrubs. **Range and habitat:** NT, Mexico to northern Argentina. Low to mid-elevations in forest and woodland close to water, including mangroves. **Similar species:** Rufous-necked and Brown Wood Rails.

Blackish Rail *Rallus nigricans* (Rallidae) 28 cm (11 in). Medium-sized rail with long yellowish bill and red legs. Rather plain plumage, dark grey below and brown-toned above. Whitish chin and black undertail-coverts and tail. Immature is browner. Shows typical rallid behaviour, generally keeping under cover in dense vegetation, rarely venturing out to edge of patches of open ground and flying just short distances. Gives short metallic call and mew. **Range and habitat:** NT, patchy distribution from Venezuela to northern Argentina in tropical and subtropical zones. Uncommon in freshwater marshes and in damp vegetation. **Similar species:** Plumbeous Rail.

Grey-breasted Crake *Laterallus exilis* (Rallidae) 15 cm (6 in). Tiny crake with grey bill, bright greenish at base. Head and foreparts greyish, darker on crown, with conspicuous rufous nape. Belly whitish with bold blackish bars on flanks. Upperparts dark brown with pale barring on wing-coverts and rump. Immatures lack rufous on nape. Denizen of dense, wet grassy vegetation. Very difficult to flush, keeping to tunnel-like runs on ground. Has trilling call and series of repeated 'keek' notes. **Range and habitat:** NT, Central America to Paraguay. Uncommon in wet grassland. **Similar species:** Paint-billed Crake.

Red-gartered Coot *Fulica armillata* (Rallidae) 44 cm (17 in). Medium-sized coot with angled profile of head from bill-tip to top of shield. Shield and bill yellowish with red on base of culmen. Legs orange with red 'garter'. Plumage dark grey with 2 white patches on undertail. Usually in groups,

but territorial when breeding. Feeds by up-ending or diving. Several call-notes including short whistles and clicks. **Range and habitat:** NT, southern Argentina and Chile, migrating to southern Brazil. Common in lowlands and mid-elevations in marshland and shallow lakes. **Similar species:** several coots.

Giant Coot *Fulica gigantea* (Rallidae) 55 cm (22 in). Huge, stocky coot with black knobs on forehead protruding either side of yellow shield. Bill mainly red with yellow sides and white culmen. Legs red. Plumage blackish grey with very indistinct white streaks on undertail. Immature is greyer below with duller bare parts. Usually in family groups; territorial disputes can be seen frequently. Adults practically flightless because of huge bulk. Calls a range of loud growls and gobbling sounds. **Range and habitat:** NT, Peru, Bolivia and Chile. High altitude lakes with extensive areas of waterplants.

Sungrebe *Heliornis fulica* (Heliornithidae) 30 cm (12 in). Small, shy bird with black and white striped head and neck and reddish bill. Male has white cheek, female buff. Rest of upperparts olive-brown. Tail tipped white, underparts whitish. Swims low in water, close to well-vegetated banks and in inundated forest. Flies short distances, low over water, when disturbed. Call a series of loud, barking honks. **Range and habitat:** NT, from southern Mexico to northern Argentina in tropical lowlands on slow-flowing rivers, flooded forest and small areas of freshwater. **Similar species:** ducks, grebes, Peter's Finfoot (see p 198).

Sunbittern *Eurypyga helias* (Eurypgidae) 48 cm (19 in). Slender, long-bodied bird with disproportionately small black and white striped head. Upperparts closely barred black, brown and grey. Vermiculated breast, rest of underparts pale buff, becoming whiter on belly. Vermiculated tail also boldly barred brown. Flight reveals bright orange patches on wings, also shown during elaborate dancing display. Moves slowly and elegantly on ground. Flies to nearby branch if disturbed. Call a high whistle which rises and falls slightly in pitch. **Range and habitat:** NT, from southern Mexico to northern Argentina in tropical lowlands, in forests close to water.

Red-legged Seriema *Cariama cristata* (Cariamidae) 69 cm (27 in). Large terrestrial bird with long red legs and raptor-like head and frontal crest. Finely vermiculated plumage is greyish-brown, with black and white barring on primaries and tail. Paler underparts are lightly streaked. Hunts various prey, including snakes, on ground and tends to run rather than fly if alarmed. Nest is placed on top of small bushes. Often give away presence by their loud call. **Range and habitat:** NT, central Brazil to northern Argentina. Lowland grassy plains with scattered bushes. **Similar species:** Black-legged Seriema.

Grey-winged (Common) Trumpeter

Grey-necked Wood Rail

Blackish Rail

Grey-breasted Crake

Red-gartered Coot

Giant Coot on nest

Sungrebe

Sunbittern

Red-legged Seriema

SHOREBIRDS, GULLS, AUKS Charadriiformes

Wattled Jacana *Jacana jacana* (Jacanidae) 25 cm (10 in). Long-legged with very long toes. Bill yellow with red basal lobes. Adult black with dark brown back and wings with yellow flight feathers. Immature white below, brown above with blackish crown, eye stripe and long white supercilium. Wades in damp vegetation or walks on floating waterplants. Noisy with loud 'kee-kick' call. **Range and habitat:** NT, common in lowlands from Panama to northern Argentina, mainly east of Andes in southern part of range. Freshwater marshes, lakes and slow-flowing rivers. **Similar species:** other jacanas (see pp 198, 248, 294).

Double-striped Stone Curlew *Burhinus bistriatus* (Burhinidae) 46 cm (18 in). Large thick-knee with white supercilium bordered black above. Large yellow eye and long yellow legs. Plumage generally brownish with paler feather edgings, giving streaked effect, and white belly. Flight feathers dark with broken white wing-bar. Spends most of day resting, active at dusk and at night. Usually solitary or in small groups. Call a rapid chattering. **Range and habitat:** NT, Central America and northern South America. Fairly common in dry open country. **Similar species:** other stone curlews (see pp 160, 198, 294).

Southern Lapwing *Vanellus chilensis* (Charadriidae) 36 cm (14 in). Large wader with black-tipped red bill and red legs. Grey head with black on face, bordered behind and above by white, extending to throat and breast. Rest of underparts white. Back brownish with bronzy shoulders. White rump and black subterminal tail band. Flight feathers black and long white wing patch. Rather slow flight. Usually in singles or small groups. Very vocal, usual call is loud 'keek-keek-keek'. **Range and habitat:** NT, common throughout South America, although largely absent from Amazonia, western Colombia, Ecuador and Peru. Open short grassland. **Similar species:** Andean Lapwing.

Collared Plover *Charadrius collaris* (Charadriidae) 15 cm (6 in). Small dainty wader with dark bill and pale yellowish legs. Black lores and bar across crown. White forehead, patch behind eye and underparts, apart from black breast band. Chestnut hindcrown and nape. Upperparts grey-brown. In flight shows narrow white wing-bar and white tail edges. Usually single or in pairs. Gives short 'cheet' call but generally silent. **Range and habitat:** NT, Mexico southwards to northern Argentina. Beside large rivers, beaches and fields. **Similar species:** Semi-palmated and Snowy Plovers (see p 62).

Tawny-throated Dotterel *Eudromias ruficollis* (Charadriidae) 27 cm (10½ in). Large plover with long, thin bill. Sides of face pale buff with narrow black line through eye. Dark crown. Neck and breast brownish grey. Throat orange-buff. Upper belly and flanks buff, black belly patch and rest of underparts white. Upperparts streaked brown and rufous. Narrow white bar on primaries. Forms flocks outside breeding season. Flight call a descending, tremulous note. **Range and habitat:** NT, Andes from Peru southwards and southern Argentina. High grasslands, arid zones, coastal areas in Peru. Some altitudinal migration and movement of southern birds northwards during austral winter. **Similar species:** American Golden Plover (see p 62), Dotterel (see p 156).

Giant Snipe *Gallinago undulata* (Scolapacidae) 40 cm (16 in). Very large snipe with rounded wings. Plumage similar to other snipes but markings more blackish than brown. Rich brown edgings on scapular and mantle feathers. Rump and uppertail cinnamon with fine dark bars. Unique among snipes in having barred flight feathers, but this is barely visible in field. Gives harsh 'kek-kek' when flushed. Display calls, probably only given at night, include a buzzing call and guttural 'HO-go go'. Very poorly known species. **Range and habitat:** NT, 2 distinct populations: Colombia and northern Brazil, and southern Brazil and Paraguay. Lowland grassland. **Similar species:** several

snipes (see pp 110, 160, 202, 250). **Not illustrated.**

Least Seedsnipe *Thinocorus rumicivorus* (Thinocoridae) 17 cm (6½ in). Small, dumpy bird. Males have grey faces, necks and breasts. White chin bordered by black, and black line runs down centre of breast to black border between breast and white underparts. Upperparts cryptically patterned buff and dark brown. Narrow white wing-bar and trailing edge. Females streaked brownish on head and breast. Ground-dwelling bird, often in large flocks. Gives a short harsh note when flushed. Song an endlessly repeated but variable series of notes, given either in flight or from perch. **Range and habitat:** NT, southern South America and coastal southern Ecuador to northern Chile. Southernmost populations migrate north in austral winter. Sandy areas with scattered vegetation at low elevations. **Similar species:** Grey-breasted Seedsnipe.

Brown-hooded Gull *Larus maculipennis* (Laridae) 36-8 cm (14-15 in). Small gull with chocolate-brown hood, replaced by dusky patches behind eye and on crown in winter. Bill and legs red. Upperwing pale grey, outer primaries white, with some black on outer web. Immatures have complex upperwing pattern and dark subterminal tail band. Rather long, narrow wings; flies with rapid wingbeat. Gregarious, noisy bird with harsh high-pitched calls. **Range and habitat:** NT, southern South America, dispersal to coasts of Brazil and northern Chile in austral winter. Breeding colonies on coast and inland lakes and marshes. **Similar species:** Andean and Grey-headed Gulls (see p 202). **Not illustrated.**

Swallow-tailed Gull *Creagrus furcatus* (Laridae) 53-8 cm (21-3 in). Distinctive gull with long wings and forked tail. Black bill, tipped yellow, and pink legs. Breeding adult has dark grey hood and white underparts. Striking upperwing pattern with black outer primaries contrasting with white triangle on hindwing and grey forewing and mantle. Nonbreeders have white heads with dark eye patch. Mainly nocturnal feeder on squid and fish, often in large groups. Calls include a harsh scream and a rattle. **Range and habitat:** NT, breeds on the Galapagos and Malpelo Island. Pelagic wanderer throughout Humboldt current region, sometimes north to Panama. **Similar species:** Sabine's Gull.

Large-billed Tern *Phaetusa simplex* (Laridae) 38 cm (15 in). Large tern with heavy yellow bill. Black crown and white underparts. Striking wing pattern with black primaries, white triangle formed across secondaries and grey tertials and forewing. Upperparts and shallowly forked tail grey. Outside breeding season, forehead is mottled white. Immatures brownish above. Seen singly or in small, dispersed groups. Breeds on sandbanks on large rivers. Noisy birds, giving a harsh 'skee-er'. **Range and habitat:** NT, rivers and lakes east of Andes.

South American Tern *Sterna hirundinacea* (Laridae) 40-4 cm (16-17 in). Medium-sized tern with red bill and legs. Breeding adults have black crowns bordered by white, pale grey underparts, whiter on belly, grey upperparts and wing with outer web of outer primary black. Rump and forked tail white. Nonbreeders have whitish foreheads. Immatures have brownish tips to back and forewing with dusky wing tips. Usually in groups. Calls include high-pitched screech and metallic note. **Range and habitat:** NT, breeding colonies on coasts south from Peru and southern Brazil. Dispersal northwards during austral winter. **Similar species:** Antarctic, Common and Arctic Terns (see pp 162, 296).

Inca Tern *Larosterna inca* (Laridae) 40-2 cm (16-17 in). Striking graceful tern with blue-grey plumage, blacker on crown, with long white 'moustache' and white trailing edge to wing, shallowly forked tail black. Large red bill with yellow base. Browner immature has shorter moustache. Like many Humboldt current seabirds, huge flocks gather to feed, often in association with marine mammals. **Range and habitat:** NT, from Ecuador to Chile along Humboldt current. Rocky coastlines, beaches and harbours.

Wattled Jacana

Southern Lapwing

Collared Plover

Double-striped Stone Curlew

Tawny-throated Dotterel

Least Seedsnipe

Swallow-tailed Gull

Large-billed Tern

South American Tern

Inca Tern

PIGEONS AND SANDGROUSE Columbiformes

Scaled Pigeon *Columba speciosa* (Columbidae) 30 cm (12 in). Medium-sized pigeon with distinctive scaled pattern on neck, upper back and breast. Bill red, tipped white. Upperparts purplish chestnut, underparts buffy, becoming paler on lower belly and vent. Tail and wings blackish brown. A fruit-eater, normally seen singly or in pairs, frequently perches on high, exposed branches. Flight fast and direct. Call a short series of deep 'whoo' calls. **Range and habitat:** NT, southern Mexico to northern Argentina. Fairly common in lowland forest and woodland. **Not illustrated.**

Ruddy Pigeon *Columba subvinacea* (Columbidae) 28 cm (11 in). Small dark pigeon with black bill and reddish legs. Plumage generally dark reddish brown, slightly paler on head, neck and underparts. Wing-linings tinged cinnamon. Fruit-eater, normally seen alone or in pairs, perched close to canopy. Calls a far-carrying, 4-syllabled coo and a harsh growl. **Range and habitat:** NT, Costa Rica to central Brazil. Generally common in lowland forest. **Similar species:** Plumbeous and Short-billed Pigeons.

Ruddy Ground Dove *Columbina talpacoti* (Columbidae) 16.5 cm (6½ in). Small, distinctive-looking dove. Male has grey cap, rufous upperparts, slightly paler below with whitish throat. Bold black spots on wing-coverts. Female quite different, with greyish brown upperparts, paler below, and rufous wings. Usually in pairs but groups aggregate on ground at rich sources of food. Call a soft double-noted hoot, repeated several times. **Range and habitat:** NT, southern Mexico to northern Argentina. Found mainly at low elevations, where common in secondary vegetation, cultivations, towns and gardens. **Similar species:** Plain-breasted Ground Dove.

Inca Dove *Scardafella inca* (Columbidae) 20 cm (8 in). Small, mainly ground-dwelling dove with long tail. Sexes similar. Plumage generally pale grey with scaly pattern. Primaries rufous with dark tips, underwing mainly rufous. Outer tail feathers edged white. Immatures are browner. Rather tame, usually in pairs or small groups, feeding on seeds and spilt grain. Males give a double-noted coo, often rendered as 'no-hope!', from exposed perches. **Range and habitat:** NT and NA, southern USA to Costa Rica. Common in open country, cultivated areas, farmyards and gardens in lowland areas. **Similar species:** Scaly Dove.

Ruddy Quail Dove *Geotrygon montana* (Columbidae) 23 cm (9 in). Small, plump, skulking dove. Bill, bare skin on lores and eye-ring red. Rufous brown above, with purplish gloss to hindneck and upper back. Cheeks and underparts pale cinnamon with rufous brown stripe across side of face. Female is darker with less pronounced facial pattern. Shy ground-dwelling bird, feeding on fallen fruit, but sometimes uses high perches. Call is soft coo, repeated after short pauses. **Range and habitat:** NT, Mexico to northern Argentina. Common in humid forest, secondary vegetation and plantations, mainly in lowlands. **Similar species:** Violaceous Quail Dove.

PARROTS AND RELATIVES Psittaciformes

Hyacinth Macaw *Anodorhynchus hyacinthinus* (Psittacidae) 100 cm (39 in). World's largest parrot. Rich cobalt-blue plumage, darker on flight feathers and underside of tail. Yellow bare skin on base of lower mandible and eye-ring. Bill greyish black. Perches on treetops, also descends to ground to collect fallen fruit. Usually in pairs or small groups. Impressive on the wing, flying high with long, streamer-like tail, constantly giving guttural contact calls. **Range and habitat:** NT, central Brazil and eastern Bolivia. Rare in lowland gallery forest, woodland, forest edge and palms. Severely threatened by hunting and trapping. **Similar species:** Indigo and Glaucous Macaws.

Little Blue (Spix's) Macaw *Cyanopsitta spixii* (Psittacidae) 54 cm (21 in). Small, beautiful macaw. Dark blue plumage with paler, greyer head and neck, underparts tinged green. Blackish bare skin on lores and around yellow eye. Seen in pairs, formerly in small groups. Very timid, readily flies off in alarm, giving a sharp 'kraa-ark' call. **Range and habitat:** NT, endemic to north-east Brazil. World's most endangered bird, virtually extinct in wild and a little more than a handful of captive birds. Occurred in gallery woodland. Population probably declined with habitat loss, then exterminated by trapping. **Not illustrated.**

Blue and Yellow Macaw *Ara ararauna* (Psittacidae) 84 cm (33 in). Large, spectacular-looking macaw, bright blue above and yellow below. Bare white skin on face, lined with black and bordered below by black bar across chin. Usually in pairs, but sometimes in broken groups of 25 or more, flying strongly with slow wingbeats. Flight call a rather nasal 'aaak'. Flights generally made early morning and late afternoon, with birds spending most of day feeding quietly in canopy, when they can be remarkably inconspicuous. **Range and habitat:** NT, Panama to southern Brazil. Lowland forest and gallery woodland.

Scarlet Macaw *Ara macao* (Psittacidae) 89 cm (35 in). Large, gaudy macaw with bicoloured bill and white bare facial skin. Scarlet body plumage, yellow greater coverts and deep blue flight feathers, rump, tail-coverts and outer tail feathers. Red inner tail. Usually in pairs or small groups, but may form larger flocks, sometimes with other species. Roosts communally. Flight call a loud, very harsh 'raaaaaak'. **Range and habitat:** NT, southern Mexico to eastern Brazil. Lowland forest and woodland. **Similar species:** Green-winged Macaw.

Dusky-headed Conure *Aratinga weddellii* (Psittacidae) 28 cm (11 in). Medium-sized parakeet with long tail. Plumage generally green with dull greyish head with conspicuous whitish bare skin around eye. Underparts have yellowish tone. Bluish flight feathers and tail tip. Usually in small noisy groups, often congregating to take nectar at flowering trees. Calls are loud harsh squawks. **Range and habitat:** NT, Amazonia. Common in riverside vegetation, forest edge and beside swamps. **Similar species:** White-eyed Parakeet.

Brown-throated Conure *Aratinga pertinax* (Psittacidae) 25 cm (10 in). Small parakeet with generally green plumage apart from pale brown face and breast. Toned yellowish on underparts and undertail. Eastern race has yellow eye-ring. Occurs in noisy flocks which fly rapidly, twisting and turning. Flight call a high-pitched screech. **Range and habitat:** NT, from southern Panama to northern Brazil. Common in lowland open country, woodland, dry scrub and mangroves. Populations may show movements, following passage of rains. **Similar species:** Blue-crowned Conure.

Ruddy Pigeon

Ruddy Ground Dove

Inca Dove

Ruddy Quail Dove

Hyacinth Macaw

Blue and Yellow Macaw

Scarlet Macaw

Brown-throated Conure

Dusky-headed Conure

Monk Parakeet *Myiopsitta monachus* (Psittacidae) 29 cm (11½ in). Medium-sized parakeet with grey forehead, face and breast. Bright green nape, duller green upperparts, blackish blue flight feathers, lower breast tinged yellowish, becoming greener on belly. Forms large noisy flocks and can be agricultural pest. Flight low and direct. **Range and habitat:** NT, mainly in lowland areas of southern Brazil, Bolivia to central Argentina. Dry open country.

Canary-winged Parakeet *Brotogeris versicolurus* (Psittacidae) 23 cm (9 in). Small parakeet with pointed tail. Plumage mainly green apart from conspicuous whitish flight feathers, with yellow on inner wing and blue outer primaries. Usually in noisy, chattering groups, often in large flocks before sunset, which fly fast above canopy, sometimes at great height. **Range and habitat:** NT, Amazon basin, south to northern Argentina. Forest edge and secondary growth beside rivers, sometimes common close to human settlements.

Cobalt-winged Parakeet *Brotogeris cyanoptera* (Psittacidae) 16.5 cm (6½ in). Small parakeet with short, pointed tail. Plumage generally green apart from yellow forehead, orange spot on chin and blue primaries. Occurs in small, noisy flocks, sometimes visiting flowering trees to feed on nectar. **Range and habitat:** NT, Amazonia. Common bird of riverbanks, secondary growth, forest edge and woodland. **Similar species:** Tui Parakeet.

Black-headed Parrot (Caique) *Pionites melanocephala* (Psittacidae) 23 cm (9 in). Small stocky parrot with bull-necked appearance. Dark crown, yellow-orange nape and face. Underparts whitish with yellow-orange thighs and vent. Upperparts bright green. Usually in small groups, feeding on fruit from mid-level to canopy. Gives a high-pitched call in flight and often noisy when perched. **Range and habitat:** NT, Venezuela to northern Brazil. Lowland forest. **Similar species:** White-bellied Parrot.

Scaly-headed Parrot *Pionus maximiliani* (Psittacidae) 29 cm (11½ in). Medium-sized parrot with black and yellow bill. Dull green plumage, with bronzy tone below. Head rather dark with dusky-edged feathers, giving scaly effect, and white-edged feathers on cheeks. Undertail-coverts and base of tail red. Bare whitish eye-ring. Usually in groups, seen flying from 1 forest patch to another. Feed in treetops, when very inconspicuous. Noisy flight call a resonant 'choik-choik'. **Range and habitat:** NT, north-east Brazil to northern Argentina. Common in low and mid-elevation humid forest and woodland. **Similar species:** several in genus.

Festive Amazon *Amazona festiva* (Psittacidae) 36 cm (14 in). Large parrot with mainly green plumage apart from narrow red line on forehead and lores and red patch on lower back which can be surprisingly difficult to see. Primaries dark blue and bluish tinge behind eye. Often forms large flocks, particularly at roosting sites when very noisy. Flight call a nasal 'wah-wah'. Feeds in canopy, often at forest edge. **Range and habitat:** NT, Venezuela to central Brazil. Lowland swampy or seasonally flooded forest, river-banks and islands. **Similar species:** Short-tailed Parrot.

Hawk-headed (Red-fan) Parrot *Deroptyus accipitrinus* (Psittacidae) 36 cm (14 in). Large, aberrant parrot with long-tailed and heavy-headed appearance. Whitish forehead, brownish streaked head, dark reddish scaly nape and underparts. Nape feathers sometimes raised to form ruff. Upperparts bright green with blackish primaries. Curious undulating gliding flight for a parrot. Normally in small groups, perched on open branches. Flight call consists of quiet 'chacks', followed by short series of loud, high-pitched calls. **Range and habitat:** NT, Venezuela to central Brazil. Lowland forest on sandy soils. **Not illustrated.**

CUCKOOS AND ALLIES Cuculiformes

Squirrel Cuckoo *Piaya cayana* (Cuculidae) 43 cm (17 in). Large, long-tailed cuckoo with yellowish green bill. Rufous brown head, breast and upperparts. Pale greyish belly, becoming darker towards vent. Underside of tail dark rufous with bold white tips. Seen singly or in pairs from mid-level to canopy, clambering through vegetation and along branches in squirrel-like manner, making short gliding flights between trees. Calls include loud nasal 'chi-kara'. **Range and habitat:** NT, Mexico to northern Argentina. Tropical to temperate areas in forest, secondary vegetation and woodland. **Similar species:** Black-bellied Cuckoo.

Groove-billed Ani *Crotophaga sulcirostris* (Cuculidae) 28-30 cm (11-12 in). Medium-sized cuckoo with rather shaggy black plumage. Upper mandible has raised compressed ridge arching from base to tip of bill and grooves on sides. Gregarious species found in small dispersed groups, moving from bush to bush with uneasy gliding flight. Calls include a dry 'hwilk' and a liquid, slurred series of 'tee-ho' notes. Communal breeder with several females sometimes laying in same bulky nest. **Range and habitat:** NT and NA, southern USA to northern Argentina. Common in lowland dry country and scrub. **Similar species:** Smooth-billed Ani (see p 70).

Pheasant Cuckoo *Dromococcyx phasianellus* (Cuculidae) 38 cm (15 in). Large parasitic cuckoo with small head and long tail. Rufous crown which is slightly crested. Narrow white line behind eye. Rest of head and breast buffy with black spots, underparts white. Dark brown above with wing-coverts appearing spotted. Dark greyish brown tail feathers narrowly tipped white. Stays low down but very skulking. Call a thin 3-note whistle, 3rd drawn out and tremulous, heard mainly at night. **Range and habitat:** NT, southern Mexico to northern Argentina. Uncommon in lowland riverside vegetation, secondary growth and forest edge. **Similar species:** Pavonine Cuckoo.

Rufous-vented Ground Cuckoo *Neomorphus geoffroyi* (Cuculidae) 46 cm (18 in). Large, terrestrial cuckoo. Bill greenish yellow. Flattened crest blackish, bluish bare skin around eye. Upperparts generally bronzy rufous, with long tail bronzy or greenish depending on race. Underparts buffish with black band across breast and rufous flanks. Very wary, but runs off rather noisily with bounding paces. Can be approachable when attracted to swarm of army ants. Has subdued, drawn-out hoot and frequently bill-snaps. **Range and habitat:** NT, Central America to Bolivia. Rare, probably overlooked, in lowland and foothill humid forest. **Similar species:** Scaled Ground Cuckoo. **Not illustrated.**

Hoatzin *Opisthocomus hoazin* (Opisthocomidae) 61-6 cm (24-6 in). Extraordinary bird with long wispy crest and hunchbacked appearance. Red eyes, bluish bare facial skin and brownish upperparts, streaked buff on nape and back. Breast buffy, rest of underparts chestnut. Wing-coverts and tail tipped pale buff, rufous primaries. Rather clumsy bird which flies weakly and tends to clamber through branches. Quite noisy, producing loud hissing sounds. Foliage-eater and only known bird to have ruminant-like system of digestion. Juvenile has clawed wing, used to clamber through trees. **Range and habitat:** NT, Guianas to Brazil. Lowland swamps, gallery forest and oxbow lakes in forested areas.

Monk Parakeet

Canary-winged Parakeet

Cobalt-winged Parakeet

Black-headed Parrot (Caique)

Squirrel Cuckoo

Scaly-headed Parrot

Festive Amazon

Groove-billed Ani

Pheasant Cuckoo

Hoatzin

OWLS Strigiformes

Spectacled Owl *Pulsatrix perspicillata* (Strigidae) 46 cm (18 in). Large, striking owl. Dark brown upperparts, head and chest with broad whitish face markings and yellow eyes. Underparts pale buff. Immature is creamy white with black, heart-shaped mask and brown wings. Largely nocturnal, but sometimes hunts in daylight. Catches prey by swooping down from a perch, normally roosts in thick cover. Call a series of deep, accelerating hoots. **Range and habitat:** NT, southern Mexico to northern Argentina. Mainly in lowlands in forest, secondary vegetation and gallery woodland. **Similar species:** Band-tailed Owl.

Tropical Screech Owl *Otus choliba* (Strigidae) 23 cm (9 in). Small owl with ear-tufts and yellow eyes. Plumage is toned brownish grey with well-spaced black streaks on underparts, finely streaked and vermiculated above. Facial disc is boldly outlined black. Scapulars and wing-coverts tipped white. Wholly nocturnal, spending day roosting in thick vegetation, often close to tree-trunk. Call is hesitant, tremulous trill, mainly at dusk and before dawn. **Range and habitat:** NT, Costa Rica to northern Argentina. Common in low to mid-elevations in forest edge, secondary vegetation, woodland and gardens. **Similar species:** Pacific Screech Owl.

Striped Owl *Asio clamator* (Strigidae) 36 cm (14 in). Medium-sized owl with brownish eyes and ear-tufts. Whitish facial disc rimmed by black, underparts pale buff boldly patterned with dark streaks. Upperparts and wings tawny brown, with dark markings. Underwing shows blackish carpel patches. Immature has darker, buffy face. Often active at dusk, hunting over open ground. Searches for prey from wing or from a low perch. Has a variety of hooting and barking calls. **Range and habitat:** NT, southern Mexico to northern Argentina. Lowland open country, fields and marshes, particularly with scattered thickets or trees. **Similar species:** Short-eared Owl (see p 78).

NIGHTJARS AND ALLIES Caprimulgiformes

Oilbird *Steatornis caripensis* (Steatornithidae) 48 cm (19 in). Curious nocturnal fruit-eating bird. Hooked bill, long wings and tail. Plumage reddish brown above, boldly spotted with black-rimmed white spots on wing feathers. Underparts buff with small white spots. Breeds in caves, only emerging at night to fly, sometimes long distances, to fruiting trees. Fruit are taken on the wing. Nests consist of mound of regurgitated seeds. Extremely noisy in cave with a range of screams and snarls. Clicking calls used for echo-location. Flight call outside is a 'kuk-kuk'. **Range and habitat:** NT, Panama south to Bolivia and east to Trinidad. Colonies are situated in caves in mountains or on coast. Undertakes seasonal movements in search of food.

Great Potoo *Nyctibius grandis* (Nyctibiidae) 51 cm (20 in). Large, pale nocturnal bird. Some variation in plumage but generally pale greyish buff above, vermiculated with black. Whiter below with fine dark vermiculations. Bolder spotting on scapulars, broken markings on breast. Wings and tail barred. Perches on high bare branches, flying out over canopy to catch flying insects and even bats, on aerial sallies. Call a drawn-out growl. **Range and habitat:** NT, southern Mexico to southern Brazil. Lowland forest and gallery woodland. **Similar species:** Common and Long-tailed Potoos.

Common Potoo *Nyctibius griseus* (Nyctibiidae) 36-41 cm (14-16 in). Large nocturnal bird with eyes which reflect orange in torchlight. Greyish brown above heavily streaked and vermiculated black and buff. Paler and greyer below with fine streaking and blotchy 'necklace' of darker markings across breast. Tail barred. Makes sallies from a perch such as an exposed branch, telegraph pole or fence-post. Spends day motionless. Call an evocative series of mournful whistles. Northern populations produce a more guttural call. **Range and habitat:** NT, Mexico to northern Argentina. Lowland and mid-elevation in forest edge and scattered woodland. **Similar species:** Great and Long-tailed Potoos.

Semi-collared Nighthawk *Lurocalis semitorquatus* (Caprimulgidae) 20 cm (8 in). Long-winged, short-tailed nighthawk. Very dark brown with white chevron on throat. Racial variation with some birds showing slightly paler, unbarred bellies. Some birds show pale band across base of upperwing. Active at dusk, flying bat-like at treetop height in pursuit of flying insects. Calls vary according to race but include an inflected 'whick', a mellow hoot and a 'tot-ta-quirrt'. **Range and habitat:** NT, Central America to northern Argentina. Widespread from lowland forest and forest edge to temperate woodland. **Similar species:** Band-tailed Nighthawk. **Not illustrated.**

Common Pauraque *Nyctidromus albicollis* (Caprimulgidae) 28 cm (11 in). Medium-sized, long-winged nightjar. Typically cryptically marked with vermiculations of brown, buff and black. White chevron on throat. Male has white band on primaries; outer 2 tail feathers black and next 2 mainly white. Female has buff wing marking and pale tips to outer 3 tail feathers. Scapulars boldly marked with black and buff. Rests on ground during day. Active at night catching insects on aerial sallies. Calls include a hollow, whistling 'ker-whee-oo' and a nasal 'waa-oo'. **Range and habitat:** NT and NA, southern USA to northern Argentina. Widespread from lowlands to subtropical zone in forest edge, secondary vegetation, and scattered woodland. **Similar species:** White-tailed and Band-winged Nightjars.

Scissor-tailed Nightjar *Hydropsalis brasiliana* (Caprimulgidae) 48 cm (19 in). Large nightjar with very long outer tail feathers, tipped white. Rest of tail feathers have white notches and edging, with central tail feathers also elongate. Otherwise plumage typically cryptic with distinctive rufous collar. **Range and habitat:** NT, Brazil to central Argentina. Low and mid-elevation woodland and savannah. **Similar species:** Swallow-tailed Nightjar.

SWIFTS AND HUMMINGBIRDS Apodiformes

White-collared Swift *Streptoprocne zonaris* (Apodidae) 20 cm (8 in). Large swift with powerful flight. Black with white collar, widest on chest, and notched tail. Immatures have broken collar. Usually in flocks, sometimes of hundreds of birds, flying at considerable height but also coming down low and sweeping down canyons at great speed. Breed on damp cliffs, sometimes behind waterfalls; forage over great distances. Call a nasal chattering. **Range and habitat:** NT, Mexico to northern Argentina. Over mountains and foothills. **Similar species:** Biscutate Swift.

Chapman's Swift *Chaetura chapmani* (Apodidae) 13 cm (5 in). Medium-sized swift with short square-ended tail and rather uniform appearance. Blackish plumage, rather glossy above, with dark brownish rump and dark grey underparts. Usually in small groups and almost always with other species of swifts. Rather poorly known. **Range and habitat:** NT, Central America to central Brazil, east to Guianas and Trinidad. Low to mid-elevation, over forest and open areas with scattered woodland. **Similar species:** Chimney Swift (see p 72). **Not illustrated.**

Lesser Swallow-tailed Swift *Panyptila cayennensis* (Apodidae) 13 cm (5 in). Slender with long forked tail, normally appearing pointed. Black with white throat, collar and patch on side of rump, visible from below. Usually seen singly or in pairs, sometimes in mixed species groups of swifts, but generally flying somewhat higher. Nest a long downward-pointing tube of plant down, fixed by saliva to tree-trunks, below branches, rock faces or even buildings. **Range and habitat:** NT, Mexico to Brazil. Widespread at low and mid-elevations over forest and wooded areas. **Similar species:** White-tipped and Fork-tailed Palm-swifts. **Not illustrated.**

Spectacled Owl
(juvenile and adult)

Tropical Screech Owl

Striped Owl

Oilbird

Great Potoo

Common Potoo

White-collared Swift

Common Pauraque

Scissor-tailed Nightjar

Long-tailed Hermit *Phaethornis superciliosus* (Trochilidae) 13 cm (5 in). Hummingbird of understorey with long, decurved bill and elongated white-tipped central tail feathers. Plumage generally brownish with bronzy green tinge to upperparts and buffy rump. Buff stripes on face. Paler below. Visits isolated flowers, travelling considerable distance, sometimes more than 1 km (over ½ mile), between feeds. Also collects small invertebrates from foliage. Nest hangs on underside of a broad leaf. Males assemble to sing in courtship leks. **Range and habitat:** NT, Mexico to central Brazil. Lowland forest understorey and nearby secondary growth, often close to streams. **Similar species:** several other hermits.

Swallow-tailed Hummingbird *Eupetomena macroura* (Trochilidae) 19 cm (7½ in). Large hummingbird with long, deeply forked dark blue tail. Bill black and slightly decurved. Head, throat and breast deep purplish blue, rest of body shining green. Agile flier which frequently glides. Call a repeated 'shep-shep-shep-shep'. **Range and habitat:** NT, Guianas and central Brazil to Bolivia. Mainly mid-elevation scrub and savannah.

White-necked Jacobin *Florisuga mellivora* (Trochilidae) 12 cm (4½ in). Medium-sized hummingbird with rather short bill. Male boldly patterned with dark blue head and throat, upperparts, breast and central tail bright green; rest of underparts, outer tail and nape white. Tail feathers narrowly edged black. Female green above, scaled bluish on throat, white belly and greenish blue tail, tipped white on sides. Mainly a canopy or forest edge species, visiting flowering trees. **Range and habitat:** NT, Mexico to central Brazil. Lowland species of forest borders and clearings.

Black-throated Mango *Anthrocothorax nigricollis* (Trochilidae) 10 cm (4 in). Medium-sized hummingbird with slightly decurved bill. Male blackish below, dark greenish above with dark maroon tail. Female quite different, having white underparts with bold black stripe passing from centre of chin to belly. Encountered at flowering trees. **Range and habitat:** NT, Central America to northern Argentina. Mainly in lowlands at forest edge, secondary vegetation, cleared areas and gardens. **Similar species:** Green-breasted Mango.

Festive Coquette *Lophornis chalybea* (Trochilidae) 7.5 cm (3 in). Tiny hummingbird with short, straight bill. Dark greenish plumage with brilliant green on side of head, extending behind as tuft with white spots. White bar across rump. Female lacks tufted appearance and whitish plumage on chin, which extends back to side of face. Feeds mainly at flowering trees, flight is bee-like. **Range and habitat:** NT, Venezuela to Brazil. Lowland species of forest edge and open areas with shrubs or trees. **Similar species:** several other coquettes.

Violet-capped Woodnymph *Thalurania glaucopis* (Trochilidae) 12 cm (4½ in). Medium-sized hummingbird with slightly decurved black bill. Male has violet crown and bronzy green back. Underparts golden green. Forked tail steel blue. Female greyish white below with green discs on sides, forked dark blue tail with white tips. Usually solitary at lower levels in forest. **Range and habitat:** NT, south-east Brazil. Very common in lowland forest and scrub. **Similar species:** Fork-tailed Woodnymph.

Green-throated Carib *Sericotes holosericeus* (Trochilidae) 12 cm (4½ in). Medium-sized hummingbird with long, black, slightly decurved bill. Mainly green with bright green throat, violet-blue breast patch and blackish belly. Tail is bluish black. Sexes alike. A pugnacious bird which defends flowering trees and shrubs from other nectar feeders. **Range and habitat:** NT, Caribbean islands. Common at low elevations, mainly in cultivated areas and secondary growth, sometimes forest. **Similar species:** Purple-throated Carib.

Green-crowned Brilliant *Heliodoxa jacula* (Trochilidae) 13 cm (5 in). Large hummingbird with straight black bill. Green with inconspicuous violet spot on throat. Long bluish black tail is forked. Female white below, heavily spotted green. Upperparts green with white spot behind eye and white malar stripe. Tail tipped white. Usually perches on inflorescence to feed. Visits flowers from mid-level to canopy, sometimes defending nectar sources. **Range and habitat:** NT, Costa Rica to Ecuador. Mainly in highland forest, although may descend to lower altitude outside breeding season. **Similar species:** Violet-fronted Brilliant.

Giant Hummingbird *Patagona gigas* (Trochilidae) 23 cm (9 in). Huge hummingbird with long wings and characteristic flight. Dull bronze-green above and buffy below, whitish rump and undertail-coverts. Female has spotted underparts. Flies like a swallow with frequent glides. Hovers slowly but often perches to feed. Aggressive, frequently seen chasing other species away from flowers. **Range and habitat:** NT, along Andean chain from Ecuador to central Chile. Mainly in dry shrubby highland areas. Some populations show altitudinal migration. **Similar species:** Great Sapphirewing.

Sword-billed Hummingbird *Ensifera ensifera* (Trochilidae) 14 cm (5½ in). Hummingbird with extraordinary long bill, about 11 cm (4½ in). Rather dark bronzy green plumage and forked tail. Female has heavily spotted and streaked underparts. Singles visit fuchsias and other tubular flowers, approaching from below. Usually perch with bill pointing upwards. **Range and habitat:** NT, along Andean chain from Venezuela to Bolivia. Widespread in humid montane forest up to treeline.

Scintillant Hummingbird *Selasphorus scintilla* (Trochilidae) 6.5 cm (2½ in). Smallest bird in the world. Tiny hummingbird with straight black bill. Male has orange-red gorget with incomplete white collar. Mainly buffish below with green spots. Upperparts bronzed green with rufous and black striped tail. Female is buffy on throat and has dark subterminal tail band. Territorial in breeding season. Usually visits small flowers on shrubs. **Range and habitat:** NT, Costa Rica and Panama. Widespread at mid-elevations in plantations, secondary vegetation and gardens. **Similar species:** Volcano Hummingbird.

Black-eared Fairy *Heliothryx aurita* (Trochilidae) 10 cm (4 in). Elegant hummingbird with short black bill. Male brilliant green above, blackish ear-coverts and spotless white below. Female has some greyish markings on throat. Very active feeder with rapid, darting flight, visiting flowers at mid-level to canopy. Also hawks insects. Sometimes apparently associated with mixed species flocks. **Range and habitat:** NT, east of Andes from Guianas to southern Brazil. Lowland humid forest. **Similar species:** Purple-crowned Fairy.

Long-tailed Hermit at Passion Flower

Swallow-tailed Hummingbird

White-necked Jacobin

Black-throated Mango

Festive Coquette

Violet-capped Woodnymph

Green-throated Carib

Green-crowned Brilliant female

Giant Hummingbird

Sword-billed Hummingbird

Scintillant Hummingbird

Black-eared Fairy

TROGONS Trogoniformes

Resplendent Quetzal *Pharomachrus mocinno* (Trogonidae) 36 cm (14 in). Spectacular-looking male has 60 cm (24 in) long streamers formed from uppertail-coverts. Male has flattened bushy green crest, extended forward over yellow bill. Bright green above and on breast. Rest of underparts red with white undertail. Female lacks crest and such extravagant streamers, has grey belly, red undertail coverts and black and white barred tail. Usually solitary or in pairs, although a group may assemble at a fruiting tree. Also feed on large insects plucked from leaves in swooping sallies. Song a series of mellow, deep notes. **Range and habitat:** NT, Central America. Quite common in humid montane forest.

Collared Trogon *Trogon collaris* (Trogonidae) 25 cm (10 in). Smallish trogon. Male has yellow bill, is green above and on chest with white chest band and red underparts. Tail narrowly barred white and black. Brown replaces green on female, which has a broken white eye-ring and bicoloured bill, she is rather pinker below with a finely vermiculated tail which appears greyish. Solitary or in pairs at low to midheight, feeding in typical trogon fashion of peering from a perch and flying out to snatch small animals such as insects or fruit from nearby trees. Song a slow, rather sad series of notes, 'cow cu-cu cu-cu...'. **Range and habitat:** NT, Mexico to Brazil. Low to mid-elevation in humid forest and secondary vegetation. **Similar species:** several other trogons.

Violaceous Trogon *Trogon violaceus* (Trogonidae) 24 cm (9½ in). Small trogon. Male has yellow eye-ring and pale greyish bill. Upperparts glossy green, becoming bluish on head and chest. Narrow white chest band and deep yellow underparts. Tail barred black and white. Female grey above with smudged boundary between chest and belly, broken eye-ring is white. Generally perches at mid-level. Song a series of even-pitched, rather fast 'kew-kew-kew' notes. **Range and habitat:** NT, Mexico to Brazil. Mainly in lowlands in forest edges, clearings and woodland. **Similar species:** Black-throated and White-tailed Trogons.

KINGFISHERS AND ALLIES Coraciiformes

Ringed Kingfisher *Megaceryle torquata* (Alcedinidae) 38 cm (15 in). Large, rather noisy kingfisher with untidy crest. Male bluish above, white on sides of neck and throat and rufous below. Female has bluish and white chest band. Usually seen alone or in pairs, on high perches. Flies high, giving a loud 'klek' call. Calls also include a hard rattle. Nests in banks, often in colonies. **Range and habitat:** NT, NA, southern USA to Tierra del Fuego. Common, mainly in lowlands, beside rivers and lakes, also in coastal areas. **Similar species:** Belted Kingfisher (see p 74).

Green Kingfisher *Chloroceryle americana* (Alcedinidae) 19 cm (7½ in). Small kingfisher with dark green upperparts, white spots on wing and white collar. Male has rufous chest band, female 2 green bands, rest of underparts white. Usually solitary, flying low or perched close to water. Flight call a low 'choot', often repeated as a chatter. **Range and habitat:** NT and NA, southern USA to Argentina. Fairly common beside rivers and lakes, marshes and woodland streams, from low to mid-elevations. **Similar species:** Amazon Kingfisher.

Puerto Rican Tody *Todus mexicanus* (Todidae) 10 cm (4 in). Small attractive bird with long, wide-based bill with red lower mandible. Emerald green above, greyish below with red throat patch and yellow flanks. Watch from exposed perches for insects which are captured, in flight, from substrates such as undersides of leaves. Whilst perched constantly move heads in jerky fashion. Rattling sound produced in flight. **Range and habitat:** NT, endemic to Puerto Rico. Rather widespread from drier coastal areas to more humid mountain slopes. **Similar species:** 4 other todies, none of which also occurs on Puerto Rico.

Rufous Motmot *Baryphthengus martii* (Momotidae) 46 cm (18 in). Large motmot with long tail, sometimes with racket-tips, which is swayed from side to side when bird is alarmed. Blackish face-mask with rest of head and underparts to lower belly rufous. Rest of underparts and upperparts greenish. Small black spot on chest. Usually solitary, perched at mid-level to subcanopy. Makes sudden flights to new perches and large insect prey are plucked off vegetation in flight. Call is series of owl-like hoots, often given before dawn. **Range and habitat:** NT, Central America to northern Argentina. Lowland and foothill species of humid forest. **Similar species:** Broad-billed Motmot.

Blue-crowned Motmot *Momotus momota* (Momotidae) 41 cm (16 in). Medium-sized with long tail, usually with rackets. Black face-mask and red eyes, with centre of crown black, broadly bordered by brilliant turquoise-blue. Greenish above and dull buff below with small black chest spots usually visible. Sits still singly or in pairs on mid-level perches and easily overlooked unless seen in flight. Call a low, double-noted hoot. **Range and habitat:** NT, Mexico to northern Argentina. Common from lowlands to mid-elevations in secondary vegetation, cacao plantations, forest edge and woodland. **Similar species:** Turquoise-browed Motmot.

WOODPECKERS AND ALLIES Piciformes

Rufous-tailed Jacamar *Galbula ruficauda* (Galbulidae) 23 cm (9 in). Medium-sized jacamar with long bill and tail. Metallic green above and on breast. White throat and rest of underparts, including tail, rufous. Female has buffy throat. Perch, usually in pairs, on exposed branch which they leave to hawk large flying insects such as butterflies. Usually return to same or nearby perch. Call a high-pitched 'peee', normally given in accelerating series. **Range and habitat:** NT, Mexico to southern Brazil. Common, mainly at low elevations, in forest edge habitats, secondary growth and shrubby vegetation. **Similar species:** White-chinned and Great Jacamars.

Great Jacamar *Jacamerops aurea* (Galbulidae) 30 cm (12 in). Stout-billed, heavily built jacamar with metallic green upperparts and chin. White throat band (buffy in female) and rufous underparts. Underside of tail blackish. Bill appears slightly decurved. Usually solitary and easily overlooked, perches for long periods at mid-levels within forest. Call a hawk-like whistle. **Range and habitat:** NT, Costa Rica to central Brazil. Uncommon in lowland and humid forest of foothills. **Similar species:** Rufous-tailed Jacamar.

White-necked Puffbird *Notharchus macrorhynchus* (Bucconidae) 25 cm (10 in). Large puffbird with heavy black bill. White forehead, throat, sides of face and collar. Underparts white apart from black breast band and some barring on flanks. Black on hindcrown, eye-stripe and upperparts. Sits on high exposed perches and, like other puffbirds, appears very lethargic. Makes short flights to collect insect prey from nearby vegetation, usually returning to original perch. Generally silent. In breeding season gives bubbling trill. **Range and habitat:** NT, southern Mexico to northern Argentina. Widespread in lowland forest edge, secondary vegetation and woodland. **Similar species:** Brown-banded Puffbird.

Spotted Puffbird *Bucco tamatia* (Bucconidae) 18 cm (7 in). Medium-sized puffbird with stout black bill and red eyes. Dull brown above, greyish on face with white malar bordered below by broad black patch and warm buff throat. Underparts whitish, heavily scalloped with black. Seen singly or in pairs, on a low perch and usually remarkably confiding. Call an accelerating series of whistles. **Range and habitat:** NT, Orinoco and Amazon basins. Fairly common in seasonally inundated forest and riverside woodland. **Similar species:** Chestnut-capped Puffbird.

Resplendent Quetzal

Collared Trogon

Violaceous Trogon

Ringed Kingfisher

Green Kingfisher

Puerto Rican Tody

Rufous Motmot

Blue-crowned Motmot

Rufous-tailed Jacamar

Great Jacamar

White-necked Puffbird

Spotted Puffbird

Black Nunbird *Monasa atra* (Bucconidae) 27 cm (10½ in). Large puffbird with long, slender, red bill, slightly decurved. Generally greyish black, paler on rump and underparts, apart from blackish throat. Wing-coverts on inner wing white. Usually in pairs or small groups, perched confidingly at low to mid-level. Plucks insects and small reptiles from vegetation on short sallies or from ground in swooping flight. Nests in burrows. Loud call, often given in choruses. **Range and habitat:** NT, north-eastern South America. Lowland humid forest, gallery forest and clearings. **Similar species:** Black-fronted Nunbird.

Swallow-wing Puffbird *Chelidoptera tenebrosa* (Bucconidae) 15 cm (6 in). Small aberrant puffbird with rather short bill and broad wings. Plumage mainly black with white rump and undertail-coverts and rufous belly. Usually in pairs or small groups perched on exposed treetop branches, making aerial sallies for flying insects. Generally silent. **Range and habitat:** NT, Venezuela to Brazil. Lowland species common along sandy river-banks, forest borders and gallery woodland.

Black-spotted Barbet *Capito niger* (Capitonidae) 19 cm (7½ in). Large barbet with stout black bill. Upperparts black with dull yellow crown and brighter yellow supercilium from behind eye extending down back to form V. Yellow wing-bar and spotting on tertials. Underparts yellow with red or orange throat (depending on race) and black spots on flanks. Female much more heavily marked black below. Usually in pairs and often in mixed species flocks from mid-level to canopy. Feeds on fruit and also searches foliage for large insects. Call a deep double hoot. **Range and habitat:** NT, Guianas to central Brazil. Common in lowland humid forest. **Similar species:** Lemon-throated Barbet.

Emerald Toucanet *Aulacorhynchus prasinus* (Ramphastidae) 33 cm (13 in). Small toucan with black bill with yellow ridge on upper mandible and white base. Plumage mainly green with chestnut vent and undertail. Throat colour variable according to race. Usually in small groups, constantly on move. Normally adopts horizontal stance, when alarmed nods head and cocks tail. Call a series of 'churt' notes. **Range and habitat:** NT, Central America and along Andean chain from Venezuela to Bolivia. Common at mid to high elevations in humid, montane forest and secondary vegetation. **Similar species:** several toucanets.

Curl-crested Araçari *Pteroglossus beauharnaesii* (Ramphastidae) 36 cm (14 in). Small slender toucan with long graduated tail. Upper mandible black with orange-brown culmen, lower mandible white. Upperparts black apart from red back and rump. Underparts yellow with red band across upper breast. Crest of curly feathers: Usually in small groups in canopy. Has rather harsh call-note. **Range and habitat:** NT, south of Amazon in Peru and Brazil. Lowland Amazon forest. **Similar species:** several araçaris.

Grey-breasted Mountain Toucan *Andigena hypoglauca* (Ramphastidae) 46 cm (18 in). Medium-sized toucan with strikingly patterned bill. Upper mandible mainly orange-red, lower mandible black. Large yellow region with black vertical bar in basal half of bill. Bluish bare skin around eye, black hood, blue-grey collar and underparts apart from red vent. Olive-green upperparts with conspicuous yellow rump and blackish tail tipped chestnut. Often in pairs or small parties, generally in canopy but visits shrubs in fruit. Mainly silent but gives loud nasal call at dawn. **Range and habitat:** NT, Andean chain from Colombia to Peru. Fairly common in montane forest to treeline. **Similar species:** Plate-billed Mountain Toucan.

Channel-billed Toucan *Ramphastos vitellinus* (Ramphastidae) 53 cm (21 in). Large toucan with bill mainly black with light blue base. Bare facial skin around eye bluish. Black upperparts apart from red uppertail-coverts. Throat white, becoming yellowish on breast. Red border between breast and black belly. Undertail-coverts red. Seen singly, in pairs or small groups, mainly in canopy but sometimes at lower levels. Hops energetically along branches. Flight is heavy and undulating. Very vocal with far-carrying yelping call. **Range and habitat:** NT, northern South America. Low to mid-elevations in humid forest and forest edge. **Similar species:** several other large toucans.

Toco Toucan *Ramphastos toco* (Ramphastidae) 62 cm (24 in). Large toucan with enormous orange bill, with black oval patch at tip of upper mandible. Black line at base of bill. Plumage black apart from white throat and rump, and red undertail-coverts. Seen singly or in small groups, usually in treetops. **Range and habitat:** NT, Guianas to northern Argentina. Mainly in lowland woodland, stands of palms and secondary vegetation.

Plain-breasted Piculet *Picumnus castelnau* (Picidae) 9 cm (3½ in). Tiny, dumpy bird with very short tail. Male has black crown with red spots, replaced by white spots on hind-crown. Female has all-black crown. Nape is scaled grey, rest of upperparts brownish olive, lightly barred yellowish with yellow edgings to primaries. Unmarked underparts have a yellowish olive wash. Black tail with sides and inner web of central feathers white. Usually solitary and unobtrusive at low to subcanopy level. Call a high-pitched trill. **Range and habitat:** NT, western Amazonia. Uncommon in lowland swamp forest, secondary vegetation close to rivers and river islands. **Similar species:** White-bellied Piculet.

Campo Flicker *Colaptes campestris* (Picidae) 29 cm (11½ in). Large woodpecker. Crown and nape black. Brown and buff barred upperparts with whitish rump. White around eye, otherwise ear-coverts golden yellow, extending on to chest. Reddish malar, throat white or black according to race. Underparts buffy white with well-spaced dark bars. Feeds, often in groups, on ants collected from ground. Nests in dead trees or in burrows in ground or termite nests. Repertoire of several calls including series of 'keep' notes. **Range and habitat:** NT, eastern South America. Lowland fields, grasslands and clearings. **Similar species:** Andean Flicker (no altitudinal overlap).

Blond-crested Woodpecker *Celeus flavescens* (Picidae) 22 cm (8½ in). Medium-sized woodpecker. Head and long crest creamy white or buff. Male has red malar, female blackish or white. Very variable species with upperparts varying from blackish barred with pale gold to creamy buff with bold black spots (depending on race). Underparts either black or brownish with buff scaling. Diet probably mainly consists of ants, collected from twigs and branches. **Range and habitat:** NT, central Brazil to northern Argentina. Lowland and foothill forest, woodland and scrub. **Similar species:** Pale-crested Woodpecker.

Helmeted Woodpecker *Dryocopus galeatus* (Picidae) 29 cm (11½ in). Large black and white woodpecker with large red crest and ivory-coloured bill. Ear-coverts finely barred black and buff, red malar (in male) and buffy throat. Upperparts and neck black with white stripe on side of neck and white rump and uppertail-coverts. Underparts barred black and white. Extremely poorly known species. **Range and habitat:** NT, southern Brazil, Paraguay and northern Argentina. Very rare species of lowland primary forest. Threatened by habitat loss. **Similar species:** several in genus. **Not illustrated.**

Black Nunbird

Swallow-wing Puffbird

Black-spotted Barbet

Emerald Toucanet

Curl-crested Araçari

Grey-breasted Mountain Toucan

Channel-billed Toucan

Toco Toucan

Plain-breasted Piculet

Campo Flicker

Blond-crested Woodpecker

PERCHING BIRDS Passeriformes

Wedge-billed Woodcreeper *Glyphorynchus spirurus* (Dendrocolaptidae) 14 cm (5½ in). Tiny woodcreeper with short bill. Brown plumage with warmer-toned rump and tail. Buff bases to flight feathers visible in flight. Buff supercilium and spotting on head and breast. Orange-buff chin. Usually solitary or in pairs, often in mixed species parties. Creeps up tree-trunk to about mid-level, thence flying to base of nearby tree. Call is short 'cheef'. **Range and habitat:** NT, Mexico to Brazil. Lowland and mid-elevation humid forest. **Similar species:** Fulvous-dotted Treerunner.

Striped Woodcreeper *Xiphorhynchus obsoletus* (Dendrocolaptidae) 20 cm (8 in). Medium-sized woodcreeper with pale buff stripes on dull brown underparts and mantle. More solidly pale buff throat. Warm brown wings, rump and tail. Bill slightly decurved. Usually solitary or in pairs, frequently in mixed species parties at low to mid-levels. Call an excited trill. **Range and habitat:** NT, lowland Amazon forest in seasonally inundated forest and swampy woodland. **Similar species:** several other woodcreepers.

Red-billed Scythebill *Campylorhamphus trochilirostris* (Dendrocolaptidae) 23 cm (9 in). Medium-sized woodcreeper with remarkable long, deeply decurved reddish bill. Plumage generally brown, warmer on wings and tail with buff streaks on head, breast and mantle. Found singly or in pairs, usually in mixed species flocks. Uses long bill to probe in bases of epiphytes and crevices in bark. Normally at low to mid-levels. Calls include a high-pitched trill. **Range and habitat:** NT, Panama to northern Argentina. Uncommon in lowland forest edge and woodland. **Similar species:** Curve-billed Scythebill.

Rufous Hornero *Furnarius rufus* (Furnariidae) 19 cm (7½ in). Large terrestrial furnariid. Slightly decurved bill and rather long dark legs. Plumage uniformly dull brown, more rufous on tail and paler on throat. Walks confidently in open areas, collecting insects and seeds from ground. Call a high-pitched chattering. Horneros build domed nests of mud on posts or on horizontal branches. **Range and habitat:** NT, southern Brazil to northern Argentina. Mainly lowlands, but range extends into highlands of Bolivia. Common in drier woodland, open country and gardens. **Similar species:** Pale-legged Hornero.

Plain-mantled Tit-spinetail *Leptasthenura aegithaloides* (Furnariidae) 15 cm (6 in). Small furnariid with very long, pointed tail and small, stubby bill. Dark crown, finely streaked brown. Unstreaked back greyish brown and rufous patch on wing. Throat whitish, rest of underparts buff or pale grey. Usually in pairs or small groups searching, in an engaging acrobatic manner, clusters of leaves and flower heads for insects and seeds. **Range and habitat:** NT, fairly common in arid zone of southern South America. **Similar species:** Brown-capped Tit-spinetail.

Dark-breasted Spinetail *Synallaxis albigularis* (Furnariidae) 15.5 cm (6 in). Small furnariid with rather short, ragged-looking tail. Olive brown above with rufous hind-crown and shoulders. Underparts dark grey with whitish belly. Throat white with bases of feathers black so that when feathers are raised throat appears blackish. Skulking, usually in pairs, in tangled vegetation and grass close to ground. Large stick nest placed low in bushes. Call a conversational 'di-dududu'. **Range and habitat:** NT, western Amazonia. Common in secondary vegetation and shrubby growth. **Similar species:** Dusky and Pale-breasted Spinetails.

Rusty-backed Spinetail *Certhiaxis vulpina* (Furnariidae) 14.5 cm (5½ in). Small spinetail with rather dingy grey underparts, slightly paler on throat. Upperparts rusty brown. Pale supercilium and narrow dark eye stripe, with fine dark streaking on sides of head. Usually in pairs, seen clambering up vine tangles and thick-stemmed tall grasses, searching for insect prey. Calls include a high-pitched whinny and a hard rattle. **Range and habitat:** NT, Amazonia. Fairly common in lowland secondary vegetation with tall grass and bushes, particularly on river islands. **Similar species:** Crested Spinetail.

Striped Woodhaunter *Philydor subulatus* (Furnariidae) 16.5 cm (6½ in). Small foliage-gleaner with quite slender, slightly decurved bill. Plumage generally dull brown, slightly more olive below, with buff throat and streaks on head, underparts and mantle. Extent of streaking depends on subspecies. Usually in mixed species flocks, keeping fairly low down and searching for insect prey in vine tangles and clusters of dead leaves. Song a series of whistles followed by trill; usual call a loud, nasal 'chook'. **Range and habitat:** NT, Central America to central Brazil. Lowland and foothill species of humid forest. **Similar species:** several foliage-gleaners.

White-browed Foliage-gleaner *Philydor amaurotis* (Furnariidae) 16 cm (6½ in). Small foliage-gleaner with bold white supercilium behind eye. Upperparts dark brown, tail rufous. Crown finely spotted white. White throat, buffy below with bold white breast markings. Usually in mixed species flocks at mid-level to subcanopy. Active forager, hopping along branches, peering and poking acrobatically into clusters of dead leaves. **Range and habitat:** NT, southeast Brazil and northern Argentina. Lowland forest. **Similar species:** several similar foliage-gleaners.

Streaked Xenops *Xenops rutilans* (Furnariidae) 11.5 cm (4½ in). Small busy bird with lower mandible upturned. Conspicuous white crescent on cheek and pale supercilium. Plumage generally brown, paler below, with narrow buff streaks on head, back and below. Bold cinnamon wing-bar. Usually in pairs, frequently in mixed species flocks. Climbs along branches, poking into rotten stumps, balancing on tiny twigs. Call a high-pitched series of shrill notes. **Range and habitat:** NT, Central America to Argentina. Widespread at mid-elevations in humid forest. **Similar species:** Slender-billed Xenops.

Plain Xenops *Xenops minutus* (Furnariidae) 12 cm (4½ in). Small with short bill with upturned appearance. Plumage dull olive-brown with buff supercilium and white crescent on side of face, sometimes slightly streaked on breast. Bold cinnamon wing-bar and black streak on side of rufous tail. Forages actively at low to mid-levels, usually in mixed species flocks. Call a high-pitched 'cheet', song a chattering trill. **Range and habitat:** NT, southern Mexico to northern Argentina. Species of lowland and mid-elevations in humid forest and secondary growth. **Similar species:** Spectacled Spinetail.

Tawny-throated Leafscraper *Sclerurus mexicanus* (Furnariidae) 15 cm (6 in). Rather plump terrestrial bird with long bill which distinctly droops at tip. Dark brown plumage with bright rufous throat and rump. Tail blackish. Usually solitary, hopping on ground and tossing over leaf litter, hence its alternative name, 'leaftosser'. Difficult to see, normally flies off if disturbed, giving a sharp 'chick' call. **Range and habitat:** NT, southern Mexico to southern Brazil. Lowland and mid-elevation humid forest. **Similar species:** several other leafscrapers.

Giant Antshrike *Batara cinerea* (Formicariidae) 32 cm (13 in). Huge antshrike with very heavy, hooked bill. Male has black crest and crown with upperparts black with fine white barring. Underparts greyish. Female has chestnut crown, black upperparts with buff barring and dull pale brown underparts. Usually in pairs in dense forest undergrowth, foraging on large arthropods and small vertebrates. **Range and habitat:** NT, eastern Brazil, Paraguay, southern Bolivia and northern Argentina. Low and mid-elevation forest. **Similar species:** Undulated Antshrike.

Black-hooded Antshrike *Thamnophilus bridgesi* (Formicariidae) 16.5 cm (6½ in). Medium-sized dark plumaged antshrike. Male black, slightly greyer on belly, with small white spots on wing-coverts, white tips to outer 3 tail feathers and white wing-linings. Female has black head with fine white streaks, blackish brown above with white tips to wing-coverts and on tail. Underparts dirty olive with white streaks. Usually in pairs, actively gleaning insects from vegetation. Call a series of nasal notes. **Range and habitat:** NT, Costa Rica and Panama. Lowlands and foothills in secondary vegetation and forest borders. **Similar species:** male similar to several antshrikes elsewhere.

Amazonian Antshrike *Thamnophilus amazonicus* (Formicariidae) 15 cm (6 in). Medium-sized antshrike with rather stout bill. Male grey with black wings boldly tipped white to coverts and white edgings to scapulars and tertials. Black tail tipped white. Crown grey or black according to race. Female has rufous buff head and breast, paler below, with dark wings and tail similarly well marked with white. Usually in pairs, foraging actively at low levels, sometimes in mixed species flocks. **Range and habitat:** NT, Guianas to central Brazil. Fairly common at edge of swampy forest and open woodland in Amazonia. **Similar species:** Slaty Antshrike.

Plain Antvireo *Dysithamnus mentalis* (Formicariidae) 11.5 cm (4½ in). Small, rather plump bird with stout bill and short tail. Male grey above, darker on crown and ear-coverts, paler below. Wing-coverts narrowly tipped white. Female has chestnut crown, olive-brown above with indistinct buff tips to wing coverts. Paler below with greyish tinge on ear-coverts and narrow white eye-ring. In pairs or family groups and sometimes in mixed species flocks. Hop along branches at low levels, gleaning prey of adjacent foliage. Song a series of rising and falling whistles. **Range and habitat:** NT, Mexico to northern Argentina. Common at mid-elevations in humid forest. **Similar species:** Spot-crowned Antvireo, some antshrikes.

Streaked Antwren *Myrmotherula surinamensis* (Formicariidae) 9.5 cm (4 in). Very small, active antbird with very short, stubby tail. Male heavily streaked black and white with 2 bold white wing-bars. Female has rufous crown, streaked black or unstreaked (depending on race). Underparts warm buff and virtually unstreaked. Wings dark with 2 white wing-bars and white edgings to scapulars. Usually in pairs, constantly moving, often hovering, as they glean foliage for small insects. Song a ringing trill. **Range and habitat:** NT, Panama to Peru and Brazil. Common in lowlands and foothills in forest edges and secondary vegetation, often close to water. **Similar species:** Cherrie's and Stripe-chested Antwrens.

Pygmy Antwren *Myrmotherula brachyura* (Formicariidae) 8 cm (3 in). Tiny antbird with very short stubby tail. Black and white streaked above, apart from greyish rump and darker tail. Pale cheeks, black malar stripe and white throat. Rest of underparts lemon-yellow. Female is buffy on head. Family parties actively search for insect prey from canopy to mid-level, incessantly giving their accelerating short trilling call. Sometimes join mixed species flocks. **Range and habitat:** NT, Panama to Brazil. Common in lowland humid forest and on forest edge. **Similar species:** Short-billed Antwren.

Banded Antbird (Antcatcher) *Dichrozona cincta* (Formicariidae) 10.5 cm (4 in). Small, terrestrial antbird. Long bill and short tail. Brown above with white bar across lower back and pale greyish rump. Wing-coverts tipped white, with greater coverts tipped golden, forming bold wing-bars. Underparts whitish with necklace of bold black spots. Female has buffier barring. Walks on forest floor in crake-like fashion. Usually solitary, giving high-pitched series of rising drawn-out whistles. **Range and habitat:** NT, Amazonia. Uncommon, but probably overlooked in lowland humid forest. **Similar species:** Spot-backed Antbird.

Dot-winged Antwren *Microrhopias quixensis* (Formicariidae) 13 cm (5 in). Large antwren with long, graduated tail. Male black with white dots on wing-coverts, bold wing-bar, and white tips to tail. Female similar to male above but underparts entirely rufous, or with black throat (depending on race). Usually in pairs and in mixed species flocks. Actively search for insect prey in clusters of dense foliage. Song a rapid series of whistles. **Range and habitat:** NT, Mexico to Brazil. Mainly in lowland humid forest and secondary vegetation.

Scaled Antbird *Drymophila squamata* (Formicariidae) 11 cm (4½ in). Small antbird with longish tail. Upperparts black with white supercilium and bold white spots on back, wing-coverts and white bars on tail. Underparts white with black spots. Female brown above with buffy spots on back, tail black with buff bars and underparts white with fine black streaking. Rather wren-like in behaviour, hopping through dense tangles of vegetation. **Range and habitat:** NT, eastern coastal Brazil. Lowland humid forest. **Similar species:** several in same genus.

Blackish Antbird *Cercomacra nigrescens* (Formicariidae) 15 cm (6 in). A slim, long-tailed medium-sized antbird. Male is uniform blackish grey with wing-coverts tipped white and white edging to bend of wing. Female is brown above with warm rufous forehead, cheeks and underparts. Pairs reside in thick tangled vegetation. Calls include a rising series of nasal notes and a trill. **Range and habitat:** NT, Guianas to Brazil. Lowland forest edge and secondary vegetation, particularly in seasonally flooded areas. **Similar species:** Dusky Antbird.

White-shouldered Antbird *Myrmeciza melanoceps* (Formicariidae) 18 cm (7 in). Large antbird with stout black bill. Male all-black apart from bluish bare skin around eye and white flash on shoulder (visible in flight). Female has black head and chest, rest of plumage rufous chestnut. Usually solitary or in pairs, sometimes attending swarms of army ants with other species. Keeps close to ground, often in thick undergrowth, 'pumping' tail when perched. Loud song a series of double notes repeated. **Range and habitat:** NT, Amazonia. Common mainly in seasonally inundated forest. **Similar species:** Sooty Antbird.

Black-spotted Bare-eye *Phlegopsis nigromaculata* (Formicariidae) 18 cm (7 in). Large, distinctive antbird with red bare facial skin. Head and underparts blackish, upperparts olive-brown heavily spotted black. Wings and tail rufous-toned. Sexes similar. Perch close to ground in typical hunchbacked fashion. Usually in family groups and associates with other species at swarms of army ants. Drop from low perches to ground to collect large arthropods flushed by ants. Call a drawn-out, nasal squeal. **Range and habitat:** NT, Amazonia. Common mainly in seasonally inundated forest. **Similar species:** Reddish-winged Bare-eye.

Giant Antshrike (female)

Black-hooded Antshrike (immature male)

Amazonian Antshrike

Plain Antvireo

Streaked Antwren

Pygmy Antwren

Banded Antbird (Antcatcher)

Dot-winged Antwren

Scaled Antbird

Blackish Antbird

White-shouldered Antbird

Black-spotted Bare-eye

Rufous-tailed Antthrush *Chamaeza ruficauda* (Formicariidae) 19 cm (7½ in). Large, plump terrestrial antbird. Dark brown above with bold white stripe behind eye. Whitish below with feathers edged black, giving strongly streaked effect. Sexes similar. Usually solitary, walks with tail cocked across forest floor and fallen branches. Song a single 'cu' repeated as a trill for almost a minute. **Range and habitat:** NT, patchy distribution in Colombia, Venezuela and southeast Brazil. Humid montane forest. **Similar species:** Shorttailed Antthrush.

Wing-banded Antbird (Wing-banded Antthrush) *Myrmornis torquata* (Formicariidae) 16.5 cm (6½ in). Large, plump terrestrial antbird with very short tail and long bill. Male warm brown above with blackish scaling. Wings dark brown with several bold bright buff bars, more extensive panel on flight feathers. Throat and chest black, bordered by fine barring of black and white. Rest of underparts dull grey with buff vent. Female has rufous throat patch. Usually in pairs, hopping on ground, tossing over leaf litter in search for prey. Song a series of loud whistles. **Range and habitat:** NT, Central America to central Brazil. Lowland humid forest. **Not illustrated.**

Chestnut-belted Gnateater *Conopophaga aurita* (Conopophagidae) 13 cm (5 in). Small plump bird with short tail and long legs. Attractive male is dark brown above with blackish scaling, has a white tuft of feathers behind eye which can sometimes be obscured. Black face and chin, rufous chestnut breast and whitish belly. Female has rufous on face. Usually in pairs perched close to ground, dropping down to collect insect prey or toss over leaves. Quiet and unobtrusive. **Range and habitat:** NT, Amazonia. Uncommon in lowland humid forest.

Rufous-vented Tapaculo *Scytalopus femoralis* (Rhinocryptidae) 14 cm (5½ in). Small mainly terrestrial bird with fairly long cocked tail. Dark grey plumage with browner tone on wings and back. Some pale scaling on belly and flanks, vent chestnut with black bars. Some individuals show diagnostic white spot on crown. Very difficult to see as it hops through tangles of vegetation close to ground. Song a long series of rapidly repeated 'chuock' or 'chu-ock' notes. **Range and habitat:** NT, Andes from Venezuela to Bolivia. Fairly common, mainly at mid-level, in undergrowth of humid forest. **Similar species:** several very similar in genus.

Purple-breasted Cotinga *Cotinga cotinga* (Cotingidae) 19 cm (7½ in). Medium-sized, plump cotinga with short black bill. Male dark blue with black wings and tail and dark purple throat and breast. Female brown with feathers edged pale buff, giving wholly scaled appearance. Usually solitary, although small groups may gather at a fruiting tree. Remain perched for long periods, making short sallies to pluck fruit. Often perch in exposed treetop branches. Silent though mechanical whirring sound produced in flight. **Range and habitat:** NT, southern Venezuela to northern Brazil. Uncommon in lowland forest on sandy soils. **Similar species:** Banded Cotinga. **Not illustrated.**

Plum-throated Cotinga *Cotinga maynana* (Cotingidae) 19 cm (7½ in). Medium-sized, stocky cotinga with yellow eyes. Male is turquoise-blue with black on primaries and underside of tail. Neat dark purple throat patch appears almost black at a distance. Female is greyish brown, wholly scaled with pale buff. Vent and underwing-coverts are cinnamon. Perches high on exposed branches. Usually solitary but several may gather at fruiting tree. Fruit are plucked on a sally, the bird normally sitting quietly in tree after each feed. In direct flight a whirring whistle is produced by wings, otherwise silent. **Range and habitat:** NT, western Amazonia. Uncommon in lowland forest and woodland. **Similar species:** Spangled Cotinga.

Black-necked Red Cotinga *Phoenicircus nigricollis* (Cotingidae) 24 cm (9½ in). Large, brilliantly coloured cotinga. Striking male is black with a scarlet crown, breast and belly, rump and tail (tipped black). Female has duller brown and reddish plumage. Usually seen alone in subcanopy and visits fruiting canopy trees. Males are smaller than females and congregate in courtship leks, about which very little is known. More often heard than seen, far-carrying call is rather crow-like. **Range and habitat:** NT, western Amazonia. Fairly common in lowland humid forest. **Similar species:** Guianan Red Cotinga.

Bare-necked Fruitcrow *Gymnoderus foetidus* (Cotingidae) 30-4 cm (12-13 in). Rather large, long-tailed, small-headed cotinga. Male black with silvery grey uppersurface to wings and bluish bare skin on neck. Female dark grey with smaller area of bare skin on neck. Usually in dispersed groups often seen flying high. Mainly silent. **Range and habitat:** NT, Amazonia. Common in lowlands, mainly at forest edge and usually close to water. **Not illustrated.**

Red-ruffed Fruitcrow *Pyroderus scutatus* (Cotingidae) 38-43 cm (15-17 in). Very large cotinga. Plumage black with extensive area of 'crinkled' red feathers on throat. Underparts show variable amount of chestnut, depending on race. Usually solitary visitors to fruiting trees. Males gather in small communal leks, producing a deep booming call. **Range and habitat:** NT; different races in northern Andes, Guianas and south-east Brazil to northern Argentina. Uncommon in low to mid-elevation cloud forest, woodland and forest edge.

Amazonian Umbrellabird *Cephalopterus ornatus* (Cotingidae) 46-51 cm (18-20 in). Large, peculiar cotinga with long feathered wattle hanging from throat and thick tuft of feathers on crown. Plumage glossy bluish black. Eyes are white. Sexes similar, although female has smaller adornments. Usually solitary or in small groups, often seen flying with deep undulating flight across rivers and swamps. Perches high in trees, hopping heavily through branches. Males give deep booming sound. **Range and habitat:** NT, Amazonia. Fairly common in lowlands and foothills, mainly at forest edge and close to water. **Similar species:** Long-wattled Umbrellabird.

Three-wattled Bellbird *Procnias tricarunculata* (Cotingidae) 25-30 cm (10-12 in). Large, stout cotinga. Male has white head, neck and chest, rest of plumage chestnut. 3 long worm-like wattles hang from base of very broad bill. Female olive-green above, yellow below with olive streaks. Usually solitary, feeding at mid-level upwards. Snatches fruit in flight or plucked from a perch. Male calls from a canopy perch, producing very loud, explosive 'BOK'. **Range and habitat:** NT, Central America. During breeding season in montane forest, then migrates to lower elevation woodland and secondary vegetation.

Bearded Bellbird *Procnias averano* (Cotingidae) 28 cm (11 in). Large cotinga with very wide-based bill. Male has brown head, black wings; otherwise whitish grey. Cluster of black wattles hangs from throat. Female olive-green above, yellowish below heavily streaked green. Males spend most of day on canopy perches giving very loud, far-carrying 'BOK' call, or occupy mid-level perches carrying out stereotyped displays to conspecifics. Fruit mainly taken in flight. **Range and habitat:** NT, northern South America. Quite common in low to mid-elevation humid forest, shows vertical migration.

Rufous-tailed Antthrush

Plum-throated Cotinga

Chestnut-belted Gnateater (female)

Rufous-vented Tapaculo

Black-necked Red Cotinga

Red-ruffed Fruitcrow

Amazonian Umbrellabird

Bearded Bellbird

Three-wattled Bellbird

Guianan Cock-of-the-rock *Rupicola rupicola* (Cotingidae) 32 cm (13 in). Large, stocky bird. Male bright orange with brownish black wings with white speculum and short dark tail. Compressed crest arches above crown and almost covers short orange bill. Female dark brown with smaller crest and greyish orange eye. Males spend most of day gathered at communal lek, displaying within individual 'courts' on or close to ground. Call a bugle-like 'ka-waooh'. **Range and habitat:** NT, southern Venezuela and Guianas. Widespread in rocky forest on Guianan shield. **Similar species:** Andean Cock-of-the-rock.

Andean Cock-of-the-rock *Rupicola peruviana* (Cotingidae) 32 cm (13 in). Very similar in appearance to Guianan species. Body plumage of male may be scarlet or orange-red (depending on race). Wings and tail black with white tertials enlarged and flattened. Female is dark reddish brown with white eyes and small crest. Male courtship leks differ from former species, being at mid-level to subcanopy. Away from leks, usually solitary, visiting fruiting trees. **Range and habitat:** NT, fairly common at low to mid-altitudes on slopes of Andes from Venezuela to Bolivia. Occurs in wooded ravines, often close to rivers. **Similar species:** Guianan Cock-of-the-rock.

Sulphur-bellied Tyrant Manakin *Neopelma sulphureiventer* (Pipridae) 12 cm (4½ in). Rather large-billed, long-tailed flycatcher-like manakin with semiconcealed yellow crown patch. Olive-green above, dirty white throat with sides of underparts greyish olive and a clean yellow belly. Usually solitary, perched at low levels and partially insectivorous, flying out on short sallies to collect airborne prey or insects from foliage. **Range and habitat:** NT, eastern Peru to northern Bolivia. Lowland forest and woodland. **Similar species:** Pale-bellied and Saffron-crested Tyrant Manakins. **Not illustrated.**

Orange-collared Manakin *Manacus aurantiacus* (Pipridae) 10 cm (4 in). Medium-sized manakin with orange legs and black bill. Dumpy, bull-necked appearance. Male has black crown, nape and upper back and wings, olive-green rump and tail. Rest of head and breast bright orange, forming a collar. Rest of underparts bright yellow. Female is dull olive above, yellower below. Usually solitary but groups of males gather in undergrowth in courtship leks. Calls include loud 'chee-pooh' and various mechanical whirring and clicking sounds. **Range and habitat:** NT, Costa Rica and Panama. Common in lowlands and foothills in humid secondary forest and forest edge. **Similar species:** White-bearded Manakin.

White-bearded Manakin *Manacus manacus* (Pipridae) 10 cm (4 in). Medium-sized manakin with orange legs and black bill. Male has black crown, nape, wings, back and tail. Otherwise white with grey rump. Female dull olive-green, paler below. Males form large lekking groups, active throughout year, with each male occupying individual court. 6 types of stereotyped display, generally involving jumps and slides along perches. Noisy, producing several calls including soft trill and mechanical sounds. **Range and habitat:** NT, Venezuela to Argentina. Common in lowland and foothill secondary vegetation and forest edge. **Similar species:** Golden-collared Manakin.

Wire-tailed Manakin *Pipra filicauda* (Pipridae) 10.5 cm (4 in). Large manakin with white eyes. Male has spectacular plumage with red crown and nape, yellow face, rump and underparts, black back, wings and tail. Wing has white flash on flight feathers. Long filaments extend from tail. Female olive-green, yellower on belly with shorter tail filaments. Perches at low levels, usually solitary although several may gather at fruiting shrub. Males form dispersed leks. Complicated display involves jumps and short flights and culminates with male brushing female's face with tail filaments. **Range and habitat:** NT, Venezuela to Brazil. Lowlands in seasonally inundated forest, woodland and secondary growth.

Black and White Monjita *Xolmis dominicana* (Tyrannidae) 19 cm (7½ in). Large, long-tailed flycatcher. White body with black wings and tail. Primaries tipped white. Females brownish tinged on crown, nape and back. Uses exposed perches as vantage posts to search for invertebrate prey, dropping from perch to collect prey from ground. **Range and habitat:** NT, south-east Brazil to north-east Argentina. Uncommon species of lowland open scrub, usually near water or swamps. Population in sharp decline. **Similar species:** several in same genus.

Fork-tailed Flycatcher *Tyrannus savana* (Tyrannidae) 28-38 cm (11-15 in). Medium-sized flycatcher with enormously long forked tail. Black hood and tail, rest of upperparts grey, wings dark grey, white underparts. Makes aerial sallies from exposed perches such as fence-posts to catch insects, also visits fruiting trees. Migrants form large communal roosts, sometimes thousands strong: a spectacular sight as flock gathers in air. Call a short 'tik'. **Range and habitat:** NT, Mexico to central Argentina. Widespread at low and mid-elevations in open country or beside large rivers in forested regions. Has several migratory populations. **Similar species:** Scissor-tailed Flycatcher (see p 76).

Tropical Kingbird *Tyrannus melancholicus* (Tyrannidae) 22 cm (8½ in) Large flycatcher with slightly forked tail. Rather heavy black bill. Grey head with concealed reddish crown stripe, dusky on ear-coverts. Olive-green back and dull brown wings and tail. Whitish throat, olive smudge on chest, rest of underparts yellowish. Perches out in open on treetops, wires and posts. Makes aerial sallies for flying insects. Has high-pitched trilling call. **Range and habitat:** NT, Mexico to Argentina. Common widespread species from lowlands to temperate zones in open terrain, woodland and forest edge. Some populations migratory. **Similar species:** several kingbirds (see p 76).

Social Flycatcher *Myiozetetes similis* (Tyrannidae) 16.5 cm (6½ in). Medium-sized flycatcher with rather stubby black bill. Head dark olive-brown apart from white throat, broad white supercilium which does not quite encircle head and concealed crimson crown patch. Upperparts olive-brown, underparts yellow. Immature has rufous edgings to wing and tail. Usually in noisy family groups, giving nasal 'teeer'. Feeds on small insects and fruit, the latter usually collected in a brief hovering flight. **Range and habitat:** NT, Mexico to Argentina. Common bird of lowland forest borders, secondary vegetation and gardens. **Similar species:** Rusty-margined Flycatcher.

Great Kiskadee *Pitangus sulphuratus* (Tyrannidae) 22 cm (8½ in). Large, noisy, rather stocky flycatcher. Quite a heavy black bill. Crown black with concealed yellow centre. Broad white supercilium completely encircling head and blackish ear-coverts. White throat, rest of underparts yellow. Brown upperparts with rufous edgings to wing and tail feathers. Perches in open, making aerial sallies or swoops to ground to collect food. Boisterous manner, calling its name triumphantly 'Kiska-dee'. **Range and habitat:** NT and NA, southern USA to Argentina. Common in tropical and subtropical zones in forest borders, secondary vegetation and gardens. **Similar species:** Lesser Kiskadee, Boat-billed Flycatcher.

Greyish Mourner *Rhytipterna simplex* (Tyrannidae) 20 cm (8 in). Large flycatcher. Plumage generally grey with slight brown tone to wings and tail. Usually solitary, but sometimes joins mixed species flocks, perched inconspicuously on branch at mid-level or subcanopy. Makes short flights to snatch insects from nearby foliage. Has loud song consisting of rising series of notes which terminate abruptly. **Range and habitat:** NT, Venezuela to southern Brazil. Uncommon in lowland humid forest. **Similar species:** Screaming Piha.

Guianan Cock-of-the-rock

Andean Cock-of-the-rock

Orange-collared Manakin

White-bearded Manakin

Fork-tailed Flycatcher

Wire-tailed Manakin on display perch

Black and White Monjita

Tropical Kingbird

Social Flycatcher

Great Kiskadee

Greyish Mourner

Royal Flycatcher *Onychorhynchus coronatus* (Tyrannidae) 16.5 cm (6½ in). Medium-sized, slender flycatcher with hammer-headed appearance at rest. Long black bill, plumage generally dull brown, more rufous on tail and pale tips to wing-coverts. Paler below with whitish chin and dark scaling on breast. Spectacular crest forms brilliant fan of red feathers (more orange in females) tipped inky blue; it is rarely seen fully expanded in field and its function is not fully understood. Usually solitary and inconspicuous at low levels. Call a low disyllabic whistle. **Range and habitat:** NT, southern Mexico to south-east Brazil. Uncommon in lowland humid forest. **Similar species:** Northern Royal Flycatcher sometimes treated as separate species.

Grey-crowned Flycatcher *Tolymomyias poliocephalus* (Tyrannidae) 11.5 cm (4½ in). Small flycatcher with pale eyes and blackish bill. Crown greyish, rest of upperparts bright olive-green with yellow edgings to wing feathers. Underparts yellowish green, slightly smudged olive on breast. Active at mid-level and subcanopy, sometimes in mixed species flocks, searching for small insects in foliage. Call a wheezy whistle. **Range and habitat:** NT, Venezuela to south-east Brazil. Uncommon in lowland humid forest and forest edge. **Similar species:** other members of same genus.

Rufous-tailed Flatbill *Ramphotrigon ruficauda* (Tyrannidae) 16 cm (6½ in). Medium-sized, large-headed flycatcher with wide-based bill. Dark olive-green above with rufous wing edgings and tail. Yellowish underparts with heavy olive diffuse streaking, buff vent. Usually solitary or in pairs, sometimes in mixed species flocks, perched at low levels. Remains motionless on perch before making short flight to new branch. Call consists of 2 or 3 drawn-out whistles. **Range and habitat:** NT, Amazonia. Uncommon in humid forest. **Not illustrated.**

Spotted Tody Flycatcher *Todirostrum maculatum* (Tyrannidae) 9.5 cm (4 in). Very small flycatcher with long tail and long flattened bill. Grey head with prominent yellow eye and whitish lores. Olive-green above with yellow wing edgings. Throat whitish with fine black spot-like streaks, rest of underparts yellowish. Solitary or in pairs, keeping low down in foliage, usually detected by loud 'peep' call. **Range and habitat:** NT, Venezuela to northern Bolivia. Common in lowland secondary vegetation and river-banks. **Similar species:** Johannes' Tody Tyrant.

Large Elaenia *Elaenia spectabilis* (Tyrannidae) 18 cm (7 in). Medium-sized flycatcher. Crown, sides of head and upperparts grey-olive. Throat and upper breast medium grey, contrasting with yellowish grey underparts. 3 whitish wing-bars. Usually solitary, perched in open situations at low to mid-levels. Call is soft 'wheeoo'. **Range and habitat:** NT, breeds in Brazil and northern Argentina, migrating to Amazonia during austral winter. Occurs in secondary vegetation, gardens and clearings, often close to rivers on wintering quarters. **Similar species:** Small-billed Elaenia.

Sooty-headed Tyrannulet *Phyllomyias griseiceps* (Tyrannidae) 10 cm (4 in). Small flycatcher with stubby black bill. Rather bushy cap is dusky, obviously darker than olive upperparts. Darker wings, flight feathers with pale edgings but no wing-bars. Narrow whitish supercilium. Underparts tinged pale olive on breast, cleaner yellow below. Usually solitary at mid-level to canopy, actively searching foliage for small insects. Call a short trill. **Range and habitat:** NT, Panama to central Peru. Fairly common at low and mid-elevations in woodland, shrubby areas and forest borders. **Similar species:** several similar tyrannulets.

White-winged Becard *Pachyramphus polychopterus* (Tyrannidae) 14 cm (5½ in). Medium-sized flycatcher, showing close affinities to cotingas. Male grey with black crown, mantle, tail and wings or all-black according to race. Wing-coverts, secondaries, scapulars and tail broadly edged white. Female quite different with olive-brown upperparts, pale below with whitish spectacles and buff wing and tail markings. Usually in pairs and often in mixed species flocks

at mid-level or subcanopy. Song a series of descending soft whistles. **Range and habitat:** NT, Central America to northern Argentina. Common in lowland and mid-elevation secondary vegetation and woodland. **Similar species:** several becards.

Masked Tityra *Tityra semifasciata* (Tyrannidae) 20 cm (8 in). Rather stout, heavy-billed flycatcher, showing close affinities to cotingas. Red bare facial skin and bill with dusky tip. Male mainly white with front of head black, black wings and tail (tipped white) and with greyish tone to back. Female has brown crown and mantle. Usually in pairs, hopping heavily in canopy and reaching out to collect fruit. Also makes short flights to seize fruit or insect prey from vegetation. Call a dry croak. **Range and habitat:** NT, Mexico to Argentina. Common at low and mid-elevations in forest edge and woodland. **Similar species:** Black-tailed Tityra.

Sharpbill *Oxyruncus cristatus* (Oxyruncidae) 16.5 cm (6½ in) Rather curious bird with sharp pointed bill and reddish eye. Black crown with concealed red crest. Olive-green above, pale yellowish below. Black scaling on face and throat, rest of underparts heavily spotted black. Generally solitary in canopy and subcanopy, searching in acrobatic fashion for insects, using sharp bill to poke into leaf clusters and buds. Also takes fruit. Males gather in small leks to produce trilling song. **Range and habitat:** NT, patchily distributed from Central America to Paraguay in foothill montane forest.

Rufous-tailed Plantcutter *Phytotoma rara* (Phytotomidae) 18 cm (7 in). Finch-like with stout bill. Male has rufous crown and streaked brown underparts. Blackish wings have white bar across median coverts and buff edgings to inner wing. Tail has black subterminal band and central feathers, with rufous on inner webs and rest of feathers, conspicuous in flight. Underparts pale rufous. Female is streaked brownish with buffy wing-bars. Usually in pairs or small groups, perching in exposed places such as telegraph wires and tops of bushes. Feed on fruit, buds and leaves and can cause damage in orchards. Has a croaking call. **Range and habitat:** NT, southern Chile and Argentina. Low to mid-elevation in scrub, orchards and gardens. **Similar species:** other plantcutters.

White-winged Swallow *Tachycineta albiventer* (Hirundinidae) 14 cm (5½ in). Medium-sized swallow. Bluish green above with white rump and patch on inner wing. Underparts white. Immatures are greyer. Usually seen in pairs flying low over water or perched on emergent bare branches. Nests in cavities close to water. Call a short trill. **Range and habitat:** NT, Venezuela to Argentina. Common in lowland lakes and large rivers. **Similar species:** Mangrove, White-rumped and Chilean Swallows.

Blue and White Swallow *Notiochelidon cyanoleuca* (Hirundinidae) 13 cm (5 in). Small slender swallow. Dark blue above, white below with black undertail-coverts. Immature browner above with buffish underparts. Often in small groups, with darting, flittering flight. Often seen perched on telegraph wires. Nests in crevices under roofs, in trees or cliffs or excavates hole in bank. Song a thin twittering trill. **Range and habitat:** NT, Central America to Tierra del Fuego. Common at mid-elevations to treeline in open areas and villages. Southern populations are migratory. **Similar species:** Tree Swallow (see p 78), Pale-footed Swallow.

Royal Flycatcher

Grey-crowned Flycatcher

Large Elaenia

Spotted Tody Flycatcher

Sooty-headed Tyrannulet

White-winged Becard

Masked Tityra

Sharpbill

Rufous-tailed Plantcutter

White-winged Swallow

Blue and White Swallow

Azure-hooded Jay *Cyanolyca cucullata* (Corvidae) 29 cm (11½ in). Small jay with stiffened feathers on forehead. Head, upper back and breast black apart from light blue crown and nape. Rest of plumage dark blue. Usually in small, noisy groups moving through trees, searching in epiphytes and crevices for large invertebrate prey. Call a loud 'jeet-jeet'. **Range and habitat:** NT, Central America. Uncommon in mid-elevation mountain forest and forest edge. **Similar species:** White-collared Jay.

Plush-crested Jay *Cyanocorax chrysops* (Corvidae) 35 cm (14 in). Large jay with rather stiff bushy crest. Head and breast black with yellow eye and bright blue patches around eye. Nape whitish blue, rest of upperparts violet-blue with white tip to tail. Underparts creamy white. Usually in small noisy groups, sometimes with other species of jay, rather inquisitive. Calls include loud 'cho-cho-cho'. **Range and habitat:** NT, Brazil to northern Argentina. Common at low and mid-elevations in forest and woodland. **Similar species:** White-naped and Black-chested Jays.

Thrush-like Wren *Campylorhynchus turdinus* (Troglodytidae) 20 cm (8 in). Large long-tailed wren. Dull brown above with narrow whitish supercilium. Underparts white, heavily spotted dark brown. Southern race unmarked below and greyer above. Family parties clamber about in epiphyte clumps and vine tangles in subcanopy, peering downwards and probing into detritus. Song, often given in duet, a loud repeated 'chooka-cookcook'. **Range and habitat:** NT, western and southern Amazonia, south-west and eastern Brazil. Lowland humid forest and forest border.

Moustached Wren *Thryothorus genibarbis* (Troglodytidae) 15 cm (6 in). Medium-sized wren. Greyish brown crown and nape. Rest of upperparts warmer brown, unmarked wings, tail brown with bold dark bars. White supercilium, ear-coverts streaked black and white, white malar, black submalar and white throat. Rest of underparts dull buff. Usually in pairs in undergrowth. Song, often delivered as duet, a series of rapid phrases. Andean populations have more liquid song and may be separate species. **Range and habitat:** NT, southern Amazonia to eastern Brazil, lowlands and foothills in bamboo thickets and forest undergrowth. Andean population south to Ecuador occurs at mid-elevations in forest borders and old clearings. **Similar species:** Coraya Wren.

Scaly-breasted Wren *Microcerculus marginatus* (Troglodytidae) 11.5 cm (4½ in). Small, terrestrial wren. Long bill and short, cocked tail. Upperparts dark, greyish brown, white underparts with flanks and vent brownish, faintly or boldly scaled black. Walks furtively, crake-like, on forest floor and over fallen logs. Memorable song consists of opening flourish followed by tantalizing series of whistles, descending in pitch and with increasing intervals. **Range and habitat:** NT, Amazonia. Common at low and mid-elevations in humid forest. **Similar species:** several, sometimes all considered to be races of single species, the Nightingale Wren.

Black-capped Mockingthrush *Donacobius atricapillus* (Troglodytidae) 22 cm (8½ in). Striking slender bird with long mobile tail. Upperparts dark brown with yellow eye. White wing flash and tips to tail. Underparts creamy with dark scaling on flanks. Usually seen in pairs, perched on tops of waterside grassy vegetation or shrubs. Often set up antiphonal duets, with pair facing each other, tails wagging from side to side. Other calls include loud whoops. **Range and habitat:** NT, Panama to northern Argentina. Common in lowlands beside rivers and lakes. **Similar species:** Bicoloured Wren.

Black-billed Thrush *Turdus ignobilis* (Turdidae) 24 cm (9½ in). Dark thrush with blackish bill. Dull brown plumage with whitish throat, streaked black. Some races have white crescent on lower throat. Underparts pale greyish brown becoming whiter on centre of belly. Usually solitary, seen hopping on open ground or perched on telegraph wires or roofs. Song a subdued series of fluty notes. **Range and habitat:** NT, Colombia to Bolivia. Common in lowland and mid-elevations in forest borders, secondary vegetation, open areas and gardens. **Similar species:** several similar thrushes.

Lawrence's Thrush *Turdus lawrencii* (Turdidae) 23 cm (9 in). Large thrush with bright orange-yellow bill and eye-ring. Dark brown above, paler below with pale throat streaked black and whitish belly and vent. Female has blackish bill and weaker eye-ring. Feeds on ground but also visits fruiting canopy trees. Song, delivered from canopy, shows remarkable mimicry of wide range of forest species. Will sing from 1 perch, then fly away to continue from another and can be very difficult to see. **Range and habitat:** NT, western Amazonia. Fairly common in lowland humid forest. **Similar species:** Pale-vented and Bare-eyed Thrushes.

Cocoa Thrush *Turdus fumigatus* (Turdidae) 23 cm (9 in). Large thrush with brown bill. Plumage generally uniform rufous brown. Underparts slightly paler with pale streaked throat and buffy centre of belly and vent. Usually in pairs, usually on or close to ground, sometimes higher. Rather shy. Gives rich, musical song consisting of repeated phrases. **Range and habitat:** NT, Venezuela to central Brazil, southeast Brazil and Caribbean. Mainly in lowlands, in humid forest and wooded areas. **Similar species:** Pale-vented Thrush. **Not illustrated.**

Clay-coloured Thrush *Turdus grayi* (Turdidae) 23 cm (9 in). Thrush with yellowish bill and reddish eye. Dull olive-brown above, buffier below with paler throat, lightly streaked. Immatures, as in most thrushes, retain some pale-tipped wing-coverts from juvenile plumage. Forages on ground, searching in leaf litter for earthworms and other small prey. Also visits fruiting trees. Pleasant musical song is rather varied with trills, whistles and warbling notes. **Range and habitat:** NT, Mexico to Colombia. Fairly common in low to mid-elevations in gardens, secondary vegetation and, especially outside breeding season, in forest. **Similar species:** Pale-breasted Thrush.

Tropical Gnatcatcher *Polioptila plumbea* (Polioptilidae) 11.5 cm (4½ in). Small bird with long, slender tail and fine bill. Male has glossy black cap, grey above with blackish wings edged white and black tail with outer tail feathers white. Underparts whitish, smudged grey on breast. Female lacks black cap, apart from in Peruvian race when similar to male. Engagingly active bird, often in mixed species flocks, which constantly twitches its tail whilst searching for insects on tips of small twigs. Call a nasal mew, song a series of thin, high-pitched notes. **Range and habitat:** NT, Mexico to Peru and eastern Brazil. Common at low and mid-elevation forest edge, secondary growth and drier woodland. **Similar species:** Slate-throated, Guianan and Blue-grey Gnatcatchers (see p 82).

Yellowish Pipit *Anthus lutescens* (Motacillidae) 13 cm (5 in). Slender, small terrestrial bird with thin bill. Brown above with buff and dusky streaks. Tail blackish with white outer tail feathers. Narrow eye-ring. Underparts washed whitish yellow with brown streaks on breast. Walks on ground and difficult to see unless in flight. Usually occur in loose groups. Sings in flight, starting with series of 'tsits', then glides back to ground with long 'zeeeeeeeeeu'. **Range and habitat:** NT, Panama, eastern South America and Pacific coast of Peru. Common in lowland grassland. **Similar species:** Chaco Pipit.

Azure-hooded Jay

Plush-crested Jay

Thrush-like Wren

Moustached Wren

Lawrence's Thrush (immature)

Scaly-breasted Wren

Black-capped Mockingthrush

Black-billed Thrush

Clay-coloured Thrush

Tropical Gnatcatcher (female)

Yellowish Pipit

Rufous-browed Peppershrike *Cyclarhis gujanensis* (Vireonidae) 14-15 cm (5½-6 in). Large, stocky vireo with heavy bill. Olive-green above, greyer on crown, with rufous supercilium and forehead, and grey cheeks with brownish eyes. Underparts yellow with white centre to belly and vent. Plumage rather variable between races. In pairs, often in mixed species flocks, rather sluggish, keeping under cover of foliage from mid-level to canopy. Short, musical song frequently repeated. **Range and habitat:** NT, Mexico to Argentina. Fairly common at low and mid-elevations in forest borders and secondary vegetation, mainly in drier areas. **Similar species:** Black-billed Peppershrike.

Rufous-crowned Greenlet *Hylophilus poicilotis* (Vireonidae) 12.5 cm (5 in). Small, warbler-like vireo. Distinctive plumage with rufous crown, extending down to nape, olive-green upperparts, whitish face and throat with fine blackish streaking on ear-coverts (depending on race). Rest of underparts yellowish with buffy tinge on breast. Often in mixed species flocks, mainly foraging at mid-level. Has loud, ringing song. **Range and habitat:** NT, south-east Brazil, northern Argentina and Paraguay. Common in low and mid-elevation humid forest, secondary vegetation and scrub.

Crested Oropendola *Psarocolius decumanus* (Icteridae) 33-43 cm (13-17 in). Large oropendola with white bill. Plumage essentially black with chestnut on rump and vent and all but central tail feathers yellow. At close quarters, wire-like crest of male visible. Usually solitary, but sometimes in flocks with other oropendolas; breed in colonies. Spectacular display by male, seemingly about to topple from perch while producing extraordinary series of loud gurgling and tearing sounds. **Range and habitat:** NT, Panama to northern Argentina. Low to mid-elevation humid forest and partially cleared areas. **Similar species:** Chestnut-headed Oropendola.

Montezuma Oropendola *Psarocolius montezuma* (Icteridae) 38-50 cm (15-20 in). Very large oropendola with black bill, tipped orange. Black head and chest, otherwise deep chestnut with black tips to flight feathers and yellow tail, apart from black central tail feathers. Bare facial skin blue, pinkish at base of bill. Usually in small groups in canopy, searching for invertebrates in crevices and also feeding on fruit. Nests in colonies in isolated trees. Male display involves deep bowing, producing loud liquid gurgling call with accompanied screeches. **Range and habitat:** NT, Central America. Common in lowlands and foothills in forest borders and cultivated areas. **Similar species:** Chestnut-mantled and Olive Oropendolas.

Yellow-rumped Cacique *Cacicus cela* (Icteridae) 24-8 cm (9½-11 in) Medium-sized icterid with whitish bill and eyes. Mainly black with yellow rump, base of tail, vent and flash on upperwing. Often in noisy groups, but sometimes solitary. Forages from canopy to low levels, searching for invertebrates and fruit. Nests colonially, invariably in isolated tree close to human habitation or surrounded by water, often associated with wasp nests. Several loud barking calls. Song of nominate race is highly mimetic. **Range and habitat:** NT, Panama to central Brazil. Common in lowlands in forest edge and partially cleared areas. **Similar species:** Band-tailed Oropendola.

Troupial *Icterus icterus* (Icteridae) 23 cm (9 in). Small icterid with pale eye. Rather variable, some races may be separate species. Essentially orange and black. Black bib always appears to have ragged border with orange underparts. Head wholly black or with orange crown, back may be black or orange depending on race. Usually some white on wing. Normally in pairs, foraging mainly at low levels. Musical song consists of several long fluid whistles delivered from high perch, usually broken off by flight. Takes over nests of other species. **Range and habitat:** NT, Colombia to Paraguay. Common in lowland forest edge, woodland and scrub. **Similar species:** several orioles (see p 90).

Giant Cowbird *Scaphidura oryzivora* (Icteridae) 28-36 cm (11-14 in). Medium-sized icterid with small-headed, bull-necked appearance. Male wholly glossy black. Female slimmer, blackish brown with pale eye. Has undulating flight and long-tailed, heavy-bodied profile is characteristic. Usually alone or in small groups, walking on ground or in attendance at an oropendola or cacique colony, the nests of which are parasitized. Apart from some grating flight calls, mainly silent. **Range and habitat:** NT, southern Mexico to northern Argentina. Uncommon at low and mid-elevations in forest borders, cleared areas and river banks. **Similar species:** Great-tailed Grackle.

Red-crowned Ant-tanager *Habia rubica* (Emberizidae: Thraupinae) 18 cm (7 in). Medium-sized tanager with blackish bill. Male reddish brown with inconspicuous scarlet crown patch, bordered by black. Paler and rosier below, with brighter throat. Female olive-brown above with orange-yellow crown patch. Dull olivaceous buff below, cleaner on throat. Family parties are noisy, chattering components of low-level mixed species flocks. **Range and habitat:** NT, Mexico to northern Argentina. Fairly common in lowland humid forest. **Similar species:** Red-throated Ant-tanager.

Silver-beaked Tanager *Ramphocelus carbo* (Emberizidae: Thraupinae) 18 cm (7 in). Medium-sized tanager with bold, swollen silvery white lower mandible. Male mainly black with velvety maroon tone to upperparts and tinged dark crimson on throat and chest. Female has duller bill, with reddish brown plumage and redder tone on rump. Immature similar to female. Noisy conspicuous groups move through bushes, feeding on fruit and insects. Call a loud, metallic 'chink'. **Range and habitat:** NT, common in lowlands east of Andes, south to southern Brazil. Forest borders and secondary vegetation. **Similar species:** female similar to Red-crowned Ant-tanager.

Blue-grey Tanager *Thraupis episcopus* (Emberizidae: Thraupinae) 16.5 cm (6½ in). Medium-sized tanager. Pale bluish grey head and underparts which contrast with darker blue back, wings and tail. Lesser wing-coverts bright blue or white with white wing-bar (depending on race). Usually in small groups which forage actively in treetops and shrubs. Has rather squeaky call. **Range and habitat:** NT, Mexico to central South America. Very common at low and mid-elevation in wide range of habitats from humid forest borders and open woodland to gardens. **Similar species:** Sayaca Tanager.

Palm Tanager *Thraupis palmarum* (Emberizidae: Thraupinae) 16.5 cm (6½ in). Rather dingy, greyish olive medium-sized tanager. Darker and glossier above, forehead tinged yellowish. Most obvious feature is 2-toned wing with paler wing-coverts contrasting with flight feathers which, apart from bases, are blackish. Pairs or small groups often seen in palms or tops of other trees, searching for insects in acrobatic fashion. Call rather squeaky. **Range and habitat:** NT, Central America to southern Brazil. Common in lowlands in forest borders, shrubby vegetation and settled areas. **Similar species:** Golden-chevroned Tanager.

Thick-billed Euphonia *Euphonia laniirostris* (Emberizidae: Thraupinae). 11 cm (4½ in). Small tanager with stubby bill and short tail. Forecrown and underparts yellow, rest of upperparts blackish blue with white spots on outer 2 tail feathers. Female is olive-green with yellower tone below. Usually in pairs or family parties, sometimes in mixed species flocks of frugivores at mid-level to canopy. A remarkable mimic of other species' calls. **Range and habitat:** NT, Costa Rica to Brazil. Common in lowland woodland, secondary growth and forest edge. **Similar species:** several euphonias.

Rufous-browed
Peppershrike

Rufous-crowned
Greenlet

Crested
Oropendola

Montezuma
Oropendola

Yellow-rumped
Cacique

Troupial

Giant Cowbird

Red-crowned
Ant-tanager

Silver-beaked
Tanager (female)

Blue-grey
Tanager

Palm Tanager

Thick-billed
Euphonia

Paradise Tanager *Tangara chilensis* (Emberizidae: Thraupinae) 15 cm (6 in). Multi-coloured gem. Apple-green head with black on nape and around eye. Black upperparts with bright red or red and orange rump. Turquoise wing-coverts and underparts, more violet on throat and black centre of belly and vent. Common in mixed species canopy flocks, visits fruiting trees and actively searches branches with epiphytes. Call a high-pitched twitter. **Range and habitat:** NT, Amazonia. Lowland humid forest and forest borders.

Opal-rumped Tanager *Tangara velia* (Emberizidae: Thraupinae) 14 cm (5½ in). Small dark tanager. Black upperparts with opal-coloured lower back and rump and shining blue uppertail-coverts. Wings, head and underparts purplish blue with chestnut belly and vent. Some races have pale blue forehead and eastern Brazilian race has pale blue, not purple-blue, on underparts. Usually in mixed canopy flocks with other tanagers. **Range and habitat:** NT, Amazonia and eastern coast of Brazil. Fairly common in lowland humid forest and forest edge. **Similar species:** Opal-crowned Tanager.

Black-faced Dacnis *Dacnis lineata* (Emberizidae: Thraupinae) 11.5 cm (4½ in). Small honeycreeper with short, pointed bill. Black mask contrasting with yellow eye and blue crown. Rest of upperparts black with blue rump, scapulars and edges to tertials. Underparts turquoise-blue with white belly and vent. Western race has yellow replacing white, also evident as tuft in front of wing. Female drab olivaceous, paler below with yellow belly in western race. Usually in pairs and often in mixed species flocks with other tanagers. Searches foliage for small insects and visits fruiting trees. **Range and habitat:** NT, Amazonia and northern Andean valleys and foothills. Common in lowland humid forest and secondary growth. **Similar species:** Turquoise Dacnis.

Bananaquit *Coereba flaveola* (Coerebidae) 10 cm (4 in). Very small bird with fine, decurved bill. Blackish brown above with bold white supercilium, white speculum and tips to outer tail. Yellow rump and underparts apart from grey throat and whitish vent. All-black forms occur on some Caribbean islands. Active, highly energetic bird usually alone or in pairs. Can be very tame. Visits flowers for nectar and takes fruit from low levels to canopy. Call a shrill twitter. **Range and habitat:** NT, Mexico to northern Argentina, throughout Caribbean. Low to mid-elevation, locally abundant in gardens, secondary vegetation and woodland.

Buff-throated Saltator *Saltator maximus* (Emberizidae: Cardinalinae) 20 cm (8 in). Large finch with heavy bill. Bright olive-green above with greyer head and short white supercilium. Black malar and white chin, bordered buffy below. Underparts greyish olive with belly and vent buff. Usually solitary or in pairs and rather wary. Mainly keeps to low or mid-levels, visiting fruiting trees or shrubs. Pleasant, repetitive song is thrush-like. **Range and habitat:** NT, Mexico to southern Brazil. Common in low and mid-elevation secondary vegetation and wooded areas. **Similar species:** Greyish Saltator.

Yellow Cardinal *Gubernatrix cristata* (Emberizidae: Cardinalinae) 20 cm (8 in). Large, boldly marked, crested finch. Male has black-fronted crest and throat, yellow supercilium and malar. Otherwise olive-green above with blackish centres to back feathers, yellower below. Female greyer with white replacing yellow on face. In pairs and small groups, perched in open on shrubs. Feeds mainly on ground. Attractive musical song. **Range and habitat:** NT, southern Brazil, Uruguay and northern Argentina. Rare in lowland woodland and scrub. Threatened by trapping for pet trade.

Blue-black Grosbeak *Cyanocompsa cyanoides* (Emberizidae: Cardinalinae) 16 cm (6½ in). Medium-sized, heavy-billed finch. Plumage of male wholly deep dark blue, rather brighter on forehead and around black bill. Female dark chocolate brown. Usually in pairs which remain furtively in undergrowth. Call a metallic 'chink'. Song a rich jumble of ascending and descending notes. **Range and habitat:** NT, Mexico to Bolivia. Fairly common, mainly in lowlands, in forest edge and secondary woodland. **Similar species:** Ultramarine Grosbeak.

Sooty Grassquit *Tiaris fuliginosa* (Emberizidae: Emberizinae) 11.5 cm (4½ in). Small finch with conical dark bill, pinkish on gape. Male uniformly dull blackish. Female drab olive-brown above, paler below. Usually in pairs or small groups, not associated with other species, foraging in grass or low down in shrubs. Builds grass domed nest, close to ground. Song a thin, high-pitched short series of notes. **Range and habitat:** NT, locally from Venezuela and Trinidad to eastern Brazil. Uncommon in lowland forest borders, drier scrub and woodland. **Similar species:** female in particular may be confused with several small finches.

Yellow-bellied Seedeater *Sporophila nigricollis* (Emberizidae: Emberizinae) 11.5 cm (4½ in). Small finch. Male has pale bill, black hood, dark olive-green above, sometimes with inconspicuous white speculum on wing, and whitish yellow below. Female has dark bill and is olive-brown, paler and warmer below. Often in small flocks mixed with other small finches. Usually on or close to ground, although male often perches high on telegraph wires. Gives short, musical song. **Range and habitat:** NT, Central America to southern Brazil. Common at low and mid-elevations in open grassy areas, woodland and forest edge. **Similar species:** Black and White and Dubois' Seedeaters.

Saffron Finch *Sicalis flaveola* (Emberizidae: Emberizinae) 14 cm (5½ in). Small bright yellow finch. Male almost wholly yellow with orange crown. Upperparts slightly olive-tinged, with faint streaking. Female similar but somewhat duller. Immature pale greyish brown, streaked above, with yellow restricted to collar and vent. Usually in flocks, foraging on ground or perched in small trees or bushes. Has musical, repetitive song. **Range and habitat:** NT, patchily distributed throughout South America and introduced to Panama and Jamaica. Common in lowland savannah, parks and gardens. **Similar species:** Orange-fronted Yellow Finch.

Grey and White Warbling-finch *Poospiza cinerea* (Fringillidae) 13 cm (5 in). Small finch. Grey upperparts with glossy black cap or dark grey cheeks (depending on race). White underparts. Flight feathers edged grey and outer tail feathers mostly white. Immature blackish only on sides of head. Usually in small groups, often in mixed species flocks, actively searching for insects in foliage. Rather inquisitive. Has chipping call. **Range and habitat:** NT, Bolivia to northern Argentina (*melanoleuca*) and southern Brazil (*cinerea*). Common at mid-elevation in woodland and shrubby areas.

Great Pampa Finch *Embernagra platensis* (Fringillidae) 20.5-23 cm (8-9 in). Large, stocky finch with orange bill with dark culmen. Dull olive-grey upperparts, slightly greyer on rump. Fine blackish streaking on crown and back (or unstreaked depending on race), yellowish wing edgings. Grey face, duskier around eye, and underparts with buffy flanks and whitish belly. Immature heavily streaked. Usually in pairs, sometimes in small groups, foraging on ground close to cover, or perched on fence-posts or bushes. Rather jerky flight with broad tail slightly cocked. Has short, spluttering song. **Range and habitat:** NT, southern Brazil to northern Argentina. Common from lowlands to mountains in cultivated areas, grasslands and marshes. **Similar species:** Buff-throated Pampa Finch. **Not illustrated.**

Paradise Tanager

Opal-rumped Tanager

Black-faced Dacnis

Bananaquit

Buff-throated Saltator

Yellow-bellied Seedeater

Sooty Grassquit

Yellow Cardinal

Saffron Finch

Blue-black Grosbeak

Grey and White Warbling-finch

THE PALAEARCTIC REGION

The Palaearctic Region comprises all of Europe, those parts of Africa north of the Sahara desert and all of Asia north of the Himalayas. Among offshore islands and island groups included in the region are Iceland, the Faroes, the Azores, the Canaries, the Selvagens and the Cape Verde Islands. The northern boundary is formed by the Arctic ocean, to the west lies the Atlantic, while the eastern boundary is the Pacific ocean. To the south, the boundaries with the Afrotropical and Oriental Regions are less well defined, though there is a nearly continuous chain of mountainous regions from the Atlas in the west to the Himalayas in the east, to the north of which lies a more or less continuous line of deserts.

Within the vast extent of this region there is, naturally enough, great diversity of geography, topography, climate and vegetation. The last great influence on all of these was the sequence of at least four Pleistocene glaciations, the final one of which ended some 10,000 years ago. The climatic regime now affecting the Palaearctic is still postglacial, though it has been described as interglacial.

During the last glaciation, ice-sheets extended quite well south in the western half of the region, but hardly at all in the eastern. The effect of the glaciation was to push the major climatic zones some thousands of miles southwards, thus completely altering the situation in the western Palaearctic, though much less so in the eastern. As the ice retreated there was a steady expansion northwards and north-westwards of the birds which had been forced south and south-east by the onset of the glaciation. This expansion is still going on with several examples of species extending their ranges during this century, the Collared Dove being the most spectacular.

As mentioned above, the boundaries between the Palaearctic and the Afrotropical and Oriental Regions are relatively poorly defined and consequently there are occurrences in the extreme south of the Palaearctic Region of what are really species of the latter more southerly regions, such as the sandgrouse. However, to the west the Atlantic ocean has formed a very effective barrier to avian spread, with no known colonizations from the Nearctic, though the Spotted Sandpiper has bred just once, and this despite hundreds of vagrants crossing from west to east every year. The Pacific ocean is also a complete barrier except in the far north, where the chain of islands stretching between Siberia and Alaska has allowed some species to colonize in either direction.

The enormous size and diversity of the region demands some method of subdivision so that the main climatic and vegetational zones can be described. An accepted scheme is through the concept of faunal type, with from five to seven zones designated. Here five have been distinguished, running from north to south, namely Arctic, boreal, temperate, Mediterranean and desert.

The northern latitudes comprising the Arctic zone are characterized by a very short summer during which large numbers of comparatively few species of birds breed, and very long winters, often with extremely cold temperatures, making the area hardly suitable for any bird life. Thus, virtually all breeding birds of the region are migrants.

Much of the Arctic zone has comparatively low precipitation. Vegetation is classified as tundra, with few if any plants reaching 50 cm (20 in) in height. Trees and shrubs are restricted to creeping

varieties. The underlying permafrost greatly impedes drainage, leading to the formation of extensive marshes, in part compensating for the low precipitation. Such marshes provide nesting and feeding areas for many birds. Among these are the Brent and White-fronted Geese, King Eider, Sanderling, Knot, Grey Phalarope and Glaucous Gull. Land birds are scarce but include the Arctic Redpoll and Snowy Owl.

The more southerly part of the Arctic zone includes a region of greater precipitation and less cold temperatures. Here there are

often patches or larger extents of scrub, growing fairly high and thus providing nest shelter and food for a greater variety of birds. Marshy areas also abound. Characteristic birds include the Willow Grouse, Lapland Bunting and Spotted Redshank.

The seas in the Arctic zone are generally shallow and fertile and, in summer, support vast numbers of seabirds which nest on coastal cliffs, including auks, the Kittiwake and the Fulmar. The conditions of weather and vegetation found on the upper regions of the higher mountains in more southerly latitudes show distinct similarities to those in the Arctic and for this reason are usually included in that zone. Some breeding birds undertake a purely vertical migration between summer and winter, breeding on the high tops and wintering lower down; examples are the Ptarmigan and Snow Bunting.

To the south of the Arctic zone lies a broad continuous belt of coniferous forest comprising the boreal zone. The northern limit of trees merges with the low Arctic scrub, with maximum

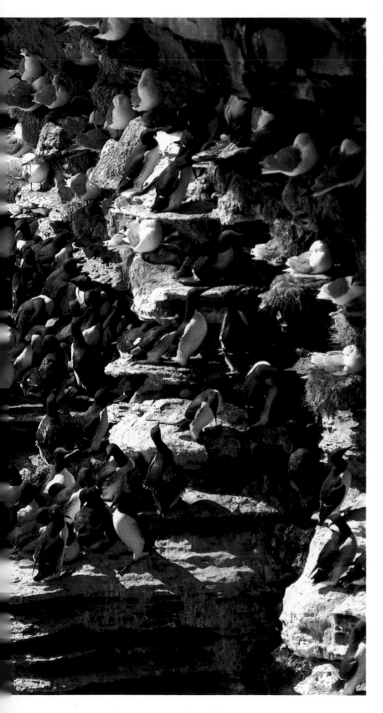

penetration of trees northwards along more sheltered river valleys. Clearings and river valleys within the forest zone are frequently wet and boggy, or filled with lakes, providing a sheltered habitat for wetland birds in addition to the forest birds which occur. Birds of this zone can be residents or migrants. They include the Smew, Spotted Redshank and Wood Sandpiper around the marshy pools, and the Goshawk, Redwing, Brambling and Siskin in the forests.

Small areas similar to the boreal zone occur around the lower slopes of the larger mountain massifs further south. The Capercaillie is a typical species.

The temperate zone was formerly a zone of deciduous forests, once more or less continuous but now much reduced and fragmented by man, with the bulk of the cleared areas converted into agricultural land, though with considerable urbanization, too. This type of habitat is more or less confined to the western Palaearctic, with isolated pockets further east and also further south on higher ground. It reappears in the eastern Palaearctic in, for example, China. Species diversity is much greater than in the coniferous forests of the boreal zone. The Green Woodpecker, various tits, Robin, Blackbird and several different finches are characteristic groups found in deciduous woodland, while ground-nesting birds such as the Lapwing and the Skylark breed in the cleared areas.

Wetlands in the temperate zone support a great number of different breeding species including many ducks – such as Mallard and Teal – as well as Coot, Moorhen, Redshank and various gulls. Seabirds are numerous on coastal cliffs, especially the Gannet, Puffin and Guillemot. Also of vital importance to birds are the many estuaries providing feeding and wintering grounds for enormous numbers of mainly Arctic breeding waders such as the Knot, Dunlin and Ringed and Grey Plovers.

The more southerly latitudes of the Palaearctic Region are characterized by a much warmer, drier climate known as the Mediterranean zone. Here evergreen trees and shrubs flourish in some areas, while in others – for example, the Russian steppes – the natural vegetation is open grassland. Several species of warblers, such as the Subalpine and Dartford, are typical of this zone, as are larger woodland birds such as the Golden Oriole and Hoopoe. Wetlands hold the Purple Heron, Little Egret and Little Bittern and the drier areas attract bustards and larks.

The desert zone contains deserts which extend not quite continuously right across the southern part of the Palaearctic from the northern half of the Sahara, through those of northern Arabia and thence eastwards into Mongolia and the Gobi. Among typical species of this zone are wheatears, larks and sandgrouse.

The avifauna of the Palaearctic Region is estimated to contain just under 950 species of which about 450 are nonpasserines and 500 passerines. Because of the effect of the last glaciations, mentioned above, species diversity increases very markedly as one goes from west to east.

A total of 71 genera are confined to the Palaearctic, including such species as the Ruff, Pheasant, Long-tailed Tit and Chaffinch.

Ornithologists in the Palaearctic Region are well served by societies. Most countries have one or more devoted to birds, some to conservation, others to research, while in western Europe the hobby of bird-watching, usually linked to conservation, has enabled some societies to attract very large and influential memberships. Further information on birds in the UK can be obtained from the Royal Society for the Protection of Birds, The Lodge, Sandy, Bedfordshire SG19 2DL.

Seabird city, Orkney, Scotland. *Here cliff ledges provide nest sites for many thousands of seabirds. Each species has slightly different nesting requirements so that almost the entire cliff face is occupied.*

DIVERS Gaviformes

Black-throated Diver *Gavia arctica* (Gaviidae) 60-70 cm (24-8 in). Large, stream-lined diving bird with slender, pointed bill. In breeding plumage, head and nape soft grey, with black throat patch. Back black, finely patterned with white. Sides of throat and chest white, streaked black. Underparts white. Winter plumage grey-black above, whitish below. Loud, eerie, wailing calls in breeding season; deep 'kwuk-kwuk' in flight. **Range and habitat:** widespread circumpolar Holarctic. Breeds on large freshwater lakes in arctic and northern temperate zones. Winters in coastal waters within breeding range and migrates to southern temperate zone. **Similar species:** Red-throated Diver.

Red-throated Diver *Gavia stellata* (Gaviidae) 50-60 cm (20-4 in). Slender diving bird with thin upturned bill. Grey head and black and white streaked neck in breeding plumage, with patch of red on throat. Back dark grey, underparts whitish. In winter, overall pale grey above, whitish below. In flight, shows typical narrow wings of diver, head, neck and legs held below line of back. Wailing calls made in breeding season; quacking 'kuk-kuk-kuk' in flight. **Range and habitat:** widespread circumpolar Holarctic. Breeds freshwater, often tiny pools, in arctic and northern temperate zones. Winters in coastal waters from breeding range south to warm temperate zone. **Similar species:** Black-throated Diver.

GREBES Podicipediformes

Great Crested Grebe *Podiceps cristatus* (Podicipedidae) 48 cm (19 in). Largest grebe with long thin neck and fine, pointed bill. In summer plumage, highly distinctive black ear-tufts and chestnut and black plumes on side of head, raised in elaborate courtship display. In winter white face and neck, cap remaining blackish, and grey-brown body. Far-carrying barking call, variety of harsh crowing and moaning sounds making up song. **Range and habitat:** locally common breeder in fertile fresh and brackish waters in temperate Holarctic, also AF (few areas) and AU, wintering within and to south of range on freshwaters, estuaries and sheltered coasts. **Similar species:** Red-necked Grebe.

Little Grebe *Tachybaptus ruficollis* (Podicipedidae) 27 cm (10½ in). Smallest grebe of WP, dumpy, very short-necked with short, stout bill. Pale patch at gape obvious from distance. In summer, chestnut throat and cheeks distinctive, rest of upperparts dark brown, paler below. Winter plumage lacks chestnut, paler all over with whitish throat and cheeks. Song a shrill trill, like horse whinnying. Calls include sharp 'whit-whit'. **Range and habitat:** widespread breeder on small ponds, lakes and slow-moving rivers of temperate Holarctic, also AF and OR, wintering mainly within breeding range on freshwaters and estuaries. **Similar species:** Black-necked Grebe, Pied-billed Grebe (see p 50).

TUBE-NOSED SWIMMERS Procellariiformes

Fulmar *Fulmarus glacialis* (Procellaridae) 47 cm (18 in). White or grey-white gull-like shearwater, with typical gliding flight on stiff wings. Upperparts – back, wings and tail – grey (dark grey in 'blue' northern colour phase), underparts pure white (pale grey in 'blue' phase). Very distinctive yellow bill, short and thick with prominent tubed nostrils on upper mandible. Rarely seen on land except on nest-ledge. Mainly glides in flight, often soaring close to cliffs, rarely flapping. Call a hoarse, chuckling grunt. **Range and habitat:** extremely numerous. Breeds on cliff ledges in north temperate and arctic WP, eastern NA, Pacific coasts of NA and EP. Winters at sea.

Manx Shearwater *Puffinus puffinus* (Procellaridae) 37 cm (15 in). Medium-sized shearwater showing contrasting black upperparts and pure white underparts. Black cap extending to below eye. Black border to underside of wings. Glides on stiffly outstretched wings, turning one way then the other

low over waves, showing alternately black and white. Mediterranean races have dark brown upperparts and brown-washed underparts. Calls only at colony with wild caterwauling and crooning. **Range and habitat:** colonial nester in underground burrows on offshore islands in temperate and southern WP; other races in PA and AU. Winters at sea, WP birds reaching coasts of NT. **Similar species:** Cory's Shearwater.

Storm Petrel *Hydrobates pelagicus* (Hydrobatidae) 15 cm (6 in). Tiny petrel, smallest seabird in WP. Blackish overall with conspicuous white rump. Square-ended tail. Faint whitish wing-bar apparent close to. Small, slightly hooked black bill and very short black legs. Latter dangle and feet patter on surface during weak, fluttering flight over surface of sea. Only comes to land to breed. Silent at sea but loud churring call in nest-burrow, rising and falling and ending with characteristic 'hiccup'. **Range and habitat:** numerous colonial breeder in rock crevices on offshore islets of temperate and Mediterranean WP. Winters at sea south to southern AF. **Similar species:** Leach's Storm Petrel.

FULL-WEBBED SWIMMERS Pelecaniformes

(Eastern) White Pelican *Pelecanus onocrotalus* (Pelecanidae) 140-80 cm (55-71 in). Huge waterbird with long bill and extensible throat pouch. Body white with black tips to very broad white wings and black trailing edges underneath. Bill yellowish, legs and feet pink. In breeding plumage, slight crest to head and rosy tinge to plumage. Flies with alternate gliding and slow wingbeats, head hunched back into shoulders. Feeds in groups. Flies in lines or Vs. Harsh growling and grunting calls at nest. **Range and habitat:** nests in colonies in shallow freshwaters and deltas of southern WP and AF. Vulnerable. Some migrate south for winter. **Similar species:** Dalmatian Pelican, Pink-backed Pelican (see p 186).

Gannet *Morus bassanus* (Sulidae) 90 cm (35 in). Large, powerful seabird. Body and wings white, apart from extensive black wing tips and yellow tinge to head of adult. Dark brown juveniles moult through stages of increasing whiteness. Large, bluish dagger-like bill, dark legs and feet. In flight, body cigar-shaped, wings narrow and pointed, flaps with short glides. Feeds by spectacular vertical dive with half-closed wings. Harsh barking calls at nest. **Range and habitat:** many large colonies on cliffs and stacks, confined to northern Atlantic coasts of WP and NA. Winters at sea, including to south of breeding range. **Similar species:** Boobies (see pp 98, 286).

Cormorant *Phalacrocorax carbo* (Phalacrocoracidae) 90 cm (35 in). Large waterbird. Breeding adult black all over, glossed purple-blue, except for white cheeks, chin and thigh patch (last-named lost in winter). Other WP races have more white on throat and head. Long, hooked bill, grey above, yellowish below, with yellow skin at base. Juveniles brownish, paler beneath. Flies with outstretched long neck. Accomplished diver. Stands upright on rocks with wings characteristically half-extended. Gutteral calls at nest, otherwise silent. **Range and habitat:** Widespread breeder on coasts and large inland freshwaters in temperate Holarctic, eastern NA, OR, AF and AU. **Similar species:** Shag.

Shag *Phalacrocorax aristotelis* (Phalacrocoracidae) 76 cm (30 in). Medium-sized waterbird. Adult in breeding plumage black, glossed bluish green, with short upright crest at front of head. Long, hooked beak dark, paler at base. Crest lost in winter, when plumage browner and throat whitish brown. Juveniles brown, paler beneath. Flies with neck outstretched, dives easily. Stands fairly upright on rocks, often with wings half-outstretched. Only vocal at nest, with croaks and grunts, also hissing if approached. **Range and habitat:** widespread colonial breeder on cliffs and among boulders of virtually all WP coasts. Winters in adjacent maritime waters. **Similar species:** Cormorant.

■□□ **Black-throated Diver**
□■□ **Red-throated Diver**
□□■ **Great Crested Grebe** (pair)

■□□ **Little Grebe** with young on nest
□■□ **Fulmar**
□□■ **Manx Shearwater**

■□□ **Storm Petrel**
□■ **(Eastern) White Pelican**

■□□ **Gannet** nesting
□■□ **Cormorant** (pair) calling at nest
□□■ **Shag** courtship

HERONS AND ALLIES Ciconiiformes

Grey Heron *Ardea cinerea* (Ardeidae) 90 cm (35 in). Very tall, long-legged, long-necked wading bird with heavy, dagger-like bill. Soft blue-grey upperparts, whitish head, neck and underparts. Black crest and streaks down neck. Bill yellow and legs brown, both turning orangey red in spring. Slow, flapping flight on very broad wings with black flight feathers. Voice a loud, harsh 'fraank', only in flight. **Range and habitat:** widespread WP, EP, AF and OR, breeding in trees and reeds or on cliffs near marshes, lakes and estuaries. Some resident, others migrate and disperse, particularly to coasts. **Similar species:** Purple Heron.

Purple Heron *Ardea purpurea* (Ardeidae) 80 cm (31 in). Tall, long-legged and long-necked wading bird with grey upperparts, glossed purple, and rufous neck and chest. Slight black crest, dark streaking to head and neck. Some rufous tips to elongated wing plumes. Flies with slow, flapping flight of broad wings, showing very dark above, lacking contrast. Calls only in flight, a high-pitched 'fraank'. **Range and habitat:** southern temperate and subtropical WP, EP, AF and OR, in wetlands with dense vegetation. Winters mainly within breeding range, some dispersal. **Similar species:** Grey Heron.

Little Egret *Egretta garzetta* (Ardeidae) 56 cm (22 in). Slightly built, slim, long-legged heron. All-white in plumage, with long, thin black bill, long black legs and unexpected bright yellow feet. Very rarely black all over instead of white. In breeding season, adult develops 2 head plumes c. 16 cm (6½ in) long, also many elongated plumes on upper breast, mantle and wings (known as 'aigrettes'). Silent away from nesting colony, where gives variety of harsh and bubbling calls. **Range and habitat:** southern temperate and subtropical WP, EP, AF, OR and AU, in marshes, lakes and estuaries, some winter dispersal. **Similar species:** Snowy Egret (see p 52).

Little Bittern *Ixobrychus minutus* (Ardeidae) 35 cm (14 in). Diminutive, black and pinky buff or streaky brown heron. Male pinky buff neck and body, with black cap and back. Large pinky buff patch on black wings. Female duller and streaked brown. Very skulking, climbing among reeds. Rapid flight with shallow wingbeats, quickly drops into cover. Croaking call in flight; male gives regular barking croak in spring. **Range and habitat:** southern temperate WP and EP; subtropical AF, OR and AU, in densely vegetated freshwater marshes. WP and EP breeding birds migrate south for winter, remainder non-migratory. **Not illustrated.**

Bittern *Botaurus stellaris* (Ardeidae) 76 cm (30 in). Bulky, thick-necked heron with dagger-like bill. Overall golden brown in colour, heavily marked with dark brown and black spots, streaks and mottles. Black cap. Flight rather owl-like, on broad, rounded wings. Principal call of adult male a far-carrying booming, preceded by a few short grunts, audible up to 3-5 km (2-3 miles). **Range and habitat:** temperate WP and EP, also AF. Mainly migratory to south for winter. Almost always in dense reed-beds, fresh and brackish. **Similar species:** American Bittern (see p 52).

Spoonbill *Platalea leucorodia* (Threskiornithidae) 86 cm (34 in). Large, white, long-legged heron with elongated black bill ending in distinctive spatulate tip. Overall white plumage with bare yellowish skin round chin, throat and eyes. In breeding season, develops crest plumes and yellow gorget. Wing tips of immature blackish. In flight, neck sags below line of back, legs trail. Mostly silent, apart from low grunts at nesting colonies. **Range and habitat:** southern temperate and subtropical WP, EP, AF and OR. Some non-migratory, others migrate south for winter. Inhabits marshes and estuaries, more marine in winter. **Similar species:** African Spoonbill (see p 188).

White Stork *Ciconia ciconia* (Ciconiidae) 102 cm (40 in). Very large black and white, long-legged and long-necked bird with long, pointed red bill. Body and inner part of wings pure white; outer part of wings, including trailing edge, black. Long red bill and legs. Graceful flight on slow wingbeats with frequent and often prolonged soaring. Silent except for noisy bill-clappering between adults at nest. **Range and habitat:** widespread breeder in WP and EP, migrating south to AF and OR for winter. Inhabits open, shallow wetlands and farmland, frequently nesting on buildings. **Similar species:** Yellow-billed Stork (see p 188).

Black Stork *Ciconia nigra* (Ciconiidae) 97 cm (38 in). Very large, mainly black stork, long neck and legs. Overall black upperparts, also head, neck and chest. Belly pure white. Long, pointed bill, eyes and legs red. Flies well on broad wings, also soars frequently. Silent away from nest, but bill-clappering between pair at nest, also variety of whistling and hissing calls. **Range and habitat:** widespread temperate and southern WP and EP, also AF, in forests with streams and pools. Migratory from WP and EP south to AF and OR for winter, in well-vegetated wetlands.

WATERFOWL Anseriformes

Mute Swan *Cygnus olor* (Anatidae) 152 cm (60 in). Huge, long-necked waterbird. Pure white plumage in adult, with enlarged black knob at base of orange-red bill, larger in male. Juvenile grey-brown with grey bill, lacking knob. Nesting pairs strongly territorial, also in flocks. Variety of grunting and hissing calls. Wings make musical throbbing in flight. **Range and habitat:** widespread in lowland WP and EP; introduced AF, AU and NA. Mainly fresh and brackish waters of all types. Partial migrant moving south for winter. **Similar species:** Trumpeter (see p 54) and Whooper Swans.

Whooper Swan *Cygnus cygnus* (Anatidae) 152 cm (60 in). Huge, long-necked waterbird. Adult plumage completely white, with extensive yellow on upper mandible of black bill. Juvenile grey, with pink and grey bill, becoming white after 1 year. Flight powerful and direct. Loud and far-carrying musical trumpeting calls, produced on ground and in flight. **Range and habitat:** breeds northern temperate and subarctic WP and EP, on shallow freshwater pools in tundra, taiga and steppe. Migrates south for winter to large freshwaters and sheltered coasts. **Similar species:** Trumpeter (see p 54) and Bewick's Swans.

Bewick's Swan *Cygnus columbianus* (Anatidae) 122 cm (48 in). Smallest all-white swan, with yellow and black bill. Smaller and shorter-necked than Whooper Swan. Adult plumage pure white, with variable yellow on upper mandible of black bill. Legs and feet grey-black. Juvenile grey with pink and grey bill, becoming like adult at about 1 year. High-pitched musical honking and baying notes given in flight and on ground. **Range and habitat:** breeds arctic regions of WP and EP on marshy tundra pools and beside rivers. Migrates south to winter on temperate marshes and estuaries. **Similar species:** Whooper Swan.

Grey Heron

Purple Heron

Little Egret at nest

Bittern

Spoonbill

White Stork (pair) at nest

Black Stork and young

Mute Swan

Whooper Swan (pair)

Bewick's Swan

Greylag Goose *Anser anser* (Anatidae) 80 cm (31 in). Large, heavy grey goose with big head, thick neck and triangular bill. Overall grey body, paler below, a few scattered black marks on belly, white undertail. Western race has orange bill, eastern has pink bill and is overall paler and larger. Legs pink. Very pale grey forewings conspicuous in powerful direct flight, often in large flocks. Loud cackling honks, particularly in flight. **Range and habitat:** breeds northern and temperate WP and EP marshes and open wetlands. Winters to south, reaching OR, often feeding on farmland. **Similar species:** Pink-footed Goose.

Pink-footed Goose *Anser brachyrhynchus* (Anatidae) 70 cm (28 in). Medium-sized compact grey goose with round head and fairly short neck. Overall pinky brown, greyer on back, paler below; white undertail and dark brown head and neck. Bill dark brown and pink, legs pink. Usually in large flocks. Powerful, rapid flight, often in V skeins. Forewing paler grey than hindwing in flight. Call a musical 'wink-wink'. **Range and habitat:** breeds on tundra of Iceland, Greenland, Svalbard; winters Britain, Netherlands, Belgium, roosting on lakes and estuaries, feeding on farmland. **Similar species:** Greylag Goose.

White-fronted Goose *Anser albifrons* (Anatidae) 70 cm (28 in). Medium to large long-necked goose, darker grey-brown on back, paler below. Adult has conspicuous white forehead and variable black barring on breast and belly, absent in juvenile, white undertail; pink bill. All-dark upperwings lack contrast in flight. Greenland race has orange bill, is larger and darker. Legs orange. Musical, laughing 'lyo-lyok' calls, especially in flight. In large and small flocks. **Range and habitat:** circumpolar Holarctic, breeding on tundra marshes, wintering lowland marshes and farmland of NA, WP and EP. **Similar species:** other grey geese.

Barnacle Goose *Branta leucopsis* (Anatidae) 63 cm (25 in). Small goose with short neck and compact, round body. Grey above, narrowly barred black on upperparts, paler whitish grey belly and white undertail. Black breast, neck and head; creamy white cheeks, thin black line from bill through eye. Small, dainty black bill. Legs black. Occurs in large dense flocks, flying rapidly, when wings show overall grey above. Yapping call, not unlike small dog's. **Range and habitat:** breeds tundra of Greenland, Svalbard, north-west Siberia; winters farmland and coasts of Britain, Ireland, Netherlands. **Similar species:** Canada Goose (see p 54).

Red-breasted Goose *Branta ruficollis* (Anatidae) 51-8 cm (20-3 in). Tiny goose boldly patterned in black, white and russet red. Crown and back of neck black, large russet cheek patch outlined in white, russet throat and chest, with narrow white borders. Belly and back black with broad white flank stripe and white undertail. 2 narrow white wing-bars. Walks fast, flight agile. High-pitched 'kee-kwa' calls. Gregarious, often flocking in winter with White-fronted Geese. **Range and habitat:** breeds tundra of extreme north-western EP, wintering on open steppes and farmland of south-eastern WP.

(Common) Shelduck *Tadorna tadorna* (Anatidae) 61 cm (24 in). Boldly marked goose-like duck. Body white, head and upper neck greenish black. Broad chestnut breast band, blackish stripe down centre of belly. Blackish scapulars, forming black bar either side of white back when wings closed. Bill red, red knob at base larger in male, absent in juvenile, which is overall duller. Legs pink. In flight, outer wings black, inner forewing white, trailing edge green and chestnut. Male whistles, female quacks. **Range and habitat:** temperate coasts and larger wetlands of WP and EP, some non-migratory, others migrate to more southern coasts and open wetlands.

Mallard *Anas platyrhynchos* (Anatidae) 58 cm (23 in). Large dabbling duck, big head and bill. Male has green head, yellow bill, white neck-ring, red-brown chest, pale grey back and flanks, black and white tail with curled central

feathers. Female overall pale brown, marked with black. Rapid flight, showing white-edged blue speculum. Male has nasal 'raab', female loud quack. Commonest duck in WP, semi-domesticated in urban areas. **Range and habitat:** breeds throughout Holarctic, except far north, on great variety of wetlands, winters within range and to south, including OR. Introduced to AU. **Similar species:** female – other female dabbling ducks.

Wigeon *Anas penelope* (Anatidae) 46 cm (18 in). Medium-sized dabbling duck. Male has chestnut head with buff crown stripe, pinkish breast, grey body with white stripe above flanks, black and white tail, conspicuous white forewings in flight. Female overall rufous or grey-brown, forewing pale grey. Highly gregarious, feeding on land in dense flocks, flight rapid and twisting. Male has loud whistling 'whee-oo' call. **Range and habitat:** breeds WP and EP in arctic and northern temperate marshes. Migrates to southern temperate coasts and marshes, including in AF and OR. **Similar species:** American and Chiloe Wigeons (see pp 54, 102).

Teal *Anas crecca* (Anatidae) 35 cm (14 in). Smallest dabbling duck, compact body and very short neck. Male has chestnut and green head, dark grey body, white flank line (absent in North American race, which has vertical white line either side of breast), creamy yellow undertail. Female brown, mottled and streaked darker. Gregarious, flight very rapid, performing aerial manoeuvres like waders. Male has high-pitched whistle, female quacks. **Range and habitat:** circumpolar Holarctic tundra and northern temperate marshes, migrating south for winter including to AF and OR. **Similar species:** female – other small female dabbling ducks.

Garganey *Anas querquedula* (Anatidae) 38 cm (15 in). Small dabbling duck; male has chestnut-brown head and neck, conspicuous white eye stripe curving down to nape. Upperparts dark with elongated white scapulars, flanks grey. Pale blue forewing prominent in flight. Female brown, whitish throat and pale spot at base of bill. Occurs in huge flocks in winter, more scattered in summer. Flight rapid and direct. Call of male harsh rattle, female quacks. **Range and habitat:** breeds temperate marshland of WP and EP. Migrates to open wetlands of AF and OR for winter. **Similar species:** female – other small female dabbling ducks.

Pintail *Anas acuta* (Anatidae) 56-66 cm (22-6 in). Long-necked slender dabbling duck. Pointed tail, c. 10 cm (4 in) long in male, which has chocolate brown head and nape, clear white neck, grey body streaked black and white on back, yellow and black undertail. Female pale brown, slightly elongated tail. Flight fast, often in Vs. Not very vocal; male has soft whistle, female soft quack. **Range and habitat:** breeds circumpolar Holarctic tundra and northern temperate marshes. Migrates south to coasts and marshes within Holarctic and to AF and OR. **Similar species:** female – other medium-sized female dabbling ducks.

Shoveler *Anas clypeata* (Anatidae) 51 cm (20 in). Very heavy-headed dabbling duck with broad, spatulate bill. Male has dark green head, conspicuous white breast, chestnut flanks with white line above, dark back, black and white undertail. In flight, shows blue forewing. Female pale brown, pale blue forewing. Very large bill immersed as bird feeds while swimming; gives top-heavy appearance in flight. Male has quiet 'took-took' call, female soft quack. **Range and habitat:** circumpolar Holarctic tundra and northern temperate marshes, migrating south to marshes within Holarctic and also AF and OR. **Similar species:** other shovelers.

Greylag Goose

Pink-footed Goose

White-fronted Goose

Barnacle Goose

Red-breasted Goose

(Common) Shelduck

Mallard

Wigeon

Teal

Garganey

Pintail

Shoveler

Tufted Duck *Aythya fuligula* (Anatidae) 43 cm (17 in). Compact diving duck with round head and drooping crest. Male mainly black, glossed purple on head, white flanks and underparts. Female dark brown, paler mottled flanks and whitish underparts, occasionally some white round base of bill. Both have white wing-bar in flight. Swims buoyantly and dives with preliminary leap. Flight rapid, wings making whistling sound. Male has soft whistle, female low growl, but not very vocal. **Range and habitat:** northern temperate and arctic tundra and marshes of WP and EP; migrates to temperate and southern open freshwaters and coasts of WP and EP, also to AF and OR. **Similar species:** other black and white or brown diving ducks (see pp 56, 288).

Greater Scaup *Aythya marila* (Anatidae) 48 cm (19 in). Bulky diving duck with large bill. Male has black head, glossed green, black breast and tail, pale grey back and white flanks. Bill blue-grey with black tip. Female mainly dark brown, paler on flanks, variable white patch at base of bill. White wing-bar shows in flight, which is strong but heavy. Swims buoyantly and dives easily. Not very vocal; female has harsh growl. **Range and habitat:** breeds circumpolar Holarctic tundra and northern marshes. Migrates to temperate WP, EP and NA, where normally on sheltered coasts and some large inland brackish or freshwater wetlands. **Similar species:** other black and white or brown diving ducks (see pp 56, 288).

Pochard *Aythya ferina* (Anatidae) 46 cm (18 in). Large, long-billed diving duck with high-domed head. Male has chestnut head and black breast, pale grey upperparts and flanks, black tail. Female yellow-brown on head and breast, grey-brown on back and flanks. In strong, fast flight, pale wing stripe shows on hindwing. Round-backed shape on water, where dives freely. Generally silent, though wings whistle in flight and female has growling note. **Range and habitat:** breeds northern temperate WP and EP marshes and pools, migrating south to marshes, open wetlands and coasts, including to AF and OR. **Similar species:** other red-headed and brown diving ducks (see pp 56, 288).

Eider *Somateria mollissima* (Anatidae) 58 cm (23 in). Heavily built, short-necked seaduck. Male has white face with pale green patches at rear, pinkish breast, white back with elongated scapulars trailing over black flanks and tail, round white patch at sides of rump. Long sloping forehead leading into long wedge-shaped greenish bill. Female cinnamon brown, paler on head, mottled darker on body. Heavy flight on broad wings, low over water, flocks usually flying in lines. During courtship display, male utters musical cooing, female has guttural 'gok-gok'. **Range and habitat:** circumpolar arctic and temperate coasts of Holarctic, wintering within range and coasts to south. **Similar species:** other eiders.

Common (Black) Scoter *Melanitta nigra* (Anatidae) 48 cm (19 in). Stockily built seaduck. Male is only all-black duck, has large head with prominent black knob at base of yellow bill. Knob is also yellow in North American race. Female dark brown, slightly paler on flanks and breast, head capped dark brown, cheeks and throat buff-white. Bill green-black with small basal knob. No wing-bars in flight, which is fast and strong and low over water. Swims buoyantly with tail occasionally cocked, dives frequently. **Range and habitat:** breeds arctic tundra and northern temperate freshwater wetlands of WP and EP, locally also in NA. Winters predominantly coasts to south. **Similar species:** other scoters (see p 56).

Goldeneye *Bucephala clangula* (Anatidae) 46 cm (18 in). Dumpy, large-headed seaduck. Male has large, round black head, glossed green, round white patch in front of eye. Back black with extensive black-outlined white panels, breast and underparts white. Female brown-headed with grey back and brown flanks, showing traces of white at top of flanks. Both sexes show double white wing-bar in flight, which is fast and direct. Swims buoyantly and dives readily and

frequently. Mostly silent, with soft, nasal calls during display. **Range and habitat:** circumpolar arctic tundra and northern temperate Holarctic lakes and rivers, wintering on coasts and freshwater wetlands to south. **Similar species:** Barrow's Goldeneye (see p 56).

Long-tailed Duck *Clangula hyemalis* (Anatidae) 41-53 cm (16-21 in). Slim, elegant seaduck with greatly elongated tail in male. Plumage of both sexes undergoes complex changes during year. Breeding male has dark brown head, with white cheeks, brown chest and black and white flanks. Winter male white-headed and white on back. Central tail feathers project c. 13 cm (5 in). Breeding female has dark brown and white head, dark brown back, whitish underparts. Winter female browner and duller. Dark wings in rapid, swinging flight. Male has yodelling display call, female soft quack. **Range and habitat:** breeds circumpolar Holarctic tundra, winters coasts to south. **Similar species:** female – female or juvenile Harlequin Duck (see p 56).

Goosander *Mergus merganser* (Anatidae) 62 cm (24 in). Large, bulky sawbill. Male has green-glossed black head, shaggy at back, with long, narrow bright red bill ending in slightly hooked tip. Back black; neck, breast and flanks white, tinged with pink or cream in front. In flight, which is powerful and fast, whole forewing is white. Female has rufous head, shaggy at back, red bill, whitish throat and underparts, grey upperparts and flanks. In flight, rear of forewing is white. Silent except in courtship when variety of harsh notes. **Range and habitat:** breeds circumpolar Holarctic by rivers and freshwaters, mainly in forests. Winters mainly to south on fresh and brackish open waters. **Similar species:** Red-breasted Merganser.

Red-breasted Merganser *Mergus serrator* (Anatidae) 55 cm (22 in). Medium-sized, slender sawbill. Male has green-black head with double crest and long, narrow red bill ending in a hook. White neck-ring above reddish breast spotted with black. Upperparts black with broad white flank stripe, underparts and flanks grey. Female has reddish head with double crest, grey-brown upperparts and grey underparts. Both sexes have white on inner wing in flight, which is fast and direct. Swims low in water and dives smoothly. Both sexes have low calls during courtship. **Range and habitat:** breeds circumpolar boreal and northern temperate Holarctic wooded lakes and rivers. Migrates south, mainly to coasts, for winter. **Similar species:** Goosander.

Smew *Mergus albellus* (Anatidae) 41 cm (16 in). Small and compact sawbill. Male mainly white on head, breast and upperparts, with fine black markings around eye, and black lines along flanks and back. Flanks and tail grey. Female has chestnut cap to white face, blackish round eye, chestnut down back of neck. Body grey, darker above. Male appears black and white in flight with white markings on inner part of dark wings. Female has less white. Flies very fast, with great agility. Swims high on water, dives frequently. Rarely calls in winter, whistles in courtship. **Range and habitat:** northern forests of WP and EP nesting beside still or running water. Winters on large freshwaters and shallow, sheltered coasts to south.

Tufted Duck

Greater Scaup

Pochard

Eider

Common (Black) Scoter

Goldeneye

Long-tailed Duck

Goosander (pair)

Red-breasted Merganser

Smew

DIURNAL BIRDS OF PREY Falconiformes

Osprey *Pandion haliaetus* (Pandionidae) 51-9 cm (20-3 in). Medium-sized raptor, normally found close to water. Small white head with dark streak through eye and down nape to dark brown back and wings. Underparts strikingly white including feathered upper legs. In flight, looks dark above, white beneath. Wings long and narrow, usually angled at carpal joint. Head protrudes quite noticeably, tail comparatively short. Call a shrill whistle. **Range and habitat:** widespread northern Holarctic, also AF, OR and AU, breeding and wintering at wetlands of all types. **Similar species:** pale morph of Booted Eagle.

Red Kite *Milvus milvus* (Accipitridae) 58-64 cm (23-5 in). Long, angular-winged medium-sized raptor with deeply forked tail. Rich chestnut overall, slightly darker above, prominent nearly white head. Flight very free with wings and tail constantly twisted and flexed. Upperwings with blackish flight feathers, blackish and chestnut shoulders. Underwings show large white patch towards tip. Long tail chestnut with white band at tip, conspicuously notched. Call a shrill mew. **Range and habitat:** open woodland and farmland of central and southern WP. Northern populations mostly migrate to within breeding range of southern populations, which are non-migratory. **Similar species:** Black Kite.

Black Kite *Milvus migrans* (Accipitridae) 53-9 cm (21-3 in). Medium-sized, rather bulky raptor. Upperparts dark dirty brown, head slightly paler and greyer. Underparts similar, toning to rufous on belly and undertail. In flight, long wings held at angle and somewhat arched. Upper- and underwings dark brown or blackish towards leading edge but with a pale panel towards tip. Dark tail slightly forked and notched. Very vocal with shrill, high-pitched trilling calls. **Range and habitat:** widespread, often near wetlands, in temperate and southern WP and EP, also almost throughout AF, OR and AU. **Similar species:** Red Kite, Marsh Harrier (see p 152).

Short-toed Eagle *Circaetus gallicus* (Accipitridae) 63-9 cm (25-7 in). Small eagle. Upperparts brown, tinged with pale buff. Underparts paler, usually with dark brown spots on breast. Head rounded, resembling owl's, often with dark hood. Wings long and broad, especially at carpal joint, light brown above, with whitish patch; underwings and underparts appear almost white, contrasting with black outer half of primaries. Buzzard-like mewing call. **Range and habitat:** open lowlands and forest edges of southern WP and western EP migrating to winter in equatorial AF. Resident in Indian subcontinent (OR). **Similar species:** Osprey and buzzards.

Sparrowhawk *Accipiter nisus* (Accipitridae) 28-38 cm (11-15 in). Small raptor. Male has slate-grey upperparts with rufous patches on cheeks, whitish below with narrow orange barring (also on underwings). Larger female has dark brown or grey-brown upperparts, barred brown on paler underparts and underwings. Tail with dark brown bands. Long tail and shortish rounded wings obvious in low, dashing flight. Chattering 'kew-kew' call. **Range and habitat:** widespread in wooded areas of WP and EP, also northern OR. EP population migratory to OR. **Similar species:** Goshawk.

Goshawk *Accipiter gentilis* (Accipitridae) 48-58 cm (19-23 in). Medium-sized raptor. Upperparts of male grey-brown with darker crown and whitish stripe over eye. Underparts paler, heavily barred brown. Female larger, upperparts browner with more obvious white stripe over eye, underparts paler and barred brown. Both sexes have barred underwings and tails. Long tail and broad, rounded wings. Flight fast and dashing. Calls shrill and screaming. **Range and habitat:** widespread in wooded areas of northern and temperate Holarctic, mostly non-migratory, but more northern breeding birds moving south for winter. **Similar species:** Sparrowhawk.

Buzzard *Buteo buteo* (Accipitridae) 50-6 cm (20-2 in). Medium-sized raptor, not unlike small eagle with rounded wings ending in 'fingers' in soaring flight. Very variable plumage, typically dark brown upperparts, slightly paler below, but often paler or even whitish on head and body and sometimes wings. In flight, always darker above and much paler below, usually showing dark carpal patches on underwings and dark trailing edge. Call a high-pitched 'mew'. **Range and habitat:** open and wooded country in temperate WP and EP, partially migratory to AF and OR. **Similar species:** other buzzards (see pp 58, 194).

Honey Buzzard *Pernis apivorus* (Accipitridae) 50-8 cm (20-3 in). Medium-sized raptor with long, pointed wings and comparatively small head. Upperparts grey-brown; male greyer, female browner. Underparts variable from dark brown to very pale, almost white, with varying amounts of darker barring. In flight wings show 3 dark bars towards trailing edge and dark carpal patches below. Tail has 2 narrow dark bars at base, 1 broad bar close to tip. Shrill whistling calls. **Range and habitat:** mainly temperate woodland and semi-open areas of WP and western EP. Wholly migratory to southern AF. **Similar species:** other buzzards (see pp 58, 194).

Booted Eagle *Hieraeetus pennatus* (Accipitridae) 45-53 cm (18-21 in). Small eagle with 2 markedly different colour phases. Commoner pale phase largely rufous above and strikingly white below on body, wings and tail. White forewing contrasts with dark flight feathers and trailing edge. Dark phase is rich dark rufous brown above and below with paler rufous tail. In flight, shows slight pale patch towards wing tips. Flight buoyant, with looping display. Shrill 'kee-kee' call. **Range and habitat:** southern woodlands of WP and EP, migrating AF and OR. Also breeds extreme south AF. **Similar species:** other small eagles (see p 242).

Golden Eagle *Aquila chrysaetos* (Accipitridae) 76-90 cm (30-5 in). Large eagle with prominently protruding head in flight and long tail. Body and wings dark red-brown to blackish above, dark brown below. Crown of head and nape golden brown. Immature shows varying amounts of white (for 1st 4-5 years) on rump and outer wings. Soars for long periods, with wing feathers extended. Not very vocal but has mewing cry and barking call. **Range and habitat:** widespread in mountainous areas and lowland woodland in WP, EP and NA; many resident, more northerly birds moving south for winter. **Similar species:** other large eagles (see pp 58, 242).

Egyptian Vulture *Neophron percnopterus* (Accipitridae) 58-66 cm (23-6 in). Small vulture. Body white with yellow tinge to head and neck and bare yellow skin on head and throat. Shaggy ruff at back of head also tinged yellow. Bill small. Main wing feathers black, so black wing tips and trailing edge in flight, rest of wings white. Wings straight, tail longish and wedge-shaped. Normally silent. In AF uses a stone to open Ostrich eggs. **Range and habitat:** open country from lowlands to mountains in southern WP and EP, also AF and OR. WP and EP populations migrate south in winter to AF and OR.

Griffon Vulture *Gyps fulvus* (Accipitridae) 97-104 cm (38-41 in). Huge vulture. Body and wing-coverts uniform ginger-buff. Flight feathers and tail dark brown or blackish; considerable contrast in flight between dark wing tips and trailing edge and pale forewings. Head and neck down-covered, whitish ruff at base of neck. Soaring flight on broad straight-edged wings with well-extended 'fingers'. Usually silent. **Range and habitat:** open, usually mountainous, areas of southern WP and south-western EP, also AF and OR, some moving short distances south for winter. **Similar species:** other large vultures (see p 192).

Osprey

Red Kite

Black Kite

Short-toed Eagle

Sparrowhawk at nest

Goshawk (female)

Buzzard at nest

Honey Buzzard in light colour phase

Booted Eagle feeding young

Golden Eagle at nest

Egyptian Vulture

Griffon Vulture

Marsh Harrier *Circus aeruginosus* (Accipitridae) 48-56 cm (19-22 in). Large, broad-winged harrier. Male's upperparts brown, shoulders creamy. Paler head and rufous underparts streaked brown. Extensive grey on rear half of wing, contrasting with dark wing tip and creamy leading edge. Female all-dark brown except for creamy head, shoulders and breast. Broad dark stripe through eye. In flight, dark except for leading edge to wing and head. Heavy flapping flight; slow glides with wings in shallow V. Shrill calls during courtship. **Range and habitat:** shallow fresh and brackish waters and lowlands of temperate WP and EP, migrating in winter to AF and OR. Also resident in AF and AU. **Similar species:** other harriers (see p 242).

Hen Harrier *Circus cyaneus* (Accipitridae) 43-51 cm (17-20 in). Medium-sized harrier. Male pale grey above, whitish below, with small white rump and black primaries which contrast sharply with all-grey upper and white underwings in flight. Much larger female dark brown above, pale streaky brown below. Large white rump and strongly barred tail. Flight buoyant with rapid beats interspersed with gliding, wings held in shallow V. Chattering and whistling calls in display. **Range and habitat:** breeds in wide variety of open habitats in circumpolar northern temperate Holarctic, mostly migrating south for winter reaching OR. Also widespread resident in NT. **Similar species:** other harriers (see p 242).

Peregrine Falcon *Falco peregrinus* (Falconidae) 38-48 cm (15-19 in). Medium to large falcon with prominent black crown and moustache. Male grey-backed with well-barred paler underparts and barred tail. Much larger female generally dark grey-brown above, paler brown below, much more heavily barred than male with large drop-shaped spots on breast. Flight very swift and powerful on pointed wings, stooping from considerable height. Also soars on stiff slightly swept-back wings. Harsh chattering and whining calls. **Range and habitat:** widespread in open habitats in northern and central Holarctic, also widespread in OR and AU, and parts of AF and NT. Some populations migratory. **Similar species:** other large falcons (see pp 58, 242).

Hobby *Falco subbuteo* (Falconidae) 30-6 cm (12-14 in). Small, slim falcon with long swept-back wings and shortish tail giving silhouette similar to Swift. Upperparts of adult very dark slate-grey, tail slightly paler, with broad buff-white collar, very dark cap and obvious blackish moustache. Underparts very pale whitish buff, heavily streaked and spotted black. Thighs and undertail rufous, conspicuous in flight. Aerially active, soaring on stiff wings, catching prey on wing by stooping and hawking. Shrill 'kew-kew' note when disturbed. **Range and habitat:** widespread in lowlands with scattered small woodlands throughout WP and EP, also parts of OR. Completely migratory, south to AF and OR. **Similar species:** other small falcons, especially Red-footed.

Merlin *Falco columbarius* (Falconidae) 27-33 cm (10½-13 in). Small falcon with few obvious markings. Male has slatey blue upperparts, broad black tip to slightly paler tail. Head with dark crown and rufous nape, moustachial streak faint. Underparts rusty, heavily streaked with black. Rusty underwings. Female mainly dark brown above, with paler streaks and brown-streaked underparts, with brown tail boldly barred with cream. Outer wing feathers of both sexes darker in flight. Shrill and chattering 'kee-kee' call. Fast and dashing flight. **Range and habitat:** widespread in open habitats throughout circumpolar Holarctic arctic and north temperate. Migrates south for winter, some reaching OR and NT. **Similar species:** other small falcons (see pp 58, 196).

Kestrel *Falco tinnunculus* (Falconidae) 33-6 cm (13-14 in). Long-winged, long-tailed falcon with characteristic prolonged hovering flight when hunting. Male has blue-grey head with narrow black moustache, chestnut upperwings and back spotted with black, blue-grey tail with black band at tip, and buff underparts, spotted and streaked with black. Female is rufous brown above, streaked and barred black,

paler below, also streaked and spotted. Both sexes show dark wing tips in flight. Hovers head to wind with sloping body and fanned tail. Shrill 'kee-kee' calls. **Range and habitat:** widespread in great variety of habitats throughout WP and EP, also widespread in AF and parts of OR. Partially migratory. **Similar species:** other falcons (see pp 58, 196).

FOWL-LIKE BIRDS Galliformes

Willow/Red Grouse *Lagopus lagopus* (Phasianidae) 38-41 cm (15-16 in). Smallish round-bodied gamebird with small head and very variably patterned plumage. Male has red wattle over eye. Summer male has dark rufous upperparts, barred with black. Both sexes underparts and most of wings white. Summer females paler rufous brown with more obvious black barring. Both sexes all-white in winter. Particoloured birds occur in spring and autumn moults. British/Irish race (Red Grouse) has no white in summer or winter. Call a loud 'kok-kok' and 'goback-goback'. **Range and habitat:** widespread in circumpolar Holarctic tundra, scrubby taiga, moorland and other open habitats. Short distance movements in winter to less exposed areas. **Similar species:** Ptarmigan.

Ptarmigan *Lagopus mutus* (Phasianidae) 33-6 cm (13-14 in). Small mountain gamebird. Male has red wattle above eye. Upperparts of summer male darkish grey-brown, mottled blackish and buff. Underparts and wings white, as are feathered legs. Summer female browner than male with golden and blackish markings and less white below; wings white. Both sexes have distinct autumn plumage, darker than in summer, though retaining white underparts and wings, before changing to all-white in winter. Flight conspicuous because of white wings. Usually very tame. Makes hoarse cackles and croaks as well as crowing song during display. **Range and habitat:** circumpolar Holarctic tundra and mountains. Non-migratory. **Similar species:** Willow/Red Grouse.

Black Grouse *Tetrao tetrix* (Phasianidae) 41-56 cm (16-22 in). Bulky gamebird with marked size difference between sexes. Much larger male has glossy black plumage with a white wing-bar and white undertail. Tail feathers extended into lyre-shaped curves. Female mostly rufous brown, barred black on upperparts and dark brown on underparts. Small white bars on wings. Male has red wattles above eyes, enlarged in spring. Flight on rapid wingbeats and then gliding. Male has sneezing call, female a harsh 'kok-kok'. Often perches on low trees or scrub. **Range and habitat:** widespread in northern wooded and scrub regions of WP and EP. Mainly non-migratory. **Similar species:** Capercaillie.

Capercaillie *Tetrao urogallus* (Phasianidae) 58-90 cm (23-35 in). Huge gamebird, much larger male reminiscent of turkey with shaggy head and broad tail. Upperparts of male blue-grey, with brown on back and wings. Darker head with throat feathers erected when calling, and red wattle over eye. Breast black with green gloss, grey underparts with pale barring. Tail dark-tipped, fanned in display. Female orange-buff to brown overall, heavily barred black above, with chestnut breast and buff-barred whitish underparts. Calls of both sexes raucous. Display song of male quickening rattle ending in pop and gurgle, like bottle being uncorked and poured. **Range and habitat:** northern forests of WP and EP where resident. **Similar species:** Black Grouse.

Marsh Harrier

Hen Harrier (pair) at nest

Peregrine Falcon at nest

Hobby at nest

Merlin

Kestrel (pair) at nest

Willow/Red Grouse

Ptarmigan

Black Grouse in male lek

Capercaillie

Pheasant *Phasianus colchicus* (Phasianidae) 52-90 cm (20-35 in). Gamebird. Male purple-chestnut all over, marked with black, except for bottle green head, short brown wings and black belly. Bare skin round eye bright red. Tail up to half overall length. Much plumage variation, some dark all over, some with white neck-ring. Female pale buff-brown all over, chestnut on back and breast, elongated tail. Flies with explosive burst of flapping followed by glide. Voice a harsh 'kok-kok' and cackles. **Range and habitat:** widespread in mixed wooded and open habitats of temperate WP, EP and OR, also introduced to NA and AU. Non-migratory. **Similar species:** other pheasants (see pp 244, 246).

Red-legged Partridge *Alectoris rufa* (Phasianidae) 30 cm (12 in). Medium-sized, bulky gamebird with upright stance. Sexes similar with olive to chestnut upperparts, white stripe above eye, black through eye, bold white cheeks and chin. Blackish collar above grey breast, flanks barred chestnut, black, grey and white. Underparts buff-orange. Legs and bill red. Main call a loud 'chuka-chuka', also wheezing notes. Usually in pairs or family groups (coveys). Runs freely, flies fast and low, gliding on down-bent wings. **Range and habitat:** confined to open habitats, including farmland, of warm temperate south-west WP where resident; introduced Britain. **Similar species:** other partridges (see pp 196, 244).

Grey Partridge *Perdix perdix* (Phasianidae) 30 cm (12 in). Medium-small stocky gamebird with round head and short neck. Male has brown upperparts, streaked buff; grey mantle, neck and breast. Face orange. Flanks grey, barred with chestnut. Prominent broad dark brown horseshoe mark on pale belly. Female duller, less well marked, paler face and indistinct horseshoe mark on belly. Lives in family groups (coveys) through autumn and winter. Flies fast and low, often gliding on wings held down-curved. Runs freely with neck outstretched. Voice a creaking 'keev-it'. **Range and habitat:** widespread in open, vegetated habitats, including farmland, of temperate WP and eastern EP, where resident. Introduced to NA. **Similar species:** other partridges (see pp 196, 244).

Quail *Coturnix coturnix* (Phasianidae) 18 cm (7 in). Very small, dumpy gamebird with comparatively long, thin wings. Male has buff upperparts marked with both black and pale streaks, particularly along flanks. Breast rufous shading to pale buff belly. Head variably chestnut with bold black throat and neck stripes, sometimes indistinct; white stripe above eye. Female much duller, black bands on throat and neck broken into spots, blackish spots on side of chest. Main call a plaintive 'wet-my-lips'. Flies reluctantly. **Range and habitat:** open vegetated habitats of central and southern WP and EP, also parts of OR and AF. Migratory in Palaearctic. **Similar species:** juvenile partridges and other quail (see pp 196, 244, 292).

CRANES, RAILS, AND ALLIES Gruiformes

Crane *Grus grus* (Gruidae) 106-14 cm (42-5 in). Tall, elegant, long-legged and long-necked bird. Body grey all over, becoming grey-brown with wear, except black upper neck and head with broad white patch round eye, trailing down hindneck. Small red patch of bare skin on crown and lores. Long, red-brown bill. Long inner wing feathers form trailing cloak over tail. In flight, dark outer wing feathers contrast with paler inner areas. Strong flight, neck and legs outstretched. Voice a loud 'kroo'. **Range and habitat:** breeds in marshy clearings among forested taiga and steppe woodlands of WP and EP, migrating south to open marshlands of AF and OR. **Similar species:** other cranes (see pp 60, 198, 246).

Great Bustard *Otis tarda* (Otididae) 75-105 cm (29-41 in). Huge running bird, male up to 50% larger than female. Male has grey head and neck with, in breeding season, long white moustachial bristles. Upperparts cinnamon barred black, underparts white with rufous breast band. In flight and display, shows much white on black-tipped wings. Female

duller, neck thinner, no breast band, less white on wings. Flight steady with slow regular wingbeats. Walks with stately gait. Mostly silent. **Range and habitat:** patchily distributed in open plains and plateaux of southern WP and EP, those in latter region migratory.

Water Rail *Rallus aquaticus* (Rallidae) 27-9 cm (10½-11½ in). Small grey and brown rail with long and slightly decurved, reddish bill. Crown and upperparts dark brown, streaked paler. Face, throat and breast blue-grey apart from tiny white chin spot and dark stripe through eye. Black flanks narrowly barred white. Whitish undertail, with buff markings. Long flesh-coloured legs dangled in weak flight. Secretive but occasionally feeds in open. Loud squealing and grunting calls, and regular 'kipp-kipp-kipp' song. **Range and habitat:** dense aquatic vegetation in fens and lakes of WP and EP. Resident in western WP, otherwise migratory to south for winter. **Similar species:** other rails (see pp 62, 108, 196, 248, 292).

Spotted Crake *Porzana porzana* (Rallidae) 22-4 cm (8½-9½ in). Medium-sized rail. Male has green-brown upperparts and olive-brown underparts strongly tinged with grey. Some black spotting above, with white spotting above and below. Flanks barred black and white. Buff undertail, longish green legs. Short bill yellow with red spot at base. Female more spotted, particularly on throat and underparts. Very secretive and skulking. Presence announced by far-carrying whiplash call 'hwit-hwit-hwit'. **Range and habitat:** marshes and bogs of central and southern WP and EP, migrating for winter to wetlands of AF and OR. **Similar species:** other rails (see pp 62, 108, 248).

Corncrake *Crex crex* (Rallidae) 25-8 cm (10-11 in). Dumpy land-based rail. Sexes similar with dull yellowy buff upperparts tinged with grey, especially on head, neck and breast, and streaked on back with black. Underparts pale red-brown, flanks barred with chestnut. Forewings chestnut, remainder red-brown. Weak, fluttering flight with dangling legs. Call a 2-note rasping 'crex-crex', heard more at night. Secretive. **Range and habitat:** lowland well-vegetated areas including pastures and crops in temperate WP and EP, migrating to eastern AF for winter. **Similar species:** other brown crakes or rails (see pp 62, 108, 248).

Moorhen *Gallinula chloropus* (Rallidae) 31-5 cm (12-14 in). Bulky waterbird with distinctive habit of flicking white undertail. Dark blackish brown upperparts, slate-grey below. White line along flanks and conspicuous white undertail. Bill red with yellow tip, extending into red frontal shield. Swims jerking head and tail, flies weakly with dangling legs. Walks readily and climbs reeds and bushes. Normally wary but tame in urban parks. Main call a croaking 'cur-ruc'. **Range and habitat:** widespread in varied freshwater habitats in temperate and southern WP and EP, in southern NA, most of OR, also parts of AF and NT. Some resident, others migratory. **Similar species:** other gallinules (see p 60).

Coot *Fulica atra* (Rallidae) 36-40 cm (14-16 in). Round-bodied waterbird. Body plumage very dark slaty black, varying from glossy black head and neck to more grey tone on flanks and underparts. Short, pointed bill and frontal shield white and very conspicuous. White-tipped inner flight feathers show as pale wing-bar in flight when grey-green legs trail behind. Gregarious, forming large winter flocks. Swims buoyantly, making frequent brief dives. Also feeds on land. **Range and habitat:** widespread on varied freshwaters with emergent or marginal vegetation in temperate and southern WP and EP, also widespread in OR and AU. Some populations migratory and found in winter on brackish and freshwater. **Similar species:** other coots (see pp 60, 108, 196).

Pheasant
Red-legged Partridge
Grey Partridge

Quail
Crane (pair)

Great Bustard
Water Rail
Spotted Crake

Corncrake
Moorhen
Coot

SHOREBIRDS, GULLS, AUKS Charadriiformes

Oystercatcher *Haematopus ostralegus* (Haematopodidae) 41-5 cm (16-18 in). Large and robust wading bird with conspicuous bill. Glossy black upperparts, head, neck and upper breast. Pure white lower breast and rest of underparts, also rump and band across trailing edge of wings. Variable white collar present in winter round front of neck. Long, stout orange-red bill. Pink legs and feet. Highly gregarious outside breeding season. Shrill 'kleep-kleep' call, loud piping display notes. **Range and habitat:** breeds northern and temperate WP and EP, particularly on coasts, also inland along river valleys and by larger freshwaters. Winters coasts of WP and EP, also AF and OR. **Similar species:** other oystercatchers (see pp 200, 294).

Avocet *Recurvirostra avosetta* (Recurvirostridae) 41-5 cm (16-18 in). Long-legged wading bird with distinctive long, thin, upturned bill. Overall body plumage pure white, with black cap to head and black nape, black wing tips, black bar across centre of wings and black bar at wing roots. Bill c. 8 cm (3 in) long, delicately upcurved and tapering to point. Used in highly distinctive side-to-side sweeping movement through shallow water. Flight graceful and fast, with legs and feet trailing behind tail. Chief call 'kloo-eet'. **Range and habitat:** shallow brackish and saline wetlands of western and southern WP, central EP and scattered through AF, wintering similar habitats in WP, AF and OR. **Similar species:** other avocets (see pp 66, 296).

Black-winged Stilt *Himantopus himantopus* (Recurvirostridae) 36-40 cm (14-16 in). Slender wading bird on exceptionally long, thin legs. Cheeks, neck, underparts and tail pure white, crown blackish, though can be white. Back black, also both wing surfaces. Needle-like black bill c. 6 cm (2¼ in) long. Bright pink legs 12.5-16.5 cm (5-6½ in) long. Rapid flight action on pointed wings, with legs trailing far beyond tail. 'Kik-kik-kik' call used mainly in alarm or when mobbing predator. **Range and habitat:** discontinuous distribution in shallow fresh and brackish open wetlands of southern WP and EP, also widespread in southern AF, and almost throughout NT, AF, OR and AU. Palaearctic populations migratory, to AF. **Similar species:** other stilts (see pp 66, 294).

Ringed Plover *Charadrius hiaticula* (Charadriidae) 18-20 cm (7-8 in). Plump, pale sandy brown and white shorebird with head and neck banded black and white. Crown and upperparts sandy brown, rest of head and underparts white. Black bands through eyes and above bill and black collar. Prominent pale wing-bar. Short, stubby orange bill tipped black. Orange legs. Slow and quick runs with very rapid steps, interspersed with pauses. Flies rapidly and freely. Territorial in breeding season, gregarious at other times. Calls a liquid 'too-ee' and sharper 'queep'. **Range and habitat:** arctic tundra and northern temperate shores of WP and EP, and also north-eastern arctic NA. Winters western coasts of WP, also widely in AF. **Similar species:** other small plovers (see pp 62, 110, 200, 294).

Little Ringed Plover *Charadrius dubius* (Charadriidae) 14-16 cm (5½-6½ in). Small brown and white plover with black and white banding on head and chest. Upperparts and crown dull brown, remainder of head and underparts pure white. Black band through eyes, joining over white forehead. Very narrow black line joining eyes to bill. Black collar below white throat. Thin dark bill, yellowish or pinkish legs. Runs freely, pausing at intervals. Flight fast and agile, showing uniform brown upper surface of wings. High-pitched 'pee-oo' call. Territorial breeder, not especially gregarious in winter. **Range and habitat:** widespread in lowland freshwater wetlands of WP, EP and OR. Wintering in AF and OR. **Similar species:** other small plovers (see pp 62, 110, 200, 294).

Golden Plover *Pluvialis apricaria* (Charadriidae) 27-9 cm (10½-11½ in). Medium to large short-billed wader with rounded head and bulky body. Crown, upperparts and upperwings blackish brown liberally speckled with gold, especially on nape. Face, chin, throat and centre of underparts jet black, with white sides to neck, shoulders, flanks and undertail. Black less extensive in southerly breeding birds. Black lost in winter. Underwings whitish. Gregarious except when breeding. Liquid 'tloo-ee' call. Rapid, agile flight; easy running gait. **Range and habitat:** arctic tundra and north temperate open mountain areas of WP and north-western EP, wintering in west and south WP where found on farmland and floodlands. **Similar species:** other golden plovers (see p 62).

Grey Plover *Pluvialis squatarola* (Charadriidae) 28-31 cm (11-12 in). Stoutly built medium to large short-billed wader. Upperparts finely barred and spotted black and white to give silvery appearance. Forehead and sides of nape and shoulders white; face, chin, throat and underparts black. Undertail white. In flight shows pale underwings with black at junction with body. Black lost in winter and upperparts paler. Call a plaintive 'tlee-oo-ee'. Gregarious, wades and runs. **Range and habitat:** breeds circumpolar Holarctic tundras, except Greenland. Migrates to winter on coasts of NA, NT, AF, OR and AU; small numbers remain on coasts of WP and EP. **Similar species:** golden plovers (see p 62).

Dotterel *Eudromias morinellus* (Charadriidae) 20-3 cm (8-9 in). Plump, small-billed plover. In summer plumage upperparts are grey and brown with blackish crown and very prominent white stripes above eyes joining to form V on nape. Grey breast separated from chestnut underparts by white line. Blackish belly and white undertail. Wings uniform brown above, pale below. Female brighter. In winter, paler and browner all over, losing chestnut on underparts. Quiet 'peep-peep' call. Wingbeats noticeably rapid in fast and strong flight. **Range and habitat:** breeds arctic tundra and northern temperate mountain uplands, and rarely lowlands, of WP and EP, migrating to winter in dry open country in southern WP. **Similar species:** other larger plovers (see pp 110, 200, 250, 294).

Lapwing *Vanellus vanellus* (Charadriidae) 29-32 cm (11½-13 in). Medium-large crested plover, black and white at distance. Upperparts dark glossy purple-green. Crown, crest and throat black, and black mark through eye. Lower face and nape whitish. Breast black, underparts white, undertail cinnamon, rump white. Rounded wings black above, black and white below. Distinctive slow erratic flopping wingbeats and very manoeuvrable tumbling flight in display making 'lapping' sound. Call a loud 'pee-wit'. Very gregarious. Runs and pauses while feeding. **Range and habitat:** widespread in mainly temperate open habitats of WP and EP, including farmland and marshes. Partially migratory, reaching OR, and often on coasts in winter. **Similar species:** other lapwings (see pp 110, 200, 250, 294).

Turnstone *Arenaria interpres* (Scolopacidae) 22-4 cm (8½-9½ in). Boldly marked small short-billed wader. Breeding plumage has white head and underparts, dark streaks on crown and black on face and breast. Upperparts chestnut with blackish stripes and mottlings. Underparts white. In winter, becomes grey; blackish on back, head and breast. In flight, highly patterned with 2 bold white wing-bars, white rump and black band to tail. Gregarious outside breeding season. Call a rapid 'kit-it-it'. Flight fast but steady. Walks among seaweed and stones flicking them over with bill. **Range and habitat:** breeds circumpolar Holarctic tundra, also northern temperate coasts of WP and NA. Winters coasts of WP, EP, NA, NT, AF, OR and AU. **Similar species:** Black Turnstone (see p 66).

Oystercatcher
Avocet
Black-winged Stilt at nest

Ringed Plover brooding young
Little Ringed Plover shading eggs

Golden Plover
Grey Plover on tundra

Dotterel
Turnstone on nest
Lapwing

Curlew Sandpiper *Calidris ferruginea* (Scolopacidae) 18-20 cm (7-8 in). Slim, upright wader with prominent white rump. Breeding plumage dark chestnut-brown above, richer chestnut-red below, with black, brown and whitish markings. Whitish wing-bar, and rump and undertail conspicuously white in flight. In winter, brown-grey above and white below, upperparts with dark streaks and pale feather edgings. Thin decurved bill. Found singly or in small groups. Flight note a soft 'chir-rup'. **Range and habitat:** breeds arctic tundra of EP, migrating, some via WP, to coastal and inland wetlands of AF, and coasts of OR and AU. **Similar species:** Dunlin.

Dunlin *Calidris alpina* (Scolopacidae) 16.5-19 cm (6½-7½ in). Small and abundant shore wader with black legs and slightly decurved bill. Breeding plumage distinctive with reddish tinged dark upperparts and black belly patch on paler underparts. Winter plumage brown-grey above, paler below. White wing-bar, outer tail and outer rump show white in flight. Highly gregarious on passage and in winter. Call a grating 'schreep'. Trilling display song. **Range and habitat:** circumpolar Holarctic breeder on arctic tundra, and north temperate moorland and saltmarsh. Migrates in winter to NA, WP, AF and OR, on tidal mudflats and margins of open freshwaters. **Similar species:** other small waders (see pp 64, 250).

Little Stint *Calidris minuta* (Scolopacidae) 14-15.5 cm (5½-6 in). Tiny, dumpy wader. Overall summer plumage dark above with rufous markings, speckled rufous chest and white belly. In winter, loses rufous tinge, becomes very pale grey-brown above, white below. Short black legs and short fine bill. Active, running and pecking, with fast flight when white wing-bar and white sides to rump show. Flight note a high 'chik'. **Range and habitat:** breeds arctic tundra of WP and EP, migrating to winter on coasts and inland freshwaters of south-east WP, and widely in AF and OR. **Similar species:** other small waders and stints (see pp 64, 250).

Knot *Calidris canutus* (Scolopacidae) 24-7 cm (9½-10½ in). Medium-small wader, uniform grey in winter. Breeding plumage chestnut above, marked with black, underparts rich cinnamon-pink. Flight feathers grey and black. Narrow white wing-bar also visible on winter plumage bird, which has all-grey body, whitish on lower belly and rump. Straight black bill, legs olive-green. Highly gregarious, often in very large flocks, wheeling and turning in unison. Call a low 'knut'. **Range and habitat:** widespread breeder on arctic tundra of NA and a few places in EP. Migrates to winter on coasts of NT, WP, AF, OR and AU. **Similar species:** Great Knot.

Sanderling *Calidris alba* (Scolopacidae) 19-22 cm (7½-8½ in). Small, very active wader of winter shorelines. Breeding plumage rufous buff on upperparts and breast, streaked and spotted white and black. White belly, short black bill and black legs. Winter plumage white over much of body, grey back and crown, black streaks on closed wings. Conspicuous white wing-bar in flight. Runs actively to and fro at tide edge. Gregarious. Calls 'twick-twick' in flight. **Range and habitat:** breeds arctic tundras of EP and NA, migrating in winter to coasts of NA, NT, WP, AF, OR and AU.

Purple Sandpiper *Calidris maritima* (Scolopacidae) 20-2 cm (8-8½ in). Small but stocky shorebird of rocky coasts. Dark uniform grey-brown above, paler beneath, with rufous spotting on upperparts in breeding plumage, more even grey in winter, mottled on breast and flanks. Belly white. Moderately long slightly decurved dark bill, yellow at base, and distinctive yellowish legs. Winters in small flocks. Tame at all seasons. Not very vocal, main call a low 'weet-wit'. **Range and habitat:** arctic tundras and upland plateaux of WP, EP and NA. Moves to coasts of WP and NA for winter.

Red-necked Phalarope *Phalaropus lobatus* (Scolopacidae) 17-19 cm (6½-7½ in). Slender, almost dainty swimming wader with needle-thin bill. Breeding female much brighter than male, slate-grey head, chest and back, last-named with buff lines. White chin, orange-red sides of neck and chest,

white underparts. Winter plumage pale blue-grey above, white below, dark markings round eye and on wings. Prominent white wing-bar. Very tame. 'Tirrick' call when alarmed. **Range and habitat:** breeds widely on circumpolar Holarctic tundra and north temperate freshwaters. Polyandrous. Winters at sea off coasts of west NT, OR and northern AU. **Similar species:** Grey Phalarope.

Redshank *Tringa totanus* (Scolopacidae) 26-30 cm (10-12 in). Medium-sized shank with noticeable red legs and noisy calls. Grey-brown above, streaked breast and whitish underparts, lacking obvious markings. Longish bill red at base. In flight appears black and white, with dark wings and broad white trailing edge, white rump. Fast, active flight with jerky wingbeats. Alarm and flight calls, musical, ringing 'tu-tu-tu' or 'tewk'. Gregarious outside breeding season. **Range and habitat:** widespread on coasts and wide variety of freshwater habitats of WP and EP, mostly migratory south to northern AF, OR and AU. **Similar species:** other shanks (see p 62).

Spotted Redshank *Tringa erythropus* (Scolopacidae) 29-32 cm (11½-13 in). Large shank, all-dark in summer. Breeding plumage black all over, relieved by white spotting on back and dusky tone to face, neck and chest. Base of bill and legs dull red. In winter, ash-grey above, paler below. In flight, wings dark, rump and lower back white. Gives diagnostic clear 2-note 'chew-it' call on being flushed. **Range and habitat:** breeds by marshes and pools among scattered trees in northern WP and EP, migrating south reaching AF and OR. **Similar species:** none in summer, other large shanks in winter (see p 62).

Greenshank *Tringa nebularia* (Scolopacidae) 29-32 cm (11½-13 in). Large, long-legged shank. Dark grey above, spotted and streaked dark grey and black in summer. Paler grey on neck and breast, streaked darker in summer. Belly white. Long bill slightly upturned. Legs green. On taking flight, gives far-carrying 'teu-teu-teu' call, and shows conspicuous white rump and white wedge up back. **Range and habitat:** breeds by marshes and pools often among trees in northern WP and EP, wintering south in WP and widely in AF, OR and AU. **Similar species:** other large shanks (see p 62).

Common Sandpiper *Actitis hypoleucos* (Scolopacidae) 19-21 cm (7½-8 in). Small, stocky wader. Upperparts olive-brown, underparts pure white with brown mark on either side of breast. White eye stripe, broad white wing-bar, barred white sides to tail. Call a shrill 'twee-twee-twee', given in flight, when wings held down in short glides. Constantly wags rather long tail when walking or still. Usually solitary, or in pairs or small groups. **Range and habitat:** wide variety of freshwaters, often running water, almost throughout WP and EP, migrating to winter widely in AF, OR and AU. **Similar species:** Spotted Sandpiper (see p 62).

Green Sandpiper *Tringa ochropus* (Scolopacidae) 22-4 cm (8½-9½ in). Medium-small sandpiper, noticeably dark above. Upperparts dark olive-brown with small white spots present in summer, soon wearing away. Neck, breast and flanks spotted and streaked brown in summer, paler in winter, white on belly. Rises abruptly when flushed and towers, giving loud 'weet-weet-weet' note before zig-zagging away. **Range and habitat:** breeds swamps and marshes often among trees in northern WP and EP, wintering south to southern WP, AF and OR. **Similar species:** other medium-small sandpipers (see pp 64, 250).

Curlew Sandpiper coming into summer plumage

Dunlin in summer plumage

Little Stint

Knot

Sanderling

Purple Sandpiper

Red-necked Phalarope

Redshank

Spotted Redshank at nest

Greenshank

Common Sandpiper

Green Sandpiper

Ruff *Philomachus pugnax* (Scolopacidae) 22-31 cm (8½-12 in). Wader with longish neck and small head, very variable in size. Larger male develops bizarre multi-coloured ruff and ear-tufts, black, white, rufous or mixture, in spring. Female grey-brown marked with sepia. In winter both grey-brown, with scaly pattern on closed wing. In flight, narrow white wing-bar and conspicuous white oval either side of dark rump. Rarely calls. Elaborate displays by groups of courting males. Often singly, small flocks at other times. **Range and habitat:** breeds lowland marshes and wet scrubby areas of temperate and northern WP and EP. Migrates to fresh and brackish waters in southern WP, AF and OR. **Similar species:** Buff-breasted Sandpiper (see p 64).

Curlew *Numenius arquata* (Scolopacidae) 51-61 cm (20-4 in). Very large, long-legged wader. Body overall pale grey-brown, variably streaked darker brown on head, neck and breast merging to produce ill-defined bib; paler on back and wings. Belly white, darker markings on flanks. Conspicuous white rump in flight. Bill 10-15 cm (4-6 in) evenly decurved; long, dark horn, very long legs blue-grey. Loud 'cour-lee' call and bubbling musical song delivered in flight. Gregarious outside breeding season. **Range and habitat:** open habitats with some water, often upland, in temperate and northern WP and EP. Migrates to winter mainly on coasts of WP, EP, AF and OR. **Similar species:** Whimbrel.

Whimbrel *Numenius phaeopus* (Scolopacidae) 39-43 cm (15-17 in). Large, long-legged brown wader with decurved bill. Brown plumage, darker above, streaked and spotted buff. Dark crown split by broad buff stripe; buff stripes above eyes. Belly white, streaked brown on flanks. White rump shows in flight, absent in NA race. Long bill bent downwards towards tip. Call a 7-note tittering trill 'ti-ti-ti-ti-ti-ti-ti'. Also bubbling song. Gregarious in small flocks. **Range and habitat:** discontinuous circumpolar breeding distribution through northern open forests, marshes and moorlands of WP, EP and NA. Winters south to coasts of NA, NT, AF, OR and AU. **Similar species:** other curlews (see p 64).

Bar-tailed Godwit *Limosa lapponica* (Scolopacidae) 36-40 cm (14-16 in). Medium to large slim wader. Summer adult has reddish chestnut underparts with fine black streaking on crown and nape and heavy black markings on rufous back. Wing-coverts grey-brown. In winter, pale grey-brown streaked above and toning to whitish below. Fine, tapering bill 8.5-10 cm (3¼-4 in) long, gently upcurved. In flight, whitish triangular rump above barred tail. Harsh 'kirrik' flight call. Gregarious. **Range and habitat:** breeds subarctic tundra and marshes of WP and EP, and north-western NA. Winters coasts of WP, AF, OR and AU. **Similar species:** other godwits (see p 64, 66).

Black-tailed Godwit *Limosa limosa* (Scolopacidae) 38-43 cm (15-17 in). Large, upright wader with fine straight bill, long legs trailing in flight. Breeding plumage pink-chestnut on head, breast and upper back; streaked and barred blackish on lower back and flanks. Belly whitish, wing-coverts grey-brown. In winter, uniform pale grey-brown above, white below, head and neck brownish. White rump and black-barred tail. Broad white wing-bar and blackish trailing edge in flight. Loud 'wicka-wicka' flight call. Gregarious. **Range and habitat:** moist lowland areas of mainly temperate WP, also scattered in EP. Migrates to inland and coastal wetlands of WP, AF, OR and AU. **Similar species:** other godwits (see p 64, 66).

Woodcock *Scolopax rusticola* (Scolopacidae) 32-6 cm (13-14 in). Large, heavily built wader with shortish legs and long, straight bill. Camouflaged plumage of reddish brown, darker above, mottled and marbled with black. Broad black bars on head and nape, face and throat buff. Rufous tail. Very broad, rounded wings, flies with bill angled downwards. Solitary and wary, flying mainly at night or at dusk, weaving between trees. Silent except in 'roding' display flight when gives grunting notes followed by 'twisick'. **Range and habitat:** widespread in moist woodlands with clearings throughout temperate and boreal WP and EP. Migrates to similar or scrub habitat in southern WP and EP, also OR. **Similar species:** other woodcocks (see p 66).

Snipe *Gallinago gallinago* (Scolopacidae) 25-8 cm (10-11 in). Medium-sized wader, bill 6-7 cm (2¼-2¾ in) long. Upperparts dark brown mottled black; cream stripes on head and back. Pale buff cheeks and neck, streaked darker, belly white, boldly barred flanks. When flushed, gives 'scaap' call and zig-zags, then towers. White trailing edge to wing. Display call a persistent 'chip-er', also 'drumming' with stiff outer tail feathers. Wary, sometimes in small groups. **Range and habitat:** very widespread in lowland wetlands and scrub of temperate and northern Holarctic, also NT and AF. Northern populations migratory to southern WP, EP, NA; also NT, AF and OR. **Similar species:** other snipe (see pp 110, 202, 250).

Stone-curlew *Burhinus oedicnemus* (Burhinidae) 38-43 cm (15-17 in). Robust, rather ungainly wader. Overall plumage sandy brown, streaked blackish on upperparts and breast, belly white. Large, staring yellow eyes, yellow and black bill, yellow legs. Slow and direct flight showing 2 white and 1 black wing-bars and white underwings. Walks or runs over ground, stopping at intervals. Has wild, wailing 'coor-lee' call. Rather shy, most active at dusk and during night. **Range and habitat:** dry heathlands and steppes of southern WP and EP, also OR; wintering south into AF more often at or near wetlands. **Similar species:** other stone-curlews (see pp 110, 198, 294).

(Collared) Pratincole *Glareola pratincola* (Glareolidae) 24-7 cm (9½-10½ in). Slender tern-like wader with long wings and short legs, and short, stubby bill. Upperparts uniform darkish brown. Narrow black line outlines cream throat. Breast and flanks fawn, rest of underparts white. White rump conspicuous in flight, chestnut underwings not obvious alongside black wing feathers. Fast and erratic tern-like flight on long pointed wings; tail deeply forked. Tern-like call, 'kir-rik'. Gregarious and colonial breeder. **Range and habitat:** scattered breeder on steppes, dry floodplains and semi-deserts in southern WP and EP and throughout AF. Palaearctic populations winter AF. **Similar species:** other pratincoles (see pp 202, 250).

Great Skua *Catharacta skua* (Stercorariidae) 56-61 cm (22-4 in). Very large and dark skua, more gull-like in shape than other skuas. Plumage dark brown on upperparts, grey-brown below, mottled with rufous. Yellowish streaks on head, lacking in winter. Stout, hooked bill. White patches towards tips of wings, both above and below, conspicuous in flight, which is fast and powerful when chasing other birds, slower and heavier at other times. Nests colonially, defending nest territory vigorously, striking intruders. Loud 'kak-kak' and 'ha-ha' calls. **Range and habitat:** breeds moorland and tundra in few localities northern WP, other races breeding NT and AN. Winters at sea. **Similar species:** South Polar Skua.

Arctic Skua *Stercorarius parasiticus* (Stercorariidae) 38-48 cm (15-19 in). Medium-sized skua with narrow, pointed wings and projecting central tail feathers; 2 colour phases plus intermediates. Dark phase dark brown all over, greyer below and with black cap. Pale phase underparts white, tinged yellow on cheeks and neck, grey-brown breast and flanks, sometimes breast band. Wings all-dark with pale patch towards tip. Graceful flight, fast and skilful when chasing other birds. Loud 'kaa-aow' call when breeding. Colonial breeder. **Range and habitat:** circumpolar Holarctic tundra and moorland. Winters at sea and round coasts of WP, EP, AF, NA, NT, OR and AU. **Similar species:** Long-tailed and Pomarine Skuas (see p 68).

Ruff

Curlew at nest

Whimbrel at nest

Bar-tailed Godwit on tundra breeding grounds

Black-tailed Godwit

Woodcock at nest

Snipe

Stone-Curlew (pair with young)

(Collared) Pratincole at nest

Great Skua

Arctic Skua

Black-headed Gull *Larus ridibundus* (Laridae) 35-8 cm (14-15 in). Small gull with chocolate brown head in summer. Nape and whole of underparts white. Back grey. Wings grey with prominent white leading edge and black tips. Winter adult white-headed with dusky black marks on head and behind eye. Thin bill and legs dark red. Flight buoyant. Gregarious, breeding colonially. Often tame. Harsh 'kraah' call. **Range and habitat:** widespread on coasts, inland marshes and pools of temperate and northern WP and EP. Winters inland and coasts of WP, EP, also NA, AF and OR. **Similar species:** other small dark-headed gulls (see pp 62, 250).

Mediterranean Gull *Larus melanocephalus* (Laridae) 37-40 cm (15-16 in). Medium-sized gull with black head in summer. All-white neck and underparts, dove-grey wings and back. Stout reddish bill, with slightly drooping tip, and red legs. Black head, except in winter, when blackish mark behind eye and dusky areas over crown. In flight, white on outer half of upperwing and white underwing. Gregarious at all seasons; breeds colonially. Harsh 'kee-ow kee-er' calls. **Range and habitat:** restricted as breeder to scattered lowland marshes of central and southern WP. Winters coasts, including to north. **Similar species:** other small dark-headed gulls (see pp 66, 250, 296).

Herring Gull *Larus argentatus* (Laridae) 53-9 cm (21-3 in). Abundant heavily built large gull. Head, neck, underparts and tail white. Back and upperwings variable slate-grey to silver-grey. Outer primaries black with white tips. White trailing edge to wings. Heavy yellow bill with red spot near tip of lower mandible. Legs pink or yellow. Gregarious, often tame. Many noisy calls, including 'kee-yow' and 'gah-gah-gah'. **Range and habitat:** widespread in coastal WP and central and northern EP and NA. Winters coasts and some inland areas of WP, EP and NA. **Similar species:** other large gulls (see pp 66, 68, 296).

Lesser Black-backed Gull *Larus fuscus* (Laridae) 51-6 cm (20-2 in). Large gull with dark upperparts, varying from dark grey to black with race, also within races. Head, neck, underparts and tail white. Wings dark above, white below. White tips to primaries and white trailing edge. Stout yellow bill with red spot on lower mandible. Legs and feet yellow. Noisy calls, including 'kyow-kyow' and 'gar-gar-gar'. Gregarious. **Range and habitat:** breeds coastal and some inland wetlands of western and northern WP and EP. Winters at sea and coasts of WP, NA and coasts and inland AF. **Similar species:** other large dark-backed gulls (see pp 68, 296).

Greater Black-backed Gull *Larus marinus* (Laridae) 64-79 cm (25-31 in). Very large bulky gull. Back and upperwings slate-black. Head, neck, underparts and tail pure white. Trailing edge to wings and tips of outermost feathers white. Heavy, deep bill yellow with red spot on lower mandible. Legs flesh-coloured. Black eye-mark gives scowling appearance. Harsh barking call, 'yowk'. **Range and habitat:** breeds temperate and northern coasts of WP and NA. Winters same coasts and to south, also at sea. **Similar species:** other large dark-backed gulls (see pp 68, 296).

Common Gull *Larus canus* (Laridae) 38-43 cm (15-17 in). Medium-sized grey-backed gull. Head, neck, underparts and tail white, back and upperwings pale blue-grey. Outer half of primaries black, white spots at tips. Bill and legs yellow-green. Easy, graceful flight. High-pitched 'kee-yaa' call. Very gregarious at all times, breeding in colonies. Often inland. **Range and habitat:** widespread breeder on coasts and inland temperate and northern WP and EP and western half of NA. Winters coasts and inland within southern part of breeding range and to south, reaching OR. **Similar species:** other medium-sized gulls (see pp 66, 110).

Kittiwake *Rissa tridactyla* (Laridae) 38-43 cm (15-17 in). Slim and dainty gull. Back and upperwings darkish grey, darker in NA race. Head, neck, underparts and tail pure white. Wings show black triangular tips and white trailing edge. Bill yellow-green, legs black. In winter, nape grey, blackish marks behind eye. Very buoyant, graceful flight. Gregarious at all times, breeding in dense, cliff-ledge colonies. Very noisy 'kittee-wayke' call when nesting. **Range and habitat:** northern and arctic coasts, especially cliffs, of WP, EP and NA. Winters at sea in northern oceans. **Similar species:** small gulls (see pp 66, 110).

Sandwich Tern *Thalasseus sandvicensis* (Laridae) 38-43 cm (15-17 in). Medium-large tern with slim yellow-tipped black bill. Crown and shaggy crest black, pale ash-grey back and upperwings, latter fading to white at edges. Forehead and crown white in winter, white streaks in black crest. Black legs and feet, with yellow soles. Flight strong and buoyant on long pointed wings, forked tail. Harsh 'kirrick' call. Gregarious, colonial breeder. **Range and habitat:** breeds on flat coasts of south and west WP, also scattered localities in NT. Winters coastal waters of WP, NA, AF and OR. **Similar species:** other medium to large terns (see pp 68, 110, 202, 250).

Common Tern *Sterna hirundo* (Laridae) 30-6 cm (12-14 in). Slender medium-small tern. Cap black, extending down nape. Upperparts and upperwings pale grey, sides of face and neck white, underparts and forked tail very pale grey-white. Underwings white. Longish bill deep red with black tip. Legs and feet red. In winter, forehead becomes white and bill black with reddish base. Harsh, grating 'kee-yar' call. Colonial breeder. **Range and habitat:** widespread inland and coastal breeder in WP, EP and NA, also scattered in NT and AF. Winters at sea and coasts of WP, EP, NA, NT, AF, OR and AU. **Similar species:** other medium-small terns (see pp 68, 202, 250, 296).

Arctic Tern *Sterna paradisaea* (Laridae) 30-9 cm (12-15 in). Slender medium-small tern; tail deeply forked, 10-18 cm (4-7 in) long. Black cap and nape. Pale blue-grey upperparts, white cheeks and pale grey underparts. Longish, pointed bill blood-red. Short legs and feet red. Wings pale grey above, white below, appear translucent from beneath and edged blackish. Forehead whitish in winter. Graceful, buoyant flight, diving vertically. Colonial breeder. Harsh, grating 'kee-errr' call. **Range and habitat:** circumpolar northern and arctic Holarctic coasts and tundra. Winters south polar seas. **Similar species:** other medium-small terns (see pp 68, 250, 296).

Little Tern *Sterna albifrons* (Laridae) 23-6 cm (9-10 in). Tiny, short-legged tern. Black cap and nape, white forehead and streak over eye. Blue-grey upperparts shading to white rump and tail. Blue-grey and white upperwings. White face, underparts and underwings. Yellow bill with black tip, legs orangey. Black cap mostly lost in winter, greyer wings. Active, jerky flight, hovering before diving. Shrill 'kik-kik' call. Colonial breeder. **Range and habitat:** widespread breeder on shingle coasts and inland in temperate and southern WP, EP and NA; also NT, AF, OR and AU. Winters coasts and shallow seas. **Similar species:** other small terns (see pp 68, 250).

Black Tern *Chlidonias niger* (Laridae) 23-6 cm (9-10 in). Marsh tern with slightly forked tail. In summer, plumage very dark slate-grey, pale grey underwings and white at base of undertail. In winter, black-capped head, white underparts and much paler underwings. Restless flight on long wings, beating to and fro, dipping to surface. Thin squeaky calls. Gregarious, colonial breeder. **Range and habitat:** widespread breeder in south temperate freshwater wetlands in WP, EP and NA. Winters estuaries and coastal lagoons of AF and NT. **Similar species:** other marsh terns.

Black-headed Gull

Mediterranean Gull coming into summer plumage

Herring Gull

Lesser Black-backed Gull

Greater Black-backed Gull

Common Gull

Kittiwake (pair) at nesting ledge

Sandwich Tern

Common Tern

Arctic Tern

Little Tern at nest

Black Tern and young at nest

Puffin *Fratercula arctica* (Alcidae) 29-31 cm (11½-12 in). Stubby auk with round head and grotesque face and bill in summer. Upperparts black, underparts white. Face grey, red eye-ring and blue appendages above and below. Greatly enlarged triangular bill striped blue-grey, yellow and orange. Much reduced by moult and mostly yellow in winter and face much greyer. Legs and feet red-orange. Rapid whirring flight. Wings pure black above, white below. Colonial breeder. Deep growling 'kaa-arr-arr' call at nest. **Range and habitat:** breeds coastal islands and stacks north-west WP and north-east NA, including arctic. Winters at sea. **Similar species:** other auks (see p 68).

Razorbill *Alca torda* (Alcidae) 39-43 cm (15-17 in). Medium to large auk with deep, vertically flattened bill. Black head, neck and upperparts. Narrow white bar across folded wings, forming white trailing edge in flight. Underparts white, inner part of underwings white, outer grey. Vertical white line on black bill and white line from upper bill to eye. In winter, cheeks and throat whitish. Flight fast and whirring. Gregarious. Colonial breeder. Has variety of growling and snoring notes at colony. **Range and habitat:** northern and arctic cliff coasts of north-west WP and north-east NA. Winters at sea. **Similar species:** other large auks (see p 68).

Guillemot *Uria aalge* (Alcidae) 40-4 cm (16-17 in). Large auk with longish, pointed black bill. Head, neck, and upperparts blackish brown, more grey-brown in southern race. Thin white bar across folded wings shows as narrow white trailing edge in flight. Dark, indented furrow behind eye. White eye-ring and white line along furrow in proportion of birds ('bridled'). Dark marks on flanks, otherwise underparts white. Black of neck and cheeks becomes white in winter. Fast, whirring flight. Colonial breeder. Loud growling and trumpeting calls at colony. **Range and habitat:** northern and arctic cliff coasts of WP, EP and NA. Winters at sea. **Similar species:** other large auks (see p 68).

Black Guillemot *Cepphus grylle* (Alcidae) 33-5 cm (13-14 in). Small, compact auk. Breeding plumage all-black except for large oval white patch on wings, highly visible at rest and in flight. Longish, pointed black bill and very obvious bright red legs and feet. In winter, becomes overall grey-white on head and body, retaining black tail and black and white wings. Rapid, whirring flight. Walks relatively easily. Solitary or in small groups. Call a shrill whistle. **Range and habitat:** northern and arctic coasts of WP, EP and NA, wintering at sea. **Similar species:** winter plumage confusable with male Smew (see p 148).

Little Auk *Alle alle* (Alcidae) 20-2 cm (8-8½ in). Tiny, stubby auk. Head, neck and breast chocolate brown, upperparts and tail glossy black. White streaks down back and across folded wings. Wings black above, with white trailing edge, underwing mainly dark. Tiny white mark above eye. Very short black bill. Highly gregarious, breeding in enormous colonies, though gale-blown vagrants often solitary. Very fast agile flight, flocks twisting and turning in unison. High-pitched laughing cries at breeding colony. **Range and habitat:** arctic coastal screes of north-west WP and extreme north-east NA. Winters at sea. **Similar species:** small auklets (see p 68).

PIGEONS AND SANDGROUSE Columbiformes

Pin-tailed Sandgrouse *Pterocles alchata* (Pteroclididae) 31-9 cm (12-15 in). Short-legged and short-necked terrestrial birds with greatly elongated tail streamers. Male has greenish upperparts, spotted yellow in spring, yellow-chestnut face, black eye-line, chin and throat. Black-outlined broad chestnut-buff breast band, pure white belly and underwing. Female duller, more buff on upperparts, more yellow on face and chest. Highly gregarious. Groups undertake close aerial manoeuvres and gather at drinking places. Ringing 'cata-cata' call in flight. **Range and habitat:** lowland plains of southern WP and south-west WP, most

resident, some winter in south-west OR. **Similar species:** other sandgrouse (see p 204).

Rock Dove *Columba livia* (Columbidae) 31-5 cm (12-14 in). Medium-sized plump dove with shortish tail, ancestor of highly variable domestic pigeons. Blue-grey head and body, with glossy purple breast and glossy green neck. Paler grey on back and closed wings, which have 2 black bars. Rump white, tail blue-grey with broad blackish band at tip. Bill dull grey with whitish cere. Dashing flight, often low with wheeling and gliding. 'Coo-roo-coo' call in display. Gregarious. **Range and habitat:** widespread in cliff caves (domestics use buildings) throughout WP, EP, NA, OR and scattered in NT and AU. Non-migratory. **Similar species:** other doves (see p 204).

Stock Dove *Columba oenas* (Columbidae) 31-5 cm (12-14 in). Medium-sized compact dove with shortish tail. Upperparts and underparts mostly grey, with blue tone on head and underbody, slatey grey on back. Glossy green sides to neck, purplish tinge to breast. Forewing pale grey with 2 short black wing-bars. Outer wing dark grey. Tail with broad black band at tip. Bill yellowish, legs and feet pink. Active flier in small groups or pairs. Call a grunting repeated double 'coo'. **Range and habitat:** widespread in open woodland of temperate and southern WP and western EP. Some resident, others winter south, mainly within breeding range. **Similar species:** other doves (see pp 204, 252).

Woodpigeon *Columba palumbus* (Columbidae) 39-43 cm (15-17 in). Large, bulky pigeon with rather small head and long tail. Head, back and upper tail blue-grey with lighter grey mantle and forewings. Glossy green sides and back of neck with bold white patch. Breast pinkish, belly blue-grey. White wing-bar shows on folded wing. Flight feathers and tail tip blackish. Fast, powerful flight, with clattering take-off. Song a 5-note rhythmic cooing. Often in large flocks. **Range and habitat:** widespread in temperate and northern woodlands of WP and western part of EP. Non-migratory or moves south within breeding range. **Similar species:** other large pigeons (see pp 204, 252).

Collared Dove *Streptopelia decaocto* (Columbidae) 29-32 cm (11½-13 in). Medium-sized long-tailed dove. Overall plumage dull grey-brown, pinkish on head and breast, becoming whitish buff on belly and undertail. Black collar round hindneck, edged white. Outer wings dark, inner wings paler underneath. Outer tail whitish above, clearer white below, contrasting with dark inner tail. Persistent 3-note cooing song. Rapid flight, often gliding or swooping. Gregarious, especially when feeding. **Range and habitat:** southern and temperate lowlands of WP, also OR and southern EP. Resident but much dispersal by juveniles, including colonization of WP this century. **Similar species:** other doves (see pp 204, 252).

Turtle Dove *Streptopelia turtur* (Columbidae) 26-9 cm (10-11½ in). Small, graceful dove. Grey-blue head and nape; pinky mauve neck and breast, former with black and white on sides, latter shading to whitish belly. Back rufous brown, mottled with black. Tail conspicuously black with white edgings above and white tip below. Wings chestnut and grey above, paler below. Some racial variation across range. Rapid and agile flight, particularly among trees. Long-drawn-out crooning 'roorr' call. **Range and habitat:** breeds in open country with scattered trees in temperate and southern WP and EP, migrating south to central AF for winter. **Similar species:** other doves (see pp 204, 252).

Puffin

Razorbill

Guillemot

Black Guillemot

Little Auk

Pin-tailed Sandgrouse

Rock Dove

Stock Dove

Woodpigeon

Collared Dove

Turtle Dove

CUCKOOS AND ALLIES Cuculiformes

Cuckoo *Cuculus canorus* (Cuculidae) 32-4 cm (12-13 in). Long-tailed, hawk-like parasitic bird. Male dull grey above, darker brown-grey on tail and flight feathers. Throat and breast grey, rest of underparts white narrowly barred brown-black except undertail. Some white spotting on tail. Adult female commonly browner with less distinct barring beneath. Rarely, female rufous above, pale below, barred all over with black. Direct flight on pointed wings. Familiar 'cuck-oo' call from male; female has bubbling trill. **Range and habitat:** widespread throughout WP and EP, also southeast OR; migrates to southern AF and OR for winter. **Similar species:** other cuckoos (see pp 204, 252, 302).

OWLS Strigiformes

Barn Owl *Tyto alba* (Tytonidae) 33-6 cm (13-14 in). Medium-sized owl. Upperparts, including crown and hind-neck, orangey buff, finely speckled and mottled brown, grey and white. Pronounced facial disc white outlined with brown. Whole of underparts pure white, with slight dark spotting and buff tone, more obvious on female, or, in other races, greyer above and more buff below. Long, white feathered legs. Slow, wavering but buoyant flight. Mainly nocturnal and crepuscular. Shrieks in flight, snores at nest. **Range and habitat:** widespread in lowlands of temperate WP and NA, and almost throughout NT, AF, OR and AU.

Little Owl *Athene noctua* (Strigidae) 21-3 cm (8-9 in). Small owl with broad head. Upperparts darkish brown (much paler in some races), flecked and spotted white. Obvious facial disc whitish with brown streaking round bright yellow, fierce-looking eyes. Underparts white, heavily spotted brown forming band on breast, but lighter on belly. Wings brown above, white below. Flight fast and bounding. Main call sharp 'kee-oo'. Solitary or in pairs. **Range and habitat:** widespread in lowland lightly wooded country, also dry uplands throughout temperate and southern WP and EP, and parts of AF. Non-migratory. **Similar species:** other small owls (see pp 72, 254).

Tawny Owl *Strix aluco* (Strigidae) 36-40 cm (14-16 in). Medium-sized owl with rounded face. Facial disc pale grey-brown with paler areas between noticeably black eyes. Upperparts variable brown, red-brown or occasionally greyish, streaked and barred dark brown and white. Underparts paler brown to whitish, streaked darker brown. Broad rounded wings barred towards tips; wingbeats slow and deliberate. Nocturnal, roosts by day in hole or against trunk, calling sharp 'kee-vick' as well as familiar long hooting song. Solitary or in pairs. **Range and habitat:** in wooded areas, less often parks and towns, of much of WP and south and east EP. Non-migratory. **Similar species:** other medium-sized owls.

Long-eared Owl *Asio otus* (Strigidae) 34-7 cm (13-15 in). Medium-sized owl with prominent ear-tufts. Overall plumage rufous brown only slightly paler below, mottled and barred grey and buff above, streaked darker below. Elongated face buff, outlined in black, with dark brown patches around eyes. Nocturnal but sometimes hunts by day, with few beats of broad rounded wings interspersed with long glides. Various sharp barking and squealing calls, also double 'oo-oo' song. **Range and habitat:** mainly wooded areas with clearings throughout temperate and northern WP, EP and NA, partially migratory to south, reaching OR. **Similar species:** other medium-sized owls.

NIGHTJARS AND ALLIES Caprimulgiformes

Nightjar *Caprimulgus europaeus* (Caprimulgidae) 26-8 cm (10-11 in). Large-headed crepuscular and nocturnal bird with huge gape for catching aerial insects. Superbly camouflaged, brown with elaborate patterns of white, silver-grey, buff, rufous, dark brown and black markings, including streaks, spots and bars. In buoyant flight, which includes swift turns and glides, prominent white spots visible near wing and tail tips (buff spots in female). Solitary or in pairs. 'Koo-ick' flight call, song a prolonged churring trill. **Range and habitat:** lowland woods, plantations with open glades in temperate and southern WP and EP, migrating to AF for winter. **Similar species:** other nightjars and nighthawks (see pp 72, 116, 212, 254, 304).

SWIFTS AND HUMMINGBIRDS Apodiformes

Swift *Apus apus* (Apodidae) 16-17 cm (6-6½ in). Very long-winged, short-tailed all-dark swift. Overall plumage colour black-brown, some green gloss on back. Forehead slightly grey, chin and throat white, sometimes brown tinged. Very faint whitish markings on feathers of rump and underparts. Completely aerial, only landing at nest site. Scythe-shaped wings held stiffly and beaten rapidly between long glides. Screaming calls during group chases around buildings. **Range and habitat:** widespread round built-up areas, also mountain and cliff terrain throughout much of WP and EP, migrating south to southern AF for winter. **Similar species:** other swifts (see pp 72, 116, 216, 254, 304).

KINGFISHERS AND ALLIES Coraciiformes

Bee-eater *Merops apiaster* (Meropidae) 27-9 cm (10½-11½ in). Brilliantly coloured bee-eater with long curved bill and extended inner tail feathers. Chestnut crown and nape, becoming golden on back. White, blue, black and green stripes on head. Throat bright yellow, separated from green-blue underparts by narrow black line. Long tail green. Wings mainly green with chestnut inner areas. Flight a mixture of wheeling and gliding, interspersed with rapid wingbeats in pursuit of aerial prey. Distinctive liquid 'quilp' call. Gregarious and colonial breeder. **Range and habitat:** varied open landscapes with soft-soiled nesting cliffs in southern WP and EP, also extreme southern AF. Migrates to AF for winter. **Similar species:** other bee-eaters (see pp 210, 256, 304).

Roller *Coracias garrulus* (Coraciidae) 29-32 cm (11½-13 in). Large, crow-like bird. Head, body and large panel on closed wing pale turquoise-blue, becoming more purplish blue on lower back and rump. Pale whitish blue streaks on forehead, throat and breast. Upperwings and mantle chestnut-brown. Paler blue tail with dark central feathers and black tips. In powerful flight, upper wings show turquoise-blue and blackish blue. Harsh 'rack-kack' calls. **Range and habitat:** breeds in open woodlands of temperate and southern WP and western EP, migrating to eastern AF for winter. **Similar species:** other rollers (see pp 208, 256).

Kingfisher *Alcedo atthis* (Alcedinidae) 15-16 cm (6-6½ in). Small, brilliantly coloured kingfisher. Crown, nape, moustache and upperparts bright blue, with paler sheen on back and darker tinge to wings and tip of short tail. Cheeks and underparts rich chestnut, whitish on throat and sides of neck. Long dagger-shaped bill black, with some orange at base (more in female). Fast, whirring flight, perches or hovers before plunging into water. Shrill 'cheee' call. **Range and habitat:** small freshwater wetlands in temperate and southern WP and EP, also OR and parts of AU, partially migratory to south. **Similar species:** other small kingfishers (see pp 120, 208).

Hoopoe *Upupa epops* (Upupidae) 27-9 cm (10½-11½ in). Boldly patterned and crested bird with long, slightly decurved bill. Body pink-brown, brighter on head and upper back, paler below, shading to whitish belly. Head surmounted by long fan-shaped crest of pink-brown feathers tipped white and black. Wings and tail boldly barred black and white. Erratic jerking flight of broad rounded wings strongly reminiscent of butterfly. Soft-toned but far-carrying song, 'poo-poo-poo'. **Range and habitat:** open country with scattered woods and trees in central and southern WP and EP, widespread throughout AF and OR, northern population migratory to south for winter.

Cuckoo

Barn Owl

Little Owl

Tawny Owl

Long-eared Owl

Nightjar

Swift

Bee-eater

Roller

Kingfisher

Hoopoe

WOODPECKERS AND ALLIES Piciformes

Green Woodpecker *Picus viridis* (Picidae) 30-3 cm (12-13 in). Large woodpecker often feeding on ground. Red crown extending down nape, mottled with grey. Upperparts dull greeny yellow, becoming bright yellow on rump, which is conspicuous in flight. Underparts pale grey-green, almost white on throat. Black patch round eye and black and red moustache. Powerful undulating flight on broad wings, outer part brownish faintly barred cream. Strong bill used for probing in ground as well as attacking trees. Hops in upright stance on ground. Loud, laughing 'plue-plue-plue' call. Solitary. **Range and habitat:** woodland and open country with trees in temperate and southern WP. Resident. **Similar species:** other large woodpeckers (see pp 74, 122, 216, 258).

Great Spotted Woodpecker *Picoides major* (Picidae) 22-4 cm (8½-9½ in). Medium-sized, boldly marked woodpecker. Glossy black cap with white forehead and cheeks, red on nape and black moustache. Black back, white shoulder patches, black tail. Underparts white or buff-white, except for red patch under tail. Some races have red and black breast band. Undulating flight with several wing-flaps alternating with complete closure of wings. White patch at base of wing. Climbs spirally round tree trunks. Drums loudly on dead wood, also abrupt, clear 'tchick' call. **Range and habitat:** forests and open woodland of all but extreme southern WP and EP, also parts of OR. Non-migratory. **Similar species:** other medium-sized woodpeckers (see pp 74, 122, 216, 258).

Lesser Spotted Woodpecker *Picoides minor* (Picidae) 14-15 cm (5½-6 in). Very small boldly patterned woodpecker. Crown of male dull red, that of female browny white and black. Back and wings black boldly marked with white transverse bars. Forehead, face and underparts whitish tinged with buff, with black moustache, and sides of breast and flanks streaked black. Undulating flight. Climbs tree-trunks agilely, also flutters and clambers among outer twigs. Main call a slow, weak 'pee-pee-pee', also quiet 'chick'. Drums with bill on dead wood. Shy and solitary. **Range and habitat:** widespread in forests and open woodlands of all but extreme southern WP and EP. Non-migratory. **Similar species:** other small woodpeckers (see pp 74, 216).

Wryneck *Jynx torquilla* (Picidae) 15-16 cm (6-6½ in). Small cryptically patterned woodland bird, rather like large warbler with a longish tail. Overall grey-brown plumage mottled and barred with dark brown and black, producing highly camouflaged effect. Darker above than below. Pale grey unmarked areas above eye and down back. Underparts pale yellowish buff, with light transverse barring. Longish tail with 3 dark bars. Hops on ground with tail raised, flies jerkily, sometimes climbs tree-trunks. At rest twists head and neck to and fro. Shrill 'quee-quee' call. **Range and habitat:** breeds open woodlands, also parks and orchards, of central and southern WP and EP, extending into eastern OR. Migrates to winter in AF and OR.

PERCHING BIRDS Passeriformes

Crested Lark *Galerida cristata* (Alaudidae) 16.5-17.5 cm (6½-7 in). Plump, medium-sized lark with pointed crest. Overall buff-brown, well streaked above, much paler below, shading to whitish on belly and undertail. Diffuse streaks and elongated spots on upper breast. Darker crest and cheek patch, latter outlined with paler buff. Whitish chin. Relatively long bill and short tail, latter with buff outer feathers. In flight, wings broad with buff undersides. Conspicuous and often tame bird with plaintive song of short phrases delivered from ground or air. Clear 'whee-whee-oo' call. **Range and habitat:** open ground, often arid, also around habitations of central and southern WP and EP, also northern AF and OR. Sedentary. **Similar species:** other larks (see pp 218, 260, 306).

Woodlark *Lullula arborea* (Alaudidae) 15-16 cm (6-6½ in). Stocky medium-sized lark with slight crest. Overall buff plumage, shading to buffish white on belly. Streaked black on head with bold whitish shapes above eye which meet behind. Cheeks darker buff. Neck, throat and breast finely streaked black. Upperparts broadly streaked black. In flight, wings rounded and tail short. Black and white marks at bend in wing. Undulating hesitant flight. Song-flight spiral ascent then circling, song liquid and fluting. Flight call 'tit-loo-eet'. Solitary or in pairs. **Range and habitat:** breeds open habitats with scattered trees of temperate and southern WP. Northern populations migratory to winter within range of southern populations. **Similar species:** other larks (see pp 218, 260, 306).

Skylark *Alauda arvensis* (Alaudidae) 17-19 cm (6½-7½ in). Medium-sized lark with prolonged aerial song. Buff-brown above, well streaked with dark brown or black. Crown streaked black ending in small erectile crest. Throat and breast warm buff, finely streaked black-brown. Rest of underparts whitish, streaked dark brown on flanks. Tail with white outer feathers. Rises vertically during song-flight, then hovers and sings for up to 5 minutes before dropping. Pointed wings and longish tail give cruciform shape when hovering. Liquid 'chirrup' call. Gregarious outside breeding season. **Range and habitat:** open habitats throughout WP and EP. Northern populations migrate south in winter to reach AF and OR. **Similar species:** other larks (see pp 218, 260, 360).

Swallow *Hirundo rustica* (Hirundinidae) 17-19 cm (6½-7½ in). Medium-sized hirundine, with outer tail feathers elongated by 2-7 cm (¾-2¾ in). Shiny blue-black upperparts, also narrow breast band. Forehead and throat chestnut-red. Underparts creamy buff to red-brown, varying with race. Underside of forewings similar, rest of underwings dusky black. White spots at tips of tail feathers. Elegant, swooping flight in long curves, but jinking after insects. Song a pleasing mix of warbles and trills. Call a shrill 'witt-witt-witt'. Gregarious, especially on migration. Perches on wires, rarely on ground. **Range and habitat:** widespread breeder throughout all but southernmost WP, EP and NA, also OR. Winters AF, NT, OR and AU. **Similar species:** other hirundines (see pp 78, 132, 224, 260, 306).

Sand Martin *Riparia riparia* (Hirundinidae) 12 cm (4½ in). Small hirundine, nesting in holes. Dark sandy brown above, including head and cheeks, extending into broad breast band. Chin and throat white, rest of underparts white with smudgy brown marks on flanks. Underwings and undertail dusky brown with faint white spotting on some wing feathers. Fast and fluttering flight on narrow pointed wings. Tail only slightly forked. Weak twittering song made up of repeated harsh chirring call notes. Gregarious and colonial breeder. **Range and habitat:** widespread breeder near freshwater wetlands throughout all but southernmost WP, EP and NA, and into OR. Migrates south to NT and AF for winter. **Similar species:** other hirundines (see pp 78, 132, 224, 260, 306).

House Martin *Delichon urbica* (Hirundinidae) 12-13 cm (4½-5 in). Medium-sized round-headed hirundine with forked tail. Upperparts metallic blue-black apart from conspicuous white rump. Tail dull black. Underparts pure white. Undersides of wings grey near body, with dark grey flight feathers. Underside of tail dark grey. Rapid but fairly steady flight, not swooping. Often high-flying. Lands on ground to collect nest mud and can walk fairly easily. Melodious chirruping song. Call a hard 'chirrup'. Gregarious, breeding colonially. Over much of range has deserted natural cliff sites for buildings. **Range and habitat:** widespread throughout WP and EP and parts of OR. Migrating to winter in AF and OR. **Similar species:** other hirundines (see pp 78, 132, 224, 260, 306).

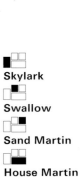

Green Woodpecker

Great Spotted Woodpecker

Wryneck

Lesser Spotted Woodpecker

Crested Lark

Woodlark

Skylark

Swallow

Sand Martin

House Martin

Tree Pipit *Anthus trivialis* (Motacillidae) 15-16 cm (6-6½ in). Medium-sized elegant pipit with quite large bill. Olive-brown upperparts streaked with black, finely on head, broader streaks on back. Rump and tail unstreaked. Underparts yellowish buff, becoming whitish on belly, with lines of bold spots on breast, fading on to flanks. Strong pinkish legs. Perches on trees. Runs and walks and wags tail on ground. Song-flight rising from tree and often dropping back in spirals. Song of several trilling phrases. High-pitched 'tseep' call. Mostly solitary. **Range and habitat:** breeds in open habitats with trees of temperate and northern WP and EP, migrating south to AF and OR for the winter. **Similar species:** other medium-sized pipits (see pp 218, 260, 306).

Meadow Pipit *Anthus pratensis* (Motacillidae) 14-15 cm (5½-6 in). Medium-sized dumpy pipit. Generally dull olive-brown upperparts and paler underparts, shading to dull grey-white. Upperparts heavily streaked blackish. Breast has small, dense blackish streaks gradually reducing on flanks. Outer tail feathers white. Walks and runs on the ground. Brief song-flight has fluttering ascent and a parachuting descent, song a series of tinkling notes followed by a trill during descent. Gregarious outside breeding season. **Range and habitat:** open areas, including moorland and bogs among scrub, in temperate and north WP including low arctic. Partial migrant, moving to south of breeding range in WP and into south-west EP. **Similar species:** other medium-sized pipits (see pp 218, 260, 306).

Rock Pipit *Anthus petrosus* (Motacillidae) 16-17 cm (6-6½ in). Medium-large dark-legged pipit. Dark olive-grey upperparts with faint darker streaking on back. Underparts dirty buff, almost white on throat, usually with some darker streaking on pink-buff breast. Fairly obvious pale stripe above eye. Outer tail feathers pale grey, rest of tail dark brown. Flight buoyant and strong, ending in curving sweep on landing. Short upward song-flight accompanying musical and trilling song. Penetrating 'feest' call. Mostly solitary or in small groups. **Range and habitat:** confined as breeder to rocky coasts of north-west WP, more northerly breeders migrating south-west for winter, mainly on coasts. Others non-migratory. **Similar species:** other medium-sized pipits (see pp 218, 260, 306).

White Wagtail *Motacilla alba* (Motacillidae) 17-18 cm (6½-7 in). Boldly marked wagtail. 2 races, 1 pale grey above, the other black (Pied Wagtail). Crown, chin and breast of both black, forehead and cheeks white. Wing feathers with prominent white edgings. Long tail black with white outer feathers. Flight bounding and easy; runs, walks and hops, head moving backwards and forwards, and tail wagging, especially after a run. Twittering song can be delivered from air. Call a shrill 'tchizzick'. Small flocks form on migration. **Range and habitat:** widespread in wide range of habitats, usually near water, in temperate and northern WP and EP, and northern OR, northern populations migratory to south, reaching AF and OR. **Similar species:** other wagtails (see pp 218, 260).

Yellow Wagtail *Motacilla flava* (Motacillidae) 16-17 cm (6-6½ in). Longish-tailed but small wagtail. Back and rump greeny yellow to olive-brown, underparts bright yellow in male, paler yellow in female. Very variable head colour: heads of males of different races have crowns of black, grey, white or yellow, often with darker area round eye, and white or yellow chin. Females more similar and mostly grey or greeny brown. Long tail dark with white outer feathers, wagged constantly. Brief warbling song. Shrill, plaintive 'schreep' call. Gregarious. **Range and habitat:** widespread in lowlands near wetlands throughout WP and EP, reaching extreme north-west NA. Migrates to AF, OR and AU for winter. **Similar species:** other wagtails (see pp 218, 260).

Grey Wagtail *Motacilla cinerea* (Motacillidae) 18-19 cm (7-7½ in). Very long-tailed and slender wagtail. Male grey above, including head, female duller grey. Whitish stripe

through eye. Male has black bib outlined in white, female chin and throat whitish. Underparts lemon-yellow, duller in female, brightest on breast and undertail. White edgings to wing feathers. Very long tail black with white outer feathers. Very active, with darting and flitting movements. Tail wagged constantly. Trilling, warbling song. Call a staccato 'tzitzi'. Solitary or in small groups. **Range and habitat:** widespread breeder beside mainly running freshwater from lowland to altitude in temperate WP and EP. Mainly migratory south to AF, OR and AU. **Similar species:** other wagtails (see pp 218, 260).

Waxwing *Bombycilla garrulus* (Bombycillidae) 17-18 cm (6½-7 in). Plump woodland bird with pronounced crest. Overall body colour vinous brown to chestnut, pinker on crest, nape and breast, paler on belly, but darker under tail. Black bib outlined with white and black stripe through eye. Wings dark grey-black with white and yellow markings, plus bright red waxy tips to inner flight feathers. Rump and tail grey with yellow band at end of tail. Female generally duller than male with smaller waxy tips. Call a weak trill 'sirrr'. Gregarious. **Range and habitat:** northern forests of WP, EP and western NA. Migratory and eruptive south to open country with scattered trees and hedges. **Similar species:** Cedar Waxwing (see p 78).

Red-backed Shrike *Lanius collurio* (Laniidae) 16-18 cm (6½-7 in). Slimly built, boldly coloured shrike. Male of western race has blue-grey head and nape contrasting with broad black mask running from base of bill through eye. Eastern race male has chestnut head. Cheeks and throat white. Back and upperwings chestnut. Underparts pale pinky white. Outer wings dark brown, tail black with white sides. Female rufous brown above, underparts paler with dark crescentic barring. Perches on bushes. Flight undulating, can hover. Warbling song with harsh 'chack' calls. **Range and habitat:** wooded and open scrub areas in most of WP and EP except extreme north, also northern OR. Migrates south to AF and OR. **Similar species:** other shrikes (see pp 78, 226, 264).

Great Grey Shrike *Lanius excubitor* (Laniidae) 23-5 cm (9-10 in). Large shrike with long tail and longish bill. Head and mantle soft grey. Broad black mask extends back from base of stout, hooked bill through eye, narrowly bordered above with white. Underparts white. Wings and long, rounded tail black, with 1 or 2 small white wing-bars and white outer tail feathers. Solitary. Perches conspicuously on top of low trees and bushes, swooping down to catch prey. Harsh 'chek-chek' call. **Range and habitat:** widespread in open areas with bushes of temperate, southern and desert regions of WP, EP, OR and northern AF, also NA. Mainly non-migratory, with some dispersal southwards of northerly breeding birds. **Similar species:** other shrikes (see pp 78, 226, 264).

Dunnock *Prunella modularis* (Prunellidae) 14-15 cm (5½-6 in). Small, drably plumaged brown and grey ground-dwelling bird. Head and breast dark grey-brown, with brown streaking on crown and ear-coverts. Upperparts and wings rufous brown, inconspicuously streaked black; 2 faint white bars on closed wings. Belly whitish, flanks grey-brown streaked black. Thin, dark bill and quite noticeable red-brown eye. Main call a thin 'tseep', song a nondescript squeaky warbling. Hops and walks on ground, creeps through bushes. Mainly solitary. **Range and habitat:** widespread in forested and open woodland, also parks and gardens, of boreal and temperate WP. Mainly non-migratory, some southward migration within WP. **Similar species:** other accentors and sparrows (see pp 88, 182, 264).

Tree Pipit

Meadow Pipit

Rock Pipit

White Wagtail

Yellow Wagtail
at nest

Grey Wagtail

Waxwing

Red-backed
Shrike

Dunnock

Great Grey
Shrike

Grasshopper Warbler *Locustella naevia* (Sylviidae) 12-13 cm (4½-5 in). Medium-sized warbler. Olive-brown above, tinged yellowish and streaked darker brown, only slightly on rump. Underparts pale buff to white, sometimes yellowish, slightly streaked darker on chest sides. Longish tail broad and noticeably round-ended. Skulks and creeps among reeds and other undergrowth. More often heard than seen. Very high-pitched, far-carrying churring, long-sustained and ventriloquial song. Sharp 'tchik' call. **Range and habitat:** temperate and southern wet or dry well-vegetated marshes, meadows and scrub of WP and western EP. Migrates to AF and OR. **Similar species:** other *Locustella* warblers (see p 268).

Reed Warbler *Acrocephalus scirpaceus* (Sylviidae) 12-13 cm (4½-5 in). Medium-sized warbler. Upperparts warm brown tinged rufous, especially on rump. Underparts paler brown, deep buff on flanks. Faint pale stripe above eye. Tail rounded. Longish, thin bill and grey-brown legs. Skulks in reeds, sidling up stems. Flight jerky and low, with spread tail. Song a prolonged rhythmic series of repeated notes, 'churr-churr-churr, chack-chack-chack'. **Range and habitat:** reed-beds in wetlands of temperate and southern WP and south-western EP, wintering in tropical AF. **Similar species:** other *Acrocephalus* warblers (see pp 222, 310).

Sedge Warbler *Acrocephalus schoenobaenus* (Sylviidae) 12-13 cm (4½-5 in). Medium-sized streaky brown warbler. Overall brown upperparts, streaked with black, most densely on crown above conspicuous creamy stripe over eye. Rump rufous, unstreaked. Tail graduated. Skulks in thick vegetation, sidles up stems. Flight weak-looking with tail fanned. Prolonged song a mixture of harsh and musical notes, including mimicry. Also harsh churring note and anxious 'tuc'. **Range and habitat:** widespread in tall vegetation and thickets usually near water throughout WP, except south, and western EP. Migrates to subtropical and tropical AF. **Similar species:** other *Acrocephalus* warblers (see pp 222, 310).

Whitethroat *Sylvia communis* (Sylviidae) 13-15 cm (5-6 in). Medium-sized grey-brown warbler. Head and cheeks of male grey, throat white. Upperparts grey-brown, streaked rufous on wings. Tail longish, with white sides. Breast pinkish buff. Head in some males and females browner, underparts less pink. Undulating flight between thick cover, but may perch conspicuously or flutter into air to deliver short, scratchy warbling song. Scolding 'chack' and 'wheet' calls. **Range and habitat:** widespread in scrub and hedgerows near cultivation throughout WP and western EP, migrating to tropical AF. **Similar species:** other *Sylvia* warblers.

Lesser Whitethroat *Sylvia curruca* (Sylviidae) 13-14 cm (5-5½ in). Small grey-brown warbler. Upperparts uniform grey-brown, greyer on head and rump, with noticeable dark mask extending below and behind eye. Throat white, breast and belly pinkish shading to buff on flanks and becoming whiter through summer. Outer tail feathers white. Legs dark grey. Solitary and skulking. Hard, rattling song preceded by soft warble. Harsh 'chack' call-note. **Range and habitat:** widespread in scrub and open woodland of WP and central and western EP. Winters in subtropical AF and OR. **Similar species:** other *Sylvia* warblers.

Blackcap *Sylvia atricapilla* (Sylviidae) c. 14 cm (5½ in). Medium-sized grey-brown warbler with black or brown cap. Male upperparts grey-brown with glossy black cap and grey nape, cheeks and underparts, becoming paler on belly. Female olive-brown above, paler grey-brown below, with red-brown cap. Undulating flight and restless, skulking behaviour. Pleasing warbling song, rich and varied, in short phrases, also a harsh 'tak-tak' call. **Range and habitat:** widespread in broad-leaved and mixed woodlands, also parks and gardens, almost throughout WP and central EP. Winters from southern WP to subtropical AF. **Similar species:** other *Sylvia* warblers.

Garden Warbler *Sylvia borin* (Sylviidae) c. 14 cm (5½ in). Medium-sized warbler. Overall grey-brown to buff plumage, greyer on back, paler below especially whitish buff throat and lower belly. Trace of pale stripe over eye. Solitary and skulking, delivering musical and sustained warbling song (similar to Blackcap) from deep in bush. Also harsh 'tak-tak' call. **Range and habitat:** widespread in mixed woodland with scrub layer in much of WP and western EP. Winters subtropical and tropical AF. **Similar species:** other *Sylvia* warblers.

Dartford Warbler *Sylvia undata* (Sylviidae) 12-13 cm (4½-5 in). Small, dark-headed warbler. Dark brown upperparts with slate-grey head, rufous underparts, small white spots on throat. Female browner above, paler pinky buff below. Eye prominently red. Short wings whirr in weak, low flight. Long tail, equals body length, often cocked. Mainly skulking but also perches on tops of bushes. Harsh, warbling song and scolding 'chirr-ik'. **Range and habitat:** confined to scrub and heath areas of south-west WP where resident. **Similar species:** other small dark-headed warblers.

Willow Warbler *Phylloscopus trochilus* (Sylviidae) c. 11 cm (4½ in). Tiny greeny yellow or grey-brown leaf warbler. Upperparts of western race greeny brown, tinged olive, underparts tinged yellow, wearing whiter by late summer. Narrow, pale stripe above eye. Other races paler, less yellow, more brown above and white below. Legs mainly pale brown, sometimes darker. Restless, flitting among vegetation, flicking wings and tail. Song a gentle descending scale. Call a plaintive 'hoo-eet'. **Range and habitat:** widespread in temperate and northern woodlands and thickets of WP and EP. Winters widely in AF. **Similar species:** other leaf warblers (see p 268).

Chiffchaff *Phylloscopus collybita* (Sylviidae) c. 11 cm (4½ in). Tiny brown and buff or grey and buff leaf warbler. Western race upperparts olive-brown with pale yellowish stripe above eye. Underparts pale buff lightly tinged yellow, particularly on breast, belly whitish. Other races much browner or greyer, lacking much or all yellow. Legs normally dark brown. Wings rounded. Restless with much wing flicking. Song a monotonous 'chiff-chaff', plus 'hweet' call. **Range and habitat:** all types of forest and open woodland areas of much of WP and EP, wintering in south of range and in AF and OR. **Similar species:** other leaf warblers (see p 268).

Wood Warbler *Phylloscopus sibilatrix* (Sylviidae) 12-13 cm (4½-5 in). Small to medium-sized leaf warbler. Upperparts greeny yellow with marked yellow stripe above eye. Throat and chest yellow, with contrasting white belly. During summer becomes duller and greyer with less yellow. Songs a repetition of call-note 'pwee-ur', also a quick series of notes ending in a trill. Arboreal. **Range and habitat:** mainly beech or part beech woods of boreal and temperate WP. Migrates to subtropical and tropical AF. **Similar species:** other leaf warblers (see p 268).

Bonelli's Warbler *Phylloscopus bonelli* (Sylviidae) 11-12 cm (4-4½ in). Small grey and white leaf warbler with some yellow. Upperparts grey-brown, but greyer on head and mantle, with very faint pale stripe above eye. Rump and outer tail feathers yellowish, also edges of wing feathers. Underparts grey-white on breast tending to white on belly. Dark pink-brown legs. Short trilling song and 'hoo-eet' call-note. **Range and habitat:** confined to broad-leaved and coniferous woodlands especially in hills of south-western WP. Migrates to winter in tropical AF. **Similar species:** other leaf warblers (see p 268).

Grasshopper Warbler at nest

Reed Warbler at nest

Sedge Warbler at nest

Whitethroat

Lesser Whitethroat at nest

Blackcap

Garden Warbler at nest

Dartford Warbler at nest

Willow Warbler

Chiffchaff at nest

Wood Warbler

Bonelli's Warbler

Goldcrest *Regulus regulus* (Sylviidae) 9 cm (3½ in). Tiny arboreal bird. Overall plumage colour dull green, paler shading to whitish below. Male has orange-yellow crown stripe, female's is yellow, both bordered with black. Pale area round eyes giving 'surprised' expression; 2 white and 1 dark wing-bars. Restless, flitting, canopy feeder. Very high-pitched 'zi-zi-zi' call and repeated 'seeda-seeda-seeda' song. **Range and habitat:** widespread in coniferous and some broad-leaved forests and woodlands of WP, EP and northern OR, including montane. Northerly populations move south in winter. **Similar species:** Firecrest and kinglets (see p 82).

Firecrest *Regulus ignicapillus* (Sylviidae) 9 cm (3½ in). Tiny arboreal bird. Upperparts greenish, underparts whitish. Male crown orange and yellow, female yellow, bordered with black above prominent white stripe over eye, black stripe through eye. Bronze patch on shoulder; 2 white and 1 dark wing-bars. Restless, flitting, among scrub and trees. **Range and habitat:** coniferous and mixed forests and scrub of south-west WP and east and west NA, including montane. Mainly resident, some southerly movement of northern breeders. **Similar species:** Goldcrest and kinglets (see p 82).

Spotted Flycatcher *Muscicapa striata* (Muscicapidae) 14 cm (5½ in). Medium-sized flycatcher. Dull brown plumage, uniform darker above, paler to whitish below. Black streaks on crown, dark brown streaks on throat and breast. Broad bill, large dark eyes. Upright stance on twigs and wires, agile twisting flight to catch insects, returning to same perch. Thin squeaky song of 6 high notes, call a shrill 'tzee'. **Range and habitat:** widespread in open woodland and forest edges, also parks and gardens of WP and EP. Winters subtropical and tropical AF and northern OR. **Similar species:** other flycatchers (see pp 220, 266).

Pied Flycatcher *Ficedula hypoleuca* (Muscicapidae) 12-13 cm (4½-5 in). Small black and white or brown and white flycatcher. Male contrastingly coloured black above with white forehead, chin and underparts, also white wing patches and outer tail feathers. Female olive-brown upperparts, white to whitish buff below, smaller white wing patch, lacks white forehead. Upright stance on perch, restless flicking wings and tail, flying up to catch insects. 'Whit' call and varied, jangling song. **Range and habitat:** deciduous and some mixed woodlands of WP and western EP, migrating to subtropical AF for winter. **Similar species:** other flycatchers (see pp 220, 266).

Stonechat *Saxicola torquata* (Turdidae) 12-13 cm (4½-5 in). Dumpy, short-winged chat. Head and upperparts of western race summer male black with contrasting white neck patches, orange-red breast and white belly. White wing patch and rump. Eastern race brighter, larger orange or white rump. In winter, black mottled with rufous. Female dull brown streaked head and upperparts, pale rufous breast. Very upright stance, bobs constantly. Hard 'tsak-tsak' call, squeaky warbling song, often in flight. **Range and habitat:** open moors, heaths and grasslands of WP, EP and AF, also northern OR. **Similar species:** other chats (see pp 222, 268).

Whinchat *Saxicola rubetra* (Turdidae) 12-13 cm (4½-5 in). Upright, brown and white chat. Male dark brown, streaked black on upperparts, orangey buff below, becoming whitish on belly. Bold white eye stripe and black cheeks. White wing-bar and white sides to tail. Female dull brown, whitish below, less obvious eye stripe and wing-bar. Perches on bushes and wires, constantly bobbing and flicking wings. Scolding 'tic-tic' call and short warbling song. **Range and habitat:** woodland clearings and open areas with bushes throughout most of WP and western EP, wintering in subtropical and tropical AF. **Similar species:** other chats (see pp 222, 268).

Wheatear *Oenanthe oenanthe* (Turdidae) c. 15 cm (6 in). Medium-sized wheatear. Head and back of male soft grey with black mask through eye, bordered above by narrow white stripe. Underparts creamy buff, becoming white on belly. Blackish wings and tail tip, conspicuous white rump. Female duller, browner above, browner mask and wings. Very active, bobbing, wagging tail. Harsh 'chack-chack' call and squeaky warbling song. **Range and habitat:** widespread in open habitats almost throughout WP and EP, including arctic, also northern NA. Winters in dry open areas of AF. **Similar species:** other wheatears.

Black Redstart *Phoenicurus ochruros* (Turdidae) 14-15 cm (5½-6 in). Small, plump thrush. Summer male black all over except for white wing patches and orange-red rump and tail. Underparts of eastern races deep chestnut. Becomes grey-black in winter. Female grey-brown, paler below, with orange-red rump and tail. Active, shivers tail. Sharp calls 'tsic', 'tic' and 'tucc-tucc', warbling song with rattling notes. **Range and habitat:** rocky and mountainous areas, also buildings, of central and southern WP and EP. Winters to south in WP, AF and OR. **Similar species:** other redstarts (see p 268).

Redstart *Phoenicurus phoenicurus* (Turdidae) 14-15 cm (5½-6 in). Slim thrush. Upperparts of summer male grey, with white forehead and black face and throat. Orange-red breast, flanks and tail, whitish on belly. White wing patch in eastern race. Winter male browner. Female grey-brown above, pale orange-buff below, orange-red tail. Restless, quivering tail. 'Hoo-eet' and 'hweet-tuc-tuc' calls, song a squeaky warble. **Range and habitat:** widespread in open broad-leaved and some conifer woodlands of most of WP and central western EP. Winters subtropical AF. **Similar species:** other redstarts (see p 268).

Robin *Erithacus rubecula* (Turdidae) c. 14 cm (5½ in). Small, dumpy thrush. Upperparts and wings uniform olive-brown. Face, throat and breast orange-red edged with grey. Belly and undertail white. Races vary in plumage tone and breast colour. Rounded appearance, lacking obvious neck. Alert, darting flight, often tame. Solitary. Repeated 'tic-tic' alarm call. Pleasing but thin warbling song. **Range and habitat:** widespread in woodland and open areas with bushes or trees throughout WP except extreme south and in north-east EP. Northern populations migrate to south WP and EP in winter. **Similar species:** other small thrushes (see pp 222, 268).

Bluethroat *Erithacus svecica* (Turdidae) c. 14 cm (5½ in). Slim version of Robin with blue throat. Upperparts dark brown, underparts buff-white. White eye stripe. Metallic blue throat and breast of male contains white or red spot, according to race, and bordered in black, white and red. Female throat and breast whitish with slight blue and reddish markings. Chestnut patches at tail base. Skulking. Rich, varied warbling song and 'hweet' and 'tac' calls. **Range and habitat:** widespread in swampy woodland of temperate and northern WP and EP and extreme north-west NA, migrating south to subtropical AF and OR.

Nightingale *Erithacus megarhynchos* (Turdidae) 16-17 cm (6-6½ in). Medium-small brown thrush with chestnut tail. Overall body plumage russet-brown above, becoming brighter chestnut on rump and brightest on long, rounded tail. Paler grey-brown below, whitish on chin and vent. Skulking, rarely seen, singing from deep cover. Song rich, varied and loud, based on repeated clear notes, some guttural. Various harsh call-notes, 'tak', 'kerr', 'charr'. Solitary. **Range and habitat:** woodlands with thick undergrowth in central and southern WP and western EP, migrating to tropical AF for winter. **Similar species:** Thrush Nightingale.

Goldcrest

Firecrest

Spotted Flycatcher

Pied Flycatcher

Stonechat

Whinchat

Wheatear

Black Redstart

Redstart

Robin

Bluethroat

Nightingale

Blackbird *Turdus merula* (Turdidae) 24-6 cm (9½-10 in). Medium-sized dark thrush. Male glossy black all over apart from very dark brown wings, some races duller black. Bill and eye-rings orangey yellow. Female dark brown above, slightly paler below, occasionally rufous tinged, and faintly streaked and mottled. Wings, rump and tail black-brown. Bill dark horn. Active on ground, hopping and running. Scolding 'tchook-tchook' and 'pink-pink'; rich, fluting song. **Range and habitat:** widespread in forests, open woodlands, parks and gardens of WP, parts of EP, OR and AU. Some movement south in winter by northerly breeding populations. **Similar species:** other dark thrushes (see pp 80, 134, 220, 266).

Ring Ouzel *Turdus torquatus* (Turdidae) 23-5 cm (9-10 in). Medium-sized dark thrush. Male dark sooty brown all over, except for broad white crescentic mark across breast and pale grey panel on wing, more conspicuously white in some races. Female dark brown, breast crescent narrower and less clear white. Underparts with pale feather edgings. Both sexes show more mottled or scaly plumage in winter due to paler feather edgings. Harsh scolding 'tac-tac' and clear, shrill song of 3 to 4 notes. **Range and habitat:** upland woods, open moorland and heaths of northern and montane WP, migratory to southern WP for winter. **Similar species:** other dark thrushes (see pp 80, 134, 220, 266).

Fieldfare *Turdus pilaris* (Turdidae) 25-7 cm (10-10½ in). Large, gregarious thrush. Head and rump grey, back and upperwings chestnut-brown. Black marks on forehead and cheeks. Dirty yellow on breast heavily spotted with black, belly and black-spotted flanks white. Female less heavily spotted, duller on back. Wings dark brown, tail black. Hops and runs on ground, pausing upright. Gregarious in autumn and winter, in fields and on trees. Loud 'chak-chak' call and squeaky, whistling song. **Range and habitat:** temperate and northern forests and scrub of WP and EP, migrating south within WP for winter. **Similar species:** other thrushes (see pp 80, 134, 220, 266).

Redwing *Turdus iliacus* (Turdidae) 20-2 cm (8-8½ in). Medium-small brown thrush. Upperparts uniform brown with bold white stripe above eye, brown cheeks bordered below with white, dark stripe running down from base of bill. Breast whitish, yellowish buff at sides, heavily spotted dark brown, belly whitish. Flanks and inner part of underwing chestnut-red. Highly gregarious in autumn and winter, in fields and berried trees and hedgerows. Thin 'tseep' call in flight. Short repetitive warbling song. **Range and habitat:** northern woods and scrub of WP and EP, migrating south to temperate and southern WP for winter. **Similar species:** other medium-small thrushes (see pp 80, 134, 220, 266).

Song Thrush *Turdus philomelos* (Turdidae) 22-4 cm (8½-9½ in). Medium-sized brown thrush. Overall warm brown upperparts and upperwings, though variable from ochre-brown to grey-brown in different races. Faint eye stripe and patterning on cheeks. Throat, breast and flanks very pale buff or whitish liberally marked with black arrowhead spots. Belly white, unspotted. Yellowy buff underwings. Loud 'chuk' alarm call, thin 'tsip' in flight. Musical song of repeated 2- to 4-note phrases. **Range and habitat:** widespread in mixed forests, parks and gardens of temperate and boreal WP and western EP. Partially migratory to southern WP. **Similar species:** other medium-sized thrushes (see pp 80, 134, 220, 266).

Mistle Thrush *Turdus viscivorus* (Turdidae) 26-8 cm (10-11 in). Large, pale thrush. Upperparts grey-brown, tinged olive, with pale edgings to back, rump and wing feathers. Underparts and underwings white with yellow-buff tinge to breast. Large, wedge-shaped black spots on throat and upper breast, rounder spots on lower breast and flanks. Upright stance on ground. In pairs or small groups. Defends fruit trees in winter. Grating chatter in flight. Loud, clear song comprises short ringing phrases. **Range and habitat:**

woodlands, parks and gardens of much of WP and western EP, avoiding warmer south. More northerly and easterly populations move south in winter. **Similar species:** other large thrushes (see pp 80, 134, 220, 266).

Long-tailed Tit *Aegithalos caudatus* (Aegithalidae) c. 14 cm (5½ in). Tiny, black and white bird with extremely long tail. Head white with or without black or grey stripe above eye and down sides of nape, depending on race. Back, wings and tail black with variable white wing patch, depending on race, and white sides to tail. Underparts white, pinkish suffusion on inner wings, rump and belly. Tiny bill. Tail occupies more than half total length. Agile, restless, frequently in small flocks. Penetrating 'zee-zee-zee' call, also 'tsirrup'. Rare song formed from call-notes. **Range and habitat:** widespread in woodland and scrub throughout temperate and southern WP and EP. Resident.

Penduline Tit *Remiz pendulinus* (Remizidae) 10-11 cm (4-4½ in). Tiny, long-tailed bird. Head and nape grey, with broad black mask across forehead and through eye. Eastern race has chestnut crown sometimes joining chestnut of back. Upperparts and inner wing chestnut, rest of wings and long tail black with white edgings. Throat pale grey, rest of underparts buff tinged pinkish on breast. Restless, agile bird, flicking tail and climbing among reed stems. Thin 'tsee' and 'tsi-tsi' calls. **Range and habitat:** reed marshes and thickets by water throughout central and southern WP and EP. Mainly resident but some dispersal southwards after breeding.

Nuthatch *Sitta europaea* (Sittidae) c. 14 cm (5½ in). Small, short-necked and short-tailed arboreal bird. Crown and upperparts blue-grey, fine black line through eye. Throat white, cheeks and rest of underparts vary: whitish all over in northern and western parts of range, deep orangey buff chest and flanks with white belly in southern and eastern races. Clambers up and down trees (head first), undulating flight. Ringing 'chwit-chwit' call, loud 'twee-twee-twee' and fast trill as song. **Range and habitat:** widespread in mainly coniferous forests and broad-leaved woods of boreal and temperate WP, EP and OR. Largely resident but some southward dispersal from northerly breeding areas. **Similar species:** other nuthatches (see pp 82, 272).

Wallcreeper *Tichodroma muraria* (Sittidae) 16-17 cm (6-6½ in). Small bird of rocky places. Crown and upperparts charcoal grey, short tail black with grey tip. Throat and breast black, female and winter male with white throat, rest of underparts pale grey. Whole of forewings bright crimson, main wing feathers black with some white spotting near tips of outer flight feathers. Fine-pointed and decurved bill. Hops up rock faces, flicking wings. Butterfly-like flight. High-pitched piping calls and song. **Range and habitat:** montane cliffs and gorges of southern WP and EP, resident but also descending to lower altitude in winter when may be found on buildings.

Treecreeper *Certhia familiaris* (Certhiidae) 12-13 cm (4½-5 in). Small brown and white arboreal bird with fine, decurved bill. Upperparts brown, heavily streaked with darker brown and buff, rump rufous. Whitish bar above eye, pale buff wing-bar and whitish underparts, buff on flanks. Northern races paler above, white on flanks. Climbs in spirals round trunks and along underside of branches, supported on tail, flitting down to base of next tree. High-pitched 'tseee' call repeated to make song. **Range and habitat:** widespread in mainly coniferous forests of WP, EP, mainly resident with only local movements in some areas. **Similar species:** other treecreepers (see p 272).

■□□ **Blackbird**
□■□ **Ring Ouzel** at nest
□□■ **Fieldfare** at nest

■□□ **Redwing**
□■□ **Song Thrush**
□□■ **Mistle Thrush** feeding on apples

■□□ **Long-tailed Tit** feeding young
□□■ **Penduline Tit** at nest

■□□ **Nuthatch**
□■□ **Wallcreeper**
□□■ **Treecreeper**

Great Tit *Parus major* (Paridae) c. 14 cm (5½ in). Large tit. Head black with conspicuous white cheeks. Bib, and broad stripe down yellow breast and belly, black; wider in male than female. Yellowish nape, greeny back, blue-grey wings and tail, white wing-bar and white outer tail feathers. White undertail. Acrobatic, often in small flocks. Often tame. Many calls including hard 'pink' and 'teacher-teacher' song. **Range and habitat:** wooded and forested areas, also parks and gardens, throughout WP, EP and much of OR. Northern breeders move south for winter, others resident. **Similar species:** other large tits (see p 82).

Blue Tit *Parus caeruleus* (Paridae) 11-12 cm (4-4½ in). Small tit. Crown blue, bordered with white. Black stripe through eye above white cheek. Small black bib. Nape and upperwings blue, back green. Whole of underparts yellow except for narrow dark line down belly. Plumage brighter in summer than winter. Very agile and acrobatic, clinging upside-down to twigs. In parties in winter, often very tame. Main call and song based on 'tsee-tsee-tsee' notes. **Range and habitat:** widespread in forests, scrub, parks and gardens throughout WP. Mainly resident. **Similar species:** other small tits and chickadees (see pp 82, 228, 272).

Coal Tit *Parus ater* (Paridae) c. 11 cm (4½ in). Small tit with large head and thin bill. Crown and bib black, cheeks and centre of nape white. Upperparts olive to slate-grey, depending on race, with 2 pale wing-bars. Underparts whitish to pale buff, darker on flanks, some races having white areas tinged yellow or even rufous. Agile and acrobatic, in small parties. High-pitched thin call, 'tsee-tsee' and louder song, 'weecha-weecha'. **Range and habitat:** widespread chiefly in coniferous but also broad-leaved woodland of WP and EP, some altitudinal movement in winter. **Similar species:** other small tits and chickadees (see pp 82, 228, 272).

Marsh Tit *Parus palustris* (Paridae) 11-12 cm (4-4½ in). Small tit. Crown, nape and bib glossy black, becoming browner during summer. Bib flecked with white, cheeks whitish. Upperparts uniform warm brown, underparts pale buff. Active and agile among trees, usually in pairs or families. Loud calls 'pitchoo' and 'tchay', with repeated 'chip-chip-chip' song. **Range and habitat:** lowland broad-leaved, often damp, woodland of boreal and temperate WP and eastern EP. Resident, with some local dispersal. **Similar species:** other small tits and chickadees (see pp 82, 228, 272).

Crested Tit *Parus cristatus* (Paridae) 11-12 cm (4-4½ in). Small tit with pronounced crest. Crown and stiff upswept pointed crest speckled black and white, becoming shorter and blacker during summer through abrasion. Black bib. Dirty white cheeks outlined in black, with narrow black collar. Upperparts including nape dull brown, underparts buff to whitish. Agile and acrobatic among twigs, also creeps up trunks. Main call thin 'tsee-tsee', also deep trill repeated to form song. **Range and habitat:** confined to tall coniferous forests, mainly upland, of WP. Resident. **Similar species:** other small tits and chickadees (see pp 82, 228, 272).

Dipper *Cinclus cinclus* (Cinclidae) 17-18 cm (6½-7 in). Plump, short-necked, short-tailed waterbird. Blackish brown above, chocolate on head. Clear white throat and breast with blackish belly. Chestnut band below white breast in 2 races. Constantly bobs, jerking tail, and flicking white inner eyelid across eye. Runs or plunges into water, including fast-flowing. Solitary. Metallic 'clink-clink' call, also loud 'zit-zit'. Song of grating and warbling notes. **Range and habitat:** widespread but scattered through mainly montane WP and EP. Resident apart from some altitudinal movements. **Similar species:** other dippers (see p 78).

Wren *Troglodytes troglodytes* (Troglodytidae) 9-10 cm (3½-4 in). Tiny, short-tailed dumpy bird. Overall rufous brown above, paler below, barred darker brown especially on back, flanks and undertail. Pale stripe above eye. Tail usually cocked. Flight whirring, rarely far. Active and creeping among stones and vegetation. Scolding 'tic-tic-tic' call and very loud (for size) warbling trilling song. **Range and habitat:** widespread in dense undergrowth of forests, parks and gardens throughout WP, eastern and southern EP and NA. Resident. **Similar species:** other wrens (see pp 78, 134).

Corn Bunting *Emberiza calandra* (Emberizidae) 17-18 cm (6½-7 in). Large, heavy-headed bunting. Overall grey-brown above, heavily streaked blackish. Streaked brown cheeks outlined in buff. Underparts whitish buff, more finely streaked on breast and flanks, belly whitish unstreaked. Bill deep and yellowish. Perches on bushes and wires, feeds on ground. Sometimes flies with legs dangling. Gregarious in winter. Harsh 'chip' and 'kwit' calls, and jangling trilling song. **Range and habitat:** open plains and agricultural land in southern WP and south-western EP, mainly resident. **Similar species:** other buntings (see pp 88, 274).

Yellowhammer *Emberiza citrinella* (Emberizidae) 16-17 cm (6-6½ in). Slim bunting. Male has bright yellow head and underparts and chestnut upperparts streaked with black. Chestnut rump unstreaked, sides of tail white. In winter, head and throat streaked brownish, yellow underparts duller. Female much less yellow than male, duller chestnut on back and more streaked and patterned on head and underparts. Gregarious in winter, hopping on ground. Perches, flicking tail. Harsh 'twit' call, high-pitched jangling song. **Range and habitat:** broad-leaved woodland edges and more open areas of temperate and northern WP. Mainly resident. **Similar species:** other buntings (see pp 88, 274).

Reed Bunting *Emberiza schoeniclus* (Emberizidae) 14-16 cm (5½-6½ in). Mainly brown bunting with conspicuous black head in male. Head, throat and bib of male black, with strongly contrasting broad white collar. Whitish moustache. Back and wings rufous brown streaked black. Underparts white, flanks slightly streaked brown. Head becomes brownish in winter. Female streaky brown, paler below, lacking marked head pattern. Perches on reeds and bushes, hops on ground. Gregarious. Varied calls including 'tseep', and squeaky 2- to 5-note song. **Range and habitat:** widespread in marshy vegetation and scrub almost throughout WP and EP. Northerly breeders migrate reaching northern OR, remainder resident. **Similar species:** other buntings (see pp 88, 274).

Lapland Bunting *Calcarius lapponicus* (Emberizidae) 14-16 cm (5½-6½ in). Boldly patterned northerly breeding bunting. Summer male has black head and bib, outlined in white and with chestnut nape. Upperparts brown, streaked black and chestnut, underparts white, flanks streaked chestnut. Becomes much paler and less coloured in winter. Female like winter male, showing little black or chestnut. Gregarious, ground feeder. Distinctive 'chew' call, also short trill and melodious warbling song. **Range and habitat:** circumpolar Holarctic breeder in tundra and scrub, wintering on temperate coasts and open plains of WP, EP and NA. **Similar species:** other buntings (see pp 88, 274).

Snow Bunting *Plectrophenax nivalis* (Emberizidae) 16-17 cm (6-6½ in). Medium-large bunting. Summer male overall white, with black back, outer wings and centre of tail. In winter, head and breast rufous buff, black areas mottled tawny and white. Summer female has grey and rufous head, brown streaks on black back, white body and large white wing patches. Becomes duller in winter. Gregarious in winter. Feeding on ground. Musical trilling song and rippling 'tirrirripp' flight call. **Range and habitat:** circumpolar Holarctic tundra and some montane areas, wintering on coasts and open plains to south.

Great Tit

Blue Tit

Coal Tit

Marsh Tit

Crested Tit

Dipper

Wren

Corn Bunting

Yellowhammer in winter

Reed Bunting

Lapland Bunting at nest

Snow Bunting

Chaffinch *Fringilla coelebs* (Fringillidae) 15-16 cm (6-6½ in). Medium-sized finch with prominent white shoulder and wing-bars. Male has soft blue-grey head with small black forehead, chestnut back and yellow-green rump. Wings and tail brown. Cheeks, throat and breast pinkish, becoming paler on belly. White on shoulder, wing-bar and outer tail. Female grey-brown above, tinged yellow, paler below, white as in male. Undulating flight, hops and walks on ground. Highly gregarious in winter. Metallic 'pink-pink' call, also 'chip' in flight. Clear, rattling song. **Range and habitat:** widespread and common in forests, woods, parks and gardens of WP, northerly breeders migrating south within WP in winter. **Similar species:** Brambling.

Brambling *Fringilla montifringilla* (Fringillidae) 14-15 cm (5½-6 in). Medium-sized finch. Summer male's head, upper cheeks and upper back black; throat, breast and shoulders orange. Wings and tail blackish with small white shoulder mark and narrow white wing-bar. Rump and underparts white. Black of head mottled buff in winter but breast and shoulders becoming richer orange. Female like winter male, greyer on head and paler on breast, with similar white markings on wings and white rump. Gregarious in winter. Call-notes 'tsweek' and 'chuc', song a drawn-out 'dzweee'. **Range and habitat:** northern scrub and birch woods of WP and EP, migrating to temperate areas for winter. **Similar species:** Chaffinch.

Greenfinch *Carduelis chloris* (Fringillidae) 14-15 cm (5½-6 in). Large, thickset finch. Olive-green above, including crown, bright yellow-green underparts. Grey panel on closed wing, with bright yellow patch on outer wings and at sides of base of tail. In winter, tinged grey or grey-brown. Female grey-brown above, grey below, tinged yellow, becoming whitish on belly, overall green tone in summer. Stout and pale bill. Strong, undulating flight, gregarious in winter. Trilling flight call, nasal 'dzwee' and twittering song. **Range and habitat:** widespread in woodlands and all open areas with trees in WP and extreme south-west EP, northerly breeders migrating south to winter within southern WP and EP. **Similar species:** other finches.

Goldfinch *Carduelis carduelis* (Fringillidae) c. 12 cm (4½ in). Small, colourful finch. Red face extending back to behind eyes, less extensive in female. White cheeks extending up to black crown and nape, latter with whitish spot. Tawny brown back, whitish rump, dark forked tail with white tips. Wings blackish with broad bright yellow wing-bar and small white tips. Small, whitish bill. Active, with dancing flight. Gregarious, feeding agilely from thistle heads. Liquid 'tswitt-twitt-twitt' flight note and trilling song. **Range and habitat:** widespread in open woodlands and open areas with trees throughout WP and western EP. Northerly breeders winter within southerly birds' breeding range. **Similar species:** other finches (see p 92).

Bullfinch *Pyrrhula pyrrhula* (Fringillidae) 14-15 cm (5½-6 in). Stout, short-necked finch. Crown and small bib black, upperparts of male blue-grey, rump white. Cheeks and underparts deep pinky red, except for white lower belly. Wings and tail black, with broad white wing-bar. Female grey-buff above, pale pinky buff below, other markings as male. Very short, stout bill and large head, lacking discernible neck. Soft, low whistling 'deu' call-note, used in short, creaky song. Secretive, in pairs or families. **Range and habitat:** widespread in coniferous forests, also some broad-leaved and more open treed areas of WP and EP. More northerly breeders move south for winter. **Similar species:** other finches (see p 92).

Hawfinch *Coccothraustes coccothraustes* (Fringillidae) 16-17 cm (6-6½ in). Stout, thick-necked finch with massive bill on large head. Head of male rufous, richer on nape, black round base of huge, stout bill. Grey collar, red-brown back, rufous rump, short rufous tail with white tip. Wings dark with broad white wing-bar. Underparts pale pinky brown. Female greyer brown on body and greyer wings. Rapid, undulating flight, perching in tree-tops. Solitary or in small parties, shy. Abrupt 'tick' call, also high-pitched 'tsip', rarely heard short scratchy song. **Range and habitat:** scattered distribution in broad-leaved woodland in WP and parts of EP, northerly breeders moving south for winter. **Similar species:** other finches (see p 92).

Siskin *Carduelis spinus* (Fringillidae) c. 12 cm (4½ in). Small, dumpy yellowish finch. Male has black crown and chin, otherwise bright yellow-green tinged grey above and bright yellow below, turning to white on belly. Dark streaking on flanks. Dark wings and short, forked tail with 2 yellow wing-bars and yellow at sides of base of tail. Female's upperparts greyer and underparts whiter, well streaked, tinged with yellow. Yellow wing-bars and tail base as male. Clear 'tsuu' call-note and warbling, twittering song. Restless and acrobatic in trees. Gregarious. **Range and habitat:** mainly coniferous forests and woods of boreal and temperate WP and eastern EP, northerly breeders moving south in winter. **Similar species:** other finches (see p 92).

Serin *Serinus serinus* (Fringillidae) 11-12 cm (4-4½ in). Tiny, yellowish finch with notched tail. Head, upperparts and breast of male yellow, streaked brownish on crown, cheeks, sides of breast and back. Rump pure yellow, unstreaked. Streaking heavier on head and back in winter, but less on flanks, overall less yellow. Belly whitish, flanks streaked brown. Wings and tail brown. Female paler brown, tinged yellow on upperparts and breast, green-yellow on head. Heavily streaked except for pale yellow rump. Trilling flight call and thin jangling song. Fast, undulating flight. Agile in trees. **Range and habitat:** forest edges, open habitats with trees throughout south-west WP. Northern breeders move south for winter. **Similar species:** other small finches (see p 92).

Redpoll *Acanthis flammea* (Fringillidae) 12-13 cm (4½-5 in). Small, dumpy finch with red forehead. Rest of head and upperparts tawny brown, heavily streaked dark brown; 2 pale wing-bars. Chin black. Rump, cheeks, throat, breast and flanks pink, brighter in summer. Some dark streaking on flanks. Forked tail. Female more streaked below, lacking much pink. Some races greyer above and with more white on wing and rump, which can be unstreaked, while flanks may be more heavily marked. Agile, acrobatic while feeding, gregarious. Hard 'chuch-uch' and 'err' flight notes, used in song. **Range and habitat:** circumpolar Holarctic tundra, open scrub and woods south to temperate zone, northern breeders winter further south. **Similar species:** other small finches (see p 92).

Crossbill *Loxia curvirostra* (Fringillidae) 16-17 cm (6-6½ in). Bulky, large-headed finch with long mandibles crossed at tip. Male crimson-orange, brighter on head, rump and breast, duller elsewhere, dark streaked on cheeks and mantle. Wings and notched tail dark brown. Female dull yellowy green tinged grey, sometimes a few golden or orange feathers. Dark streaking on head, upperparts and flanks. Crossed bill varies in size according to race. Fast, undulating flight. Agile, clambers about branches, clings to cones with 1 foot. Metallic 'jip-jip' call in flight. Song of short, staccato trills. **Range and habitat:** in conifers in northern, temperate and montane WP, EP, NA and parts of OR. Resident or irruptive migrant southwards. **Similar species:** other crossbills (see p 92).

Chaffinch
Brambling at nest
Greenfinch

Goldfinch
Bullfinch male and female feeding young

Hawfinch
Siskin
Serin bathing

Redpoll
Crossbill

House Sparrow *Passer domesticus* (Ploceidae) 14-15 cm (5½-6 in). Small sparrow. Male has grey crown, white cheeks and large black bib. Dark stripe through eye. Mantle chestnut streaked black, rump grey. Wings and tail brown, former with short white wing-bar. Underparts dirty white. Duller in winter with white streaks in black bib. Female brown all over, paler beneath, buff stripe behind eye, 2 pale wing-bars. Very gregarious, lives in human settlements; rapid flight, hops and flicks tail. Many chirrups and cheeps, strung together into song. **Range and habitat:** widespread throughout WP, EP and OR, introduced to NA, NT, AF and AU. Mostly non-migratory. **Similar species:** other sparrows (see pp 88, 232, 234, 276).

Tree Sparrow *Passer montanus* (Ploceidae) c. 14 cm (5½ in). Small sparrow. Crown and nape chestnut-brown. Black spot on white cheeks, small black bib. White collar round sides of neck. Back brown, streaked black, rump yellowish brown. Wings and tail brown, with 2 thin white wing-bars. Underparts whitish, buff under tail. Hard, high-pitched 'chip-chip' call, 'teck-teck' in flight. Song of rapid chirping. Gregarious with other sparrows and finches in winter, but shy. **Range and habitat:** widespread throughout lightly wooded and cultivated areas of WP, EP and OR. Mainly resident though some movement south by northerly breeders. **Similar species:** other sparrows (see pp 88, 232, 234, 276).

Starling *Sturnus vulgaris* (Sturnidae) 21-2 cm (8-8½ in). Male in summer black, glossed purple and green, brownish wings and tail with pale feather edgings. Female tends to have more pale spots on body, while both sexes in winter are well-spotted buff above, whitish below. Longish, pointed bill yellow in summer, brown in winter. Rapid direct flight, walks and runs on ground. Highly gregarious, often roosting in immense numbers. Song a mimetic combination of whistles, chatters, screams and clicks. **Range and habitat:** widespread in woodlands, cultivated land and urban areas of WP, south-western EP and parts of OR, introduced NA, AF and AU. **Similar species:** other starlings (see pp 228, 276).

Golden Oriole *Oriolus oriolus* (Sturnidae) 23-5 cm (9-10 in). Large, brilliantly coloured bird. Male bright yellow all over except for black wings and black base and centre to tail and small black stripe from eye to base of reddish bill. Small yellow wing-bar. Female greenish yellow above, greyish below streaked with brown. Wings and most of tail dark brown. Shy, secretive birds, more often heard than seen. Squealing cat-like call. Melodious and fluting song 'weela-weelo'. **Range and habitat:** widespread in broad-leaved woodlands and orchards of temperate and southern WP and EP, also OR. Non-migratory in OR, remainder winter in tropical AF. **Similar species:** African Golden Oriole (see p 228).

Jay *Garrulus glandarius* (Corvidae) 33-6 cm (13-14 in). Large, boldly coloured bird. Pinky brown body, brighter on nape and back, paler below. Forehead and crown whitish, streaked black. Black moustache, white throat. Pure white rump and long black tail conspicuous in flight. Black wings, with pinky brown shoulders, white patch to rear and speckled blue and white forewing. Stout bill. Undulating flight, hops. Harsh scolding 'skaak-skaak', also 'kaa' and 'keu'. Soft, scratchy warbling song. Very wary. **Range and habitat:** in most types of woodland in temperate and southern WP and EP, also OR. Mainly non-migratory, but irruptive movements southward by northern breeders. **Similar species:** other jays (see pp 92, 134).

Magpie *Pica pica* (Corvidae) 46-50 cm (18-20 in). Large, very long-tailed black and white crow. Head, breast and back black, with purplish gloss. Wings black, glossed blue and green, with much white in black-tipped outer feathers. Large white patches at wing bases in flight. Very long, tapering tail (c. half overall length) black, glossed green to purple. Lower breast and belly pure white. Weak-looking flight, often gliding. Walks and hops. In pairs or small flocks. Harsh chattering call and musical 'chook-chook'. No real song. **Range and habitat:** woodland and cultivated areas of WP, EP, OR and NA. Non-migratory apart from some postbreeding dispersal. **Similar species:** Yellow-billed Magpie.

Carrion Crow *Corvus corone* (Corvidae) 45-9 cm (18-19 in). Large all-black or black and grey crow. Carrion Crow black all over, with green and purple gloss, former especially on head and neck. Hooded Crow has black head, throat, wings and tail, and grey body, very finely streaked black. Hybrids occur. Stout black bill with feathered base. Powerful, deliberate flight. Walks and hops. Solitary or in small flocks. Thrice-repeated calls of 'kaaah' or 'keerk'. No song. **Range and habitat:** widespread wherever there are trees; Carrion Crow confined to south-west WP, Hooded Crow throughout WP and EP. Northern breeders winter to south. **Similar species:** other crows (see pp 92, 228, 278, 322).

Rook *Corvus frugilegus* (Corvidae) 44-7 cm (17-18 in). Medium-sized crow. Black all over, glossed purple and green, mainly purple on head and neck. Adult has bare whitish skin round base of black bill including chin and below eyes. Shaggy feathers on upper leg. Highly gregarious, nesting in colonies. Flight rapid. Waddles on ground, also hops. No song, calls include 'kaah', 'kaa-aah' and higher-pitched 'ki-ook'. **Range and habitat:** widespread in broad-leaved and mixed woodlands and stands of trees in grassy and cultivated land in most of WP and EP. Some non-migratory, but northern breeders move south for winter, reaching OR. **Similar species:** other crows (see pp 92, 228, 278, 322).

Jackdaw *Corvus monedula* (Corvidae) 32-4 cm (12-13 in). Small and compact crow. Black all over except for ash-grey nape, which contrasts with glossy black of cap and throat. Northern race has paler silvery grey nape and some whitish feathers at base of neck. Short, stubby black bill. Gregarious, often mixing with Rooks. Nests in holes. Flight rapid on rather narrow, pointed wings, includes diving and tumbling. Brisk walk on ground. No song. High-pitched 'chack' and shrill 'kyow' calls. **Range and habitat:** open wooded areas almost throughout WP and western EP. Northern and eastern breeders move south and west for winter. **Similar species:** other small crows (see pp 92, 228, 278, 322).

Raven *Corvus corax* (Corvidae) 60-7 cm (24-6 in). Very large crow with very stout bill. Completely black all over, glossed green and purple. Large head emphasized by shaggy throat feathers and heavy, deep bill. Long tail wedge-shaped. Wings long and narrow, outer part slightly swept back in powerful flight. During display becomes aerobatic, tumbling, nose-diving, soaring, even flying upside-down. Walks on ground. Usually in pairs, but in small groups at roosts. Loud, deep croaking call 'pruk-pruk'. No song. **Range and habitat:** circumpolar Holarctic in wide variety of habitats almost throughout WP, EP and NA. Non-migratory, but some wandering in winter by northern breeders. **Similar species:** other ravens (see pp 92, 228).

Chough *Pyrrhocorax pyrrhocorax* (Corvidae) 36-41 cm (14-16 in). Medium-sized all-black crow with bright red bill and legs. Glossy blue-black all over, except for wings and tail which are glossed greenish. Long red bill pointed and decurved. Legs red. Very buoyant flight, often undulating, on broad wings with tips of flight feathers spread. Quite aerobatic, soaring and diving with closed wings. Agile on ground. Normally in pairs or small flocks. Main call a loud, far-carrying 'chee-ow', also quieter 'chuff' and deeper 'kaa'. No song. **Range and habitat:** coasts and mountains of mainly south-western and southern WP and EP. Non-migratory. **Similar species:** Alpine Chough.

House Sparrow
Tree Sparrow
Starling feeding on meat

Golden Oriole at nest
Jay

Magpie
Carrion Crow
Rook

Jackdaw
Raven calling
Chough

THE AFROTROPICAL REGION

African White-backed Vultures *attend a wildebeest carcass in Kenya. Different vulture species feed on separate parts of the carcass so that it is swiftly consumed.*

Africa was formed about 100 million years ago during the fragmentation of the southern supercontinent of Gondwana. It is the second largest of the world's land masses after Eurasia, and unique in completely spanning the tropics, from the Mediterranean far into the southern oceans.

Although one land mass, Africa is not a single zoogeographical unit because its northern reaches are occupied by the Sahara, the greatest of the world's deserts. This immense wilderness, over 1,500 km (930 miles) wide north–south, has long acted as a barrier to the movement of many lifeforms, and now forms the boundary between two of the earth's major zoogeographical realms, the Palaearctic Region to the north, and the Afrotropical (formerly known as Ethiopian) Region to the south. The Sahara is not a total barrier, however; for example, it is crossed or skirted twice a year by millions of migrant birds from the Palaearctic.

Although the bulk of the Afrotropics is within continental Africa, this region is often considered to include peninsular Arabia and the island of Madagascar.

Like sub-Saharan Africa, Arabia is divided from Eurasia by wide deserts, but its avifauna shows both Palaearctic and Afrotropical affinities. Madagascar separated from Africa's eastern coast over 30 million years ago and its fauna and flora possess numerous endemic elements, including many lower primates (dominated by the lemurs) and five families of birds, such that it is often considered a zoogeographical realm, the Malagasy Region, in its own right.

After extended periods as a stable land mass, Africa's surface has been eroded into vast peneplains. In the absence of mountains large enough to influence climate on a continental scale, the distribution of Africa's rainfall and vegetation is mainly decided by the annual movements of the Inter-Tropical Convergence Zone (ITCZ), a belt of low atmospheric pressure and resulting high rainfall that follows the sun on its annual progression north and south of the equator.

Areas within about 10° of the equator receive the ITCZ (and thus wet seasons) at two well-separated times of the year. This produces high, evenly distributed rainfall, which results in the growth of forest; Africa's tropical rainforests are second in extent only to those of South America. They contain a wealth of birds, including the Congo Peacock (*Afropavo congensis*) and many members of such families as hornbills, barbets and greenbuls.

Regions more than 10° from the equator receive a single wet season each year, followed by a prolonged drought. Depending on the total amount of rainfall, which decreases away from the equator into the tropics, this type of regime produces savannah woodland, semi-arid bush or, at its driest, desert. For these reasons, Africa's basic types of vegetation are arranged in parallel east–west belts of decreasing luxuriance, on either side of the equator.

The savannah woodlands consist of deciduous trees of moderate height, which provide numerous nest-holes for such groups as starlings, rollers and the smaller, open country hornbills. The arrival of the rainy season produces tall, lush growths of grass alive with displaying bishops and widowbirds. The semi-arid bush is also largely deciduous, and is dominated by various acacias. This dry country supports a great variety of weavers, and the vegetation is often festooned with their hanging, grassy nests. Numerous ground-dwelling birds include the bustards, sandgrouse and pipits. Beyond these semi-arid lands, the rainfall is insufficient to support much vegetation and the land is open desert, the Sahara to the north and the Kalahari and Namib to the south. The diversity of birds in these two southern deserts is distinctly lower than that in the moister Afrotropical regions – hence the appearance of the phrase 'apart from the far south-west' in many of the species' summaries given here.

Two other distinctive Afrotropical areas should be mentioned.

Africa's southernmost tip has a subtropical climate like that of the Mediterranean; this produces a low, mixed scrub known as fynbos, which is very similar to the maquis of the north. The many seabirds around these shores include the Jackass Penguin (*Spheniscus demersus*).

The rift valley is the surface expression of weaknesses in Africa's crust. The soda-lakes found on the rift's floor at such sites as Nakuru and Natron accommodate enormous concentrations of flamingos which are among the world's foremost wildlife spectacles. Dramatic uplifts that accompanied rifting produced the Ethiopian and East African highlands. The latter are of relatively restricted extent but they contain close conjunctions of diverse ecosystems that, in Kenya and Uganda, support the richest local avifaunas on the continent.

The Afrotropical avifauna consists of approximately 1,700 species in 95 families, and is exceeded in richness only by that of the Neotropics. One order (the mousebirds) and 12 families are endemic to the Afrotropical Region. Five of these families contain unique, single species: these are the Ostrich, the Whale-headed Stork or Shoebill, the Hammerkop, the Secretary Bird and, in the Malagasy area, the Cuckoo-roller. The Ostrich, a relic of Africa's Gondwanan days, has become confined to sub-Saharan Africa through human agencies, having been hunted to extinction in the Middle East, and virtually so in North Africa only recently.

The largest of the Afrotropical's unique families is the 20 turacos, an ancient group possibly related to owls and nightjars. The wood hoopoes and helmet shrikes are also endemic to the Afrotropical mainland, while the mesites, ground-rollers, asities and vanga shrikes are confined to the Malagasy area. The guineafowl are endemic to Africa as a whole, though fast declining north of the Sahara. Other uniquely Afrotropical groups, of but two species each, are the sugarbirds, the oxpeckers and the picathartes or rockfowl.

Families at their most diverse in the Afrotropics include the bustards, coursers, bee-eaters, rollers, barbets, honeyguides, larks, shrikes, sunbirds and weavers. The prolific speciation of the cisticola warblers (40 taxa), sunbirds (74) and weavers (126) is especially notable. The Afrotropical Region is also exceptionally rich in large diurnal raptors.

Families poorly represented in the Afrotropical Region relative to other areas include the doves, parrots, trogons, kingfishers, woodpeckers, broadbills, pittas, babblers, cuckoo shrikes and white-eyes. Most of these attain their greatest diversity further east, in the Oriental and Australasian Regions. The parrots are also diverse in South America and, like the Ostrich, probably represent Gondwanan stock.

Hundreds of millions of migrant birds of over 200 species breed in the Palaearctic and spend the northern winter in the Afrotropical Region. The aquatic species utilize all types of waters, but the land birds favour Africa's savannah woodlands and bush and make very little use of the vast tropical rainforests.

Many of these birds, like the Common Pochard (*Aythya ferina*) and Turtle Dove (*Streptopelia turtur*), penetrate south only as far as Africa's northern tropics, but others like the Garganey (*Anas querquedula*) and Swallow (*Hirundo rustica*) reach the far south.

Much of the migration is on broad fronts but some takes place in bands as narrow as 30 km (19 miles) across, particularly where the birds' progress is channelled around highlands. One such stream passes over Ngulia Lodge in Kenya, where spectacular 'falls' of up to 100,000 birds have occurred.

Relatively small numbers of these Palaearctic migrants regularly remain in sub-Saharan Africa for the northern summer. The White Stork (*Ciconia ciconia*), Black Stork (*C. nigra*). Booted Eagle (*Hieraaetus pennatus*), Eurasian Bee-eater (*Merops apiaster*) and House Martin (*Delichon urbica*) have established resident breeding populations in the far south, where the annual variations in day-length and season most resemble those of the Palaearctic.

Most Afrotropical birds were formerly considered nonmigratory, but in fact many frequently perform movements of some kind, ranging from local wandering to regular, transequatorial migration. Many dry country species are nomadic, moving erratically to avoid adverse climatic conditions and breeding opportunistically on entering areas that are green after recent rains. Intra-African migrants make regular movements between breeding and off-season ranges within Afrotropical Africa. This type of migration has been rather arbitrarily defined as involving journeys of more than 400 km (250 miles) in either direction, but species like Abdim's Stork (*Ciconia abdimii*) cover many thousands of miles each year.

Afrotropical birds are threatened on a continental scale by such factors as desertification, deforestation and the destruction of wetlands, which are a consequence of the southwards spread of the Sahara, and of the unsustainable exploitation of natural habitats by the large and rapidly increasing human populations. The vast international trade in captive birds, including bustards, is another serious problem.

Birds that benefit from the spread of man include those capable of utilizing dry, degraded environments (e.g. larks, sandgrouse), those that scavenge around habitation (e.g. crows, vultures) and those that have taken to nesting on buildings (e.g. swifts, swallows). Artificially introduced species like the Feral Pigeon (*Columba livia*), Indian House Crow (*Corvus splendens*) and House Sparrow (*Passer domesticus*) have also spread widely.

Various Afrotropical ornithological societies exist, and many produce publications. Further information can be obtained from: the East African Natural History Society (OS-C), PO Box 48019, Nairobi, Kenya; the West African Ornithological Society, 1 Fishers Heron, East Mills, Fordingbridge, Hampshire FP6 2JR, UK; and the Southern African Ornithological Society, PO Box 87234, Houghton, Johannesburg 2041, Republic of South Africa.

OSTRICHES Struthioniformes

Ostrich *Struthio camelus* (Struthionidae) 190-240 cm (75-94 in). Flightless, unmistakably huge, the largest living bird. Adult male has mainly black plumage, skin of neck and legs pinkish in most races, bright blue-grey in Somali Ostrich of north-eastern AF. Female smaller, plumage mainly grey-brown, neck and legs greyer. Immature as female, plumage increasingly black in male. Usually silent; breeding male produces deep booming, like lion roaring. Found in pairs and small groups in open country; conspicuous, but resembling bushes at long range. **Range and habitat:** recently hunted to extinction in Arabia, almost so in North Africa; remains in 3 regions of AF: at 10-20° N; south from Sudan, Ethiopia to Tanzania; south of 15° S. Locally common. Resident; seasonal and nomadic movements known. **Similar species:** Rhea (see p 96), Emu (see p 282).

PENGUINS Sphenisciformes

Jackass Penguin *Spheniscus demersus* (Spheniscidae) 63 cm (25 in). Rather small penguin, black above and on face, with conspicuous broad white eyebrow extending back to broader white stripe down side of neck; pinkish skin above and around eye, bill black with subterminal pale band; underparts white, with black band across upper breast and down on to flanks. Immature has head and neck sooty, bare skin as adult. Call, frequently at night, is loud braying. Found singly or in flocks, foraging at sea. **Range and habitat:** only resident penguin in Africa, from Namibia to southern Mozambique. Locally common offshore, especially islands and on coasts. Resident, also wandering. **Similar species:** Rockhopper and Macaroni Penguins.

FULL-WEBBED SWIMMERS Pelecaniformes

Long-tailed (Reed) Cormorant *Phalacrocorax africanus* (Phalacrocoracidae) 60 cm (24 in). Rather small, long-tailed, blackish or dark brown cormorant with red eyes and yellow bill. Breeding adult has short crest on forehead and contrasting glossy feathers on upperwing; nonbreeding adult duller, lacking crest, rather blotchy below. Immature has ventral torso dusky whitish. Usually silent; occasionally croaking sounds; more vocal at nest. Found singly or in small parties, on open shores, perches or water; sometimes approachable. **Range and habitat:** widespread in AF south of 17° N. Locally common on all types of water except open sea, particularly near fringing vegetation. Resident, moving locally. **Similar species:** Great Cormorant, African Darter.

African Darter *Anhinga rufa* (Anhingidae) 90 cm (35 in). Medium-sized waterbird recalling slender cormorant, distinguished by long, thin, kinked neck, equally slim head and pointed (not hook-tipped) bill. Adult has black torso streaked white on wings, and bright chestnut-brown neck with white lateral stripe. Crown black in male, browner in female. Immature buff on neck and underparts. Usually silent; various grunts, croaks at nest. Found singly or in small groups, on water or nearby prominent perch, often swimming with only neck showing; often unobtrusive. **Range and habitat:** widespread in AF south of 15° N. Locally common on fresh, sluggish waters with fringing vegetation, also coastal inlets on occasion. Resident, moving locally. **Similar species:** other darters (see pp 52, 238, 286).

Pink-backed Pelican *Pelecanus rufescens* (Pelecanidae) 140 cm (55 in). Rather small, grey pelican, washed pinkish on torso when breeding; bill pouch pinkish or greyish, small spot of naked black skin in front of eye. In flight, slight contrast between underwing-coverts and browner flight feathers. Immature with head and wings brownish, otherwise whitish. Usually silent; harsh croaks at nest. Singly or in small groups, not fishing in coordinated parties; on water or open shores, often with Eastern White Pelican *P. onocrotalus*; fairly approachable. **Range and habitat:** widespread in AF south of 15° N, apart from much of far south.

Locally common on any freshwater with sufficient fish. Resident, but dispersing widely after breeding. **Similar species:** Eastern White Pelican (see p 142).

HERONS AND ALLIES Ciconiiformes

Black-headed Heron *Ardea melanocephala* (Ardeidae) 95 cm (37 in). Large grey heron with top of head, hindneck and legs black, and white throat. In flight, contrast between pale underwing-coverts and black flight feathers. Immature has hindneck and top of head greyish; foreneck white, tinged buff or rufous. Call is loud 'oo-arrk'; also harsh sounds at nest. Found singly or in small groups, foraging on open ground or shores; moderately approachable. **Range and habitat:** widespread in AF south of 15° N. Common but seldom numerous, in variety of inland wetlands and also far from water in open country or cultivated land. Resident, with post-breeding dispersal; moves into arid country after rains. **Similar species:** Grey Heron (see p 144).

Goliath Heron *Ardea goliath* (Ardeidae) 140 cm (55 in). Very large, heavily built heron with massive bill, slate-grey wings and tail, and rich chestnut head, neck and underparts. Combination of great size and slow, deliberate wingbeats distinctive even at long range. Immature paler, grey-brown on back, underparts whitish with dark streaking. Call is loud 'woork', often given when flushed. Found singly or in pairs, occasionally loose flocks, foraging in open; wary. **Range and habitat:** WP in Kuwait, Iraq, Red Sea; otherwise AF, where widespread south of 15° N, apart from far south-west. Rather uncommon around permanent waters, both fresh and saline, including running tide. Resident, dispersing after breeding. **Similar species:** Purple Heron (see p 144)

Black Heron *Egretta ardesiaca* (Ardeidae) 66 cm (26 in). Medium, slaty black heron with conspicuous yellow feet that turn reddish during breeding. Immature duller, browner. Usually silent; threat is raucous scream. Found singly or in small parties, seasonal flocks up to hundreds in some areas; fishes in open water, holding wings in canopy above head; bold. **Range and habitat:** widespread in AF south of 15° N, apart from far south-west and much of Congo basin. At best locally common, along open margins of lakes, marshes, coastal creeks, sometimes rivers. Some populations resident, dispersing after breeding; others possibly intra-African migrants. **Similar species:** Slatey Egret, dark phases of Little Egret and Reef Heron (see p 144).

Cattle Egret *Bubulcus ibis* (Ardeidae) 52 cm (20 in). Rather small, short-necked egret, nonbreeding adult (most of year) white with yellowish bill and legs. Breeding adult has bill and legs more orange, with crown, breast and back buff-orange. Immature white, with bill, legs and feet black. Usually silent; occasional croaking calls. Found in small flocks, foraging on open ground; habitually takes insects flushed by large mammals, often approachable. **Range and habitat:** OR, southern WP, EP; widespread in AF south of 15° N, apart from far south-west; has recently spread to NA, NT, AU. Common, locally numerous, in open grassy country, often away from water. Some populations resident, others regular intra-African migrants. **Similar species:** other white egrets (see pp 52, 144, 286).

Hammerkop *Scopus umbretta* (Scopidae) 56 cm (22 in). Unique brown waterbird, recalling short, squat stork, with black legs, stout, conical black bill and thick drooping crest on back of head. In buoyant flight, can be mistaken for ibis or raptor until longish neck and heavy bill are seen. Immature resembles adult. Call is series of loud, shrill whistles, again suggesting raptor, often rising to strained crescendo. Found singly or in pairs, sometimes small, loose flocks, foraging in shallow open water; massive nest on tree or rock face; bold, sometimes approachable. **Range and habitat:** widespread in AF south of 15° N, also south-west Arabia. Fairly common around slow inland waters, also estuaries. Basically resident, colonizing seasonal waters, nomadic in arid lands. **Similar species:** Hadada (see p 188).

Ostrich in display

Jackass Penguin

Long-tailed (Reed) Cormorant drying wings

African Darter

Pink-backed Pelican

Black-headed Heron

Black Heron fishing

Goliath Heron

Cattle Egret showing breeding plumes

Hammerkop

Whale-headed Stork (Shoebill) *Balaeniceps rex* (Balaenicipitidae) 150 cm (59 in). Very large, dark grey, stork-like bird with unmistakable huge, pale, shoe-shaped bill and short tuft-like crest. Immature browner, with smaller bill. Mainly silent; occasional chuckling and shrill whistles; clatters bill. Found singly or paired, very rarely small, loose groups; shy, often in dense vegetation, can soar at great heights. **Range and habitat:** central AF, from south Sudan and Central African Republic south to Zambia, Botswana. Very local, often obscure, in *Miscanthidium* and papyrus swamps. Resident, possibly wandering very locally. **Similar species:** Marabou high overhead.

Woolly-necked Stork *Ciconia episcopus* (Ciconiidae) 86 cm (34 in). Rather small stork with glossy brownish black body and wings; contrasting white neck has peculiar, short 'woolly' feathering, white patch at base of bill. In flight, shows conspicuous white abdomen. Immature browner, lacking patch at base of bill. Usually silent; makes harsh sounds and clatters bill at nest. Found singly, paired, or in small loose groups; not secretive. **Range and habitat:** east in OR from India to Philippines, Indonesia; widespread in AF south of 15° N, apart from far south-west. Local, often uncommon, in vicinity of water, usually in wooded country. Some populations resident, others intra-African migrants.

African Open-bill Stork *Anastomus lamelligerus* (Ciconiidae) 92 cm (36 in). Rather small, blackish brown stork; bill pale, with narrow gap between closed mandibles. Immature browner, gap reduced or absent. Usually silent; occasional honking or croaking sounds. Usually gregarious, foraging in open; bold, fairly approachable. **Range and habitat:** widespread in AF south of 15° N, apart from far south-west. Locally common, usually near wetland, where specialized bill extracts molluscs from shells. Some resident; most intra-African migrants, which can temporarily appear in any habitat. **Similar species:** Asian Open-bill Stork (see p 238).

Saddle-bill Stork *Ephippiorhynchus senegalensis* (Ciconiidae) 145 cm (57 in). Africa's largest stork, with black and white plumage and unmistakable huge, black and red bill with yellow 'saddle' at base of upper mandible. Eye dark in male, yellow in female. In flight, white flight feathers conspicuous even at great range. Immature grey and mottled, bill black. Silent. Found singly or paired, foraging in open; shy. **Range and habitat:** widespread in AF between 15° N and 20° S. Local, thinly distributed, around inland wetlands. Resident.

Abdim's Stork *Ciconia abdimii* (Ciconiidae) 76 cm (30 in). Small stork with dark upperparts and white ventral torso; facial skin blue and red, bill tawny; legs pinkish grey with contrasting red tarsal joints and feet. In flight, large white patch on back and rump visible at great range. Immature browner, bill redder. Usually silent; occasional weak whistles; clatters bill. In flocks up to thousands, foraging in open; moderately approachable, nesting even in towns. **Range and habitat:** intra-African migrant that breeds during the rains in northern AF, from Senegal east to Ethiopia and extreme south-west Arabia, then migrates into southern AF for southern rainy season, south to South Africa. Seasonally common in open country, often far from water. **Similar species:** Black Stork (see p 144).

Marabou (Marabou Stork) *Leptoptilos crumeniferus* (Ciconiidae) 152 cm (60 in). Huge, unmistakable stork, dark above and white below, with bare pinkish head and neck, and similarly naked, pendulous air sac dangling over breast. In flight readily identified by massive bill and contrast between white belly and black wings and tail. Immature has whitish, 'woolly' feathering on head. Usually silent, sometimes clattering bill; snores and grunts at nest. Single or in small parties, flocks at food or on migration; bold, moderately approachable. **Range and habitat:** widespread in AF between 15° N and 25° S. Common in small numbers in open, dry or wetland habitats, also scavenging at carrion and refuse. Resident or intra-African migrant.

Yellow-billed Stork *Mycteria ibis* (Ciconiidae) 97 cm (38 in). Medium-sized, mainly white stork with orange-yellow, slightly decurved bill, bare red facial skin and red legs; plumage tinged pinkish when breeding. In flight, white underwing-coverts contrast with black flight feathers. Immature grey-brown, bill and face duller, legs brown. Generally silent; clatters bill; various screams at nest. Single or in small groups, foraging in open; bold, moderately approachable. **Range and habitat:** widespread throughout AF between 15° N and 30° S. Fairly common at inland waters; local on coasts. Few entirely resident, substantial movements occur. **Similar species:** White Stork (see p 144).

Sacred Ibis *Threskiornis aethiopicus* (Threskiornithidae) 89 cm (35 in). Mainly white with elongate black plumes drooping over tail, decurved black bill and naked black head and neck; legs blackish. Immature lacks plumes; head and neck feathered. Usually silent; squeals and croaks at nest. Found singly or in flocks up to hundreds, foraging in open; not shy. **Range and habitat:** locally Middle East, in Iraq and Iran; widespread in AF south of 17° N, apart from far south-west. Common in wetlands, including sea coast; also on cultivated land, and scavenging at habitation. Some resident, others making regular movements.

Hadada *Hagedashia hagedash* (Threskiornithidae) 76 cm (30 in). Medium-sized grey-green ibis with metallic purplish green wash (often inconspicuous) on closed wing, and whitish stripe on cheek. In flight, wings distinctively broad and rounded, recalling Hammerkop. Immature duller, lacking metallic wash. Call (origin of name) is very loud, harsh 'ark-ha, haaaaa', most frequent at twilight. Single, in pairs or small parties, larger flocks known, foraging in open; often approachable. **Range and habitat:** widespread in AF south of 15° N, apart from far south-west. Locally common on damp ground or near water, usually near woodland or forest. Basically non-migratory. **Similar species:** Green and Glossy Ibises, Hammerkop (see pp 100, 186).

African Spoonbill *Platalea alba* (Threskiornithidae) 91 cm (36 in). White waterbird with long, grey and red spatulate bill, bare red facial skin and red legs. Immature with blackish wing tips, yellow bill and blackish legs. Usually silent; occasional grunts and croaks; clatters bill. Single or in small flocks, foraging in open, shallow water; rather retiring. **Range and habitat:** widespread in AF south of 17° N, apart from far south-west. Locally common at open waters. Few truly resident; regular intra-African movements. **Similar species:** other spoonbills (see pp 144, 288).

Greater Flamingo *Phoenicopterus ruber* (Phoenicopteridae) 142 cm (56 in). Tall, very pale flamingo with black-tipped pink bill and pink legs. In flight, scarlet wing-coverts contrast with black flight feathers. Call is loud, goose-like honk; flocks grunt and murmur. Very gregarious, often flocking in thousands, frequently alongside Lesser Flamingos. **Range and habitat:** NT, southern NA and Palaearctic; in AF, migrants from Palaearctic in north, AF-resident populations in east, and south of 15° S. Locally abundant on open, shallow waters, particularly if saline or brackish. Substantial dispersal from few major AF breeding sites; also erratic, nomadic. **Similar species:** Lesser Flamingo.

Lesser Flamingo *Phoeniconaias minor* (Phoenicopteridae) 101 cm (40 in). Medium-sized, distinctly pink flamingo with black-tipped, very dark maroon bill and red legs. In flight, pink and red wing-coverts contrast with black flight feathers. Immature distinct. Call is goose-like honk; flocks grunt and murmur. Extremely gregarious, in concentrations that can exceed a million, often alongside Greater Flamingos, wary, walking away into water. **Range and habitat:** Persian Gulf and north-west India, but mainly AF, especially on rift valley soda-lakes from Ethiopia to southern Tanzania; also west coast, and south of 15° S. Locally extremely abundant on saline and brackish waters; moribund wanderers elsewhere. Resident, dispersing after breeding, and erratic nomad. **Similar species:** Greater Flamingo.

■□□
Whale-headed Stork (Shoebill)
□■□
Woolly-necked Stork
□□■
African Open-bill Stork

■□□
Saddle-bill Stork
□■□
Abdim's Stork
□□■
Marabou (Marabou Stork)

■□□
Yellow-billed Stork
□■□
Sacred Ibis
□□■
Hadada

■□□
African Spoonbill
□■□
Greater Flamingo
□□■
Lesser Flamingo
in territorial display

WATERFOWL Anseriformes

Maccoa Duck *Oxyura maccoa* (Anatidae) 45 cm (18 in). Rather small diving duck, sole AF stifftail; breeding male with unmistakable combination of black head, vivid cobalt-blue bill and bright chestnut body. Nonbreeding male, female and immature brown, with conspicuous dark horizontal streak across white face. Usually silent; grunts in threat and courtship. In pairs, or small groups with fewer males than females; swims low in water, often with only head showing; rather retiring, dives if alarmed. **Range and habitat:** 2 areas of AF: north-east, in Tanzania, Kenya and Ethiopia, and south of 20° S. Rather local, on shallow inland waters with fringing vegetation. Resident. **Similar species:** African Pochard, other stifftails (see pp 56, 102).

White-backed Duck *Thalassornis leuconotus* (Anatidae) 42 cm (17 in). Rather small diving duck, mottled brown, fulvous and white, with conspicuous white spot on either side of face near base of bill. White back visible only in flight. Immature as adult. Usually silent; occasional soft whistles. In pairs or small parties, often swimming low in water, with tail submerged; unobtrusive, often close-sitting, inconspicuous when motionless among emergent vegetation. **Range and habitat:** widespread in southern AF, apart from far south-west, and north to Kenya, Ethiopia; also from Senegal to Chad in north. Local, on quiet fresh waters with fringing and floating vegetation. Non-migratory, moves in response to changing water levels.

African (Southern) Pochard *Netta erythrophthalma* (Anatidae) 50 cm (20 in). Medium-sized diving duck, distinctive male very dark brown, paler on flanks, with blue-grey legs and bill, and red eyes. Female paler brown, with curved white stripe on side of head. Immature as female, but paler. In flight, shows conspicuous white bar running length of dark upperwing. Usually silent; occasional purrs, quacks. Found paired, or in thousands; sometimes shy. **Range and habitat:** local tropical NT, declining significantly; widespread in AF south of 10° S, north to Uganda, Kenya, Ethiopia. Common on large, deep fresh waters. Resident or intra-African migrants, with annual nonbreeding influx September-October to northern AF, departs south in March. **Similar species:** female Maccoa Duck, African Black Duck.

Cape Wigeon (Cape Teal) *Anas capensis* (Anatidae) 46 cm (18 in). Rather small, pale, speckled brown and grey duck with pink bill, black at base. In flight, shows contrasting white and green speculum on dark brown wings. Immature as adult. Usually silent; occasional whistles. In pairs or small parties, flocks of hundreds known. Fairly approachable. **Range and habitat:** widespread in AF south of 15° S, and north through East Africa to Ethiopia; also scattered areas of north-west. Locally common, usually on shallow brackish or saline waters. Resident, but wandering and also making nomadic and possibly more regular movements. **Similar species:** Red-billed Pintail.

Hottentot Teal *Anas punctata* (Anatidae) 35 cm (14 in). Small, yellowish brown duck with dark speckling on breast and mantle, and distinctive dark cap above pale face; bill blue-grey. In flight, shows green speculum with white trailing edge, on dark brown wings. Immature browner, bill greyer. Usually silent; occasional low quacks, thin whistles. In pairs or small parties, often unobtrusive. **Range and habitat:** widespread in AF south of 15° S, apart from arid south-west, and north through East Africa to Ethiopia; also Chad. Locally common on sluggish, fresh or brackish waters, especially with surface vegetation. Resident or intra-African migrant. **Similar species:** Red-billed Pintail.

Red-billed Pintail *Anas erythrorhynchos* (Anatidae) 48 cm (19 in). Medium-sized, speckled brown and white duck, readily identified by red bill and dark cap above white face. In flight, shows conspicuous creamy buff speculum on dark brown wings. Immature duller, bill pinkish brown. Usually silent; occasional barely audible whistles. In pairs, or flocks up to hundreds; not shy. **Range and habitat:** widespread in AF south of 10° S and north through East Africa to Ethiopia, Sudan. Common on fresh and alkaline waters, including temporary pools. Resident or intra-African migrant. **Similar species:** Hottentot Teal, Cape Wigeon.

White-faced Whistling Duck *Dendrocygna viduata* (Anatidae) 48 cm (19 in). Medium-sized duck with white face (often stained by muddy water), bright chestnut lower neck, barred flanks and, when out of water, distinctly erect stance on rather long legs. Immature has face pale brown. Call is loud, shrill whistle 'seeeee-ooooooooo', uttered in flight. In pairs, or flocks up to hundreds; often noisy, not shy. **Range and habitat:** tropical NT, south to Uruguay, northern Argentina; widespread in AF south of 15° N, except far south-west and Congo basin. Common, locally numerous, on variety of quiet waters. Resident, flocking and moving in nonbreeding season; probably intra-African migrant. **Similar species:** female South African Shelduck.

Pygmy Goose *Nettapus auritus* (Anatidae) 34 cm (13 in). Small duck with dark green upperparts, white face and conspicuous bright rufous orange breast and flanks. Male has black-bordered green patch on side of head; female and immature have this area speckled greyish. In flight, short white speculum contrasts with dark upperwing. Male's call is soft whistle, female's is soft quack. Found in pairs or small parties, rarely in large flocks; not shy, but often still, unobtrusive, easily overlooked among emergent vegetation. **Range and habitat:** widespread in AF between 15° N and 20° S. Local on quiet fresh waters, especially with much floating vegetation. Resident, probably also intra-African migrant. **Similar species:** other pygmy geese (see p 240).

Comb (Knob-billed) Duck *Sarkidiornis melanotos* (Anatidae) 62-79 cm (24-31 in). Large duck with speckled head and neck, and glossy blue-black upperparts contrasting with white underparts. Male has dark knob, enlarged when breeding, at base of upper mandible; female smaller, lacking knob. In flight, black underwings contrast with white body. Immature buffish below, more speckled above. Usually silent. Found in pairs or flocks of up to hundreds, perching on shores or trees; sometimes shy. **Range and habitat:** tropical NT, from Venezuela south to Paraguay; OR, from India to China; widespread in AF south of 15° N, apart from far south-west. Locally or seasonally common on wide variety of fresh waters. Resident or intra-African migrant. **Similar species:** Spur-winged Goose.

Egyptian Goose *Alopochen aegyptiacus* (Anatidae) 72 cm (28 in). Large, goose-like duck, brown above, greyer below, with prominent patches of dark chestnut around eyes and in centre of breast. In flight, shows conspicuous white wing-coverts. Immature duller, more uniform, chestnut patches reduced or absent. Frequent loud wheezing and honking calls. In pairs or family parties, moulting flocks of hundreds, on open ground or water; noisy, bold. **Range and habitat:** in WP, resident Egypt, occasional Tunisia, Algeria, feral UK; in AF, 1 of commonest and most widespread of anatids south of 15° N, on all types of water except sea. Basically resident, with moult migrations and wandering. **Similar species:** Ruddy and South African Shelducks (see p 240).

Spur-winged Goose *Plectropterus gambensis* (Anatidae) 80-101 cm (31-40 in). Largest AF anatid, readily identified by reddish bill and facial skin, dark glossy upperparts and white belly. Mature male has fleshy knob on forehead, variably white face. Female smaller, browner, less facial white. In flight, white wing-coverts contrast with dark flight feathers. Immature as female but browner, face feathered. Usually silent; occasional soft whistles. Found in pairs, or in flocks of up to hundreds during off-season; rather shy and retiring. **Range and habitat:** widespread in AF south of 15° N, apart from Congo basin and far south-west. Locally common on fresh waters, particularly inundated areas around wetlands. Resident, or intra-African migrant. **Similar species:** Comb Duck.

Maccoa Duck

White-backed Duck

African (Southern) Pochard

Cape Wigeon (Cape Teal)

Hottentot Teal

Red-billed Pintail

White-faced Whistling Duck

Spur-winged Goose

Pygmy Goose

Comb (Knob-billed) Duck

Egyptian Goose

DIURNAL BIRDS OF PREY Falconiformes

Secretary Bird *Sagittarius serpentarius* (Sagittariidae) 120-50 cm (47-59 in). Very large, black and pale grey raptor with bare red facial skin, lax crest of long black feathers on back of head, and long pinkish legs. In occasional soaring flights, recognized by contrast between pale wing-coverts and black flight feathers, long neck, and elongated central tail feathers. Immature has yellow facial skin. Generally silent; various coughs, croaks at nest. Singly or paired, or in rare small groups, walking purposefully; rather wary and shy. **Range and habitat:** widespread in AF south of 17° N. Common in open country. Resident, occasionally wandering.

Rüppell's Griffon *Gyps ruppelli* (Accipitridae) 100 cm (39 in). Large, speckled, brown plains vulture with yellow eyes and cream bill. In flight, shows slender pale bars on underwing-coverts, coarsely speckled ventral torso. Immature with underwing barring reduced. Harsh squeals and hisses when disputing carcasses. Highly gregarious; moderately approachable. **Range and habitat:** northern AF, from Senegal east to Ethiopia, Kenya, northern Tanzania. Common over open dry country with good numbers of large mammals; roosts and nests on cliffs. Basically resident, making daily foraging flights of up to 150 km (95 miles). **Similar species:** African White-backed and Cape Vultures.

African White-backed Vulture *Gyps africanus* (Accipitridae) 97 cm (38 in). Large, rather plain, brown plains vulture with dark eyes and black bill. In flight, shows white back, and contrast between pale underwing-coverts and blackish flight feathers. Immature lacks white back, and has faintly streaked undertorso. Harsh squeals when disputing carcasses. Highly gregarious; quite approachable. **Range and habitat:** widespread in AF between 15° N and 30° S, apart from equatorial forest. Common over open country with good numbers of large mammals; roosts and nests on trees. Basically resident, but making long daily foraging flights and following migrations of wild ungulates, e.g. Wildebeest. **Similar species:** Rüppell's Griffon, Cape Vulture, Griffon Vulture (see p 150).

Lappet-faced Vulture *Aegypius tracheliotus* (Accipitridae) 105 cm (41 in). Largest and most powerful AF plains vulture, identified by naked crimson head with lateral skin folds and massive yellowish bill. In flight, white under-forewing-bar and thighs contrast with dark underparts. Immature has head duller, less white on thighs. Generally silent; harsh guttural sounds when disputing carcasses. Single or paired, sometimes 4 at carcass; moderately approachable. **Range and habitat:** widespread in AF at 15-30° S, and north through East Africa to areas of northern tropics and Arabia, reaching WP in Israel. Moderately common, never numerous, in open country with large numbers of big mammals. Resident, territorial, foraging locally. **Similar species:** White-headed Vulture.

Hooded Vulture *Necrosyrtes monachus* (Accipitridae) 65-70 cm (26-8 in). Small, dark brown vulture with bare pink facial skin, white downy feathering on hindneck and head, and white legs. In flight, overall brown coloration broken by pale feet and thighs, and pale area at base of flight feathers. Immature has brown down on head and neck, face as adult. Generally silent; squeals at carcasses and when nesting. Single or paired, occasionally small parties; sometimes approachable. **Range and habitat:** widespread in AF between 17° N and 22° S. Common in many habitats, from desert to forest edge; often around habitation in northern AF. Resident, making only local foraging flights. **Similar species:** immature Egyptian Vulture (see p 150).

Palm-nut Vulture *Gypohierax angolensis* (Accipitridae) 60 cm (24 in). Small, black and white vulture with yellowish bill and bare reddish facial skin; legs yellowish orange or pinkish. In flight, black secondaries, tail and primary tips contrast with white underparts. Immature brown, with yellowish facial skin and whitish legs. Generally silent;

occasional guttural and crowing calls. Found singly or in pairs; sometimes approachable. **Range and habitat:** widespread in AF south from Senegal, Sudan and Kenya to Botswana and Mozambique. Locally common in wooded areas, particularly near *Raphia* and *Elaeis* palms, whose fruit it eats. Non-migratory, but wanderers known. **Similar species:** Egyptian Vulture (see p 150).

Black-shouldered Kite *Elanus caeruleus* (Accipitridae) 32 cm (13 in). Small, graceful raptor, pale grey above with red eyes, white tail and black carpal patches visible on closed wing; underparts white. One of very few AF raptors to hover frequently. Immature browner above, washed warm buff on breast, eyes yellowish. Usually silent; occasional weak whistles. Single or paired, perching prominently; moderately approachable. **Range and habitat:** OR and WP, including Spain, Portugal, North Africa; widespread in AF south of 16° N. Common, never numerous, in many open habitats. Non-migratory or moving locally, e.g. to concentrations of rodent prey. **Similar species:** African Scissor-tailed Kite, Australian Black-shouldered Kite (see p 290).

African Scissor-tailed Kite *Chelictinia riocourii* (Accipitridae) 37 cm (15 in). Small, elegant, tern-like raptor with deeply forked tail; upperparts pale grey, underparts white with black carpal patch on underwing. Flight buoyant and graceful, hovering frequently. Immature browner above, streaked below, tail less deeply cleft. Usually silent; low mewing calls at nest. Single or in small flocks; rather shy. **Range and habitat:** mainly at 8-16° north in AF, east from west coast to Ethiopia, Somalia, Kenya. Locally common in dry, open country. Intra-African migrant, breeding March-June during northern tropics' rains and then moving south to avoid ensuing dry season. **Similar species:** black-shouldered kites (see p 290).

Bat Hawk *Macheiramphus alcinus* (Accipitridae) 45 cm (18 in). Medium-sized, very dark brown raptor with palish throat, large yellow eyes and whitish legs and feet. In flight, resembles thickset falcon *Falco* species, with narrow tail, tapered wings. Immature whitish on belly. Generally silent; occasional high whistles. Found singly; not especially shy, but spends day in dense canopies, emerging at dusk to hunt bats and small birds on wing. **Range and habitat:** OR east to New Guinea; widespread in AF south of Senegal and Somalia. Rather uncommon in densely wooded habitats and town centres, always near open areas to hunt. Basically non-migratory, possibly dispersing after breeding.

Verreaux's Eagle *Aquila verreauxii* (Accipitridae) 84 cm (33 in). Very large, black eagle with boldly contrasting white V on back. In flight, shows distinctively narrow, 'pinched' bases to wings, and white flash at base of underwing primaries. Immature rufous on mantle, nape and crown, otherwise heavily speckled and streaked buffish, blacker on breast and face. Generally silent; occasional yelping and mewing. In pairs, singly or with immature; often inconspicuous high on rock faces, not shy. **Range and habitat:** widespread in all but interior areas of AF south of 10° S, and north through East Africa to Ethiopia, southern Arabia; also Sudan, Chad. Reaches WP in southern Egypt, Sinai, Israel. Locally common around rocky hills, gorges, cliffs with hyrax, the main prey. Basically resident, immatures wander.

Tawny Eagle *Aquila rapax* (Accipitridae) 65-75 cm (26-9 in). Medium-sized eagle, varying from creamish to very dark brown, with blackish flight feathers and tail; long, yellow gape, extending back to level with centre of eye. Immature has 2, diffuse, white wing-bars, and white crescent on rump. Usually silent, but various yelping and barking calls. Single, paired; sometimes in small parties; can be approachable. **Range and habitat:** widespread in AF south of 17° N, and north to south-western Arabia. Also Morocco and Algeria in WP, east to India, Burma. Locally common in scrub or woodland or scavenging at habitation. Some are resident, others local migrants. **Similar species:** Steppe Eagle, other brown eagles (see p 150).

□■□ **Secretary Bird**
□■□ **Rüppell's Griffon**
□□■ **Lappet-faced Vulture**

■■□ **African White-backed Vulture**
□□■ **Verreaux's Eagle**

■□□ **Hooded Vulture**
□■□ **Palm-nut Vulture**
□□■ **Black-shouldered Kite**

■□□ **Bat Hawk**
□■□ **African Scissor-tailed Kite**
□□■ **Tawny Eagle**

Crowned Solitary Eagle *Harpyhaliaetus coronatus* (Accipitridae) 81-90 cm (32-5 in). Very large, thickset eagle, dark above with crest on nape and barred tail; undertorso buffish or white, blotched and barred dark. In flight, shows underwings with rufous coverts, white flight feathers barred black, and broad black trailing edge. Immature grey above, white on head and ventral torso; underwing-coverts faintly rufous. Male loudly whistles 'ki-wii ki-wii ki-wii . . .' during high undulating display flight; female a lower 'kuu-kuu-kuu'. Single or paired, shy. **Range and habitat:** AF from Senegal and Ethiopia south to South Africa. Sometimes common, often local, in forest and woodland. Non-migratory. **Similar species:** immature Martial Eagle.

Martial Eagle *Hieraaetus bellicosus* (Accipitridae) 76-83 cm (30-3 in). Large, powerful eagle with dark brown upperparts, head and breast; below white, rather inconspicuously spotted brown; short crest on nape. In flight shows contrast between white belly and dark head, tail and underwings. Immature grey above, white below; in flight, whitish underwing-coverts contrast with dark flight feathers. Generally silent; ringing 'klow-eee', also barking. Single or paired; sometimes wary and shy. **Range and habitat:** widespread in AF south of 15° N, apart from forest regions and many areas of west. Moderately common in open habitats. Non-migratory, making long daily foraging flights. **Similar species:** Short-toed Eagle (see p 150).

African Hawk Eagle *Hieraaetus spilogaster* (Accipitridae) 60-70 cm (24-8 in). Medium-sized eagle, blackish above, white below, with short dark streaks on breast. In flight, black underwing-coverts contrast with white flight feathers; black terminal band to tail. Immature browner above, with rufous head, undertorso and underwing-coverts. Usual call is 'kloo-kloo-kloo', also various yelping and barking sounds. Found singly or in pairs; moderately approachable. **Range and habitat:** widespread in AF south of 15° N, apart from far south-west and equatorial forests. Locally common, never numerous, in scrub and woodland, often around rocky prominences. Mainly resident. **Similar species:** Ayres' Hawk Eagle, Booted Eagle (see p 150).

Long-crested Eagle *Spizaetus occipitalis* (Accipitridae) 52-8 cm (20-3 in). Small, blackish eagle with long, loosely erect crest on rear of crown. In flight, shows black and white barred tail, white patch at base of upperwing primaries, and black underwing-coverts contrasting with white and grey flight feathers. Immature browner, crest shorter, some white speckling on neck and wings. Various high, sharp, repeated whistles, 'ki-ki-ki-ki. . .' and drawn-out screams, 'keeee-eeeer'. Single or sometimes in pairs, perching conspicuously; often approachable. **Range and habitat:** widespread in AF south of 15° N, apart from far south-west. Locally common in wooded country, including forest edge and farmland. Basically resident, some wandering.

Lizard Buzzard *Kaupifalco monogrammicus* (Accipitridae) 37 cm (15 in). Small raptor with grey upperparts, head and breast, and white throat crossed by vertical black streak; belly finely barred grey on white; cere and legs red. In flight, shows 2 (sometimes 1) conspicuous white bars on dark tail. Immature buffer than adult, with less well-defined throat streak. Call is far-carrying, ringing whistle 'keee-oo keeee-oo', also 'doo-doo-doo. . .'. Single, sometimes paired, still-hunting from high perches; sometimes inconspicuous in dense canopies, sometimes shy. **Range and habitat:** AF within 15° of equator, reaching 30° S in east. Quite common in well-wooded areas. Basically resident, moving locally or nomadically. **Similar species:** Gabar Goshawk.

Brown Harrier (Brown Snake) Eagle *Circaetus cinereus* (Accipitridae) 71 cm (28 in). Medium-sized eagle, darkish brown at rest, with large yellow eyes, large owl-like head and bare whitish legs. In flight, easily identified by heavily barred tail, and contrast between dark body plumage and silvery flight feathers. Immature paler, may have whitish mottling on undertorso, and streaking on crown. Call is

drawn-out 'zeee-oooo' whistle; also guttural 'tok'. Singly, usually perching conspicuously; sometimes retiring. **Range and habitat:** widespread in AF south of 15° N, except western equatorial forests and far south. Quite uncommon in dry, open scrub and woodland. Basically resident. **Similar species:** immature Bateleur, Short-toed Eagle.

Bateleur *Terathopius ecaudatus* (Accipitridae) 60-70 cm (24-8 in). Striking, medium-sized eagle, mainly black at rest, with large owl-like head, bright red legs and facial skin, and bright chestnut tail and mantle. In flight, readily identified by long pointed wings, virtual lack of tail and habitual sideways rocking action; underwings blazing white, with narrow black trailing edge in female and much broader edge in male. Immature brown, tail longer than adult's. Generally silent, sometimes barking. Single or paired, usually in conspicuous low gliding flight; not shy. **Range and habitat:** widespread in AF south of 15° N, except Congo forests and far south. Common in open habitats, including woodland. Basically resident. **Similar species:** Brown Harrier Eagle.

African Fish Eagle *Haliaeetus vocifer* (Accipitridae) 66-75 cm (26-9 in). Large eagle with white head, mantle, breast and tail; shoulders and belly deep chestnut, flight feathers blackish. In flight, distinctive broad-winged, short-tailed silhouette. Immature initially brown, developing white areas of adult plumage with brown mottling and streaking. Call is loud, 3- or 4-part, ringing gull-like scream. Paired, sometimes single, perching prominently, noisy, quite approachable. **Range and habitat:** widespread in AF south of 16° N. Locally common around fish-bearing waters. Some resident, most disperse after breeding to even small seasonal waters. **Similar species:** Palm-nut Vulture, American Bald Eagle (see pp 58, 192).

Augur Buzzard *Buteo augur* (Accipitridae) 45-53 cm (18-21 in). Large buzzard with blackish upperparts, rather short reddish tail, and barred panel in upperwing secondaries; 2 distinct colour phases occur, commoner pale morph mainly white below, rarer dark phase mainly black. Distinctive flight silhouette, with broad wings often held above body, and short, broad tail. Immature brown above and on tail, with dark markings below. Call is far-carrying high crowing. Single or paired; not shy. **Range and habitat:** eastern and southern AF, from Ethiopia to East Africa and scattered areas at 10-25° S. Common in open habitats, usually near hills and rocky outcrops. Basically non-migratory. **Similar species:** Red-backed and Jackal Buzzards.

African Little Sparrowhawk *Accipiter minullus* (Accipitridae) 23-5 cm (9-10 in). Very small sparrowhawk, size only of oriole; grey above, with 2 conspicuous white spots on black tail; below white, finely barred rufous on breast and belly; eye, cere and legs yellow. White rump visible in flight. Immature brown above, with tail spots and narrow white rump; heavily spotted below. Often silent; call is repeated high 'kee'. Found singly, sometimes in pairs; small, unobtrusive, overlooked in dense canopies, often retiring. **Range and habitat:** AF, widespread in moist areas to south of Congo basin, and north to East Africa, Ethiopia. Usually uncommon, in dense woodlands. Basically non-migratory. **Similar species:** Red-thighed Sparrowhawk.

Somali Chanting Goshawk *Melierax poliopterus* (Accipitridae) 50-60 cm (20-4 in). Large goshawk with upright stance and long orange or red legs; upperparts and breast pale grey, belly finely barred grey on white; cere yellow to orange-red. In flight, shows black wing tips and tail and white rump; shallow wing-beats. Immature browner, with streaked breast and more heavily barred belly. Chanting is loud, melodious piping, accelerating to crescendo then fading, given from perch or in flight in breeding season. Single or paired, perching prominently; not shy. **Range and habitat:** north-eastern AF, in East Africa, Ethiopia, Somalia. Common in open scrub and woodland. Basically non-migratory. **Similar species:** Dark Chanting Goshawk.

Crowned Solitary Eagle

Martial Eagle

African Hawk Eagle

Long-crested Eagle

Brown Harrier (Brown Snake) Eagle

Lizard Buzzard

Bateleur

African Fish Eagle

Augur Buzzard

Somali Chanting Goshawk

African Little Sparrowhawk
female at nest

African Harrier Hawk *Polyboroides typus* (Accipitridae) 61-6 cm (24-6 in). Quite large, blue-grey raptor with bare yellow facial skin, lax crest on rear of small head, and finely barred abdomen. In flight, shows black tips and trailing edges to broad wings, and white bar on black tail. Immature uniformly brown or mottled paler, sometimes warm buff below. Generally silent; long quavering whistle 'keeeeeeeee' in flight. Single or sometimes paired, foraging on tree-trunks and cliff faces, thrusting long legs into holes and nests; quite retiring, unobtrusive. **Range and habitat:** widespread in AF south of 15° N, apart from dry south-west. At best locally common, in well-wooded areas. Basically non-migratory. **Similar species:** Madagascar Harrier Hawk.

Red-headed (Red-necked) Falcon *Falco chicquera* (Falconidae) 30-6 cm (12-14 in). Small falcon with rufous crown and nape, and warm buff wash on cream breast; remainder of plumage blue-grey above and white below, with fine dark barring. In flight, shows dark subterminal bar to pale tail. Immature browner, some streaking below. Call is shrill 'kee-kee-kee-kee. . .'. Single, paired or in family parties; sometimes unobtrusive, shy. **Range and habitat:** widely distributed in India; widespread in AF south of 15° N, apart from Congo forests of west. Uncommon in low, open, mainly dry country, often near palm trees. Basically resident, and wandering. **Similar species:** Lanner.

African Pygmy Falcon *Polihierax semitorquatus* (Falconidae) 19 cm (7½ in). Tiny falcon, only size of shrike, male grey above with white face and underparts, female with chestnut mantle. In flight, conspicuous white rump; tail and flight feathers black, spotted white. Immature as female, washed warm buff on breast. Call sharp and high-pitched 'kee-kee-kee. . .', or 'koo-koo-koo. . .'. Single or paired, or in small, loose parties, perching conspicuously, mostly taking prey from ground; moderately approachable. **Range and habitat:** 2 regions of AF: in north-east, from Sudan, Ethiopia and Somalia south to Tanzania; in south, in Namibia, Botswana, South Africa. Fairly common in open, dry bushed and wooded country, breeding in old weaver nests. Basically resident. **Similar species:** White-rumped Pygmy Falcon.

FOWL-LIKE BIRDS Galliformes

Stone Partridge *Ptilopachus petrosus* (Phasianidae) 25 cm (10 in). Unique medium-sized gamebird, resembling partridge; mainly dark brown inconspicuously speckled buff, with pale buff or cream lower breast and belly; legs red; walks and runs with tail cocked upwards like bantam. Rarely flies. Immature as adult, barred above and below. Noisy; characteristic high, piping call 'ooeeet-ooeeet-ooeeet'. In small parties, running on rocks and cliffs; very shy, betrayed by calls. **Range and habitat:** northern AF, at 6-16° N, from Senegal east to Ethiopia, north-west Kenya. Locally common on rocky hills and cliffs in dry country. Very sedentary. **Similar species:** Crested Francolin.

Swainson's Francolin *Francolinus swainsonii* (Phasianidae) 34-9 cm (13-15 in). Rather large francolin, best identified by combination of streaked brown plumage, conspicuous red bare facial skin and neck, black legs and bill. Female smaller. Immature has face and neck feathered buffy white with white spotting elsewhere. A noisy bird, producing a loud crowing 'krraa krraa' repeated several times and usually delivered at dusk from a branch or anthill. Otherwise shy, found on ground in pairs or small groups. **Range and habitat:** southern AF. Common on cultivated land, woodland edges and open secondary growth. Non-migratory. **Similar species:** 34 other AF francolins.

Yellow-necked Francolin *Francolinus leucoscepus* (Phasianidae) 33-6 cm (13-14 in). Large francolin with upright stance, grey-brown above and buff below, finely streaked white; identified by bare yellow skin on throat, bare red skin around eye, and black legs. In flight, open wing shows large pale patch at base of primaries. Immature duller, bare skin paler, legs brown. Call is loud and frequently uttered,

retching, grating 'gruaaaark', often at twilight. Paired or small flocks, foraging openly on ground; moderately approachable, bold. **Range and habitat:** north-eastern AF, in Ethiopia, Somalia, Kenya, Tanzania. Common in many less open habitats. Non-migratory. **Similar species:** Grey-breasted and Red-necked Spurfowls.

Helmeted Guineafowl *Numida meleagris* (Phasianidae) 52-7 cm (20-2 in). Large, plump gamebird with fine white spotting on dark grey plumage, rendered unmistakable by combination of horny casque protruding from crown, and naked red and blue head and neck, with pendulous wattles near base of bill. Immature browner, casque reduced or absent, downy feathering on neck. Loud, frequent call is grating 'keerrrrr'. In flocks of up to hundreds, foraging on open ground; sometimes approachable, bold. **Range and habitat:** WP in Morocco, where fast declining; widespread in AF south of 15° N (including south-west Arabia), apart from Congo forests and extreme south-west. Common in many wooded habitats, usually near water. Resident. **Similar species:** other guineafowl in AF.

Vulturine Guineafowl *Acryllium vulturinum* (Phasianidae) 58-61 cm (23-4 in). Long-tailed guineafowl, with grey head and neck naked save for tuft of rich brown, downy feathering on nape; feathers of mantle and breast elongated and striped black, white and bright cobalt-blue; remainder of plumage dark grey, finely spotted white. Immature browner, with more downy feathering on head. Often silent; loud, high-pitched alarm 'chink chink. . .chickerr'. Flocks on ground; not shy. **Range and habitat:** north-eastern AF, in Somalia, Ethiopia, Kenya, Tanzania. Common in open bushed and wooded areas, often far from water. Resident. **Similar species:** Helmeted Guineafowl.

CRANES, RAILS AND ALLIES Gruiformes

Little Button-quail (Andalusian Hemipode) *Turnix sylvatica* (Turnicidae) 14-15 cm (5½-6 in). Small, quail-like bird with pale eyes, female brown above with rufous breast, white belly and heart-shaped spots on flanks. Male duller. In brief low flights, shows brown rump and pale upperwing-coverts. Immature duller, breast spotted. Call is ventriloquial hooting, 'hoom'. Single, paired or in small parties, foraging on ground; skulking and often overlooked. **Range and habitat:** WP in Portugal, Spain, North Africa, declining rapidly; also OR east to Philippines. Widespread in AF south of 15° N, apart from Congo forests and far south-west. Locally common in open grassy habitats. Resident or intra-African migrant. **Similar species:** Black-rumped Button-quail, true quails (see pp 154, 244, 246). **Not illustrated.**

African Black Crake *Porzana flavirostra* (Rallidae) 20-3 cm (8-9 in). Medium-sized black crake with red eyes, legs and feet, and greenish yellow bill. Immature greyer with pale throat, duskier bill and legs. Call is deep, straining, growling grunt, repeated 2 or 3 times; also various clucking notes, and sharp 'chik' alarm. Single, paired or in small, loose parties, foraging on ground; not secretive, sometimes approachable. **Range and habitat:** widespread in AF south of 15° N. Common, most easily seen crake, on swamp with open water and some vegetation. Basically non-migratory. **Similar species:** Blackish Rail (see p 108).

Crested (Red-knobbed) Coot *Fulica cristata* (Rallidae) 42 cm (17 in). Medium-sized, thickset black waterbird with conspicuous white bill and frontal shield, and 2 dark red knobs (often very inconspicuous) on forecrown; legs and lobed feet greenish to greyish. Immature grey-brown above, off-white below. Call is resonant, penetrating 'kronnk'. Single, paired or in flocks up to thousands, swimming, diving, seldom flying; not shy. **Range and habitat:** WP in Spain, Morocco, Algeria, Tunisia, but declining. Widespread in AF south of 10° S, and north through East Africa to Ethiopia. Common, seasonally abundant, on inland waters. Moves extensively when not breeding. **Similar species:** Moorhen, Red-fronted Coot (see p 154).

African Harrier Hawk

Stone Partridge

Red-headed (Red-necked) Falcon

African Pygmy Falcon female

Swainson's Francolin

Yellow-necked Francolin

Helmeted Guineafowl

Vulturine Guineafowl

African Black Crake

Crested (Red-knobbed) Coot

Crowned Crane *Balearica pavonina* (Gruidae) 105 cm (41 in). Medium-sized crane with golden crest on rear of crown, and bright red and white throat and face wattles; rest of plumage mainly grey, with large white patch in closed wing. In flight, distinctively drooping neck, and white wing-coverts. Immature smaller, browner, lacking wattles. Call is 2-note honk, 'graa-oww'. Paired when nesting, or flocks up to 200, foraging in open; moderately approachable. **Range and habitat:** widespread in AF south of 16° N, apart from Congo forests and far south-west. Locally common on marshes and other damp habitats. Resident, moving locally. **Similar species:** South African Crowned Crane.

Blue (Stanley) Crane *Anthropoides paradisea* (Gruidae) 105 cm (41 in). Medium-sized, pale blue-grey crane with pinkish bill, white crown and elongated dark grey tertials curving down behind wings; legs black. In flight, dark flight feathers contrast with pale wing-coverts. Immature paler, lacking tertial elongation; crown warm brown. Call is loud, high-pitched 'craaaaaarrk'. Paired when breeding, or in flocks up to hundreds, foraging in open; not shy. **Range and habitat:** 2 areas of southern AF: Etosha, northern Namibia; South Africa, from Cape Province to Transvaal. Common in grasslands. Resident, moving outside breeding season. **Similar species:** Demoiselle Crane.

Wattled Crane *Bugeranus carunculatus* (Gruidae) 120 cm (47 in). Large crane, grey above and black below, with white breast and neck, dark grey crown, and reddish and white wattle hanging from naked, dark reddish face; legs black. Flies reluctantly. Immature has wattles reduced or absent, crown white. Generally silent; occasional loud, ringing 'horoonnk'. Found in pairs, or in flocks up to hundreds, foraging in open; shy. **Range and habitat:** 3 areas of AF: Ethiopian highlands; central south, from Namibia to Zimbabwe; eastern South Africa. Uncommon around open lakes and marshes, and on damp grasslands. Resident.

African (Peters') Finfoot *Podica senegalensis* (Heliornithidae) 46-53 cm (18-21 in). Cormorant-like waterbird with conspicuous red bill, legs and lobed feet, whitish stripe on side of neck and dark, white-spotted wings and back; underparts white. Male has head and neck dark blue-grey. Female smaller, chin and foreneck whitish. Immature as female but browner, less spotted. Generally silent; occasional hard 'bak', weaker 'kiiii'. Single or paired, foraging beneath overhanging riverine vegetation; secretive, often overlooked. **Range and habitat:** widespread in AF south of 10° N. Uncommon on quiet inland waters with dense overhanging vegetation. Non-migratory, moving locally. **Similar species:** other finfoots, Sungrebe (see p 108).

Kori Bustard *Ardeotis kori* (Otididae) 120-50 cm (47-59 in). Large, crested bustard, grey-brown above with black and white tips to wing-coverts; neck finely vermiculated grey-white, with peculiarly thick appearance due to loose feathering; underparts white. Female smaller. Flies reluctantly but strongly. Immature spotted above. Generally silent; displaying male produces deep grunting, 'vum, vum, vum, vummmmmm'. Found singly, in pairs or small, loose parties, walking in open areas; quite retiring. **Range and habitat:** 2 regions of AF: widespread south of 15° S; also north-east, from Tanzania to Ethiopia. Locally common in open country. Mainly resident. **Similar species:** Arabian Bustard.

White-bellied Bustard *Eupodotis senegalensis* (Otididae) 53 cm (21 in). Small, handsome bustard, brown on back and wings, with white belly. Male has dark cap, white face, black throat and blue-grey foreneck; female has brown cap, with face, throat and foreneck white or pale brown. Immature resembles female. Call is far-carrying squawk, 'aarwarka, aawarka'. Paired or in family parties, walking; shy, betrayed by call. **Range and habitat:** many areas of AF within 17° of equator, also eastern South Africa. Rather local and uncommon in grassland and open bush. Resident. **Similar species:** 3 similar small bustards in AF.

Hartlaub's Bustard *Eupodotis hartlaubii* (Otididae) 53 cm (21 in). Small bustard. Male has black underparts, foreneck, cheek patch and line behind eye; back and wings brown; in flight, wings white, rump and tail blackish. Female brown above, underparts and foreneck white, rump and tail greyish. Immature undescribed. Call is quiet 'pop... hooooomm'. Single, paired, occasionally loose groups; sometimes approachable. **Range and habitat:** north-eastern AF, from Tanzania to Sudan, Ethiopia. Local or uncommon in grassland and open bush, usually in dry areas. Resident. **Similar species:** 3 similar AF bustards.

(Buff-)Crested Bustard *Eupodotis ruficrista* (Otididae) 53 cm (21 in). Small bustard. Male brown above with whitish V markings; belly black; reddish pink crest on nape. Female similar, lower breast white; crest less developed. Immature not described. Male's call is series of piercing, mournful whistles; also croaking duet. Single or paired, on ground; secretive, retiring, betrayed by call. **Range and habitat:** 3 regions of AF: north-east, from Tanzania to Ethiopia; northern tropics, from Senegal to Sudan; at 15-30° S. Common in dry, densely bushed areas. Resident. **Similar species:** 3 similar AF bustards.

SHOREBIRDS, GULLS, AUKS Charadriiformes

Spotted Stone-curlew (Cape Dikkop) *Burhinus capensis* (Burhinidae) 43 cm (17 in). Large stone-curlew, upperparts buff with profuse dark spotting; underparts white, washed buff on breast; legs, large eyes and base of blackish bill yellow. In flight, wings conspicuously black and white. Immature streaked dark above. Call, usually at night, is series of shrill whistles 'chi-weee', increasingly loud then fading. Found paired on ground; mainly nocturnal, resting in shade by day; unobtrusive. **Range and habitat:** widespread in AF south of 16° N, apart from western forests. Sometimes common, never numerous, in open scrub and woodland. Resident, moving locally. **Similar species:** other stone-curlews (see pp 110, 160, 294).

Senegal Stone-curlew (Thick-knee) *Burhinus senegalensis* (Burhinidae) 37 cm (15 in). Medium-sized, upperparts grey-brown, thinly streaked dark, with greyish panel in closed wing; underparts white, streaked on breast; legs, large eyes and base of bill yellow. In flight, black flight feathers contrast with pale wing-coverts. Immature more finely streaked. Call is series of high 'pee' notes. Single, in pairs or small flocks, active by day and night; unobtrusive. **Range and habitat:** at 5-16° N in AF, reaching WP along Nile, in Egypt. Locally common near water in bushed or grassy habitats. Resident, and local migrant. **Similar species:** Other stone-curlews (see p 110, 160, 294).

African Jacana *Actophilornis africanus* (Jacanidae) 28 cm (11 in). Striking gallinule-like bird, with chestnut-brown body, white foreneck and face; bill and frontal shield bright bluish white; long legs and elongated toes, greyish. Immature brown above, white below, tinged golden on breast; crown and hindneck blackish. Strident screeching, trilling and grating calls. Single, in pairs or flocks; noisy, somewhat retiring. **Range and habitat:** widespread in AF south of 16° N. Common on fresh waters with floating vegetation where they forage. Resident, or local migrant. **Similar species:** other jacanas (see pp 110, 248, 294).

Lesser (Smaller) Jacana *Microparra capensis* (Jacanidae) 15 cm (6 in). Sparrow-sized, brown above with black rear-neck, chestnut cap and white eyebrow; underparts white, long legs, toes greenish. In short flights, shows white trailing edge to blackish wings. Immature has crown dark, nape golden. Often silent; low grunt, 'pruk'. Found singly, in pairs or loose parties, foraging on floating vegetation or swampy shores; shy. **Range and habitat:** scattered areas of AF from Niger to Ethiopia, and through East Africa to Angola, South Africa. Local, usually uncommon, on fresh waters with floating vegetation. Resident or local migrant. **Similar species:** other jacanas (see pp 110, 248, 294).

■□□
Crowned Crane
□■□
Blue (Stanley) Crane
□□□
Wattled Crane

■□□
African (Peters') Finfoot
□■□
Kori Bustard
□□■
White-bellied Bustard

■□□
Hartlaub's Bustard male displaying
□■□
(Buff-)Crested Bustard
□□□
Spotted Stone-curlew (Cape Dikkop)

■□□
Senegal Stone-curlew (Thick-knee)
□■□
African Jacana
□□■
Lesser (Smaller) Jacana

Painted Snipe *Rostratula benghalensis* (Rostratulidae) 24-8 cm (9½-11 in). Medium-sized, snipe-like wader; bill long, with slightly drooping tip. Female grey-green above with dark rufous head and neck, and conspicuous broad white stripes through eye and centre-crown, and on mantle; underparts white with black breast band. Male smaller, greyer, lacking rufous coloration. Flight low, rail-like, with dangling legs and showing dark, rounded, buff-spotted wings. Immature as male, lacking breast band. Often silent; sharp 'chek'. Found singly or in pairs, unobtrusive in marshes; shy and often overlooked. **Range and habitat:** WP in Egypt; east through OR to Philippines, also AU. Widespread in AF south of 15° N, apart from Congo forests and far south-west. Local and uncommon along quiet swampy margins where vegetation provides refuge. Resident or migrant. **Similar species:** South American Painted Snipe.

African Black Oystercatcher *Haematopus moquini* (Haematopodidae) 45 cm (18 in). Large, thickset, glossy black wader with red eyes, long straight orange-red bill, and purplish red legs. Immature browner, legs greyish. Loud piping calls 'tee-weeep, tee-weeep', with sharper 'tik tik' alarm notes. Found singly or in small parties on open shores; somewhat shy. **Range and habitat:** far southern AF, on coasts of Namibia and South Africa. Locally common along coastline of mainland and offshore islands. Basically non-migratory; breeds in more remote areas. **Similar species:** other black oystercatchers (see p 294).

White-fronted Sandplover *Charadrius marginatus* (Charadriidae) 18 cm (7 in). Small plover, sandy brown above with narrow white collar, white forehead and thin dark eye stripe; underparts white, sometimes washed buff on breast; bill and legs black. Male has darker crown than female. Immature paler on crown, all white below. Flight call is softly piped 'twi-it', also sharp alarm. Found singly or in pairs, flocking in off-season, on open shores; sometimes approachable, can be unobtrusive. **Range and habitat:** widespread in AF south from Kenya, Zaire, Niger and Senegal. Common or locally so on open, sandy seashores, also around large inland waters. Basically resident. **Similar species:** other *Charadrius* plovers (see pp 62, 110, 156, 294).

Kittlitz's Sandplover *Charadrius pecuarius* (Charadriidae) 16 cm (6½ in). Small plover, grey-brown above with conspicuous white stripe above eye extending back to form complete white collar around hindneck; bill, legs and eye stripe black; underparts white with buff or orangish breast. Immature has head brownish, underparts white. Flight call is plaintive piping, 'chip-eeep'. Single, paired or in loose flocks on open flats; can be approachable, sometimes overlooked. **Range and habitat:** WP in Egypt; widespread in AF south of 16° N, with exception of Congo forest regions. Locally common on bare or short grass flats adjacent to inland waters, sometimes marine coast. Resident or local migrant. **Similar species:** White-fronted Sandplover.

Three-banded Plover *Charadrius tricollaris* (Charadriidae) 18 cm (7 in). Small plover, brown above with white forehead, eyebrow and collar, red eye-ring, and black-tipped red bill; underparts white, with 2 black breast bands; legs pinkish. Immature speckled buff above, with brown head and incomplete breast bands. Flight call is plaintive piping. Single or paired, loosely flocking in some areas, on open shorelines; moderately approachable, can be unobtrusive. **Range and habitat:** widespread in southern AF, and north to East Africa, Sudan, Ethiopia; also Mali, Chad, Cameroon. Common but never numerous on shores of inland waters. Resident or local migrant. **Similar species:** other *Charadrius* plovers (see pp 62, 110, 156, 294).

Crowned Plover *Vanellus coronatus* (Charadriidae) 30 cm (12 in). Large plover, identified by conspicuous white ring encircling black crown; upperparts and breast greyish brown, latter with dark lower margin, below which belly is white; black-tipped red bill, conspicuous long red legs. Immature speckled buff above, lacking dark breast margin.

Call is strident scream 'streeep', rapidly repeated in flight. In pairs and small flocks on open ground; diurnal, but often active and calling at night; noisy, aggressive when nesting, otherwise somewhat retiring. **Range and habitat:** widespread in AF south of 10° S, and through East Africa to Ethiopia, Somalia. Common on open and bushed grasslands, mainly in drier country. Basically resident. **Similar species:** other *Vanellus* plovers (see pp 110, 156, 250, 294).

Spur-winged Plover *Vanellus spinosus* (Charadriidae) 26 cm (10 in). Medium-sized plover, grey-brown above with black cap, and conspicuous vertical black stripe from chin to centre of white breast; upper belly black, abdomen white; bill and legs black; wing spur not fieldmark. In flight, prominent white rump and wing-bar. Immature speckled buff above. Call is loud screech, 'dee-dee-durrr-it'; sharp 'teck' alarm. Found in pairs or small, loose flocks on open ground; rather retiring. **Range and habitat:** WP in Egypt and Near East; widespread in AF at 5-16° N, from Senegal east to Somalia, Kenya. Common on open flats and short grassland near water; rare on exposed seashores. Resident or local migrant. **Similar species:** Long-toed Lapwing.

Blacksmith Plover *Anitibyx armatus* (Charadriidae) 29 cm (11½ in). Large plover, black with contrasting white forehead, hindneck and belly, and pale grey mantle and wings; eye reddish brown, bill and legs black. Immature has conspicuous broad, vertical, dark brown stripe on side of white neck; head brown. Call is distinctive, hard, repeated 'tink', often blending into scream. Found singly or in pairs, sometimes loose flocks, on open ground; noisy, rather retiring. **Range and habitat:** widespread in AF south of 10° S, and north through Tanzania to Kenya. Common on open flats and grassland near water, but rarely coastal. Resident, or wandering locally. **Similar species:** Long-toed Lapwing.

Senegal Wattled Plover *Vanellus senegallus* (Charadriidae) 34 cm (13 in). Large, grey-brown plover with conspicuous red and yellow wattles at base of black-tipped yellow bill, prominent white forehead and streaked neck; long yellow legs. In flight, shows prominent white patch on inner wing. Immature has wattles and forehead white reduced. Call is sharp, shrill, repeated 'kweep', often ascending and accelerating; frequently at night. Found singly or in pairs, flocking in off-season; somewhat retiring. **Range and habitat:** widespread in AF south of 15° N, apart from Congo forests, far south-west and much of far east. Locally common on bare ground or grassland, often near water. Resident or local migrant. **Similar species:** White-crowned Plover.

Black-headed Plover *Vanellus tectus* (Charadriidae) 25 cm (10 in). Medium-sized plover, grey-brown above and white below, with upswept black crest on rear of black crown, red wattles at base of black-tipped red bill, vertical black median stripe on white breast, and long maroon legs. In flight, shows white wing-coverts, rump and black-tipped tail. Immature speckled buff and black above. Call is shrill, rasping 'kerr, kerr, kerr', also 'kerr-vick'. Found in pairs or small flocks, on open ground; sometimes approachable. **Range and habitat:** widespread in AF at 10-17° N, from Senegal to Somalia, Kenya. Locally common in dry, open and bushed habitats. Non-migratory, or moving locally. **Similar species:** Spur-winged Plover.

Long-toed (Lapwing) Plover *Vanellus crassirostris* (Charadriidae) 30 cm (12 in). Large plover, brown above, with striking white face and throat contrasting with black hindneck and breast; eyes red, bill pinkish with dark tip; long, dull red legs with elongated toes. In flight, wings appear mostly white. Immature not described. Call is plaintive whistle, 'wheeeet'; sharper 'tickkk' alarm. Found in pairs or small groups, foraging on floating vegetation or open ground; wary, often retiring. **Range and habitat:** mainly eastern AF, from Mozambique and Zambia north to East Africa, south Sudan; also Angola, Chad. Uncommon to locally common around permanently swampy areas with floating vegetation. Non-migratory or local migrant.

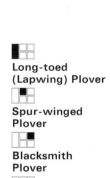

Painted Snipe

African Black Oystercatcher

White-fronted Sandplover

Kittlitz's Sandplover

Three-banded Plover

Crowned Plover

Long-toed (Lapwing) Plover

Spur-winged Plover

Blacksmith Plover

Senegal Wattled Plover

Black-headed Plover

African Snipe *Gallinago nigripennis* (Scolopacidae) 30 cm (12 in). Medium-sized wader with long, straight bill; upperparts blackish brown with buff streaking, underparts white. Flight fluttering. Immature less boldly marked. Generally silent; characteristic 'ts-uck' when flushed; deep 'drumming' produced by tail feathers during aerial display. Single, in pairs or loose parties, on mud in marshy vegetation; somewhat nocturnal; secretive and often overlooked. **Range and habitat:** widespread in AF south of 10° S, apart from dry south-west; also East Africa, Ethiopia. Locally common in marshes, inundated grassland. Resident **Similar species:** other snipe (see pp 110, 160, 250).

Crab Plover *Dromas ardeola* (Dromadidae) 38 cm (15 in). Large, thickset, white wader with black mantle, long greyish legs and heavy black bill. In flight, contrasting black flight feathers. Immature browner, with dark streaking on crown. Call is sharp 'cheerook', also constant chatter and barking. In small parties, flocks reaching hundreds at times, on open shores; diurnal, active at night; retiring. **Range and habitat:** breeds in WP on shores of Arabia, then migrates August-April to western India, Indian ocean islands and east coast of AF south to Mozambique and (stragglers) South Africa. Locally common on sandy and muddy marine shores. **Similar species:** Eurasian Avocet (see p 156).

Temminck's Courser *Cursorius temminckii* (Glareolidae) 20 cm (8 in). Rather small, brown courser with bright chestnut cap, black and white eye stripes extending to nape, rufous upper belly, black central belly and white vent; legs long and grey-white. Immature streaked black on crown, speckled buff above, less black on belly. Generally silent; grating, metallic flight call 'eerrrr-errr'. In pairs or small flocks, on open ground; rather retiring. **Range and habitat:** widespread in AF south of 15° N, except equatorial forests and far south-west. Locally common on short grassland, including recently burnt ground. Mainly resident or nomadic. **Similar species:** other coursers (see p 248).

Two-banded Courser *Rhinoptilus africanus* (Glareolidae) 21 cm (8 in). Small, pale courser, mottled buff and black above, pale buff below with 2 black breast bands, long whitish legs. In flight, shows white rump and rufous patch in wing. Immature as adult, initially lacking breast bands. Generally silent; low piping 'peee-peeee'. Single, in pairs or scattered groups, on open ground; often active at night; unobtrusive. **Range and habitat:** 3 areas of AF: Ethiopia and Somalia; southern Kenya and Tanzania; far south-west. Locally common in dry, open or sparsely bushed plains. Resident. **Similar species:** Heuglin's and Jerdon's Coursers.

Violet-tipped (Bronze-winged) Courser *Rhinoptilus chalcopterus* (Glareolidae) 27 cm (10½ in). Large courser, brown above and white below, with broad dark streak across white face, brown breast band with blackish lower margin, and dull red legs. In flight, shows white rump, black flight feathers and violet primary tips. Immature buff above, washed buff below. Call is piping 'jee-koo-eet'. Single, in pairs or small parties; nocturnal, spending day under bushes; retiring, often overlooked. **Range and habitat:** widespread in AF south of 15° N, except Horn, Congo forests and far south-west. Uncommon in bushed and wooded country. Resident and intra-African migrant. **Similar species:** Crowned Plover.

Egyptian Plover *Pluvianus aegyptius* (Glareolidae) 21 cm (8 in). Squat, short-legged courser, grey above with black mantle and cap, and white eyebrows joining on nape; underparts buffish white with thin black breast band; legs greyish. In flight, black bar on pale wing. Immature as adult, speckled brown above, wing-coverts rusty. Call is loud, rapidly repeated 'churrsk'. Found in pairs or flocks on open ground; usually tame. **Range and habitat:** widespread in AF at 5-15° N from Senegal to Ethiopia, also south to Congo, Zaire. Locally common around rivers and lakes with emergent bars and short vegetation. Resident and intra-African migrant.

White-collared Pratincole *Glareola nuchalis* (Glareolidae) 19 cm (7½ in). Small pratincole, dark brown above, with conspicuous white collar (rufous some regions) from cheeks around hindneck; grey breast merges with white abdomen; legs red. In flight, shows stubby forked tail, white rump. Immature speckled, legs black, lacking collar. Call is shrill 'pee-pee-pee-. . .'; alarm a faint 'tip-tip-tip'. In small flocks on open shores and rocks; approachable, can be unobtrusive. **Range and habitat:** widespread in AF from Sierra Leone and Angola east to Uganda, Zaire, Mozambique, also Niger, Chad, Ethiopia. Locally common on rivers and lakes with exposed rocks. Resident and intra-African migrant. **Similar species:** other pratincoles (see pp 160, 250).

Grey-headed Gull *Larus cirrocephalus* (Laridae) 41 cm (16 in). Medium-sized gull, breeding adult with pale grey hood and red bill and legs; upperwing grey with black tip and white flash towards base of primaries. Nonbreeding adult has hood indistinct. Immature has white head with dark spot on ear-coverts; dark tail band. Call is loud, raucous 'kraaarr'. In flocks up to thousands; not shy. **Range and habitat:** widely distributed NT; widespread in eastern AF from South Africa to Sudan, Ethiopia; also western coasts, Niger river. Common on coasts and inland waters; scavenger at harbours, rubbish tips. Resident and intra-African migrant. **Similar species:** 3 similar red-billed gulls in AF.

Sooty Gull *Larus hemprichii* (Laridae) 44 cm (17 in). Medium-sized gull, breeding adult grey-brown above with blackish brown hood; collar, rump and tail white; bill yellow-green with black band and red tip; legs greenish yellow. Nonbreeding adult has paler hood, more diffuse collar. Immature has black tail, lacks collar. Call is loud scream, 'kee-aawr'. In flocks, up to hundreds, on open coasts; scavenges at harbours; not shy. **Range and habitat:** breeds coasts of Arabia and north-east AF south to Kenya, before moving down east AF coast to Tanzania October-May. Seasonally common. **Similar species:** White-eyed Gull.

White-cheeked Tern *Sterna repressa* (Laridae) 32-7 cm (13-15 in). Medium-sized tern, breeding adult dark grey, with black cap above conspicuous broad white streak extending from lores to nape; bill black with reddish base, legs red. Nonbreeding adult has whitish speckled forehead and whitish underparts. Immature as nonbreeding adult, dark leading edge to upperwing; underparts white. Call is harsh, loud 'kee-errr' or 'kee-leeek'. In flocks on open coasts; retiring. **Range and habitat:** breeds coasts of western India, Arabia and north-east AF June-October then disperses, some south to Tanzania. Seasonally common. **Similar species:** Antarctic and Whiskered Terns (see p 296).

Greater Crested Tern *Thalasseus bergii* (Laridae) 50 cm (20 in). Breeding adult pale or dark grey above with white forehead, black cap and yellow bill; white below, legs black. Nonbreeding; cap flecked white, neck speckled greyish. In flight, shows grey rump and deeply forked tail uniform with wings. Immature heavily marked above, white below; bill yellowish. Often silent; hard 'tik-tik', screamed 'kreeee'. Found in small parties, sometimes singly, on open coasts. **Range and habitat:** coasts of Arabia, east to OR, AU; eastern and far south-western AF coasts. Common. Resident or disperse after breeding. **Similar species:** Lesser Crested Tern, Sandwich Tern (see p 162).

African Skimmer *Rynchops flavirostris* (Rynchopidae) 38 cm (15 in). Medium-sized tern-like species, breeding adult blackish above, white below, long red bill with yellow tip and elongated lower mandible; legs red. Nonbreeding adult has pale hindneck collar. In flight, shows white rump and forked tail, dark wings. Immature speckled buff above, bill and legs yellow. Calls hard 'tik', screaming 'kreeep'. Found in small parties, flocks of 1,000 known, on open shores; retiring. **Range and habitat:** widespread in AF from Senegal, Ethiopia south to Angola, Mozambique. Locally common on inland waters. Intra-African migrant, dispersing after breeding. **Similar species:** other skimmers (see p 68).

■□□
African Snipe
leaving nest
□■□
Crab Plover
■■□
Temminck's Courser

■□□
Two-banded Courser
□■□
Violet-tipped (Bronze-winged) Courser
□□■
Egyptian Plover

■□□
White-collared Pratincole
□■■
Grey-headed Gull
□□■
Sooty Gull

■□□
White-cheeked Tern
□■□
Greater Crested Tern in courtship display
□□■
African Skimmer

PIGEONS AND SANDGROUSE Columbiformes

Chestnut-bellied Sandgrouse *Pterocles exustus* (Pteroclididae) 30 cm (12 in). Medium-sized sandgrouse with long, central tail feathers. Male sandy, with black breast band and chestnut belly. Female vermiculated and streaked, with face and breast band plain buffish; belly dark. In flight, shows dark belly and underwings. Immature as female but greyer, no breast band. Call is musical 'gutta gutta...'. Found in small parties, larger drinking flocks, on open ground, but shy and often overlooked. **Range and habitat:** Egypt, east to Middle East, India; widespread in AF at 10-19° N, from Mauritania east to Somalia, southern Arabia. Locally common in dry, open country. Resident. **Similar species:** Spotted and Black-bellied Sandgrouse.

Yellow-throated Sandgrouse *Pterocles gutturalis* (Pteroclididae) 33 cm (13 in). Large sandgrouse, male grey-brown with custard yellow face and throat bordered by black gorget; belly chestnut-brown. Female with yellowish face, mottled buff and brown above and on breast, belly as male. In flight, shows dark underwing and wedge-shaped tail. Immature as female, more finely mottled above. Call is guttural 'gutta'; also 'weet-weet'. In pairs or small parties, in hundreds when drinking or migrating, on open ground; shy. **Range and habitat:** 3 regions of AF: interior south of 14° S; from Zambia to Tanzania, southern Kenya; Ethiopia. Common or locally so on open grass plains. Resident; intra-African migrant in south.

Red-eyed Dove *Streptopelia semitorquata* (Columbidae) 30 cm (12 in). Large, grey-brown dove with black collar on hindneck; forecrown pale grey; eye appears dark at any distance. In flight, tail shows terminal band greyish white above, white below. Immature speckled buff above, warm buff below, collar reduced or absent. Distinctive, far-carrying call 'coo coo, coo-coo, coo-coo'; shrill 'zeeee' when landing. Single or in pairs, small parties foraging on ground, perching in canopies; rather shy. **Range and habitat:** widespread in AF south of 12° N, except far south-west and Congo forests. Common in well-wooded country. Resident. **Similar species:** other *Streptopelia* doves (see pp 164, 252).

Laughing Dove *Streptopelia senegalensis* (Columbidae) 24 cm (9½ in). Small dove with pinkish head, rusty back and blue-grey closed wing; breast pinkish rufous, speckled black at base of neck; abdomen white. In flight, shows rump and wing-coverts blue-grey, flight feathers black; white sides and tip to tail. Immature browner above, greyer below. Call is soft chuckling, 'ooooo coo-coo-oo'. Single, in pairs or feeding flocks, foraging on ground; unobtrusive, often approachable. **Range and habitat:** widespread India, Middle East, North Africa; AF south of 20° N except Congo forests and arid south-west. Common in reasonably open habitats, including cities. Resident or intra-African migrant. **Similar species:** other *Streptopelia* doves (see pp 164, 252).

(Madagascar) Green Pigeon *Treron australis* (Columbidae) 30 cm (12 in). Large, thickset, grey-green or yellow-green pigeon, with mauve carpal patch, red bill and red or yellow legs. Plump, bull-necked silhouette during fast, very direct flight. Immature duller, lacking carpal patch. Call is shrill whistling, 'teea-weet teea-weet...', ending with quiet 'tok' notes. Found in small parties, foraging parrot-like in dense canopies; camouflaged, often silent and unobtrusive, betrayed by call. **Range and habitat:** widespread in AF south from Senegal, northern Zaire, Ethiopia, Kenya, except far south-west. Common or locally so in wooded country. Resident, moving locally to fruiting trees. **Similar species:** 3 other green pigeons in AF.

Speckled Pigeon *Columba guinea* (Columbidae) 33 cm (13 in). Large grey pigeon with pale chestnut back and wings, latter spotted white; also bare red skin around eye. In flight, prominent pale grey rump. Immature brown, including facial skin. Call is rapid series of high notes, 'coo-coo-coo...', becoming louder then fading. Found singly or in

flocks, foraging on ground, often around habitation; can be approachable. **Range and habitat:** widespread in AF south of 16° N, apart from lowland forest and country between Congo basin and Botswana. Common or locally so near cliffs or buildings, including city centres; foraging in open country. Resident, making long foraging flights. **Similar species:** Rock Dove (see p 164).

Namaqua Dove *Oena capensis* (Columbidae) 27 cm (10½ in). Small size and long graduated tail unique among AF doves. Male grey above with black face, throat and breast, purple bill with yellow tip, and white underparts. Female with black bill, face to breast greyish. In rapid, direct flight, shows reddish flight feathers and black and white bars on lower back. Immature as female, heavily spotted above. Often silent; call is soft 'cuh-hoooo'. Single or in pairs, sometimes flocks, foraging on open ground; often approachable. **Range and habitat:** WP in Near East; widespread in AF (including Arabia) south of 20° N. Common in scrub or open woodland. Resident, local or intra-African migrant.

Emerald-spotted Wood Dove *Turtur chalcospilos* (Columbidae) 20 cm (8 in). Small, short-tailed, grey-brown dove, paler below, with metallic green spots (often appearing black) visible in closed wing; bill blackish or reddish. In brief low flights, shows reddish flight feathers, black and white bands on back. Call is soft cooing, starting hesitantly, accelerating, dying: 'coo-coo...coo...coo-coo...coo...coo...coocoocoocoocoocoocoocoocoocoo'. Single or in pairs, foraging on ground, often in shade; shy, betrayed by call. **Range and habitat:** widespread in AF south of 15° S, apart from far south-west; and north to East Africa, Ethiopia, Somalia. Common in dry, bushed and wooded country. Resident. **Similar species:** 4 other AF wood doves.

Tambourine Dove *Turtur tympanistria* (Columbidae) 23 cm (9 in). Small dove, male dark brown above with white face, supercilium and underparts. Female greyer about face and breast. In low flight, shows reddish primaries. Immature barred above, grey-brown below with white belly. Call is prolonged, soft cooing, starting deliberately, accelerating, dying: 'coo-coo...coo-coo...coo-coo...coocoocoocoocoo....'. Single or in pairs, foraging on ground in shade, unobtrusive, shy. **Range and habitat:** widespread in AF between 9° N and 15° S, also south to eastern South Africa. Common in dense scrub or woodland. Non-migratory.

CUCKOOS AND ALLIES Cuculiformes

Didric Cuckoo *Chrysococcyx caprius* (Cuculidae) 19 cm (7½ in). Small cuckoo, male metallic green above with white eyebrow and red eye; underparts white, barred green on flanks. Female more coppery above, breast washed buff. Immature dull green or rusty above with coral bill; underparts white, streaked or spotted greenish. Call is shrill whistle, 'dee-dee-dee deederick'. Single, sometimes pairs, perching prominently; brood parasite of, e.g., various weavers; bold, moderately approachable, unobtrusive if not calling. **Range and habitat:** widespread in AF south of 16° N, apart from extreme south-west; summer visitor, southern Arabia. Common in open scrub and woodland. Resident equatorial areas or intra-African migrant. **Similar species:** African Emerald and Klaas' Cuckoos.

Black and White Cuckoo *Oxylophus jacobinus* (Cuculidae) 33 cm (13 in). Large, long-tailed crested cuckoo, black above with white patch in closed wing; white or pale grey below; dark morph black apart from wing patch. Immature brownish above, buff below. Call is loud piping 'tleu' or 'tleeu-ip'. Single, in pairs or small groups; brood parasite of bulbuls; often noisy. **Range and habitat:** widespread in OR, from India to Burma; widespread in AF south of 16° N, except Congo forests and far south-west. Common in open scrub and woodland. Some resident; most migrant, either intra-African or from OR. **Similar species:** Levaillant's and Great Spotted Cuckoos.

■□□ **Chestnut-bellied Sandgrouse**
□■□ **Yellow-throated Sandgrouse** two males in foreground
Laughing Dove

■■□ **Red-eyed Dove**
□■□ **Speckled Pigeon**

■□□ **(Madagascar) Green Pigeon**
□■□ **Namaqua Dove**
□□■ **Emerald-spotted Wood Dove**

■□□ **Tambourine Dove**
□■□ **Didric Cuckoo** female
□□■ **Black and White Cuckoo**

Red-chested Cuckoo *Cuculus solitarius* (Cuculidae) 29 cm (11½ in). Medium-sized cuckoo, dark grey above with yellow eye-ring, rufous breast and whitish, thinly barred belly. Immature blackish above and about head, with white feather-edgings and nape patch; underparts whitish, thinly barred dark. Call, given day and night, is far-carrying whistle, 'wheet wheet whou', rendered 'it will rain'. Single, sometimes several in same area, usually high in canopies; brood parasite of thrushes; inconspicuous, shy, sometimes approachable, betrayed by call. **Range and habitat:** widespread in AF south of 10° N, apart from Somali and far south-west arid areas. Common or seasonal in wooded country. Resident equatorial areas or intra-African migrant. **Similar species:** red-chested race of Black Cuckoo.

White-browed Coucal *Centropus superciliosus* (Cuculidae) 44 cm (17 in). Medium-sized coucal with red eye and long, broad, black tail; wings and back rufous red, underparts white. Some in southern AF have crown to upper mantle bluish black; rest have brownish crown and nape finely streaked white, and broad white eyebrow. Flight low, floundering. Immature has eyebrow and underparts buff. Call is distinctive melodic bubbling, falling in pitch then rising, like water poured from bottle. Single or paired, in dense herbage; skulking, sometimes approachable, betrayed by call. **Range and habitat:** widespread in AF south of 5° S except south-west; and through East Africa to Ethiopia, south-west Arabia. Common in dense grass and thickets, often near water. Mainly non-migratory. **Similar species:** other coucals (see pp 254, 302).

Senegal Coucal *Centropus senegalensis* (Cuculidae) 40 cm (16 in). Rather small coucal with red eye, black crown and nape, rufous back and wings, white underparts and long, broad, black tail; dark morph in west AF has brownish throat, rufous underparts. Immature brown and closely barred above with fine white streaking; buffish below. Call is melodic bubbling, descending in pitch then rising, like water poured from bottle. Single or in pairs, low in dense herbage; shy, betrayed by call. **Range and habitat:** southern AF from Botswana to Zambia, Malawi; also between 5° S and 15° N, east to Ethiopia, Kenya; also Nile delta in WP. Local, in dense grass, often away from water. Mainly non-migratory. **Similar species:** other coucals (see pp 254, 302).

Hartlaub's Turaco *Tauraco hartlaubi* (Musophagidae) 41 cm (16 in). Medium-sized, green turaco with blue-black, subtly crested crown bordered below by thin white line extending from gape to ear-coverts; also prominent white spot in front of eye, and bare red circumorbital skin. Brilliant crimson flight feathers conspicuous during brief, flapping flights. Immature unknown. Call is loud, hoarse, high-pitched croaking, 'caw-caw-caw-caw...', also sharp 'chik', 'chukkerrrr' and growling. Single or in pairs, small parties at fruiting trees, usually high in canopies, running silently along branches; cryptic, often unobtrusive, betrayed by call. **Range and habitat:** eastern AF, in Kenya, eastern Uganda, north-eastern Tanzania. Locally common in highland forest and adjacent well-timbered areas. Non-migratory. **Similar species:** 10 similar green turacos in AF.

Lady Ross's Turaco *Musophaga rossae* (Musophagidae) 51 cm (20 in). Large, deep violet-black turaco with short erect crimson crest and conspicuous yellow bill, frontal shield and circumorbital skin. Crimson flight feathers conspicuous during usually brief and clumsy flapping flights. Immature duller, crimson only in centre of crown, forehead feathered, eye-ring black. Call, sometimes in chorus, is hoarse, quite high-pitched crowing, 'caw caw caw...'. In pairs or small parties, usually high in canopy, running silently along branches; dark, often unobtrusive, betrayed by call. **Range and habitat:** central AF around periphery of Congo basin forests from Angola to Zambia, Kenya, Central African Republic, Cameroon. Locally common in woodland. Non-migratory. **Similar species:** Violet Turaco.

Great Blue Turaco *Corythaeola cristata* (Musophagidae) 75 cm (29 in). Largest and most spectacular turaco, bright grey-blue above with thick black crest, bright yellow bill with red tip, long broad tail with subterminal black band; breast blue merging to greenish yellow below, abdomen chestnut. Brief flapping, gliding flights, often losing height; no red in wings. Immature duller, crest reduced, breast greyish yellow. Loud, far-carrying call begins with deep bubbling trill, then rapid series of croaking 'tok' notes. In small parties, high in canopy, climbing and running on branches; noisy or silent, unobtrusive. **Range and habitat:** west and central AF, east from Sierra Leone to the Congo basin and Uganda. Locally common in forest, sometimes near cleared areas. Non-migratory.

White-bellied Go-away-bird *Corythaixoides leucogaster* (Musophagidae) 51 cm (20 in). Large, slim, grey turaco with long crest, white abdomen and long, broad tail. Brief flapping flight, shows conspicuous broad white bar on tail, also white wing-bar; can momentarily resemble raptor. Immature browner, with rather fluffy plumage on belly, head and neck. Call is loud, barking 'back, back, go-back', also nasal 'wayeerrr, go-wayeerrr...'. Single, in pairs or small parties, often on elevated, prominent perches; noisy, not skulking but retiring. **Range and habitat:** north-eastern AF, from Ethiopia and Somalia to Kenya, Tanzania. Common in dry, open scrub and woodland. Resident or wandering. **Similar species:** 4 other grey turacos.

PARROTS AND RELATIVES Psittaciformes

African Grey Parrot *Psittacus erithacus* (Psittacidae) 30 cm (12 in). The largest African parrot, grey with boldly contrasting bright scarlet tail; bare whitish facial skin around yellow eyes; pale feather-edgings give overall scaly appearance to plumage. Flight fast, direct, often very high. Immature as adult but eyes grey, tail darker near tip. Calls are high clear whistles and screams, also raucous, screeching alarms. Found in flocks, foraging in canopies; shy. **Range and habitat:** central and western AF, from Kenya and Uganda, through Congo basin, to forests around Gulf of Guinea. Common, locally abundant, in well-wooded areas, including gardens. Non-migratory, moving only to forage and roost.

Brown Parrot *Poicephalus meyeri* (Psittacidae) 23 cm (9 in). Medium-sized parrot, brown above with contrasting yellow shoulders and (sometimes) patch on crown; breast brown, abdomen bluish green. In low fast flight, shows conspicuous blue-green rump and yellow underwing-coverts. Immature as adult, but yellow absent on crown, reduced in wing. Calls are raucous screams and shrieks, including distinctive 2-note sound, apparently given in duet by pair. Found in pairs or small parties, foraging in dense canopies, sometimes perching conspicuously; wary, shy, often retiring, betrayed by calls. **Range and habitat:** interior AF, from Angola and Botswana north-east to south-eastern Zaire, East Africa and Ethiopia; also Sudan, Chad. Common or locally so in wooded areas and open country with scattered large trees. Resident. **Similar species:** 5 AF parrots.

Fischer's Lovebird *Agapornis fischeri* (Psittacidae) 14 cm (5½ in). Sparrow-sized parrot, green above, paler below, with dull yellow collar, brown cap and conspicuous white eye-ring; bill, throat and forehead orange-red. Contrasting blue rump seldom seen in fast, direct, high flight. Immature as adult, duller. Calls are high shrill whistles and twittering. In small parties, occasionally larger numbers, foraging on ground and vegetation; rather wary and retiring, betrayed by calls. **Range and habitat:** eastern AF, in interior northern Tanzania, recently Rwanda, Burundi; also feral populations in Kenya and Tanzania, often hybridizing with Yellow-collared Lovebird *A. personata*. Common in open woodland and cultivation. Non-migratory; apparently extending range in Rwanda, Burundi. **Similar species:** 7 other AF lovebirds and hybrids.

White-browed Coucal

African Grey Parrot

Senegal Coucal

Red-chested Cuckoo

Hartlaub's Turaco

Lady Ross's Turaco

White-bellied Go-away-bird

Great Blue Turaco

Brown Parrot

Fischer's Lovebird

KINGFISHERS AND ALLIES Coraciiformes

Lilac-breasted Roller *Coracias caudata* (Coraciidae) 36 cm (14 in). Medium-sized roller, readily identified by contrast between lilac throat and breast and azure blue abdomen (north-eastern race *lorti* has breast blue, lilac only on throat); forehead and eyebrow whitish, back brown; blue tail has outer 2 feathers elongated to form pointed streamers. In flight, bright blue coverts contrast with dark flight feathers. Immature duller, throat and breast buffish. Call is loud harsh squawk, 'zaaak'. Single or paired, perching prominently, takes prey from ground; aggressive, often noisy. **Range and habitat:** widespread in AF south of Congo basin, apart from extreme south; also north to East Africa, Ethiopia, Somalia. Common in open scrub and woodland. Resident or intra-African migrant. **Similar species:** 4 blue rollers in AF, European Roller (see p 166).

Rufous-crowned Roller *Coracias naevia* (Coraciidae) 36 cm (14 in). Rather large and thickset, square-tailed roller, with brown or greenish crown, white forehead and eyebrow, and maroon-brown underparts profusely streaked white; back and wings olive-grey or greenish, tail dark blue. In flight, shows glossy dark blue and purple wings. Immature browner, underparts more broadly streaked white. Call is loud but hardly strident 'grerrr', also 'ka-kaaaa'. Single or paired, perching prominently, takes prey from ground. Still and quiet or noisy, active. **Range and habitat:** widespread in AF between 15° N and 30° S, apart from Congo forest areas. Rather uncommon or local in open, usually dry, scrub and woodland. Resident or intra-African migrant.

African Broad-billed Roller *Eurystomus glaucurus* (Coraciidae) 27 cm (10½ in). Small, stocky roller, brown above, lilac below, with azure-blue tail and vent, and broad, bright yellow bill. In flight, brown wing-coverts contrast with dark blue flight feathers; overall appearance like falcon. Immature brownish above and on upper breast, belly and lower breast grey-blue, bill yellowish. Call is nasal 'ke-aiow'; also chattering 'tikk-k-k-k-r-r-r'. Single or paired, sometimes flocks when foraging or migrating, hawking insects on wing, rather crepuscular; perching prominently, sometimes aggressive, noisy. **Range and habitat:** widespread in AF from 15° N south to Angola, eastern South Africa. Locally common in wooded areas. Resident or intra-African migrant. **Similar species:** Dollarbird (see p 256).

Giant Kingfisher *Megaceryle maxima* (Alcedinidae) 46 cm (18 in). Crow-sized, by far the largest African kingfisher, with massive black dagger-like bill, shaggy crest and black upperparts speckled white. Male has chestnut breast, white belly with black bars; female has heavily spotted breast, chestnut abdomen. Immature male as adult, speckled black on chestnut breast; immature female as adult, breast less heavily spotted. Call is loud, raucous, laughing or shrieking 'tek-tek-tek...'. Single or in pairs, often unobtrusive in leafy canopy, shy, betrayed by call. **Range and habitat:** widespread in AF south of 15° N, apart from far south-west and Somalia. At best locally common, along wooded edges of rivers and lakes, also coastal lagoons. Resident.

Lesser Pied Kingfisher *Ceryle rudis* (Alcedinidae) 28 cm (11 in). 2nd largest African kingfisher, strikingly black and white with substantial black bill and shaggy crest. Male has white underparts with thick black breast band above thinner; female similar, with single thick band, usually incomplete. Immature as female, throat and breast speckled dark brown. Call is high-pitched, squeaky chattering, 'kweek-kweek-kweek...', often in chorus. In pairs or small, loose parties, perching prominently by water, hovering high and plunging for fish; noisy, often approachable. **Range and habitat:** from Near and Middle East to India, China; also Nile in North Africa; widespread in AF south of 15° N, apart from far south-west. Common, or locally numerous, around open fresh or sea water. Resident or intra-African migrant.

Malachite Kingfisher *Alcedo cristata* (Alcedinidae) 14 cm (5½ in). Small kingfisher, ultramarine above with often inconspicuous crest, dark cap extending down to eyes, red bill and white patch on neck; underparts rufous orange with white throat. Immature darker above, browner below, speckled dark on breast, bill black. Flight call is sharp 'teek'; chittering song, often in duet. Single or paired, perching openly at waterside, but often quiet, unobtrusive; moderately approachable. **Range and habitat:** widespread in AF south of 15° N, apart from far south-west and parts of Congo basin. Common in open fringing vegetation around water bodies. Resident or local migrant. **Similar species:** 5 similar kingfishers in AF, Kingfisher (see p 166).

Pygmy Kingfisher *Ceyx picta* (Alcedinidae) 13 cm (5 in). Small kingfisher, ultramarine above with rufous orange eyebrow and hindneck, diffuse mauve wash on ear-coverts, white spot on side of neck, and red bill; underparts rufous orange with white chin and throat. Immature as adult but speckled blackish above, paler below, bill blackish. Often silent; call, mainly in flight, is high squeak, 'seep'. Single or paired, often perched in low and inconspicuous situations; quiet, shy, overlooked. **Range and habitat:** widespread in AF between 15° N and 20° S, reaching 30° S along south-east coast. Locally common in variety of dense habitats, including forest far from water. Resident, or intra-African migrant. **Similar species:** 5 AF kingfishers.

Woodland Kingfisher *Halcyon senegalensis* (Alcedinidae) 23 cm (9 in). Medium-sized kingfisher, pale blue above with pale grey hood (latter can be washed pale blue or brown according to race) and extensive black 'shoulder' in closed wing; large bill has upper mandible red, lower black; underparts whitish, legs black. In flight, striking black and brilliant azure contrast in open wings. Immature has breast and flanks washed buffish with dusky vermiculation, lower mandible sometimes patchily reddish. Loud distinctive call is high note followed by descending trill, 'chi, chiiiiiirrrr'. Single or paired, often perching prominently, or unobtrusive. **Range and habitat:** widespread in AF between 15° N and 22° S, east to Ethiopia, Kenya and range of allopatric Mangrove Kingfisher *H. senegaloides*. Common in scrub and woodland, often near water. Mainly resident near equator or intra-African migrant. **Similar species:** Mangrove, Blue-breasted and Smyrna Kingfishers (see p 256).

Grey-headed Kingfisher *Halcyon leucocephala* (Alcedinidae) 20 cm (8 in). Medium-sized, red-billed kingfisher, with grey or light brown head and breast, black back and shoulders, brilliant cobalt-blue rump, tail and flight feathers; belly rich chestnut. Immature is rather duller, with buffy neck, face and breast mottled with dark grey, belly is buff and bill dusky red. Calls include a clear, repeated 'piuu' and some chattering notes. Song is a rather weak trill. Single or paired, perches prominently, takes prey from the ground; sometimes quiet and unobtrusive. **Range and habitat:** AF, widespread but largely absent from Congo basin, South Africa and parts of East Africa. Common in woodland, riverine thickets, wooded grassland and forest edge. Resident and local migrant. **Similar species:** 6 AF kingfishers.

Striped Kingfisher *Halcyon chelicuti* (Alcedinidae) 18 cm (7 in). Rather dull smallish kingfisher, with streaked, grey-brown cap divided from brown back by whitish collar; bill black on upper mandible, dull reddish below; underparts white, finely streaked dark on breast and flanks. During wing-opening display, male shows black band on underwing. In flight, shows bright, pale blue rump. Immature buff below, speckled on breast. Call is ringing, far-carrying trill, 'ki-kirrrir', often in series and duet, accompanied by wing-opening display. Single or paired, perches conspicuously, takes prey from ground; moderately approachable, located by call. **Range and habitat:** widespread in AF between 15° N and 22° S, except equatorial forest. Common in wooded areas, often far from water. Resident or migrant. **Similar species:** Brown-hooded Kingfisher.

Lilac-breasted Roller

Rufous-crowned Roller

Giant Kingfisher

African Broad-billed Roller

Lesser Pied Kingfisher

Malachite Kingfisher

Pygmy Kingfisher

Woodland Kingfisher prospecting for nest-hole

Grey-headed Kingfisher

Striped Kingfisher

Carmine Bee-eater *Merops nubicus* (Meropidae) 36-8 cm (14-15 in). Unmistakable, large, bright carmine-red bee-eater with elongated central tail feathers. North tropical race *nubicus* has vent and head dark greenish blue; southern tropics *nubicoides* (may be separate species) has chin and throat carmine. In flight, shows contrasting pale blue rump. Immature browner above, pinky brown below, lacking tail elongation. Call is deep, husky 'turk'. Found in flocks up to hundreds, perching prominently, hawking flying insects; conspicuous, moderately approachable. **Range and habitat:** widespread in AF from 15° N to Angola, Botswana, eastern South Africa, except equatorial forest. Seasonally common in open lowland scrub and woodland. Intra-African migrant; converges on equator after breeding.

Little Bee-eater *Merops pusillus* (Meropidae) 16 cm (6½ in). Smallest AF bee-eater, green above with black mask, and yellow throat bordered below by black gorget; remainder of underparts buffish ochre. In low brief dashing flights, wings and tail appear mainly rufous orange. Immature washed greenish below, lacking gorget. Often silent; call is quiet 'chip'. In pairs or family parties, sometimes singly, perching low in open, hawking insects; sometimes unobtrusive, moderately approachable. **Range and habitat:** widespread in AF south of 15° N, apart from far south-west and Congo forest regions. Commonest AF bee-eater, in open scrub and woodland. Resident, moving locally. **Similar species:** 5 similar AF bee-eaters, Little Green Bee-eater.

White-fronted Bee-eater *Merops bullockoides* (Meropidae) 23 cm (9 in). Medium-sized, square-tailed bee-eater with black mask, white forehead, chin and cheeks, and bright red throat; upperparts green, with cinnamon nape; underparts cinnamon with dark blue vent. In flight, shows contrasting blue lower rump. Immature greenish below; paler red on crown and throat. Call is deep squeak, 'qrrrrt'. Single or in flocks, perches prominently, hawking insects from air or vegetation; conspicuous, sometimes approachable. **Range and habitat:** southern AF, east from Angola to Tanzania and north-east South Africa; also Kenya. Locally common in open country, often near gullies. Resident or wandering. **Similar species:** Red-throated Bee-eater.

Black Bee-eater *Merops gularis* (Meropidae) 19 cm (7½ in). Small, square-tailed, blackish bee-eater with scarlet throat and azure rump and belly. Immature lacks red throat and is generally duller. Normally silent, but gives a high-pitched 'tssssp' and a liquid trill. Perches prominently in pairs or small parties but sometimes unobtrusive, not particularly tame. Forages at canopy height. **Range and habitat:** lowland forests of west and central AF, east from Guinea to Uganda, south to northern Angola. Uncommon or local in rainforest glades, streams and clearings. Probably mainly sedentary. **Similar species:** Blue-headed Bee-eater.

Trumpeter Hornbill *Ceratogymna bucinator* (Bucerotidae) 59-66 cm (23-6 in). Large, glossy black hornbill with white belly and bare purplish pink skin around eye. Male has casque running length of blackish bill; female smaller, casque reduced. In flight, shows white rump, white tips to secondaries and long, mainly black tail. Immature as female, brownish on face. Loud call is distinctive wailing, like crying of child,'waaaaaaaa, waaaaaaaa...'. Usually in small flocks, foraging in canopy; bold, moderately approachable. **Range and habitat:** south-eastern AF, from Kenya and Tanzania to southern Zaire, Zambia, eastern South Africa. Locally common in wooded and forested areas. Basically resident. **Similar species:** 4 AF hornbills.

Silvery-cheeked Hornbill *Ceratogymna brevis* (Bucerotidae) 75-80 cm (29-31 in). Large, glossy black hornbill with white lower abdomen and bare blue skin around eye; silvery-tipped feathers on face. Male has brownish bill with pale basal band, surmounted by huge creamish casque; female smaller, casque reduced and darker. In flight, shows white rump; wings appear entirely dark. Immature has face brownish, casque reduced or absent. Call is loud, quacking

or braying; also sharp barks, grunts. In pairs or flocks, foraging in canopies; often noisy, bold; quite approachable. **Range and habitat:** eastern AF, from north-east South Africa to Kenya, Ethiopia. Locally common in woodland. Apparently resident, but moving seasonally when foraging. **Similar species:** 4 AF hornbills.

African Grey Hornbill *Tockus nasutus* (Bucerotidae) 44-9 cm (17-19 in). Smallish hornbill, grey-brown above, white below, with white eyebrow and greyish upper breast. Male has black bill, cream at base of upper mandible; female smaller, bill creamy above, black below, with reddish tip. In flight, shows whitish streak on back, and white tip to tail. Immature as male, greyer; bill grey. Call is series of whistles, 'peee peee peee peeeoow peeeooooow...'. In pairs or flocks, bold, quite approachable. **Range and habitat:** at 7-17° N in AF, east to Ethiopia, south-west Arabia; also south through East Africa to Namibia, northern South Africa. Common, never numerous, in open wooded country. Resident, local migrant. **Similar species:** Pale-billed Hornbill.

Red-billed Hornbill *Tockus erythrorhynchus* (Bucerotidae) 43-50 cm (17-20 in). Smallish hornbill with slender red bill; crown and upperparts black, profusely spotted white on closed wing; face and underparts white. Immature has smaller, browner bill, wing-spots buffish. Call is loud high clucking, 'tok-tok-tok...'. In pairs or small to large flocks, foraging on open ground; conspicuous, often noisy, usually approachable. **Range and habitat:** at 9-17° N in AF east to Ethiopia, and thence south through East Africa to Namibia, Botswana, north-eastern South Africa. Common, sometimes local, in open scrub and woodland. Resident, moving locally. **Similar species:** 5 AF hornbills.

Yellow-billed Hornbill *Tockus flavirostris* (Bucerotidae) 48-58 cm (19-23 in). Medium-sized hornbill, black or dark brown above, boldly spotted white on closed wing, with unmistakable large yellow or orangish bill; underparts white. Male has massive bill, gular skin rosy; female smaller, bill reduced, gular skin black. Immature as female, bill mottled dusky; dark streaking on breast. Call is low staccato clucking, 'tok-tok-tok...', often accelerating, then fading. Single, in pairs or small parties, foraging openly on ground or vegetation; moderately approachable. **Range and habitat:** 2 regions of AF: north-east from Tanzania to Ethiopia, Somalia; also widespread at 15-30° S (sometimes considered separate species). Common or local in open scrub and woodland. Resident. **Similar species:** Red-billed Hornbill.

Crowned Hornbill *Tockus alboterminatus* (Bucerotidae) 50-4 cm (20-1 in). Medium-sized, blackish brown hornbill with white belly, deep orange-red bill and white flecking on nape. Male has long casque-ridge, gular skin black; female smaller, casque reduced, gular skin greenish yellow. In flight, shows white-tipped outer tail feathers. Immature as adult, bill yellow, no casque. Call is loud piping, often rising and falling, 'pee-pee-pee...'. In pairs or small parties, foraging openly on vegetation; moderately approachable. **Range and habitat:** widespread in southern AF, from Angola to Tanzania and South Africa, avoiding south-west; also Kenya, Ethiopia. Common, local in wooded and forest edge habitats. Resident. **Similar species:** 4 AF hornbills.

Southern Ground Hornbill *Bucorvus leadbeateri* (Bucerotidae) 90-100 cm (35-9 in). Huge, black, terrestrial hornbill with black bill surmounted by low basal casque-ridge. Male has bright red wattles on face and throat; female smaller, wattles bluish on throat. White primaries conspicuous during occasional low flights. Immature brownish, bill reduced and grey; wattles greyish or yellowish. Call is deep, rhythmic, lion-like grunting, 'oo-oo-oo...', often in duet. In pairs or small parties, sedately walking; rather retiring. **Range and habitat:** widespread in AF south of Congo basin, apart from far south-west; also north to Uganda, Kenya. Locally common in open country. Resident, moving locally. **Similar species:** Abyssinian Ground Hornbill.

Carmine Bee-eater

Little Bee-eater

White-fronted Bee-eater

Black Bee-eater

Silvery-cheeked Hornbill

Trumpeter Hornbill

African Grey Hornbill

Red-billed Hornbill

Yellow-billed Hornbill

Crowned Hornbill

Southern Ground Hornbill

Green Wood Hoopoe *Phoeniculus purpureus* (Phoeniculidae) 30-6 cm (12-14 in). Large wood hoopoe with conspicuous long, decurved, red bill, red legs and long graduated tail; plumage generally blackish, glossed bright metallic green on head, mantle and underparts, deep violet on wings and tail. In flight, shows conspicuous white wing-bar and tail-spots. Immature duller, throat buffish; legs and short, straight bill black. Call is high hysterical cackling, accelerating to crescendo, usually in chorus. In small parties, foraging agilely on bark of tree-trunks and branches; active, often noisy, quite approachable. **Range and habitat:** widespread in AF south of 15° N, except equatorial forests, arid Horn and far south-west. Locally common in wooded country. Resident, wandering. **Similar species:** 6 other wood hoopoes in AF.

Scimitarbill *Rhinopomastus cyanomelas* (Phoeniculidae) 25-8 cm (10-11 in). Medium-sized wood hoopoe with thin, black, sharply decurved bill, black legs and long tail. Male generally black, glossed violet-blue; female with face and ventral torso dull dark brown. In flight, shows conspicuous white wing-bar and tips to outer 3 tail feathers. Immature as female, bill shorter, less curved. Call is low, mournful, repeated whistle, 'phooee, phooee. . .'. Single or paired, sometimes in small parties, foraging on bark of tree-trunks and branches; active but often quiet, quite approachable. **Range and habitat:** widespread in AF between 16° N and 30° S, apart from equatorial lowland forests. Moderately or locally common in open scrub and woodland. Resident. **Similar species:** 6 other wood hoopoes in AF.

Abyssinian Scimitarbill *Rhinopomastus minor* (Phoeniculidae) 23 cm (9 in). Small wood hoopoe with long, conspicuous, sharply decurved, bright orange-red bill, black legs and long graduated tail. Male generally black, glossed purple-blue and violet; female browner below. In flight, shows conspicuous white wing-bar (absent in southern race). Immature as female, browner below, bill brownish. Often silent; call is harsh chatter, 'ki-ki-ki-ki. . .', rising and falling. Single or paired, occasionally in small groups, foraging agilely on tree-trunks, branches, flowers; active but sometimes unobtrusive, shy. **Range and habitat:** north-eastern AF, from Somalia, Ethiopia, southern Sudan to Tanzania. Sometimes common, never numerous, in mainly dry, open scrub and woodland. Non-migratory. **Similar species:** 6 other AF wood hoopoes.

OWLS Strigiformes

Pearl-spotted Owlet *Glaucidium perlatum* (Strigidae) 16-18 cm (6½-7 in). Small owl with large yellow eyes, no 'ear' tufts, and striking 'false face' on nape; upperparts brown, profusely spotted white, especially boldly on relatively long tail; underparts white with coarse, irregular, dark brown streaking. Immature less spotted on crown and mantle. Call is loud distinctive series of whistles, very far-carrying and increasing in pitch and volume, culminating in several very extended descending notes: 'keee, keee, keee. . .peeeooooo, peeeoooo. . .'. Single or paired, often roosts unseen by day but frequently betrayed by combination of partly diurnal habits, mobbing by other birds and unique calls; often approachable. **Range and habitat:** widespread in AF south of 15° N, except Congo forests and far south-west. Common in open scrub and woodland. Non-migratory. **Similar species:** 4 owls in AF, Spotted Owlet (see p 254).

African Marsh Owl *Asio capensis* (Strigidae) 36 cm (14 in). Medium-sized owl, dark brown above, pale buff below, with conspicuous dark eyes set in pale facial disc; short 'ear' tufts sometimes visible. In flight, appears long-winged, with prominent reddish buff primaries. Immature has facial disc darker, buff speckling on back. Call is loud tearing croak, 'zirkk'. Single, in pairs or flocks, nocturnal and crepuscular, often emerging on cloudy days to quarter open country like harrier, otherwise roosting unseen on ground; rather shy. **Range and habitat:** uncommon in Morocco; widespread in AF south of Congo basin, apart from south-west, also north to Ethiopia, Sudan, scattered parts of west. Locally common in open marshy and grassy areas. Resident, nomadic. **Similar species:** Cape Grass and Short-eared Owls (see p 70).

Verreaux's Eagle Owl *Bubo lacteus* (Strigidae) 61-5 cm (24-6 in). Very large, finely vermiculated, milky grey owl with rather small (often inconspicuous) 'ear' tufts and prominent black border to pale grey facial disc; large dark eyes with striking pink lids. Immature browner, more barred. Call is distinctively deep and gruff, often irregular, grunting, 'hoo. . .hoo hoo. . .hoo. . .'; also extended, shrieking whistle, 'seeeeeeeeeerrr'. Single or paired, sometimes with immature, nocturnal, roosting motionless in large shady tree by day; often unseen, sometimes approachable. **Range and habitat:** widespread in AF south of 14° N, apart from lowland equatorial forests and far south-west. Quite common in wooded habitats, particularly near water. Resident, immatures disperse locally. **Similar species:** 7 other large owls in AF, Great Horned Owl (see p 70).

Pel's Fishing Owl *Scotopelia peli* (Strigidae) 63-5 cm (25-6 in). Very large owl, barred rufous brown above, paler buff with fine streaking below; large dark eyes in poorly defined facial disc; crown flattish, lacking 'ear' tufts, feathering fluffed out when excited to give rounded head; legs bare, whitish, often concealed by body plumage. Immature whitish with rufous wash, becoming darker. Call is low booming grunt, 'hoooommmmm-ut', far-carrying and often repeated for minutes at a time. Single or paired, sometimes with immature, taking surface fish at night, roosting unseen high in dense waterside canopy by day; moderately approachable. **Range and habitat:** widespread in AF from Senegal, Nigeria, Ethiopia south to Congo, Botswana, eastern South Africa. Local common to rare in riverine forest and swamps. Non-migratory. **Similar species:** Rufous Fishing Owl, 6 AF eagle owls, Malay Fish Owl (see p 254).

NIGHTJARS AND ALLIES Caprimulgiformes

Abyssinian Nightjar *Caprimulgus poliocephalus* (Caprimulgidae) 24 cm (9½ in). Medium-sized, square-tailed nightjar, streaked and mottled grey, brown and black above with diffuse rufous collar; underparts barred and vermiculated grey, brown and buff, with small white patch on side of throat. In flight, male shows white outer tail feathers and flash in primaries; female less white in tail, wing flash buff. Immature as female, more rufous brown. Call, at night, is loud whistle, 'phee-oo-weet', followed by wavering, descending trill. Single or paired, nocturnal, roosting unseen on ground by day; betrayed by call. **Range and habitat:** mainly east AF, from Ethiopia to Malawi; also Angola. The only AF montane nightjar, locally common above 1,000 m (3,300 ft) in woods and forest edge. Resident. **Similar species:** 17 other square-tailed nightjars in AF, European Nightjar (see p 166).

Standard-winged Nightjar *Macrodipteryx longipennis* (Caprimulgidae) 23 cm (9 in). Medium-sized nightjar, breeding male with unmistakable elongated 2nd primaries forming standards over twice body length, consisting of bare feather shaft with flag-like vane at tip, giving impression of 2 small birds following nightjar in flight; shed on reaching 'wintering' range. Nonbreeding male brownish, throat pale buff, no standards. Female and immature as male, paler, no standards. Male's song is soft, high stridulation. In pairs or loose groups, active dusk and night, roosting unseen on ground by day; close-sitting. **Range and habitat:** widespread in AF at 5-15° N, from Senegal, Liberia to Ethiopia, Kenya. Seasonally common in bushed and open wooded country. Intra-African migrant, breeds December-June, then moves north. **Similar species:** when without standards, several nightjars (see pp 72, 116, 166, 254, 304).

Green Wood Hoopoe

Scimitarbill

Abyssinian Scimitarbill

Pel's Fishing Owl

Standard-winged Nightjar

Pearl-spotted Owlet

African Marsh Owl

Abyssinian Nightjar

Verreaux's Eagle Owl

COLIES OR MOUSEBIRDS Coliiformes

Speckled Mousebird *Colius striatus* (Coliidae) 30-5 cm (12-14 in). Medium-sized mousebird, brown above with greyish face, short crest and long, broad pointed tail; bare white skin behind eye in some races; underparts paler, barred on breast; legs red or purplish. In flight, short rounded wings, long tail and fast gliding like small pheasant. Immature duller, with slight crest, stubby tail. Call is soft 'sioo' or 'tooo'; sharper alarm 'swik'. In small parties, foraging openly and agilely on vegetation; usually conspicuous and approachable. **Range and habitat:** widespread in AF between 10° N and 10° S, apart from central Congo basin; also south in east to Cape. Common in fairly open scrub and woodland, gardens and forest edge. Non-migratory. **Similar species:** 5 other mousebirds.

Blue-naped Mousebird *Urocolius macrourus* (Coliidae) 30-5 cm (12-14 in). Slender-tailed mousebird, grey-brown above with bare red skin around eye, red base to black bill and bright turquoise nape patch; underparts paler, legs pinkish. In flight, shows pale rump and long tail. Immature paler, lacking nape patch, facial skin pink. Call is thin but far-carrying whistle, 'seeeee seeeeeeeeee'; also 'treee-treee'. In small parties, foraging openly on vegetation; rather shy. **Range and habitat:** widespread in AF at 10-20° N from Senegal to Ethiopia, Somalia, and south to East Africa. Common in dry, bushed and wooded country. Resident, wandering locally. **Similar species:** 5 other mousebirds.

TROGONS Trogoniformes

Narina's Trogon *Apaloderma narina* (Trogonidae) 30 cm (12 in). Striking, long-tailed, pigeon-sized bird, male bright green with greenish yellow bill and scarlet abdomen; ventral tail white. Female pinkish brown on face and upper breast, grading to grey lower breast; abdomen pinker. Immature duller above with white-spotted wing-coverts; greyish below, lightly barred. Call is series of far-carrying ventriloquial hoots, 'woo-hoo, woo-hoo...'. Single or paired, occasionally in 3s, motionless in canopy, unseen except during brief flights; moderately approachable, betrayed by call. **Range and habitat:** widespread in AF between 10° N and 20° S, also eastern South Africa. Moderately common, often local, in well-wooded areas. Resident, or local migrant. **Similar species:** 2 other AF trogons.

WOODPECKERS AND ALLIES Piciformes

White-headed Barbet *Lybius leucocephalus* (Capitonidae) 16 cm (6½ in). Medium-sized barbet, blackish brown above and white below, with white hood and heavy blackish bill. Of several distinctive races, 1 with white tail, another with blackish belly. In flight, white rump contrasts with dark back and wings. Immature has bill, and white of plumage, brownish. Call is loud, variable 'chek-chek-shee-chip-chip...'. Single, paired or in small parties, foraging on fruiting trees; sometimes unobtrusive, moderately approachable. **Range and habitat:** 3 areas of AF: from Nigeria to Chad, Central African Republic; from southern Sudan to Kenya, Tanzania; Angola. Quite common in open scrub and woodland. Resident, moving to fruiting trees.

Spotted-flanked Barbet *Tricholaema lacrymosa* (Capitonidae) 13 cm (5 in). Smallish, black above with white eyebrow, yellow edging in wing, and black bill; underparts white, with dark spots on flanks and dark stripe from chin to breast. Male has eye pale yellowish, female dark brown. Immature duller, greyer above, eye dark. Call is loud hoot, 'poop...poop'; also short croaks. Single or paired, foraging on bushes and trees; sometimes unobtrusive, not shy. **Range and habitat:** eastern AF, from southern Sudan and eastern Zaire to Uganda, Kenya, Tanzania, Zambia. Locally common, seldom numerous, in moist, bushed and wooded country. Resident, moving locally to fruiting trees. **Similar species:** Black-throated and Pied Barbets.

Red and Yellow Barbet *Trachyphonus erythrocephalus* (Capitonidae) 23 cm (9 in). Large 'ground barbet'; male has speckled black and yellow head with black crown and throat, and red, black and white flash on ear-coverts; rest of upperparts black, spotted white or yellowish; underparts yellow, with narrow black- and white-spotted breast band and red vent. Female has cap and ear-coverts orange-red, throat yellow. In brief low flights, shows red rump. Immature duller, browner above, male with crown darkening. Call is loud duet, 'ka-shee-ka-chao, ka-shee-ka-chao...'. In pairs or small parties, foraging on low vegetation; noisy, active, approachable. **Range and habitat:** north-east AF, from Somalia, Ethiopia to Kenya, Tanzania. Locally common in dry scrublands, near termite hills. Non-migratory. **Similar species:** 3 other 'ground barbets'.

d'Arnaud's Barbet *Trachyphonus darnaudii* (Capitonidae) 16 cm (6½ in). Medium-sized 'ground barbet' with erectile crest on black or grey crown; rest of head yellow, finely speckled with black; remainder of upperparts dark brown, spotted white; underparts yellowish with variable black throat, merging to whitish belly and scarlet vent; bill pale, but blackish in west Kenya. Immature browner on crown and throat, greyer below. Call is prolonged loud 'kerr-ta-te-doodle, kerr-ta-te-doodle...'. In pairs, or 3 or 4 together, foraging on ground or low vegetation; noisy but sometimes shy, quite approachable. **Range and habitat:** north-eastern AF, from Somalia and Ethiopia to East Africa. Common in dry scrublands. Resident. **Similar species:** 3 other 'ground barbets' in AF.

Red-fronted Tinkerbird *Pogoniulus pusillus* (Capitonidae) 10 cm (4 in). Small barbet, black upperparts streaked and speckled white and yellowish, with red forehead, black stripe through eye and thinner black malar stripe; underparts pale yellowish. In flight, shows yellowish rump. Immature has forehead black. Call is far-carrying hollow note, given for long periods, 'tonk-tonk-tonk...'; also shrill trill. Single, or paired, foraging openly on bushes and trees; busy, active, sometimes retiring. **Range and habitat:** 2 areas of AF: eastern South Africa; north-east, from Tanzania and Kenya to Ethiopia, Somalia. Common in scrublands or forest edge. Resident. **Similar species:** Yellow-fronted Tinkerbird, Red-fronted Barbet.

Golden-rumped (Yellow-rumped) Tinkerbird *Pogoniulus bilineatus* (Capitonidae) 10 cm (4 in). Small barbet, black above with yellow edge on wing, thin white eyebrow, white stripe from forehead to neck, and black malar stripe; bill black; underparts greyish, yellower on belly. In flight, shows yellow rump. Immature speckled yellow on mantle, bill paler. Call is far-carrying, hollow note, uttered for long periods in groups of 4 to 6 with brief pauses, 'tonk-tonk-tonk-tonk ... tonk-tonk-tonk-tonk-tonk ...'; also short series of croaks. Single or paired, foraging openly on bushes and trees; busy, active, sometimes approachable. **Range and habitat:** widespread within 10° of equator, also south to Mozambique, eastern South Africa. Common in dense woodland, moist thicket. Non-migratory. **Similar species:** 3 other 'black' tinkerbirds in AF.

Black-throated (Greater) Honeyguide *Indicator indicator* (Indicatoridae) 20 cm (8 in). Large honeyguide, male brown above, white below, with pink bill, black throat and whitish cheeks. Female with dark bill. In flight, shows white outer tail feathers and small yellow 'shoulder' on wing. Immature brown above, bill black, with yellow throat and breast, belly white. Male's call is far-carrying 'wheet-terrr...wheet-terrr...'; guides with chattering, 'tik-tik-ik-ik-tik-ik...'. Found singly on bushes and trees, often near bees' nests; brood parasite of e.g., bee-eaters, barbets; unobtrusive, can be shy. **Range and habitat:** in AF south of 15° N, except Congo forests, arid Horn, far south-west. Fairly common, often local, in scrublands. Resident. **Similar species:** 3 other large honeyguides in AF, Indian Honeyguide (see p 258).

Speckled Mousebird

Blue-naped Mousebird

Narina's Trogon

White-headed Barbet

Spotted-flanked Barbet

Red and Yellow Barbet

d'Arnaud's Barbet

Red-fronted Tinkerbird

Golden-rumped (Yellow-rumped) Tinkerbird

Black-throated (Greater) Honeyguide immature

Ground Woodpecker *Geocolaptes olivaceus* (Picidae) 26 cm (10 in). Large, terrestrial woodpecker, male with grey hood, pinkish white eye, and dark, faintly reddish moustachial stripe; rest of upperparts dark greenish brown with fine pale spots and bars; pinkish red breast merging to crimson on abdomen, with barred flanks. Female lacks reddish tinge to moustache. In flight, shows prominent red rump. Immature paler, pinker below, eye white, often some red on nape. Call is harsh scream, 'keeearrrgh'; also rhythmic 'tick-dee'. In pairs or small parties, foraging on ground; noisy, sometimes aggressive, can be inconspicuous. **Range and habitat:** far southern AF, in South Africa. Locally common on reasonably open, rocky or uneven ground, up to 2,100 m (6,900 ft). Resident, moving locally, seasonally.

Golden-tailed Woodpecker *Campethera abingoni* (Picidae) 20-2 cm (8-9 in). Medium-sized woodpecker, male with black and red speckled forecrown, red hindcrown and moustache; upperparts otherwise olive or grey-brown, barred and spotted white, and tail yellowish. Underparts whitish or buffy, streaked black (some populations spotted on breast, throat blackish). Female has black forecrown speckled white, lacks moustache. Immature as female, duller. Call is wailing 'waaaa', occasional soft drumming. Found in pairs or singly, foraging on trunks and branches. Approachable, can be unobtrusive. **Range and habitat:** widespread in southern AF apart from far south-west, and north to East Africa, Sudan, Somalia. Locally common, never numerous, in bush, woods, forests. Resident. **Similar species:** 6 similar AF woodpeckers.

Grey Woodpecker *Dendropicos goertae* (Picidae) 18 cm (7 in). Medium-sized woodpecker, male with grey head and red hindcrown. Upperparts otherwise green, with red rump; underparts grey, usually with red patch on belly. Female has head entirely grey. Immature faintly barred below, rump paler, abdominal red reduced. Call is distinctive, high rattling 'wikk-wikk-wikk....', rarely drums. Found in pairs or singly, foraging on trunks and branches. Moderately approachable, betrayed by call. **Range and habitat:** mainly at 5-17° N in AF east to Ethiopia, and south to East Africa. Also Congo, Zaire, Angola. Common or locally so in many wooded habitats. Resident. **Similar species:** Olive and Ground Woodpeckers.

Bearded Woodpecker *Dendropicos namaquus* (Picidae) 24 cm (9½ in). Large woodpecker, male with red hindcrown, and white face cut by broad dark stripe behind eye and similar black malar stripe; upperparts otherwise olive or grey-brown, barred and speckled white, tail yellowish; underparts yellowish olive, finely barred whitish. Female has hindcrown black. Immature as male, crown blackish red, speckled white. Call is loud, high chattering, 'kik-kik-kik...'; also frequent loud, far-carrying, decelerating drumming, 'trrrrrrrrr-tap-tap-tap'. Single or paired, foraging on large tree-trunks and branches; noisy, often conspicuous, quite approachable. **Range and habitat:** widespread in AF south of 10° S, apart from far south-west, and north through East Africa to Ethiopia, Central African Republic. Somewhat local, never numerous, in open woodland. Non-migratory. **Similar species:** 3 similar AF woodpeckers.

Red-throated (Rufous-necked) Wryneck *Jynx ruficollis* (Picidae) 18 cm (7 in). Resembling smallish woodpecker, but grey-brown nightjar-like upperparts with fine mottling, speckling and vermiculation, and broken, black median stripe from crown to mantle; below, reddish rufous throat and breast, with whitish, finely streaked belly and rufous-washed vent (Ethiopian race *aequatorialis* has underparts mostly rufous). Immature darker above; less rufous, more barred, below. Call is loud series of high shrieks, 'keek-keek-keek...'. Single or paired, foraging on ground, tree-trunks and branches; unobtrusive, quite approachable. **Range and habitat:** fragmented distribution in AF, from Cameroon, southern Sudan, Ethiopia south to Angola, Zambia, eastern South Africa. Locally common in wooded

areas. Resident, some dispersal; local migrant extreme south. **Similar species:** Wryneck (see p 168).

SWIFTS AND HUMMINGBIRDS Apodiformes

Nyanza Swift *Apus niansae* (Apodidae) 15 cm (6 in). Medium-sized, dark brown swift with whitish throat, forked tail and in good light shows pale panel on upperwing secondaries. Flight fluttering, not powerful. Immature not described. Call is very thin, shrill screech. In flocks, foraging high, lower down around hills, cliffs, buildings, and after rain and storms; often with other swift species. **Range and habitat:** north-eastern AF, from Ethiopia and northern Somalia south to north-eastern Zaire, northern Tanzania. Locally common around cliffs, rocks and buildings in highland areas, also foraging over lowlands. Resident or intra-African migrant. **Similar species:** other dark swifts (see pp 72, 116, 166, 254, 304). **Not illustrated.**

Little (House) Swift *Apus affinis* (Apodidae) 13 cm (5 in). Rather small, thickset, blackish swift with square tail, broad white rump extending down to flanks, and white throat. Flight fluttering, not strong and direct. Immature browner, washed olive. Call is frequent loud, twittering scream, 'tri-i-i-i-i-i...'. In flocks up to hundreds, foraging high, also low around cliffs, buildings and during storms; often with other swift species; noisy, often nesting on buildings, approachable. **Range and habitat:** WP in North Africa and Arabia, Near East, and east through OR to China, Borneo. Widespread in AF south of 16° N, except Horn and far south-west. Common away from arid country, particularly near buildings (in cities), rock faces. Resident in many areas, also local, partial and WP migrant. **Similar species:** other white-rumped swifts (see p 254).

Böhm's Spinetail *Neafrapus boehmi* (Apodidae) 9 cm (3½ in). Very small, almost tail-less swift, black above with white rump; throat and breast pale brownish, merging in white abdomen. Like torso-less 'flying wing' in erratic slow, bat-like, often low flight; wings very broad-based, tapering sharply to tips. Immature duller. Often silent; call is shrill twittering, 'tee-tee-tee-deep'. Single, paired or in small parties, foraging low, often with swallows. **Range and habitat:** mainly southern AF, in scattered areas from Angola, Zambia, Mozambique to Tanzania, Kenya. At best locally common, over woodland and forest edge. Resident or moving locally. **Similar species:** 3 other white-bellied spinetails in AF, White-throated Spinetail (see p 254). **Not illustrated.**

White-rumped Swift *Apus caffer* (Apodidae) 15 cm (6 in). Medium-sized, slim black swift with narrow white rump, deeply forked tail with slender outer feathers and white throat. Immature speckled white on wings. Often silent. Call is low, twittering scream. Usually in pairs, sometimes loose groups, foraging low, often with other swift species; not shy. **Range and habitat:** WP in Morocco, southern Spain; widespread in AF south of Ethiopia, Chad, Senegal, apart from southern interior and equatorial forests. Locally common around buildings and cliffs or over open country, often near old nests of swallows, which it mainly uses. Resident and intra-African migrant. **Similar species:** Horus and Little Swifts.

African Palm-swift *Cypsiurus parvus* (Apodidae) 14-17 cm (5½-6½ in). Small, slender, pale grey-brown swift with long, slender and tapering wings, and long, cleft tail with thin swallow-like streamers. Tail usually held closed during fast veering flight. Immature speckled warm buff, with shorter streamers. Call is muted, thin and high twitter; also quiet 'si-ip'. In pairs or loose flocks, foraging low; noisy, distinctive, often aggressive. **Range and habitat:** widespread in AF between 15° N and 20° S, apart from arid Horn; also south to eastern South Africa, north to south-west Arabia. Locally common over open, low-lying country with native or exotic palms, also nesting/roosting under bridges. Resident; wanders. **Similar species:** Asian Palm-swift.

Ground Woodpecker pair at nest-hole in bank

Red-throated (Rufous-necked) Wryneck

White-rumped Swift

Golden-tailed Woodpecker

Bearded Woodpecker

African Palm-swift

Grey Woodpecker

Little (House) Swift at colonial nest site

PERCHING BIRDS Passeriformes

Rufous-naped Bush Lark *Mirafra africana* (Alaudidae) 18 cm (7 in). Medium-sized lark, brown streaked with black above, pale buff below, with fine black streaking on breast; some rufous in closed wing, but rufous nape not infallible fieldmark. In flight, shows short, rounded, conspicuously rufous wings, and short tail. Immature has breast streaking more diffuse, upperparts rather scaly. Call is mournful, clear whistle, 'chi-wee', from perch. Single, sometimes paired, perching prominently on, e.g., bushes, termite hills; not shy. **Range and habitat:** widespread in AF south of 10° S, except far south-west, and north through east to Kenya; also west. Locally common on open and bushed grassland. Non-migratory. **Similar species:** Red-winged Bush Lark, Fawn-coloured Bush Lark.

Fawn-coloured Bush Lark *Mirafra africanoides* (Alaudidae) 14 cm (5½ in). Rather small lark, rufous brown streaked with black above, with conspicuous whitish eye stripe; underparts pale buff, streaked reddish to brown on breast. In flight, shows prominent white outer tail feathers. Immature spotted dark above, more extensively white in tail. Song is brief, staccato series of 'chip' and 'cheree' whistles, from perch or flight. Single or paired, perching prominently on trees and bushes; not shy. **Range and habitat:** widespread in AF south of 15° S; also north-east, from Tanzania to Ethiopia and Somalia. Locally common on open and bushed grasslands, mainly in areas with dry, sandy soils. Non-migratory. **Similar species:** Rufous-naped Bush Lark, Singing Bush Lark (see p 306).

Red-capped (Short-toed) Lark *Calandrella cinerea* (Alaudidae) 15 cm (6 in). Medium-sized lark, palish brown above with diffuse darker streaking and conspicuous bright orange rufous pectoral patches and crown; underparts unmarked white. Immature darker brown above with some white spotting; dark spotting on breast. Call is brief twittering; also trilling song, given in flight. In small flocks, or loose gatherings of hundreds when not breeding; perching and foraging on open ground; not shy. **Range and habitat:** widespread in AF south of 5° S, and north through East Africa to Ethiopia, Somalia. Locally common on open, short grassland and bare or burnt ground. Resident or nomadic. **Similar species:** Lesser Short-toed Lark.

Chestnut-backed Finch Lark *Eremopterix leucotis* (Alaudidae) 13 cm (5 in). Small, thickset lark with heavy bill, male black with white ear coverts, hindneck collar and thighs, and chestnut back. Female mottled, with narrow white collar and blackish central belly. Immature as female, paler. Call is sharp 'chirrup-chew', musical song. Found in flocks on open ground, often with Chestnut-headed Finch Lark *E. signata* in north-eastern AF. Rather shy. **Range and habitat:** two regions of AF at 10-17° N and south of 15° S, apart from far south-west. Common to scarce on open plains with scattered bushes and trees. Resident, local migrant, nomad. **Similar species:** 4 other AF *Eremopterix*.

African Pied Wagtail *Motacilla aguimp* (Motacillidae) 20 cm (8 in). Conspicuously pied wagtail, mainly black above with prominent white eyebrow and much white in closed wing; underparts white, with prominent black breast band. Immature duller, browner and greyer. Call is frequent brisk 'tissip'; also sustained jumbled song. Single or paired, sometimes small, loose parties, walking and running on open ground, habitually wagging tail; often tame. **Range and habitat:** widespread in AF south of 10° N, apart from far south-west. Commonest AF wagtail, found on flat, open ground, usually near water or habitation, including city centres. Resident; local migrant in subtropics. **Similar species:** 5 other wagtails in AF, White Wagtail (see p 170).

Cape Wagtail *Motacilla capensis* (Motacillidae) 19 cm (7 in). Shorter-tailed wagtail, greyish or olive-brown above with white eyebrow and outer tail. Underparts whitish with dark breastband (vestigial or absent in some). Immature

browner, tail shorter. Call is loud 'tseep' or 'tsee-eep', also piping song. Found singly or in pairs, sometimes flocks, walking or running on open ground, often wagging tail. Can be tame. **Range and habitat:** south of 10° S in AF, and north to East Africa. Common to scarce in open, waterside habitats, forest clearings, also near habitation. Resident, moving locally, seasonally. **Similar species:** 5 other wagtails in AF.

Dark Plain-backed Pipit *Anthus leucophrys* (Motacillidae) 17 cm (7 in). Medium-sized pipit, brown above with buffy eyebrow, virtually unmarked mantle and buffy outer tail, no moustachial stripe. Underparts cinnamon, extremely faintly streaked on breast. Bill dark with yellowish base, legs pinkish. Immature mottled above. Flight call is low 'chiz-zick', sparrow-like, chirping song, from perch. Found singly, in pairs or flocks, foraging on open ground. Moderately approachable. **Range and habitat:** widespread south of 15° N in AF apart from much of Congo basin and south-west. Common to local in short grasslands, often near rocks, termite hills. Resident, nomad, migrant. **Similar species:** 5 similar AF pipits.

Yellow-throated Longclaw *Macronyx croceus* (Motacillidae) 20 cm (8 in). Large pipit, dark and mottled above, bright yellow below, with broad black breast band; hind claw elongate. White corners of tail conspicuous during low, jerky flight. Immature buffish below, breast band diffuse. Call is repeated loud, mewing whistle, 'toooeeee', from perch; also vigorous whistled song. Usually in pairs, on ground or low perches; often approachable. **Range and habitat:** northern AF, from Gulf of Guinea east to Kenya, and thence south through eastern areas to eastern South Africa. Locally common in open, bushed and wooded grassland, also swampy areas. Resident with erratic local movements. **Similar species:** Golden Pipit, and 7 other longclaws.

Rosy-breasted Longclaw *Macronyx ameliae* (Motacillidae) 20 cm (8 in). Large pipit, dark and mottled above, male deep salmon-red below with prominent black breast band. Female paler, more buffy, washed reddish below. Note white corners of tail in low flight. Immature as female but duller, scalloped buff above. Call is plaintive whistle, 'chooitt', from perch; series of squeaking whistles during song-flight. Single or paired, on ground; shy, often unobtrusive. **Range and habitat:** mainly southern areas of AF, from eastern South Africa north to Angola, southern Zaire, Tanzania, Kenya. Local, never numerous, in open, often damp, grassland. Non-migratory, but local movements frequent in some areas. **Similar species:** 7 other longclaws.

Arrow-marked Babbler *Turdoides jardineii* (Timaliidae) 23 cm (9 in). Medium-sized, rather dark grey-brown babbler with orange or yellow eyes and very small pale, arrow-shaped markings (produced by feather tips) on throat and breast. Immature lacks arrow-marks, has eyes dark. Call, often 1st indication of presence, is loud 'churrr churrr churrr...', increasing to crescendo, usually in chorus. Almost invariably in small parties, flying low 1 after the other between bushes and trees; often noisy and extrovert, then quiet, shy. **Range and habitat:** widespread in AF south of 5° S, north to Uganda, Kenya. Locally common in bushed and wooded country. Resident. **Similar species:** Brown and Black-lored Babblers, other babblers (see p 270).

Rufous Chatterer *Turdoides rubiginosus* (Timaliidae) 19 cm (7½ in). Slender, rather long-tailed babbler, brown above, warm cinnamon below, with pale yellow eyes. Immature duller below, eyes dark. Loud churring and chattering calls, also whistles, usually in chorus, often 1st indication of presence. Usually in small parties, flying low 1 after the other between patches of low cover, also hopping on ground with tail cocked; sometimes approachable, often rather shy. **Range and habitat:** north-eastern AF, from Tanzania to Sudan, Ethiopia, Somalia. At best locally common, in dense low cover, mainly in drier areas. Resident. **Similar species:** Fulvous and Scaly Chatterers.

Rufous-naped Bush Lark

Fawn-coloured Bush Lark

Chestnut-backed Finch Lark

Red-capped (Short-toed) Lark

Cape Wagtail

African Pied Wagtail

Dark Plain-backed Pipit

Yellow-throated Longclaw

Rosy-breasted Longclaw

Arrow-marked Babbler

Rufous Chatterer

Garden (Yellow-vented) Bulbul *Pycnonotus barbatus* (Pycnonotidae) 20 cm (8 in). Jaunty, restless, thrush-like species, brown above, darker on head, with short erectile crest; below, paler brown breast grades to whitish abdomen, vent is bright yellow. Immature duller, paler, buffish above in some races. Call is frequent twittering 'tchit...tchit...'; song is brief chattering 'quick-doctor-quick'. In pairs or loose parties foraging on ground or in vegetation; extrovert. **Range and habitat:** widespread North Africa and Middle East; outside forests, probably most ubiquitous bird in AF, in most areas south of 20° N except Cape and arid south-west. Common in scrub/wooded areas, even city centres. Resident. **Similar species:** other bulbuls (see p 262).

Yellow-whiskered Greenbul *Pycnonotus latirostris* (Pycnonotidae) 18 cm (7 in). Medium-sized greenbul, dark olive-green above, washed brownish on tail, and paler below, merging to yellowish on belly; easily identified by bright yellow moustaches on either side of throat. Immature lacks moustaches. Loud distinctive song, throughout day, from cover, is monotonous 'chip chup chip chip chup', interspersed with high trill. Single or paired, in dense cover; shy. **Range and habitat:** mainly west and central AF, from Gulf of Guinea, east through Congo basin, to Sudan, Kenya, Tanzania. Common in forest, bamboo, even large urban gardens. Resident, moving to favoured food sources. **Similar species:** 46 AF greenbuls, other bulbuls (see p 262).

Dusky Flycatcher *Muscicapa adusta* (Muscicapidae) 12 cm (4½ in). Small, plump, short-tailed, grey-brown above, dusky brown below merging to whitish on throat and belly. Immature speckled and spotted buff above, brown spotting and streaking below. Generally silent; occasional long, sibilant, 'tseeeet'. Single or paired, hawking insects from low perches; often unobtrusive; usually approachable. **Range and habitat:** mainly eastern AF, from Ethiopia, south through East Africa to Cape; also Angola, Cameroon. Common in forest edge and moist well-wooded habitats, gardens. Non-migratory; local migrant in southern subtropics. **Similar species:** other flycatchers (see pp 174, 266).

White-eyed Slatey Flycatcher *Melaenornis fischeri* (Muscicapidae) 15 cm (6½ in). Plump, slate-grey above with white eye-ring; paler grey below merging to whitish on abdomen. Immature spotted white above, dark spotting below. Usually silent; call is sharp 'tzit', also chattering, descending trill. Found singly or in pairs on low perches, taking insects from air and ground, most active at twilight; often approachable. **Range and habitat:** mainly highlands of eastern AF, from Malawi and Zambia north to East Africa and Ethiopia; also Angola. Common in forest edge and moist well-wooded sites, cultivation, gardens. Non-migratory. **Similar species:** Abyssinian Slatey Flycatcher.

Southern (South African) Black Flycatcher *Melaenornis pammelaina* (Muscicapidae) 20 cm (8 in). Large, glossy bluish black flycatcher with dark brown eyes and slightly notched tail. Wings palish in flight. Immature blackish above with buff spots; profusely spotted and barred rusty below. Usually quiet; persistent 'sweer' in twilight; also faint piping song. Single, paired or in small, loose parties, on exposed perches, taking insects from air or ground; frequently near very similar Drongo; often unobtrusive, rather shy. **Range and habitat:** widespread in AF south of 5° S, apart from dry south-west, north to Tanzania, Kenya. Common, often local, in bushed and wooded habitats. Resident. **Similar species:** Black Flycatcher, 2 drongos, Black Cuckoo Shrike in AF (see p 224).

Silverbird *Melaenornis semipartitus* (Muscicapidae) 18 cm (7 in). Slender, quite long-tailed flycatcher with silvery grey upperparts and bright rufous underparts. Immature spotted black and buff above, mottled black on buff below. Usually silent; soft thrush-like song. Found singly or in pairs, using perches and taking insects from air and ground; shy. **Range and habitat:** north-eastern AF, from interior northern Tanzania to Uganda, western Kenya, south Sudan and

western Ethiopia. Locally common but never numerous in many open, bushed and wooded habitats, in both dry and moist country, but absent from many apparently suitable areas. Non-migratory; wandering.

Chin-spot Puff-back Flycatcher *Batis molitor* (Platysteiridae) 12 cm (4½ in). Small, short-tailed flycatcher, male grey, white and black above with thick black eye stripe and contrasting pale yellow eye; underparts white, with conspicuous broad black breast band. Female similar, with chestnut breast band and chin-spot. Immature resembles female, washed buff or rufousish above. Song is soft but far-carrying, descending piping, 'chin-spot' or 'chin...spot...fly...cat-cher'; also harsh alarms. Single or paired, foraging in bushes and canopy; bold. **Range and habitat:** widespread in AF south of 5° S, apart from far south-west, and north through East Africa to south Sudan. Common but never numerous in wooded habitats. Generally non-migratory. **Similar species:** other puff-back flycatchers.

Blue Flycatcher *Elminia longicauda* (Monarchidae) 14 cm (5½ in). Small, strikingly beautiful flycatcher with bright blue upperparts, paler breast and white abdomen; tail long, habitually fanned. Immature distinctly paler, with some buff spotting above. Call is loud chirp; also sunbird-like twittering song. Single or paired, foraging from ground level to tree-tops, also pursuing insects in flight; conspicuous, not shy. **Range and habitat:** mainly west and central AF, from Senegal to Angola; also east to Sudan, Uganda, north-west Tanzania, western Kenya. Locally common, never numerous, in moist scrublands and forest edge. Resident. **Similar species:** White-tailed Blue Flycatcher.

African Paradise Flycatcher *Terpsiphone viridis* (Monarchidae) 16-40 cm (6½-16 in). Medium-sized, male unmistakable with rusty orange back, wings and long (over twice body length) tail, glossy blue-black head and breast, white abdomen; uncommon colour phase, males only, has orange replaced by white. Female as male, tail shorter. Immature as female, less glossy on head. Call is sharp 'schveit'; song a ringing whistle, 'pee-pee, pee-pee-pee-pee, pee, pee, pee'. Single or paired, hawking insects; not shy, sometimes unobtrusive. **Range and habitat:** widespread in AF south of 15° N (including southern Arabia), except far south-west. Locally common in scrub and woodland. Non-migratory or intra-African migrant. **Similar species:** other paradise flycatchers (see pp 264, 308, 310).

Olive Thrush *Turdus olivaceus* (Turdidae) 23 cm (9 in). Large, dark olive-brown thrush with rufous orange belly and orange-yellow bill and legs. Immature spotted black on breast and flanks. Call is thin 'weeeet', sharper alarm; loud melodious song, from perch. Single or paired, foraging on ground, perching in bushes, trees; generally shy. **Range and habitat:** eastern and southern AF, from Ethiopia and Kenya south to Mozambique and South Africa. Common in wide range of wood and forest habitats, including alpine heath, gardens and plantations. Basically resident, but moving locally, e.g. in postreproductive flocking or altitudinal migration. **Similar species:** 5 similar thrushes in AF; other *Turdus* thrushes (see pp 80, 134, 176, 266).

Groundscraper Thrush *Turdus litsitsirupa* (Turdidae) 22 cm (8½ in). Large thrush, grey-brown above, heavily spotted black on buff or white below, with curved black mark on ear-coverts. In flight, shows orange-buff patch in wing. Immature spotted whitish or buff above, finely spotted below. Alarm is chuckling 'chee'; song a varied succession of shrill whistles, from perch. Single or paired, foraging on ground, perching in bushes, trees; usually approachable. **Range and habitat:** 2 populations in AF: Ethiopian highlands; from Angola and Tanzania south to northern South Africa. Generally common, sometimes local, in open scrublands, including near habitation. Basically resident. **Similar species:** Spotted Ground Thrush.

Garden (Yellow-vented) Bulbul

Yellow-whiskered Greenbul

Dusky Flycatcher

White-eyed Slatey Flycatcher

Southern (South African) Black Flycatcher

Silverbird

Chin-spot Puff-back Flycatcher

Blue Flycatcher

African Paradise Flycatcher male at nest

Olive Thrush

Groundscraper Thrush

Anteater Chat *Myrmecocichla aethiops* (Turdidae) 20 cm (8 in). Medium-sized, rather thickset and erect thrush, entirely brownish black at rest, appearing black from any distance. In flight, open wings show conspicuous white patch on primaries. Immature brownish on wing-coverts. Call is piping whistle; song a strong and melodic whistling, from perch. In pairs or family parties, foraging on ground, sometimes perching prominently on low bushes, trees or termite hills; not skulking, approachable. **Range and habitat:** mainly at 10-15° N in AF, from Senegal east to southern Sudan; also Kenya, north-east Tanzania. Common in open grassy, bushed or wooded country, often near termite hills. Non-migratory. **Similar species:** Anteating Chat, Sooty Chat, female White-winged Cliff Chat.

White-browed Robin Chat *Cossypha heuglini* (Turdidae) 20 cm (8 in). Medium-sized thrush with black cap cut by white eyebrow; underparts, rump and outer tail feathers bright rufous orange, back and wings greyish. Immature spotted buff above, mottled black and buff below, lacking eyebrow. Call is monotonously repeated whistle, 'doo-doo, swee. . .doo-doo, swee. . .'; song, in duet from perch, most often at twilight, is loud and beautiful, often ventriloquial, whistled crescendo, usually incorporating much mimicry. Singly or paired, foraging on ground, perching in cover; shy, retiring. **Range and habitat:** widespread in AF at 10-25° S, north to East Africa, Chad, Somalia. Common in low cover of moist well-vegetated country. Resident. **Similar species:** 16 other robin chats in AF.

Red-backed (White-browed) Scrub Robin *Erythropygia leucophrys* (Turdidae) 15 cm (6 in). Small thrush, pale brown to rufous brown above, with bold white eyebrow and variable amount of white in wing; tail mainly rufous, white-tipped; underparts whitish, streaked dark on throat and breast. Immature mottled buff and dark brown. Harsh, chattering alarm calls; song is loud, penetrating, often ventriloquial series of whistles with much repetition, from perch; also mimicry. Single, sometimes paired, foraging on ground, perching in cover; shy, retiring. **Range and habitat:** widespread in AF south of 5° S, apart from far south-west; and north to East Africa, Sudan, Ethiopia, Somalia. Common in great variety of bushed and wooded habitats where some ground cover is present. Non-migratory or moving locally. **Similar species:** other scrub robins.

(White-) Starred (Forest) Robin *Pogonocichla stellata* (Turdidae) 15-16 cm (6-6½ in). Beautiful small thrush with slate-blue head, greenish back and wings, yellow-edged black tail and contrasting bright yellow underparts; the 'stars' are small white spots in front of each eye and in the centre of the breast. Immature initially dark, profusely spotted yellow; later unspotted, greenish above, lemon-yellow below. Harsh, grating alarm and 2-note piping call; song is soft fluting, 6-note whistle. Usually found singly, foraging on ground or vegetation; unobtrusive, retiring, in cover. **Range and habitat:** eastern AF, north from Cape to Malawi and East Africa. Locally common in low cover of forests (including bamboo) and dense woodland. Non-migratory; regular altitudinal migration in far south. **Similar species:** Swynnerton's Bush Robin.

Banded Tit Warbler *Parisoma boehmi* (Sylviidae) 11 cm (4½ in). Medium-sized warbler, grey-brown above with pale yellow eye, white feather-edging in wing and white sides to black tail; underparts white with fine black spotting on throat, conspicuous black breast band and rufous brown flanks and vent. Immature spotted with buff above, lacking breast band. Call is loud squeaking 'tick-warra, tick-wurrr'; also brief repeated, trilling song. Single, sometimes paired, foraging openly on bushes and trees, especially acacia; not shy. **Range and habitat:** north-eastern AF, from eastern Tanzania to Kenya, Ethiopia, Somalia. Locally common, never numerous, in dry, bushed or wooded country. Basically resident, wandering to avoid drought. **Similar species:** Black-collared Apalis. **Not illustrated.**

African Reed Warbler *Acrocephalus baeticatus* (Sylviidae) 12 cm (4½ in). Small warbler, unmarked brown above, darker on wings and tail, with faint pale stripe above pale brown eye; lower mandible yellowish; underparts white or buff-white, tinged warmer buff on breast and flanks; legs dark brown. Immature as adult, generally buffer. Call is sharp 'tickk', also harsh 'churrrrr' alarm; song is slow, scratchy, grating, with much repetition, resembling Eurasian Reed Warbler *A. scirpaceus*. Single or paired, foraging in dense vegetation; unobtrusive, shy, retiring. **Range and habitat:** widespread in AF south of 15° S, north to East Africa; also Senegal, Cameroon, Chad, Sudan, Somalia. Local, seldom common, in dense, moist, low vegetation. Resident or regular intra-African migrant. **Similar species:** other *Acrocephalus* warblers (see pp 172, 310).

African Yellow Warbler *Chloropeta natalensis* (Sylviidae) 14 cm (6 in). Medium-sized warbler, male greenish brown above with yellower rump, darker crown and dull yellow edging in wings and tail; underparts bright yellow. Female duller below. Immature as female, more buff below and in wings and tail. Hard 'churr' alarm, shrill, hoarse song with brief, rapidly repeated phrases. Found singly or in pairs, foraging low in dense herbage but occasionally perching prominently. Generally unobtrusive, often retiring. **Range and habitat:** widespread in AF at 5° N-15° S; also Ethiopia, eastern South Africa. At best locally common, in moist, well vegetated hollows and forest edge, often near water. Resident. **Similar species:** Mountain Yellow Warbler and Papyrus Yellow Warbler.

Bar-throated Apalis *Apalis thoracica* (Sylviidae) 12-13 cm (4½-5 in). Small, long-tailed warbler, dark above with pale eye and white outer tail. Pale underparts cut by black breastband in male, band narrower or absent in female. Great racial variation, with upperparts greyish, greenish or brownish, belly white or yellow. Immature duller, breastband much reduced. Male's song is ringing clicking with galloping rhythm; female replies in duet, with high 'cheek' calls. Found in pairs or singly, foraging in bushes and trees. Often tame. **Range and habitat:** eastern AF, from Kenya south to eastern South Africa. Common in bush, woodland and forests. Resident, some altitudinal migration. **Similar species:** 19 other *Apalis* in AF.

Red-faced Crombec *Sylvietta whytii* (Sylviidae) 10 cm (4 in). Tiny, plump, short-necked, short-tailed warbler, silvery-grey above, with face, ear coverts and underparts rufous. Immature browner above with some fawn speckling. Call is high, whistling 'see-see-see-seeee', also sharp, far-carrying 'chick' alarm, often rapidly repeated. Found in pairs or singly, foraging busily on bushes and trees, systematically searching bark and leaves for invertebrates. Bustling, not skulking or retiring. **Range and habitat:** eastern AF, from Sudan and Ethiopia south to East Africa, Zimbabwe, Mozambique. Locally common in diverse bushed and wooded habitats, including forest edge. Resident, moving locally. **Similar species:** 8 other crombecs in AF.

Grey-backed Camaroptera *Camaroptera brevicaudata* (Sylviidae) 11 cm (4½ in). Plump, restless little warbler, grey above with dull green wings and moderately short tail, frequently cocked over rump; below, throat and breast grey merging to whitish belly, but underparts whitish to buffish in some races. In some areas, nonbreeding plumage is browner above, whiter below. Immature usually yellowish below, sometimes browner or greener above. Call is loud bleating 'squeeeeeeeb'; song a loud deep clicking. Single or paired, foraging on shady ground or in dense vegetation; skulking, retiring. **Range and habitat:** widespread in AF south of 15° N, apart from far south-west. Common in dense, low vegetation. Resident, moving very locally. **Similar species:** Green-backed Camaroptera.

White-browed Robin Chat

Anteater Chat

African Reed Warbler

Red-backed (White-browed) Scrub Robin

(White-)Starred (Forest) Robin

Red-faced Crombec

African Yellow Warbler

Bar-throated Apalis

Grey-backed Camaroptera

Rattling Cisticola *Cisticola chiniana* (Sylviidae) 12-13 cm (4½-5 in). Large cisticola, streaked black or greyish on brown back and wings, with warmer brown cap (sometimes streaked black) and panel in closed wing; tail brown, tipped whitish, with subterminal black spots; underparts whitish. Male larger. Immature yellowish on breast; some races rustier above. Alarm is loud 'swearing', 'cheeeee-cheeee. . .'; song a similar 'cheee-cheee-cheee-churrrrrr', often from prominent perch. In pairs or family parties, foraging in grass, bushes, trees, often in open; noisy. **Range and habitat:** widespread in AF south of 5° S, except far south-west, and north to Sudan, Ethiopia. Common in open bushed and wooded country. Non-migratory. **Similar species:** other cisticolas, e.g. Fan-tailed Warbler (see p 310).

Singing Cisticola *Cisticola cantans* (Sylviidae) 13 cm (5 in). Large, unstreaked cisticola, brown to greyish above with contrasting warmer brown cap and panel in closed wing; dusky subterminal band and whitish tip to tail; underparts whitish, washed buff on breast. Some races have nonbreeding plumage, rustier or streaked darker above. Immature duller with faint streaking, or yellowish wash below. Alarm is extended 'cheeeerr'; song a loud stuttering 'pit-choo' or 'pit-choo-chi'. Single, in pairs or family parties, foraging in undergrowth; shy. **Range and habitat:** mainly eastern AF, north from Mozambique to East Africa and Ethiopia; also northern tropics, west to Senegal. Common in low, mainly moist, dense vegetation. Non-migratory. **Similar species:** Fan-tailed Warbler (see p 310).

Tawny-flanked Prinia *Prinia subflava* (Sylviidae) 12 cm (4½ in). Small, slim, restless warbler, brown above with white eyebrow, and white below with warm buffish flanks; long tail frequently cocked vertically and wagged from side to side. Immature washed yellow below, bill yellowish. Call is sharp, far-carrying 'sip, sip. . .', also harsh alarm 'shzbeeeee'; piping song. In pairs and small parties, foraging low in vegetation but not skulking; often approachable. **Range and habitat:** east through OR to Java; widespread in AF south of 15° north, apart from far south-west. Locally common at low to medium levels in moist, dense scrubland. Non-migratory. **Similar species:** other prinias (see p 268).

Mosque Swallow *Hirundo senegalensis* (Hirundinidae) 24 cm (9 in). Large, thickset swallow, glossy blue-black above with rufous rump and long tail streamers. Underparts rufous with white face and upper breast. Displays blazing white underwing coverts during falcon-like flight. Immature duller, streamers shorter, brownish speckling on upperwing. Call is deep, nasal 'harrrrp'. Found singly or in pairs, occasionally small parties, hawking aerial insects. Not shy. **Range and habitat:** widespread at 15° N to 20° S in AF apart from equatorial forests. Locally common, never numerous, in open wooded areas. Resident, moving locally. **Similar species:** 14 similar swallows in AF.

Wire-tailed Swallow *Hirundo smithii* (Hirundinidae) 13 cm (5 in). Small swallow, glossy violet-blue above with rufous orange crown and long, thin tail streamers; underparts white. Immature grey-brown above, washed buff below, streamers very reduced. Usually silent; occasional low twitterings. Single or paired, sometimes loose parties while foraging, hawking aerial insects; often tame. **Range and habitat:** widespread in OR, east to Laos, Vietnam; widespread in AF south of 10° N, apart from far south-west and Congo basin forests. Locally common, never numerous, in open areas near water; often around habitation. Basically non-migratory. **Similar species:** other swallows in AF.

Lesser Striped Swallow *Hirundo abyssinica* (Hirundinidae) 17 cm (6½ in). Medium-sized swallow, blue-black above with rufous chestnut crown and rump, and long tail streamers; easily identified by coarse dark streaking on white ventral torso. Immature duller, washed buff below, tail streamers reduced. Call consists of loud, distinctive discordant squeaks; also discordant piping song. In pairs or small flocks, hawking aerial insects; often approachable,

nesting on buildings. **Range and habitat:** widespread in AF south of 15° N, apart from far south-west and Congo forest regions. Common in many open habitats, including forest edge; often around habitation. Non-migratory, or local or intra-African migrant. **Similar species:** other AF swallows.

African Sand Martin *Riparia paludicola* (Hirundinidae) 13 cm (5 in). Small martin with slightly forked tail; most brown with white abdomen, but a few entirely brown; north African race greyish white below with darker throat. Immature speckled dark buff above, washed buff on abdomen. Harsh alarm call, also weak twittering. In flocks up to thousands, hawking aerial insects; not shy, can be unobtrusive. **Range and habitat:** widespread resident in Morocco, and OR east to Philippines; widespread in AF south of 15° N, apart from Congo forests. Locally common, usually near inland wetlands. Resident; postbreeding dispersal and other movements. **Similar species:** 6 other brown martins in AF, Sand Martin, Dusky Crag Martin (see pp 168, 260).

Banded Sand Martin *Riparia cincta* (Hirundinidae) 17 cm (6½ in). Large, robust, sluggish martin, brown above with small white streak in front of eye, and square or slightly notched tail; ventral torso and underwing-coverts white, with brown breast band. Immature speckled buff above. Often silent; various twittering calls. In pairs or small parties, larger flocks after breeding season, hawking low-flying insects, perching on grass stems; not shy. **Range and habitat:** widespread in AF south of 10° N, except far south-west. Rather local, foraging low over open or lightly bushed grasslands. Sometimes resident, but mainly local or intra-African migrant. **Similar species:** 7 other brown martins in AF, Sand Martin, Dusky Crag Martin (see pp 168, 260).

Black Saw-wing *Psalidoprocne pristoptera* (Hirundinidae) 17 cm (7 in). Medium-sized, slightly glossy, black rough-wing swallow with long, deeply forked tail. In flight, white-winged birds show conspicuous white or pale underwing coverts. Immature as adult, duller. Weak twittering calls, but often silent. Found singly or in small flocks, sometimes with White-headed Rough-wings, hawking aerial insects. Not shy. **Range and habitat:** widespread in AF south of Cameroon and Ethiopia, apart from arid Horn and far south-west. Locally common, sometimes numerous, in moist, open, bushed, wooded and forest edge habitats. Resident, intra-African migrant in south. **Similar species:** immature White-headed Rough-wing.

Black Cuckoo Shrike *Campephaga sulphurata* (Campephagidae) 21 cm (8 in). Male is medium-sized glossy, faintly bluish black, somewhat cuckoo-like bird with rounded tail, orange-yellow gape wattles and, in some individuals, conspicuous yellow shoulders. Female brown with dark barring above, much bright yellow edging in wings and tail, and black, yellow and white barring below. Immature as female, more heavily barred. Generally silent; call is sharp 'chup', also insect-like trill. Single or paired, foraging unobtrusively in mid-storey of vegetation; not shy. **Range and habitat:** widespread in AF south of 5° S, apart from far south-west; also north to Ethiopia. Seasonally common in bushed and wooded habitats. Some equatorial populations non-migratory, others intra-African migrants. **Similar species:** 3 other cuckoo shrikes (see p 262).

Fork-tailed Drongo *Dicrurus adsimilis* (Dicruridae) 25 cm (10 in). Medium-sized glossy black, rather shrike-like bird with reddish eye and deeply forked, diverging 'fish-tail'. In flight, wings show pale greyish wash. Immature greyer above and below, tail less deeply cleft. Call is loud, strident 'waank'; loud, strident song, also mimicry. Single or paired, on exposed perches, hawking insects, killing small birds; noisy, aggressive, extrovert; also unobtrusive forest race, the Velvet-mantled Drongo. **Range and habitat:** widespread in AF south of 15° N. Common in many nonforest habitats; Velvet-mantled is rarer, in interior of Congo forests. Basically non-migratory. **Similar species:** other drongos (see pp 278, 322).

Rattling Cisticola

Singing Cisticola

Tawny-flanked Prinia

Mosque Swallow

Black Saw-wing

Wire-tailed Swallow

Lesser Striped Swallow

African Sand Martin

Banded Sand Martin

Black Cuckoo Shrike

Fork-tailed Drongo

Long-crested Helmet Shrike *Prionops plumata* (Laniidae) 20 cm (8 in). Medium-sized shrike, black and white above with yellow eyes and surrounding wattles, and bristly crest on forecrown; below white, with orange-yellow legs. Immature browner above, spotted on wing-coverts, lacking eye-wattles. Frequent loud, chattering calls and whistling, often in chorus; also bill-snapping. Habitually in small flocks, foraging through bushes and trees; often approachable. **Range and habitat:** widespread in AF south of 15° N, apart from far south-west and Congo forests. Locally common in generally dry scrublands. Resident, making extensive erratic, foraging movements; irrupts in south. **Similar species:** Grey-crested Helmet Shrike.

White-crowned Shrike *Eurocephalus ruppelli* (Laniidae) 24 cm (9½ in). Distinctive, noisy, rather large shrike, brown above with white crown and nape; underparts white. White rump prominent during characteristically stiff-winged, gliding flight. Immature speckled buffish above, crown and nape greyish brown. Calls are sharp, high-pitched, chattering squawks and whistles. Found in pairs or small parties, perching prominently, high on bushes and trees; often retiring. **Range and habitat:** north-eastern AF, from Tanzania to Kenya, Ethiopia, Somalia, southern Sudan. Common in generally dry scrublands. Basically non-migratory. **Similar species:** Rüppell's White-crowned Shrike.

Brubru Shrike *Nilaus afer* (Laniidae) 13 cm (5 in). Small bush shrike, male boldly patterned black and white above with prominent white eyebrow; below white, with conspicuous rufous chestnut flanks. Female similar, black replaced by dark brown. Immature mottled dark above, with some irregular dark barring or streaking below. Male gives high, far-carrying trill, 'trreeeeeeee'; female often makes wheezing sounds in duet. Single or paired, restlessly foraging at mid-levels in vegetation; unobtrusive, not retiring. **Range and habitat:** widespread in AF south of 17° N, except Cape and Congo basin areas. Fairly common in scrublands, especially in acacia canopies. Non-migratory. **Similar species:** *Batis* flycatchers (see p 220).

Grey-backed Fiscal *Lanius excubitoroides* (Laniidae) 25 cm (10 in). Medium-sized shrike with pale grey upperparts and bold black streak across forehead and through eye, extending to side of neck. Wings black. Long tail is black with white basal region and tip. Underparts of male white, female has some chestnut on flanks. Immature is lightly barred dusky brown. Calls are varied, may be musical or harsh. Very gregarious, sometimes in large groups feeding on emerging flying ants; noisy, conspicuous, aggressive. **Range and habitat:** AF in southern Sudan, Ethiopia, Uganda, western Kenya and Tanzania. Fairly common in open bush country. Resident and local migrant. **Similar species:** in AF, other fiscals, puffback shrikes, also boubous.

Yellow-billed Shrike *Corvinella corvina* (Laniidae) 30 cm (12 in). Large, very long-tailed shrike, brown streaked black above, with dark facial mask and conspicuous chrome-yellow bill; ventral torso buffish white, thinly streaked black. Male has pale chestnut flank patch. In flight, shows conspicuous rufous primaries. Immature mottled blackish above, barred blackish on breast. Typical call is monotonously repeated 'scis-scis'; also shrill whistles, shrieks. In small parties, perching prominently, following each other like babblers; often approachable. **Range and habitat:** at 5-15° N in AF, east to southern Sudan and Kenya. Locally common in open scrublands, particularly near water and dense undergrowth. Resident.

Mufumbiri Shrike *Laniarius mufumbiri* (Laniidae) 20 cm (8 in). Medium-sized, but robust bush shrike mainly black above with golden-yellow crown and nape with white tips to wing coverts. Underparts bright crimson with buff vent. Immature has a yellower crown. Call is a loud, mellow 'yo-yo'. Single or paired; shy, retiring in dense cover, usually betrayed by calls. Often rather inquisitive to strange sounds. **Range and habitat:** AF in south and west Uganda, western Kenya and north-western Tanzania. Restricted to papyrus swamps, where locally common. Non-migratory. **Similar species:** Black-headed Gonolek, Common Gonolek.

Slate-coloured Boubou *Laniarius funebris* (Laniidae) 18 cm (7 in). Medium-sized, rather rotund, dark slate to blackish bush shrike. Female slightly paler below. Immature indistinctly barred brown above; greyer below with fine blackish barring. Various whistles, harsh 'churrs' and softer 'plops'; but mainly loud, rapid, precisely synchronized duets between male and female, typically 'hoo-hoo, hoooeeet', often repeated. Single or paired, skulking and retiring in or under low dense vegetation, betrayed by calls. **Range and habitat:** north-eastern AF, from Tanzania and Kenya to Sudan, Ethiopia, Somalia. Common in drier, dense scrublands. Non-migratory. **Similar species:** other black boubous and black helmet shrikes.

Southern Boubou *Laniarius ferrugineus* (Laniidae) 23 cm (9 in). Large, rather robust bush shrike, glossy black above, sometimes with white wing-bar and/or whitish rump; below white, washed pink or peach, especially on vent. Immature spotted buffish above; washed buff below, sometimes with brown barring. Various whistling, snarling and churring calls, but characteristically an extensive vocabulary of loud, precisely synchronized duets, often incorporating pure, bell-like notes. Found in pairs, skulking in vegetation or in open, betrayed by calls; sometimes tame. **Range and habitat:** widespread in AF south of 10° N, apart from far south-west and Congo forests. Common in moist, dense cover. Non-migratory, perhaps moving locally. **Similar species:** 2 similar boubous.

Black-backed Puffback *Dryoscopus cubla* (Laniidae) 16 cm (6½ in). Small, black and white bush shrike with crimson eyes; male glossy blue-black above, with downy white rump feathers puffed out during courtship. Female duller above, forehead and eyebrow white, rump grey. In flight, wings make buzzing 'purrrrp'. Immature as female, with pale plumage washed buff. Various clicks, whistles and bill-snapping; typical call is high shrill whistle, preceded by click. In pairs, foraging in canopies for insects; not retiring. **Range and habitat:** widespread in AF south from Angola, southern Zaire and Kenya, apart from far south-west. Common in moist woods and forest edges. Non-migratory. **Similar species:** in AF 4 other puffbacks and black and white shrikes.

Black-headed Tchagra *Tchagra senegala* (Laniidae) 21 cm (8 in). Medium-sized bush shrike with black cap cut by striking white eyebrow, rusty red wings, and longish black tail with white tip; underparts white, washed grey on breast. Immature has brownish crown. Usual call is far-carrying piping whistle, from flight or perch; also harsh alarm, 'churrr'. Single or paired, foraging in or under low cover but also reaching higher, more exposed situations; shy. **Range and habitat:** WP in Morocco, Algeria, Tunisia; widespread in AF south of 15° N (including southern Arabia), apart from far south-west and Congo forests. Common in low cover and thickets, in scrublands. Non-migratory. **Similar species:** other tchagras in AF.

Sulphur-breasted Bush Shrike *Telophorus sulphureopectus* (Laniidae) 18 cm (7 in). Medium-sized, slim bush shrike, grey and yellow-green above with yellow forehead and eye stripe; underparts bright yellow, merging to bright orange on breast. Immature barred black, lacking orange breast and yellow forehead. Call is far-carrying ringing whistle, including characteristic 'tee-tee too-toooo' and 'tee-tee-tee-teeeeee' phrases; also harsh alarms. Single or paired, foraging at medium to high levels in vegetation; sometimes skulking and retiring. **Range and habitat:** widespread in AF south of 10° N, apart from Congo basin and far south-west. Locally common in many scrub habitats. Apparently non-migratory. **Similar species:** other bush shrikes.

Long-crested Helmet Shrike

Black-headed Tchagra

White-crowned Shrike

Brubru Shrike

Grey-backed Fiscal

Yellow-billed Shrike

Mufumbiri Shrike

Slate-coloured Boubou

Southern Boubou

Sulphur-breasted Bush Shrike

Black-backed Puffback

White-bellied (White-breasted) Tit *Parus albiventris* (Paridae) 14 cm (5½ in). Noisy, active, mainly black tit, with much white edging in wing and conspicuous unmarked white belly. Male is glossy black; female duller, sooty black. Immature has some yellow in flight feathers. Characteristic call is loud, sharp 'see see see'; song a repeated softer 'dee dee dee durrrr'. In pairs, family groups or single, restlessly foraging at medium to high levels in bushes and trees; conspicuous, bold. **Range and habitat:** mainly eastern AF, in Tanzania, Kenya, Uganda, southern Sudan; also Cameroon in west. Fairly common but never numerous in generally moist bush and woodland. Non-migratory. **Similar species:** other tits (see pp 82, 272).

African Golden Oriole *Oriolus auratus* (Oriolidae) 24 cm (9½ in). Medium-sized oriole, male golden yellow with dark pink bill and black eye stripe extending back to ear-coverts; much yellow edging in black wings. Female greener above, paler yellow with olive streaking below; eye stripe similar. Immature as female, greenish edging in wings. Characteristic call is liquid, melodious whistle, 'weela-wooo'; also cat-like mewing. Single, paired or in small parties, foraging high in dense canopy; shy, betrayed by calls. **Range and habitat:** widespread in AF between 15° N and 22° S, apart from Congo forests. Uncommon in woods. Intra-African migrant converging on equator after breeding. **Similar species:** Golden Oriole (see p 182). **Not illustrated.**

African Black-headed Oriole *Oriolus larvatus* (Oriolidae) 24 cm (9½ in). Medium-sized, bright golden yellow oriole, greenish on back and tail, with black hood; coral-pink bill, white and yellow edging in wing. Immature has blackish bill; streaked yellowish on head, black on breast and belly. Characteristic call is melodious, liquid whistle, 'weeela-weeeooooo'; also harsh and mewing sounds. Single or paired, foraging high in dense canopy; shy, betrayed by calls. **Range and habitat:** widespread in AF south of 5° S, except far south-west; also north through East Africa to Sudan, Ethiopia. Common in well-wooded areas. Basically resident. **Similar species:** 6 similar orioles; various black-headed weavers (see pp 182, 232, 278).

African White-necked Raven *Corvus albicollis* (Corvidae) 54 cm (21 in). Large, thickset, short-tailed, glossy black raven with massive black bill tipped with white, and white crescent on hindneck and upper mantle. Immature browner, white patch streaked dark; sometimes thin white breast band. Call is distinctively high croak, 'kraaak'. Single or paired, up to hundreds at good feeding sites; highly aerial, foraging on ground; approachable, extrovert. **Range and habitat:** highlands of eastern AF, from South Africa to eastern Zaire and East Africa. Locally common around cliffs and rocky hills, often scavenging boldly at carcasses and habitation. Non-migratory, making long foraging flights, wandering. **Similar species:** other ravens.

Pied Crow *Corvus albus* (Corvidae) 46 cm (18 in). Large, glossy black crow with conspicuous white breast and collar. Immature duller, white feathering speckled dusky. Call is loud, raucous, frequently uttered 'caww'; also subdued grunts, snoring. In pairs, also flocks up to hundreds, foraging boldly on ground; occasionally shy, usually approachable, opportunistic. **Range and habitat:** widespread in AF south of 20° N, apart from far southern interior; vagrants north to Algeria and Libya in WP. Common, omnivorous scavenger, mainly in and around settlements where sympatric with the Brown-necked Raven, more widespread south of latter's range. Basically resident.

Piapiac *Ptilostomus afer* (Corvidae) 35 cm (14 in). Small, slender, noisy, glossy black crow, browner on wings and on long graduated tail, with black bill and violet eye. Immature duller, with pinkish violet bill tipped with black. Shrill, frequently uttered, squeaking calls; also harsh, croaking alarms. In small parties, foraging on open ground; active, often shy. **Range and habitat:** northern AF, from Senegal and Guinea east to southern Sudan, Uganda, western Kenya.

Common in open wooded grasslands, especially near *Borassus* palms; often seizing insects flushed by large mammals; in towns in western AF. Resident or moving extensively but erratically. **Similar species:** Rüppell's Long-tailed Glossy Starling.

Cape Rook (Black Crow) *Corvus capensis* (Corvidae) 48 cm (19 in). Large but not heavily built, glossy black crow with noticeably slender black bill. Immature duller, washed brownish. Call is frequently uttered, loud high-pitched 'kraaah'; also lower liquid and growling sounds. Single or paired, more often small flocks, occasionally thousands, foraging on ground; not shy. **Range and habitat:** 2 separate areas of AF: widespread south of 10° S except the interior; and north-east, from East Africa to southern Sudan, Ethiopia, Somalia. Sometimes common or locally so in open country with scattered trees and cultivated land. Non-migratory, moving locally to feeding sites, occasionally wandering. **Similar species:** Brown-necked Raven.

Wattled Starling *Creatophora cinerea* (Sturnidae) 21 cm (8 in). Medium-sized, drab grey-brown starling with black wings and tail, and pale or white rump. Breeding male unmistakable with bare yellow and black skin and long black wattles on head. Nonbreeding male has head feathered, without wattles; female similar, with black wing-coverts. Immature as female, browner. Often silent; alarm is nasal 'raaah'; also squeaky, high-pitched song. In flocks up to thousands, foraging on ground; not shy. **Range and habitat:** widespread in AF south of 5° S, and north to Ethiopia, Somalia and sometimes south Arabia. Locally common, irregularly abundant, in open scrublands, often foraging around large mammals. Resident, but frequently nomadic in search of food.

Greater Blue-eared Glossy Starling *Lamprotornis chalybaeus* (Sturnidae) 23 cm (9 in). Medium-sized, thickset, metallic blue-green starling with bluish ear-coverts and conspicuous orange-yellow eyes; central tail feathers green, washed with blue. Immature sooty black, glossed green; eyes grey. Typical call is harsh 'skeea-eeeear'; song a mixture of similar squawking and chattering notes. In pairs during breeding, then flocks up to hundreds, foraging on ground; often tame. **Range and habitat;** at 10-15° N in AF from Senegal to Ethiopia, and south through East Africa to Angola, eastern South Africa. Common in open scrublands, often around habitation. Resident or moving extensively when not breeding. **Similar species:** other glossy starlings, New World grackles (see pp 90, 182, 276).

Violet Starling *Cinnyricinclus leucogaster* (Sturnidae) 18 cm (7 in). Small starling, male unmistakable with brilliant, metallic crimson-violet head and upperparts, white underparts. Female brown above; below white or buffish, with dusky spotting and streaking; eyes yellow. Immature as female, eyes brown. Often silent; twitters in flight, also twanging, whining song. In pairs when breeding; then mainly small flocks, often of 1 sex; foraging high in trees or on ground; often shy. **Range and habitat:** widespread in AF between 25° S and 15° N, apart from Congo forest areas; also north to south-west Arabia in northern summer. Seasonally common in wooded and forested country. Intra-African migrant converging on equator after breeding. **Similar species:** female resembles spotted thrushes.

Rüppell's Long-tailed Glossy Starling *Lamprotornis purpuropterus* (Sturnidae) 35 cm (14 in). Large, metallic blue and violet starling with long graduated tail and white eyes. Immature duller, blacker. Harsh alarm, 'skaaaaa'; also squeaky chattering and whistling, and short, musical song. In pairs or small groups, foraging in trees or on ground; often tame. **Range and habitat:** north-eastern AF, from eastern Zaire and northern Tanzania to Somalia, Ethiopia, Sudan. Common in moist open bushed and wooded country; more local in arid areas, usually near water; often around habitation. Basically resident, wandering. **Similar species:** other glossy starlings in AF, New World grackles (see p 90).

White-bellied (White-breasted) Tit

African Black-headed Oriole

African White-necked Raven

Pied Crow

Piapiac

Cape Rook (Black Crow)

Greater Blue-eared Glossy Starling

Wattled Starling
display calling

Violet Starling

Rüppell's Long-tailed Glossy Starling

Golden-breasted Starling *Cosmopsarus regius* (Sturnidae) 35 cm (14 in). Slender starling with metallic turquoise-green head and neck, white eyes, blue-violet gloss on wings and long tail; underparts golden yellow. Immature duller, blackish above, brownish on face and breast. Call is loud chattering whistle. In pairs or small parties, foraging high in vegetation; shy. **Range and habitat:** north-eastern AF, from northern and eastern Tanzania to Kenya, Somalia, south-east Ethiopia. Locally common in open scrublands. Resident or moving locally, e.g. dispersing during rains. **Similar species:** structure as Ashy Starling.

Superb Starling *Lamprotornis superbus* (Sturnidae) 18 cm (7 in). Distinctive small starling, metallic blues and greens above, with blackish head and contrasting yellowish white eyes; below, conspicuous thin white band divides metallic blue breast from rufous chestnut abdomen; vent white. Immature duller, blacker; eyes dark; lacks breast band. Loud whining alarm, 'weeeaaaaa'; also varied warbling song, with mimicry. In small parties, foraging on ground; tame. **Range and habitat:** north-eastern AF, from Tanzania to south Sudan, Ethiopia, Somalia. Common in many bush and woodland habitats and city centres. Basically resident. **Similar species:** 3 AF starlings (see p 228).

African Red-winged Starling *Onychognathus morio* (Sturnidae) 28 cm (11 in). Slender, moderately long-tailed starling; male glossy blue-black; female grey on head and neck, streaked grey on breast. In flight, rufous primaries. Immature as male, lacking gloss. Call is loud, melodious, liquid whistle; harsher alarm. Nesting in pairs, otherwise flocks up to hundreds, foraging on trees or ground; often shy, but nests on buildings. **Range and habitat:** mainly eastern AF, from Cape to Ethiopia; also patchily in northern tropics. Locally common in wooded country, often around cliffs and buildings, including city streets. Resident. **Similar species:** other *Onychognathus* starlings.

Yellow-billed Oxpecker *Buphagus africanus* (Sturnidae) 22 cm (9 in). Slender, medium-sized starling, darkish brown on head and upperparts with red-tipped yellow bill and narrow yellow eye-ring. Underparts pale buff. In flight, prominent pale buff rump. Immature duller, bill dusky. Call is penetrating hiss 'shssssss', also cackling, chattering. In small parties, gleaning parasites, flesh and blood from hides of large mammals, often with Red-billed Oxpecker. Shy. **Range and habitat:** widespread in AF between 15° N and 20° S apart from Congo forests. Common in open country. Resident. **Similar species:** Red-billed Oxpecker.

Cape Sugarbird *Promerops cafer* (Meliphagidae) 24-44 cm (9½-17 in). Distinctive, bulbul-sized species, like sunbird; male with long decurved bill, brown upperparts and long, graduated brown tail; underparts whitish, russet on breast, vent bright yellow. Female similar, tail shorter. Immature with tail shorter still, lacking yellow vent. Harsh alarm; song a sustained jumble of discordant twanging and grating notes, from perch. Single or paired, rarely in small parties, foraging at flowers; bold. **Range and habitat:** extreme southern AF, in south Cape Province, South Africa. Common in fynbos vegetation, often near proteas. Resident, moves seasonally to feed at protea flowers. **Similar species:** Gurney's Sugarbird, various sunbirds (see p 274).

African Yellow White-eye *Zosterops senegalensis* (Zosteropidae) 10.5 cm (4 in). Small white-eye, yellowish-green above with thin white eye-ring and bright yellow forehead. Underparts bright canary-yellow. Immature duller, initially lacking eye-ring. Usual call is frequent, high peeping whistle; also a soft, low warbling song. Nests in pairs, otherwise found in small parties, restlessly foraging through cover. Not shy, often tame. **Range and habitat:** widespread throughout most of AF, south to north-east South Africa. Common in scrubland, forest edge and gardens. **Similar species:** other white-eyes (see pp 272, 318).

Yellow-tufted Malachite Sunbird *Nectarinia famosa* (Nectariniidae) 14-24 cm (5½-9½ in). Large sunbird, breeding male unmistakable bright emerald green with inconspicuous yellow pectoral tufts, and blackish wings and tail, latter with greatly elongated central feathers. Nonbreeding male speckled green, yellow below, often lacking elongated tail. Female and immature greyish, with yellow malar stripe; white-edged tail. Shrill 'tseee' and 'siip' notes; also rapid twittering, jingling song, from perch. Single or paired, foraging at flowers; active, bold. **Range and habitat:** eastern AF, from South Africa to Ethiopia. Locally common, visiting flowers in many habitats. Resident or local migrant. **Similar species:** Scarlet-tufted Malachite Sunbird.

Mariqua Sunbird *Nectarinia mariquensis* (Nectariniidae) 13 cm (5 in). Medium-sized, square-tailed sunbird, male with metallic green head and mantle, dark blue breastband shading downwards to maroon, and black belly, wings, tail. No non-breeding plumage. Female greyish above, yellowish below with dark streaking. Immature as female, male with throat blackish. Call is sharp 'sip-sip'; high, rapid song. Found singly or in pairs, foraging at flowers. Not shy. **Range and habitat:** two regions of AF, north-east, in Ethiopia, East Africa, and at 15-25° S. Locally common in open, bushed and wooded habitats. Resident. **Similar species:** 14 AF sunbirds.

Variable Sunbird *Nectarinia venusta* (Nectariniidae) 11 cm (4½ in). Small sunbird, male metallic blue-green above with metallic purple breast and nonmetallic white, yellow or orange belly. Female grey-brown above, whitish to yellowish below. Immature as female, throat blackish in male. Harsh 'cheer' alarm, softer 'tsip' calls; short tuneful, twittering song, from perch. Usually found in pairs, small groups at favoured feeding sites, foraging at flowers and leaves; often tame. **Range and habitat:** widespread in AF between 10° N and 17° S, apart from Congo basin. Common in bushed and wooded areas. Resident, some seasonal movement. **Similar species:** Collared and White-bellied Sunbirds.

Scarlet-chested Sunbird *Nectarinia senegalensis* (Nectariniidae) 15 cm (6 in). Medium-sized sunbird, male always black with brilliant scarlet breast and metallic turquoise-green crown and throat. Female brown above, yellowish with dark mottling below. Immature as female, male developing scarlet breast. Harsh 'chaak' alarm, squeaky song; characteristic call deliberate, monotonously repeated 'sheep, ship, shop', from high perch. Single or paired, small groups at feeding sites, foraging at flowers; active, sometimes approachable. **Range and habitat:** widespread in AF between 15° N and 20° S, apart from Congo forests. Common in moist scrub and woodland. Resident, moving locally, seasonally. **Similar species:** Hunter's Sunbird.

Lesser Double-collared Sunbird *Nectarinia chalybea* (Nectariniidae) 12.5 cm (5 in). Medium-sized, sunbird with long, decurved bill. Male has metallic green head, throat and mantle, narrow purple chest-band followed by broader red band. Rest of underparts dirty grey. Rump blue, wings and tail blackish. Female grey below with slight yellowish wash, dull brownish grey above. Calls include a harsh 'zzik'. Song a jumble of notes which rise and fall in pitch. Usually in pairs, sometimes in groups, foraging on nectar or insects. **Range and habitat:** southern AF, South Africa to southern Angola. Fairly common in forest and gardens. **Similar species:** several 'double-collared' sunbirds.

Beautiful Sunbird *Nectarinia pulchella* (Nectariniidae) 12-15 cm (4½-6 in). Rather small sunbird, breeding male metallic green with elongated central tail feathers, and bright red breast bordered yellow on flanks; belly black in some races. Nonbreeding male metallic green on rump and tail, otherwise greyish above, whiter below. Female similar, yellowish below, with whitish eyebrow. Immature similar, throat black in male. Monotonous 'sip' calls, soft song. Single, in pairs or small groups, feeds at flowers; not shy. **Range and habitat:** mainly at 8-17° N in AF, from Senegal to Ethiopia; also East Africa. Locally common in open bushed and wooded areas. Resident. **Similar species:** Olive-bellied and Smaller Black-bellied Sunbirds.

Golden-breasted Starling

Superb Starling

African Red-winged Starling

Yellow-billed Oxpecker

Cape Sugarbird

African Yellow White-eye

Yellow-tufted Malachite Sunbird

Mariqua Sunbird

Variable Sunbird

Scarlet-chested Sunbird

Lesser Double-collared Sunbird

Beautiful Sunbird

Red-billed Buffalo Weaver *Bubalornis niger* (Ploceidae) 25 cm (10 in). Large, stocky weaver, male black with white on flanks, and conspicuous large red bill. Female brown above, mottled and streaked dusky on white below, bill as male. White wing patch visible in flight. Immature as female, bill yellowish or brown. Loud, garrulous, falsetto chattering and croaking calls. In pairs or small parties, foraging on ground; not shy. **Range and habitat:** north-east AF, from Tanzania to Ethiopia, Somalia; south AF, from Angola, Zambia to South Africa. Common in dry, open scrublands. Non-migratory, wandering; migrant far south. **Similar species:** White-billed Buffalo Weaver.

White-browed Sparrow Weaver *Plocepasser mahali* (Ploceidae) 16 cm (6½ in). Medium-sized, noisy weaver, brown above with broad white eyebrow, white edging in blackish wings; underparts white, sometimes spotted on breast. In flight, shows white rump. Immature with pinkish bill. Sharp 'chuck' alarm; loud, chattering, liquid song with 'chop' and 'cheep' notes. In pairs or small parties, feeds on ground; not shy. **Range and habitat:** 2 regions of AF: north-east, in Kenya, Sudan, Ethiopia; south of 15° S, except far south-west. Locally common in dry, open, wooded country, near habitation. Resident. **Similar species:** Donaldson-Smith's Sparrow Weaver.

Grey-headed Social Weaver *Pseudonigrita arnaudi* (Ploceidae) 13 cm (5 in). Rather small and short-tailed greyish buff weaver, blacker on wings, with conspicuous grey or pale grey cap; tail black or dark, broadly tipped dirty grey. Immature buffer, including cap. Brief, piping flight call; also chattering and squeaking at nest. In small parties, feeds on ground; busy, often tame. **Range and habitat:** north-eastern AF, from Tanzania north to Uganda, Sudan, Ethiopia. Locally common in mainly dry bushed and open wooded country, often on sandy soils; sometimes around habitation. Basically non-migratory or wandering.

Chestnut Sparrow *Sorella eminibey* (Ploceidae) 11 cm (4½ in). Small sparrow, breeding male warm chestnut-brown, wings and tail darker with some pale edging; bill black. Nonbreeding male, and female, greyer-brown above with pale brown eyebrow and some dark streaking; buffish or greyish below; bill horn. Immature similar, still greyer above. Low chirping calls. In small parties or larger flocks, feed on ground; sometimes shy. **Range and habitat:** north-eastern AF, from Tanzania to Sudan, Ethiopia, southern Somalia. Locally common in open scrublands, often near habitation or water. Resident, wandering widely. **Similar species:** Chestnut Weaver.

White-headed Buffalo Weaver *Dinemellia dinemelli* (Ploceidae) 23 cm (9 in). Large, stocky weaver with white head and underparts, striking bright reddish orange rump and vent, brown back, wings and tail. In flight, white patch at base of upperwing primaries. Immature has orange rump. Call is harsh squeal, also loud chattering. Found in pairs and small parties, foraging openly on ground; can be wary. **Range and habitat:** north-eastern AF, in Ethiopia, Somalia, East Africa. Common in dry, open bushed and wooded country. Resident. **Similar species:** White-crowned Shrike.

Golden Palm Weaver *Ploceus bojeri* (Ploceidae) 15 cm (6 in). Medium-sized weaver, male always bright yellow, greener above, with head and breast washed chestnut-orange; mantle unmarked; dark eye contrasts with pale face. Female yellow, dusky to greenish above, brighter below. Immature as female, paler below. High-pitched chattering and sizzling calls, especially around nest. In small parties; feeds on ground; noisy, active, fairly tame. **Range and habitat:** north-east AF, in Tanzania, Kenya, southern Somalia. Common or locally so in open bushed and wooded areas, often around habitation or water. Non-migratory, wandering. **Similar species:** 4 other golden weavers in AF.

Reichenow's (Baglafecht) Weaver *Ploceus baglafecht* (Ploceidae) 15 cm (6 in). Medium-sized weaver, male black above (some green) with pale eyes, yellow face cut by black mask (face black in some), and golden yellow edging in wing; underparts yellow (white in some). Female as male, face and crown all black. Immature as female, streaked dark on olive above. Call is sharp 'chirp'; short chattering, sizzling song. In pairs or family parties; feeds on ground and vegetation; active, often tame. **Range and habitat:** mainly eastern AF, from Malawi to Ethiopia; also Cameroon. Common in moist woodland; often around habitation. Non-migratory. **Similar species:** 5 similar AF weavers.

Black-headed Weaver *Ploceus melanocepalus* (Ploceidae) 18 cm (7 in). Large weaver with greatly varying races, breeding male yellow above with black hood or face, variably chestnut nape, and yellow edging in black wings and tail; underparts yellow, sometimes washed chestnut. Nonbreeding male grey-brown, with yellow breast, white belly. Female similar, greener and yellower when breeding. Immature as breeding female, browner above. Loud sizzling and chattering at nest. Found in flocks, foraging on ground and vegetation; noisy, approachable. **Range and habitat:** widespread in AF south of 15° N, apart from far south-west. Common in open bushed and wooded habitats, often around habitation or water. Resident and local migrant. **Similar species:** many similar yellow and black weavers (see p 276).

Red-headed Weaver *Anaplectes melanotis* (Ploceidae) 15 cm (6 in). Medium-sized weaver, breeding male with bright red head and breast, white belly; some races red on mantle and wings. Nonbreeding male greyer above, some races dull orange on head, with yellow breast; female as latter races. Immature as female, buffish below. Often silent; calls 'tink', also sizzling, squeaky song. Single or paired; feeds in trees; unobtrusive, not shy. **Range and habitat:** widespread in AF south of 13° N, apart from far south-west and much of west. Locally common, never numerous, in scrub and woodland. Mainly resident; local migrant in far south. **Similar species:** Cardinal and Red-headed Queleas.

Red Bishop *Euplectes orix* (Ploceidae) 14 cm (5½ in). Medium-sized bishop. Breeding male bright red above with pale brown wings and tail. Face (including forehead), lower breast and belly black. Nonbreeding male, female and immature are brown, sparrowlike, heavily streaked black, buff and brown above, buff below with fine streaking on breast and flanks. Sharp 'chiz' flight call, rather wheezy twittering song. Found in flocks, foraging on the ground, busy and noisy. **Range and habitat:** widespread throughout AF. Common in bushed or open moist grassland, also cultivations and gardens. Basically resident, wandering. **Similar species:** 3 similar bishops in AF.

Yellow (-rumped) Bishop *Euplectes capensis* (Ploceidae) 15 cm (6 in). Large bishop, breeding male black with conspicuous, bright lemon-yellow shoulders and rump. Nonbreeding male retains yellow but otherwise sparrow-like, heavily streaked on upperparts and breast, belly white. Female and immature as nonbreeding male, rump and shoulders dull yellow-olive. Often quiet; sharp 'tzeet' call; subdued twittering in courtship. Paired when breeding, flocks in off-season, foraging on, e.g., standing sedges; active, not shy. **Range and habitat:** local in AF, at 5-30° S north to Ethiopia; also Cameroon. Locally common in tall, bushed grassland and woodland edge. Basically resident. **Similar species:** Yellow-mantled Wydah, Golden Bishop.

Jackson's Wydah *Euplectes jacksoni* (Ploceidae) 14-35 cm (9-14 in). Medium-sized widowbird, breeding male black with brown shoulders; in flight, long tail is distinctively droop-tipped. Nonbreeding male sheds tail, and is brown, streaked, sparrow-like. Female and immature similar. Call is soft 'cheeee', song 'see-see-chip-chip-chip'. In flocks, females outnumber males, foraging on ground. Unique courtship display; male performs jumping 'dance' around grass tuft. Unobtrusive out of season. **Range and habitat:** eastern AF, in Kenya and Tanzania. Locally common in moist, open grassland. Resident, moving locally outside breeding season. **Similar species:** 6 other AF widowbirds.

Red-billed
Buffalo Weaver

White-browed
Sparrow Weaver

Grey-headed
Social Weaver

Chestnut
Sparrow

White-headed
Buffalo Weaver

Golden Palm
Weaver male
nest-weaving

Reichenow's
(Baglafecht)
Weaver

Black-headed
Weaver

Red-headed
Weaver

Red Bishop

Jackson's
Whydah

Yellow
(-rumped)
Bishop in display

Bronze Mannikin *Lonchura cucullata* (Estrildidae) 9 cm (3½ in). Small mannikin, brown above with head, neck and breast blacker and inconspicuously glossed bottle-green; bill black above, blue-grey below; underparts white with fine dark barring along flanks. Immature brown, paler below, with blackish bill and black tail. Alarm call is hard 'stik'; also low twittering song. Found in small parties, foraging on ground or standing grasses; busy, often tame. **Range and habitat:** widespread in AF south of 10° N, apart from far south-west. Common in open grassy scrub and woodland, often near habitation. Resident, wandering. **Similar species:** 5 other mannikins and silver-bills in AF.

Red-billed Firefinch *Lagonosticta senegala* (Estrildidae) 10 cm (4 in). Male is pinkish red, browner on back, wings and tail, with pinkish tinge to grey bill; small white spots on breast; pale vent. Female has pinkish red lores and rump, otherwise brown above, buffish below, bill as male. Immature as female, lores grey-brown, bill dark. Alarm is sharp 'tzit'; also softer 'hueet', and low fluting song. In pairs or small groups, foraging on ground; active, quite tame. **Range and habitat:** widespread in AF south of 20° N, except Congo forests and far south-west. Common in open scrublands, often around habitation. Basically resident, local migrant far south. **Similar species:** 7 other firefinches in AF.

Purple (Village) Indigobird *Vidua chalybeata* (Ploceidae) 11 cm (4⅓ in). Breeding male glossy bluish, purplish or greenish black, with coral-pink legs and (depending on race) coral or white bill. Nonbreeding male, and female, sparrow-like, streaked above, paler below, with especially conspicuous buff streaks on crown. Immature as female, browner below, no crown streaks. Call is sharp 'tikk'; mimics song of Red-billed Firefinch, of which it is brood parasite. Single in breeding season, otherwise in small flocks, foraging on ground; active, sometimes tame. **Range and habitat:** widespread in AF south of 20° N, apart from Congo forests and far south-west. Somewhat uncommon in open scrub or woodland. Resident, regular local movements. **Similar species:** 10 other AF indigobirds.

Purple Grenadier *Uraeginthus ianthinogaster* (Estrildidae) 14 cm (5½ in). Large, unmistakable waxbill, male with red bill, bright cinnamon head and neck, darker brown wings and tail, and conspicuous bright violet-blue rump, underparts and eye patch. Female with conspicuous white eye-ring and violet-blue rump, otherwise brown above, pale cinnamon below, merging to whitish on belly. Immature as female, lacking eye-ring. Usually quiet; low tinkling, chirping, trilling notes. In pairs or small parties, foraging on ground; sometimes unobtrusive, tame. **Range and habitat:** north-eastern AF, from Tanzania north to Ethiopia, southern Somalia. Locally common in open bush and wooded areas. Non-migratory. **Similar species:** Common Grenadier.

Red-cheeked Cordon-bleu *Uraeginthus bengalus* (Estrildidae) 13 cm (5 in). Male unmistakable bright sky blue, with red patch on ear-coverts; bill pinkish; crown, back and wings brown, belly buffish. Female as male but duller blue, lacking red ear-coverts. Immature as female, less extensively blue below. Usually quiet; low piping and squeaking; also subdued, monotonously repeated song, 'tz, tz, seeee'. In pairs or family groups, feeds on ground; busy, usually tame. **Range and habitat:** mainly at 10-15° N in AF, from Senegal east to Ethiopia; also south through East Africa to southern Zaire and Zambia. Common in scrub and woodland, often around habitation. Resident. **Similar species:** 2 other cordon-bleus in AF.

Red-billed Quelea *Quelea quelea* (Ploceidae) 13 cm (5 in). Rather small weaver, breeding male with red bill, and black face surrounded by buff or rosy wash; upperparts heavily streaked buff and black; underparts whitish, lightly spotted on breast and flanks. Nonbreeding male, and female, lack black face and wash. Immature as female, bill pinkish grey. Often silent; sharp 'chak' alarm, loud chattering at nest. In flocks up to millions, foraging on ground or on, e.g., cereal crops, hence Africa's main avian agricultural pest; bold. **Range and habitat:** widespread in AF south of 17° N, apart from western forests and extreme south. Common, seasonally abundant, in dry, open, bushed grasslands. Few resident; most breed opportunistically, move in response to rain. **Similar species:** nonbreeding quelea resemble sparrows, weavers (see p 276).

Pin-tailed Whydah *Vidua macroura* (Ploceidae) 12-34 cm (4½-13 in). Breeding male has red bill, black cap and upperparts, white collar and underparts, and long, slender, black tail. Nonbreeding male short-tailed, sparrow-like, with red bill and especially broad black and buff streaks on crown. Female and immature similar, bill pinkish. Sharp 'tseep' flight call; shrill, twittering song. In small parties with more females, foraging on ground; brood parasite of Common Waxbill; active, bold. **Range and habitat:** widespread in AF south of 15° N, except arid south-west. Common in open bushed and wooded habitats. Resident, moving locally. **Similar species:** Shaft-tailed Whydah.

Fischer's Whydah *Vidua fischeri* (Ploceidae) 10-29 cm (4-11 in). Breeding male has black head and upperparts, red bill, pale buff crown and underparts and straw-coloured, slender, elongated, central tail feathers. Nonbreeding male, and female, brown, streaked above, sparrow-like, with red bill. Immature browner, less streaked, bill blackish. Call is sharp 'sip'; short, much-repeated song. Found in small parties, females outnumbering males, foraging on ground. Brood parasite of Purple Grenadier. Active, sometimes shy. **Range and habitat:** north-eastern AF, from Tanzania to Ethiopia, Somalia. Locally common in dry bush. Resident, moving locally. **Similar species:** Shaft-tailed Whydah.

Brimstone Canary *Serinus sulphuratus* (Fringillidae) 15 cm (6 in). Large, thickset canary, greenish yellow streaked with black above, brighter yellow below, with diffuse, dusky moustachial stripe and, in some, dusky breast; heavy conical, greyish bill. Immature duller, some breast and flank streaking. Call is harsh 'chirrup'; jumbled, deep, harsh, husky, rather tuneless song. Single or paired, sometimes small parties, foraging on ground and vegetation; unobtrusive, not shy. **Range and habitat:** widespread in AF south of 10° S, apart from arid south-west; also north through east to Kenya, Uganda. Sometimes common in open, moist bushed and wooded country. Resident or moving locally. **Similar species:** 14 similar canaries in AF, Serin (see p 180).

African Citril *Serinus citrinelloides* (Fringillidae) 11 cm (4½ in). Small, rather slender canary, yellowish green above with dark streaking on mantle and, sometimes, prominent yellow eyebrow; below yellow, sometimes lightly streaked. Male has black or greyish face; female lacks this coloration and is more distinctly streaked below. Immature as female but browner above, paler below. Low cheeping calls; song is sustained, far-carrying, shrill, tuneful whistle. In small parties, feeds on ground and vegetation; sometimes unobtrusive, but not shy. **Range and habitat:** eastern AF, from Malawi north to Ethiopia. Common in many moist, fairly open habitats, including gardens and forest edge. Basically resident, wandering. **Similar species:** 14 similar AF canaries, Serin (see p 180).

Cinnamon-breasted Rock Bunting *Emberiza tahapisi* (Fringillidae) 14 cm (6 in). Medium-sized bunting, male streaked black on brown above, cinnamon below, with white-striped black hood. Female similar, hood browner, striped grey. Immature as female, duller. Call is 2-note 'see-eeeeee'; song a high 'chee-trrr, chee chee'. Found singly or in pairs, sometimes small parties, foraging openly on ground. Often unobtrusive. **Range and habitat:** widespread south of 15° N in AF (including southern Arabia), apart from equatorial forests and far south-west. Locally common on open, rocky ground with some grass or bush. Resident and local migrant. **Similar species:** 8 *Emberiza* in AF.

Bronze Mannikin

Purple (Village) Indigobird

Purple Grenadier

Red-cheeked Cordon-bleu

Pin-tailed Whydah

Red-billed Quelea

Red-billed Firefinch

African Citril

Fischer's Whydah
nonbreeding male

Brimstone Canary

Cinnamon-breasted Rock Bunting

THE ORIENTAL REGION

The tropical mainland of Asia, together with the islands, or island groups, of Sri Lanka, Sumatra, Java, Borneo, Taiwan, Hainan and the Philippines, constitutes the Oriental Region. The most clearly defined boundary runs along the narrow arc of the Himalayan mountain chain, which forms both a physical and a climatic barrier between the subtropical forests on the south side and the arid Tibetan plateau to the north. The valley of the river Indus forms a convenient, if somewhat arbitrary, boundary on the western flank, in an area where the number of truly Oriental species diminishes, and Palaearctic forms become more noticeable. There is, however, a much broader and more complex area of transition to a Palaearctic fauna through the mountainous country of south-eastern Tibet and south-western China. The boundary skirts the edge of the Tibetan plateau, runs north and east to include Szechuan, and then follows a latitude of between 27° and 30°N across China to the China sea. South of this line, the Philippine Islands, with their somewhat specialized fauna, are included, though the whole archipelago of Wallacea, lying between Borneo and New Guinea, has a strong Australo-Papuan element, is highly transitional in nature, and is not incorporated in the Oriental Region in this analysis.

The total land area of some 9 million sq km (3.5 million sq miles) is only half that of South America, or about one-third of the African land mass, but is characterized by an astonishing variety of climate, vegetation and scenery. In the south-east of the region, the humid tropics, with their very uniform annual temperature, support one of the three largest areas of primary rainforest in the world, with an immensely rich tree fauna, and an abundant though specialized avifauna with many striking endemic species such as the Great Argus Pheasant and the Rhinoceros Hornbill. There are two much smaller patches of rainforest, one in south-west India and the other in the slopes and foothills of the eastern Himalayas and Assam, which, despite the discontinuity, hold many bird species in common.

Tropical semi-evergreen forest covers a much larger total area in the region than true rainforest, and is equally rich in bird species, if not more so in some respects. Though very variable according to altitude and rainfall, especially as between the dry evergreen forest of the lowlands and wet hill evergreen, it is the home of many of the babblers and pheasants for which the region is notable. There is a very marked contrast in the vegetation in the west of the region, in north-west India and Pakistan, with its semi-desert conditions and dry thorn scrub supporting birds more characteristic of the arid lands which run from Morocco to the steppes of central Asia. However, it is worth noting that in quite recent times, even up to 200 years ago, this area was much lusher, with subtropical vegetation in Pakistan capable of supporting mammals such as rhinoceros. Human activity has been a major factor in modifying landscape and vegetation across the Oriental Region, and is now a more potent and threatening force for change than ever before. Apart from the complete destruction of the natural vegetation in some areas, which of course eliminates the existing fauna, there are significant wholesale changes which are less apparent at first sight. For instance, clearings in primary forest regenerate with a lush second growth which can look like a

Painted Storks *feed against the backdrop of an Indian sunset.*

virgin habitat to the unpractised eye, but in fact holds quite different species.

Over most of the region outside the rainforest areas, the marked dry seasons constitute a monsoon regime, though there are very marked variations locally in the pattern and extent of rainfall. Assam, in north-east India, has the highest annual rainfall in the world at over 400 in, while the hills of south-west India receive over 130 in per annum. The monsoon is heavier and comes earlier in the year in the eastern half of the Himalayas, and decreasing rainfall westwards is associated with a less rich flora and fauna, and a diminishing number of species typical of the eastern Himalayas and South-East Asia. Many species whose ranges extend the length of the Himalayan chain have evolved darker forms in the eastern, wetter section, and paler forms in the west, and these are given subspecific status.

The total number of bird species in the region, at around 1,900, compares very favourably with any similar-sized tropical area in the world, although there is in fact only one endemic family, the leafbirds (Irenidae) with 14 species. Many bird families generally typical of the Old World tropics are well represented, such as trogons with 11 species (two in Africa but 23 in South America), hornbills, barbets, broadbills, pittas, bulbuls and sunbirds. However, the outstanding features of Oriental bird-life are the richness of the pheasant family, with 89 species, and the great array of babblers, flycatchers, warblers and thrushes, totalling 525 species.

The distribution of birds in the region falls very clearly into three subregions, to such an extent that only 14% of the total number of species occur in all three of these. The richest subregion in terms of species, which is probably best called Sino-Himalayan (the alternative name of Indo-Chinese sounds better but is less exact), comprises continental South-East Asia, Taiwan, Hainan, southern and western China, and the southern faces and foothills of the Himalayas west to Kashmir. Around two-thirds of the total number of Oriental species, some 1,200, are found in this area of nearly 4 million sq km (1.5 million sq miles), containing many of the highest mountains and the most spectacular scenery in the world. Fifty-seven species of pheasants, many in China, are the jewel of the avifauna, though many are gravely threatened by habitat destruction. Even more striking is the number of babblers, with 153 species, though there is great diversity too in many other groups. For example, there are 24 species of owls, 26 bulbuls, 79 warblers, 75 thrushes, 45 flycatchers and 33 woodpeckers.

A feature which lends both interest and charm to birding in these hilly tracts is the perceptible vertical zonation of forest types and bird-life. A few hours' walk in Nepal can take one from dry evergreen on the valley floor, through wet hill evergreen, bamboo, broad-leaved deciduous woods with much oak and chestnut and finally pine forests before the snowy peaks come clearly into view.

It was in this subregion that a fascinating zoogeographical puzzle came to light in 1966 with the chance discovery of a species of swallow new to science. This was the White-eyed River Martin, a very distinctive swallow, having only one close relative and that a very similar-looking bird confined to the rivers of West Africa.

Around 38% of the species found in the Sino-Himalayan subregion penetrate south through peninsular Thailand to another clearly defined faunal area, the Malaysian subregion. This comprises the southern tips of Burma and Thailand, Malaysia, and the great islands of Sumatra, Java and Borneo lying on the Sunda shelf, which are separated from the mainland by a uniformly shallow sea about 36 m (120 ft) deep, and which in recent geological time were part of continental Asia. Some 916 bird species occur in this mosaic of islands and rainforest, including perhaps the most typical Oriental birds. Of the 14 leafbirds, 13 are found here, and among the 25 species of pheasant are the quite spectacular Great Argus and a clutch of rare peacock-pheasants. All 10 Oriental broadbills occur here, and 24 of the 27 flowerpeckers, which, with eight of the region's 10 frogmouths, gives an Australo-Papuan flavour to the avifauna. Also well represented are pittas, with 12 species, and there are 34 cuckoos, 36 bulbuls, 74 babblers and 60 flycatchers.

The great rainforests which are the glory of the subregion are fragile and highly threatened, and yet have huge potential for research. Bird-life can seem almost absent at times in the forest, until one suddenly comes upon a 'bird-band' or hunting party of many different species, which forage actively through the vegetation at all levels from the canopy to the forest floor. Often, before one can see and note down the birds present, they have moved on and all is quiet again.

The Philippines, though included in this analysis in this subregion, have a specialized avifauna with some Bornean and also some Australo-Papuan elements. Endemic and rare in the Philippines is the magnificent Monkey-eating or Philippine Eagle, arguably the most spectacular bird of prey in the world. Other endemic species in this group of over 7,000 large and small islands are a small parrot, the Guaiabero (*Bolbopsittacus lunulatus*) whose nearest relative is Malaysian, and the aberrant large tree-creepers (*Rhabdornis*) of which there are two species remarkable for their brush-tipped tongues, plumage pattern and heavy, curved bills. The Philippine avifauna, with over 300 breeding species, is not as rich as that of Borneo or Java, but has half again as many breeders as Sri Lanka.

The third subregion, which within itself has several markedly differing faunal areas, comprises the Indian subcontinent south of the Himalayan foothills and west of the Brahmaputra river, Pakistan and Sri Lanka. In general, it is characterized by having many families of dry-country birds more typical of North Africa or South-West Asia, such as bustards, coursers, larks, chats and sandgrouse. The total species list of the subregion at about 700 is markedly lower than that of the other two, and it is of interest to note that only 25% of the species in the Malaysian subregion also occur here. However, there are many interesting endemic species such as a flycatcher, a thrush, some bulbuls, barbets, a malcoha and a coucal, especially in the humid hills of south-west India and in Sri Lanka.

There are some striking similarities between birds which occur in the Sino-Himalayan subregion and again in the wet hill forests of south-west India, despite the gap of some 2,500 km (1,550 miles), indicating the loss of suitable habitat in the intervening area, probably with the recession of the last Ice Age. Notable species in this context include the Great Hornbill, Fairy Bluebird and Broad-billed Roller, and groups such as laughing thrushes, trogons and frogmouths have evolved distinct species. There are also interesting species in northern India, possibly relics of the ancient pre-Himalayan drainage systems, such as the recently extinct Pink-headed Duck, the Bristled Grass Warbler and the Sind Jungle Sparrow. Jerdon's Courser, thought to be extinct as it had not been sighted since 1900, was re-discovered in central India in 1986 – a spectacular event in the history of Indian ornithology.

Ornithology in the Oriental Region has had many illustrious students over the last century and a half, and a vast amount of information has been published in such journals as *Stray Feathers*, which appeared from 1873–99, and that of the Bombay Natural History Society, founded in 1886 and still in the forefront of research. A more recent development has been the establishment of the Oriental Bird Club, which collates and publishes news and research from the whole region in its excellent journal *Forktail*.

FULL-WEBBED SWIMMERS Pelecaniformes

Spot-billed (Grey) Pelican *Pelecanus philippensis* (Pelecanidae) 150 cm (59 in). Greyish upperparts and the pinkish tinge below give this species a dingier appearance than the White Pelican. Nonbreeding and young birds are markedly browner. Very buoyant in the air for such a heavy bird, whether in low flap-and-glide flight over the water or soaring in thermals. A colonial breeder, formerly in huge numbers in the region in swamps long since drained, though there are still large pelicanries in a few places. Voice a deep croak. **Range and habitat:** OR. Resident and locally migratory, in well-watered plains. Outside breeding season on large rivers and creeks. **Similar species:** White and Dalmatian Pelicans (see p 142).

Indian Cormorant *Phalacrocorax fuscicollis* (Phalacrocoracidae) 62 cm (24 in). A slim, medium-sized cormorant with a white tuft on the sides of the head in breeding plumage; yellow throat and slender bill distinguish it from Javanese Cormorant. Often occurs in large flocks and breeds colonially, often in company with other species, the timing of breeding dependent on the rains and water conditions. Mostly silent. **Range and habitat:** OR. Common resident, wandering locally, on lakes, rivers, reservoirs or on the coast. **Similar species:** Great and Javanese Cormorants.

Javanese Cormorant *Phalacrocorax niger* (Phalacrocoracidae) 50 cm (20 in). A small, stocky cormorant with a thickish bill, long tail and dark throat which becomes white outside the breeding season. The thicker neck and head help to distinguish it from similar species. Gregarious and often in huge numbers in suitable localities, though equally likely to be only 1 or 2 birds on a tiny patch of marsh or village pond. Shares tree nesting sites with other cormorant and heron species. Largely silent. **Range and habitat:** OR. Common resident in the plains in marshland, by lakes, ponds, rivers and reservoirs and at the coast. **Similar species:** Pygmy and Indian Cormorants.

Oriental (Indian) Darter *Anhinga melanogaster* (Anhingidae) 90 cm (35 in). The long, wedge-shaped tail is obvious when this cormorant-like species is seen soaring over marshes, though a more usual view just shows the thin, snaky head and neck as the bird swims semi-submerged. Fish are caught after an underwater chase with a spear-like thrust of the strong neck and sharp bill. The twig nest is in trees, often with other species nesting close by. **Range and habitat:** OR. Resident on lowland lakes, marshes and rivers, also on the coast in mangroves. **Similar species:** other darters (see pp 52, 186, 286).

Christmas Island Frigate Bird *Fregata andrewsi* (Fregatidae) 100 cm (39 in). Breeding only on Christmas Island in the southern Indian ocean, this species has one of the more restricted ranges of its family, and is certainly the smallest in terms of total population. Nests are placed in tall, isolated trees from which the bird has a clear, wind-assisted take-off; rising off a calm sea appears to be a major problem. In the air, though, frigate birds are masters of the element, swooping to gather fish or offal from the surface. The voice is an occasional, harsh croak. **Range and habitat:** OR and PA, Christmas Island, dispersing presumably over adjacent seas: occurs off Javan and Bornean coasts. **Similar species:** Great, Least and Ascension Frigate Birds (see pp 52, 286). **Not illustrated.**

HERONS AND ALLIES Ciconiiformes

Pond Heron *Ardeola grayii* (Ardeidae) 52 cm (20 in). A strikingly plumaged small heron in its breeding plumage of maroon and buff, though virtually indistinguishable from related species in the streaky brown winter dress. Common at the water's edge everywhere, whether a village pond or in mangrove swamps, the white wings suddenly become obvious as it flaps up. Nesting is in colonies in trees which are often, though not necessarily, near water. The note is a heron-like croak, more varied and continuous during nesting activities. **Range and habitat:** OR, Indian subcontinent and Sri Lanka. Common resident, wandering locally, in open country by ponds, paddyfields or by the coast.

Asian Openbill Stork *Anastomus oscitans* (Ciconiidae) 80 cm (31 in). One of the smaller storks, though common in its range, and sharing with a close relative in Africa the unique oval gap in the closed bill. The purpose of this is evidently to facilitate coping with the snails on which the birds feed. A strong flier and partial to soaring, but agile on the wing and capable of rushing, tumbling descents under complete control. Breeding colonies are sometimes very large, with thousands of birds standing on or by their tree nests, and a constant coming and going of birds feeding in nearby waterways. It makes a typical stork-like clattering with the mandibles. **Range and habitat:** OR, except Indo-Malayan subregion, resident but also dispersing long distances, by lakes, rivers and marshes in the lowlands. **Similar species:** African Openbill Stork (see p 188).

Black-necked Stork *Ephippiorhynchus asiaticus* (Ciconiidae) 130 cm (51 in). A majestic and striking species with its black dagger-like bill and scarlet legs – the female differs only in having a staring yellow eye as opposed to black in the male. Like many large, shy birds, it is not numerous, and is usually seen singly or in pairs. The large stick nest is placed high up in a tree, and the display is a ceremonial face-to-face affair, with wing-flapping and clattering of bills. **Range and habitat:** OR to north AU. Resident in lowland areas in marshes or by rivers.

Greater Adjutant Stork *Leptoptilos dubius* (Ciconiidae) 160 cm (63 in). This huge, ugly scavenger is the Asian counterpart of the African Marabou, and is perhaps seen at its best when soaring majestically, often in company with vultures. Its powers of flight render it capable of wandering great distances, for although not known to breed now in India, it occurs regularly across northern India as far west as Sind. Burma and Indo-China are the main breeding areas. Nests are in tall trees, and breeding colonies were sometimes very large. Numbers are now thought to be dramatically reduced, though a colony has recently been located in Assam. **Range and habitat:** OR; except southern India, Sri Lanka and Indo-Malayan subregion. **Similar species:** Lesser Adjutant Stork.

Black Ibis *Pseudibis papillosa* (Threskiornithidae) 68 cm (27 in). A somewhat solitary ibis which appears to be more tolerant of dry habitats such as stubble fields and cultivated areas than related species. The black plumage, bare scarlet skin on the head and red legs are distinctive, and unlike the White Ibis there are no plumes on neck or back in the breeding season. Often seen singly or in small parties at best, it shuns the noisy, crowded colonies of breeding waterbirds, and it is only rarely that more than 2 or 3 pairs nest in the same tree. It has a loud call when taking wing but is mostly a quiet species. **Range and habitat:** OR; northern India to Indo-China, in open country near rivers, lakes or marshes. **Similar species:** Glossy Ibis.

Spot-billed (Grey) Pelican

Indian Cormorant with (left) Javanese Cormorant

Javanese Cormorant

Oriental (Indian) Darter

Pond Heron

Black-necked Stork

Asian Openbill Stork

Greater Adjutant Stork

Black Ibis

WATERFOWL Anseriformes

Indian Whistling Duck *Dendrocygna javanica* (Anatidae) 42 cm (17 in). Deriving its Whistling name from the constantly uttered flight call, it is one of the relatively few ducks which nest in cavities in trees, though occasionally nests are placed in waterside vegetation on the ground. A surface feeder in water with plenty of emergent vegetation, it has an unusually upright posture when standing, and walks well, though a poor flier on its rounded wings. In this connection, it is of note that Whistling Ducks have curious notches on the inner vane of the primary feathers. Usually found in flocks, sometimes large. The sexes are alike. **Range and habitat:** OR. Resident, wandering locally, on ponds, rivers or marshes on the plains.

Barheaded Goose *Anser indicus* (Anatidae) 75 cm (29 in). A breeding bird in Central Asia, this goose migrates over the Himalayas, and the only wintering areas lie in the Indian subcontinent. Like all geese, it occurs in family parties which aggregate to form large flocks on stretches of water or rivers adjacent to cultivated areas where the birds can feed. Grazing the crops is a nocturnal activity, and by day the wary geese rest around river or lake shores. Said to be remarkably tame when on the remote and undisturbed breeding grounds, in contrast to alert behaviour in winter. Call a musical honking. **Range and habitat:** EP, in winter to OR. Open country, lakes, rivers, crops.

Ruddy Shelduck *Tadorna ferruginea* (Anatidae) 66 cm (26 in). Orange plumage and pale head distinctive, the female lacking the black collar of breeding males; black flight feathers and a white wing patch are prominent when the birds are on the wing. More goose- than duck-like, it grazes on crops and grass as well as taking aquatic animal food. Not as gregarious as some relatives, it is usually seen in small parties which haunt rivers, sandbanks and lakes in the winter. It has a loud, goose-like honking call. **Range and habitat:** southern WP and EP, wintering to OR, occasional in AF. Open country, mountains, steppes, lakes and rivers. **Similar species:** Common Shelduck (see p 146).

Cotton Teal *Nettapus coromandelianus* (Anatidae) 33 cm (13 in). A diminutive duck, with a stubby bill and short legs, consequently a poor walker (and does not graze), though strong and agile in flight. It feeds on the surface on waterplants and small animal life, and is often to be seen in pairs or small groups. Prefers quite reedy ponds or even very small patches of water, and nests in tree-holes at heights from 2 m (6½ ft) to 30 m (98 ft). The female lacks the iridescence of the dark parts of the male's plumage, being browner, and the white underparts are dingier. **Range and habitat:** OR. Locally common and resident by ponds, marshes and wet paddyfields in the plains. **Similar species:** African Pygmy Goose (see p 190).

Falcated Teal *Anas falcata* (Anatidae) 50 cm (20 in). The name is a reference to the sickle shape of the inner secondaries, which fall in a curve over the primaries on the folded wing. The drake is also adorned with a full, drooping crest behind the beautifully glossed head, and the duck has a vestigial crest, giving but a trace of the heavy-headed look. A wary bird, it mingles with the other wintering species in small numbers on lakes and rivers, occasionally on the sea. Females quack; males make a short, whistling note. **Range and habitat:** EP, in winter to OR, occasionally to WP. Rivers, lakes, flooded land. **Similar species:** Gadwall, Pintail, Wigeon (see p 146).

Spotbill Duck *Anas poecilorhyncha* (Anatidae) 60 cm (24 in). The bright red spot at the base of the bill is nearly lacking in the female, and totally so in the Chinese subspecies, which also differs in being much browner, without such well-defined pale feather edges. Similar in lifestyle and voice to the Mallard, it is rather heavier and slower in flight. Frequenting shallow waters in which it can up-end for food, it is mostly resident, and both local

movements and timing of breeding are governed by water conditions. **Range and habitat:** EP, OR in marshland, by ponds and reservoirs. **Similar species:** Pacific Black Duck (see p 288).

White-winged Wood Duck *Cairina scutulata* (Anatidae) 81 cm (32 in). A rare, forest-haunting duck, and a close relative of the Muscovy Duck of South America. It has a bony knob on the wing similar to that of the Comb Duck, and the male has a swollen and reddish base to the bill when breeding, though not exaggerated as in the Muscovy. In flight, the large white wing patch is striking and prevents confusion with the Comb Duck. It is a little-known species as it is partly nocturnal and inhabits the sort of terrain which is very difficult to work. The call is a mournful honking, which is most often heard as the birds fly to and from their feeding grounds at dawn and dusk. **Range and habitat:** OR. Resident around boggy pools and streams in evergreen forest. **Similar species:** Comb Duck, Muscovy Duck (see p 102).

DIURNAL BIRDS OF PREY Falconiformes

Jerdon's Baza *Aviceda jerdoni* (Accipitridae) 48 cm (19 in). A sluggish, somewhat buzzard-like hawk with long wings and a crest which is raised vertically when the bird is sitting. Tending to be crepuscular, it is not one of the more obvious hawks in its habitat, and feeds on small prey such as insects and frogs, though quite a strong flier, especially in the soaring display above the tea-gardens or hill forest. There are several subspecies which vary in the saturation of colour and the weight of the barring; there is a small variation in size in some races. Juveniles lack the broad barring on the underparts, which are much purer white. The call is a buzzard-like mewing. **Range and habitat:** OR to the Philippines and Celebes. Resident in forests up to 1,800 m (5,900 ft). **Similar species:** Crested Goshawk. **Not illustrated.**

Oriental Honey Buzzard *Pernis ptilorhynchus* (Accipitridae) 68 cm (27 in). Closely related to the Palaearctic Honey Buzzard and just as variable in plumage, though the populations of tropical forests are crested and sedentary. In flight, the long tail and slender head and neck are distinctive when the bird is seen soaring over a patch of forest, and the whistling call is frequently uttered. Although taking honeycombs and bee larvae, prey such as small mammals, reptiles and birds are also recorded. Quite a sociable hawk, gathering in numbers at roosts and, especially as regards the northern populations, when on migration. **Range and habitat:** EP and OR; in open country, deciduous woods and evergreen forests, to 1,700 m (5,500 ft) in the hills. **Similar species:** Common Buzzard, Jerdon's Baza. **Not illustrated.**

Brahminy Kite *Haliastur indus* (Accipitridae) 48 cm (19 in). The immature bird is quite unlike the handsome adult, being mainly brown with a pale patch under the wing at the base of the primaries, so somewhat buzzard-like. Always in the vicinity of water, whether rivers, reservoirs or coastal harbours, where it wheels about, flying with great ease and buoyancy, swooping to snatch food from the surface, or pick up a dead fish from the shore. When quartering flooded paddyfields, it behaves and flies much as a harrier. It is a gregarious bird, and large numbers congregate in good feeding areas and at roosts. As a species, it undoubtedly benefits from human activities such as rice-growing and fisheries, and is a common sight round towns and villages near water. **Range and habitat:** OR and AU. Resident in suitably watered areas and coasts, and up to 1,800 m (5,900 ft) in the hills. **Similar species:** Whistling Kite (see p 290).

Indian Whistling Duck

Barheaded Goose

Ruddy Shelduck

Cotton Teal

Falcated Teal

Spotbill Duck

White-winged Wood Duck

Brahminy Kite

White-bellied Sea-eagle *Haliaeetus leucogaster* (Accipitridae) 70 cm (28 in). A familiar sight around the coasts of tropical Asia, this is an impressive bird in flight on its broad white wings with black borders. The tail is short, and this with the heavy build gives it a somewhat vulturine aspect. The immature bird is dark brown, and becomes progressively lighter with age. Like many eagles, it spends much time perched, often in a prominent position commanding a wide view, which it will return to after a successful foray with a fish or snake. The prey is caught in a powerful dive to the surface either off the perch or more usually in flight, like an Osprey. In display, the pair soar high in the air, and call frequently with their loud, clanging, goose-like notes. **Range and habitat:** OR and AU, along sea coasts and inland, but not far, along major waterways. **Similar species:** Pallas's Fish Eagle, White-tailed Eagle.

Indian White-backed Vulture *Gyps bengalensis* (Accipitridae) 90 cm (35 in). The blackish skin on the bare head and neck resembles that of the Indian Griffon, though this species lacks the white back patch. In flight the Indian White-backed Vulture can be recognized by the conspicuous white bar below the wing. Commonly associating with other vultures, this is a medium-sized species often seen perched hunched-up on treetops or scrambling for food at carcasses or at slaughter houses. In the morning it remains perched on trees until about three hours after dawn and takes wing rather reluctantly. Once up however, it soars majestically on thermals as it searches for food. **Range and habitat:** OR. India to Indo-China; in cultivated and well-wooded country. **Similar species:** other griffons (see pp 150, 192).

Crested Serpent Eagle *Spilornis cheela* (Accipitridae) 75 cm (29 in). The thick erectile crest, yellow eyes and bare yellow legs are constant features in all the races – over 20 – of this variable species. Plumage differences concern size and clarity of the spots and barring, overall coloration and body size. The wings are noticeably rounded and heavily barred, making the bird easy to recognize in flight, when it is apt to be quite noisy, with a variety of whistling calls. It is a woodland bird, and when sitting quietly in some large leafy tree is inconspicuous, though soaring and calling above the canopy draw attention to its presence. It drops on its prey, mainly reptiles, from a perch, rather than in flight like some close relatives of open country. **Range and habitat:** OR; resident in wooded areas in the plains and hills up to 3,000 m (9,850 ft). **Similar species:** Short-toed Eagle, Mountain Hawk-eagle (see p 150).

Pied Harrier *Circus melanoleucos* (Accipitridae) 49 cm (19 in). The male is one of the most strikingly marked of the harriers, with the black head, breast and back contrasting with pure white underparts, and silvery grey wings and tail. The female is a typical brownish hawk with streaked underparts and a barred tail. In low, though buoyant and airy, flight, it quarters the paddyfields on the look-out for frogs, lizards or small rodents, but does not apparently pose much of a threat to birds, though taking munias regularly in Borneo. It is a quiet bird in its winter quarters in tropical Asia, though on its breeding grounds makes a repeated alarm call and a chattering note. **Range and habitat:** EP, wintering across OR, resident in northern Burma and Assam. Open country, marshes, paddyfields, hill grasslands. **Similar species:** Pallid Harrier, Marsh Harrier (see p 152).

Grey Goshawk (Frog Hawk) *Accipiter soloensis* (Accipitridae) 30 cm (12 in). The plain breast of the male, washed with rufous, is unusual among the sparrowhawks, as is the habit of feeding largely on frogs, on which it often stoops from a perch. As a result of specializing on this prey, it is a bird of more open country than its relatives, both in breeding quarters and when wintering southwards through South-East Asia. The tail lacking barring above, and white underside of the flight feathers renders it easier to tell in the field than some small hawks. It migrates northwards in parties, but is more solitary and so less conspicuous on the return journey in the autumn. **Range and habitat:** EP, in winter to Indo-China, Borneo and the Greater Sundas. **Similar species:** Shikra. **Not illustrated.**

White-eyed Buzzard *Butastur teesa* (Accipitridae) 42 cm (17 in). One of a small group of 4 closely related hawks, 1 in Africa and the other 3 Asian, with rather pointed wings and a long, square-cut tail. Like the others, this species is a somewhat unspectacular bird, often walking about to pick up small prey off the ground, though also fond of using low perches, such as fence-posts, as a look-out. The sexes are alike, and immature birds are paler and streakier below, with pale edges to wing feathers. Not a gregarious bird, it is usually seen singly or in pairs. The call is a buzzard-like double note. **Range and habitat:** OR; Indian subcontinent to Burma. Resident in open country or sparse woodland in the plains to 1,200 m (3,900 ft) in the hills. **Similar species:** Rufous-winged and Grasshopper Buzzards.

Monkey-eating (Philippine) Eagle *Pithecophaga jefferyi* (Accipitridae) 95 cm (37 in). One of the most spectacular of the world's eagles, and unfortunately high on the list of the world's most threatened species. Like several other large, rare eagles, it is a forest bird, and takes mammal prey such as flying lemurs and monkeys. The feet and claws are very powerful, and the bill laterally compressed and deeply arched. This, together with the full crest and pale blue eyes, gives the bird an impressive appearance. It builds a huge nest high in a forest tree and lays a single egg, which takes about 2 months to hatch. **Range and habitat:** OR, Mindanao Island, Philippines. Resident in heavy forest up to 1,300 m (4,300 ft).

Indian Black Eagle *Ictinaetus malayensis* (Accipitridae) 75 cm (29 in). A somewhat aberrant eagle, adapted to a diet which largely features nestlings and eggs. The claws are notably long but only shallowly curved, which may well facilitate the snatching of nests, and the bird is capable of flying extremely slowly and buoyantly on its broad wings. It is particularly adept at following the contours of foliage and small clearings while making a close inspection of the surroundings for prey – which also includes small rodents and mammals on occasion. The sexes are alike in their black plumage, and the bright yellow feet are easily seen when the bird is soaring overhead. **Range and habitat:** OR. Resident in hill forests up to 2,700 m (8,900 ft). **Similar species:** Crested Hawk-eagle.

Crested Hawk-eagle *Spizaetus cirrhatus* (Accipitridae) 72 cm (28 in). Variable in plumage, this eagle has a somewhat hawk-like profile with its broad wings and long tail, which must aid manoeuvrability in its forest habitat. There is a dark plumage phase, but more commonly the bird is brown above and white, heavily streaked, below. The Indian race has a long, floppy crest. Like many related species, it can become a determined killer of poultry, but usual prey is small mammals and birds. It is not a conspicuous species, though calls noisily during the soaring display flight above the forest. The nest is placed high in a forest tree, and a single egg is laid. **Range and habitat:** OR. Resident in open woodland and forest in the plains and foothills. **Similar species:** Mountain Hawk-eagle, Indian Black Eagle. **Not illustrated.**

Laggar Falcon *Falco jugger* (Falconidae) 45 cm (18 in). Somewhat smaller than either Lanner or Saker Falcons, the 2 common, large, resident falcons of India. Distinct from the local race of the Peregrine, the Shaheen Falcon, by having white rather than cinnamon underparts. Though an accomplished flier, it is not as dashing as the Peregrine, and on occasion picks up dead birds, though quite capable of catching small mammals and birds with ease. Often seen in pairs which stay together the year round, and not infrequently in towns, where it perches conspicuously on buildings. **Range and habitat:** OR; Indian subcontinent to Burma. Resident in open or rocky country. **Similar species:** Lanner and Saker Falcons, Peregrine (see p 152).

White-bellied Sea-eagle

Indian White-backed Vulture

Crested Serpent Eagle

Pied Harrier

White-eyed Buzzard

Monkey-eating (Philippine) Eagle

Indian Black Eagle

Laggar Falcon

FOWL-LIKE BIRDS Galliformes

Painted Partridge *Francolinus pictus* (Phasianidae) 30 cm (12 in). The yellowish legs and chestnut face (though the female has a whiter throat) are quick pointers to identity, but otherwise the black back and underparts with white spotting are very like that of the Black Francolin. More active at dawn and dusk than in the heat of the day, it is a difficult bird to watch, though a pair may sometimes be seen foraging in cultivated fields or in thin bush. It feeds on seeds, shoots and insects, and nests on the ground in a patch of concealing vegetation. It has a typical partridge-type call in the breeding season, when the male ascends some small mound or other vantage point to proclaim his whereabouts. **Range and habitat:** OR; peninsular India and Sri Lanka. Resident in the plains in thin scrub, grasslands and also cultivation. **Similar species:** Black Francolin, Grey Francolin. **Not illustrated.**

Chinese Francolin *Francolinus pintadeanus* (Phasianidae) 33 cm (13 in). The male is best identified by the face pattern, with a black stripe dividing the white cheeks and throat. The black and white spotting on back and underparts recalls several related species, and the female is duller and more buff. It is the only species of francolin in China. A shy and rarely seen bird in grassy or lightly bushed country, though, like its relatives, noisy in the breeding season. It is not gregarious, and does not form coveys, though quite numerous in some localities. It runs with head lowered and tail cocked up when alarmed. **Range and habitat:** OR and south EP. Resident in dry, rocky or grassy hills, scrub jungle and cultivation. **Similar species:** Black Francolin. **Not illustrated.**

Indian Grey Francolin *Francolinus pondicerianus* (Phasianidae) 31 cm (12 in). More reminiscent of the European Grey Partridge, with its rufous tail feathers, than the other Asian francolins with black in the tail. However, like the francolins, it has a spur above the hind toe. The tail colour is noticeable when the bird is put up, though often a covey will escape danger by splitting up and running in different directions. Tolerant of most types of habitat except forest, it is often in coveys which break up with the onset of the breeding season. Commonly roosts in trees, and will also take refuge in a tree or bush when disturbed. **Range and habitat:** OR; India and Sri Lanka. Resident in scrub jungle and cultivation in the plains. **Similar species:** Chinese Francolin.

Jungle Bush Quail *Perdicula asiatica* (Phasianidae) 17 cm (6½ in). The bush quails are a group of 4 species confined to the Indian subcontinent, differing from the true quails in having a heavier and deeper bill, and a spurred tarsus. In this and 1 other species, the male is closely barred on the underparts whereas the female is plain rufous. Keeps in small coveys which, when put up, rise with a startling fluster and disperse at once in all directions. The species is noted for only making very short flights when disturbed, and is in general very sedentary. It has developed several well-marked subspecies. The call in the breeding season is a somewhat partridge-like repetitive grating note. **Range and habitat:** OR; peninsular India and Sri Lanka. Resident in grasslands, cultivation and thin forest. **Similar species:** Rock Bush Quail.

Common Hill Partridge *Arborophila torqueola* (Phasianidae) 28 cm (11 in). Typical of this group of 14 species of the Himalayas and South-East Asia, though this and 1 other are only 2 species whose ranges extend to the western Himalayas. As with many Himalayan species, the race of the western, drier region is paler in coloration than those occurring further east. Keeping in family parties, it forages in dense undergrowth in ravines and hill gullies, often in deciduous forest. Roosts in trees, into which it will often fly when disturbed, and has a long, low whistle repeated at intervals. **Range and habitat:** OR; western Himalayas to

Vietnam. Resident in broad-leaf or evergreen hill forest. **Similar species:** Rufous-throated Hill Partridge.

Crested Wood Partridge *Rollulus rouloul* (Phasianidae) 25 cm (10 in). The remarkable fan-like crest and dark green plumage are unique among gamebirds. Both male and female have a hairy crest growing from the crown, though the female lacks the bushy fan, is much lighter green in colour than her glossy mate, and has a grey head. Like many of its relatives, it is often to be seen in coveys or family parties, which scratch about on the ground, often in clearings, in lowland forest. It appears to prefer fairly dry forest with plenty of bamboo or palm. The diet includes fallen fruits, seeds and insects, and the note is a repeated, low whistle. **Range and habitat:** OR. Resident in heavy forest in Malaysia, Sumatra and Borneo.

Red Spurfowl *Galloperdix spadicea* (Phasianidae) 37 cm (15 in). The spurfowl are a group of 3 species confined to India and Sri Lanka, somewhat resembling a domestic fowl in shape, but with a rather long, rounded and not arched tail, and 2 or 3 spurs on the leg in the male. Not a gregarious bird, it is usually seen singly or in small family parties, though it is in any case difficult to see, and a swift runner when disturbed. The dense cover which the birds frequent does not permit long flights, and sometimes they will take to the trees for refuge and for roosting purposes. The note is a hen-like cackling. **Range and habitat:** OR. Resident in peninsular India in thick scrub and near cultivation in the lowlands and lower hills. **Similar species:** Painted and Sri Lankan Spurfowl.

Blood Pheasant *Ithaginis cruentus* (Phasianidae) 46 cm (18 in). A short-tailed, bushy-crested pheasant of high altitudes in the Himalayas. The female, in her brown plumage, is much dowdier than the male with his scarlet throat and eye-ring, and scarlet-streaked, lime-green breast. There are no fewer than 14 subspecies of this sedentary bird, differing mainly in the extent of red on the face and breast and in overall size. Usually in parties except in the breeding season, this pheasant lives at or near the snow-line, and is a notoriously tame bird. Roosts and finds shelter in the fir woods or bamboo, but feeds in the open, digging into the snow for seeds and leaves. **Range and habitat:** EP and OR, in southern Tibet, the Himalayas and north-west China, Resident in alpine scrub or forest above 3,500 m (11,500 ft).

Satyr Tragopan *Tragopan satyra* (Phasianidae) 68 cm (27 in). The 5 species of tragopan are bulky pheasants with a rounded but not elongated tail, and the males are remarkably ornamented for display with erectile fleshy horns and brilliantly patterned and inflatable throat lappet. The Satyr Tragopan is typical, being a sedentary species, not gregarious, and very shy and elusive. Entirely at home in trees, though also ground feeding, it is a mountain bird of steep, forested slopes, and, unlike most pheasants, is a tree-nester. It is a popular and confiding bird in captivity. **Range and habitat:** OR, central Himalayas, in hill forests up to 4,250 m (13,900 ft).

Sclater's Monal Pheasant *Lophophorus sclateri* (Phasianidae) 72 cm (28 in). The 3 species of monal are heavy, brilliantly metallic pheasants, with an unusual tail for a pheasant, being flat and square. The lengthened upper mandible is effective in digging up roots and bulbs. Living at up to the tree limit in fir or rhododendron scrub, it is often solitary or in very small groups. Being a very sedentary bird, it is unfortunately all too easily snared, though little is known about its biology in the wild. It has a loud, whistling call, often uttered by a sentinel bird from its viewpoint on a rocky hillside. **Range and habitat:** OR; northern Burma and adjacent parts of Tibet and China. Resident in forest up to 4,000 m (13,100 ft). **Not illustrated.**

Indian Grey
Francolin

Jungle Bush
Quail

Common Hill
Partridge

Red Spurfowl

Satyr Tragopan

Crested Wood
Partridge

Blood Pheasant

Red Junglefowl *Gallus gallus* (Phasianidae) 65 cm (26 in). The unmistakable ancestor of domestic fowl, it has 2 close relatives in India and Sri Lanka and 1 in Java. Unlike the domestic variety, the wild Junglefowl is noteworthy for having an eclipse plumage into which it moults after breeding, losing the neck hackles and long tail feathers. Gregarious birds, they have a similar crow to that of the barnyard cockerel, though shorter, and are difficult to observe, being very alert and shy. The usual view is of the backs of 3 or 4 birds as they jump up at the side of a track. **Range and habitat:** OR; India throughout South-East Asia to the Philippines. Resident in light forest and scrub. **Similar species:** Grey Junglefowl, Green Junglefowl.

Silver Pheasant *Lophura nycthemera* (Phasianidae) 67 cm (26 in). A sedentary species dwelling in thick evergreen forest, it has developed no fewer than 14 subspecies, which differ in the intensity and pattern of the markings and also in size. The 'true' Silver Pheasant, the race originally described from China, is the largest and also the whitest above, with very long, almost unmarked central tail feathers. It lives in small family parties in hill forests, especially bamboo, and though not shy and quite a noisy bird, is not easy to see. The striking appearance of the male has for centuries made them desirable birds in captivity. **Range and habitat:** southern EP and OR in south and west China and continental South-East Asia, in hill forests. **Similar species:** Kalij Pheasant.

Great Argus Pheasant *Argusianus argus* (Phasianidae) Male up to 190 cm (75 in), female 63 cm (25 in). The striking feature of this species is the great development and beautiful patterning of the secondaries and the immensely long central tail feathers. These are shown to advantage in display in which the male faces the female with wings spread and arched, hiding the head, and the tail standing vertically behind. For this purpose the cock maintains a scrupulously cleared patch on the forest floor. He spends much time here, advertising his presence vocally, ready to display to any female ready to breed. The call is a commonly heard, far-carrying and repeated bark. **Range and habitat:** OR; in Malaysia, Borneo and Sumatra. Resident in lowland forest. **Similar species:** Malay Crested and Rheinart's Crested Arguses.

Common Peafowl *Pavo cristatus* (Phasianidae) Male is up to 200 cm (79 in), female 85 cm (33 in). The decorative splendour of this bird has made it a favourite in captivity and a familiar sight for more than 30 centuries, though presumably for all that time people have winced at the harsh double call, especially irritating at night. Persecution and hunting renders it a shy bird in the wild, though it is tame around villages. It is surprising how easily a cock with full train can fly up out of tall grass to perch in an isolated tree at the approach of danger. Always extremely alert, peafowl are particularly wary of the large cats which share their jungle home. **Range and habitat:** OR; in India and Sri Lanka. Resident in forests in the plains and foothills. **Similar species:** Green Peafowl.

CRANES, RAILS AND ALLIES Gruiformes

Bustard Quail *Turnix suscitator* (Turnicidae) 15 cm (6 in). This species is representative of the family of 14 species, which differ from true quails by lacking the hind toe and the crop, and also in that females are larger and more brightly coloured, and leave incubation and care of the young to the male. The hen makes a loud drumming or booming call to attract a mate. Usually seen singly, it is an unobtrusive bird of grassy undergrowth, difficult to watch or even put up, though when it flies it looks like a typical quail, small rounded wings beating rapidly until it abruptly lands in cover and vanishes. **Range and habitat:** OR; throughout India and Sri Lanka. Resident in grassy places, cultivation or thin forest. **Similar species:** Little Button-quail. **Not illustrated.**

Japanese White-naped Crane *Grus vipio* (Gruidae) 127 cm (50 in). Of the 8 species of crane recorded from the OR, all but 1 are winter visitors. The White-naped Crane migrates as far south as Taiwan, on the north-eastern limits of the region, though the main wintering areas are on the lower Yangtse in eastern China, in Korea and Japan. It is one of the most beautiful of the family with its red face and legs, and dark stripe down the neck. Within the winter flocks, family groups tend to preserve their identity, and males attack other males which come too close to their own group. By the time the birds are back on their breeding grounds in Manchuria they are already paired up. **Range and habitat:** EP; wintering OR south to Taiwan, on marshes, crops and by rivers. **Similar species:** Common and Sarus Cranes. **Not illustrated.**

Sarus Crane *Grus antigone* (Gruidae) 150 cm (59 in). A huge, pale grey crane, almost always to be seen in pairs, and very familiar to, and protected by, the country people generally in India. It is unusual among the cranes in being sedentary, and in its tameness, presumably a result of not being molested. Like all its relatives, though, it has a dancing display with much wing-flapping, bowing and noisy trumpeting. The pair also call together, with necks stretched up, when greeting or as a warning. The nest is a big heap of reeds and rushes built on swampy ground in a marsh or paddyfield. **Range and habitat:** OR; India to South-East Asia. Resident in well-watered plains country. **Similar species:** Common Crane (see p 154).

Great Indian Bustard *Choriotis nigriceps* (Otididae) 125 cm (49 in). One of the rarest of the world's large bustards, and gravely threatened by hunting and habitat destruction. Very shy and wary and unapproachable on foot, it can however be easily shot from a vehicle. The male is a massive bird weighing up to 18 kg (40 lbs), though able to run with facility, and a powerful flier. Moves about in small groups which wander in search of suitable grasslands according to weather conditions, rendering demarcated reserves a problematical solution to the hunting threat. The clutch is a single egg, and the breeding season extended. **Range and habitat:** OR; north, west and central India. Rare resident and nomad in sparse grasslands and light scrub. **Similar species:** Kori Bustard, Arabian Bustard (see p 198).

Bengal Florican *Houbaropsis bengalensis* (Otididae) 66 cm (26 in). This species and the Lesser Florican are very reminiscent in appearance of several African bustards, and are among the most threatened species in the family. Highly valued for its eating qualities and hence still a target for the pot, the male is a conspicuous object during display when he flutters high into the air above the tall grass, making a humming noise while beating the air. Unusually in the family, he is slightly smaller than his mate. Usually found solitarily, these bustards forage in open, grassy areas at dusk and in the early morning, and keep to thicker cover during the day. **Range and habitat:** OR; Himalayan foothills and north-east India, possibly still in Kampuchea. Resident in areas of tall grassland and scattered bushes. **Similar species:** Lesser Florican.

Lesser Florican *Sypheotides indica* (Otididae) 46 cm (18 in). Smaller and more slender than the Bengal Florican, it has particularly long legs, and the male is unusual in having a form of eclipse plumage following the breeding season. The white inner wing and shoulder contrast in flight with the black neck and underparts, particularly in the jumping display flight, which is repeated with extraordinary regularity and persistence. When disturbed, prefers running to flying, when its wingbeats are rapid for a bustard, but is generally less shy than its relatives. Successful breeding very dependent on the rains allowing good grass growth. **Range and habitat:** OR; Pakistan and India. Resident and nomadic in grassy areas. **Similar species:** Bengal Florican. **Not illustrated.**

Silver Pheasant

Great Argus Pheasant

Red Junglefowl

Common Peafowl (pair)

Bengal Florican

Sarus Crane (pair)

Great Indian Bustard

Blue-breasted Banded Rail *Rallus striatus* (Rallidae) 28 cm (11 in). The chestnut crown and nape contrasting with the greyish breast, and especially the finely banded upperparts, distinguish this species from related rails in its range. Like many other sedentary birds, it has developed a number of subspecies which differ both in size and in intensity of coloration. It is a difficult subject to study in its wetland habitat, taking cover at a quick run at any sign of danger, and is rarely seen in the open. In the highlands of Borneo it has been caught in numbers in rodent snares. It is generally solitary, and makes a sharp double call-note. **Range and habitat:** OR. Resident in swamps, mangroves or flooded fields from India to the Philippines. **Similar species:** Water Rail, Slaty-legged Crake (see p 154). **Not illustrated.**

Ruddy (Ruddy-breasted) Crake *Porzana fusca* (Rallidae) 23 cm (9 in). A small, rich brown crake with noticeably red legs. Like many crakes, it has been able, despite apparently weak flight, to colonize quite remote islands, in this case the Andamans and Christmas Island. In the Andamans it has been found nesting in dense, wet forest, though in other parts of its range, a more usual site would be in waterside vegetation. It is particularly secretive and also rather quiet, and so a very unobtrusive species. It is apparently more migratory than many of its relatives, being a summer visitor to much of eastern China, and a winter visitor (as well as being resident) in Sri Lanka. **Range and habitat:** EP, OR; resident and migratory to some extent. Inhabits marshes and swamps. **Similar species:** Red-legged Crake.

White-breasted Waterhen *Amaurornis phoenicurus* (Rallidae) 30 cm (12 in). Not only one of the commonest of waterside birds in the OR, it also happens to be, with the Moorhen, one of the species in the family most adapted to feeding in the open. Although not particularly shy, it is nevertheless very wary, and runs rapidly into cover if disturbed. It is a noisy bird with a sharp double note which is often repeated for minutes on end. Like many of the family, it is pugnacious and quarrelsome. Breeding is during the rains in the monsoon areas, the nest on or by water, sometimes 1 m (39 in) or more above the ground in a bush or tree. **Range and habitat:** OR; India and Sri Lanka to the Philippines. Resident and local migrant, by ponds, marshes or wet areas in the plains and hills up to 2,400 m (7,900 ft). **Similar species:** Moorhen (see p 154).

Watercock *Gallicrex cinerea* (Rallidae) 42 cm (17 in). Owes its name to the pugnacious nature of the males, who battle furiously in the breeding season, and develop at this period a fleshy red horn above the frontal shield. Resembles an outsize Moorhen with red legs, but lacks the white undertail-coverts. Actions similar too, as it prospects with jerking tail round the margin of a pond, running into cover when alarmed. It is more active at dawn and dusk, and does not feed in the open, preferring to keep among lush waterside vegetation. The call, uttered for long periods at a stretch, is a rhythmic series of booming notes. **Range and habitat:** OR, and in summer to EP. Resident and migrant in well-watered lowlands.

SHOREBIRDS, GULLS, AUKS Charadriiformes

Pheasant-tailed Jacana *Hydrophasianus chirurgus* (Jacanidae) 30 cm (12 in). A striking and elegant adornment to any lotus-covered pond, over which the enormously elongated toes enable it to walk with ease. In the breeding season the male grows the 15 cm (6 in) arching tail, to which the bird owes its name. Often surprisingly inconspicuous when feeding among the tangle of waterplants, it is easily seen in flight when the white wings are striking and the long toes dangle awkwardly. Often congregates in quite large flocks in winter. The female is larger than the male, who incubates the eggs and rears the chicks. **Range and habitat:**

OR, in summer to EP. Largely resident on lakes, marshes and ponds. **Similar species:** Bronze-winged Jacana.

Bronze-winged Jacana *Metopidius indicus* (Jacanidae) 28 cm (11 in). Unlike the Pheasant-tailed Jacana, this bird has no special breeding plumage, and the tail is short and rounded. It is even more addicted to ponds where the emergent vegetation is really luxurious and there are bordering reeds and sedges. It is a rather shy bird, and can often be difficult to see among the curling lily leaves, and is an unwilling flier; it seems to be more adept at swimming and diving than its longer tailed relative, sinking the body to avoid being seen. The male incubates the eggs and rears the chicks. **Range and habitat:** OR. Resident on lakes and ponds in the lowlands. **Similar species:** African Jacana (see p 198).

Ibisbill *Ibidorhyncha struthersii* (Ibidorhynchidae) 40 cm (16 in). A strange shorebird with no obviously close relations, notable for its red, down-curved bill, rounded wings, and red legs and feet which lack a hind toe. The shape of the bill is well adapted for probing between the rounded boulders along the Himalayan rivers it frequents, and the bird also has a habit of wading in the water and tilting the whole head and neck underneath in the manner of a dipper. It has rather slow, lapwing-like wingbeats, and utters a clear, double whistle in flight, when the white patch on the wing is noticeable. Nowhere particularly common, usually seen in pairs or small parties. Nests in early spring before the snow has melted. **Range and habitat:** EP and OR. Resident by rocky rivers in Tibet and the Himalayas.

Great Stone Plover *Esacus recurvirostris* (Burhinidae) 50 cm (20 in). The heavy bill of this outsize stone curlew enables the bird to cope with the crabs and shellfish on which it feeds, and this feature is even further developed in the closely related species in Australia. The staring yellow eyes are a clue to the bird's nocturnal habits; the daytime is spent roosting on a river sand-bar or saltpan. Despite being nocturnal, it is accustomed to brilliantly sunny, hot conditions in the day, and nests on the bare ground completely in the open. Not a gregarious bird, though often in pairs. It has a loud, mournful cry like others of its family. **Range and habitat:** OR. Resident by rivers, saltpans and on the coast. **Similar species:** Stone-curlew (see p 160).

Indian Courser *Cursorius coromandelicus* (Glareolidae) 26 cm (10 in). Until recently, thought to be the only breeding courser in India, the family being more typical of dry grasslands in Africa. Usually in pairs or small groups, it is a bird well adapted for life on sandy, barren areas, being very fast on its legs. The feet lack a hind toe, in common with many other ground-dwelling species. It is also, however, strong and quick on the wing when forced to fly. Takes food from the ground by tilting the body down in a mechanical way, then runs a metre or two before stopping stock still, when it is often difficult to see. It nests in the hot weather before the ground is saturated by the monsoon. **Range and habitat:** OR; India and Sri Lanka. Resident in open country in the plains. **Similar species:** Cream-coloured Courser.

Ruddy (Ruddy-breasted) Crake

White-breasted Waterhen

Watercock

Pheasant-tailed Jacana

Bronze-winged Jacana

Ibisbill

Great Stone Plover

Indian Courser

Little Pratincole *Glareola lactea* (Glareolidae) 17 cm
(6½ in). This dainty little pratincole is a common sight as it
wheels about over the big rivers of southern Asia, congregating to rest and roost on the sandbanks. It is constantly
active in the air, flying perhaps more erratically than its
larger relatives, perhaps because of taking smaller, more
agile flying insects. The sexes are alike, the black underwing
and shallowly forked tail giving a distinctive appearance.
Nests colonially on bare ground or sand along river-banks,
dispersing after the breeding season to suitable stretches of
water or coastal areas. It has a low, harsh note. **Range and
habitat:** OR; the Indus river and Sri Lanka to South-East
Asia. Resident and nomadic by marshes and rivers in the
lowlands. **Similar species:** Oriental Pratincole.

Yellow-wattled Lapwing *Vanellus malabaricus* (Charadriidae) 25 cm (10 in). The plumage and yellow facial wattle
recall a very similar African species, though in this species
the wattle lies along the throat as opposed to being pendant.
It is quieter and less frantic than its larger red-wattled
relative in the OR, and prefers a much drier type of habitat,
especially ploughed fields and arid or stony ground. Often to
be seen foraging in small groups, which rise with a double
piping call, showing the prominent white wing-bar, though
when standing motionless, often difficult to detect against a
greyish brown dried-up field. **Range and habitat:** OR; India
and Sri Lanka. Resident in open country in the lowlands.
Similar species: Red-wattled Lapwing.

Red-wattled Lapwing *Vanellus indicus* (Charadriidae)
33 cm (13 in). A noisy, demonstrating and vigilant bird,
particularly relentless when mobbing predators near a nest
or chicks. The heavy, floppy flight is transformed when the
bird corkscrews down with great agility to discomfort its
target, swooping up immediately to a safe height. Usually in
pairs or small groups, it feeds in typical plover fashion,
tilting down to take an insect, then running on a little way.
The sexes are alike, and there is no seasonal change in the
plumage as with the Lapwing. There are several subspecies,
differing in size, coloration and extent of white on the neck.
Range and habitat: WP and OR; South-West Asia to Sri
Lanka and South-East Asia. Resident near water, marshes or
cultivation in the plains and lower hills. **Similar species:**
Yellow-wattled Lapwing.

Grey-headed Lapwing *Vanellus cinereus* (Charadriidae)
35 cm (14 in). Strikingly patterned in flight, with the rather
pointed, black-tipped wings having a broad white band
along the trailing edge. The long, yellow legs project well
beyond the tail tip. The muted plumage colouring renders
the bird often difficult to see when standing quietly in the
grass or on a paddyfield. Very noisy and demonstrative on
the breeding grounds, mobbing intruders with a sharp,
short note. Sociable in winter, when small parties gather by
rivers or in the fields. The food is largely aquatic insects.
Range and habitat: EP, in winter across OR, on marshes,
grasslands and cultivation, river margins. **Similar species:**
Yellow-wattled Lapwing.

Swinhoe's Snipe *Gallinago megala* (Scolopacidae) 27 cm
(10½ in). Identification of Asiatic snipes in the field has been
hindered by lack of critical data until the last few years, and
museum criteria such as this species having 20 tail feathers
compared with 28 in the Pintail Snipe are of little use to the
modern birder who rarely has a bird in the hand. Swinhoe's
may be commoner in the region in winter than previously
thought, though already known to be common in the
Philippines. Larger and with a longer bill than Pintail Snipe,
it is also a duller and paler bird, and not as heavily striped on
the back as Common. **Range and habitat:** EP, in winter to OR
in marshes or wet paddyfields. **Similar species:** Pintail
Snipe, Common Snipe (see p 160). **Not illustrated.**

Long-toed Stint *Calidris subminuta* (Scolopacidae) 15 cm
(6 in). A bird of wet paddyfields and swampy or marshy
ground rather than the open shore, and rather prone to
stretch up while standing, maybe to get a clearer view over
surrounding vegetation. A dark, well-streaked stint, it does
in fact have a middle toe about 15% longer than that of the
Little Stint. In its winter quarters in South-East Asia, the
birds tend to feed in loose groups through the marsh
vegetation, and it is often surprising how many can be
flushed from a comparatively small wet area. It rises with a
dry trilling call. **Range and habitat:** EP, in winter to OR; in
swampy places and wet paddyfields. **Similar species:** Red-
necked Stint, Little Stint (see p 158).

Spoon-billed Sandpiper *Eurynorhynchus pygmeus* (Scolopacidae) 17 cm (6½ in). In winter, when this species consorts
with other small sandpipers on the mudflats, it is by no
means easy to pick it out by the curious bill shape, although
this can be obvious enough if the bird is close and head-on to
the observer. Very active and sprightly, it dabbles the bill
from side to side as it moves through shallow water,
changing direction without lifting the head up. It also takes
insects off the ground, these forming the main diet in the
breeding season. The greyish upperparts of winter become
more rufous in summer, as do the face and breast, much as in
the Red-necked Stint. **Range and habitat:** EP, in winter to
OR, on shores and mudflats. **Similar species:** Little Stint,
Red-necked Stint (see p 158).

Brown-headed Gull *Larus brunnicephalus* (Laridae)
46 cm (18 in). Somewhat larger than the similar Black-headed
Gull, and with a white spot in the black wing tip, it is
common in winter around the coasts of southern Asia. At
this season, the brown hood shrinks to a dark patch on the
side of the whitish head. Often in large flocks which
scavenge around harbours, lakes or estuaries in typical gull
fashion. Also takes insect food from the surface while
swimming. It is not a particularly noisy bird, but has various
strident notes, and is especially vocal at breeding sites. The
brown hood of summer is showing on most birds in late
spring before they migrate north. **Range and habitat:**
central PA, in winter to OR; on rivers, lakes and sea-coasts.
Similar species: Black-headed Gull (see p 162).

Indian River Tern *Sterna aurantia* (Laridae) 40 cm (16 in).
It is interesting that 2 species of tern, 1 of them this species,
are almost exclusively inland birds of rivers in southern
Asia, and 2, also with yellow bills, in South America, but
that no counterpart exists in Africa. The River Tern is among
the largest of the 'black-capped' terns and has a heavy
yellow bill which becomes duller, with a dark tip, in winter.
At this time, the crown is heavily streaked with white. It
nests fairly early in the year before the monsoon rains swell
the river levels and cover the sandbanks which are a usual
nest-site. **Range and habitat:** OR, but not in Sri Lanka or the
Indo-Malayan subregion. Resident and locally migratory, on
rivers, reservoirs and sometimes sea coasts. **Similar species:**
Black-bellied Tern.

Black-bellied Tern *Sterna melanogaster* (Laridae) 33 cm
(13 in). A long-tailed, more slender tern with a lighter bill
than the River Tern, and with a conspicuous blackish belly
in the summer plumage. Traces of the black underparts are
often retained in winter. It is even more restricted to fresh
water, living by large rivers or open stretches of water in the
lowlands. A sociable bird, it is usually in flocks which rest
on the river-bank or on sandy islands when not beating the
water in their search for fish. This tern also hawks for insects
on occasion, taking them off the surface or in aerial chase.
Range and habitat: OR, from Pakistan to South-East Asia.
Not in Sri Lanka or the Indo-Malayan subregion. Resident
along rivers in the plains. **Similar species:** Indian River
Tern. **Not illustrated.**

Little Pratincole

**Yellow-wattled
Lapwing**

**Red-wattled
Lapwing**

**Grey-headed
Lapwing**

Long-toed Stint

**Spoon-billed
Sandpiper**

**Brown-headed
Gull**

Indian River Tern

PIGEONS AND SANDGROUSE Columbiformes

Yellow-legged Green Pigeon *Treron phoenicoptera* (Columbidae) 33 cm (13 in). Often concealed among dense foliage, it is often surprising how many birds can be beaten out of a tree, all bursting out with loud wing-clapping. Forages quietly but voraciously on ripe fruit, especially figs. The feet of green pigeons have swollen, padded soles as an adaptation for grasping twigs, and the bird is very nimble and agile when feeding, despite being quite bulky. Does not feed on the ground. Several races are distinguished by size and amount of yellow in plumage. **Range and habitat:** OR; India and Sri Lanka to Indo-China. Resident in forest, groves and gardens. **Similar species:** 23 species of green pigeon are broadly similar.

Rufous (Eastern) Turtle Dove *Streptopelia orientalis* (Columbidae) 33 cm (13 in). Very similar to Turtle Dove, and its eastern counterpart, but larger and heavier and without white tail tips. Rufous, well-patterned upperparts distinguish it from other Asian doves and pigeons. Breeding pairs gather together in parties to migrate, and in winter resident birds of Indian peninsula share habitat with more northern breeders. Often seen along paths or forest tracks where it picks up grain or seeds, flying up with expanded tail. Several races over total range differ largely in richness of rufous and colour of tail tip. **Range and habitat:** EP, OR, in winter to south of range, in forest, light woodland or cultivation. **Similar species:** Turtle Dove (see p 164).

Little Cuckoo-dove *Macropygia ruficeps* (Columbidae) 28 cm (11 in). The long, graduated tail gives this rufous little dove its name. The sexes are alike, but the female lacks the metallic reflections. Like other doves, it is a ground feeder, coming into clearings and small patches of cultivation in the forest to forage in parties, which coo as they move about. It also takes berries, searching forest slopes for ripe fruit. It is a rapid flier when alarmed, showing the chestnut underwing. **Range and habitat:** OR, in Indo-Malayan and Sino-Himalayan subregions. Resident in hill evergreen forest. **Similar species:** Barred Cuckoo-dove.

Spotted (-necked) Dove *Streptopelia chinensis* (Columbidae) 28 cm (11 in). Very familiar throughout its range, often around villages and towns, it walks about fearlessly on tracks and paths, constantly searching for grain and seeds. If sufficiently alarmed, it jumps into the air almost vertically, with the widely fanned tail showing the white tips. Often in pairs, though will congregate in large numbers at suitable sources of food, such as reaped paddyfields. The pleasant cooing and gentle disposition make it a favourite cagebird everywhere. Plumage variations are responsible for the separation into 8 subspecies across the wide range. **Range and habitat:** OR; resident and common around villages, gardens and cultivation in the plains. **Similar species:** Rufous (Eastern) Turtle Dove.

Emerald Dove *Chalcophaps indica* (Columbidae) 26 cm (10 in). A fast, low-flying dove, often seen hurtling across forest paths or clearings, for although quite familiar, it is a shy bird. Looks rather dark in flight, and shows the 2 white rump bars and short tail. The female is duller and browner than the male. Very much a ground bird, it is rarely far from cover, and the deep cooing may often be heard from amongst dense foliage. It is not a sociable species, usually being seen singly or in pairs. **Range and habitat:** OR, AU. Resident in mixed or evergreen forest in the hills and plains. **Similar species:** Green Imperial Pigeon.

PARROTS AND RELATIVES Psittaciformes

Vernal Hanging Parrot (Indian Lorikeet) *Loriculus vernalis* (Psittacidae) 15 cm (6 in). Tiny, and difficult to see in thick foliage when climbing about searching for fruit or nectar. Flies from 1 tree to another with an undulating flight on whirring wings which are rather short and triangular. Often in small parties which keep in touch with a continuous

distinctive squeaking which often draws attention to their presence. Remarkable for sleeping while hanging upside-down from a twig. Typically adept at clinging and clambering at all angles while feeding. **Range and habitat:** OR. Resident in forested areas. **Similar species:** Sri Lanka and Malay Lorikeets. **Not illustrated.**

Alexandrine Parakeet *Psittacula eupatria* (Psittacidae) 52 cm (20 in). A popular cagebird whose numbers may have been reduced in the wild by nestlings being taken for pets. A heavy-headed, stream-lined bird in flight, it is destructive to crops and fruit, and keeps in small parties which congregate at roosts. It makes a loud scream when on the wing, but is not much of a talker in captivity. The nest-hole, in a tree, is carved out with the powerful bill, or an old hole is suitably enlarged. The birds pair for life, and are very affectionate. **Range and habitat:** OR; India and Sri Lanka to Indo-China. Resident in wooded country in the plains. **Similar species:** Blossom-headed Parakeet.

Blossom-headed (Plum-headed) Parakeet *Psittacula cyanocephala* (Psittacidae) 36 cm (14 in). The female lacks the beautiful rosy colour of the male's head, and the black collar. A small, slim species, though with a long, narrow tail. The rapid wingbeats enhance the speedy impression in flight, and the birds are very adept at swerving and dodging in thick woodland. Like other parakeets, it is noisy in flight, with a scream which is less harsh than that of some relatives. The nest is a hole, often high in a dead tree, several pairs often choosing adjacent trees and forming a small colony. **Range and habitat:** OR; India and Sri Lanka. Resident in forests and cultivation in the plains and lower hills. **Similar species:** Alexandrine Parakeet.

CUCKOOS AND ALLIES Cuculiformes

Common Hawk Cuckoo *Cuculus varius* (Cuculidae) 33 cm (13 in). Like many cuckoos, almost totally unobtrusive and silent in the colder months. It makes up for this by the endless and irritating repetition of its call in the hot weather and the rains – a 3-syllabled, frenzied and accelerating shout. It is parasitic on various babbler species and lays an egg of a beautiful turquoise colour. Resembles a sparrow-hawk, even to the immature being streaked rather than barred on the underparts. The food is largely caterpillars. **Range and habitat:** OR; India and Sri Lanka. Resident and nomadic in wooded areas and cultivation or gardens. **Similar species:** Large Hawk Cuckoo.

Oriental Cuckoo *Cuculus saturatus* (Cuculidae) 31 cm (12 in). Tends to parasitize even smaller host species than does the European Cuckoo, though both are much the same size, commonly laying in the nests of tiny warblers. It is very similar in plumage, though the barring on the underparts is broader and more widely spaced. The 4 notes of the call, however, all at the same pitch, render it easily identifiable. Movements outside the breeding season are difficult to study owing to its unobtrusive habits at that time, though some populations are certainly migratory. **Range and habitat:** EP, OR in central Asia, the Himalayas to southern China and the Greater Sundas. Resident or migratory in wooded hill country. **Similar species:** European Cuckoo (see p 166). **Not illustrated.**

Plaintive Cuckoo *Cacomantis merulinus* (Cuculidae) 23 cm (9 in). An attractively coloured, small cuckoo with its tawny underparts, and as with many other cuckoos, a reddish brown, barred phase is common among females. The name refers to the quality of a commonly heard call, though the song is a multi-note whistle. It has a rapid and agile flight on pointed wings, and is often quite easy to see as it haunts open woodland and gardens, and tends to call from treetops rather than in deep foliage. It parasitizes small warblers. **Range and habitat:** OR; from south China to Burma, Java and the Philippines. Resident or local migrant, in open wooded country. **Similar species:** Banded Bay Cuckoo, Lesser Cuckoo.

■□□ **Yellow-legged Green Pigeon**
□■□ **Rufous (Eastern) Turtle Dove**

■□□ **Little Cuckoo-dove**
□■□ **Spotted (-necked) Dove**

■□□ **Emerald Dove**
□■□ **Alexandrine Parakeet**
□□■ **Blossom-headed (Plum-headed) Parakeet**

■□□ **Common Hawk Cuckoo**
□■□ **Plaintive Cuckoo**

Small Green-billed Malcoha *Rhopodytes viridirostris* (Cuculidae) 40 cm (16 in). The dark grey plumage and skulking habits render this an inconspicuous bird in the tangled, dense foliage it frequents. It clambers, climbs or hops about rather like a squirrel, the long, white-tipped tail sticking out behind. It is a poor flier, soon diving into cover. The bright green bill and blue eye patch soon identify it when the head is pushed briefly through the leaves. Unlike many cuckoos, it builds its own nest and rears the young. **Range and habitat:** OR; Peninsular India and Sri Lanka. Resident in woodland in the plains and foothills. **Similar species:** Lesser Green-billed Malcoha.

Common Crow Pheasant (Coucal) *Centropus sinensis* (Cuculidae) 48 cm (19 in). Coucals are unlike other cuckoos in having a long, straight claw on the inner hind toe, and are much less arboreal. Rather ungainly birds, they clamber through dense grass or brushwood, the big tail flopping about or flicked or expanded, and sometimes perch in the open on a small bush to call. They forage for lizards, nestlings or small mammals on the ground, pushing through thick cover or under bushes, and are reluctant fliers. The large nest is made of twigs, and the call is a resonant, repeated booming. **Range and habitat:** OR. Resident in grassy, bushed areas, gardens and scrub. **Similar species:** Lesser Coucal.

OWLS Strigiformes

Malay Fish-owl *Bubo ketupu* (Strigidae) 41 cm (16 in). Roosting in the daytime on a branch of a tall tree, often by the junction with the main trunk, it emerges in the late afternoon to patrol the ricefields or forest streams. The food is largely amphibians, fish or small mammals, and grasping slippery prey is facilitated by the prickly scales on the soles of the feet. The fish-owls enjoy bathing, and splash about with their wings at the edge of a shallow pool or stream. The call is a soft whistle or trilling sound. **Range and habitat:** OR; in Indo-Malayan subregion. Resident in lowland forest near streams or ricefields or at the coast. **Similar species:** Brown Fish-owl.

Spotted Owlet *Athene brama* (Strigidae) 20 cm (8 in). Much more likely to be seen in daylight than some other owls, when its presence is often indicated by the noisy mobbing of various small birds. One of the most familiar owls because of its preferred habitat of groves or gardens, often around houses, but avoiding thick forest. The yellow eyes and heavily white-spotted plumage render identification easy. Often lives in small groups, usually at least 2 birds roosting together. When disturbed, bobs up and down and stares at intruder before flying off with undulating flight. Call a series of screeches. **Range and habitat:** OR; except Sri Lanka and Indo-Malayan subregion. Resident in light woodland, cultivation, villages. **Similar species:** other owlets.

Jungle Owlet *Glaucidium radiatum* (Strigidae) 20 cm (8 in). A much less familiar bird than its spotted cousin, as it is both more strictly nocturnal, and also haunts much denser vegetation in woods and thick forest. The lack of ear-tufts distinguishes it from other small owls of the same habitat. Spends the daytime roosting in a tree-hole, but is sometimes discovered and mobbed by other birds, and then flies off to deeper cover. It preys mainly on large insects and small animals and birds. The call is a repeated, musical hoot. **Range and habitat:** OR; India and Sri Lanka. Resident in forests up to 2,000 m (6,500 ft). **Similar species:** other owlets.

NIGHTJARS AND ALLIES Caprimulgiformes

Hodgson's Frogmouth *Batrachostomus hodgsoni* (Podargidae) 26 cm (10 in). The family name refers to the enormously wide gape, such that, seen from above, the bill is a rounded triangular shape, though the base is mostly covered with bristles. The feet are extremely small and weak. The food, mostly moths and insects, is probably taken as much from branches as in the air. Difficult to see, frogmouths spend the day perched on a dead branch or stump in thick forest, and have mottled, highly protective plumage. The call is a soft, descending whistle. **Range and habitat:** OR; Himalayas to Indo-China. Resident in evergreen hill forest. **Similar species:** other frogmouths (see p 304). **Not illustrated.**

Indian Nightjar *Caprimulgus asiaticus* (Caprimulgidae) 24 cm (9½ in). Most usually seen at night, either hawking noiselessly for insects, or resting on a path or track, when the eyes shine red in a car's headlights. The nightjars, of which there are 11 species in the OR, are best identified by the call, which in this bird is a repetition of 3 notes running into a throaty trill. Spends the day squatting quietly on the ground, almost invisible owing to the protective plumage, but starts to feed and call at dusk. Seasonal movements are difficult to study, as calling ceases after the breeding season. **Range and habitat:** OR; India and Sri Lanka to Indo-China. Resident in dry scrub, gardens, wasteland. **Similar species:** other nightjars (see pp 72, 116, 166, 212, 304).

SWIFTS AND HUMMINGBIRDS Apodiformes

White-throated Spinetail Swift *Hirundapus caudacuta* (Apodidae) 20 cm (8 in). The name is a reference to the shafts of the tail feathers, which project beyond the vane as a short spike. A large swift and a very powerful flier, which makes a loud, whooshing noise as it sweeps by. Often flies very high in the air, but when hawking insects at low levels the great size, with a wing spread of up to 53 cm (21 in) is apparent, as is the bulky body with a white patch at the rear. Often in small groups, the birds scream as they chase each other. **Range and habitat:** EP, OR, in winter to AU. Resident and migratory, wandering great distances. **Similar species:** Eurasian Swift (see p 166).

Northern White-rumped (Fork-tailed) Swift *Apus pacificus* (Apodidae) 18 cm (7 in). A widely distributed swift in eastern Asia, wandering widely southwards in winter. The forked tail and gleaming white rump patch facilitate identification even when the bird is flying at a great height, as it often does in fine weather. In Tibet, nests under eaves and in stonework of buildings, but outside breeding season almost continuously on the wing. Has a screaming note much as in Swift. **Range and habitat:** EP, OR, in winter to AU. Largely migratory. **Similar species:** Eurasian Swift (see p 166).

Edible Nest Swiftlet *Aerodramus fuciphagus* (Apodidae) 12 cm (4½ in). One of a group of about 20 species, some of which make the nests which are used for birds's-nest soup. The best nests for this purpose consist entirely of coagulated saliva, looking a clear white in colour, and the present species builds one of the purest and therefore most favoured. In appearance it is a tiny, slim swift with a well-forked tail and a fluttery, bat-like flight, which hawks for minute insects over beaches and around coconut trees. The nests are built in cliff or cave crevices. **Range and habitat:** OR. Resident around beaches, plantations or forest. **Similar species:** Eurasian Swift (see p 166). **Not illustrated.**

Asian Palm-swift *Cypsiurus balasiensis* (Apodidae) 13 cm (5 in). A slim, small swift with a long, deeply forked tail which distinguishes this species from swiftlets. The name is derived from its habit of nesting in palm trees. Unlike swiftlets, palm swifts construct their nests from feathery seeds, often those of the kapok tree. These are glued together with saliva and stuck onto the upper surface of older, hanging palm fronds. Colonial breeder. Hawk flying insects on the wing, and congregate around the palms in which they breed. **Range and habitat:** OR; resident in open country with palm trees. **Similar species:** swiftlets.

Small Green-billed Malcoha feeding young

Common Crow Pheasant (Coucal) feeding young

Malay Fish-owl

Spotted Owlet

Jungle Owlet

Indian Nightjar

White-throated Spinetail Swift

Northern White-rumped (Fork-tailed) Swift

Asian Palm-swift

KINGFISHERS AND ALLIES Coraciiformes

White-breasted (Smyrna) Kingfisher *Halcyon smyrnensis* (Alcedinidae) 28 cm (11 in). The heavy red bill, and white breast contrasting with chocolate head and underparts, are diagnostic of this familiar bird – one of the most noticeable kingfishers throughout the whole OR. In flight, shows a large, white wing patch. Perches in a characteristically exposed position, often on telephone wires or roadside posts, and equally at home in wet paddyfields or forest clearings, for it eats more insects and small animals than fish. Bobs nervously on its perch, and has a loud, rippling whistle. The 6 subspecies vary in saturation of plumage colours and size of the white bib. **Range and habitat:** WP, EP, OR; Iran to the Philippines. Resident in open country, thin forest, sea coasts, at up to 2,500 m (8,200 ft) in the Himalayas. **Similar species:** Black-capped Kingfisher.

Black-capped Kingfisher *Halcyon pileata* (Alcedinidae) 30 cm (12 in). Much more restricted to the neighbourhood of water than the similar white-breasted bird, and easily distinguished by the black head, pale buffy underparts and blackish violet back. Also shows a large white wing patch in flight. Often seen on a post sticking out of a swampy area or in mangroves, it is mainly a fish- and crab-eater, well equipped for dealing with such prey with its formidable bill. A solitary and rather shy bird, not allowing a close approach. It has a shrill, whistling call. **Range and habitat:** OR; India to eastern China and the Philippines. Resident, wandering locally, on rivers, lakes and sea coasts. **Similar species:** White-breasted Kingfisher.

Chestnut-headed (Bay-headed) Bee-eater *Merops leschenaulti* (Meropidae) 20 cm (8 in). The green back, and lack of elongated central tail feathers, distinguish it from the European Bee-eater, and it is also a much smaller bird. Young birds have heads the same colour as the back and are duller. Often seen in small groups, it perches on exposed branches in forest glades or by the side of a track, swooping out now and then to chase a passing insect. Several birds sit side by side and change positions as 1 after another circles out and then returns to the perch. It is very vocal when on the wing, making a musical, liquid, rippling note. **Range and habitat:** OR; India to Indo-China and Java. Resident and migratory in wooded areas. **Similar species:** European Bee-eater (see p 166).

Blue-tailed Bee-eater *Merops philippinus* (Meropidae) 30 cm (12 in). The bright blue tail and rump distinguish it from the otherwise similar Blue-cheeked Bee-eater, and it shows the same beautiful, coppery underwing in flight. It is a highly gregarious species, many birds nesting together and migrating in parties. It is addicted to the neighbourhood of water, whether forest streams, paddyfields or coastal estuaries, and spends much time on the wing in such areas hawking for food. The prey features the larger flying insects such as dragonflies, wasps and bees. The call is a soft, rippling whistle, often uttered in flight. **Range and habitat:** OR; India to Borneo. Migratory in northern parts of range. Open country, light woodland, seashores. **Similar species:** Blue-cheeked Bee-eater.

Blue-bearded Bee-eater *Nyctyornis athertoni* (Meropidae) 36 cm (14 in). The largest of the bee-eaters, and, with its red-bearded relative, a somewhat unusual one in behaviour. It is entirely a forest bird, and forages through the foliage and branches for insects as well as catching them in flight. It seems possible that beetles form a good part of its diet. It is a shy bird, and does not gather in the large, noisy groups characteristic of some of the others of the family, usually being seen singly. The call is a loud, harsh croak, uttered with the long throat feathers fluffed forwards. **Range and habitat:** OR; Himalayas and India to South-East Asia. Resident in evergreen forest. **Similar species:** Red-bearded Bee-eater. **Not illustrated.**

Red-bearded Bee-eater *Nyctyornis amicta* (Meropidae) 28 cm (11 in). A sedentary species which takes the place of its blue-bearded relative in the Indo-Malayan subregion of the OR. The male has much more extensive lilac colouring on the crown, but otherwise the female is similar, as with most bee-eaters. An unobtrusive and solitary bird, which, however, draws attention to its presence by the throaty, descending chuckle, which is often answered by a 2nd bird. Insects are caught in flight from perches often concealed in foliage, though it is a sluggish bird and is inactive for long periods. **Range and habitat:** OR; southern Burma to Borneo. Resident in forest. **Similar species:** Blue-bearded Bee-eater.

Great Indian Hornbill *Buceros bicornis* (Bucerotidae) 130 cm (51 in). Like all large birds, now much threatened by habitat destruction and disturbance. A splendid hornbill, very impressive in flight, when the wingbeats make a loud throbbing noise, easily heard even when the bird is invisible over the forest canopy. Very buoyant and agile on its feet, bouncing along large boughs and nimble when foraging for fruit. Also takes small animals and reptiles. Needs a large tree cavity in which to nest, so selects only the biggest trees, which unfortunately are most likely to be felled for their timber. Quite a noisy bird with a variety of loud barking and bellowing noises. **Range and habitat:** OR; western India and Himalayas to Malaya. Resident in tall evergreen forest. **Similar species:** other hornbills (see p 210).

Helmeted Hornbill *Rhinoplax vigil* (Bucerotidae) 125 cm (49 in). Remarkable for its solid, rather than hollow, casque, and also for its call. This is a familiar sound of the forest, a series of loud barks which gradually accelerate until they culminate in a cackling laugh. The casque, which is hard like ivory, used to be in great demand for carving into ornaments, and the long central tail feathers are still valued as ornaments. Hornbills play a large part in the folklore of Bornean tribes. Usually seen singly, these large birds have a flap-and-glide flight over the forest canopy, though often difficult to see through the dense vegetation. **Range and habitat:** OR; Malaysia, Sumatra and Borneo. Resident in tall primary forest. **Similar species:** other hornbills (see p 210). **Not illustrated.**

Indian Roller *Coracias benghalensis* (Coraciidae) 30 cm (12 in). The flash of brilliant blue on the wings as it drops down from a perch is always surprising, as, at rest, this roller often looks a rather drab bird. It is a very familiar sight in open country, both for choosing conspicuous look-outs from which to see grasshoppers or other prey, and also for its noisy and acrobatic display. This takes the form of excited flapping and throwing itself about in the air, diving and rolling and swooping up again, with constant loud, harsh calls. The sexes are alike, and the several subspecies vary somewhat in colour. **Range and habitat:** OR; except Greater Sunda, Borneo and the Philippines. Resident and locally migratory in open country in the plains. **Similar species:** European Roller (see p 166).

Eastern Broad-billed Roller *Eurystomus orientalis* (Coraciidae) 30 cm (12 in). Often known as the Dollarbird, in reference to the circular silvery blue patch visible below the expanded wing. Most often seen sitting motionless at the extremity of some bare twig, it watches for insects which it catches in flight, only occasionally dropping to the ground like the Indian Roller. The bill is particularly short and broad, though the feet are small and weak. It is more of a forest bird, preferring clearings among tall trees, often in heavy jungle in the Himalayan foothills. The many subspecies vary in both size and intensity of the plumage colours. **Range and habitat:** EP, OR and AU; northern birds wintering southwards. **Similar species:** African Broad-billed Roller (see p 208).

White-breasted (Smyrna) Kingfisher

Black-capped Kingfisher

Chestnut-headed (Bay-headed) Bee-eater

Blue-tailed Bee-eater

Great Indian Hornbill

Red-bearded Bee-eater

Indian Roller

Eastern Broad-billed Roller

WOODPECKERS AND ALLIES Piciformes

Crimson-breasted Barbet *Megalaima haemacephala* (Capitonidae) 17 cm (6½ in). Commonly known as the Coppersmith in many places, owing to the metallic note, which is repeated with great regularity and persistence. Although brightly coloured, its small size and habit of sitting quietly amongst foliage render it unobtrusive, but for the note. It feeds voraciously on ripe fruit, gathering in some numbers in suitable trees, often in company with other species. Often, however, it is quite solitary. The young bird is much greener overall, lacking the bright patches of colour. Several subspecies occur, 1 in the Philippines having the throat and facial patches yellow. **Range and habitat:** OR; except southern China. Resident in light woodland in the plains and up to 2,000 m (6,500 ft) in the hills. **Similar species:** Blue-throated Barbet.

Small Green Barbet *Megalaima viridis* (Capitonidae) 23 cm (9 in). Typical of the family in structure, though it lacks brightly coloured facial or breast patches, and in this resembles several larger, mainly green, barbets. It also has a similar call to these closest of its relatives, a resonant double note preceded by a metallic trill, and is extremely noisy in the breeding season. However, although the call is familiar, the bird itself, being leaf-green and staying mainly in the foliage, is not much seen. It is a fruit- and insect-eater. **Range and habitat:** OR; south-west India, resident in evergreen forest in the plains and hills. **Similar species:** Large Green Barbet.

Great Barbet *Megalaima virens* (Capitonidae) 32 cm (13 in). In spite of its bright colours, it usually looks a drab bird in the field, though the large, pale bill is noticeable. The mournful, double whistle is a very familiar sound, echoing across all the valleys of the hill country across southern Asia, though sighting the bird is not easy. In winter, there is a general move to lower levels, and the birds gather in small flocks. Although mainly a fruit-eater, insects and even small birds are sometimes taken. The powerful bill is useful for excavating the nest-hole, while the bird supports itself on the tail like a woodpecker. **Range and habitat:** OR; Himalayas to Indo-China and eastern China. Resident in hill forests. **Similar species:** Oriental Green Barbet.

Blue-throated Barbet *Megalaima asiatica* (Capitonidae) 22 cm (8½ in). A widely distributed bird, with several subspecies differing in plumage details. Often common around villages, especially in the Himalayan foothills, though it keeps entirely in the trees, moving around in search of fruit and sometimes large insects. When calling from an exposed perch, often high up, the bright blue of the throat is easily seen. A sedentary species, it is most vocal, and so most noticed, in the spring, and the constantly and rapidly repeated call is a hard, triple, rolling note, which soon becomes very familiar. **Range and habitat:** OR; Himalayas to Indo-China and Borneo. Resident in the hills up to 2,000 m (6,500 ft) in open or secondary forest. **Similar species:** Oriental Green Barbet.

Indian Honeyguide *Indicator xanthonotus* (Indicatoridae) 15 cm (6 in). A particularly unobtrusive bird, it sits quietly in the foliage by some shady path, perching upright, and often with the wings drooped so exposing the brilliant yellow rump. It is neither very lively nor noisy, the only note being a short, quiet call uttered on the wing, so it is often overlooked. It is the only representative in India of a family of 14 characteristic of Africa and unique in their diet of honeycomb and wax. Parasitic in breeding habits, though it is not yet known which species, possibly barbets, the Indian bird chooses as a foster parent. **Range and habitat:** OR; in the Himalayas up to 3,500 m (11,500 ft). Resident in forests. **Similar species:** other honeyguides (see p 214). **Not illustrated.**

Speckled Piculet *Picumnus innominatus* (Picidae) 10 cm (4 in). A tiny woodpecker in habits, though instead of having stiff tail feathers to use as a support while climbing, it has a soft, rounded tail, and often sits across a branch like a perching bird. Usually, though, it is seen climbing around small branches or twigs, or foraging up bamboo stems, tapping away in its search for insects. Often in company with other species in hunting parties. It makes a short, sharp call, and also drums like larger woodpeckers. Makes a tiny hole in a rotten branch to excavate the nest. **Range and habitat:** OR; wide ranging but not in Sri Lanka. Resident in hill forests. **Similar species:** other piculets. **Not illustrated.**

Brown-capped Woodpecker *Picoides moluccensis* (Picidae) 14 cm (5½ in). The red streak on either side of the crown is lacking in the female, but both birds are in general quite typical in plumage of the pied woodpeckers, with white barring on the back and streaking on the underparts. This species has a wide geographical range, and a number of subspecies have been described, varying in the tone of the plumage and extent of the white marks. It is a tiny but quite common woodpecker, not unduly shy, and often seen in company with other foraging species such as tits and nuthatches. It makes a muted drumming, and has a quiet call-note. **Range and habitat:** OR; India to Borneo. Resident in thin forest and bamboo. **Similar species:** other pied woodpeckers (see pp 74, 168).

Greater Yellow-naped Woodpecker *Picus flavinucha* (Picidae) 33 cm (13 in). A splendid, bright green woodpecker, especially when the golden crest is expanded as the bird clings to a branch when calling. At this moment, the head is thrown back so that the bill points up vertically, and the throat is expanded. The female differs in having the throat brown, instead of bright yellow. Many subspecies have been listed, varying in both size and plumage details. It ascends tree-trunks jerkily, and has a bounding flight, but often forages at low levels or on the ground. **Range and habitat:** OR. Wide ranging except Sri Lanka, Java, Borneo and the Philippines. Resident in evergreen or deciduous woodland. **Similar species:** Lesser Yellow-naped Woodpecker.

Lesser Golden-backed (Flame-backed) Woodpecker *Dinopium benghalense* (Picidae) 30 cm (12 in). A widely distributed and common woodpecker, which shows a good deal of variation in the colour of the upperparts and also the streaking on the throat and breast, and has been split into several subspecies accordingly. The female has a black, rather than crimson, crown. Feeds as much on the ground as in trees, where it is often accompanied by other species in hunting parties, and is more familiar than some woodpeckers owing to its preference for gardens, avenues and scattered trees round cultivation and houses. It has a loud, harsh scream. **Range and habitat:** OR; India and Sri Lanka. Resident in lightly wooded areas in the plains and foothills. **Similar species:** Crimson-backed Woodpecker.

Bay Woodpecker *Blythipicus pyrrhotis* (Picidae) 28 cm (11 in). A large woodpecker, the rufous plumage and yellow bill being distinctive in the field. It is a noisy bird, with a loud, descending series of raucous notes. Often in groups of 2 or 3 birds, which chatter away until aware of an intruder, when they disappear silently. Favourite foraging places are low down in bamboo or dense vegetation in evergreen forest, which makes observation difficult. It less often ascends large trees in the usual woodpecker fashion. Several subspecies have been described, but the sexes differ only in the female lacking the crimson on the neck. **Range and habitat:** OR; Himalayas to south China and Malaysia. Resident in evergreen hill forest. **Similar species:** Rufous Woodpecker. **Not illustrated.**

Crimson-breasted Barbet

Great Barbet

Small Green Barbet

Blue-throated Barbet

Brown-capped Woodpecker

Greater Yellow-naped Woodpecker

Lesser Golden-backed (Flame-backed) Woodpecker

PERCHING BIRDS Passeriformes

Long-tailed Broadbill *Psarisomus dalhousiae* (Eurylaimidae) 27 cm (10½ in). A somewhat unobtrusive bird, and often difficult to see in the heavy foliage it frequents, as the bright colouring of the plumage blends exactly with the light and shade on the leaves. Often sits quietly, perched upright, for long periods, but also flutters about catching insects, and is more easily seen then or when flying through a clearing. Solitary or in small groups which are usually quiet, though the bird has a loud whistle which draws attention to its presence. Sexes are alike in this broadbill. **Range and habitat:** OR; Himalayas to Indo-China and Borneo. Resident in evergreen forest in the plains and hills. **Similar species:** other broadbills.

Indian (Blue-winged) Pitta *Pitta brachyura* (Pittidae) 20 cm (8 in). Gaudy bird with long, strong legs and short tail. Lives on the ground in forest or thick scrub. Not easy to see or study, though it is not especially shy, and has a loud, musical whistle, often heard at dusk. When calling, the head is thrown back, and the tail is slowly wagged if the bird is nervous. In the deep woodland shade, when the colours cannot be seen, the dumpy shape and upright posture are diagnostic. The sexes are alike. **Range and habitat:** OR; India to eastern China and Borneo. In evergreen or deciduous forest, the southern populations resident, northern ones migrate southwards in winter.

Hooded Pitta *Pitta sordida* (Pittidae) 20 cm (8 in). The large, white wing patch is noticeable in flight, which is rapid, on whirring wings. The sexes are alike, though the young bird is paler and duller than the adults. Often quite common in parts of its range, may haunt quite open habitat, such as the margins of mangrove swamps. The food consists mainly of worms, caterpillars and small insects. It has a double, whistled note which is frequently uttered. Sedentary bird with 13 subspecies which vary in size and intensity of coloration. **Range and habitat:** OR. Widespread resident, wandering locally, in forest, scrub and mangroves. **Similar species:** other pittas.

Gurney's Pitta *Pitta gurneyi* (Pittidae) 20 cm (8 in). A pitta of very restricted distribution, and hence very vulnerable to habitat destruction, especially as its preferred lowland jungle has been logged even more extensively than the hill forests. Hardly seen in the wild for 50 years, it was rediscovered in 1986 as a result of investigating birds illegally held for sale in Bangkok, and the 1st nest was found since 1915. The female has the crown brown instead of brilliant blue. The call is a short, mellow, double note, and it has another, single, contact note. **Range and habitat:** OR, in southern Burma and southern Thailand. Resident in dense, semi-evergreen rain forest. **Similar species:** other pittas. **Not illustrated.**

Ashy-crowned Finch-lark *Eremopterix grisea* (Alaudidae) 13 cm (5 in). A small, nondescript little lark which gathers in large flocks outside the breeding season. Even when large numbers are feeding over a dry field, they are quite inconspicuous, holding themselves very low to the ground as they search for seeds. The female lacks the black eye-line and underparts. In the breeding season, which can extend over many months, the male has a conspicuous song-flight, notable for the regular earthward dives followed by swoops upwards to gain altitude and resume the song once more. **Range and habitat:** OR; India, Pakistan and Sri Lanka. Resident in the plains in dry, open countryside. **Similar species:** other larks.

Oriental Skylark *Alauda gulgula* (Alaudidae) 18 cm (7 in). Very reminiscent of the European Skylark, though not as sweet a singer. Character of the song and type of song-flight extremely similar, however. It is a slightly smaller bird, with a shorter tail and more rounded wings, and sandy rather than white outer tail feathers, and has a quite different call-note in flight. Often in small parties which may mingle with wintering skylarks. Many subspecies have been described of this widely distributed lark. **Range and habitat:** EP and OR; northern birds wintering southwards. Resident and migratory in open country and grasslands. **Similar species:** Skylark (see p 168). **Not illustrated.**

Dusky Crag Martin *Hirundo concolor* (Hirundinidae) 13 cm (5 in). A sooty little swallow, with white spots on the square tail which show up as the bird banks in the air. Slow and buoyant flight, sailing to and fro around old buildings or rocky crags. The birds call with a low, soft, chipping note. The sexes are alike. Usually found in pairs or small groups, it is not a colonial nester and does not occur in large parties like some of the migratory swallows or the Cliff Swallow. **Range and habitat:** OR; India to Indo-China. Resident around cliffs, buildings or towns in the plains or lower hills. **Similar species:** Crag Martin.

Indian Cliff Swallow *Hirundo fluvicola* (Hirundinidae) 13 cm (5 in). The shallowly forked tail and rather fluttery flight recall the Sand Martin, with which this species often occurs, for it often nests colonially in the vicinity of water, over which the 2 species hawk together. The dull chestnut cap, streaked underparts and pale rump are diagnostic of this species; the sexes are alike, the young bird duller. Nesting colonies are often large, many nests being clustered together, the site selected as likely to be on a building as a cliff-face. **Range and habitat:** OR; Himalayas and northern and peninsular India. Resident but migratory in some areas, in the plains and lower hills. **Similar species:** Swallow (see p 168). **Not illustrated.**

Indian Tree Pipit *Anthus hodgsoni* (Motacillidae) 18 cm (7 in). The softly streaked, olive upperparts and white stripe over the eye, and the dark streaking on the very pale underparts, are good pointers to its identity. Commonly seen in small parties in shady places and clearings in woodland, the birds fly up into trees when disturbed, and sit there with tails wagging nervously. Like the Tree Pipit, has a song-flight which starts from, and returns to, a perch in a tree. The call-note resembles that of the Tree Pipit. **Range and habitat:** EP and OR; northern populations wintering southwards. Inhabits open woodland, grassy places, cultivation. **Similar species:** Tree Pipit (see p 170).

Nilghiri Pipit *Anthus nilghiriensis* (Motacillidae) 18 cm (7 in). A pipit of very limited distribution, and like many birds of high rainfall areas, of rather dark appearance, with heavy streaking on the underparts. Usually seen singly or in pairs in the breeding season, it is typical of grassy areas and coffee plantations in the hills, and though familiar in its range, little has been published about its habits or song. **Range and habitat:** OR. Resident above 1,000 m (3,300 ft) in the hills of south-west India. **Not illustrated.**

Large Pied Wagtail *Motacilla maderaspatensis* (Motacillidae) 22 cm (8½ in). One of the largest of the wagtails, and in effect an Indian version of the large Pied Wagtail of Africa. A common and conspicuous bird, with its pied plumage, wagging tail and constant presence in the open along streams or around ponds. Often perches on houses or boats, where it sometimes nests, being quite a confiding bird. Usually in pairs, though sometimes parties gather outside the breeding season. **Range and habitat:** OR; Pakistan and India. Resident in the plains and hills by water. **Similar species:** White Wagtail (see p 170).

Forest Wagtail *Dendronanthus indicus* (Motacillidae) 17 cm (6½ in). A somewhat aberrant wagtail in habits, living in often dense evergreen forest, though always by streams, waterfalls or very wet paths and tracks. Easily identified by the double breast band, and 2 prominent white wing-bars. The sexes are alike. Has a curious trick of swaying the body and tail, as well as normal tail-wagging. Partial to perching in trees, in which it takes refuge if disturbed. The note is unusual for a wagtail – a sharp, single call. **Range and habitat:** EP, in winter to OR, in dry or wet forest and bamboo. **Similar species:** White Wagtail (see p 170).

Hooded Pitta

Long-tailed Broadbill

Indian (Blue-winged) Pitta

Ashy-crowned Finch-lark

Dusky Crag Martin

Indian Tree Pipit

Large Pied Wagtail

Forest Wagtail

Large (Black-faced) Cuckoo-shrike *Coracina novae-hollandiae* (Campephagidae) 28 cm (11 in). The female lacks the black eye patch, and has closely barred underparts. There are nearly 20 subspecies of this very widely distributed bird, varying in details of plumage and in size. Often seen in pairs or small parties which keep to the treetops, though conspicuous on account of the loud, shrill whistle which is frequently uttered. It is an active bird, constantly on the move in the search for the insects, caterpillars and fruit on which it feeds. **Range and habitat:** OR and AU. Resident, wandering locally, in open forest in the plains and lower hills. **Similar species:** Black-headed Cuckoo-shrike.

Black-headed Cuckoo-shrike *Coracina melanoptera* (Campephagidae) 20 cm (8 in). Female and young birds lack the black head of the male, and are browner, with barred underparts. Actively searches through foliage singly or in little parties, and often in company with other species. Keeps mostly in the trees, where it feeds on insects and berries. It has a variety of whistling notes, often 1 note being repeated several times at the same pitch, and is especially noisy in the breeding season. **Range and habitat:** OR; in India and Sri Lanka. Resident in open woodland in the plains and lower hills. **Similar species:** Black-winged Cuckoo-shrike.

Common Wood-shrike *Tephrodornis pondicerianus* (Campephagidae) 16 cm (6½ in). An unobtrusive and quietly coloured bird, not perching conspicuously like the true shrikes to watch for prey, but moving rather slowly and methodically through the foliage as it searches for insects. The sexes are much alike, the eye stripe being less well marked in the female. Usually in pairs, and difficult to see as it is often in thick foliage, but betrays its presence with repeated, descending whistle. The male also has a warbling song. **Range and habitat:** OR; India to Indo-China. Resident in open deciduous forest in the plains. **Similar species:** Brown-tailed Wood-shrike. **Not illustrated.**

Rosy Minivet *Pericrocotus roseus* (Campephagidae) 18 cm (7 in). The more northern populations of this species, in China, lack the rosy underparts and rump which make the Himalayan race such an attractive bird. A forest bird which keeps to the treetops, often in flocks in winter, when it moves well to the south of its breeding range in China. Usually seen perching in tall trees, or moving through the forest, the birds following each other over the canopy or across the clearings. The whistling calls are less musical than those of some of its relatives. **Range and habitat:** EP and OR. Resident or migratory, in hill forests. **Similar species:** Scarlet Minivet. **Not illustrated.**

Small Minivet *Pericrocotus cinnamomeus* (Campephagidae) 15 cm (6 in). Differs somewhat from its relatives in being a plains bird of quite open scrub and sparsely wooded places. However, similar in habits in its lively manner of hunting through foliage in parties. The birds keep in touch with a constant, quiet whistling. It has a very long breeding season, and builds a particularly beautiful and tiny nest about 25 mm (1 in) deep. The female lacks the dark throat, is browner above, and has paler, duller underparts. More than 10 subspecies have been described. **Range and habitat:** OR; India and Sri Lanka to Borneo. Resident in open woodland in the plains. **Similar species:** Scarlet Minivet.

Scarlet Minivet *Pericrocotus flammeus* (Campephagidae) 20 cm (8 in). The largest, best known and one of the most brightly coloured of the minivets. It is a common bird of hill forests in southern Asia, conspicuous when flying over the treetops, the birds flashing red, yellow and black against the green foliage. Flocks keep in touch with a constant, liquid whistling, and the birds are very active as they chase insects among the leaves or in short, aerial flutters. Many subspecies occur, with variations in plumage intensity and pattern, and in size. **Range and habitat:** OR. A very widely distributed resident in hill forests. **Similar species:** Small Minivet.

Red-vented Bulbul *Pycnonotus cafer* (Pycnonotidae) 20 cm (8 in). One of the most familiar garden birds in all India, it is a noisy, friendly and active favourite, though capable of doing much damage to flowers and fruit. The sexes are alike, and the white tail tip and rump are seen even more easily than the red patch below the tail. The chirpy call is constantly uttered, but the bird is not much of a songster. Generally keeps in the trees, and often sits conspicuously on the top of a bush but rarely descends to the ground. The food is mainly insects, fruit and berries. **Range and habitat:** OR; India and Sri Lanka to Burma. Resident in the plains and hills, not in desert or dense forest. **Similar species:** Red-whiskered Bulbul.

Red-whiskered Bulbul *Pycnonotus jocosus* (Pycnonotidae) 20 cm (8 in). The long, pointed crest and red patch behind the eye are common to both sexes. A very common and well-known bird, gregarious and obvious around villages, gardens and neighbouring scrub. It is cheerfully noisy, the calls being louder and more musical than those of its red-vented relation. On the whole, more partial to areas of higher rainfall, with a more easterly general distribution. Several subspecies occur, and the bird has been introduced in Australia and Florida. Some subspecies lack the white tail tip, and there are differences in back colour and crest length. **Range and habitat:** OR; India to Hong Kong. Resident in open woodland and forest edge. **Similar species:** Red-vented Bulbul.

Ashy Bulbul *Hypsipetes flavalus* (Pycnonotidae) 20 cm (8 in). The short, rounded crest and light brown cheeks are distinctive; the subspecies occurring in Malaya lacks the conspicuous yellow wing patch of the Himalayan and Indo-Chinese birds. A sociable and noisy bird outside the breeding season, the flocks foraging actively through the treetops for berries and insects. Has a musical, fluty call-note and a short song. Birds breeding at higher altitudes move to lower levels in the winter and to less densely forested areas. **Range and habitat:** OR; Himalayas to Indo-China and Borneo. Resident in hill forests. **Similar species:** Chestnut Bulbul.

White-browed Bulbul *Pycnonotus luteolus* (Pycnonotidae) 17 cm (6½ in). A dull-coloured bird, though the white stripe over the eye and the red eye can be noticed if one can get a good look, for it is a very skulking species. It inhabits dense shrubbery and thickets in predominantly dry country. It is a noisy bird, so that it is much more often heard than seen, and the song is a sudden jumble of loud notes. It also has a churring alarm note. Not one of the more sociable bulbuls, usually found in pairs. **Range and habitat:** OR; peninsular India and Sri Lanka. Resident in dry scrub. **Similar species:** Yellow-browed Bulbul. **Not illustrated.**

Black Bulbul *Hypsipetes madagascariensis* (Pycnonotidae) 25 cm (10 in). A widely distributed bird with many subspecies, some of which have the head and breast white. The red bill and legs are unique among bulbuls. A restless bird of treetops, highly gregarious and lively and often in large parties. The birds do not stay long in 1 place, and follow each other from tree to tree making a variety of squeaky whistles. It is a strong flier. It is attracted to flowering and fruiting trees, and also makes short flights after insects. **Range and habitat:** OR, where widespread except Malaysian subregion. Also AF (Madagascar and the Comoros). Resident in hill forests. **Similar species:** Ashy Bulbul.

Large (Black-faced) Cuckoo-shrike

Black-headed Cuckoo-shrike

Small Minivet

Scarlet Minivet

Red-vented Bulbul

Red-whiskered Bulbul

Ashy Bulbul

Black Bulbul

Common Iora *Aegithina tiphia* (Irenidae) 15 cm (6 in). The most widely distributed member of its family, which is notable for being the only family peculiar to the OR. A small, active, tree-loving bird, with short wings and tail. The feathers on the rump are very thick and soft, and are fluffed out in display. A very variable species, especially in the amount of black on the upperparts. Females have greener wings and a green tail. Often in pairs, searching through the foliage for insects, sometimes in company with other species in a hunting party. The calls feature a variety of sweet whistles. **Range and habitat:** OR; from Pakistan to Borneo. Resident in the plains and lower hills in woodland, gardens and scrub. **Similar species:** Marshall's Iora.

Golden-fronted Leafbird *Chloropsis aurifrons* (Irenidae) 19 cm (7½ in). The shining orange forehead distinguishes this species from several similar-looking relatives. The sexes are much alike, the female just lacking the yellow border round the black chin patch. Always keeps in bushes and trees, where it searches for insects and probes flowers for nectar with the delicately curved and pointed bill. It is an active and agile bird, almost tit-like when foraging, and fast and confident on the wing. It has a musical, whistling call-note and a pleasant song, but is also a versatile and inveterate mimic of other birds. **Range and habitat:** OR; Himalayas and India to Sumatra. Resident in woodland in the plains and hills up to 1,800 m (5,900 ft). **Similar species:** Orange-bellied Leafbird.

Orange-bellied Leafbird *Chloropsis hardwickii* (Irenidae) 19 cm (7½ in). The only leafbird with orange underparts. The female lacks the black and blue face-mask, and the young bird is entirely green. The commonest of its family in hill forests, ascending to 2,300 m (7,500 ft) in the Himalayas, though moving to lower levels with the onset of winter. Often to be seen at flowering trees with other nectar-seeking species, it is an important pollinator and also responsible for spreading parasitic plants of the mistletoe family, whose berries it eats. It has a fine song, and is also a great mimic. **Range and habitat:** OR; Himalayas to Hainan and Malaysia. Resident in hill evergreen forest. **Similar species:** Golden-fronted Leafbird.

(Blue-backed) Fairy Bluebird *Irena puella* (Irenidae) 27 cm (10½ in). The brilliant metallic blue of the male's plumage is not always easy to see in the field while the bird is moving through dense evergreen forest in the shade, but becomes startlingly apparent once it shifts into the sunlight. The female is a duller, greener blue. The rich red eyes are obvious in a good view. Despite apparently liking the densest cover, it is quite a noisy and conspicuous bird, with a sharp, double whistle, constantly repeated and unmistakable. Numbers often gather in a fig tree to gorge on the fruit in company with such birds as barbets and green pigeons. **Range and habitat:** OR; south-west India, eastern Himalayas to Java. Resident in evergreen forest.

Black-throated Accentor *Prunella atrogularis* (Prunellidae) 15 cm (6 in). The black throat at once distinguishes it from the other members of the family, though the streaked plumage is otherwise similar. The sexes are alike. Usually in small groups, it is only a winter visitor to the Himalayas from its breeding grounds in central Asia. Keeps mostly to the ground, where it forages for insects, turning over the leaves and twiglets. Often perches in bushes and trees, where it usually flies on being disturbed. It is a quiet bird, but sometimes makes a short, soft, contact note. **Range and habitat:** EP, in winter to OR in the Himalayas. Frequents scrubby or grassy hillsides. **Similar species:** Rufous-breasted Accentor.

Ashy Woodswallow *Artamus fuscus* (Artamidae) 19 cm (7½ in). The most widespread of only 2 representatives in the OR of an Australo-Papuan family. A thickset bird with a strong bill, very long wings and a short tail. Most often seen above woodland glades where it wheels about, much like a bee-eater, on widely spread, triangular wings. Perches on bare or exposed twigs while on the look-out for the flying insects on which it feeds. It is a social bird, and frequently calls with a loud, harsh cry. **Range and habitat:** OR; India and Sri Lanka to Indo-China. Resident in lowland wooded areas. **Similar species:** White-breasted Woodswallow (see p 320).

Rufous-backed (Black-headed) Shrike *Lanius schach* (Laniidae) 25 cm (10 in). A large and impressive shrike, it is somewhat variable in plumage. Some races have a grey, not rufous, back, and the head may be all-black, or grey with a black eye-stripe. The sexes are alike. Often conspicuous as it perches at the top of a bush, watching the surrounding grass for insects, small animals or fledgelings, which are killed by blows from the heavy bill. An aggressive and noisy bird, it makes a variety of harsh calls, but also mimics the calls of other birds. **Range and habitat:** OR; Himalayas to the Philippines. Resident and migratory, in open or bushy country in the plains and hills. **Similar species:** Bay-backed Shrike.

Bay-backed Shrike *Lanius vittatus* (Laniidae) 18 cm (7 in). A smart little shrike, the deep, rich colour of the back at once distinguishing it. The sexes are alike. It does not have the plumage variation of the wider-ranging Rufous-backed Shrike. A common bird around field edges or cultivated patches, it watches from the edge of a bush or a fence for its prey, taking small items such as insects, sometimes mice or nestlings. Like many shrikes, it has a rather pretty warbling song, in contrast to the harsh call-note. **Range and habitat:** EP and OR; South-West Asia, Pakistan and India. Resident and locally migratory, in lightly wooded or open country. **Similar species:** Red-backed Shrike (see p 170).

Brown Shrike *Lanius cristatus* (Laniidae) 19 cm (7½ in). A rather dull-coloured shrike which can be easily confused with some races of the Red-backed Shrike. As with other species, the young bird has wavy concentric barring on the underparts. A noisy bird with a harsh, chattering note, uttered as it sits on a bush or telephone wire, the tail switching about as it maintains its balance. It feeds voraciously on insects, especially large grasshoppers, and has been seen to catch frogs. **Range and habitat:** EP, wintering widely across OR. Frequents woodland glades, gardens, grassy scrub. **Similar species:** Red-backed Shrike (see p 170).

Asiatic Paradise Flycatcher *Terpsiphone paradisi* (Monarchidae) 20 cm (8 in). Only males over the age of 4 years acquire the beautiful white plumage, and even then retain a glossy black head. Younger males resemble females, though these have shorter tails. A very familiar bird owing to its striking appearance and conspicuous habits. The long tail floats behind, fluttering and twisting, as the bird pursues its aerial prey; when on a perch, the stance is upright and the tail hangs down, curving gracefully. The call is a short, grating note. **Range and habitat:** EP and OR, Arabia and South-West Asia to Java. Resident and migratory in woodland, gardens and scrub. **Similar species:** Japanese (Black) Paradise Flycatcher.

Black-naped Blue Monarch *Hypothymis azurea* (Monarchidae) 17 cm (6½ in). The female is duller and lacks the black nape patch and breast band. Many subspecies are listed, showing small plumage differences. An active and sprightly bird, pirouetting on tree-trunks and fanning the tail, only to stop suddenly and flutter out after an insect, though food is taken as much off twigs and branches. Often seems to keep in quite shady places in trees, tall hedges or bamboo groves. Often joined in its foraging activities by other species in hunting parties. The food consists of small insects, and the call is a short, sharp little note. **Range and habitat:** OR; India and Sri Lanka to the Philippines. Resident in forests and groves. **Similar species:** other monarch flycatchers.

Common Iora at nest

Golden-fronted Leafbird

Orange-bellied Leafbird

(Blue-backed) Fairy Bluebird

Black-throated Accentor

Ashy Woodswallow

Rufous-backed (Black-headed) Shrike

Bay-backed Shrike

Brown Shrike

Asiatic Paradise Flycatcher

Black-naped Blue Monarch

Grey-headed Canary Flycatcher *Culicicapa ceylonensis* (Muscicapidae) 11 cm (4½ in). A very common little flycatcher in woodland in southern Asia. Rarely high in the trees, it forages very like a warbler, flicking and fluttering among the leaves and darting out to take an insect, or hovering to take a morsel off a leaf. It is very vocal, with a characteristic double call-note and a trilling song. It is often a member of the hunting parties which move swiftly through the hillside trees in the Himalayas, where it breeds at up to 3,000 m (9,900 ft). **Range and habitat:** OR. Widely distributed in hill forests, wintering at lower altitudes. **Similar species:** Grey-faced Willow Warbler.

Tickell's Niltava *Niltava tickelliae* (Muscicapidae) 14 cm (5½ in). A persistent songster as it moves about in the shady foliage above a stream gully, with a series of rather metallic notes. It also makes a ticking alarm note. Very similar in colour and pattern to several other Asian flycatchers, though in this case the female is duller but similarly coloured, while females of other species are brown. Often difficult to see in deep forest when it is perching, which it does very upright, the head turning as it scans the surroundings. **Range and habitat:** OR. Widespread resident in thick woodland. **Similar species:** Hill Blue Niltava. **Not illustrated.**

Rufous-bellied Niltava *Niltava sundara* (Muscicapidae) 17 cm (6½ in). The shining blue patch on the sides of the neck is virtually the only distinguishing feature from a related, extremely similar species. The female also has this patch, but otherwise is brown with a white chest gorget. Despite the brilliance of the plumage when the bird is in a good light, it is by no means so obvious in the preferred habitat of a damp gully deep in evergreen forest. Keeps low down in the foliage, often taking food off the forest floor. One of those species with a discontinuous range in south-western India and again in the Himalayas. **Range and habitat:** OR; Himalayas to southern China. Resident in hill forests. **Similar species:** Vivid Niltava.

White-browed Fantail *Rhipidura aureola* (Monarchidae) 17 cm (6½ in). The very broad bill, surrounded by long bristles, and the long graduated tail are typical of this group of flycatchers. A very sprightly and restless bird, constantly twitching and pirouetting around tree-trunks, expanding and closing the tail all the time. It has a short, attractive little song, and with its posturing and indifference to man, is a very pleasing bird to watch. The food consists of small insects, which are snapped up continuously, caught in the air or picked off vegetation. The sexes are alike. **Range and habitat:** OR; India and Sri Lanka to Burma. Resident in bushy places, gardens and woodland. **Similar species:** other fantail flycatchers.

White-rumped Shama *Copsychus malabaricus* (Turdidae) 27 cm (10½ in). Probably more familiar as a cagebird than in the wild, it is highly prized for its beautiful, fluty song. Often quite difficult to see, as not only is it a shy bird, but it inhabits much denser cover than the garden-loving Magpie Robin. Its favourite habitat is among the undergrowth in fairly moist forest, often in ravines with small streams or in bamboo, and it feeds to a large extent on the ground. The female is like the male but duller, and in the race occurring in the Andamans, the male has a white, not chestnut, breast. **Range and habitat:** OR. Widespread resident in forests. **Similar species:** Magpie Robin.

Magpie Robin *Copsychus saularis* (Turdidae) 22 cm (8½ in). A very common and familiar bird almost throughout the whole region. It is a familiar sight in gardens where it hops tamely about, cocking the tail jauntily over the back. Although particularly favouring areas around human habitation, also occurs in open woodland or scrub almost anywhere. Feeds on the ground on insects, earthworms or vegetable matter. The song is an attractive whistling, very pleasing but not as rich as that of the Shama. The female is duller and greyer than the male. **Range and habitat:** OR.

Widespread resident except in dense forest or semi-desert. **Similar species:** White-rumped Shama.

Blue Whistling Thrush *Myiophoneus caeruleus* (Turdidae) 30 cm (12 in). The deep blue-back of the plumage is suddenly seen to be brilliantly spangled with a shining, paler blue when the bird moves into a sunlit patch. Most of the time it is in dark shadow in the ravines and stream-sides it haunts. Hops surefootedly about on rocks in the tumbling water on its strong legs, or forages among the dead leaves on a nearby path, when it looks very like a Blackbird. Also has the habit of raising and flirting its tail. The call is a piercing, repeated whistle. **Range and habitat:** OR. Widespread resident by hill streams in forest. **Similar species:** Blue Rock Thrush.

White-throated Rock Thrush *Monticola gularis* (Turdidae) 18 cm (7 in). The female, though lacking the bright colours of the male, also has a conspicuous white throat patch. The white wing patch and rufous rump on the male distinguish it from similar related species. A beautiful little thrush, though somewhat inconspicuous in its usual haunts on the forest floor. Sometimes visits gardens and more open ground, especially on migration. Usually solitary and somewhat shy, it flies up into low branches if disturbed. **Range and habitat:** EP and OR, wintering widely in South-East Asia, in woodland. **Similar species:** Blue Rock Thrush. **Not illustrated.**

Siberian (Ground) Thrush *Zoothera sibirica* (Turdidae) 23 cm (9 in). A large white patch underneath the wing is noticeable in flight, and the male's slaty plumage with white eyebrow is diagnostic. The female is brown, barred darker below, with a buffy eyebrow. It is a shy bird, often in small groups in winter which quickly take alarm and fly up into trees. Much time is spent on the ground, turning over the leaf debris in the search for insects, worms and berries. It makes a quiet twittering. **Range and habitat:** EP, in winter to OR. Haunts shrubbery, thickets and woods, often near water.

White's Thrush *Zoothera dauma* (Turdidae) 26 cm (10 in). A beautiful and prominently marked thrush, showing a large pale patch under the wing in flight. A very widely distributed bird which has been divided into some 16 subspecies. Many of the races are sedentary birds in isolated areas of hill forest throughout the region, and even vary as to the number of tail feathers. Usually seen singly or in pairs, often in a clearing or by a small stream, it is a quiet bird and rather shy. Feeds on the ground, on insects, fallen berries or worms. **Range and habitat:** EP, OR and AU. Northern populations migratory. Haunts woodland and forested hillsides. **Similar species:** Long-tailed Mountain Thrush.

Black-throated Thrush *Turdus ruficollis* (Turdidae) 25 cm (10 in). Closely related to the Red-throated Thrush, but lacks chestnut in the tail apart from the obvious difference in the throat. Both birds mix in winter flocks. Feeds on ground in usual manner of family, but also forages in bushes and trees for berries and also in flowering trees, possibly for insects. Has a loud, abrupt contact note and a chuckling alarm call. Early autumn migrants in the Himalayas move to lower levels with onset of winter. **Range and habitat:** EP, in winter to northern parts of OR, in woods, scrub and cultivation. **Similar species:** Red-throated Thrush.

Grey-headed Canary Flycatcher

Rufous-bellied Niltava

White-browed Fantail at nest

White-rumped Shama

Magpie Robin

Blue Whistling Thrush

Siberian (Ground) Thrush

White's Thrush at nest

Black-throated Thrush

Grey Bushchat *Saxicola ferrea* (Turdidae) 15 cm (6 in). The female is browner than her mate, with a less marked eye patch and a brown tail. A common bird on grassy or scrubby hillsides where it perches conspicuously on a bush or bracken frond. It prefers rather taller and thicker scrub than do some of its relatives, even bamboo thickets, and quite often perches in trees. It has the usual hard alarm note of the stonechat family, but also has quite a pretty little song. It feeds on small insects which are caught after a short sally from the look-out perch, in the manner of a shrike. **Range and habitat:** OR; Himalayas east through southern China, wintering at lower altitudes. Scrub and cultivation in the hills. **Similar species:** Whinchat (see p 174).

Pied Bushchat (Stonechat) *Saxicola caprata* (Turdidae) 13 cm (5 in). The female is brown and buff; the male, smart black and white. Sometimes when the male is sitting the white wing patches are concealed, but become obvious when he drops to the grass for an insect. A typical stonechat, liking cultivated terraces or hillsides dotted with bushes on which it perches, sometimes fluttering up to take an insect or in brief song-flight. Short, quiet song, and a hard alarm note. This species and the Grey Bushchat are often parasitized by cuckoos. **Range and habitat:** south EP, OR and AU, to New Guinea. Resident and migrant in open, bushy country. **Similar species:** Stonechat (see p 174).

Indian Bluechat *Erithacus brunneus* (Turdidae) 15 cm (6 in). The female is a brown bird, paler on the underparts, quite unlike the colourful male. One of a group of 5 species of similar appearance, it is a skulking, unobtrusive bird of deep undergrowth. It keeps much to the ground or lowest levels of bushes, foliage and tangle of fallen branches and ferns in moist, shady forest. Hops about quietly, now and then flicking the wings or flirting its tail. Feeds on insects. Has a brief song, starting with 3 notes, which often draw attention to its presence. **Range and habitat:** OR. Breeds in south-west China and the Himalayas, wintering in south-west India and Sri Lanka. **Similar species:** Red-flanked Bluetail. **Not illustrated.**

Brown Rockchat *Cercomela fusca* (Turdidae) 17 cm (6½ in). Modest appearance, but confiding behaviour. Often in pairs, seen in old forts and temples, where it nests in crevices in the masonry, but also inhabits rocky hillsides away from habitation. Feeds on insects caught on the ground, often pouncing from a look-out on a wall or roof. Flies directly and strongly like a wheatear. Has a warbling song and a harsh alarm note. **Range and habitat:** OR, in Pakistan and northern India. Resident in dry countryside. **Similar species:** Northern Wheatear (see p 174).

Plumbeous Water Redstart *Phoenicurus fuliginosus* (Turdidae) 14 cm (5½ in). The female is much browner than the male, and instead of a chestnut tail, has this brown with a broad white base. Flits around boulders in rushing Himalayan rivers and streams, the tail constantly being wagged and fanned, as it searches for flies. Sometimes makes a little fluttering jump in the air or forages on the bank, soon returning to the centre of the stream. Always on the move, it is a delightful bird to watch. It has a short song and a sharp little alarm note. **Range and habitat:** OR; Himalayas to western China. Resident by hill streams. **Similar species:** White-capped Redstart. **Not illustrated.**

Daurian Redstart *Phoenicurus auroreus* (Turdidae) 15 cm (6 in). Both sexes share the white wing patch, by which the female can be told from the female Common Redstart – the males, apart from this feature, are very alike. Typical redstart behaviour, with constant shivering of the tail, and active foraging in trees and bushes for insects. Sometimes flies down to the ground from a low perch, but is soon up in the foliage again. **Range and habitat:** EP and OR. Breeds in north and east China and the eastern Himalayas, wintering to South-East Asia and the Himalayan foothills. Haunts groves, cultivation, open woodland. **Similar species:** Hodgson's Redstart.

Bar-winged Prinia *Prinia familiaris* (Sylviidae) 13 cm (5 in). Small greenish-brown bird with 2 pronounced yellowish-white wing bars, a grey throat and lemon-yellow belly. Bill medium length and slightly decurved. Tail long and graduated with white tips to the feathers. Within its rather limited range it is a familiar bird with a remarkably loud and explosive song. **Range and habitat:** endemic to Java, Sumatra and Bali where it frequents a variety of open grassy habitats below 1,500 m (4,900 ft), including gardens around habitation. Especially common on Java. **Similar species:** other Prinias.

Grey-headed Flycatcher-warbler *Seicercus xanthoschistos* (Sylviidae) 10 cm (4 in). One of a group of warblers closely related to the willow-warblers, but with a broader bill surrounded with long bristles – hence the reference to flycatchers in the name. A tiny bird, constantly flitting about in foliage, flicking its wings, or hovering for a moment to pick an insect off a leaf. It is often in company with other warblers and babblers in the mixed hunting parties which forage through the woods. The call-note, which is frequently heard, is a thin, high note. **Range and habitat:** OR. Resident in hill forests in the Himalayas. **Similar species:** other flycatcher-warblers. **Not illustrated.**

Tickell's Willow-warbler *Phylloscopus affinis* (Sylviidae) 11 cm (4½ in). The Himalayas and South-East Asia are the headquarters of the leaf- or willow-warblers, with some 23 breeding species. This is easier than some to identify, with its bright yellow underparts, and dark, olive upperparts without a pale wing-bar. Typical of the group, it is an active little insectivorous bird, which searches the foliage in the constant quest for food. Often in the lower levels of trees or in low bushes and scrub near the ground. It is a particularly high altitude nester, at up to 4,500 m (14,800 ft). **Range and habitat:** OR; Himalayas and south-west China, wintering south in India, in woodland and scrub. **Similar species:** other willow-warblers (see p 173). **Not illustrated.**

Blyth's Crowned Willow-warbler *Phylloscopus reguloides* (Sylviidae) 11 cm (4½ in). A distinct yellowish line along the crown and 2 white wing-bars, together with white borders to the tail, are the field-marks of this pretty little warbler. There are, however, several very similar species. A common and active small bird which hunts through the foliage, often with other species. It swings through the smallest twigs or makes brief flycatching sorties and never seems to be still, except perhaps when singing, though even this is often done on the move. The song is a short warble. **Range and habitat:** OR; Himalayas to Indo-China. Resident in evergreen forests. **Similar species:** other leaf-warblers (see p 173). **Not illustrated.**

Mountain Leaf-warbler *Phylloscopus trivirgatus* (Sylviidae) 10 cm (4 in). A sedentary species of montane forest-dwellers, with a very wide distribution from Malaysia through all the islands south and eastwards to New Guinea. The isolation of all these many populations has led to the evolution of no fewer than 34 subspecies, which are differentiated by variations in the plumage colour and markings. All races share the distinctly banded head pattern in black and yellow. A bird of the forest canopy, often feeding on the ferns and orchids which festoon the larger trees. It has a variety of call-notes, and a high-pitched song. **Range and habitat:** OR and AU. Resident in montane forest. **Similar species:** other leaf-warblers (see p 173).

Pallas's Grasshopper Warbler *Locustella certhiola* (Sylviidae) 11 cm (4½ in). An extremely skulking bird in thick waterside vegetation. Usually only seen briefly as it flies away, before dropping down, showing reddish brown rump and a very rounded tail with whitish tips to the feathers. Runs along the ground like a rodent, and can slip with ease through the tangled stems and debris. Churring alarm note. **Range and habitat:** EP, in winter across OR, in wet paddyfields, marshes and wet thickets. **Similar species:** Grasshopper Warbler (see p 172).

Grey Bushchat

Pied Bushchat (Stonechat)

Brown Rockchat

Daurian Redstart

Pallas's Grasshopper Warbler

Mountain Leaf-warbler

Bar-winged Prinia

Long-tailed Tailorbird *Orthotomus sutorius* (Sylviidae) 13 cm (5 in). A dull-looking little bird famed for its remarkable nest. Rufous on the crown. It has a rather long, sharp bill and a spiky tail, which is often carried cocked over the back. Some races develop longer tail feathers in the breeding season. Usually seen foraging for insects in shrubs in gardens or low scrub. Makes a loud, sharp, double call-note, which is more familiar to most people than its appearance. The nest is in a large leaf, or several smaller leaves, the edges of which are sewn together with fibres. **Range and habitat:** OR. Widespread in scrub or cultivation in the lowlands. **Similar species:** Rufous-fronted Prinia.

Rufous-fronted Prinia *Prinia buchanani* (Sylviidae) 13 cm (5 in). The dark brown tail with conspicuous white tips, together with the rufous crown, distinguish it from several similar species. A slender little warbler with a conspicuously long tail, which shows the white tips when it is fanned. Often seen in small parties in low scrub, constantly active as they creep and flutter about through the stems, twitching or cocking the tail, or hop about on the ground below in their search for insects. Makes a double call-note, and has a twittering song. **Range and habitat:** OR; in Pakistan and India. Resident in dry scrub and semi-desert. **Similar species:** other prinias. **Not illustrated**.

Olivaceous Bearded Bulbul *Criniger pallidus* (Pycnonotidae) 23 cm (9 in). The puffed-out pure white beard, greyish breast and primrose yellow belly distinguish this species from similar bearded bulbuls. Live in pairs or small groups throughout the forests of South-east Asia. Shyer than most bulbuls, they frequent the dark forest understorey, feeding on a variety of small fruits and insects. Often associate with other birds in mixed flocks, especially babblers and flycatchers. **Range and habitat:** OR. Hainan, Indo-China, northern Thailand and Burma; resident in undergrowth of evergreen forest. **Similar species:** Ashy-fronted Bearded Bulbul.

Himalayan Streaked Laughing Thrush *Garrulax lineatus* (Timaliidae) 22 cm (8½ in). A combination of the greyish body with rufous wings, tail and ear patch, and whitish shaft streaks on its back and breast distinguish this from other Laughing Thrushes. Small groups forage around scrub and rocks on open hillsides and around mountain villages. Rarely flying far, these noisy birds tend to glide downhill with bowed wings when disturbed, although around villages they are often tame and merely hop to cover. They are agile and strong on their feet as they hop around searching for berries and insects. **Range and habitat:** OR. Himalayas from Afghanistan to Bhutan, on sparsely vegetated hillsides and around villages to 3050 m (10,000 ft). **Similar species:** Striped Laughing Thrush.

White-crested Laughing Thrush *Garrulax leucolophus* (Timaliidae) 30 cm (12 in). A typical species inhabiting dry deciduous woods and bamboo thickets in the lower foothills, it moves in parties through the undergrowth and lower branches, the bushy white crests held vertically. After a period of silence or quiet chattering, the flock may burst into wild, cackling laughter, repeating this at intervals until all the birds have left the vicinity. Forages vigorously among the fallen leaves like a thrush, bounding along on its strong legs, but keeps well under cover. **Range and habitat:** OR; Himalayas to Indo-China and Sumatra. Resident in forest in the lower hills. **Similar species:** White-throated Laughing Thrush. **Not illustrated**.

White-headed Jungle Babbler *Turdoides striatus* (Timaliidae) 25 cm (10 in). A common and familiar bird, sociable and of undistinguished and untidy appearance. Brown, with a long, scrawny tail, and pale yellowish eyes and bill. The birds keep together in flocks of up to a dozen, hopping about below hedges or bushes, or flying jerkily after each other across a clearing, all the time maintaining a low conversation. When alarmed, the calls suddenly increase in volume to a noisy chatter. The flocks have a complex social structure, with nonbreeding birds assisting in the rearing of nestlings, and the same individuals staying together in the flock for long periods of time. **Range and habitat:** OR. Resident in India and Pakistan in light woodland and cultivation or gardens. **Similar species:** Common Babbler.

Common Babbler *Turdoides caudatus* (Timaliidae) 23 cm (9 in). Sociable like others of its family, and very much a ground-dwelling bird, the flocks foraging fussily about under bushes or scrub. The birds often cock up their straggly tails and bounce about after each other, now and then taking to wing with a low flap-and-glide flight to another patch of cover. The birds keep up a conversation of low whistles which become louder and shriller when they are alarmed. They feed on insects, berries and seeds, and also nectar. **Range and habitat:** OR. Resident in India in dry scrub or bushy country. **Similar species:** Jungle Babbler. **Not illustrated**.

Oriental Yellow-eyed Babbler *Chrysomma sinense* (Timaliidae) 18 cm (7 in). A common babbler of grassy places and thickets in southern Asia, where it climbs about through the tangle of stems, now and then emerging into the open to utter a snatch of song, when the yellow eye and white eye stripe are noticeable. Flutters across open spaces with a weak, jerky flight and immediately vanishes into cover again. Often in small, loose groups in company with other small birds such as munias and warblers. It makes a loud, cheeping call-note which often advertises its presence when invisible in the vegetation. **Range and habitat:** OR. Widespread in the lowlands in grassy and bushy places. **Similar species:** Chestnut-capped Babbler. **Not illustrated**.

Chestnut-backed Scimitar-babbler *Pomatorhinus montanus* (Timaliidae) 21 cm (8 in). The name refers to the long, strongly curved bill which is characteristic of this little group of babblers. Very distinctive when once seen, its black crown and cheeks contrast with white supercilium and yellow bill. The mantle, rump and flanks are bright chestnut. However, it is a skulking bird of dense undergrowth which rarely allows itself to be watched in the open. Often, its presence in some jungly overgrown gully is given away by the beautiful mellow, fluty note often repeated several times. Also has a coarse, bubbling call. **Range and habitat:** OR, Greater Sundas, Malaysia. Resident in hill forest. **Similar species:** other scimitar-babblers.

Striped Tit-babbler *Macronous gularis* (Timaliidae) 14 cm (5½ in). A very widely distributed babbler in southern Asia, with some 25 subspecies arising from the many isolated populations on islands. It is an easily distinguished species in the field with its rufous crown and dark streaking on the yellow underparts. A noisy little bird with a repeated mellow note, constantly uttered as it forages through the undergrowth. Often in small parties, which are often joined by various other kinds of small birds in hunting parties or 'bird waves', which move rapidly onwards in the forest at all levels in their quest for food. **Range and habitat:** OR. Resident in undergrowth and forest in the lowlands from south-west India to the Philippines. **Similar species:** Red-fronted Tree Babbler. **Not illustrated**.

Quaker Babbler (Brown-cheeked Fulvetta) *Alcippe poioicephala* (Timaliidae) 16 cm (6½ in). The name is an allusion to the plain colours of this small babbler; the grey head and brown wings are dull. Common inhabitant of 2nd-growth and woods from western India to China in a number of subspecies, usually seen in small parties, foraging actively in the foliage of bushes and trees. The call-note is a repetition of double notes at alternately higher and lower pitches. Populations in some areas have a black band on the sides of the crown, and are confusingly like a related species. **Range and habitat:** OR; southern India to Indo-China. Resident in forest in the plains and foothills. **Similar species:** Grey-cheeked Fulvetta. **Not illustrated**.

Long-tailed Tailorbird

Himalayan Streaked Laughing Thrush

Olivaceous Bearded Bulbul

White-headed Jungle Babbler
feeding young

Chestnut-backed Scimitar Babbler

Yellow-naped (Whiskered) Yuhina *Yuhina flavicollis* (Timaliidae) 15 cm (6 in). The short, black moustachial stripe is a more constant plumage feature in the various races of this bird than the collar, which may be yellow, rufous or almost absent. A characteristic of all yuhinas is the jaunty, pointed crest, and in its acrobatic foraging this bird resembles the familiar Crested Tit of Europe. It is always in the trees, swinging from a spray or fluttering out after an insect, moving through the foliage in company with other small insectivorous birds. **Range and habitat:** OR; Himalayas to Indo-China. Resident in hill forests, moving to lower altitudes in winter. **Similar species:** White-naped Yuhina.

Brown-capped Tit-babbler (Gould's Fulvetta) *Alcippe brunnea* (Timaliidae) 15 cm (6 in). An active and sprightly babbler of tangled vegetation and dwarf bamboo in wet hill forests, often foraging among the leaf litter on the ground. Difficult to watch in these dense and dank surroundings, it does not make observation any easier by being quite shy. It makes a constant churring contact note and also has a short, jingling song in the breeding season. Often in small groups, and sometimes with other species hunting through the bushes in winter. **Range and habitat:** OR; in the Himalayas and southern China. Resident in forest and forest edge up to 2,400 m (7,900 ft). **Similar species:** Rufous-throated Fulvetta. **Not illustrated.**

Pekin Robin *Leiothrix lutea* (Timaliidae) 14 cm (5½ in). A loud burst of musical, rich song from the depths of a shady gully often betrays the presence of this beautifully coloured but retiring babbler. It is, on account of the song, greatly valued as a cagebird, when the bright plumage is certainly easier to see than is often the case in the wild. The female bird lacks the crimson on the wing, this patch being all yellow. Outside the breeding season, behaves much as other babblers, hunting through foliage or rummaging about in the debris on the forest floor, usually in small parties. It makes a variety of harsh or piping call-notes. **Range and habitat:** OR; Himalayas to China. Resident in hill forest, moving to lower altitudes in winter. **Similar species:** Silver-eared Mesia.

Hoary Barwing *Actinodura nipalensis* (Timaliidae) 20 cm (8 in). The barwings are a small group of 6 species, intermediate in size between the small babblers and the laughing thrushes, characterized by mop-like crests and barred tails and wings. In winter, they forage in the higher levels of the hillside trees, hopping actively along the branches and searching the epiphytic ferns and mossy stems for insects. Their size, crests and long tails make them easy to identify among the mixed-species hunting parties so evident in the woods at this time of year. **Range and habitat:** OR, confined to the Himalayas. Resident in higher-level forests. **Similar species:** other barwings. **Not illustrated.**

Bar-throated Siva (Chestnut-tailed Minla) *Minla strigula* (Timaliidae) 15 cm (6 in). One of the more colourful of the smaller babblers, with orange-yellow patches on the crown, wings and tail. It is a particularly noticeable species in the mixed hunting parties which roam through the woods in winter, in company with tits, nuthatches, creepers and other small babblers. It keeps in the trees all the time, hunting actively through the foliage for insects, berries and seeds. It has a short, 3- or 4-note song, and a simple, piping call-note. **Range and habitat:** OR; Himalayas to South-East Asia and Malaya. Resident in hill forest. **Similar species:** Red-tailed Siva (Red-tailed Minla).

Black-headed Sibia *Heterophasia melanoleuca* (Timaliidae) 23 cm (9 in). Common in the hills in South-East Asia, in loose bands which roam through the forest, exploring the branches and clumps of ferns, mosses and orchids which festoon the trees at higher altitudes. Active and restless, the birds flirt their tails and constantly call to each other with a variety of notes. A very characteristic song, evocative of these misty evergreen forests, is a sad, descending cadence of 6 notes. There are several subspecies in which the back

colour varies from black to brown and grey. **Range and habitat:** OR. Resident in hill evergreen forest. **Similar species:** Long-tailed Sibia. **Not illustrated.**

Chinese Yellow Tit *Parus spilonotus* (Paridae) 14 cm (5½ in). A noisy and sociable tit, easily recognized by the pointed yellow-tipped black crest, yellow cheeks, eyebrow and lores and streaked green back. Often in parties with other species, in particular babblers, when foraging through the trees. Typical in their actions, swinging upside-down at the tip of a twig, or searching around the bole of a tree near the ground. Like other tits, they nest in holes, and constantly repeat their short, cheery songs whilst searching for food. Food consists of insects, grubs and berries. **Range and habitat:** OR. Resident in pine and mixed woodlands from the eastern Himalayas to Indo-China, usually above 1520 m (5000 ft). **Similar species:** Yellow-cheeked Tit.

Red-headed Tit *Aegithalos concinnus* (Aegithalidae) 10 cm (4 in). A tiny, long-tailed tit, delightful to watch as it is often quite oblivious of the human observer. Always on the move, swinging and flitting through the bushes, clinging upside-down as it examines a leaf or fluttering to the next bush as the flock moves along the hillside. Often forages near the ground amongst the low vegetation in forest clearings, but sometimes high in the trees. Usually in company with a variety of other species of small insectivorous birds which keep a loose association as they hunt for food. The sexes are alike. **Range and habitat:** OR; Himalayas to Indo-China. Resident in hill scrub and forest edge. **Similar species:** Long-tailed Tit (see p 176).

Himalayan Tree-creeper *Certhia himalayana* (Certhiidae) 12 cm (4½ in). Distinguished from the familiar creeper of Europe, which also occurs in the Himalayas, by the barring on the tail. Habits typical of those of the family. Usually seen ascending the trunk of a tree in jerky, hesitant fashion, probing into the cracks and crevices, or slipping round to the other side, only to reappear higher up, moving outwards underneath a large branch. The food consists of tiny insects, grubs and spiders. The call is a low, thin note, and it has a short, twittering song. **Range and habitat:** OR. Resident in woods in the Himalayas. **Similar species:** Common Tree-creeper (see p 176). **Not illustrated.**

Velvet-fronted Nuthatch *Sitta frontalis* (Sittidae) 11 cm (4½ in). A very widely distributed bird, in a number of races. The red bill and violet upperparts are diagnostic. A small nuthatch, and quite sociable, often being seen in 2s and 3s, and very often participating in the mixed-species hunting flocks. Usual nuthatch actions, hitching jerkily around branches and stems, stopping to tap here and there, or exploring a clump of fern. It is highly adaptable as regards habitat, and occurs in all kinds of woodland from dense evergreen to tea plantations or mangroves. It has a repeated, whistling call-note. **Range and habitat:** OR. Widespread resident in woodland in the hills and plains. **Similar species:** Eurasian Nuthatch (see p 176).

Oriental White-eye *Zosterops palpebrosa* (Zosteropidae) 10 cm (4 in). A tiny, green, warbler-like bird, with sharply pointed bill and a ring of white feathers around the eye. A sedentary species which has developed many subspecies across its wide range, especially in the numerous islands south and east of Malaya. Common in most types of woodland, where it forages, often high in the canopy, in company with other small birds. It feeds on small insects and nectar, so is often seen probing into the blossoms in flowering trees. It has a quiet, chirping call-note and a jingling song. **Range and habitat:** OR. Widespread resident in woodland in the hills and plains. **Similar species:** other white-eyes (see pp 230, 318).

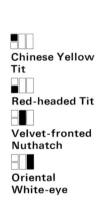

Yellow-naped (Whiskered) Yuhina

Bar-throated Siva (Chestnut-tailed Minla)

Pekin Robin

Chinese Yellow Tit

Red-headed Tit

Velvet-fronted Nuthatch

Oriental White-eye

Purple Sunbird *Nectarinia asiatica* (Nectariniidae) 10 cm (4 in). An all-dark sunbird, often looking quite black until it is seen in a good light, with a short, square tail. Constantly active around flowering shrubs and trees, where it probes into the blossoms with its strongly curved bill. Although nectar is an important food item, many small flies and insects are taken as well. It has a loud, shrill call-note, and is usually seen singly, although sometimes consorting with other nectar-seeking species. Following the breeding season, the males moult into a duller plumage similar to that of the female, except for some blue patches. **Range and habitat:** EP and OR, from Arabia to Indo-China. Resident in light woodland and gardens. **Similar species:** Purple-rumped Sunbird.

Fork-tailed Sunbird *Aethopyga christinae* (Nectariniidae) 10 cm (4 in). The finely pointed central tail feathers on the male extend to 13 mm ($\frac{1}{2}$ in) beyond the rest, so the total tail length is markedly less than on some of its close relatives. The female has a rounded tail, and is a dull olive bird, yellower on the underparts. A beautiful little sunbird, which haunts the topmost boughs of tall trees. It has the active foraging habits of the family, seeking nectar or insects among blossom, and has a loud, metallic note and a quieter chirp. **Range and habitat:** OR; south China and Hainan. Resident in woodland in hill country. **Similar species:** Yellow-backed Sunbird.

Yellow-backed Sunbird *Aethopyga siparaja* (Nectariniidae) 10 cm (4 in). The scarlet back and breast and green tail readily separate the male bird from other sunbirds, though the female is similar to those of other species, being olive above and yellower below. Constantly fluttering around blossoms in the forest or in gardens, hovering to take nectar or insects for a moment, then darting off to explore an adjacent flower. It is often seen singly, and has a typical sunbird sharp call-note. There are several races of this bird, with minor plumage variations. **Range and habitat:** OR; India to the Philippines. Resident in moist evergreen forest. **Similar species:** Mrs Gould's Sunbird. **Not illustrated.**

Plain Flowerpecker *Dicaeum concolor* (Dicaeidae) 9 cm (3$\frac{1}{2}$ in). A dully plumaged, tiny little bird, often seen in pairs which hunt through foliage and flowering trees in their search for nectar and insects. Flowerpeckers are important in assisting the spread of parasitic plants such as mistletoe, as the fruits are swallowed, and the undigested, sticky seeds deposited on branches wherever the birds happen to leave them. The birds make a loud, chipping note as they fly, often at considerable heights, from 1 treetop to another. **Range and habitat:** OR. Widespread resident in hill forests. **Similar species:** Tickell's Flowerpecker. **Not illustrated.**

Orange-bellied Flowerpecker *Dicaeum trigonostigma* (Dicaeidae) 9 cm (3$\frac{1}{2}$ in). The male is a tiny gem of a bird, bright blue head and wings contrasting with an orange rump and belly, but the female has only a vestige of these colours, being mostly a dull green-grey. It is an active bird, often frequenting the extreme tops of trees and turning with an almost mechanical flicking movement from side to side. Like other flowerpeckers, it feeds on the berries of parasitic plants, the seeds subsequently adhering to branches within the bird's droppings, and thereby colonizing fresh sites. Many subspecies have been described from the bird's wide range. **Range and habitat:** OR; from eastern India through south-east Asia to the Philippines. Resident in forests in the plains and lower hills. **Similar species:** Crimson-breasted Flowerpecker.

Thick-billed Flowerpecker *Dicaeum agile* (Dicaeidae) 9 cm (3$\frac{1}{2}$ in). A widely distributed bird in southern Asia, occurring in a number of subspecies which differ in plumage details. The swollen bill and red eyes are distinctive. A somewhat solitary species, sometimes in pairs, which hunts actively through foliage and flowers for nectar and insects. As with other flowerpeckers, also takes berries of tree parasites. However, instead of swallowing them whole, it strips the flesh off the berry, and discards the sticky seeds by wiping them off on the branch. It thus helps to concentrate the parasite on its existing host-tree. **Range and habitat:** OR. Widespread resident in hill forests. **Similar species:** Tickell's Flowerpecker. **Not illustrated.**

Crested Bunting *Melophus lathami* (Emberizidae) 15 cm (6 in). A soberly plumaged bunting with chestnut wings and tail and black head and body, but distinguished by the pointed crest. The female is a brown, sparrow-like bird with a vestigial crest. In the winter, often seen in small flocks which scratch about in the dust on jungle tracks or bare ground on hillsides, flying into nearby bushes when disturbed. It has a brief call-note and a typically dry and short song phrase. It is a resident species, and is not seen in large numbers like the migratory buntings in their winter quarters. **Range and habitat:** OR; Himalayas and north India to eastern China, on hillside clearings and open ground. **Not illustrated.**

Black-headed Bunting *Emberiza melanocephala* (Emberizidae) 18 cm (7 in). The large flocks of this handsome bunting which visit the Indian plains in winter do a good deal of damage to the crops, particularly when the numbers are concentrated in the spring, before the return passage to their breeding areas. As with most buntings, the female is quite unlike the male, having a streaky brown, sparrow-like plumage. The call is a sweet, rather husky note, and the song a short jingle, sometimes heard just as the birds are about to depart. **Range and habitat:** EP, in winter to OR in Pakistan and western India. Open country and cultivation in the plains. **Similar species:** Red-headed Bunting.

Yellow-breasted Bunting *Emberiza aureola* (Emberizidae) 15 cm (6 in). One of the better songsters among the buntings. This fact, together with the beautiful plumage of the male, makes it a favourite cagebird in China. The song can be heard from males in the spring, when they are gathering for their northward migration. Often in very large numbers in the winter quarters in South-East Asia, where they congregate in the ricefields and around villages, taking a great quantity of rice and grass seed. The bright, clean colours of the summer plumage appear as the buff feather tips of the winter dress wear off. **Range and habitat:** EP, in winter to OR, occasionally to WP, in open country and cultivation. **Similar species:** Black-headed Bunting.

Black-faced Bunting *Emberiza spodocephala* (Emberizidae) 15 cm (6 in). The female lacks the grey head and black mask, and has a pale yellow eyebrow, but is otherwise much like the male. It is scarcer than other buntings in winter, appearing in small groups, which frequent hedgerows, scrub and damp, grassy places. The birds feed on the ground, flying up into nearby bushes if disturbed, calling with a ticking note. Being rather shy, and not as conspicuous as those of its relatives which congregate in large numbers, it is probably often overlooked. There are several races of this bunting in China, varying somewhat in plumage colouring. **Range and habitat:** EP, wintering to northern parts of South-East Asia and southern China. Frequents open country, wet scrub and crops. **Similar species:** Yellow-breasted Bunting.

Red-headed Bunting *Emberiza bruniceps* (Emberizidae) 17 cm (6$\frac{1}{2}$ in). A popular cagebird on account of the striking plumage and pleasant little song. Closely related to the Black-headed Bunting, with which it occurs in very large numbers in winter, feeding in crops. When scared off or otherwise disturbed, flies up in dense flocks which colour the trees yellow. It is a variable species, the males sometimes having yellow crowns or heads, and hybridizes with the Black-headed Bunting where the breeding ranges overlap. **Range and habitat:** EP, in South-West Asia, wandering to western Europe, and in winter to OR. **Not illustrated.**

Purple Sunbird

Fork-tailed Sunbird

Orange-bellied Flowerpecker

Black-headed Bunting

Yellow-breasted Bunting

Black-faced Bunting

Red-headed Bullfinch *Pyrrhula erythrocephala* (Fringillidae) 17 cm (6½ in). A quiet and unobtrusive species, often announcing its presence with a soft, low whistle. In small parties in winter, feeding slowly and methodically in bushes and trees, nipping off buds, or sometimes picking up seeds from the ground, where they favour patches of tall herbs under the trees. When disturbed, they flit off instantly into cover, showing the white rump patch. Female's head is yellowish green instead of red. **Range and habitat:** OR. Resident in the Himalayas at up to 4,000 m (13,100). **Similar species:** Bullfinch (see p 180). **Not illustrated.**

Red Avadavat (Red Munia) *Amandava amandava* (Estrildidae) 10 cm (4 in). A favourite cagebird, being easy to keep and active and cheerful in captivity. In the wild, a sociable little bird, gathering in large numbers in grassy places, twittering as they cling to the seeding heads and stems. The tall grasses of moist ground and swamp margins are particularly favoured. The birds roost communally in such places, and also in reed-beds. The female lacks the red underparts with white spotting, and the male has a similar plumage outside the breeding season. **Range and habitat:** OR; India to China and Java, in open country in the plains and lower hills. **Similar species:** Green Munia.

Spotted Munia (Nutmeg Mannikin) *Lonchura punctulata* (Estrildidae) 10 cm (4 in). Like the avadavats, all the munias are familiar as cagebirds. This species has a very wide distribution in southern Asia, with many subspecies, although, very curiously, it is absent from Borneo. Always in flocks which feed in crops and grass, flying about in groups from place to place, or perching in nearby bushes or trees. The sexes are alike in this munia, though outside the breeding season the plumage is duller, lacking the speckled underparts. **Range and habitat:** OR; India to the Philippines. Resident in open country in the lowlands and foothills. **Similar species:** White-backed Munia. **Not illustrated.**

Pin-tailed Parrot-finch *Erythrura prasina* (Estrildidae) 15 cm (6 in). A colourful, small, finch-like bird, typical of this small group of mainly green and red members of the family. This is the only species with, in the male, an elongated and pointed tail. The sexes are similar apart from the tail shape, the female being duller. Sometimes feeds on bamboo seed-heads, but chiefly a rice-eater, and a serious pest in the paddyfields in the uplands of Borneo. Either migratory, or else wanders widely, though little is yet known about its movements. It has a short, high-pitched call-note. **Range and habitat:** OR; South-East Asia to Malaya and Borneo. Resident or nomadic in cultivation, thickets and on the forest edge. **Similar species:** other parrot-finches.

Baya Weaver *Ploceus philippinus* (Ploceidae) 15 cm (6 in). A common bird of grassy places in the plains. The nesting colonies, with their woven grass nests hanging from the branches of bushes, are a familiar sight. When in breeding plumage, the male is easily told from other Asian weavers by its black mask and yellow breast. However, outside the breeding season, both sexes resemble sparrows. Timing of breeding is related to the onset of the monsoon, and males initiate the colony by building the nests. The females only appear later to approve a nest and mate with its builder. The nest has a long, downward-pointing entrance tunnel, up which the birds fly without a pause. **Range and habitat:** OR; Pakistan to Sumatra. Resident in cultivation and scrub in the plains. **Similar species:** Streaked Weaver.

Streaked Weaver *Ploceus manyar* (Ploceidae) 15 cm (6 in). The male in breeding plumage is easily distinguished by the heavy streaking on the breast and the yellow crown contrasting with the black head. The female is difficult to tell from other species, and both sexes look alike outside the breeding season. The construction of the nest differs from that of the Baya, in that it is supported from a bunch of stems, rather than tapering at the top to a single support point. Also, the entrance tube is much shorter. The nesting

colonies often consist of only a few pairs, and are always located in tall reeds over water. **Range and habitat:** OR; Pakistan to Java. Resident in open, well-watered country in the plains. **Similar species:** Baya Weaver.

Sind Jungle Sparrow *Passer pyrrhonotus* (Ploceidae) 13 cm (5 in). For many years after its initial discovery, this species was not recorded again, probably owing to confusion with the common House Sparrow. It is, however, a smaller and slimmer bird, and the male's black bib is much narrower and does not extend beyond the throat. Females are extremely similar. Its identity as a good species was not confirmed until it was found nesting close to House Sparrows. A bird of tall vegetation along watercourses in the region of the Indus river, it occurs in small groups, and is not as obviously gregarious as its larger relative. It nests in loose colonies in trees standing in or by water. **Range and habitat:** OR. Resident in waterside scrub in semi-desert in Pakistan. **Similar species:** House Sparrow (see p 182). **Not illustrated.**

Brahminy Myna (Black-headed Starling) *Sturnus pagodarum* (Sturnidae) 22 cm (8½ in). A strikingly plumaged starling, with its long black crest, pink underparts and white-bordered tail. Sexes alike, but the female has a shorter crest. Often in parties which frequent flowering and fruiting trees, it is a common and familiar bird in India. It feeds on berries, nectar or insects, which it searches for in typical starling fashion on the ground, walking or running about in the grass or amongst grazing animals. It has a musical song and various other notes, and also mimics other bird calls. **Range and habitat:** OR; Pakistan, India and Sri Lanka. Resident in light woodland and cultivation in the plains and foothills. **Similar species:** Rose-coloured Starling.

Asian Pied Myna *Sturnus contra* (Sturnidae) 23 cm (9 in). A smart black and white starling, the plumage set off by an orange bill and bare skin around the eye. The sexes are alike. A gregarious, noisy bird, often around gardens, fields and villages. It feeds on the ground in loose flocks, taking both insects and grubs as well as vegetable food, especially grain. It associates with other species such as the Common Myna, and shares the same noisy, packed roosts. Like other starlings, it is very vocal, with a variety of whistling notes, and is also a good mimic. **Range and habitat:** OR; India to South-East Asia and Java. Resident in open, cultivated country. **Similar species:** Black-collared Starling. **Not illustrated.**

Common Myna *Acridotheres tristis* (Sturnidae) 22 cm (8½ in). Perhaps the most familiar bird in southern Asia, certainly in towns and around all human habitation. A bold, perky and yet wary bird, highly successful at taking advantage of the feeding and nesting opportunities which man's activities afford. It is as happy walking around small patches of grass in busy cities as foraging amongst cattle by a remote forest settlement. Often in noisy, active flocks, and roosts in huge numbers in groves of trees, reed-beds or on buildings. It takes a wide variety of food, both animal and vegetable, and is highly vocal, with a mixture of harsh and liquid notes. **Range and habitat:** OR, introduced into AF and Arabia. Resident except at the highest altitudes or in thick forest. **Similar species:** Bank Myna.

Hill Myna *Gracula religiosa* (Sturnidae) 30 cm (12 in). A striking species, the glossy black plumage set off by large white wing patches, and the bright yellow wattles and bare skin on the head and nape. The sexes are alike. A most familiar cagebird, owing to its wide vocal repertoire, which is as noticeable in the wild. Characteristic penetrating, short, descending whistle from which it derives its Bornean name of 'Tiong'. Often singly or in small groups in forest treetops, or gathering in fruiting trees with such birds as barbets and green pigeons. Insects and small animals are taken as well as berries and fruit. The nest is placed in a hole in a forest tree. **Range and habitat:** OR. Widespread resident in hill forests. **Similar species:** Ceylon Grackle.

Red Avadavat (Red Munia)

Pin-tailed Parrot-finch

Baya Weaver

Brahminy Myna (Black-headed Starling)

Streaked Weaver building nest

Common Myna

Hill Myna

Red-billed Blue Magpie *Urocissa erythrorhyncha* (Corvidae) 70 cm (28 in). The long, gracefully curving tail is seen to advantage when several birds fly, after each other, across a Himalayan hillside. The flight is rather jerky, several flaps followed by a glide. The birds live in small groups in the hill woods, feeding around cultivated terraces or glades. Though feeding mainly in the trees, they do forage on the ground, and as they hop about, the tail is elevated, though drooping at the tip. They are noisy birds, with a variety of harsh and chattering calls, and are also good mimics. The attractive appearance makes them popular birds in aviaries. **Range and habitat:** OR; Himalayas to southern China. Resident in hill forests. **Similar species:** Yellow-billed Blue Magpie.

Green Hunting Crow (Green Magpie) *Cissa chinensis* (Corvidae) 39 cm (15 in). A beautiful crow, the white tips to the tail and bright rufous wings noticeable in flight. The vivid green of the plumage, often highly cryptic in its forest environment, fades to a dull turquoise in museum specimens. Like many other crows, it is a noisy bird as well as being shy, so is often heard more than it is seen. It has a variety of loud whistles and raucous, chattering notes. Rarely more than 2 or 3 birds together, it stays well concealed in dense vegetation, where it forages for insects or small animals, sometimes in company with other species such as laughing thrushes. **Range and habitat:** OR; Himalayas to Borneo. Resident in hill forest. **Similar species:** Short-tailed Green Magpie.

Black Racket-tailed Treepie *Crypsirina temia* (Corvidae) 34 cm (13 in). One of 2 similar species, both having a curiously spatulate tail measuring up to 20 cm (8 in). Hunts through the forests singly or in small groups, almost always in company with other species such as laughing thrushes and other babblers or drongos. It has a variety of harsh notes, including a characteristic swearing note of 3 syllables. It is shy and unobtrusive but for the note, and always stays in the bushes or trees, appearing not to take food on the ground. **Range and habitat:** OR; southern Burma to Java. Resident in woodland. **Similar species:** Hooded Racket-tailed Treepie.

Indian Treepie *Dendrocitta vagabunda* (Corvidae) 50 cm (20 in). A typical magpie in habits, wary and inquisitive, adaptable, noisy and omnivorous. It is, as its name indicates, a woodland bird, moving about in small parties which attract attention with their distinctive, fluty, 3-note call. They climb about actively among the branches, or hop lightly up big boughs with the tail cocked, searching for insects or fruit. The range of food taken is very wide, as with most crows. A somewhat reluctant flier, it has a jerky flap-and-glide progress, 1 bird following another when the group is on the move, showing the bold white wing patch. It can be extremely shy and inconspicuous around human habitations, though daring in snatching food when opportunity offers. **Range and habitat:** OR; India to Indo-China. Resident in lightly wooded country. **Similar species:** Himalayan Treepie.

Asian Black-headed Oriole *Oriolus xanthornus* (Oriolidae) 25 cm (10 in). A common and beautiful bird, which keeps to leafy trees in gardens or light woodland, attracting attention with its melodious, fluty whistles. The sexes are alike. Largely a fruit-eater, though insects form a substantial part of the diet. Usually seen singly, sometimes in pairs, and not easily noticed in thick foliage despite the brilliant yellow plumage. However, it is not especially shy, and is conspicuous when flying from 1 tree to another. Often consorts with other species such as drongos, barbets, treepies and babblers at fruiting or flowering trees. **Range and habitat:** OR. Widespread resident in wooded areas in the plains and foothills. **Similar species:** Golden Oriole (see p 182). **Not illustrated.**

Jungle Crow *Corvus macrorhynchos* (Corvidae) 48 cm (19 in). Replaces the Raven in much of tropical Asia, being more of a forest species where the 2 birds overlap. However, also occurs in eastern Asia as far north as Manchuria. There are several subspecies, which vary in size, bill shape and details of gloss on the plumage. Often solitary or in small parties, only gathering in larger numbers at a suitably abundant food supply. It takes a wide variety of animal food and carrion, and scavenges habitually around villages and towns, though always very wary and quick to flap away if alarmed. It has the typical harsh cawing of the larger crows. **Range and habitat:** EP and OR. Widespread resident in wooded areas, cultivation or around towns and villages. **Similar species:** Carrion Crow (see p 182).

(Indian) House Crow *Corvus splendens* (Corvidae) 43 cm (17 in). An abundant and successful species, with a distinctly urban lifestyle. It is, perhaps, the most specialized of all its family in profiting from man's activities, and is to be found around every village, city or settlement in large numbers. A highly gregarious bird, roosting in huge numbers, the locality being marked by the conspicuous morning and evening flight paths to and from feeding and foraging areas. Despite its boldness and confidence, it remains quick to take alarm, but as soon as any perceived danger is past, is back immediately to snatch a morsel of food. It is frequently parasitized by a large black cuckoo, the Koel. **Range and habitat:** OR, introduced into Arabia and AF. Widespread resident in the lowlands. **Similar species:** Jackdaw (see p 182).

White-bellied Drongo *Dicrurus caerulescens* (Dicruridae) 23 cm (9 in). The combination of the black plumage, long forked tail and white belly (the white restricted to the vent in some races) is diagnostic. The shrike-like habit of perching on a conspicuous look-out, from which it can launch a chase after insect prey, makes it an easily noticed bird. It is particularly attracted to a food source such as a swarm of termites. It is a pugnacious bird, freely attacking much larger species which trespass on its nesting area, and can harass very effectively, being a superbly agile flier. It has a large variety of attractive, whistled calls, and is also a good mimic of other birds' calls. **Range and habitat:** OR. Resident in forests and forest edge in India and Sri Lanka. **Similar species:** Black Drongo.

Greater Racquet-tailed Drongo *Dicrurus paradiseus* (Dicruridae) 35 cm (14 in). With a tail measuring up to another 30 cm (12 in), this is a conspicuous species in flight. The 2 spatulate barbs at the ends of the extended outer tail feather shafts look like 2 small birds fluttering along just behind, and make a distinct thrumming noise. Often seen in flight in open parts of forest, but easily located when perched in dense evergreen foliage by the constant babble of metallic and whistling calls. It is constantly calling, and when not practising its own repertoire, is mimicking other birds with great fidelity. Though not especially gregarious, a few birds at most being seen together, it is often in company with other species at flowering trees, where it takes nectar. Flying insects form the main part of the diet. **Range and habitat:** OR. Widespread resident in forest in the plains and foothills. **Similar species:** Lesser Racquet-tailed Drongo.

Black Drongo *Dicrurus macrocercus* (Dicruridae) 30 cm (12 in). A very familiar bird around compounds or clearings, which swoops with great agility on flying insects or, like a shrike, pounces on a grasshopper or beetle on the ground. It chooses conspicuous perches, such as telephone wires, bare boughs or fences, and happily rides on the backs of cattle or buffaloes, which helpfully disturb insects from the grass. Usually in pairs, which are ferocious in defence of their nesting territory, unmercifully harrying any crow or cat which strays too close for comfort. Although it has several musical notes, and a good variety of harsh, chattering noises, it is not as accomplished vocally as some of its relatives. **Range and habitat:** OR, marginally EP in South-West Asia. Resident in wooded country in the plains from Iran to China. **Similar species:** Ashy Drongo.

Red-billed Blue Magpie

(Indian) House Crow

Green Hunting Crow (Green Magpie)

Black Racket-tailed Treepie

Indian Treepie

Jungle Crow

White-bellied Drongo

Greater Racquet-tailed Drongo

Black Drongo

THE AUSTRALASIAN REGION

The Australasian Region consists of the continental mass of Australia, along with the large islands of New Guinea and New Zealand and many smaller islands at the eastern end of Indonesia and throughout the western Pacific. It is clearly separated from the Oriental Region by Wallace's Line between Bali and Lombok and Borneo and Sulawesi, with marsupials dominant on the Australian side and placental mammals in the Orient. Birds are of course more mobile than mammals and Malaysia and Indonesia are regions of gradual change-over. Antarctica is sometimes considered to be part of the Australasian Region.

Australia and Antarctica belonged to the massive southern continent of Gondwana. Africa and India separated from Gondwana about 100 million years ago, and South America somewhat later. Antarctica and Australia split some 45 million years ago, as Australia drifted northwards. New Zealand and some of the larger Pacific islands were also separate by this time. The links between Australia and the other Gondwanan land masses can be seen in the animals living in the southern continents today. For instance the ratites, large flightless birds, are found in South America (rheas), Africa (Ostrich), Australia (Emu and cassowaries) and New Zealand (kiwis). The extinct elephant-birds of Madagascar and the Moa of New Zealand also belonged to this group.

Australia is an ancient continent, characterized by poor soils and a mostly arid climate. Rainfall increases from the centre to the coasts and the vegetation changes accordingly. Much of the interior is desert, or covered by grassland and scrub, whereas woodland and forest are found in the north, east and south-west.

Stony deserts or *gibbers* have a sparse cover of saltbush, while the sandy deserts are often well vegetated with shrubs and scattered small trees. Native chats (probably most closely related to honeyeaters), grasswrens and some kinds of parrots and pigeons are the most obvious birds in desert areas. Two widespread forms of semi-arid woodland are the *mulga* and *mallee*. Mulga is acacia woodland and mallee consists of multi-stemmed eucalypts. Both have a rich bird fauna, including many kinds of parrots, and the Mallee Fowl (*Leipoa ocellata*).

Taller woodlands and forests are dominated to a remarkable degree by one genus of trees, *Eucalyptus*, and many types of birds have adapted to live in eucalypt woodlands and forests. Eucalypts come in many forms, differing most conspicuously in the type of bark they possess: smooth-barked gums; peppermints and boxes with flaky bark; and the aptly named stringybarks and ironbarks. Many insectivorous birds search for food on the leaves and bark of eucalypts, honeyeaters and lorikeets visit the flowers for nectar and pollen, parrots consume the seeds and a variety of species from tiny pardalotes to massive black-cockatoos nest in the holes that are frequent in old eucalypts. Eucalypt woodland also has a remarkably large number of cooperatively breeding species, in which nonbreeding birds help to rear the young of others.

Heathy vegetation predominates in parts of the south-west and east coast on rocky or sandy soils. There is a diverse flora with many of the species producing masses of nectar, which the honeyeaters harvest while pollinating the flowers.

Perhaps surprisingly for such a dry continent, there are significant areas of rainforest, in north Queensland, the central

The Rainbow Lorikeet *is a common occupant of various arboreal habitats in eastern Australia.*

east coast and Tasmania. The temperate rainforests in the south are remnants from the time when great moist forests of southern beeches (*Nothofagus*) clothed much of Gondwana. This kind of tree is now also found in New Zealand and South America, and used to occur in Antarctica. The tropical rainforests are very rich in tree species, and have an avifauna more akin to that of New Guinea, including bowerbirds and fruit-doves.

Australia is famed for its great sandy beaches, but these are mostly poor bird habitat. There are estuaries and mudflats, and those near Melbourne and Adelaide, and at Broome in the north of Western Australia, provide a home for waders from the northern hemisphere in the nonbreeding season. Extensive mangroves are found across the north, and these have their own group of bird species. The Great Barrier Reef which extends along most of the coast of Queensland is the breeding ground for boobies, frigate birds, tropicbirds and terns, whereas shearwaters (muttonbirds), albatrosses and gannets breed in the islands of the Bass Strait.

People arrived in Australia at least 40,000 years ago and had a major impact on the land. Their use of fire encouraged the spread of eucalypts at the expense of rainforest and fire-sensitive trees. They apparently exterminated many of the large mammal species: giant kangaroos, diprotodonts and the marsupial lion (*Thylacoleo*). It is probable that many large birds disappeared at this time. Others would of course have been favoured by the environmental changes, especially those inhabiting grassland or profiting from eucalypts.

In the short time since Europeans settled Australia, a little over 200 years ago, they have cleared half of the remaining forests and woodlands and degraded much of the rest. The sheep, cattle and rabbits they introduced devastated the vegetation in semi-arid regions, which had never before felt the impact of hard hooves, leading to the loss of many medium-sized marsupials. Foxes and cats now prey on ground-dwelling birds, which have become scarce in many areas. Australia has a small population (17 million) but people are concentrated near the coast, which is now feeling the effects of development and tourism.

Despite the damage caused by European settlement, probably only one species of bird, the Paradise Parrot (*Psephotus pulcherrimus*) has become extinct on the Australian mainland. Many other species are endangered, rare or poorly known. Australians have recently started to identify with their native flora and fauna, rather than simply replacing them with European equivalents. Conservation now has a large popular following and the last decade has seen a proliferation of national parks. Many of these are in areas that have become part of the World Heritage scheme, including the rainforests of eastern Australia, the wilderness of south-western Tasmania, the Kakadu wetlands in the north and the central desert. The large cities all have areas of bush near or within them, such as King's Park in Perth, Sherbrooke Forest near Melbourne and Royal National Park (the second proclaimed in the world, in 1879) on the outskirts of Sydney. Also many birds, such as Kookaburras (*Dacelo novaeguineae*) and lorikeets, have become common in urban areas.

Australia's birds fall into three main groups in terms of their origin. A few are ancient survivors from Gondwana days, including the ratites mentioned above, and also the megapodes or mound-builders and probably the parrots. Most other groups evolved within Australia during the period of 30 million years or so when it was isolated from the other continents. The honeyeaters and Australian warblers, flycatchers, robins and wrens all have a long history in Australia. A few species are clearly recent arrivals, such as the Grey-backed White-eye (or Silvereye) (*Zosterops lateralis*) and a single bee-eater, pipit and lark, as their nearest relatives are found in Asia.

Altogether Australia has some 575 breeding species, somewhat over half being passerines. Only a handful of nonbreeding land birds reach the country, such as the White-throated Spinetailed Swift (*Hirundapus caudacuta*), which breeds in Japan. Numerous waders from the north and seabirds from the south also migrate to Australia. To an outsider the most obvious land birds are the parrots and honeyeaters, whereas the lyrebirds, bowerbirds and mound-builders are perhaps the most peculiar.

New Guinea originated from the leading edge of the Australian plate as it collided with the Asian plate. For much of its history it has been joined to northern Australia. It is a land of rainforest, with smaller areas of woodland and grassland. The richness of the forests, and perhaps lack of certain mammals, along with the great altitudinal range, has led to a highly diverse avifauna, with almost as many breeding birds as Australia. The parrots range from the pygmy parrots, scarcely larger than a wren, to the huge Palm Cockatoo (*Probosciger aterrimus*). Fruit-doves abound and New Guinea has a greater variety of kingfishers than anywhere else. The two most remarkable groups of birds are the birds of paradise and the bowerbirds. These are not closely related to each other, but both feed on rich fruits, which has freed the males of many species from parental care. Sexual selection imposed by the females for males with the best genes has led to spectacular plumages and bizarre displays.

New Zealand has an even longer history of isolation than Australia. During this period it suffered volcanism, glaciation and more recently the arrival of people, and consequently has a depauperate avifauna, with only 42 breeding species of land bird. Some, like the kiwis and tiny Rifleman (*Acanthisitta chloris*) are survivors from Gondwana; others, like the wattlebirds (Callaeidae), are of uncertain origin. Many species are clearly similar to Australian birds, suggesting a more recent origin. New Zealand was mostly forested, but burning by the Maoris and clearing by European settlers has replaced much forest by grassland. Sadly, many birds, like the Huia (*Heterolochia acutirostris*), have become extinct recently and many others teeter on the brink, like the Chatham Island Robin (*Petroica traversi*), which was once down to seven individuals. New Zealand now has many national parks, encompassing volcanoes, snowy mountains, glaciers, fjords and forests, and has an excellent recent history of research on endangered bird species. However, native birds are hard to find and species introduced from Europe, such as Chaffinches (*Fringilla coelebs*), often dominate, even in native forests.

New Caledonia has been a separate land mass for a long time, and although many of its birds are related to Australian species it has one oddity, the Kagu (*Rhynochetos jubatus*). The other islands of the Pacific show a decline in diversity of birds the further one travels from Australia and South-East Asia. Fiji is quite rich, with a mixture of Australian and Asian birds, whereas Tahiti has few birds. Parrots and rails have been successful in reaching even the most distant islands, but many of these have become extinct following human settlement and the introduction of predatory mammals.

Antarctica is now almost covered in snow and ice and has no true land birds. The surrounding seas are rich in plankton and other sealife, so that huge colonies of seabirds nest on the Antarctic mainland. Penguins are the best known of these, but there are skuas, terns and petrels as well. The subantarctic islands also harbour vast seabird colonies, particularly of albatrosses, petrels and shearwaters.

The Royal Australasian Ornithologists' Union (21 Gladstone Street, Moonee Ponds, Victoria 3039, Australia) has a growing membership and now has bird observatories in Western Australia, Victoria and New South Wales for visiting bird-watchers. New Zealand and New Guinea have ornithological societies, as do most of the Australian states.

CASSOWARIES AND EMUS Casuariiformes

Emu *Dromaius novaehollandiae* (Dromaiidae) 2 m (6½ ft). Huge flightless bird with long neck, long powerful legs. Long, thick hair-like plumage appears soft and shaggy. Grizzled greyish brown, with short, black hair-like feathers on head and neck, bare blue skin on face and throat. Breeding birds have pale neck ruff. Downy chick striped brown and white, immature smaller and darker brown than adult. Runs with a bouncy, swaying motion. Call a deep, resonant booming or grunting. **Range and habitat:** AU. Common nomad in open habitats from woodland to desert in Australia, occasionally near coast. In southern inland, large groups make nomadic or migratory seasonal movements. Extinct populations on Kangaroo and King Island, may have been distinct species. **Similar species:** immature Southern Cassowary, introduced Ostrich (see p 186).

Southern Cassowary *Casuarius casuarius* (Casuariidae) 1.5-2 m (5-6½ ft). Huge flightless bird with long neck, long legs, very long inner claw. Black plumage long and hair-like. Tall bony casque on head, bare blue skin on face and neck, red wattles on neck. Downy chick striped brown and yellow, immature brown with rusty head. Runs with head lowered, using blade-like casque to assist passage through vine tangles. Solitary. Call a loud, hollow booming or deep rumbling; also roaring, hissing and shrill whistle. **Range and habitat:** AU. Uncommon resident of tropical rainforest, often near streams and clearings, in New Guinea and northeast tip of Australia. **Similar species:** Emu.

KIWIS Apterygiformes

Brown Kiwi *Apteryx australis* (Apterygidae) 50 cm (20 in). Fowl-sized, dumpy, flightless bird with hairy plumage, long decurved bill. Grizzled grey-brown with rusty tinge, greyer on head. Pale to dark brown feet. No tail, short legs. Nocturnal, powerful runner. Call a shrill double whistle, repeated sounds like 'ki-wi', or a guttural hoarse cry. Snuffling noises when feeding, by probing into ground in search of earthworms. **Range and habitat:** AU. Uncommon resident of forest remnants and edge in New Zealand. **Similar species:** Little Spotted Kiwi, Greater Spotted Kiwi, both of which are barred.

Little Spotted Kiwi *Apteryx oweni* (Apterygidae) 40 cm (16 in). Fowl-sized, dumpy, flightless bird with hairy plumage, moderately long, slightly decurved bill. Irregularly banded and mottled brownish black on pale buff-grey background. Head grey, feet pale brown to flesh-white. No tail, short legs. Nocturnal, powerful runner. Call a shrill, even trill or low rolling 'churr'. **Range and habitat:** AU. Rare resident of forest remnants on South Island of New Zealand. Main population on Kapiti Island where it was introduced. **Similar species:** Brown Kiwi, Greater Spotted Kiwi, which is more chestnut.

PENGUINS Sphenisciformes

Adelie Penguin *Pygoscelis adeliae* (Spheniscidae) 70 cm (28 in). Medium-sized black and white penguin with stubby bill, peaked crown, long stiff tail of pointed feathers. Head, chin and upperparts black, underparts and prominent eyering white. Short beak partly feathered. Call a trumpeting and braying, or guttural squawk. **Range and habitat:** AN. Oceanic and circumpolar, abundant breeder in coastal and peninsular Antarctica, South Shetland, Orkney and Sandwich Islands, Bouvet and Peter I Island. **Similar species:** other *Pygoscelis* penguins.

Chinstrap (Bearded) Penguin *Pygoscelis antarctica* (Spheniscidae) 76 cm (30 in). Medium-sized pied penguin with distinctive face pattern. Upperparts black, face and underparts white with prominent black line encircling throat from ear to ear. Short unfeathered beak. Long tail. Call a loud braying, also low growling and hissing. **Range and habitat:** AN. Oceanic and circumpolar, common breeder in peninsular Antarctica, South Shetland, Orkney and Sandwich Islands, Bouvet and Peter I Islands, occasionally reaching more northerly land masses. **Similar species:** other *Pygoscelis* penguins.

Emperor Penguin *Aptenodytes forsteri* (Spheniscidae) 112 cm (44 in). Large, fat grey and white penguin with brightly coloured head pattern. Heaviest seabird at 25-45 kg (55-99 lb). Upperparts blue-grey, underparts white, slightly yellow on breast. Head black with orange-yellow throat and ear patch. Juvenile grey-brown. Rather small, slightly decurved bill with narrow pink gape and stripe on lower mandible. Greyish chicks with black and white head. Call is a resonant squawk, or shrill trumpeting or braying. **Range and habitat:** AN. Oceanic and circumpolar, breeds in winter on snow and sea-ice in Antarctica. **Similar species:** King Penguin.

King Penguin *Aptenodytes patagonicus* (Spheniscidae) 95 cm (37 in). Large grey and white penguin with brightly coloured head pattern. Upperparts blue-grey, underparts white with yellow sides to breast. Head black with golden orange ear patches and throat. Long bill with prominent pink gape and broad stripe along lower mandible. Woolly brown juveniles. Call a resonant squawk, or shrill trumpeting or braying. **Range and habitat:** AN. Oceanic and circumpolar, breeds on Macquarie, Marion, Prince Edward, Heard, Kerguelen and Crozet, South Sandwich Islands, South Georgia. **Similar species:** Emperor Penguin.

Fiordland Crested (Victoria) Penguin *Eudyptes pachyrhynchus* (Spheniscidae) 60 cm (24 in). Medium-sized pied penguin with brightly coloured bill and eye stripe. Head and upperparts glossy black with faint white streaks on cheeks, underparts white. Underflipper white with black border. Bill red, without bare pale skin at base. Black eye. Long, thin eyebrows of yellow filamentous feathers, ending in downward-drooping crest. Call a loud braying. **Range and habitat:** AU. Oceanic, breeds in south-western South Island, Stewart Island, New Zealand in coastal forest, small caves and overhangs; postbreeding movement to southern Australian seas. **Similar species:** other *Eudyptes* penguins, which have bare skin at base of bill, different underflipper patterns, differently shaped crests.

Little (Little Blue) Penguin *Eudyptula minor* (Spheniscidae) 35 cm (14 in). Small, plain penguin. Upperparts blue-grey to blue-black, face grey, underparts white. Flippers have white trailing edge. Small beak. Call is a yapping, braying and wailing. **Range and habitat:** AU. Common on the southern coasts of Australia, Tasmania, New Zealand. Some colonies are a major tourist attraction. Oceanic, coming ashore to moult and breed among rocks, dense shrubs or grass tussocks often on islands or base of cliff. **Similar species:** Yellow-eyed Penguin of New Zealand is larger, with pink bill, yellow eye and band across crown and very rare.

GREBES Podicipediformes

Hoary-headed Grebe *Podiceps poliocephalus* (Podicipedidae) 27 cm (10½ in). Small grey grebe. Upperparts dark grey, breast buff, underparts white. Broad white band through flight feathers. Breeding plumage distinctive, with thin, white hair-like plumes on black head and face. In nonbreeding plumage, dark cap extends below eye, hindneck black. Nonbreeding birds form tight flocks, flee from danger with long, pattering take-off across water. Usually silent; call near nest is a soft churring. **Range and habitat:** AU. Common nomad on lakes, swamps and estuaries in Australia, New Zealand. **Similar species:** nonbreeding Australasian Grebe is darker with fleshy yellow or cream gape, usually dives to escape danger; New Zealand Dabchick is darker and glossier. **Not illustrated.**

■□□
Emu
□■□
Southern Cassoway with young
□□■
Brown Kiwi

■□□
Little Spotted Kiwi
□■□
Adelie Penguin
□□■
Chinstrap (Bearded) Penguin

■□□
Emperor Penguin
□■
King Penguin

■□□
Fiordland Crested (Victoria) Penguin
□■
Little (Little Blue) Penguin on nest

TUBE-NOSED SWIMMERS Procellariiformes

Yellow-nosed Albatross *Diomedea chlororhynchos* (Diomedeidae) 78 cm (31 in). Small, slender albatross with yellow bill. White with black back, wings and tail. Underwing white with black border. Race in AU has grey cheeks, Atlantic race has grey head and neck, white cap. Bill black with yellow stripe from forehead to tip. Juvenile has white head, no yellow on bill. Most small albatrosses glide alternately showing pale underparts and dark wing and mantle. Call a croak at sea, braying at nest. **Range and habitat:** oceanic, breeding on subantarctic islands in Atlantic and Indian oceans. Common winter visitor off all southern continents. **Similar species:** other small albatrosses (mollymawks), especially Grey-headed and Buller's, both of which have yellow stripe on bottom of bill as well as top.

Grey-headed Albatross *Diomedea chrysostoma* (Diomedeidae) 88 cm (35 in). Small albatross with grey head, black and yellow bill. White with black back, wings and tail. Underwing white with black border. Head grey, bill black with yellow stripe along top and bottom, pink tip. Juvenile has black bill, grey underwing. Call a croak at sea, braying whistle at nest. **Range and habitat:** oceanic and circumpolar, breeding on subantarctic islands in all southern oceans. Uncommon winter visitor off all southern continents. **Similar species:** Yellow-nosed Albatross, Black-browed Albatross, Buller's Albatross which has dark grey head, white cap.

Royal Albatross *Diomedea epomophora* (Diomedeidae) 125 cm (49 in). Large, mainly white albatross with black tips and hind border on underwings. Upperwings black in northern race, white with black tips and hind border in southern. Tail white. Juvenile has black upperwings, mottled back. Bill yellowish pink with black cutting edge. Call a guttural croaking, braying whistle and bill-clappering. **Range and habitat:** oceanic, breeding on islands off New Zealand, small colony on mainland near Dunedin can be visited. Uncommon winter-spring visitor to seas off AU, South America. **Similar species:** Wandering Albatross.

Wandering Albatross *Diomedea exulans* (Diomedeidae) 135 cm (53 in). Large albatross, variable in colour but mostly white with black tips and hind edges to wings, dark mottling on upperwings and fine vermiculations on underparts. Tail white with black edges. Juvenile dark brown with white face. Immatures paler and mottled with dark collar and breast band, becoming increasingly white. Bill pink, without black cutting edge. Soars gracefully with little flapping. Call a guttural croaking, hoarse braying whistle and bill-clappering. **Range and habitat:** oceanic and circumpolar, breeding on subantarctic islands in all southern oceans. Fairly common winter-spring visitor off all southern continents. **Similar species:** Royal Albatross, Giant Petrel.

Light-mantled Sooty Albatross *Phoebetria palpebrata* (Diomedeidae) 70 cm (28 in). Small, slender, all-dark albatross with long pointed tail. Brown with white eye-ring, grey nape and back. Bill black with light blue line near cutting edge. Soaring flight elegant and graceful, with little flapping. Ghostly trumpeting at sea; call at nest is a shrill 'pee-ow', the second part deeper. **Range and habitat:** oceanic and circumpolar, breeding on antarctic and subantarctic islands in all southern oceans. Rare winter visitor to AU and South America. **Similar species:** Sooty Albatross, which has dark upperparts, yellow line along bill; Giant Petrel.

Giant Petrel *Macronectes giganteus* (Procellariidae) 90 cm (35 in). Huge, dumpy petrel with bulbous horn-coloured bill. Brown, mottled paler and greyer on head, face and underparts. Juvenile all-dark brown. White form is all-white with scattered brown spots. In flight shows short tail, ragged wing outline, more flapping and less graceful gliding than albatrosses. Call is a croaking at sea, hoarse whinnying and mewing at nest. Feeds on offal at sea and on land near breeding colonies. Occasionally kills petrels, penguins and albatrosses. **Range and habitat:** oceanic and circumpolar, breeding on Antarctica, antarctic and subantarctic islands. Common winter visitor off all southern continents. **Similar species:** juvenile Wandering Albatross, sooty albatrosses. 2 species now recognized – Northern and Southern, with pink and green bill tips respectively.

Short-tailed Shearwater *Puffinus tenuirostris* (Procellariidae) 42 cm (17 in). All-dark shearwater with dark feet, slender dark bill. Brown, with grey underwings, short rounded tail. Flies with a series of quick flaps then a long banking glide low over the water surface. Often in large flocks. Silent at sea; call at nest is a crooning, rising to crescendo, and harsher noises. Also called muttonbird, as young are still harvested for food. **Range and habitat:** oceanic. Abundant breeder on islands of southern Australia and Tasmania, migrates in winter to north Pacific ocean. **Similar species:** other dark *Puffinus* shearwaters (see p 50); Sooty has white underwing-coverts.

Pale-footed Shearwater *Puffinus carneipes* (Procellariidae) 45 cm (18 in). Large all-dark shearwater with pink feet, large pale bill. Entirely brown, tail short and fan-shaped, feet do not project beyond tail in flight. Flies with a series of stiff, deep wingbeats and long banking glides on straight wings, low over the water surface. Silent at sea; call at nest is a mewing, crooning and cackling. **Range and habitat:** oceanic. Common breeder on islands off southern Australia, New Zealand, migrates north in winter to tropical and subtropical Indian and Pacific oceans. **Similar species:** other all-dark *Puffinus* shearwaters (see p 50).

Cape (Pintado) Petrel *Daption capense* (Procellariidae) 40 cm (16 in). Large, distinctively pied petrel. Head and upperparts black with white patches in upperwings, white chequering on back and rump, white base to tail. Underparts white, underwings white with black border. Flies with stiff-winged fluttering and gliding, follows ships. Call a raucous chattering, trilled 'cooo'. **Range and habitat:** oceanic and circumpolar, breeding in Antarctica, antarctic and subantarctic islands, islands off New Zealand. Common winter-spring visitor off all southern continents. **Similar species:** Antarctic Petrel is browner, more clear-cut dark and white without chequering.

Soft-plumaged Petrel *Pterodroma mollis* (Procellariidae) 33 cm (13 in). Medium-sized grey and white petrel with face pattern, breast band, dark upperwings. Head and upperparts grey with white face and throat, black eye patch. Upperwings with dark M pattern. Underparts white with grey breast band, underwings dark with pale leading edge near body. Feet pink. Occasional all-dark individuals occur. Flight rapid, with wheeling arcs on long, backswept swift-like wings. Often in flocks. Call a shrill, trilling whistle or a moaning cry. **Range and habitat:** oceanic. Breeds on Atlantic islands off southern Africa and in Indian and south Pacific oceans, common winter visitor to seas off southern Australia, southern Africa. **Similar species:** other *Pterodroma* petrels (see p 98). **Not illustrated.**

Fairy Prion *Pachyptila turtur* (Procellariidae) 25 cm (10 in). Small blue-grey petrel. Upperparts with faint white eyebrow, dark M pattern across spread wings, broad dark tail tip. Underparts white. Bill narrow, with large nail at tip. Often in flocks. Flight swift and erratic, dips to 'hydroplane' along water surface. Call a guttural cooing. **Range and habitat:** oceanic. Abundant breeder on islands off southern Australia and New Zealand, and some subantarctic islands in all southern oceans. Common winter visitor off southern Australia, southern Africa. **Similar species:** other prions; distinguished by bill shape and width, amount of black on tail. **Not illustrated.**

Yellow-nosed Albatross

Grey-headed Albatross

Royal Albatross

Wandering Albatross

Light-mantled Sooty Albatross

Giant Petrel

Short-tailed Shearwater

Pale-footed Shearwater

Cape (Pintado) Petrel

FULL-WEBBED SWIMMERS Pelecaniformes

Australian Pelican *Pelecanus conspicillatus* (Pelecanidae) 150-180 cm (59-71) in). Large, distinctively pied pelican. Head, neck and underparts white, with grey crested nape, yellow eye-ring. Wings black with white shoulders, tail black. Large pink bill and pouch. Feeds by plunging head below surface. Swimming flocks herd schools of fish. Call a deep grunt. **Range and habitat:** AU. Common resident or visitor on lakes, rivers and estuaries in Australia, New Guinea. Irruptive, breeding in large numbers when inland lakes fill, then disperses coastwards when they dry out. Vagrants occasionally reach eastern Indonesia, New Zealand, islands in western PA. **Similar species:** none in AU.

Australian Gannet *Morus serrator* (Sulidae) 90 cm (35 in). Large, slender diving seabird with long bill, long pointed wings, long tail. White with golden-buff head, black tips and hind border to wings, black central tail feathers. Juvenile mottled dark brown, paler on underparts. Flies strongly and heavily with series of quick beats and long glides, dives with closed wings into sea. Call a croak at sea, and a nasal chuckling and cackling at nest. **Range and habitat:** oceanic. Abundant breeder on islands of New Zealand, smaller colonies off southern Australia. Common in offshore waters, especially in winter. **Similar species:** other gannets and boobies; albatrosses (see pp 98, 142).

Australian Darter *Anhinga novaehollandiae* (Anhingidae) 90 cm (35 in). Large black or grey-brown fishing bird with long pointed bill, long snake-like neck. Male black with white neck stripe, rusty patch on foreneck, white patches and streaks on wings. Female paler, grey-brown; with white underparts. Bill yellow, slender. Flight laboured, with series of quick beats and long glides, showing pale upperwing band and ample tail; soars well with distinctive 'flying cross' shape. Swims with only head and neck above water, often submerges. Perches beside water with wings open, snake-like movements of head. Call a strident, descending and accelerating croaking. **Range and habitat:** common resident or visitor on mainly freshwater wetlands. **Similar species:** other *Anhingas* (see pp 52, 186, 238).

Pied Cormorant *Phalacrocorax varius* (Phalacrocoracidae) 75 cm (29 in). Large black and white cormorant with long, prominently hooked bill. Upperparts black, face and underparts white with black thighs. Yellow, blue and orange facial skin, bone-coloured bill. Swims with bill slightly uptilted. Flies heavily with series of quick beats and glides, holding head slightly above plane of body, flocks often in V formation. Perches with wings open. Call a loud guttural cackling, croaking or screaming. **Range and habitat:** AU. Common resident of coastal waters and large lakes in Australia and New Zealand. **Similar species:** Black-faced Cormorant has black face and bill, Little Pied Cormorant has a short yellow bill.

Little Black Cormorant *Phalacrocorax sulcirostris* (Phalacrocoracidae) 65 cm (26 in). Small, entirely black cormorant with slender bill. Upperparts have bronze-green, scaled appearance. Flies with quick beats and glides, often in flocks in V formation. Often in large, tight fishing flocks which swim and submerge together. Perches in dead trees with wings open. Call a guttural croaking, ticking or whistling, mainly near nest. **Range and habitat:** OR, AU. Common resident or visitor to freshwater wetlands and estuaries in Indonesia, New Guinea, Australia and New Zealand. **Similar species:** Great Cormorant (see p 142).

Lesser (Least) Frigatebird *Fregata ariel* (Fregatidae) 75 cm (29 in). Large dark seabird with long pointed wings, forked tail. Male glossy black with inflatable red throat skin, small white patch at base of underwing. Female has white collar and breast, brown mottling on shoulders, no throat sac. Juvenile mottled brown with rusty head, white belly. Pale bill long and slender, with hook at tip. Soars effortlessly on large wings, harries other seabirds to give up food. Call a grunting, cackling and bill rattling. **Range and habitat:** tropical coastal seas. Abundant breeder on tropical islands, islets and reefs in all oceans. **Similar species:** other frigatebirds (see pp 52, 238).

HERONS AND ALLIES Ciconiiformes

Pacific (White-necked) Heron *Ardea pacifica* (Ardeidae) 90 cm (35 in). Large pied heron. Upperparts sooty black glossed blue, with maroon lanceolate hackles on back in breeding plumage. Head, neck and breast white with line of dark spots down foreneck. White 'headlights' on leading edge of wings in flight. Juvenile duller, with fawn-tinged, more heavily spotted neck and breast. Flies with slow, sweeping beats, neck retracted. Call a harsh croak. Sometimes soars. Uses stand-and-wait method to capture prey. **Range and habitat:** AU. Common resident or nomad on shallow freshwater wetlands in Australia; occasional vagrants reach New Guinea and New Zealand. **Similar species:** juvenile Pied Heron is much smaller, lacks wing 'headlights'.

White-faced Egret *Egretta novaehollandiae* (Ardeidae) 65 cm (26 in). Medium-sized, all-grey heron with white face and throat, grey hackles on back, rusty hackles on lower neck. Juvenile duller, lacking white face and hackles. Call a deep, hoarse cry, repeated. Commonest heron through most of range, in all kinds of habitats. **Range and habitat:** AU, PA. Abundant resident or nomad on shallow, fresh or saline wetlands and wet grasslands in eastern Indonesia, Australia, New Zealand, islands in western Pacific. **Similar species:** grey morph of Reef Heron.

Intermediate Egret *Egretta intermedia* (Ardeidae) 65 cm (26 in). Medium-sized, all-white egret with stout bill, neck same length as body. In breeding plumage has long filamentous plumes on breast and back, green facial skin, red or orange bill, red upper legs. Nonbreeding birds lack plumes and have orange-yellow bill and face, black legs. Stalks prey slowly with neck straight, upright. Flies with slow beats, neck retracted; in flight looks broad-winged, with slight neck bulge. Call a deep, rattling croak. **Range and habitat:** common resident, nomad or migrant on shallow, fresh and saline wetlands in AF, OR, AU. **Similar species:** other white egrets, spoonbills.

Rufous (Nankeen) Night Heron *Nycticorax caledonicus* (Ardeidae) 59 cm (23 in). Squat, dumpy heron with stout bill, short legs. Upperparts rufous with black cap, long white plumes on crown. Underparts pale rufous, shading to white on belly. Juvenile streaked and mottled brown, with white spots on back and wings. Noctural, roosting by day in hunched posture in leafy waterside trees. If disturbed, circles about before landing in another tree. Call a hollow, squeaky 'kyok'. **Range and habitat:** OR, AU, PA. Common resident, migrant or nomad of shallow, fresh and saline wetlands with nearby trees in Indonesia, New Guinea, Australia, islands in western Pacific. **Similar species:** rufous form of Green-backed (Striated) Heron, lacks cap and plumes, more rufous and spotted ventrally; Australian Bittern is yellowish brown with dark streaks and bars, lives in reeds.

Black Bittern *Ixobrychus flavicollis* (Ardeidae) 60 cm (24 in). Medium-sized, all-dark bittern. Dark brown to sooty black with yellow plumes down sides of neck, pale streaks from throat to breast. Juvenile paler, browner. Seldom seen unless flushed from trees. In flight looks compact, all-dark. Call a deep croak, cooing and booming. **Range and habitat:** OR, AU, PA. Uncommon resident or migrant of mangroves, trees fringing streams and wetlands in India, eastern Asia, New Guinea, coastal Australia, islands in western Pacific. **Similar species:** Green-backed (Striated) Heron. **Not illustrated.**

Australian Pelican

Australian Gannet colony

Australian Darter

Pied Cormorant

Little Black Cormorant

Pacific (White-necked) Heron

Lesser (Least) Frigatebird

White-faced Egret

Intermediate Egret

Rufous (Nankeen) Night Heron

Straw-necked Ibis *Threskiornis spinicollis* (Threskiornithidae) 67 cm (26 in). Pied ibis with straw-like plumes on neck. Upperparts black glossed blue-purple and green, with bald black head, white neck ruff, long yellow plumes on foreneck. Underparts and tail white. Juvenile duller with dusky feathered head, shorter bill, no plumes. Flies with quick, rhythmic beats and long glides, neck outstretched. Large flocks soar in circles or fly in V formation. Call a deep grunt. Often feeds well away from water, patrolling pasture in search of large insects. **Range and habitat:** AU. Abundant resident of grasslands and shallow freshwater wetlands in Australia. Partly nomadic or migratory, large flocks gathering at insect plagues and southern juveniles wintering in tropical northern Australia and occasionally New Guinea. **Similar species:** Pacific (White-necked) Heron, Pied Heron (see p 286).

Royal Spoonbill *Platalea regia* (Threskiornithidae) 78 cm (31 in). All-white spoonbill with black face, bill and legs. Long plumes on nape in breeding plumage. Flies with quick, rhythmic beats and long glides, neck outstretched. Call a deep grunt, bill-clappering. **Range and habitat:** OR, AU. Common resident of shallow, fresh and saline wetlands in Indonesia, New Guinea, Australia and New Zealand. **Similar species:** Yellow-billed Spoonbill has yellow bill and legs; white egrets (see pp 52, 144, 286).

WATERFOWL Anseriformes

Magpie (Pied) Goose *Anseranas semipalmata* (Anseranatidae) 70-90 cm (28-35 in). Distinctively pied goose with knob on head, long unwebbed toes. Head, neck, wings and tail black. Back, shoulders and underparts white. Pink bill and face, long yellow or pink legs. Immature has white areas mottled brown. Large flocks gather on feeding grounds. Call a resonant honking. **Range and habitat:** AU. Abundant breeding resident or local nomad of swamps and flooded grasslands and crops in New Guinea and tropical Australia. Formerly abundant breeder in south-eastern Australia but now rare, though recovering slowly. **Similar species:** none in AU.

Black Swan *Cygnus atratus* (Anatidae) 100-40 cm (39-55 in). All-black swan with red bill, white wing tips revealed in flight. Immature grey-brown, white wing feathers mottled and tipped brown. Downy young grey. Call a plaintive bugling. Feeds from surface, by submerging bill and neck or upending. **Range and habitat:** AU. Abundant resident or nomad on both fresh and brackish wetlands in Australia, vagrant to New Guinea. Introduced to New Zealand, where now abundant.

Cape Barren (Cereopsis) Goose *Cereopsis novaehollandiae* (Anatidae) 75-100 cm (29-39 in). Large all-grey goose with bulbous yellow-green cere, white crown, black spots on wings, pink legs. In flight shows black wing tips and tail. Call is a deep resonant grunting and honking, repeated. **Range and habitat:** AU. Uncommon breeder in heath and tussock-grass on islands off southern Australia, dispersing to coast and pastures of southern Australian mainland and Tasmania. **Similar species:** none in AU.

Pacific Black Duck *Anas superciliosa* (Anatidae) 55 cm (22 in). Medium-sized, dark duck with striped face. Mottled dark brown with black crown, pale and dark horizontal lines through eye and face, buff throat, iridescent patch in wing which varies from green to purple depending on angle of light. In flight shows glossy green patch on rear of inner upperwing, large white patch on inner underwing. Call a deep, hoarse quack, sometimes repeated as a descending 'laugh'. **Range and habitat:** OR, AU, PA. Abundant resident or nomad of mainly freshwater wetlands in Indonesia, New Guinea, Australia, New Zealand and Pacific islands. Declining in New Zealand through genetic swamping by introduced Mallard, with which it hybridizes. **Similar species:** Grey Teal, female Mallard (see p 146).

Grey Teal *Anas gibberifrons* (Anatidae) 42 cm (17 in). Small duck, mottled grey-brown with pale face and white throat. In flight shows small green wing patch at rear of upperwing, white stripe along centre of upperwing, small white patch on inner underwing. Call a descending 'laugh' of soft cooing quacks, or whistled 'peep'. **Range and habitat:** OR, AU, PA. Abundant nomad of fresh and brackish wetlands from Andaman Islands and Indonesia to New Guinea, Australia, New Zealand and western Pacific islands. Irruptive in Australia, congregating on ephemeral inland wetlands after rain to breed and dispersing when they dry out. **Similar species:** Pacific Black Duck; female Chestnut-breasted Teal is darker without pale throat.

Pink-eared Duck *Malacorhynchus membranaceus* (Anatidae) 40 cm (16 in). Small, boldly barred duck with long bill. Upperparts brown, underparts white with broad dark bars, buff undertail. Distinctive head pattern of white face, black eye patch, small pink ear spot. Bill has flap of skin at each side of tip. Feeds on surface by filtering out small animals. Flight rapid and whirring, showing prominent bill, white crescent on rump. Often in flocks. Call is a liquid chirrup, often given in flight. **Range and habitat:** AU. Common nomad of freshwater wetlands in Australia. Breeds on ephemeral inland waters, disperses coastwards during drought. **Similar species:** Grey Teal; female Australian Shoveller is plainer, has blue shoulders in flight.

Australian White-eyed Duck *Aythya australis* (Anatidae) 55 cm (22 in). Dark brown duck with white wing patch and undertail. Male with white iris, female dark iris. White underwing, stripe along upperwing, broader at tip. Smooth outline with no obvious tail in water. Has grey beak with blue tip. Dives. **Range and habitat:** Australia, deep lakes, flooded rivers, permanent swamps. **Similar species:** Australian Blue-billed Duck has no white on wings.

Maned Goose (Australian Wood Duck) *Chenonetta jubata* (Anatidae) 47 cm (18 in). Medium-sized, grey perching duck with small bill, brown head, white wing flashes in flight. Male grey with brown head, speckled neck and breast, dark plumes on back, black undertail. Female has white eyebrow and cheek stripe, underparts mottled brown and white. In flight shows black wing tips, broad white rear edge on inner upperwing, white underwing. Often perches on trees or stands on land. Nests in tree-holes. Call a nasal, querulous ascending mew. **Range and habitat:** AU. Abundant resident or local nomad of freshwater wetlands and pastures in Australia. **Similar species:** none in AU.

Musk Duck *Biziura lobata* (Anatidae) 47-73 cm (18-29 in). Dark, elongated diving duck that sits low in the water. Entirely dark brown with fine, pale vermiculations. Bill short and stout, head large, tail long and stiff. Male has large leathery flap below bill and throat. Female smaller, no leathery flap. Swims partly submerged, with only head and back above water, often dives. Rarely flies, walks with difficulty. Call by male in display is a shrill whistle and 'plonk' made by feet in water, while head raised and tail cocked. **Range and habitat:** AU. Common resident of deep freshwater wetlands and coastal waters in southern Australia. **Similar species:** female/eclipse Australian Blue-billed Duck is smaller with rounded head, flat bill.

Straw-necked Ibis

Royal Spoonbill

Magpie (Pied) Goose

Black Swan

Cape Barren (Cereopsis) Goose

Pacific Black Duck

Grey Teal

Pink-eared Duck

Australian White-eyed Duck

Maned Goose (Australian Wood Duck)

Musk Duck

DIURNAL BIRDS OF PREY Falconiformes

Australian Black-shouldered Kite *Elanus notatus* (Accipitridae) 35 cm (14 in). Small, long-winged white hawk with pale grey back and wings, black shoulders, black spot near bend of underwing and dark wing tips. Juvenile washed rusty on head and breast, mottled on back and wings. Winnowing and hovering flight, glides on raised wings. Call a weak whistling note and harsh wheezing. **Range and habitat:** AU. Common in open woodland, grassland and farmland in Australia. Irruptive and nomadic, following mouse plagues. **Similar species:** Letter-winged Kite has black line along inner underwing, pale wing tips.

Black-breasted Buzzard Kite *Hamirostra melanosternon* (Accipitridae) 51-61 cm (20-4 in). Robust, soaring hawk with white panels in wings, short pale tail. Adult black, with rufous nape, shoulder mottling, thighs and undertail. Juvenile/immature rufous to sandy brown with black streaks. Flight eagle-like gliding and soaring on raised wings, with sideways rocking motion. Call a repeated short, hoarse yelp. **Range and habitat:** AU. Uncommon resident of woodland in arid inland and tropical northern Australia. **Similar species:** Wedge-tailed Eagle; Square-tailed Kite is slender with long tail, barred flight feathers. **Not illustrated.**

Whistling Kite *Haliastur sphenurus* (Accipitridae) 51-9 cm (20-3 in). Medium-sized, untidy, soaring hawk. Mottled or streaked sandy brown with dark flight feathers, underwing pattern of pale leading edge and inner primaries, dark secondaries and outer primaries. Slow circling flight on bowed wings with splayed primaries, showing large pale area on upperwing and long rounded tail. Often gregarious. Call a loud whistle sliding down scale, accompanied by rapid up-scale chatter. **Range and habitat:** AU, PA. Common resident, migrant or nomad of open and lightly wooded country, often near water, in New Guinea, Australia and New Caledonia. **Similar species:** juvenile Brahminy Kite (see p 240); Little Eagle is chunkier with narrow pale upperwing band, short tail, flat wings in flight, feathered legs.

Brown (Australian) Goshawk *Accipiter fasciatus* (Accipitridae) 37-55 cm (15-22 in). Fierce, active, powerful hawk with rounded wings, long rounded tail, long yellow legs. Upperparts slate-grey washed brown, with rufous half-collar. Underparts finely barred dull rufous and white. Juvenile brown, with white underparts coarsely streaked and barred brown. Flight heavy, beats wings in rapid bursts, glides on flat wings. Call a rapid, shrill chatter or slower mewing notes. **Range and habitat:** AU, PA. Common but secretive resident and partial migrant of woodlands and open forests in eastern Indonesia, New Guinea, Australia and New Caledonia. Some southern birds winter in tropical AU. **Similar species:** Australian Collared Sparrowhawk, Brown Falcon (see p 292).

Australian Collared Sparrowhawk *Accipiter cirrhocephalus* (Accipitridae) 29-38 cm (11½-15 in). Small, fierce hawk with rounded wings, long square-tipped tail, long spindly legs, long middle toe. Upperparts slate-grey washed brown, with rufous half-collar. Underparts finely barred rufous and white. Juvenile brown, with white underparts coarsely streaked and barred brown. Flight agile and winnowing or flickering, beats wings in rapid bursts, glides on flat wings. Call a very rapid chittering or slower mewing. **Range and habitat:** AU. Uncommon, secretive resident of forests and woodlands in Australia and New Guinea. **Similar species:** Brown (Australian) Goshawk.

Grey (White) Goshawk *Accipiter novaehollandiae* (Accipitridae) 33-55 cm (13-22 in). Striking hawk with short rounded wings, long tail, long orange-yellow legs. Upperparts grey, underparts either white (Australia) or rufous (New Guinea and islands). Also an all-white colour form in both regions, and a dark grey form (sometimes with chestnut underparts) in New Guinea. Juveniles in New Guinea and islands are brown, with buff underparts streaked and barred brown. Flight powerful and agile, beats wings in rapid bursts, glides on bowed wings. Call a series of upslurred, ringing whistles. **Range and habitat:** AU, PA. Common but secretive resident of wet forests in Australia, New Guinea and islands from eastern Indonesia to Solomon Islands. **Similar species:** Brown (Australian) Goshawk in New Guinea, Black-shouldered Kite, Grey Falcon in Australia.

Wedge-tailed Eagle *Aquila audax* (Accipitridae) 85-104 cm (33-41 in). Large dark eagle with lanky build, long wedge-shaped tail, baggy feathered legs. Sooty black with tawny nape, mottled grey-brown upperwing-bar, brown undertail, pale bases to flight feathers. Juvenile/immature dark brown with golden to reddish brown nape, back and upperwing-bar, pale undertail. Stable and controlled soaring flight on long, upswept wings. Often gregarious at carrion. Call a weak yelp or squeal. **Range and habitat:** AU. Common resident or visitor, most habitats except intensively cultivated or settled areas, in Australia and New Guinea. **Similar species:** Black-breasted Buzzard Kite, juvenile White-bellied Sea Eagle; in New Guinea Gurney's Eagle has rounded tail, flat wings in flight.

Pacific (Crested) Baza *Aviceda subcristata* (Accipitridae) 35-46 cm (14-18 in). Medium-sized crested hawk with long wings and tail, short legs. Upperparts slate-grey, underparts and flight feathers boldly barred. Underwing-coverts, thighs and undertail-coverts pale rufous. Juvenile much browner. Flight buoyant and agile, with slow beats, glides on flat wings. Active and acrobatic in tree canopy, sometimes gregarious. Call a reedy 2-note whistle, rising and falling. **Range and habitat:** AU, PA. Common resident or altitudinal migrant of forests and woodlands in Australia, New Guinea and islands from eastern Indonesia to Solomon Islands. **Similar species:** Brown (Australian) Goshawk, Australian Collared Sparrowhawk.

Black Falcon *Falco subniger* (Falconidae) 45-56 cm (18-22 in). Large, sleek, fierce falcon with small head, powerful shoulders, pointed wings, short legs. Dark brown to sooty grey or sooty black with inconspicuous black cheek stripe, white chin, and dark underwings. Flight is a kestrel-like winnowing or slower, crow-like flapping, glides on bowed wings. Often soars, showing long, square-tipped, usually furled tail. Fast and agile in stoop or pursuit. Usually silent; call a deep harsh chatter, slow whining or soft whistle. **Range and habitat:** AU. Uncommon resident or nomad of plains in inland Australia. Irruptive, into southern coastal areas. **Similar species:** Black Kite, Brown Falcon (see p 292).

Australian Hobby *Falco longipennis* (Falconidae) 30-6 cm (12-14 in). Small, slender, dusky falcon with small bill and feet, long scythe-shaped wings. Upperparts slate-grey with black mask, pale forehead and half-collar. Underparts rufous with fine dark streaks. Juvenile browner. Flight often low, fast and flickering, glides on flat wings with tips swept back. Aerobatic in pursuit of small aerial prey. Call a reedy chatter. **Range and habitat:** AU. Common resident or partial migrant of open woodland, towns and cities in Australia, winter migrant to eastern Indonesia and New Guinea. **Similar species:** Peregrine Falcon, Oriental Hobby (see p 152).

Grey Falcon *Falco hypoleucos* (Falconidae) 33-43 cm (13-17 in). Medium-sized, pale grey falcon with long pointed wings, short tail, short legs. Upperparts grey with faint cheek stripe, black wing tips. Underparts white with faint shaft streaks, wings and tail finely barred. Juvenile darker and more boldly marked. Flight is a kestrel-like winnowing, glides on flat wings. In pursuit of birds, flies low and fast or stoops from a height. Call a hoarse chattering, clucking and whining. **Range and habitat:** AU. Rare in shrubland, grassland and woodland of arid inland Australia; some birds winter in northern Australia and rarely New Guinea. **Similar species:** Australian Black-shouldered Kite, Grey (White) Goshawk, Australian Hobby, Peregrine Falcon (see p 152). **Not illustrated.**

Australian Black-shouldered Kite

Whistling Kite

Brown (Australian) Goshawk

Australian Collared Sparrowhawk

Grey (White) Goshawk

Pacific (Crested) Baza

Wedge-tailed Eagle

Black Falcon

Australian Hobby

Brown Falcon *Falco berigora* (Falconidae) 41-51 cm (16-20 in). Medium-sized, scruffy, loose-plumaged falcon with large head, round shoulders, long grey legs. Variable in plumage, from rufous brown on the upperparts and white on the underparts, with dark streaks and brown thighs, to uniformly dark brown. Double cheek stripe distinct on paler birds, barred underwings and tail visible on dark birds. Often perches sluggishly. Flight slow, heavy and erratic, glides on raised wings with rounded tail often fanned, sometimes hovers. Very vocal; call is a raucous cackling, chattering or screeching. **Range and habitat:** AU. Abundant resident and partial migrant in most open and lightly wooded habitats in Australia and New Guinea; some southern juveniles winter in northern Australia. **Similar species:** Brown (Australian) Goshawk, Black Falcon (see p 290), Australian Kestrel.

New Zealand Falcon *Falco novaeseelandiae* (Falconidae) 41-8 cm (16-19 in). Medium-sized, dark, dashing and powerful falcon. Upperparts blue-black with pale bars, dark cheek stripe. Underparts cream to rufous with dark streaks, flanks barred. Juvenile duller and browner, without barring on flanks. Flight rapid and powerful, glides on flat wings. Soars, sometimes hovers or wind-hangs, makes fast, slanting dives at prey. Call a chattering, clucking or whining. **Range and habitat:** AU. Uncommon resident of wooded and open habitats in New Zealand. **Similar species:** none in New Zealand.

Australian Kestrel *Falco cenchroides* (Falconidae) 30-5 cm (12-14 in). Small, pale falcon with slender build, long tail. Upperparts pale rufous with indistinct cheek stripe, fine black markings, dark wing tips. Underparts buff with fine dark streaks. Flight winnowing, glides on flat wings. Often hovers, showing black band near tail tip. Call a shrill chatter. **Range and habitat:** AU. Abundant resident or partial migrant of most open and lightly wooded habitats in Australia and New Guinea. **Similar species:** pale-coloured Brown Falcon.

FOWL-LIKE BIRDS Galliformes

Mallee Fowl *Leipoa ocellata* (Megapodiidae) 60 cm (24 in). Beautifully mottled fowl-like bird with large feet. Upperparts grey-brown with chestnut, black and white mottling on back and wings. Underparts pale grey-brown with buff throat, black streak down breast. Walks or runs powerfully, flies heavily if pressed. Presence indicated by large nesting mounds of sand and leaf litter. Call a booming, grunting or soft cluck. Chicks are independent immediately after emerging from mound and can fly when very small. **Range and habitat:** AU. Uncommon, localized and declining resident of remnant 'mallee' in southern Australia – low, dense, multi-stemmed *Eucalyptus* scrub, acacia scrub. **Similar species:** none in AU.

Brush Turkey *Alectura lathami* (Megapodiidae) 70 cm (28 in). Turkey-like bird with vulturine head, large feet, tent-shaped or vaulted tail. Black, with bare red head, yellow neck wattles, grey scalloping on belly. Walks or runs powerfully, flies heavily if pressed. Call a deep grunt or loud booming. Male constructs huge mound by raking up leaf litter. Chicks independent on emerging. **Range and habitat:** AU. Common resident of wet forest, thick scrub and well-vegetated tropical gardens (even in suburbs), in coastal eastern Australia. **Similar species:** none in AU.

Stubble (Pectoral) Quail *Coturnix pectoralis* (Phasianidae) 18 cm (7 in). Secretive, dumpy ground bird. Upperparts grey-brown with fine black and white vermiculations, prominent cream streaks. Underparts white with brown streaks. Male has rufous throat, dense black streaks on chest. Feet flesh-pink. Readily detected by call, seldom seen unless flushed. Flies suddenly with rapid whirring of pointed wings, in flight appears grey-brown with white streaks. Call a liquid 3-note whistle 'pip-pee-whit' or a sharper 2-note 'pip-pip' repeated. **Range and habitat:** AU. Common nomad of grassland and crops in Australia; extinct in Tasmania and New Zealand. **Similar species:** other quail and button-quail (see pp 154, 196, 244, 246). **Not illustrated.**

Tasmanian Brown Quail *Coturnix ypsilophora* (Phasianidae) 18 cm (7 in). Secretive, dumpy ground bird. Upperparts brown with fine chestnut, grey and black marbling, fine white streaks. Underparts light brown with fine, wavy black barring. Some birds with blue-grey suffusion. Female sandier brown and more coarsely marked than male. Feet orange-yellow. Readily detected by call, seldom seen unless flushed. Flies suddenly with rapid whirring of rounded wings, in flight appears rich brown. Call a mournful rising whistle of 2 notes 'tu-whee' or shorter 'whee-whit', not rising. **Range and habitat:** AU. Common resident or nomad of rank grass, swamps and heath in Indonesia, New Guinea, Australia; introduced to New Zealand and Fiji. **Similar species:** other quail and button-quail (see pp 154, 196, 244, 246). **Not illustrated.**

CRANES, RAILS AND ALLIES Gruiformes

Little Quail *Turnix velox* (Turnicidae) 13-15 cm (5-6 in). Tiny quail-like bird with short, stout bill. Upperparts rufous, mottled grey-brown and black, streaked white on back and wings. Breast rufous, belly white. Male smaller, greyer and more mottled. Creeps hesitantly between clumps of grass, rocking backwards and forwards on toes, seldom seen unless flushed. Flies suddenly with rapid whirring of pointed wings, in flight appears rufous with white flanks and undertail. Call a low, disyllabic cooing. **Range and habitat:** AU. Common nomad of dry grassland and grassy woodland in Australia. **Similar species:** quail (see p 154); Red-chested Button-quail appears grey with rufous underparts. **Not illustrated.**

Buff-banded Rail *Rallus philippensis* (Rallidae) 31 cm (12 in). Small, secretive waterbird of thick vegetation. Upperparts brown, mottled black and white. Head rufous, with white eyebrow. Flight feathers barred chestnut. Underparts barred black and white, with grey throat and buff patch on breast. Short, red-brown bill. Skulks through ground vegetation, runs into cover if startled. Calls varied – croaks, clicks, rattles and squeaky 'swit-swit'. **Range and habitat:** OR, AU, PA. Common but seldom-seen resident or partial migrant of swamps and rank grass in Indonesia, New Guinea, Australia and islands of Indian and western Pacific oceans. **Similar species:** Lewin's Rail has longer bill, no eyebrow or breast band, and is rarer.

Black-tailed Native Hen *Gallinula ventralis* (Rallidae) 35 cm (14 in). Bantam-like waterbird with cocked, fan-shaped tail. Upperparts dark olive-brown, head and underparts dark blue-grey with white spots on flanks, black tail. Bill green with orange base, legs red. Runs rapidly with tail depressed. Often in large flocks. Call a sharp 'kek' or low cackle. **Range and habitat:** AU. Abundant breeding visitor to ephemeral wetlands of inland Australia; highly irruptive and nomadic, following rains and dispersing when wetlands dry out. **Similar species:** Dusky Moorhen and Purple Swamphen have red bills and white undertails; Coot (see p 154). **Not illustrated.**

Brolga *Grus rubicunda* (Gruidae) 140 cm (55 in). Large grey crane. Crown grey; face and band around nape scarlet, with grey ears; black dewlap under chin. Legs dark grey-brown to black. Performs bowing and leaping 'dances'. Flies slowly with neck extended, shallow flicking beats; soars well. Call a wild gurgling and trumpeting. **Range and habitat:** AU. Common breeding resident of shallow freshwater wetlands in northern Australia; uncommon and declining in the south. **Similar species:** Sarus Crane (see p 246) has more extensive red on head and upper neck, no dewlap, pink legs.

Brown Falcon

New Zealand Falcon

Australian Kestrel (female)

Mallee Fowl

Brush Turkey

Buff-banded Rail

Brolga (immature)

Australian Bustard *Choriotis australis* (Otididae) 75-100 cm (29-39 in). Large, heavily built ground bird with long neck and legs. Crown black, neck white finely vermiculated grey. Upperparts brown, with black and white chequering near bend of wing. Underparts white to pale grey with narrow black breast band. Stalks imperiously with bill uptilted, sometimes 'freezes' in cryptic stance. Flight powerful and heavy with flicking beats, neck extended. Call a resonant grunting or croaking; in display male gives guttural roar with neck feathers fanned, tail raised. **Range and habitat:** AU. Uncommon nomad of plains and dry woodland in northern and inland Australia, now rare in south. **Similar species:** none in AU. **Not illustrated.**

SHOREBIRDS, GULLS AND AUKS
Charadriiformes

Comb-crested Jacana *Irediparra gallinacea* (Jacanidae) 23 cm (9 in). Small waterbird with very long, thin toes. Upperparts brown, underparts white with broad black breast band. Distinctive red chicken-like comb on head; face and throat buff. Immature duller, lacks comb and breast band. Walks on waterlilies and other floating vegetation without sinking. Flight low and fluttering, with long toes trailing. Call a weak piping whistle. **Range and habitat:** OR, AU. Common resident or nomad of freshwater wetlands with floating vegetation in Indonesia, New Guinea and Australia. **Similar species:** none in AU; other jacanas in OR (see p 248).

Bush Thick-knee (Australian Stone-curlew) *Burhinus magnirostris* (Burhinidae) 55 cm (22 in). Large, long-legged ground bird with large yellow eyes. Upperparts mottled grey with pale shoulders. White eyebrow, forehead and throat. Underparts buff with dark streaks. Active at night; during day usually squats in cryptic posture, but if disturbed walks with stiff-legged gait or flies with stiff-winged, shallow beats, showing white patches in wings. Call a mournful wailing 'wee-loo' repeated to crescendo, then wavering and dying away. **Range and habitat:** AU. Common resident or nomad of open forest to woodland with abundant ground litter in northern Australia and New Guinea; rare and declining in south, extinct in Tasmania.

Sooty Oystercatcher *Haematopus fuliginosus* (Haematopodidae) 50 cm (20 in). Robust, all-black shorebird with long red bill, red eye-ring, red legs. Clambers expertly over wave-washed rocks. Call a melodious piping 'pip-peep' and longer series of notes. **Range and habitat:** AU. Uncommon resident of rocky shores, reefs and islands of Australian coastline. **Similar species:** Sooty Oystercatcher of New Zealand almost identical.

Wrybill *Anarhynchus frontalis* (Charadriidae) 20 cm (8 in). Small, squat grey and white plover with slender bill turned to the right at the tip. Head and upperparts grey, underparts white. In breeding plumage has white forehead and eyebrow, bordered by thin black line on forehead, and black breast band. Forms flocks and coordinated mass flights in autumn and winter. Call a short clear 'weet', short harsh note or rhythmic churring. **Range and habitat:** AU. Uncommon breeder on shingle of large river-beds in South Island of New Zealand. Migrates in winter to open seashores of North Island. **Similar species:** migratory *Charadrius* plovers in nonbreeding plumage (see pp 62, 110, 156, 200).

Banded Lapwing (Plover) *Vanellus tricolor* (Charadriidae) 28 cm (11 in). Large, boldly patterned plover of dry land. Upperparts soft brown with black cap, white stripe from eye to nape. Underparts white, with broad V-shaped black breast band extending up neck to face. Small red wattles on forehead. Flies with shallow, stiff-winged beats showing broad white band in wings. Call a strident but plaintive series of 3 or 4 high-pitched notes, slightly descending. **Range and habitat:** AU. Common resident or nomad of dry grassland and bare plains in Australia. **Similar species:** larger Masked Plover lacks breast band, has large yellow facial wattles.

Red-kneed Dotterel *Charadrius cinctus* (Charadriidae) 18 cm (7 in). Small, boldly marked plover with dumpy, front-heavy appearance and long legs. Upperparts brown with black cap. Underparts white with broad black breast band, chestnut flanks. Bill red, knees red on otherwise dark legs. Juvenile duller and browner, without breast band or coloured flanks. Sometimes wades. In flight shows white edges to tail, broad white trailing edge to wings. Call a liquid 'wit-wit' or low trilling. **Range and habitat:** AU. Common resident or nomad on margins of shallow wetlands in Australia and New Guinea. **Similar species:** more slender Black-fronted Plover has brown cap, white eyebrow, narrower breast band, white flanks.

Red-capped Dotterel *Charadrius ruficapillus* (Charadriidae) 15 cm (6 in). Small, pale plover of sandy areas. Upperparts fawn, with rufous cap and half-collar, white forehead bordered black. Underparts white. Juvenile lacks rufous cap. Flight rapid and darting, showing narrow white wing-bar. Call a sharp 'wit' or trill. **Range and habitat:** AU. Common resident or nomad of margins of saline wetlands and beaches in Australia; vagrant to New Zealand. **Similar species:** migratory *Charadrius* plovers in nonbreeding plumage, nonbreeding Rufous-necked Stint.

Australian Dotterel (Courser) *Peltohyas australis* (Charadriidae) 20 cm (8 in). Small, sandy plover of dry country. Upperparts mottled brown, with black bar across crown to below eyes. Underparts fawn with chestnut belly, Y-shaped black breast band from nape to centre of belly. Juvenile lacks black markings. Turns back to intruder, bobs head. Flight low and swift on long wings, showing buff upperwing patches. Call is a single metallic note or trill. **Range and habitat:** AU. Uncommon nomad of stony plains and bare ground in inland Australia. **Similar species:** Plains Wanderer is more quail-like, lacks black markings.

Banded Stilt *Cladorhynchus leucocephalus* (Recurvirostridae) 41 cm (16 in). Elegant brown and white wading bird with needle-like bill, long legs. White with blackish brown wings, chestnut breast band extending into black line along belly. Legs pink. Juvenile has no breast band. Broad white trailing edge to wings in flight. Swims; occurs in large flocks. Call is a puppy-like yapping. **Range and habitat:** AU. Locally abundant nomad of saline wetlands, salt works and estuaries in inland and southern Australia. Highly irruptive, breeding in large colonies when inland salt lakes fill. **Similar species:** Black-winged Stilt, Red-necked Avocet (see p 156, 296).

New Zealand Stilt *Himantopus novaezealandiae* (Recurvirostridae) 38 cm (15 in). Medium-sized, slender, all-black wader with needle-like bill, long pink legs. Upperparts glossed green. Juvenile white or mottled on head and underparts. In flight, long legs trail beyond tail. Call a puppy-like yapping. **Range and habitat:** AU. Rare and endangered resident of dry shingle-beds and adjacent swamps in South Island, New Zealand. Under 100 birds remain. Hybridization with Black-winged Stilt could eliminate the species. Some winter dispersal northwards within New Zealand. **Similar species:** Black-winged Stilt; Sooty Oystercatcher is dumpier with stout red bill (see p 156).

Comb-crested Jacana

Bush Thick-knee (Australian Stone-curlew)

Sooty Oystercatcher

Wrybill

Banded Lapwing (Plover)

Red-kneed Dotterel

Red-capped Dotterel

Australian Dotterel (Courser)

Banded Stilt

New Zealand Stilt

Red-necked Avocet *Recurvirostra novaehollandiae* (Recurvirostridae) 43 cm (17 in). Elegant pied wading bird with long, slender upturned bill, long legs. White with dark chestnut head and neck, broad black bands in wings. White ring round eye. Legs blue-grey, partially webbed. Juvenile duller and paler. Legs extend beyond tail in floppy flight. Call a liquid 'toot-toot', also yelps and wheezes. Gregarious. Swims or wades, sweeps beak from side to side, taking prey from surface of water or mud. **Range and habitat:** AU. Widespread and locally abundant nomad of shallow saline and freshwater wetlands in Australia. Breeds in Lake Eyre in vast numbers, when it fills. Vagrant to New Zealand. **Similar species:** Banded Stilt (see p 294).

Australian Pratincole *Stiltia isabella* (Glareolidae) 23 cm (9 in). Elegant dry-country 'wader' with very long wings, short unforked tail, long legs. Orange-buff head, breast and back, with black eye stripe and wing tips, dark chestnut belly patch, white undertail. Tail square, white with black tip, white rump. Bill red with black tip. Juvenile duller, more mottled, all-black beak. Stands erect, runs gracefully, bobs head, teeters rear end. Wings extend well beyond tail. Flight swift, erratic and graceful, swallow-like or tern-like. Call is a plaintive chirrup. **Range and habitat:** OR, AU. Common breeding migrant to lagoon edges and open plains near water in eastern, inland and northern Australia; occasionally near south coast in fields; winters in northern Australia, New Guinea and Indonesia. **Similar species:** Eastern Collared Pratincole has a forked tail.

Snowy Sheathbill *Chionis alba* (Chionidae) 41 cm (16 in). Plump all-white gull-like bird with short, stout, pinkish yellow bill with dark tip, hard sheath over nostrils, flesh-pink warty skin on forehead and face. Feet black. Walks with a pigeon-like bobbing of head. Call a harsh, guttural caw or low muttering. Feeds on seashore or scavenges around seal and penguin colonies. **Range and habitat:** AN, oceanic and circumpolar. Common resident of seabird colonies on rocky coasts of peninsular Antarctica, South Georgia, South Orkney and Shetland Islands. Partial migrant, some birds wintering off South America. Generally not far from land. **Similar species:** Black-faced Sheathbill.

Silver Gull *Larus novaehollandiae* (Laridae) 41 cm (16 in). Small grey and white gull with red bill, eye-ring and legs. White with grey back and wings, black wing tips with white leading edge and white spots on trailing edge. Some African birds have faint lavender collar. Immature mottled brown on wings, with brown bill, eyes and legs. Call a variety of harsh, raucous notes, short and sharp or prolonged. Gregarious, colonial. **Range and habitat:** AF, AU, PA. Abundant and increasing resident or nomad of beaches, coastal and inland waters, parks, airfields, cricket pitches and rubbish dumps in southern Africa, Australia, islands in south-west Pacific. Breeding on islands, occasionally near freshwater, salt lakes. **Similar species:** Grey-headed Gull (see p 202); immature Black-billed Gull has more white in wing tips.

Kelp (Southern Black-backed) Gull *Larus dominicanus* (Laridae) 58 cm (23 in). Large black and white gull with slender yellow bill. White with black back and wings, broad white hind border to wings and row of white spots along wing tips. No or slight black band on tail. Bill has red spot on lower mandible only. Juvenile mottled brown, with black bill. Call a prolonged yelping. **Range and habitat:** Circumpolar. Common resident of temperate, coastal and inland waters, rubbish dumps and farmland in AF, NT, AU, subantarctic islands; increasing in AU. **Similar species:** Pacific Gull, other black-backed gulls elsewhere (see p 162).

Pacific Gull *Gabianus pacificus* (Laridae) 63 cm (25 in). Large black and white gull with massive bill. White with black back and wings, black band near tail tip, narrow white hind border to wings, no white spots on wing tips. Deep yellow bill with prominent red tips on both mandibles. Juvenile brown, mottled on back and wings, with pink-based bill. Call is a deep barking 'owk-owk', or 'kiaw'.

Strictly marine, may follow ships. **Range and habitat:** AU. Uncommon resident of coastal waters in southern Australia, perhaps declining. Breeds on offshore islands in small colonies. **Similar species:** Kelp (Southern Black-backed) Gull.

Antarctic Skua *Catharacta antarctica* (Stercorariidae) 58 cm (23 in). Large, heavily built, brown gull-like seabird with conspicuous white patches at primary bases, visible in flight. Central tail feathers project slightly. Upperparts uniform dark brown, with some light streaking. Underparts variable, dark to pale brown. Bill strongly hooked. Flight swift and powerful on broad, pointed wings. Predatory and piratical, chasing other seabirds. Scavenges, follows ships. Call a gull-like barking. Sometimes treated as a subspecies of the Great Skua (see p 160). **Range and habitat:** oceanic and circumpolar, breeds on southern oceanic islands and on the Antarctic peninsula. Winter visitor to all southern continents and into north Pacific ocean. **Similar species:** Great (see p 160) and South Polar Skuas, juvenile Pacific and Kelp (or Southern Black-backed) Gulls.

Antarctic Tern *Sterna vittata* (Laridae) 38 cm (15 in). Medium-sized marine tern with long tail streamers. Breeding plumage grey with black cap, white cheek stripe, rump and tail, red bill and feet. Nonbreeding birds have white forehead, mottled crown, paler underparts, dark bill and feet. Immature mottled, with white underparts. Flight graceful and undulating, often hovers. Call a shrill, high-pitched chatter or rattle. **Range and habitat:** oceanic and circumpolar. Common breeder on Antarctica, antarctic and subantarctic islands, also small islands near New Zealand, some winter movement northwards to temperate waters. **Similar species:** many other terns, especially nonbreeding Arctic Tern (see p 162); Kerguelen Tern is darker grey, with grey tail.

Sooty Tern *Sterna fuscata* (Laridae) 43 cm (17 in). Medium-sized, pied marine tern. Upperparts blackish brown with white forehead and outer edges to forked tail, underparts white. Juvenile dark brown with pale spots on upperparts, pale underwings, white undertail. Flight graceful, buoyant and swooping; soars. Flocks skim and dip to water surface, often far from land. Occasionally feeds at night. Call a high-pitched, nasal 'ker-wack-wack'. Sometimes in huge colonies. **Range and habitat:** oceanic and pantropical. Breeds on islands in Atlantic, Indian and Pacific oceans off equatorial AF, OR, AU, eastern NT and NA. **Similar species:** Bridled Tern has more extensive white forehead, noddies; Spectacled Tern has white eyebrows, greyer upperparts.

Black (White-capped) Noddy *Anous minutus* (Laridae) 35 cm (14 in). Small, dark brown seabird with white cap, black face, white crescent below eye, slender bill, short forked tail. Flight swift and erratic, low over water, dips to surface. May patter feet on surface. Generally close to shore. Often in large flocks. Call is a querulous, rattling 'kirr', cackling 'krik-krik-krik'. **Range and habitat:** tropical coastal seas of AU, PA and NT, breeding on islands in equatorial Pacific and western Atlantic oceans. **Similar species:** other noddies, juvenile Sooty Tern. **Not illustrated.**

Red-necked Avocet

Australian Pratincole

Snowy Sheathbill

Silver Gull

Kelp (Southern Black-backed) Gull

Pacific Gull

Antarctic Skua

Antarctic Tern

Sooty Tern

PIGEONS AND SANDGROUSE Columbiformes

New Zealand Pigeon *Hemiphaga novaeseelandiae* (Columbidae) 51 cm (20 in). Large, brightly coloured fruit-pigeon, only native pigeon in New Zealand. Upperparts and breast iridescent green, bronze and purple, sometimes with blue-grey cast in certain lights. Underparts white. Eye, bill and feet red. Much swishing and clapping of wings as birds fly or move about tree-branches. Call a soft, penetrating coo. **Range and habitat:** AU. Common resident of forest and exotic fruiting trees and shrubs in New Zealand. **Similar species:** none in New Zealand.

Rose-crowned (Pink-capped) Fruit Dove *Ptilinopus regina* (Columbidae) 22 cm (8½ in). Small, brightly coloured pigeon of the forest canopy. Upperparts bright green, with rose-pink cap bordered yellow, grey nape, broad yellow tail tip. Throat and breast blue-grey, spotted white, with lilac patch on lower breast, belly and undertail orange. Juvenile green with mottled yellow belly. Difficult to see; detection aided by falling fruits, rapid flight with whistling wings. Call is a double coo, and a long series of accelerating 'hoo' notes. **Range and habitat:** AU. Common resident and partial migrant of tropical and subtropical rainforest in north-western and north-eastern Australia to mid-New South Wales, vagrant further south, eastern Indonesia and New Guinea. **Similar species:** other *Ptilinopus* fruit doves.

Pink-spotted Fruit Dove *Ptilinopus perlatus* (Columbidae) 26 cm (10 in). Medium-sized, brightly coloured pigeon of the forest canopy. Upperparts bright green, with pink spots on shoulders, grey tail tip. Head all-grey or yellow-brown with grey collar, brown breast band. Underparts olive-green. Gregarious. Call is an accelerating, rising and falling series of 'hoo' notes, or a series of coos alternating in pitch. **Range and habitat:** AU. Common nomad of rainforest and edge in New Guinea and satellite islands. **Similar species:** Ornate Fruit Dove. **Not illustrated.**

White-breasted (White-bibbed) Fruit Dove *Ptilinopus rivoli* (Columbidae) 25 cm (10 in). Medium-sized, brightly coloured pigeon of the forest canopy. Green with purple cap and belly, white breast band with yellow wash. Female green with yellow undertail. Island males have yellow belly and undertail. Solitary and inactive. Call is an accelerating and descending series of 'hoo' notes, and single or repeated 'hoo' notes. **Range and habitat:** AU. Common nomad of hill and montane rainforest in New Guinea and satellite islands. **Similar species:** other *Ptilinopus* fruit doves. **Not illustrated.**

Ornate Fruit Dove *Ptilinops ornatus* (Columbidae) 26 cm (10 in). Medium-sized, brightly coloured pigeon of the forest canopy. Upperparts bright green, with purple and grey shoulder patch, yellow tail tip. Head yellow-brown or purple with grey collar, brown breast band, olive-green underparts. Gregarious. Call is a series of ascending 'hoo' notes, or a series of coos alternating in pitch. **Range and habitat:** AU. Common nomad of rainforest and moss forest in New Guinea. **Similar species:** Pink-spotted Fruit Dove. **Not illustrated.**

Wompoo (Magnificent) Fruit Dove *Ptilinopus magnificus* (Columbidae) 30-40 cm (12-16 in). Large, brightly coloured, long-tailed pigeon of the forest canopy and subcanopy. Upperparts green with blue-grey head, yellow shoulder flash, orange-red bill. Throat and breast purple, belly and undertail yellow. Difficult to see; detection aided by falling fruits, calls. Flies swiftly, low to canopy, showing yellow in underwings. Call a deep, resonant far-carrying, 'wom-poo' or 'wollack-a-woo'. Has declined in southern part of range due to loss of rainforest. **Range and habitat:** AU. Uncommon resident or partial migrant of tropical and subtropical rainforest in north-east Australia to central New South Wales, New Guinea and satellite islands. **Similar species:** none in AU.

Peaceful (Gould's Zebra) Dove *Geopelia placida* (Columbidae) 22 cm (8½ in). Tiny long-tailed dove. Upperparts grey-brown, barred darker. Head and breast finely barred grey and white, belly pale fawn. Pale blue-grey eye-ring. Feeds on ground. Flight rapid and undulating, with whistling wings; appears grey with white fringes to tail. Call is a melodious coo 'doodle-doo' and low rattle. **Range and habitat:** OR, AU. Common resident of woodland in South-East Asia, New Guinea and satellite islands, north and east Australia. Has declined in some areas. Introduced Hawaii. **Similar species:** Diamond Dove.

Diamond Dove *Geopelia cuneata* (Columbidae) 20 cm (8 in). Tiny long-tailed dove. Upperparts grey-brown with white spots on wings, red eye-ring. Head and breast blue-grey, belly white. Juvenile browner, upperparts barred. Flight rapid and undulating, with whistling wings; shows rufous in spread flight feathers. Call is a soft, mournful cooing. Feeds on grass seeds on ground. Visits desert water holes. **Range and habitat:** AU. Common nomad of dry woodland in Australia. **Similar species:** Peaceful (Gould's Zebra) Dove.

Common Bronzewing *Phaps chalcoptera* (Columbidae) 32 cm (13 in). Medium-sized, plump, brown pigeon with pale forehead and chin, iridescent patch on wings. Upperparts brown with cream forehead, lilac neck, bronze-green bands on wings. Underparts pinkish fawn. Female, immature duller, browner. Feeds on ground, inconspicuously, then rockets up noisily. Often seen along roadsides in early morning. Flight swift and direct, showing buff in spread flight feathers. Call is a resonant 'oom', repeated. **Range and habitat:** AU. Common resident of open forest and woodland with moderate understorey, occasionally farmland, crops, throughout Australia. **Similar species:** Brush Bronzewing is more blue-grey and chestnut with buff forehead, chestnut throat, rufous in spread flight feathers.

Crested Pigeon *Ocyphaps lophotes* (Columbidae) 31-5 cm (12-14 in). Medium-sized, pinkish grey pigeon with conspicuous vertical crest. Grey head, brownish grey back, light grey beneath, pink on sides of neck and breast. Wings barred with black, prominent green or purple patches. Eye-ring and legs red. Tall pointed crest is black. Longer tail than other crested pigeons in Australia. Raises tail on landing. Rapid flight makes loud whistling noise. Simple, low, soft cooing. In pairs or small flocks, often in farmland, along roads. Feeds on ground on seeds, green leaves and insects. **Range and habitat:** open habitats in most of Australia, except for forested coastal areas, has extended range with forest clearing.

Wonga Pigeon *Leucosarcia melanoleuca* (Columbidae) 37 cm (15 in). Large grey and white ground pigeon. Upperparts blue-grey with white forehead. Breast blue-grey with V-shaped band of white, belly and undertail white spotted black. Bill and feet red. Feeds on seeds on ground in cover. Flies suddenly with very loud clatter of wings, lands on tree-branch with back to observer. Call a fast, penetrating, monotonous 'wonk-wonk-wonk…' repeated for long periods. **Range and habitat:** AU. Common resident of dense, damp forest, or edge of rainforest in Australia from central Queensland to eastern Victoria.

New Zealand Pigeon

Rose-crowned (Pink-capped) Fruit Dove

Wompoo (Magnificent) Fruit Dove

Peaceful (Gould's Zebra) Dove

Diamond Dove

Crested Pigeon

Common Bronzewing

Wonga Pigeon

PARROTS AND RELATIVES Psittaciformes

Kea *Nestor notabilis* (Psittacidae) 46 cm (18 in). Large, short-tailed grey-green parrot with slender, elongated upper mandible. Scaly olive-green with red underwing patches, dull red rump. Inquisitive and approachable, attracted to humans and buildings. Call is a loud screech 'keaa' given in flight. Normally feeds on plant material, but occasionally on carrion. **Range and habitat:** AU. Common resident of mountain forests, scrub and alpine grassland on South Island, New Zealand. **Similar species:** Kaka.

Kaka *Nestor meridionalis* (Psittacidae) 45 cm (18 in). Large, short-tailed bronze-brown parrot with heavy bill. Scaly olive-brown to olive-green with red collar, rump, belly and underwings, pale crown, orange cheek patch. Flies low and fast through bush or high in the air, showing large head and bill. Call is a harsh 'ka-aa' or ringing, musical whistle, given in flight. Tears up rotten wood in search of insects. **Range and habitat:** AU. Uncommon resident of native forest in New Zealand. **Similar species:** Kea. **Not illustrated.**

Red-tailed Cockatoo *Calyptorhynchus magnificus* (Cacatuidae) 50-63 cm (20-5 in). Large black cockatoo with red or orange in tail. Male black with large crest, broad scarlet band in tail. Female spotted yellow on head and barred yellow on breast, with fine black bars across orange tail band, white bill. Flight slow and laboured, showing long wings and tail. Gregarious, wary. Call is a grating, discordant metallic screech. **Range and habitat:** AU. Common, but patchily distributed, resident of open forest, woodland, semi-arid scrub in Australia. **Similar species:** smaller, tamer Glossy Cockatoo is duller with shorter crest, more bulbous bill; female has yellow blotches on head but little or no yellow spotting and barring, grey bill. It occurs in sheoak woodland near east coast.

Galah *Eolophus roseicapillus* (Cacatuidae) 36 cm (14 in). Small pink and grey cockatoo. Crest pale pink, head and underparts deep rose-pink. Upperparts grey, paler on rump. Eyes brown in male, pink in female. Commonest cockatoo in southern Australia, gathers in large, noisy, wheeling flocks. Flies strongly with deep, rhythmic beats. Call is a shrill disyllabic screech, also harsh alarm squawks. **Range and habitat:** AU. Abundant resident or nomad of most open habitats, including parks and towns, in Australia; increasing and has spread with clearing of forests near coast. Sometimes a pest in crops. **Similar species:** Gang-gang Cockatoo is darker, all-grey with red head in males.

Little Corella *Cacatua sanguinea* (Cacatuidae) 37 cm (15 in). Small white cockatoo with short pale bill, short white crest, pink face. Bulbous blue-grey skin around and below eye, yellow wash on underwings and undertail. Gregarious, forms large, noisy wheeling flocks, often in crops. Flight strong and direct. Call is a wavering 2-note cry, harsh alarm shrieks. **Range and habitat:** AU. Abundant resident or nomad of woodland, open plains and farmland in Australia and New Guinea; increasing. **Similar species:** Long-billed Corella has pink forehead and throat, long upper mandible (but Little Corella in south-west has long mandible); larger Sulphur-crested Cockatoo has long yellow crest. **Not illustrated.**

Major Mitchell's Cockatoo *Cacatua leadbeateri* (Cacatuidae) 36 cm (14 in). Pink and white cockatoo with multicoloured crest. Upperparts white, head and underparts salmon pink. Long, filamentous crest shows bands of red and yellow when raised. Eyes brown in male, pink in female. Flight is a hesitant, shallow flapping and gliding, showing pink underwings. Call is a plaintive chuckling cry, harsh alarm screeches. Usually in smaller flocks than other cockatoos. **Range and habitat:** AU. Uncommon resident of woodland and plains in inland Australia. Declining locally. **Similar species:** corellas; larger Sulphur-crested Cockatoo is all-white with yellow crest; Galah which is pale in Western Australia.

Rainbow Lory *Trichoglossus haematodus* (Loridae) 28 cm (11 in). Medium-sized, spectacularly coloured lory with long pointed tail. Upperparts bright green with blue head, yellow collar. Breast orange, belly blue, undertail yellow-green. Bill red. Gregarious, forming noisy flocks, among flowering eucalypts and other nectar-bearing trees and shrubs. Flight swift and whirring, showing yellow band across wing. Call a rolling or discordant screech, mellow whistle. **Range and habitat:** AU, PA. Abundant nomad or partial migrant of open forest, woodland, orchards and towns in eastern Indonesia, New Guinea, northern and eastern Australia and western Pacific islands. **Similar species:** Red-collared Lorikeet (probably a geographical race) has orange collar, blue nape, blue-black belly; Scaly-breasted Lorikeet is green with yellow scalloping on breast, red underwings.

Dusky Lory (Dusky-orange Lorikeet) *Pseudeos fuscata* (Loridae) 25 cm (10 in). Medium-sized, short-tailed, dull lory sometimes appearing all-dark. Scaly brown with orange cap, bill, breast and belly patches, white rump. Yellow and orange in underwings and tail. Gregarious and vocal, forming huge communal roosts. Call a loud, harsh grating screech. **Range and habitat:** AU. Abundant nomad or migrant of forest, edge and plantations in New Guinea. **Similar species:** Rainbow Lory; Duyvenbode's Lory is larger with black bill, longer tail.

Goldie's Lorikeet *Trichoglossus goldei* (Loridae) 19 cm (7½ in). Small green lory with long tail. Upperparts green, underparts yellower with dark green streaks. Cap red, face purple with blue streaks. Yellow band through flight feathers. Flight rapid with backswept wings, pointed tail. Gregarious. Call a high-pitched shriek. **Range and habitat:** AU. Common nomad of montane forest in New Guinea. **Similar species:** Striated Lorikeet is darker, lacking red and purple on head. **Not illustrated.**

Papuan (Green-winged) King Parrot *Alisterus chloropterus* (Psittacidae) 38 cm (15 in). Large, brightly coloured, long-tailed parrot. Head and underparts red, upperparts green with blue nape and rump, yellow-green shoulder flash, black tail. Female and juvenile have green head, no shoulder flash. Flight agile and graceful with deep wing-beats. Call a sharp, metallic 'keek' in flight or a high-pitched, longer whistling note 'eeek' repeated when perched. **Range and habitat:** AU. Common resident of forest in New Guinea.

Red-winged Parrot *Aprosmictus erythropterus* (Psittacidae) 32 cm (13 in). Medium-sized green parrot with red wing flashes. Male bright green with black back, blue rump, red bill and shoulders. Female and juvenile duller, greener, with smaller red shoulder patches. Tail moderately long, square, with yellow tip. Strong, erratic flight on long wings, with deep, leisurely beats. Feeds on fruit and seeds of trees, sometimes on ground and in crops. Usually pairs or small flocks. Call a sharp, metallic 'kik' in flight, or rolling 'crillik'. **Range and habitat:** AU. Common resident or nomad of woodland and mangroves in northern and eastern Australia to central New South Wales and southern New Guinea. **Similar species:** Superb Parrot is green with long pointed tail, yellow and red throat in males; Australian King Parrot is red below, has long dark tail.

Kea

Red-tailed Cockatoo

Galah

Major Mitchell's Cockatoo

Rainbow Lory

Dusky Lory (Dusky-orange Lorikeet)

Papuan (Green-winged) King Parrot

Red-winged Parrot

Cockatiel *Nymphicus hollandicus* (Cacatuidae) 32 cm (13 in). Tiny, long-tailed cockatoo. Grey with yellow face and crest, orange ear patches, white shoulders. Female duller, tail finely barred, outer tail feathers yellow. Flight strong but erratic, with leisurely beats, showing pointed wings and long pointed tail. Call a rolling, upslurred 'queel', softer plaintive wavering notes. **Range and habitat:** AU. Common migrant within Australia, moving north in autumn and south in spring, in woodland, farmland and plains.

Budgerigar *Melopsittacus undulatus* (Psittacidae) 18 cm (7 in). Tiny green and yellow parrot with long, pointed tail. Head yellow finely barred black, with black and blue spots on cheeks. Back and wings barred yellow, green and black. Rump and underparts green, tail blue. Cere blue in male, brown or pale blue in female. Flight rapid, whirring and undulating, showing yellow bands in wings and tail. Forms noisy, wheeling flocks. Call a warbling and chirruping, harsh scolding and alarm screeches. **Range and habitat:** AU. Abundant nomad or migrant of dry woodlands and grasslands in inland Australia. Breeds prolifically after rain.

Crimson Rosella *Platycercus elegans* (Psittacidae) 35 cm (14 in). Medium-sized, long-tailed red and blue parrot. Rich red with blue cheeks, wings and tail, black scalloping on back. Juvenile green with red on forehead, breast and undertail. Geographical races vary in colour: 'Yellow Rosella' is yellow with tiny red patch on forehead, 'Adelaide Rosella' is orange to brick red. Flight undulating. 3-note bell-like whistle, middle note highest, and harsh alarm notes. **Range and habitat:** AU. Common resident of forest and woodland in Australia. **Similar species:** Australian King Parrot is larger and greener, lacks blue cheeks.

Mulga Parrot *Psephotus varius* (Psittacidae) 28 cm (11 in). Small, brightly coloured parrot with long tail. Male bright green with blue wings and tail, red nape, belly and rump patches, yellow forehead, shoulder and undertail. Female duller, with red shoulder. Flight low and undulating. Short, soft whistling notes. **Range and habitat:** AU. Uncommon resident of woodland in inland Australia. **Similar species:** Red-rumped Parrot is more simply coloured, green and yellow with red rump; female is dull green without patches of colour.

Bourke's Parrot *Neophema bourkii* (Psittacidae) 19 cm (7½ in). Very small, softly coloured parrot with long tail. Upperparts brown with blue forehead and wing edges, cream scalloping on wings. Underparts pink, undertail pale blue. Female and juvenile lack blue forehead. Flight low and fluttering, showing white edges to tail. Call is a chirrup, soft twitter or shrill alarm note. **Range and habitat:** AU. Uncommon resident or nomad of dry woodland and scrub in inland Australia. **Similar species:** Diamond Dove (see p 298). **Not illustrated.**

CUCKOOS AND ALLIES Cuculiformes

Pallid Cuckoo *Cuculus pallidus* (Cuculidae) 33 cm (13 in). Medium-sized grey cuckoo with long, barred tail. Upperparts grey with yellow eye-ring, pale eyebrow and nape, black eye stripe. Underparts pale grey. Juvenile mottled and barred black and white, immature mottled brown and buff. Perches in exposed positions with tail drooping. Flight undulating on pointed wings, flicks tail on alighting. Call a series of whistles ascending in semitones by male, harsh single whistle by female. **Range and habitat:** AU. Common resident and migrant of woodland and scattered trees in farmland in Australia; occasionally Timor, New Guinea. **Similar species:** Oriental Cuckoo (see p 252). **Not illustrated.**

Horsfield's Bronze Cuckoo *Chalcites basalis* (Cuculidae) 17 cm (6½ in). Tiny, glossy cuckoo with barred underparts. Upperparts dull bronze-green with white eyebrow, dark stripe through eye and ear, rufous in spread tail. Underparts and undertail boldly barred brown and white, bars broken on belly. Juvenile duller, less barred. Flight

rapid and undulating on pointed wings, showing white wing-bar. Call a high-pitched, descending whistle 'see-oo', sparrow-like chirrup. **Range and habitat:** OR, AU, PA. Common resident or partial migrant of open forest and woodland in Australia; some birds winter in New Guinea, Indonesia and islands in eastern Indian ocean. **Similar species:** other bronze cuckoos. **Not illustrated.**

Koel *Eudynamys scolopacea* (Cuculidae) 39-46 cm (15-18 in). Large, long-tailed cuckoo. Male all-black with red eye, pale (AU) or green bill (OR). Female brown, barred and spotted buff and white; head rufous (New Guinea) or black (Australia) with white cheek stripe, finely barred underparts, heavily barred tail. Juvenile mottled rusty brown and buff, barred black. Flight shallow and fluttering. Far-carrying call a shrill 'coo-ee' and rollicking notes by male, sometimes at night, series of piercing whistles by female. **Range and habitat:** OR, AU, PA. Common resident and migrant in India, east Asia, New Guinea, eastern and northern Australia, western Pacific islands. Southern birds winter in islands north of Australia. **Similar species:** Long-tailed Koel (AU, PA) is streaked on head and underparts.

Pheasant Coucal *Centropus phasianinus* (Cuculidae) 60-80 cm (24-31 in). Huge, ungainly ground cuckoo with barred plumage, long tail. In breeding plumage head and underparts black, upperparts rusty brown finely barred black and white, streaked cream, tail barred. In nonbreeding plumage head and underparts rusty with cream streaks. Eyes red. Flight low and laboured on short, rounded wings, tail fanning. Call a series of rapid, resonant hoots, falling then rising; also sharp, hissing alarm notes. **Range and habitat:** AU. Common resident of thick undergrowth in forest, woodland, farmland, swamps in Timor, New Guinea and eastern and northern Australia. **Similar species:** immature Greater and Lesser Coucals.

Brush Cuckoo *Cacomantis variolosus* (Cuculidae) 23 cm (9 in). Small, dull cuckoo. Head and eye-ring grey, upperparts grey-brown. Breast buff, belly and undertail pale rufous. Juvenile richly mottled and barred brown and rufous, paler on underparts. Flight rapid and undulating on pointed wings, showing pale wing-bar and barred tail. Call a mournful series of downslurred whistles descending by semitones; urgent series of 3-note whistles, phrase repeated at increasing pitch; single loud, tremulous whistle. **Range and habitat:** OR, AU, PA. Common resident and migrant of forest and woodland in South-East Asia, New Guinea and Australia; southern birds winter in tropics. **Similar species:** Plaintive Cuckoo, Banded Bay Cuckoo in OR (see p 252); Fan-tailed Cuckoo has yellow eye-ring, more tapered tail.

Greater Coucal *Centropus menbecki* (Cuculidae) 64 cm (25 in). Huge, pheasant-like ground cuckoo with stout, arched bill. All-black with short wings, long broad tail, pale bill, red eyes. Immature has barred tail. Secretive in dense vegetation, where it hops clumsily. Flies infrequently and heavily on rounded wings. Call a deep series of descending hooting notes, also croaks and rattles. **Range and habitat:** AU. Common resident of undergrowth in forest and edge in New Guinea and satellite islands. **Similar species:** smaller Lesser Coucal has dark eye and bill, immature barred brown. **Not illustrated.**

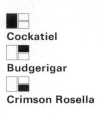

Cockatiel

Budgerigar

Crimson Rosella

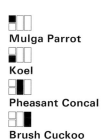

Mulga Parrot

Koel

Pheasant Concal

Brush Cuckoo

OWLS Strigiformes

Boobook Owl *Ninox novaeseelandiae* (Strigidae) 25-35 cm (10-14 in). Small brown hawk-like owl. Dark to light brown with white spots on wings, barred wings and tail. Underparts streaked, marbled or spotted white on brown or rufous. 'Mask' of dark eye patches on white face. Juvenile has downy white underparts, faintly streaked. Eyes greygreen or yellow. Falsetto double hoot 'boo-book', 2nd note lower, and soft continuous hooting. **Range and habitat:** AU. Common resident of woodland in eastern Indonesia, New Guinea, Australia, New Zealand (sometimes considered a separate species). **Similar species:** Rufous Hawk Owl, Little Owl (see p 166); Sumba Boobook, Barking Owl.

Australian Masked Owl *Tyto novaehollandiae* (Tytonidae) 37-50 cm (15-20 in). Large owl with prominent heartshaped facial disc, feathered legs, large feet and claws. Upperparts mottled dark brown to grey-brown, speckled white, strongly barred on wings and tail. Facial disc has crisp black border, dark zone around black eyes. Face chestnut, underparts rufous, coarsely spotted black in darkest birds; face and underparts white, breast sparsely spotted in palest birds. Very loud, deep rasping screech or shrill tremulous scream. **Range and habitat:** AU. Uncommon in forest, woodland, caves in New Guinea and Australia. **Similar species:** Barn Owl (see p 166); sooty owls.

NIGHTJARS AND ALLIES Caprimulgiformes

Tawny Frogmouth *Podargus strigoides* (Podargidae) 34-46 cm (13-18 in). Owl-like nocturnal bird with broad bill, large yellow eyes, weak feet, ragged tail. Large head, tuft of feathers above bill. Mottled grey, browner on shoulders; wings and tail barred. Less common rufous or brown forms. Perches by day on branches, cryptic. At night perches sluggishly, flies with quick shallow beats and long glides, appears pale. Monotonous, resonant 'oom-oom-oom . . . .'. **Range and habitat:** AU. Common resident of forest and woodland in Australia, rare in New Guinea. **Similar species:** smaller Marbled Frogmouth has banded plumes, long tail; larger Papuan Frogmouth has red eyes, long tail.

(Australian) Owlet-nightjar *Aegotheles cristatus* (Aegothelidae) 23 cm (9 in). Small nocturnal bird with fluttering, moth-like flight. Soft grey, finely barred, paler on belly; rufous colour form occurs. Face mammal-like, with large dark eyes, black crown stripe, eye stripes and collar. Wings rounded, long tail and legs. Roosts in tree-hollows, often at entrance; if flushed, flies with undulating flight to another hole. At night active in trees, in air and on ground. Shrill, strident churring, often during the day. **Range and habitat:** AU. Common resident of forest and woodland in New Guinea and Australia. **Similar species:** other owlet-nightjars.

Spotted Nightjar *Eurostopodus argus* (Caprimulgidae) 30 cm (12 in). Delicately mottled, well-camouflaged nocturnal bird with long wings, strong graceful flight. Mottled and spotted buff, grey and black with white throat, barred tail. Juvenile more rufous. Rounded wing tips, large white spot in wings. Roosts on ground, flies suddenly and silently. At night active in the air; red eye-shine. Call a hoarse 'haw-haw-haw, chokka-chokka-chokka' on same pitch. **Range and habitat:** AU. Open woodland in Australia; some north-south migration. **Similar species:** White-tailed Nightjar; Whitethroated Nightjar has small white spot in wing, pointed wing tips. **Not illustrated.**

SWIFTS AND HUMMINGBIRDS Apodiformes

Mountain Swiftlet *Aerodramus hirundinaceus* (Apodidae) 13 cm (5 in). Small uniform swift. Upperparts blackish brown, underparts silvery grey-brown, legs feathered. Flight rapid and fluttering, with long glides. Gregarious, flies high in the air over cave entrances. Call a rapid twittering. **Range and habitat:** AU. Montane New Guinea

and satellite islands, aerial over hill and mountain forest to above treeline. **Similar species:** lowland Uniform Swiftlet has browner underparts, pale throat, unfeathered legs. **Not illustrated.**

KINGFISHERS AND ALLIES Coraciiformes

Laughing Kookaburra *Dacelo novaeguineae* (Alcedinidae) 45 cm (18 in). Giant brown and white kingfisher with loud 'laughing' call. Head cream with brown streaks and ear patch. Upperparts brown with pale blue mottling on shoulders, rufous tail barred black. Rump pale blue in male, brown in female and juvenile. Underparts cream, faintly barred brown. Perched profile distinctive, with large head and bill, dropped tail. In flight shows white patches in wings, white tail tip; on alighting raises tail. Loud, rollicking and hooting 'laughter', throaty chuckling. **Range and habitat:** AU. Common resident of forest and woodland in Australia. **Similar species:** Blue-winged Kookaburra.

Red-backed Kingfisher *Halcyon pyrrhopygia* (Alcedinidae) 20 cm (8 in). Small, pale, dry-country kingfisher. Head streaked grey-green and white, with black eye stripe encircling nape, white collar. Upperparts grey-green, bluer on wings in male, with rufous rump. Underparts white. Call a single mournful whistle, or chattering and rattling. **Range and habitat:** AU. Common resident and also partial migrant of dry, open woodland in Australia; southern birds winter in northern and coastal areas. **Similar species:** Sacred Kingfisher.

Sacred Kingfisher *Halcyon sancta* (Alcedinidae) 21 cm (8 in). Small, bright woodland kingfisher. Upperparts glossy blue-green, bluer in male, with black eye stripe encircling nape, small buff 'headlights' on forehead, buff to white collar. Underparts white to rich buff. Juvenile duller, faint buff scalloping on breast and wings. Call a strident 4-note squeak, slightly descending; also trilling, rattling and harsh scolding. **Range and habitat:** AU, PA. Common resident and migrant of open forest and woodland in eastern Indonesia, New Guinea, Australia, New Zealand, western Pacific islands. Southern birds winter in tropics. **Similar species:** Red-backed Kingfisher, White-collared Kingfisher; Forest Kingfisher has large white spot in wing.

Australian Bee-eater *Merops ornatus* (Meropidae) 23-8 cm (9-11 in). Brightly coloured, with slender curved bill. Iridescent blue and green with orange nape and throat, black eye stripe and throat band, orange in wings revealed in flight. Tail black with thin, projecting central feathers. Juvenile duller, without throat band or tail wires. Flight erratic and undulating with extended wheeling glides; soars. Call a short trilling note repeated, also rippling and ticking notes. **Range and habitat:** AU. Common resident and migrant of woodland in Australia and New Guinea, southern birds winter in tropics, sometimes islands in western Pacific. **Similar species:** Blue-tailed Bee-eater (see p 256).

Boobook Owl

Australian Masked Owl (dark phase)

Tawny Frogmouth

(Australian) Owlet-nightjar

Red-backed Kingfisher

Laughing Kookaburra

Sacred Kingfisher

Australian Bee-eater

PERCHING BIRDS Passeriformes

Rifleman *Acanthisitta chloris* (Xenicidae) 8 cm (3 in). Tiny, with very short tail and long thin bill. Male bright yellowish green above, buff below, female browner and streaked above, pale eyebrow and wing-bars. Family parties, with immatures like female. Usually located by high-pitched call 'zipt'. Flicks wings while feeding by creeping along bark and moss on tree-trunks and branches. Apparently polygamous; young of first broods may help to feed subsequent young. **Range and habitat:** all main islands of New Zealand, still fairly common in native forests and also occupies exotic vegetation. **Similar species:** Bush and Rock Wrens.

Superb Lyrebird *Menura novaehollandiae* (Menuridae) 74-98 cm (29-38 in). Very large ground-dwelling passerine, with strong legs and feet. Usually glimpsed rushing off along the ground giving a loud 'whisht' alarm call, though sometimes quite tame. Dark brown above, grey-brown below with rufous throat. Male has spectacular lyre-shaped tail, with large patterned outer feathers and fine filamentous central ones. The tail is spread and arched over the body during a courtship display on a cleared arena. Expert songster with powerful melodious phrases and complex and accurate mimicry of many other birds, and perhaps mechanical noises. Polygamous, with female caring for single young alone. **Range and habitat:** fairly common in damp forests of south-eastern Australia, even in outskirts of Sydney and Melbourne. **Similar species:** Prince Albert's Lyrebird.

Singing Bushlark *Mirafra javanica* (Alaudidae) 12-15 cm (4½-6 in). Rather undistinguished smallish brown bird, heavily streaked above and on breast. Rufous wing patch and deep finch-like bill help to separate it from similar species. Song is melodious, though less impressive than that of Skylark, may include mimicry, uttered in flight. Numerous subspecies, differing in colour to match background. **Range and habitat:** AF, OR and AU, fairly common in open grassland, but has also adapted well to farmland and crops. **Similar species:** other larks, and also pipits (see pp 168, 170, 218, 260). **Not illustrated.**

White-backed Swallow *Cheramoeca leucosternum* (Hirundinidae) 15 cm (6 in). Striking black and white swallow, usually nesting colonially and roosting communally in holes in sandy banks or small cliffs. Black wings, tail and lower body, white back and breast, with greyish crown. Deeply forked tail and fluttering flight. **Range and habitat:** most of Australia, except for north and coastal areas. Typically found in open country, it originally nested in the burrows of small marsupials. These have since declined, but clearing of forests and activities related to roads and building have allowed the species to spread. Apparently not migratory but may enter torpid state in cold weather. **Not illustrated.**

Welcome Swallow *Hirundo neoxena* (Hirundinidae) 15 cm (6 in). A typical swallow, very like the common Swallow. Blue-black back, wings and tail, orange throat and forehead and white belly, but no dark breast band, very long tail streamers. A common and popular bird throughout its range, coping well with human activities. Twittering song. **Range and habitat:** eastern and southern Australia, Tasmania and New Zealand. Most habitats, except for deep in forest; nests commonly in human structures, but also in old Fairy Martin nests. **Similar species:** Swallow (see p 168).

Tree Martin *Hirundo nigricans* (Hirundinidae) 13 cm (5 in). Glossy blue-black above and dirty white below and on rump, with tiny rufous patch on forehead. Tail barely forked. In loose groups, but gathers in larger flocks on telegraph wires and fences to migrate. Twittering call. **Range and habitat:** throughout Australia, mostly nesting south of the Tropic of Capricorn, migrating north in winter, also in New Caledonia, New Guinea and Indonesia. Common in open woodland, nesting and roosting in holes in large trees. **Similar species:** Fairy Martin. **Not illustrated.**

Fairy Martin *Hirundo ariel* (Hirundinidae) 12 cm (4½ in). Small martin commonly seen feeding in groups along roadsides in open country. Black wings, tail and back, white underparts and rump, with rufous head. Square tail. Weak twittering call, and louder chirruping alarm call. **Range and habitat:** Australia, except Tasmania, in open habitats. Nests colonially, often with Welcome Swallow, placing its bottle-shaped mud nests under overhanging rocks and cliffs, but more typically now under bridges and road culverts. Some north-south migration. **Similar species:** Tree Martin.

Richard's Pipit *Anthus novaeseelandiae* (Motacillidae) 17 cm (6½ in). Large pipit, dark brown above, buff below, with heavy streaking on back and breast and pale eyebrow. Legs and beak are also pale. Some local variation in colour of this widespread species. Wags tail up and down. Abrupt flight call 'shreep', and monotonous song delivered from the ground or in flight. **Range and habitat:** EP, OR, AF and AU, uncommon visitor to WP. Common in short grassland, low heath, steppe, farmland, up to about 2,000 m (6,500 ft), also semi-arid areas in Australia. **Similar species:** many larks and other pipits (see pp 168, 170, 260).

Black-faced Cuckoo Shrike *Coracina novaehollandiae* (Campephagidae) 33 cm (13 in). Smart-looking, large grey bird, with black face and upper breast and flight feathers. Juvenile has small black face patch, extending behind the eye. Small legs, rather heavy beak, dark eyes. Undulating flight, shuffles wings after landing. Call is a harsh 'kaark', occasional sweeter notes. Also known as Grey Jay or Blue Jay. A sit-and-wait predator, snatching large larvae and beetles from the foliage. **Range and habitat:** AU and OR, widespread in all types of wooded country, even among scattered trees and in towns, some migration in Australia, with birds crossing Torres Strait. **Similar species:** Papuan Cuckoo Shrike.

Ground Cuckoo Shrike *Pteropodocys maxima* (Campephagidae) 33-7 cm (13-15 in). Largest cuckoo shrike. Generally grey, with black tail and wings, lower breast, belly, rump and lower back barred dark grey. Tail slightly forked, bright yellow iris. Immature is barred all over body and has dark face and eye. Unlike other cuckoo shrikes it feeds on the ground, walking like a pigeon. Flight direct. Metallic 'ti-yew' call. Lives in family groups, with all birds helping to rear the young. **Range and habitat:** widespread, though rather uncommon in dry woodland in inland Australia. Somewhat nomadic. **Similar species:** other cuckoo shrikes (see p 262).

White-winged Triller *Lalage sucrri* (Campephagidae) 19 cm (7½ in). The male is a striking pied bird, with black cap, back, tail and wings, clean white underparts and wing patches, and grey rump. Female grey-brown above and buff below. Male like female in nonbreeding season. Usually detected in late spring by loud trilling song, often given in an energetic song-flight. Loosely colonial, though males chase each other a lot early in breeding season. Also called Caterpillar-eater, as takes larvae snatched from foliage or pounced upon on the ground. **Range and habitat:** throughout Australia, except Tasmania, but avoids heavy forest, also New Guinea and Indonesia. Regular north-south migration, with late arrival and early departure in the south. **Similar species:** Varied Triller.

Rifleman

Superb Lyrebird

Welcome Swallow

Fairy Martin in nest

Richard's Pipit

White-winged Triller

Black-faced Cuckoo Shrike

Red-capped Robin *Petroica goodenovii* (Eopsaltriidae) 11 cm (4½ in). Male has black head, back, wings and tail, white belly and wing patch, and red breast and diagnostic red cap. Female very dull, grey-brown above and buff below, with hint of reddish on forehead. Song a rattling trill. Perches on a low branch, flycatching and pouncing on insects on the ground. Flicks wings and tail. Australian robins are remarkably similar to Palaearctic flycatchers in appearance and behaviour, but are not closely related. Red-capped Robin is the semi-arid representative of its genus; 4 other red-breasted robins occupy progressively wetter and denser habitats. **Range and habitat:** inland Australia, in eucalypt, acacia and native pine woodland. Partly migratory, more common near coast during droughts. **Similar species:** other *Petroica* robins.

Scarlet Robin *Petroica multicolor* (Eopsaltriidae) 13 cm (5 in). Attractive red, black and white robin. Male has black head, throat, back, wings and tail, scarlet breast outlined in white, and white forehead, wing-bars and tail edges. Female is grey-brown with pale red breast. Rather weak twittering song. Perches on low branch, pouncing on ground prey, also hawks and snatches insects from tree-trunks in summer. **Range and habitat:** Southern Australia, Tasmania, and many Pacific islands, dry forest, with some understorey. **Similar species:** Red-capped Robin, Flame Robin has more orange breast to throat.

White-winged Robin (Thicket Flycatcher) *Peneothello sigillatus* (Eopsaltriidae) 15 cm (6 in). Black robin, with conspicuous white wing patch. Birds from the western part of distribution have another white patch on the shoulder. Juveniles are mottled brown and black. Perches on low mossy branches and pounces on ground prey. Calls a variety of piping trills. **Range and habitat:** forests of the central ranges of New Guinea, above 2,500 m (8,200 ft). **Similar species:** White-rumped Thicket Flycatcher has a white rump.

Jacky Winter (Australian Brown Flycatcher) *Microeca leucophaea* (Eopsaltriidae) 13 cm (5 in). A dull brown flycatcher, with white outer tail feathers and eyebrow. Wags tail from side to side. Juvenile spotted with white above and brown below. Perches on trees and fence-posts, darting out after flying insects or pouncing on them on the ground. Loud song, described as 'Jacky Winter', or more accurately as 'Peter Peter'. Tiny, inconspicuous nest. **Range and habitat:** woodland and scrub throughout Australia, but not Tasmania. Also southern New Guinea. Will occupy farmland, where few trees remain, and particularly on edge of woodland or forest.

Golden Whistler *Pachycephala pectoralis* (Pachycephalidae) 17 cm (6½ in). Usual male plumage is spectacular, with black head and gorget, white throat, bright yellow underparts and collar, green back and black tail. Female is greyish brown above and buff below. Immature has rufous patch on wings. Numerous races of this widespread species, with great variety of plumage. Some females are brighter, some males duller. Males may take 2 or more years to attain bright plumage, but may breed in female-like plumage. Powerful, melodious whistling song, sometimes ending in a whipcrack. Whistlers snatch and glean large insects from foliage. As sit-and-wait predators they can be unobtrusive in nonbreeding season. **Range and habitat:** in forests in southern and eastern Australia, moving into more open habitat in winter and numerous Pacific islands. **Similar species:** Mangrove Golden Whistler.

Rufous Whistler *Pachycephala rufiventris* (Pachycephalidae) 17 cm (6½ in). One of the most ubiquitous songsters in Australia. Male has grey back, wings, tail and head, white throat, outlined in black and pale rufous underparts. Female is grey above, with white throat and buff abdomen with vertical dark streaks. Immature and second year male resemble female. Loud whistling song, usually finishing with 'ee-chong' or a series of 'chews'. Snatches and gleans large insects from foliage. **Range and habitat:** woodland and open forest, throughout Australia, except Tasmania and Kangaroo Island. **Similar species:** other whistlers.

Grey Shrike-thrush *Colluricincla harmonica* (Pachycephalidae) 24 cm (9½ in). Well-named bird; grey, thrush-like with a slightly hooked beak, but really a large whistler. Eastern birds have a brownish back and western birds brownish underparts. Male has white lores with black eye-ring, female grey lores and white eye-ring. Immatures streaked on breast and some rufous on wings and face. Feeds by fossicking on ground, on horizontal branches and from foliage, for large insects. Occasionally takes small lizards and baby birds. Beautiful, melodious and whistling song, with neighbouring birds frequently replying to each other with matched phrases. **Range and habitat:** New Guinea and Australia, wherever there are trees or shrubs. **Similar species:** other shrike-thrushes, which are more localized and generally browner.

Crested Bellbird *Oreoica gutturalis* (Pachycephalidae) 22 cm (8½ in). Striking bird with white face, black erectile crest and bib, grey-brown back and tail and white belly. Female is duller than male, has no crest and dark eyes, whereas male has orange eyes. Typically detected by haunting, ventriloquial bell-like call, and can usually be found hopping inoffensively along the ground, feeding amongst leaf litter. Attaches hairy caterpillars to nest, perhaps as a food supply or deterrent to predators. **Range and habitat:** throughout inland Australia, in dry woodland and scrub. Has declined markedly with clearing of native vegetation for growing crops. **Similar species:** none, though Bell Miner, a honeyeater, is usually called Bellbird. **Not illustrated.**

Restless Flycatcher *Myiagra inquieta* (Monarchidae) 18 cm (7 in). Black above and white below, long tail and slight crest. Female has pinkish breast. Usually seen hovering above the ground before diving into long grass after insects. Also feeds by hawking, snatching insects from leaves and bark and gleaning on the ground. It utters a strange grinding, wheezing noise, whence its common name of Scissors-grinder, also grating sounds and loud whistling song. **Range and habitat:** open, grassy, eucalypt and paperbark woodland in northern, eastern and south-western Australia. Birds in north are much smaller. **Similar species:** (Black and White Fantail) Willie Wagtail (see p 310), Satin and Leaden Flycatchers, which have dark or orange throats.

Black-faced Monarch *Monarcha melanopsis* (Monarchidae) 17 cm (6½ in). A grey flycatcher with bright rufous underparts. Adult has black forehead and throat, juvenile no black face. Calls range from loud whistles to harsh nasal rasps. Monarchs typically pursue active prey by hawking or snatching insects from foliage and branches. This species though spends much time searching leaves and twigs for insects. **Range and habitat:** Rainforest and eucalypt forest in eastern Australia. Some birds migrate over the Torres Strait in autumn. **Similar species:** Black-winged Monarch, Spectacled Monarch.

Mountain Peltops Flycatcher *Peltops montanus* (Monarchidae) 20 cm (8 in). A spectacular bird of the treetops. Black, with white patches on back and cheeks, and red rump and undertail-coverts. Typically sallies for insects from emergent trees, or at the edge of clearings, and roadsides. Call is a rapid, descending 'tit tit tit . . .', reported to be like the sound of a finger being run over a comb, also 'schweep'. Although like a flycatcher, related to the Australian butcher birds. **Range and habitat:** rainforest in foothills and mountains of New Guinea. **Similar species:** Lowland Peltops Flycatcher, which replaces it below 550 m (1,800 ft). **Not illustrated.**

Red-capped Robin at nest

Scarlet Robin at nest

White-winged Robin (Thicket Flycatcher)

Jacky Winter (Australian Brown Flycatcher) at nest

Golden Whistler at nest

Grey Shrike-thrush

Rufous Whistler at nest

Black-faced Monarch at nest

Restless Flycatcher at nest

Collared Grey Fantail *Rhipidura fuliginosa* (Monarchidae) 16 cm (6½ in). A hyperactive grey flycatcher, with white throat, eyebrow and buff underparts. Long tail, frequently fanned while perching. Never still for long. Delicate legs and feet, short broad beak with obvious bristles around gape. Juvenile more buff on head and wings. Numerous races, vary in shade of grey and amount of white on tail. Pursues small flying insects aerobatically, occasionally feeds on ground. High-pitched twittering calls and song. **Range and habitat:** Australia, in woodland and forest, Pacific islands. Mangrove Fantail of northern Australia and New Guinea may be a distinct species. Partial migrant in southern parts of range, on migration in towns and open habitat. **Similar species:** other fantails (see p 266).

Friendly Fantail *Rhipidura albolimbata* (Monarchidae) 15 cm (6 in). Sooty grey fantail, with white chin, belly, eyebrow and spots on wings and tail. Repeated, staccato whistles. Like other fantails, it willingly approaches humans, perhaps attracted to insects associated with or flushed out by them. Hawks for flying insects in middle or lower levels of forest. **Range and habitat:** rainforest in New Guinea from 1,750 m (5,700 ft) to timberline. **Similar species:** other fantails (see p 266).

(Black and White Fantail) Willie Wagtail *Rhipidura leucophrys* (Monarchidae) 20 cm (8 in). Striking fantail. White eyebrow, can be raised when alarmed. Juvenile brownish on wings and eyebrow. One of Australia's best-known and popular birds. Bold and aggressive, even attacking much larger birds. Sweet, chattering song; harsh alarm notes; sometimes sings at night. Feeds on beetles, grasshoppers, on ground, cocking and wagging long tail from side to side, or by pursuing flying insects. Perches on sheep and takes ticks from cattle. **Range and habitat:** Australia, New Guinea, parts of Indonesia, in open woodland, grassland, farmland, gardens, city parks. **Similar species:** Restless Flycatcher (see p 308).

Wedgebill *Sphenostoma cristatum* (Orthonychidae) 20 cm (8 in). Light brown, medium-sized bird, with white outer tail feathers. Pointed crest, rather heavy black beak and long wedge-shaped tail. Feeds in bushes and on ground on seeds and insects. Wedgebills have been split into 2 distinct races (sometimes treated as species) by their distribution and calls. Chiming Wedgebill calls 'why did you get drunk', and Chirruping Wedgebill 'tootsie cheer'. Otherwise appear identical. **Range and habitat:** Chiming Wedgebill in dense acacias and other shrubs in western central Australia, often along watercourses; Chirruping Wedgebill in eastern inland, in more open habitat. **Not illustrated.**

Clamorous Reed Warbler *Acrocephalus stentoreus* (Sylviidae) 17 cm (6½ in). Clean-looking pale brown bird, warm brown above, buff below, white throat and eyebrow. Loud, continuous song with melodious quality, but some harsh notes. Usually heard from dense reed-beds, but will perch in view and sometimes pursues flying insects in the open. During droughts in bushes, gardens, away from water. **Range and habitat:** north-east Africa, southern Asia, New Guinea, eastern and south-west Australia, reed-beds. **Similar species:** other *Acrocephalus* warblers (see pp 172, 222).

Little Grassbird (Little Marshbird) *Megalurus gramineus* (Sylviidae) 14 cm (5½ in). Inconspicuous small bird, dark brown above, grey-brown below, heavily streaked. Long graduated tail. Usually detected by high-pitched, 2- or 3-noted whistling call, then seen flitting from tussock to tussock at the edge of water. **Range and habitat:** eastern and south-western Australia, in grassy swamps, lake shores, saltmarshes, artesian bores and farm dams, vagrant to New Guinea. Migratory, or perhaps only silent and overlooked in winter. **Similar species:** Tawny Marshbird is larger, more rufous, in rank grassland or heath. **Not illustrated.**

Golden-capped Cisticola *Cisticola exilis* (Sylviidae) 10 cm (4 in). Male has golden head and upperparts, with dark streaking on wings, white underparts and very short tail.

Female and nonbreeding male less golden and streaked head, longer tail. Buzzing calls and songs. Can occur at high density and be conspicuous, at other times hard to see. **Range and habitat:** AU and OR, India, China, South-East Asia, New Guinea, north and east Australia. Swamps, wet grassland, crops, canefields, irrigated regions. May be migratory or irruptive. **Similar species:** other cisticolas (see p 224).

Rufous Songlark *Cinclorhamphus mathewsi* (Sylviidae) 16-20 cm (6½-8 in). Brown above, lightly streaked with obvious rufous rump, underparts pale with some darker smudging on breast, pale eyebrow. Male performs lively, melodious song-flight in breeding season, usually starting from and finishing on high branch. Female much smaller than male and less conspicuous. May be polygamous, but nests in dense grass are hard to observe. **Range and habitat:** Australia, except Tasmania. Open woodland and grassland with scattered trees, typically long grass, especially roadsides. Migratory. **Similar species:** female Brown Songlark, Richard's Pipit (see p 306). **Not illustrated.**

Brown Songlark *Cinclorhamphus cruralis* (Sylviidae) 18-25 cm (7-10 in). Conspicuous dark brown bird of open country. Male is dark chocolate brown, streaked above, but uniform below, slight eyebrow. Female streaked above and pale, slightly marked below. Usually seen in creaky, clattering song-flight descending vertically with legs dangling, or perched on a fence-post, tail cocked. Male much bigger than female. Feeds on insects and seeds on ground. **Range and habitat:** Australia, generally not near coast, partly migratory, moves out of centre during droughts. Grassland, crops. **Similar species:** female like Rufous Songlark. **Not illustrated.**

Splendid (Banded) Fairy-wren *Malurus splendens* (Maluridae) 14 cm (5½ in). Fairy-wrens are small, lively ground feeding birds, with very long tails, often cocked. Male Splendid Fairy-wren is brilliant iridescent blue, paler in the face, with black eye stripe, nape and breast band, dark beak. Female grey-brown, rufous face patch and beak and blue tail. Male outside breeding season is brown with blue wings and tail, dark face and beak. Breeds in pairs or groups, with extra birds helping to rear nestlings and fledglings. Helpers are often young that have not dispersed, usually males. Voice high-pitched squeaks and trills. **Range and habitat:** 3 isolated populations in south-west, central and inland east Australia, perhaps separate species, eucalypt forest and woodland, arid scrub. **Similar species:** Blue Wren has white belly.

Variegated Fairy-wren *Malurus lamberti* (Maluridae) 15 cm (6 in). Male has light blue or purple crown, face, upper back and tail, black throat, breast and nape patch, chestnut saddle across back and white belly. Female is grey-brown above, pale below, with dark brown face and beak and bluish tail. Male also brown in nonbreeding season. Some races have much bluer female. Social, usually seen feeding among low shrubs. Metallic squeaks and trills. **Range and habitat:** distinct races occupy central, north-west, north-east and central east Australia. Heathlands, forests with dense understorey. **Similar species:** Blue-breasted and Red-winged Fairy-wrens.

White-winged Fairy-wren *Malurus leucopterus* (Maluridae) 13 cm (5 in). Male of race *leuconotus* is a brilliant deep blue with white wings. Female dull grey-brown above, white below with blue tail and no red eye-ring. Race *leucopterus* is velvety-black in male, with blue tail and white wings. Calls are high-pitched trills. Feeds on tiny insects taken from bushes and ground. Cooperative breeder, in groups of up to 10 birds. **Range and habitat:** inland Australia, reaching coast in west and south. Race *leucopterus* on Dirk Hartog and Barrow Islands off Western Australia. Shrub steppe with low, often scattered bushes. **Similar species:** other fairy-wrens.

Collared Grey Fantail at nest

Friendly Fantail

(Black and White Fantail) Willie Wagtail

Clamorous Reed Warbler

Golden-capped Cisticola

Splendid (Banded) Fairy-wren

Variegated Fairy-wren at nest

White-winged Fairy-wren

Striped Grass Wren *Amytornis striatus* (Maluridae) 15-17cm (6½-7in). Brown above, heavily streaked with white, black moustache, orange eyebrow and buff beneath. Female has chestnut patches on belly. Plumage varies between races, to match background. Long tail, frequently cocked. Thin squeaking call. Usually keeps close to the ground, often under grass thickets. Runs rather than flies. May shelter in burrows when hot. **Range and habitat:** patchy through inland Australia, in porcupine grass on sandy soil and scrub in more rocky areas. Range has contracted probably because of grazing by sheep and changes in fire frequency. **Similar species:** other grass wrens, which differ in shade of brown, facial markings, range and habitat.

Redthroat *Pyrrholaemus brunneus* (Acanthizidae) 13 cm (5 in). Only small brown bird in Australia with pale chestnut throat. Female has white throat, both sexes grey underparts, slightly scaly crown and white lores. Tip of tail white. Rich song, often with mimicry. Active birds, gleaning insects and seeds from ground and foliage of low shrubs. **Range and habitat:** inland Australia, in mallee, mulga, shrub-steppe and deserts. Has declined where habitat damaged by grazing. **Similar species:** female like thornbills. **Not illustrated.**

Striated Field Wren *Calamanthus fuliginosus* (Acanthizidae) 13 cm (5 in). Varies from greyish brown, through olive-brown to reddish brown. Whitish belly. Heavily streaked black above and below. Usually glimpsed perched with tail cocked on top of a shrub, before darting under cover. Musical, twittering song. Sedentary, usually in pairs or small groups. **Range and habitat:** patchily distributed across southern Australia. Habitat varies from wet coastal and mountain heaths to saltbush and spinifex of arid plains, even edges of salt lakes. Perhaps 2 species, and many races. **Similar species:** heath wrens, Little Field Wren. **Not illustrated.**

Weebill *Smicrornis brevirostris* (Acanthizidae) 8 cm (3 in). Smallest Australian bird. Olive-brown above, buff below in south, paler in centre, but northern race very yellow. Pale eyebrow and iris and very short, dumpy, pale beak. Clear high-pitched calls, loud for such a tiny bird, may resemble 'I'm a Weebill'. Social, often with other small birds, tame. Feeds on outer foliage, often hovering near or clinging to leaves to take scale insects. **Range and habitat:** most of Australia and Tasmania, in dry eucalypt forest and woodland and mallee. **Similar species:** thornbills, but identified by beak and call.

White-throated Flyeater *Gerygone olivacea* (Acanthizidae) 10 cm (4 in). Bright lemon-yellow underparts with white throat, grey above with white forehead and patches on tail. Eye bright red. Immatures are all-yellow below. Beautiful high-pitched wavering, descending song. Sometimes called Bush Canary, due to song and colour. Australian warblers sometimes called Gerygones. Feeds by gleaning and dashing energetically amongst leaves for active insects. Nest constructed of shredded bark and spiders' webs, is pendant with long tail extending downwards and small hood at side. **Range and habitat:** northern and eastern Australia, in open eucalypt forest and woodland, generally not far inland. Migratory, arriving in south in October, departing in April. **Similar species:** other *Gerygone* warblers, most of which are whiter below.

Broad-tailed Thornbill *Acanthiza apicalis* (Acanthizidae) 10 cm (4 in). Thornbills are tiny birds, filling the role of warblers in Australia. They have a reputation for being hard to identify, but calls and habits help. Grey-brown above, some populations with dark rufous rump, pale below with fine striations. Scalloped forehead, reddish brown eyes. Unlike very similar Brown Thornbill, often cocks tail. Rich warbling call, harsh churring. Feeds and nests in shrub layer. Usually in pairs or family groups. **Range and habitat:** inland Australia, typically replaces Brown Thornbill west and north of Great Dividing Range, dry woodland and

scrub. Reaches coast in south-west, where also in eucalypt forest. **Similar species:** other thornbills. **Not illustrated.**

Brown Thornbill *Acanthiza pusilla* (Acanthizidae) 10 cm (4 in). Small, brown bird, identified by strong calls. Grey-brown above, with rich brown rump. Scalloped forehead and streaked throat, buff underparts. Usually solitary or in pairs, but will sometimes join other thornbills in mixed species feeding flocks. Warbling call, harsh chattering alarm call. Gleans small insects from shrubs. **Range and habitat:** south-eastern Australia and Tasmania, in rainforest and eucalypt forest with understorey. **Similar species:** other thornbills.

Southern Whiteface *Aphelocephala leucopsis* (Acanthizidae) 10 cm (4 in). Small, dull brown bird with white face. Grey-brown above, brown tail with white tip, whitish below, with buff flanks. White face bordered with black on crown and below eye. Pale iris. Tinkling, wistful calls and tuneful bell-like song. Sedentary, in small groups, often with thornbills. Feeds on insects and seeds on the ground, often turning over objects in its search. **Range and habitat:** inland Australia, near coast in south, in semi-arid and arid scrub and grassland. Still found in farmland, though has declined in some regions. **Similar species:** other whitefaces. **Not illustrated.**

New Zealand Grey Flyeater *Gerygone igata* (Acanthizidae) 11 cm (4½ in). Small greenish grey bird of the outer foliage. Pale grey throat, white belly, yellowish flanks, pale eyebrow and patches on tail. Attractive trilling song. Searches for insects and spiders by gleaning and hovering among leaves. Pendant pear-shaped nest, frequently parasitized by Shining Bronze-cuckoo (*Chrysococcyx lucidus*). **Range and habitat:** throughout New Zealand, forests and woodland, one of the few native species to adapt well to exotic vegetation in farmland, parks and gardens.

Banded Whiteface *Aphelocephala nigricincta* (Acanthizidae) 10 cm (4 in). Chestnut-brown back and black breast band distinguish it from other whitefaces. White face bordered black, grey tail and wings, rufous flanks and pale iris. Musical, weak calls. Sometimes feeding on ground with other small birds. May be a cooperative breeder. **Range and habitat:** central Australia, in saltbush and spinifex in stony and sandy deserts with scattered trees. May be nomadic. **Similar species:** Southern Whiteface is more widespread, Chestnut-breasted Whiteface much rarer. **Not illustrated.**

Crimson Chat *Ephthianura tricolor* (Acanthizidae) 10-12cm (4-4½in). Striking red, black and white bird of the Australian outback. Male has red cap, breast, belly and rump, black eye stripe, nape, back, wings and tail and white throat. Female is greyish brown above, buff below, with red smudges, also red rump. Slightly decurved beak. Bouncy flight, high-pitched calls. Short-tailed ground feeder taking mostly insects, but also visits flowers for nectar. Can be common, occurring in flocks with other chats. **Range and habitat:** inland and western Australia, nomadic and irruptive, sometimes well outside normal range, acacia scrub, shrub-steppe often near salt-lakes.

White-faced Chat *Ephthianura albifrons* (Acanthizidae) 11-12cm (4-4½in). Male has white face, throat and upper breast, with thick black border. Grey back, white belly, black wings and tail. Female generally greyer, with narrow breast band, juvenile has faint breast band. Call 'tang', rather like a Zebra Finch. Jerky flight. Wags short tail up and down. Strictly ground feeding, on insects. Loosely colonial in breeding season, flocks at other times. Commonest chat in southern Australia. **Range and habitat:** southern Australia and Tasmania, in damp grassland, edges of lagoons and salt-lakes, coastal saltmarshes, sometimes crops.

Striped Grass Wren

Weebill at nest

White-throated Flyeater

Brown Thornbill

New Zealand Grey Flyeater

Crimson Chat

White-faced Chat at nest

White-browed Treecreeper *Climacteris affinis* (Climacteridae) 14 cm (5½ in). Greyish brown bird, heavily streaked below and in face. White throat and eyebrow, female with orange line over eyebrow. Flight somewhat undulating, but also glides showing conspicuous pale orange stripe across wing feathers. Longish decurved beak and long toes. Australian treecreepers are not closely related to treecreepers on other continents, but are an endemic group perhaps related to lyrebirds and bowerbirds. Typically work their way steadily up trunks and major branches in search of insects, especially ants. Also feed on ground. Hole-nesting and probably cooperative. Calls high pitched, somewhat shrill, less vocal than other treecreepers. **Range and habitat:** central Australia, uncommon in woodland dominated by sheoak (*Casuarina*) and trees other than eucalypts. **Similar species:** other treecreepers, which typically inhabit eucalypt woodland or forest. **Not illustrated.**

Red Wattle Bird *Anthochaera carunculata* (Meliphagidae) 31-9 cm (12-15 in). Large, raucous, grey-brown honeyeater. White streaks over greyish brown body, darker head but whitish face. Yellow belly, wattle of red skin dangling from chin, which is small in juvenile and may be hard to see in adult. Longish beak and strong pink legs. Range of harsh rattling and clanking calls. Often in groups feeding in eucalypt flowers, with much noise and chasing. Highly aggressive to other honeyeaters and lorikeets. In territorial pairs in breeding season. **Range and habitat:** southern Australia, in woodland, open forest, heaths, common in town parks and gardens. Some local movements. **Similar species:** Little Wattle Bird has no wattle or yellow patch on belly.

Spiny-cheeked Honeyeater *Acanthagenys rufogularis* (Meliphagidae) 22-7 cm (8½-10½ in). Medium-sized greyish honeyeater. Scalloped or streaked above and below, with pale rufous throat and upper breast. Bare pink skin around eye, stretching to pink and black beak. Black through eye, with white spiny feathers below eye. Attractive warbling call, excellent songster at dawn. Feeds in flowers of trees and shrubs, on berries and hawks for insects. **Range and habitat:** inland Australia, occasionally up to coast, in dry woodland, mallee, scrub, stony shrub-steppe. **Similar species:** none, though can look like Yellow-throated Miner in flight.

Stitch-bird *Notiomystis cincta* (Meliphagidae) 15 cm (6 in). Male a striking honeyeater, female duller. Black head, upper back and breast bordered yellow, white tuft behind ears, greyish belly. Grey, yellow and white wings and tail in male. Female generally brown, more olive above, with white wing patch. Dark iris, prominent bristles, sometimes tilts tail. Calls 'tzit', also 'pek pek pek . . .'. Feeds on nectar, fruit and insects from foliage, at all levels in the forest. Nests in tree-holes. **Range and habitat:** now restricted to forests of Little Barrier Island, where fairly common, formerly more wide-spread on North Island, New Zealand. **Similar species:** female like Bellbird.

Little Friarbird *Philemon citreogularis* (Meliphagidae) 27 cm (10½ in). Grey-brown above, silvery white below, with bald bluish facial skin. Longish tail and beak. Immature has yellow chin and throat. Raucous cackling calls. A noisy, active, aggressive bird, often feeding in groups in flowering trees. Also takes fruit and large insects. **Range and habitat:** in open, eucalypt forests and woodland, in northern and eastern Australia. **Similar species:** other friarbirds have knobs on top of their beaks.

Yellow-throated Miner *Manorina flavigula* (Meliphagidae) 23-8 cm (9-11 in). Medium-sized grey bird, with contrasting white or pale grey rump, conspicuous in flight. Black face, with yellow on forehead and side of throat and bare yellow patches on gape, chin streak and behind eye. White tips to tail. Beak and legs bright yellow. Social, living and breeding in groups. Loud ringing calls. **Range and** habitat: open woodland and scrub, inland Australia, reaching coast in south and west. Likes partly cleared country, where it may have replaced or hybridized with endangered Black-eared Miner. **Similar species:** other miners none of which has pale rump.

Noisy Miner *Manorina melanocephala* (Meliphagidae) 24-8 cm (9½-11 in). Bold, clamorous, grey bird with black on face and crown. No contrast between rump and back. Yellow beak, legs and bare patch behind eye. Slight yellow wash on wings and tail. Range of piping calls, with some harsh elements. Highly social, living in groups or colonies. Numerous individuals may help at one nest. Performs 'corroboree' display where up to 20 individuals gather, call loudly and flutter wings. Vigorously drives away other birds, from tiny insectivores to large raptors. Loud alarm calls lead to common name of Soldierbird. Takes insects, lerps (scale insects), honeydew and some nectar. **Range and habitat:** eastern Australia, in open woodland, particularly where trees are scattered and unhealthy, also gardens. **Similar species:** other miners.

Tui (Parson Bird) *Prosthemadera novaeseelandiae* (Meliphagidae) 28-30 cm (11-12 in). Only large dark honeyeater in New Zealand. Iridescent green, with dark bronzy reflections on back and belly. Unusual white curls hanging from chin and filamentous feathers around neck, also white shoulder patch. Immature duller. Attractive bell-like calls, also strange harsher wheezes and chuckles. Mostly nectarivorous, but takes some insects and also fruit in autumn, usually in canopy. May feed lower on native flax and on garden flowers. **Range and habitat:** native forests in most of New Zealand, but will venture into settled districts. **Similar species:** Kokako, Saddleback (see p 320).

White-eared Honeyeater *Meliphaga leucotis* (Meliphagidae) 20 cm (8 in). Dark bottle-green honeyeater, with white facial patch surrounded by black. Grey crown, yellow belly and wash on wings and tail. Juvenile has yellowish face, less black. Birds in west and drier country are small and pale. Loud 'chok chok' call, also 'pick-em-up'. A rather solitary bird, typically probing amongst fibrous bark for insects and honeydew. Sometimes visiting flowers of eucalypts and shrubs. **Range and habitat:** central Queensland through south-east Australia, perhaps continuously to south-west Australia, in rather scrubby forest and mallee. **Similar species:** juvenile can be confused with other honeyeaters with yellow faces.

Common Smoky Honeyeater (Common Melipotes) *Melipotes fumigatus* (Meliphagidae) 22 cm (8½ in). Nondescript grey bird, with large yellow patch of bare skin in face, which may flush bright red. Rather sluggish for a honeyeater, but sometimes aggressive. Short beak. Quiet 'swit swit' call. Feeds mostly on fruit in taller shrubs and trees. **Range and habitat:** rainforests in mountain ranges of New Guinea. **Similar species:** subspecies Western Smoky Honeyeater has different range.

Timor Helmeted Friarbird *Philemon buceroides* (Meliphagidae) 30-7 cm (12-15 in). Very large, grey, conspicuous honeyeater. Black bare face, silvery feathers on crown and chin, forming ruff on nape. Large, gently sloping knob where beak joins forehead. Brownish throat and breast. Harsh metallic cackling calls, e.g. 'poor devil poor devil' 'sergeant major'. Noisy, aggressive flocks gather at flowering trees; also eats fruit, insects and spiders. **Range and habitat:** edge of monsoon forest, eucalypt woodland, mangroves, gardens in northern Australia, New Guinea and southern Indonesia. Somewhat nomadic. **Similar species:** other friarbirds, but note knob shape and pattern of silvery feathers.

Red Wattle Bird

Spiny-cheeked Honeyeater

Stitch Bird

Little Friarbird

Yellow-throated Miner

Noisy Miner at nest

Tui (Parson Bird)

White-eared Honeyeater

Common Smoky Honeyeater (Common Melipotes)

Timor Helmeted Friarbird

Brown Honeyeater *Lichmera indistincta* (Meliphagidae) 12-16 cm (4½-6½ in). Small, dull honeyeater, with yellow or black gape and yellow patch of skin behind eye. Olive-brown above, buff grading to white below, yellow wash on wings. One of the best songsters among the honeyeaters, melodious and chattering notes resemble a reed-warbler. Active in treetops, sometimes lower, visiting nectar from flowers such as mistletoes, mangroves and paperbarks. Also captures small insects in flight. **Range and habitat:** most of Australia, but lacking from south and south-east. New Guinea, southern Indonesia. A wide range of habitat from mangroves, forest, heath to arid scrub. **Similar species:** like female *Myzomela*, but song is diagnostic.

Golden-backed Honeyeater *Melithreptus laetior* (Meliphagidae) 15 cm (6 in). Clean-looking gold and white bird, with distinctive head pattern. Head black, with white crescent extending across nape and broad white streak from beak on either side of small black chin. Bare greenish skin above eye. Juvenile has brown head, pale beak. Usually considered race of Black-chinned Honeyeater, which has olive back and blue eye patch. Feeds on insects and exudates in canopy of eucalyptus trees, often in small groups. Sometimes visits flowering trees and shrubs. Loud, churring and ringing call. **Range and habitat:** across northern Australia in eucalypt and paperbark woodland, Black-chinned in open woodland in south-east Australia, generally not near coast. **Similar species:** other *Melithreptus* honeyeaters, but note colour of eye patch and calls. **Not illustrated.**

White-plumed Honeyeater *Meliphaga penicillata* (Meliphagidae) 15-19 cm (6-7½ in). Lively, short-beaked, yellowish honeyeater. White line on side of neck, faintly bordered with black, is distinctive, but can be hard to see. Olive-grey with yellow in face and on wings and tail. Local races differ, very yellow in west, greener in south-east. Usual call 'chick-a-wee', sometimes repeated and delivered in rising and descending song-flight. Takes insects and scales from leaves, and nectar from eucalypt flowers. Very common in suburbs, town parks in south-east Australia. **Range and habitat:** Australia, except south-west and north, and islands. Eucalypt woodland and especially red gums (*Eucalyptus camaldulensis*) along creeks in the outback. Occasionally in mallee or other scrub. **Similar species:** other *Meliphaga* honeyeaters, but yellow face, white plume and call diagnostic.

White-throated Honeyeater *Melithreptus albogularis* (Meliphagidae) 13-15 cm (5-6 in). Black head with white crescent across nape, olive-green back and very clean white underparts. White throat extending to chin and bluish white skin above eye. Very similar White-naped Honeyeater in eastern Australia has tiny black patch on chin and red above eye. A social bird that gleans among eucalypt foliage, and less often takes nectar. Call 'tserp tserp'. **Range and habitat:** northern Australia and southern New Guinea, in eucalypt and paperbark woodland. Overlaps with White-naped in south Queensland, north-east New South Wales, but usually nearer coast. **Similar species:** White-naped Honeyeater.

White-fronted Honeyeater *Phylidonyris albifrons* (Meliphagidae) 16-18 cm (6½-7 in). Fairly inconspicuous black and white honeyeater with yellow on wing. White forehead and chin stripe, black cheeks, throat and breast merging into blotches on white belly, grey back and yellow on primaries. Tiny red spot of skin behind eye. Juvenile washed-out, browner version of adult. Tuneful song. Visits flowers of eucalypts and desert shrubs like *Eremophila*. **Range and habitat:** inland southern Australia, in acacia and eucalypt scrub and woodland. Unpredictable and nomadic species. Sometimes very common in a locality, then not seen for years. **Not illustrated.**

Black Honeyeater *Certhionyx niger* (Meliphagidae) 10-12 cm (4-4½ in). Very small black and white honeyeater. Male has black back, head and stripe down centre of otherwise white breast. Female and nonbreeding male dull brown. Longish decurved beak. Thin 'peee peee' call, often delivered in buoyant flitting flight. Feeds on flowers of desert shrubs, especially *Eremophila*. **Range and habitat:** sparsely through inland Australia, with numbers fluctuating from year to year. Generally in south in spring and summer, a few records from north in autumn and winter. Arid woodland and scrub. **Similar species:** Pied Honeyeater, female like several other small honeyeaters. **Not illustrated.**

Pied Honeyeater *Certhionyx variegatus* (Meliphagidae) 15-18 cm (6-7 in). Black and white honeyeater. No black stripe down breast, but white on rump, side of tail and wings. Blue patch of skin below eye, bluish decurved beak. Female dull brown above, paler mottled below, conspicuous white scalloping on wings. Takes nectar of mistletoes and shrubs, also insects and saltbush fruit. Voice 'tee titee tee tee', diving display flight. **Range and habitat:** sparse in central Australia, complex movements, in desert scrub and acacia and eucalypt woodland. **Similar species:** Black Honeyeater is much smaller, also plumage like Hooded Robin (see p 308), though behaviour different. **Not illustrated.**

Mistletoe Flowerpecker *Dicaeum hirundinaceum* (Dicaeidae) 10-11 cm (4-4½ in). Tiny brilliant bird, often located by high-pitched call. Male is glossy blue-black on top, red on throat, breast and undertail, white belly with black stripe along keel. Female is grey above, off-white below, with pink undertail. Very short tail, but long wings and strong flight. Short, broad beak, consumes mistletoe fruits leaving skin attached to plant. Seeds pass through in 20-30 minutes and are wiped onto a horizontal branch, on which they grow. As well as loud call-note, has a complex, varied song, with mimicry. Delicate hanging nest, with side entrance. **Range and habitat:** Australia, except Tasmania, Aru Island off New Guinea, associated with parasitic mistletoe in any habitat with trees or shrubs, migratory or nomadic. **Not illustrated.**

Spotted Pardalote *Pardalotus punctatus* (Pardalotidae) 9 cm (3½ in). Tiny brightly coloured bird with very short tail. Male has bright yellow throat and undertail coverts, red rump, black crown, wings and tail spotted white and white eyebrow. Female duller with cream throat and yellow spots on crown. Very short stumpy beak. Ringing two or three note call, quite loud for such a small bird. Feeds on small insects and scale in the treetops. Nests in holes in banks and slopes. **Range and habitat:** eastern Australia, Tasmania and extreme south-western Australia. Some movement north and inland in autumn. Eucalypt forest and woodland. **Similar species:** Yellow-rumped Pardalote.

Red-browed Pardalote *Pardalotus rubricatus* (Dicaeidae) 10-12 cm (4-4½ in). Rather dull pardalote of northern Australia. Fawn above and below, black crown with white spots, yellow eyebrow with small red spot near beak. Orangey wing patch and yellow on breast. Short beak and tail, but long wings; 5-note call. Mostly sedentary, in pairs or small groups. **Range and habitat:** inland and northern Australia, in eucalypt and paperbark woodland, scattered trees in deserts. **Similar species:** other pardalotes.

Brown Honeyeater

White-plumed Honeyeater

White-throated Honeyeater

Spotted Pardalote

Red-browed Pardalote

Grey-backed White-eye (Silvereye) *Zosterops lateralis* (Zosteropidae) 12 cm (4½ in). Small greenish bird, with conspicuous white eye-ring. Olive-green head, wings, rump and tail, ring of white feathers around eye, partly outlined in black. Generally grey to white below. Many races, which vary in details of plumage. Tasmanian race has very rufous flanks, western race green on back, some eastern races very yellow on throat and undertail. Sharp black beak, brown eyes. Call high-pitched weak 'tsee-oow', lively warbling song. Flocks in winter. Very versatile in foraging behaviour, taking insects, fruit and nectar. Tame, approachable. Successful in towns, suburbs, disturbed vegetation. **Range and habitat:** western, southern and eastern Australia, including Tasmania, southern islands, Barrier Reef islands, New Zealand, islands through south-west Pacific. Southern races migrate, partial migrant elsewhere. Rainforest, eucalypt forest and woodland, scrub, heath, gardens, parks. **Similar species:** white-eyes (see pp 230, 272), but other species have different range or are much rarer.

Painted Finch *Emblema picta* (Estrildidae) 10-12 cm (4-4½ in). Small brown, red, black and white finch. Red face in male, red rump and mid-belly in both sexes. Brown back, and face in female, black underparts, boldly spotted white. Longish deep beak, tipped red. Call a loud 'trut'. Feeds on ground among rocks and tussocks, mostly in small groups. Visits water-holes, scooping up water. **Range and habitat:** patchily through central Australia, occasionally south or east of normal range. Arid, stony hills and gorges with acacia and spinifex. **Similar species:** other firetails in wetter habitats.

(Spotted-sided) Zebra Finch *Poephila guttata* (Estrildidae) 10 cm (4 in). Common, attractive, greyish brown finch. Male has grey head, orange cheek patch, bold black-white-black stripes between beak and cheeks, fine transverse black and grey stripes across breast. Chestnut sides with white spots, white belly, grey-brown back, black and white striped tail, white rump. Female more uniform greyish brown, with same facial stripes. Beak orange in both sexes, black in immature. Call a nasal 'tang', twittering song. Feeds on seeds on ground in flocks, may pull down grass stems to reach seed-heads. Large flocks may visit water-holes, to suck up water. Popular cage bird and experimental subject. **Range and habitat:** most of Australia, though not generally near coast or islands. Open habitats, breeds in introduced hawthorn scrub.

Double-barred Finch *Poephila bichenovii* (Estrildidae) 10-11 cm (4-4½ in). Small brown and white finch. White face and underparts with 2 well-spaced black bands across throat and breast. Dark grey-brown back, black wings with fine white speckles, black undertail and tail. Rump white in south, black in north. Greyish beak and legs. Immature has faint breast bars. Call 'tiaah', similar to Zebra Finch, but weaker, soft nasal song. **Range and habitat:** northern and eastern Australia, grassy woodland near water, disturbed areas, e.g. blackberry tangles. **Similar species:** Plum-headed Finch has many fine breast bands.

Streak-headed Mannikin *Lonchura tristissima* (Estrildidae) 10 cm (4 in). Dull blackish brown, with conspicuous yellow rump, black tail. Heavy grey beak. Call a buzzy 'tseed'. Occurs in flocks. Spherical nest of grass. **Range and habitat:** New Guinea lowlands to 1,400 m (4,600 ft), forest edge, secondary growth. **Similar species:** other mannikins differ in colour of head, back and tail.

Olive-backed Oriole *Oriolus sagittatus* (Oriolidae) 25 cm (10 in). Medium-sized speckled greenish bird. Head and back olive-green, finely streaked, underparts white with bold spots. Tail and wings grey. Bill and eye red. Female somewhat duller than male. Immature more rufous with pale eyebrow. Strong, undulating flight. Call far-carrying 'orry-orry-ole', complex song with mimicry, can only be fully appreciated at close quarters. Feeds in treetops on large insects and fruit. **Range and habitat:** northern and eastern

Australia, New Guinea in open forest and woodland, rainforest, occasionally in towns. Southern birds migrate north in autumn. **Similar species:** Australian Yellow and Brown Orioles, Figbird, can look like friarbirds in flight.

Brown Oriole *Oriolus szalayi* (Oriolidae) 27 cm (10½ in). Dull brown, streaky, medium-sized bird. Grey-brown, darker in face, heavily streaked on head and underparts. Red beak and eye. Immature has dark beak and eye and pale eyebrow. Loud musical warbling calls. Feeds in canopy on fruit, often with friarbirds. **Range and habitat:** New Guinea, up to 1,400 m (4,600 ft), in forest. **Similar species:** other orioles, also remarkably like friarbirds (see p 314), possibly as a result of convergence to avoid aggression. **Not illustrated.**

Great (Grey) Bowerbird *Chlamydera nuchalis* (Ptilonorhynchidae) 35 cm (14 in). Large grey bowerbird. Pale grey below, grey with brownish scalloping above. Tail heavily spotted. Heavy, decurved, black beak. Harsh, grating, hissing calls, mimicry of predators' calls. Males build a large bower of sticks in a double avenue, decorated with various objects, such as snail-shells, pale stones, bleached bones and dung. Male displays at bower, revealing lilac crest on nape. Feeds on large insects and fruit, sometimes from the ground. Often in groups. **Range and habitat:** northern Australia, inhabits drier habitat than most bowerbirds; riverine forest and thickets and open woodland. Occasionally in towns. **Similar species:** Spotted Bowerbird.

Fawn-breasted Bowerbird *Chlamydera cerviniventris* (Ptilonorhynchidae) 25-30 cm (10-12 in). Fawn and grey bowerbird. Grey-brown on back, head and throat, with feathers tipped white, buff-fawn below. Heavy black beak. Shortish tail, thick neck and heavy flight, typical of most bowerbirds. Harsh, rasping, churring calls, mechanical sounds, mimicry. Builds a large, 2-walled bower of sticks on a raised platform of sticks. Gregarious, can be noisy. **Range and habitat:** New Guinea, extreme north-east Australia, edge of rainforest, thickets in savannah. **Similar species:** Yellow-fronted Gardener Bowerbird.

Satin Bowerbird *Ptilonorhynchus violaceus* (Ptilonorhynchidae) 27-33 cm (10½-13 in). Heavily built, greenish or glossy black bird. Males black with blue sheen, violet eyes, yellow beak and legs. Female and male up to age of 5-6 years dull green above, with rufous on wings and tail, pale underparts, heavily scalloped with olive-grey. Thick beak, short tail, rounded wings. Variety of whistling, wheezing, mechanical buzzing calls, with mimicry. Males, including some green birds, display at bowers consisting of a double avenue of sticks, typically decorated with blue objects (e.g. parrot feathers, drinking straws, bottle tops, clothes pegs) but also yellow and sometimes other colours. May paint bower with mixture of charcoal and saliva. **Range and habitat:** eastern Australia, in rainforest and wet eucalypt forest, quite tolerant of human activity.

Regent Bowerbird *Sericulus chrysocephalus* (Ptilonorhynchidae) 24-8 cm (9½-11 in). Spectacular male black, golden yellow on crown, nape, shoulders, most of wings; yellow beak and eyes. Female and young male dark greyish brown, scalloped white on back, pale on head and below, scalloped brown. More slightly built than most bowerbirds, with slender beak. Male displays at bower with double avenue of sticks, usually with few decorations. May paint bower yellow with saliva and juice from crushed leaves. Fairly quiet, scolding, chattering calls. **Range and habitat:** central eastern Australia, in rainforest and dense undergrowth.

Grey-backed White-eye (Silvereye)

Painted Finch

(Spotted-sided) Zebra Finch

Double-barred Finch

Streak-headed Mannikin

Olive-backed Oriole

Great (Grey) Bowerbird

Fawn-breasted Bowerbird

Satin Bowerbird

Regent Bowerbird

White-breasted Woodswallow *Artamus leucorhynchus* (Artamidae) 17 cm (6½ in). Clean, grey and white, aerial bird. White rump and underparts, black head, upper breast, back and tail. Juvenile mottled with buff. Call 'pert pert'. Woodswallows are smallish, dumpy birds, with short tails and long wings. They feed mostly by pursuing insects high in the air, in a manner similar to bee-eaters. **Range and habitat:** northern and eastern Australia, woodland, usually near water, swamps, mangroves. Some years move south of normal range. Also from Malaysia, through Indonesia, New Guinea to Fiji. **Similar species:** other woodswallows.

Masked Woodswallow *Artamus personatus* (Artamidae) 19 cm (7½ in). Black face and throat, pale grey underparts and slate-grey crown, nape and back. White tips to tail. Female with less clear-cut division on throat. Blue beak with black tip. Immature mottled. Calls 'chip chip'. Occasionally feeds on ground or in trees. **Range and habitat:** throughout Australia, more common in west, generally not near coasts. Regular north-south migration, sometimes irrupts beyond normal range. Woodland, scrub and desert. **Similar species:** other woodswallows (see p 264). **Not illustrated.**

White-browed Woodswallow *Artamus superciliosus* (Artamidae) 19 cm (7½ in). Spectacular, gregarious bird, most distinct of woodswallows. Male is blue-grey on upper breast and upperparts, head black with a distinct white eyebrow and deep chestnut belly. Female similar, but head grey, less sharply marked. Beak blue with black tip and slightly decurved. Sparrow-like calls and chattering. Often heard passing over in flocks high in the air. Highly social, breeding in loose colonies. Numbers vary in a locality from year to year. Sometimes feeds on ground on locusts and snails, or in trees and shrubs on nectar. Often with Masked Woodswallow, with which it may hybridize. **Range and habitat:** common, inland Australia, especially in east, erupting towards coast in some years as well as regular migration. Woodland, scrub, desert.

Black-faced Woodswallow *Artamus cinereus* (Artamidae) 18 cm (7 in). Rather dull woodswallow. Smoky grey, paler below, with distinct black face. Undertail-coverts white in east, black elsewhere. White tip to tail. Juvenile mottled. Similar Dusky Woodswallow is browner, with white leading edge to wing. Calls 'chep chep' and chattering. Breeds opportunistically after, sometimes with helpers. **Range and habitat:** inland Australia, spreading to coast in west. Also vagrant to Timor and New Guinea. More sedentary than most woodswallows. Open habitat, often on telegraph wires in treeless farmland. **Similar species:** Dusky and Little Woodswallows. **Not illustrated.**

Dusky Woodswallow *Artamus cyanopterus* (Artamidae) 18 cm (7 in). Greyish-brown woodswallow, rather darker in face, underparts fawnish. Wings and tail dark grey with white leading edge to wing and tip of tail. Slate-blue, rather deep, bill with black tip. Juvenile speckled brown. Breeding pairs may have helpers. In cold weather flocks may cluster on branches to roost. Feeds in flight, and sometimes on the ground. Call 'sweep sweep', and harsh chattering when nest disturbed. **Range and habitat:** open woodland, throughout Australia, except for far north. Migratory in south. **Similar species:** Black-faced Woodswallow, Little Woodswallow has no white on wings.

Kokako *Callaeas cinerea* (Callaeidae) 37 cm (15 in). Large dark, somewhat crow-like bird. Dark bluish grey, washed with olive-brown. Black lores and eye-ring. Circular wattle hanging down from gape, is bright blue on North Island and rich orange on South Island. Heavy, hooked beak. Wide variety of calls; song described as like 2 long rich organ notes, interspersed with whistles, can be heard far away. Hops up tree-branches, gliding flight is weak. **Range and habitat:** New Zealand, has declined since European settlement, quite widespread in small numbers on North Island (total population about 1,000), very rare or possibly extinct

on South Island, in native forests. **Similar species:** Saddleback, Tui (Parson Bird) (see p 314). **Not illustrated.**

Saddleback (Tieke) *Creadion carunculatus* (Callaeidae) 25 cm (10 in). Dark, somewhat starling-like bird. Glossy black with iridescence on head and breast. Bright chestnut saddle across back, scapulars, wing-coverts and rump. Orange wattle hanging down from gape. Long slender beak. Calls 'cheep te-te-te-te' and 'che-che-u-che'. Very vocal, also other musical and clicking calls. **Range and habitat:** New Zealand, used to be widespread on all islands, now restricted to forest on a small number of offshore islands that are free of cats and rats (Hen Island, islets off Stewart Island), where it is fairly common. **Similar species:** Tui (Parson Bird) (see p 314) and Kokako are larger, different calls. **Not illustrated.**

Magpie Lark *Grallina cyanoleuca* (Grallinidae) 27 cm (10½ in). Boldly pied, ground-dwelling bird. Both sexes black back, wings and tail tip, with white underparts, rump, base of tail and shoulders. Male has black face, throat and breast, and white eyebrow and cheeks. Female has black line extending from eye downwards to meet on chest, with white forehead, chin and throat and white from breast extending up to behind eye. Call a loud 'pee-wee', giving common name of Peewee, often pairs call antiphonally. Builds large mud nest (another name is Mudlark). Walks strongly, bobbing head back and forth. Direct flapping flight. Often feeds along edge of water, taking snails. **Range and habitat:** throughout Australia, vagrant in Tasmania, also southern New Guinea. Open woodland, farmland with scattered trees, near water. **Similar species:** Black-backed Magpie, Black-throated Butcher Bird.

Apostle Bird *Struthidea cinerea* (Grallinidae) 29-32 cm (11½-13 in). Medium-sized grey bird with long tail. Slate-grey with fine white striations, brown on wings. Heavy, finch-like black beak. Harsh chattering calls. Lives in tight groups of about 12 (hence name). All members contribute to building the mud nest and feeding the young. Feeds on ground on large insects, seeds and occasionally mice. Generally sedentary, but groups wander in nonbreeding season. **Range and habitat:** inland eastern Australia, open woodland near water.

Grey Butcher Bird *Cracticus torquatus* (Cracticidae) 24-30 cm (9½-12 in). Medium-sized grey, black and white bird. Black cap, face, wings and tail, white lores, rump and underparts, grey back. Juvenile brownish. Heavy, hooked grey and black beak, strong claws. Fast, direct flight, often calling. Loud, musical call, with some harsh notes. Will vigorously attack intruders in breeding season, pecking humans on head. Young in 1st year occasionally help parents. Pounces on large insects, lizards and small mammals on ground, occasionally taking birds. Will wedge food in forks or impale prey on thorns. **Range and habitat:** most of Australia, scarce in northern inland, in woodland, scrub. **Similar species:** other butcher birds.

Black-throated (Pied) Butcher Bird *Cracticus nigrogularis* (Cracticidae) 32 cm (13 in). Medium-sized black and white bird. Black head, throat and upper breast, back, wings, tail. White belly, collar across back, wing patches, rump and tail tips. Juvenile browner. Upright stance on wires, posts, branches. Beautiful flute-like song, resembling opening bars of Beethoven's 5th Symphony. Behaviour similar to Grey Butcher Bird, but more often in communal groups. **Range and habitat:** most of Australia, but absent from cooler parts of south-east, in woodland, mallee, scrub. **Similar species:** other butcher birds, Magpie Lark.

White-breasted Woodswallow

White-browed Woodswallow (male)

White-browed Woodswallow (female)

Dusky Woodswallow

Magpie Lark

Apostle Bird

Grey Butcher Bird

Black-throated (Pied) Butcher Bird

Pied Currawong *Strepera graculina* (Cracticidae) 40-50 cm (16-20 in). Large, dark, long-tailed bird. Black, with white patches on wings, base and tip of tail. Long heavy beak, with slight hook. Yellow eye. Loud, ringing call 'curra-wong', musical whistles and fluting. Flocks in winter come into towns to feed on berries. Major nest predator. **Range and habitat:** eastern Australia, in forest and woodland in breeding season, more open habitat in winter. **Similar species:** other currawongs, differ in amount of white and calls.

Spangled (Hair-crested) Drongo *Dicrurus hottentottus* (Dicruridae) 30-2 cm (12-13 in). Glossy black bird with long fish tail. Body black with iridescent spangles, white edging to undertail feathers. Bright red eye. Heavy black beak. Noisy, rasping, hissing calls. Agile flight rather like bee-eater in pursuit of insects. Also takes nectar and occasionally other birds. **Range and habitat:** southern Asia, New Guinea, northern and eastern Australia. Forest, particularly edges, along streams or on emergent trees.

Magnificent Riflebird *Ptiloris magnificus* (Paradisaeidae) 26-33 cm (10-13 in). Male is black showing iridescent blue-green on crown, breast and tail. Filamentous plumes hanging from flanks. Female is grey-brown above, cinnamon on wings, pale below with fine dark bars. Long, curved beak. Male has spectacular display in which breast shield is spread. Growling call. Takes insects from bark, often chiselling into dead bark, also fruit. **Range and habitat:** rainforest in New Guinea and extreme north-east Australia. **Similar species:** other riflebirds, with different ranges.

Brown Sicklebill *Epimachus meyeri* (Paradisaeidae) 100 cm (39 in). Extremely long-tailed bird of paradise, with long slender, decurved beak. Male is grey-brown, with dark head and blue eyes. Also flank plumes of blue and brown. Female has chestnut crown, with finely barred underparts, light blue eyes. Male call is a loud, unmusical 'tat at tat at tat at' or 'tat at at at tat at at at at', described as having the quality of a pneumatic drill or machine gun. Also liquid gurgling notes. Display, in which breast feathers are spread, is from a traditional arboreal perch. Males occupy large, exclusive territories. Use long, curved beaks to extract insects from wood and occasionally fruits from capsules. **Range and habitat:** mid-mountain forest in Central Ranges of New Guinea, most common around 2,000 m (6,500 ft). **Similar species:** Black Sicklebill is darker, has red eyes.

Superb Bird of Paradise *Lophorina superba* (Paradisaeidae) 25 cm (10 in). Black or brown smallish bird of paradise. Male is black, with iridescent blue breast shield, which he spreads during display, also a velvety cape, which can be erected. Female brown above and barred below, like many other birds of paradise, but has rufous patch on wings, pale eye stripe and brown eyes. Males hold exclusive territories. Mostly insectivorous. **Range and habitat:** common, montane forests in New Guinea. **Similar species:** other birds of paradise, especially parotias.

King Bird of Paradise *Cicinnurus regius* (Paradisaeidae) 16 cm (6½ in). Male is a spectacular, small, red bird of paradise. White belly, dark wings, breast band and spot above eye, otherwise bright ruby-red. Very short tail with 2 long thin streamers ending in green discs. Female a more normal brown and barred plumage, with pale yellow beak. Male displays and calls 'kraan kraan . . .' from vine tangles high in the canopy. Often display in pairs. May join mixed flocks. **Range and habitat:** New Guinea and neighbouring islands up to 300 m (1,000 ft). **Similar species:** female like Magnificent Bird of Paradise.

Magnificent Bird of Paradise *Diphyllodes magnificus* (Paradisaeidae) 18 cm (7 in). Male is small and compact, generally dark, with golden back. Blue beak and streak behind eye, long, curved tail wires, from very short tail. Female brown above, barred below, with distinctive blue beak, legs and line of bare skin behind eye. Male sings from a cleared arena in a thicket to attract a female. Then he displays by raising feathers around his neck to form a shield and dancing mechanically, while clinging to a vertical stem, just above the ground. Calls range from trills to hisses. Feeds on complex fruits, like nutmeg. **Range and habitat:** forested foothills and lower mountain-sides in New Guinea and neighbouring islands. **Similar species:** female like many other birds of paradise.

Raggiana Bird of Paradise *Paradisaea raggiana* (Paradisaeidae) 33 cm (13 in). One of the most familiar red and yellow birds of paradise. Male has yellow head and back, iridescent green throat and around ivory beak, chocolate brown underparts. Long red or orange flank plumes are spread in spectacular, inverted display by males at communal leks. Females are dark brown above and below, with yellow crown and nape. Promiscuous, females attracted to display grounds by loud 'wau wau wau . . .' call of males. Mostly frugivorous, joins mixed-species flocks. **Range and habitat:** eastern New Guinea in rainforest. **Similar species:** other *Paradisaea* differ in colour of plumes, breast in males and head and underparts in females.

Australian (Black-backed) Magpie *Gymnorhina tibicen* (Cracticidae) 36-44 cm (14-17 in). Large, crow-like, pied bird. Black head, underparts, wings and tip of tail. Males in south-east (race *hypoleuca*) have white on shoulders and from nape to tail, females greyish on back. Western Magpie (race *dorsalis*) has male like *hypoleuca*, female heavily marked with black on back. Birds through rest of range (*tibicen*) have broad black band across back, females light grey on nape. Black-backed and white-backed birds hybridize to produce offspring with narrow or incomplete black bands. Lively carolling song and harsh squawking. Attack and sometimes terrify and injure humans during breeding season. Live in small breeding groups or larger nonbreeding flocks. **Range and habitat:** grassy forest, open woodland, farmland with scattered trees, towns, in most of Australia, southern New Guinea. **Similar species:** Magpie Lark, Pied Currawong. **Not illustrated.**

Little Crow *Corvus bennetti* (Corvidae) 48 cm (19 in). Smallest Australian crow. Glossy black with white eyes. Bases of feathers white, short, hackle feathers pointed. Call fairly low, monotonous 'ca ca ca . . .'. Social, rapid, agile flight, often soaring. Tame. **Range and habitat:** inland Australia, reaching coast in south and west. Woodland, arid and semi-arid scrub, towns. **Similar species:** other crows and ravens (see pp 92, 182, 228, 278), distinguished by call, size and throat feathers. **Not illustrated.**

Torresian (Australian) Crow *Corvus orru* (Corvidae) 50 cm (20 in). All-black, white eyes. Feather bases white, hackle feathers rounded. Slightly more heavily built than Little Crow, with shorter, broader tail. Shuffles wings on landing. Call is high-pitched, rapid cawing. Sometimes longer notes. **Range and habitat:** northern Australia, New Guinea, Moluccas, in most habitats. **Similar species:** other crows and ravens (see pp 92, 182, 228, 278). **Not illustrated.**

Australian Raven *Corvus coronoides* (Corvidae) 52 cm (20 in). Large, black bird, with white eyes. Long, pointed throat hackles, grey bases to feathers. High-pitched, wailing calls, last note sounding as if strangled. Fans throat when calling. **Range and habitat:** southern and eastern Australia, most habitats, but not in dense forests. **Similar species:** other ravens and crows (see pp 92, 182, 228, 278), distinguished by long call and throat hackles. **Not illustrated.**

Pied Currawong

Magnificent Riflebird

Spangled (Hair-crested) Drongo

Brown Sicklebill

Magnificent Bird of Paradise

King Bird of Paradise

Superb Bird of Paradise

Raggiana Bird of Paradise

WORLD CHECKLIST OF BIRDS

KEY

- ●   Rare or endangered species
- WP   Western Palaearctic including N Africa & Middle East
- EP   Eastern Palaearctic including Japan & N China
- AF   Afrotropical region including Red Sea
- OR   Oriental region including Himalayas, Philippines and up to Wallace's Line
- AU   East of Wallace's Line to Eastern New Guinea, Australia, New Zealand
- PA   Pacific including Solomon Islands and Hawaii
- NA   Nearctic, south to Rio Grande
- NT   Neotropical, Latin and S America, West Indies
- AN   Antarctica

Islands in Tristan da Cunha and St Helena groups are assigned to Afrotropical region and Galapagos to Pacific region.

OSTRICHES Struthioniformes

1 OSTRICHES Struthionidae

Ostrich *Struthio camelus* WP AF

RHEAS Rheiformes

2 RHEAS Rheidae

Greater Rhea *Rhea americana* NT
Lesser Rhea *Pterocnemia pennata* NT

CASSOWARIES AND EMUS Casuaraiiformes

3 CASSOWARIES Casuariidae

Double-wattled Cassowary *Casuarius casuarius* AU
Dwarf Cassowary *Casuarius bennetti* AU
One-wattled Cassowary *Casuarius unappendiculatus* AU

4 EMUS Dromaiidae

Emu *Dromaius novaehollandiae* AU

KIWIS Apterygiformes

5 KIWIS Apterygidae

Brown Kiwi *Apteryx australis* AU
Little Spotted Kiwi *Apteryx owenii* ● AU
Great Spotted Kiwi *Apteryx haastii* AU

TINAMOUS Tinamiformes

6 TINAMOUS Tinamidae

Grey Tinamou *Tinamus tao* NT
Solitary Tinamou *Tinamus solitarius* ● NT
Black Tinamou *Tinamus osgoodi* ● NT
Great Tinamou *Tinamus major* NT
White-throated Tinamou *Tinamus guttatus* NT

Highland Tinamou *Nothocercus bonapartei* NT
Tawny-breasted Tinamou *Nothocercus julius* NT
Hooded Tinamou *Nothocercus nigrocapillus* NT

Cinereous Tinamou *Crypturellus cinereus* NT
Berlepsch's Tinamou *Crypturellus berlepschi* NT
Little Tinamou *Crypturellus soui* NT
Tepui Tinamou *Crypturellus ptaritepui* NT
Brown Tinamou *Crypturellus obsoletus* NT
Undulated Tinamou *Crypturellus undulatus* NT
Pale-browed Tinamou *Crypturellus transfasciatus* NT
Brazilian Tinamou *Crypturellus strigulosus* NT
Grey-legged Tinamou *Crypturellus duidae* NT
Red-footed Tinamou *Crypturellus erythropus* NT
Yellow-legged Tinamou *Crypturellus noctivagus* ● NT
Black-capped Tinamou *Crypturellus atrocapillus* NT
Thicket Tinamou *Crypturellus cinnamomeus* NT
Slaty-breasted Tinamou *Crypturellus boucardi* NT
Choco Tinamou *Crypturellus kerriae* NT
Variegated Tinamou *Crypturellus variegatus* NT

Rusty Tinamou *Crypturellus brevirostris* NT
Bartlett's Tinamou *Crypturellus bartletti* NT
Small-billed Tinamou *Crypturellus parvirostris* NT
Barred Tinamou *Crypturellus casiquiare* NT
Tataupa Tinamou *Crypturellus tataupa* NT

Red-winged Tinamou *Rhynchotus rufescens* NT

Taczanowski's Tinamou *Nothoprocta taczanowskii* ● NT
Kalinowski's Tinamou *Nothoprocta kalinowskii* ● NT
Ornate Tinamou *Nothoprocta ornata* NT
Chilian Tinamou *Nothoprocta perdicaria* NT
Brushland Tinamou *Nothoprocta cinerascens* NT
Andean Tinamou *Nothoprocta pentlandii* NT
Curve-billed Tinamou *Nothoprocta curvirostris* NT

White-bellied Nothura *Nothura boraquira* NT
Lesser Nothura *Nothura minor* ● NT
Darwin's Nothura *Nothura darwinii* NT
Spotted Nothura *Nothura maculosa* NT

Dwarf Tinamou *Taoniscus nanus* ● NT

Elegant Crested-Tinamou *Eudromia elegans* NT
Quebracho Crested-Tinamou *Eudromia formosa* NT

Puna Tinamou *Tinamotis pentlandii* NT
Patagonian Tinamou *Tinamotis ingoufi* NT

PENGUINS Sphenisciformes

7 PENGUINS Spheniscidae

King Penguin *Aptenodytes patagonicus* NT
Emperor Penguin *Aptenodytes forsteri* AN

Gentoo Penguin *Pygoscelis papua* NT
Adelie Penguin *Pygoscelis adeliae* AN
Bearded Penguin *Pygoscelis antarctica* AN

Victoria Penguin *Eudyptes pachyrhynchus* AU
Snares I.Penguin *Eudyptes robustus* AU
Big-crested Penguin *Eudyptes sclateri* AU
Rockhopper Penguin *Eudyptes chrysocome* AF AU NT
Royal Penguin *Eudyptes schlegeli* AU
Macaroni Penguin *Eudyptes chrysolophus* NT

Yellow-eyed Penguin *Megadyptes antipodes* ● AU

Little Penguin *Eudyptula minor* AU
White-flippered Penguin *Eudyptula albosignata* AU

Jackass Penguin *Spheniscus demersus* ● AF
Humboldt Penguin *Spheniscus humboldti* ● NT
Magellanic Penguin *Spheniscus magellanicus* NT
Galapagos Penguin *Spheniscus mendiculus* NT

DIVERS Gaviformes

8 DIVERS Gaviidae

Red-throated Diver *Gavia stellata* WP EP NA
Black-throated Diver *Gavia arctica* WP EP
Pacific Diver *Gavia pacifica* NA
Great Northern Diver *Gavia immer* WP NA
White-billed Diver *Gavia adamsii* EP NA

GREBES Podicipediformes

9 GREBES Podicipedidae

Little Grebe *Tachybaptus ruficollis* WP EP AF
Australian Dabchick *Tachybaptus novaehollandiae* AU
Madagascar Little Grebe *Tachybaptus pelzelnii* ● AF
Delacour's Little Grebe *Tachybaptus rufolavatus* ● AF

Pied-billed Grebe *Podilymbus podiceps* NT
Atitlan Grebe *Podilymbus gigas* ● NT

White-tufted Grebe *Rollandia rolland* NT
Short-winged Grebe *Rollandia micropterum* NT

Great Grebe *Podiceps major* NT
Hoary-headed Grebe *Podiceps poliocephalus* AU
New Zealand Dabchick *Podiceps rufopectus* AU
Least Grebe *Podiceps dominicus* NA NT
Red-necked Grebe *Podiceps grisegena* WP EP NA
Great Crested Grebe *Podiceps cristatus* WP EP AF
Slavonian Grebe *Podiceps auritus* WP EP NA
Black-necked Grebe *Podiceps nigricollis* WP AF NA

Silvery Grebe *Podiceps occipitalis* NT
Puna Grebe *Podiceps taczanowskii* ● NT
Hooded Grebe *Podiceps gallardoi* ● NT

Western Grebe *Aechmophorus occidentalis* NA
Mexican Grebe *Aechmophorus clarkii* NT

TUBE-NOSED SWIMMERS Procellariiformes

10 ALBATROSSES Diomedeidae

Wandering Albatross *Diomedea exulans* AF AU NT
Royal Albatross *Diomedea epomophora* AU NT
Waved Albatross *Diomedea irrorata* NT
Short-tailed Albatross *Diomedea albatrus* ● PA
Black-footed Albatross *Diomedea nigripes* PA
Laysan Albatross *Diomedea immutabilis* PA
Black-browed Albatross *Diomedea melanophris* AF AU NT
Buller's Albatross *Diomedea bulleri* AU
Amsterdam Island Albatross *Diomedea amsterdamensis* ● AF
Shy Albatross *Diomedea cauta* AU
Yellow-nosed Albatross *Diomedea chlororhynchos* AF AU NT
Grey-headed Albatross *Diomedea chrysostoma* AF AU NT

Sooty Albatross *Phoebetria fusca* AF AU
Light-mantled Sooty Albatross *Phoebetria palpebrata* AF AU NT

11 PETRELS, AND SHEARWATERS Procellariidae

Giant Petrel *Macronectes giganteus* AF AU
Hall's Giant Petrel *Macronectes halli* AF AU NT

Fulmar *Fulmarus glacialis* WP NA
Southern Fulmar *Fulmarus glacialoides* AU NT

Antarctic Petrel *Thalassoica antarctica* AN

Pintado Petrel *Daption capense* AF AU NT AN

Snow Petrel *Pagodroma nivea* AN

Kerguelen Petrel *Lugensa brevirostris* AF AU

Great-winged Petrel *Pterodroma macroptera* AF AU
Mascarene Black Petrel *Pterodroma aterrima* ● AF
White-headed Petrel *Pterodroma lessonii* AF AU NT
Black-capped Petrel *Pterodroma hasitata* ● NA NT
Cahow *Pterodroma cahow* ● NT
Schlegel's Petrel *Pterodroma incerta* ● AF NT
Solomon Is Petrel *Pterodroma becki* PA
Phoenix Petrel *Pterodroma alba* PA
Peale's Petrel *Pterodroma inexpectata* AU PA
Solander's Petrel *Pterodroma solandri* AU PA
Murphy's Petrel *Pterodroma ultima* PA
Kermadec Petrel *Pterodroma neglecta* AU PA NT
Taiko *Pterodroma magentae* ● AU
Trinidade Petrel *Pterodroma arminjoniana* AF NT
Tonga Petrel *Pterodroma heraldica* AU PA
Soft-plumaged Petrel *Pterodroma mollis* AF AU NT
Cape Verde Petrel *Pterodroma feae* ● WP
Madeira Petrel *Pterodroma madeira* ● WP
Barau's Petrel *Pterodroma baraui* AF
Hawaiian Petrel *Pterodroma phaeopygia* ● PA
White-necked Petrel *Pterodroma externa* AU PA NT
Kermadec White-necked Petrel *Pterodroma cervicalis* AU
Cook's Petrel *Pterodroma cookii* ● PA
Juan Fernandez Petrel *Pterodroma defilippiana* ● NT
White-winged Petrel *Pterodroma leucoptera* AU PA
Collared Petrel *Pterodroma brevipes* AU PA
Bonin Petrel *Pterodroma hypoleuca* EP PA
Black-winged Petrel *Pterodroma nigripennis* AU PA
Chatham Is Petrel *Pterodroma axillaris* ● AU
Stejneger's Petrel *Pterodroma longirostris* PA NT
Pycroft's Petrel *Pterodroma pycrofti* ● AU

Macgillivray's Petrel *Pseudobulwaria macgillivrayi* ● PA
Tahiti Petrel *Pseudobulwaria rostrata* AU PA

Blue Petrel *Halobaena caerulea* AF AU NT

Broad-billed Prion *Pachyptila vittata* AF AU
Salvin's Prion *Pachyptila salvini* AU
Dove Prion *Pachyptila desolata* AU NT
Slender-billed Prion *Pachyptila belcheri* AU NT

Aerid–

Fairy Prion *Pachyptila turtur* AF AU NT
Thick-billed Prion *Pachyptila crassirostris* AU

Bulwer's Petrel *Bulweria bulwerii* AF AU PA NT
Jouanin's Petrel *Bulweria fallax* AF

Grey Shearwater *Procellaria cinerea* AF AU NT
White-chinned Petrel *Procellaria aequinoctialis* AF
AU NT
Black Petrel *Procellaria parkinsoni* ● AU
Westland Petrel *Procellaria westlandica* ● AU PA

White-faced Shearwater *Calonectris leucomelas* EP
PA
Cory's Shearwater *Calonectris diomedea* WP AF NA
NT
Pink-footed Shearwater *Puffinus creatopus* ● PA NT
Pale-footed Shearwater *Puffinus carneipes* AU PA
Greater Shearwater *Puffinus gravis* WP AF NT
Wedge-tailed Shearwater *Puffinus pacificus* AF AU
PA
Grey-backed Shearwater *Puffinus bulleri* AU
Sooty Shearwater *Puffinus griseus* AU PA NT
Short-tailed Shearwater *Puffinus tenuirostris* AU PA
Heinroth's Shearwater *Puffinus heinrothi* ● AU
Christmas Island Shearwater *Puffinus nativitatis* PA
Manx Shearwater *Puffinus puffinus* WP NA
Fluttering Shearwater *Puffinus gavia* AU
Hutton's Shearwater *Puffinus huttoni* AU
Black-vented Shearwater *Puffinus opisthomelas* NA
Townsend's Shearwater *Puffinus auricularis* ● PA
NA
Little Shearwater *Puffinus assimilis* WP AF AU PA
Audubon's Shearwater *Puffinus lherminieri* AF AU
PA NT

12 STORM PETRELS Hydrobatidae

Wilson's Storm Petrel *Oceanites oceanicus* NT AN
Elliot's Storm Petrel *Oceanites gracilis* NT

Grey-backed Storm Petrel *Garrodia nereis* AF AU
NT

White-faced Storm (Frigate) Petrel *Pelagodroma
marina*

White-bellied Storm Petrel *Fregetta grallaria* AF AU
NT

Black-bellied Storm Petrel *Fregetta tropica* AF AU
NT

White-throated Storm Petrel *Nesofregetta fuliginosa*
AU PA

Storm Petrel *Hydrobates pelagicus* WP

Least Storm Petrel *Halocyptena microsoma* NA

Galapagos Storm Petrel *Oceanodroma téthys* NA NT
Madeiran Storm Petrel *Oceanodroma castro* AF PA
Leach's Storm Petrel *Oceanodroma leucorhoa* WP EP
PA NA
Sooty Storm Petrel *Oceanodroma markhami* ● NA NT
Matsudaira's Storm Petrel *Oceanodroma matsudairae*
AU PA
Tristram's Storm Petrel *Oceanodroma tristrami* PA
Swinhoe's Storm Petrel *Oceanodroma monorhis* EP
Ashy Storm Petrel *Oceanodroma homochroa* NA
Ringed Storm Petrel *Oceanodroma hornbyi* ● NT
Fork-tailed Storm Petrel *Oceanodroma furcata* PA
NA
Black Storm Petrel *Oceanodroma melania* NA

13 DIVING PETRELS Pelecanoididae

Peruvian Diving Petrel *Pelecanoides garnotii* ● NT
Magellan Diving Petrel *Pelecanoides magellani* NT
Georgian Diving Petrel *Pelecanoides georgicus* AF
AU NT
Common Diving Petrel *Pelecanoides urinatrix* AF
AU NT

FULL-WEBBED SWIMMERS
Pelecaniformes

14 TROPICBIRDS Phaethontidae

Red-billed Tropicbird *Phaethon aethereus* AF OR NT
Red-tailed Tropicbird *Phaethon rubricauda* EP AF
AU PA
White-tailed Tropicbird *Phaethon lepturus* AF AU
PA NT

15 PELICANS Pelecanidae

Eastern White Pelican *Pelecanus onocrotalus* WP AF
White Pelican *Pelecanus roseus* EP OR
Pink-backed Pelican *Pelecanus rufescens* AF
Grey Pelican *Pelecanus philippensis* ● OR
Dalmatian Pelican *Pelecanus crispus* ● OR
Australian Pelican *Pelecanus conspicillatus* AU
American White Pelican *Pelecanus erythrorhynchos*
NA
Brown Pelican *Pelecanus occidentalis* NA NT

16 GANNETS AND BOOBIES Sulidae

Gannet *Morus bassanus* WP NA
Cape Gannet *Morus capensis* AF
Australian Gannet *Morus serrator* AU

Abbott's Booby *Papasula abbotti* ● OR

Blue-footed Booby *Sula nebouxii* NT
Peruvian Booby *Sula variegata* NT
Blue-faced Booby *Sula dactylatra* AF OR AU PA NT
Red-footed Booby *Sula sula* AF OR AU PA NT
Brown Booby *Sula leucogaster* AF OR AU PA NT

17 CORMORANTS Phalacrocoracidae

Double-crested Cormorant *Phalacrocorax auritus* NA
Olivaceous Cormorant *Phalacrocorax olivaceus* NT
Little Black Cormorant *Phalacrocorax sulcirostris* AU
Cormorant *Phalacrocorax carbo* WP EP AF OR AU
White-breasted Cormorant *Phalacrocorax lucidus* AF
Indian Cormorant *Phalacrocorax fuscicollis* OR
Cape Cormorant *Phalacrocorax capensis* AF
Socotra Cormorant *Phalacrocorax nigrogularis* OR
Bank Cormorant *Phalacrocorax neglectus* AF
Japanese Cormorant *Phalacrocorax capillatus* EP
Brandt's Cormorant *Phalacrocorax penicillatus* NA
Shag *Phalacrocorax aristotelis* WP
Pelagic Cormorant *Phalacrocorax pelagicus* EP NA
Red-faced Cormorant *Phalacrocorax urile* EP NA
Magellan Cormorant *Phalacrocorax magellanicus* NT
Guanay Cormorant *Phalacrocorax bougainvillei* NT
Chatham Cormorant *Phalacrocorax featherstoni* AU
Pied Cormorant *Phalacrocorax varius* AU
Black-faced Cormorant *Phalacrocorax fuscescens* AU
Rough-faced Cormorant *Phalacrocorax carunculatus*
● AU
Campbell Is Cormorant *Phalacrocorax campbelli* AU
Kerguelen Cormorant *Phalacrocorax verrucosus* AF
Red-legged Cormorant *Phalacrocorax gaimardi* NT
Spotted Cormorant *Phalacrocorax punctatus* AU
Blue-eyed Cormorant *Phalacrocorax atriceps* AF NT
S Georgia Cormorant *Phalacrocorax georgianus* NT
King Cormorant *Phalacrocorax albiventer* NT
Little Pied Cormorant *Phalacrocorax melanoleucos*
AU
Reed Cormorant *Phalacrocorax africanus* AF
Crowned Cormorant *Phalacrocorax coronatus* AF
Javanese Cormorant *Phalacrocorax niger* OR
Pygmy Cormorant *Phalacrocorax pygmeus* ● WP

Flightless Cormorant *Nannopterum harrisi* ● NT

18 ANHINGAS Anhingidae

African Darter *Anhinga rufa* AF
Indian Darter *Anhinga melanogaster* OR
Australian Darter *Anhinga novaehollandiae* AU
American Darter *Anhinga anhinga* NA NT

19 FRIGATEBIRDS Fregatidae

Ascension Frigatebird *Fregata aquila* ● AF
Christmas I Frigatebird *Fregata andrewsi* ● OR
Magnificent Frigatebird *Fregata magnificens* NT
Great Frigatebird *Fregata minor* OR AU PA
Lesser Frigatebird *Fregata ariel* OR AU PA

HERONS AND ALLIES Ciconiiformes

20 HERONS AND BITTERNS Ardeidae

Whistling Heron *Syrigma sibilatrix* NT

Capped Heron *Pilherodius pileatus* NT

Grey Heron *Ardea cinerea* WP EP AF OR
Great Blue Heron *Ardea herodias* NA NT
Cocoi Heron *Ardea cocoi* NT
White-necked Heron *Ardea pacifica* AU

Black-headed Heron *Ardea melanocephala* AF
Malagasy Heron *Ardea humbloti* ● AF
Imperial Heron *Ardea imperialis* ● OR
Dusky-grey Heron *Ardea sumatrana* OR AU
Goliath Heron *Ardea goliath* AF
Purple Heron *Ardea purpurea* WP AF OR

Great Egret *Egretta alba* WP EP AF OR NA NT
Reddish Egret *Egretta rufescens* NA NT
Pied Heron *Egretta picata* AU
Slaty Egret *Egretta vinaceigula* ● AF
Black Heron *Egretta ardesiaca* AF
Louisiana Heron *Egretta tricolor* NA NT
Intermediate Egret *Egretta intermedia* AF OR AU
White-faced Egret *Egretta novaehollandiae* AU
Little Blue Heron *Egretta caerulea* NA NT
Snowy Egret *Egretta thula* NA NT
Little Egret *Egretta garzetta* WP EP AF OR AU
Swinhoe's Egret *Egretta eulophotes* ● OR
Reef Heron *Egretta sacra* OR AU

Cattle Egret *Bubulcus ibis* WP AF OR NA NT

Squacco Heron *Ardeola ralloides* WP AF
Indian Pond Heron *Ardeola grayii* OR
Chinese Pond Heron *Ardeola bacchus* OR
Javanese Pond Heron *Ardeola speciosa* OR
Malagasy Pond Heron *Ardeola idae* AF
Rufous-bellied Heron *Ardeola rufiventris* AF

Green-backed (Striated) Heron *Butorides striatus* EP
AF OR AU NA NT

Chestnut-bellied Heron *Agamia agami* NT

Yellow-crowned Night Heron *Nycticorax violaceus*
NA NT
Black-crowned Night Heron *Nycticorax nycticorax*
WP EP AF OR NA NT
Nankeen Night Heron *Nycticorax caledonicus* OR AU
White-backed Night Heron *Nycticorax leuconotus* AF

Magnificent Night Heron *Gorsachius magnificus* ●
OR
Japanese Night Heron *Gorsachius goisagi* ● EP
Malaysian Night Heron *Gorsachius melanolophus* OR

Boat-billed Heron *Cochlearius cochlearius* NT

Bare-throated Tiger Heron *Tigrisoma mexicanum* NT
Fasciated Tiger Heron *Tigrisoma fasciatum* NT
Rufescent Tiger Heron *Tigrisoma lineatum* NT

Forest Bittern *Zonerodius heliosylus* AU

White-crested Tiger Heron *Tigriornis leucolophus* AF

Zigzag Heron *Zebrilus undulatus* ● NT

Stripe-backed Bittern *Ixobrychus involucris* NT
Least Bittern *Ixobrychus exilis* NA NT
Little Bittern *Ixobrychus minutus* WP EP AF OR AU
Chinese Little Bittern *Ixobrychus sinensis* EP OR
Schrenk's Bittern *Ixobrychus eurhythmus* EP OR
Cinnamon Bittern *Ixobrychus cinnamomeus* OR
African Dwarf Bittern *Ixobrychus sturmii* AF
Black Bittern *Ixobrychus flavicollis* OR AU

Pinnated Bittern *Botaurus pinnatus* NT
American Bittern *Botaurus lentiginosus* NA NT
Bittern *Botaurus stellaris* WP EP AF
Australian Bittern *Botaurus poiciloptilus* AU

21 HAMMERKOP Scopidae

Hammerkop *Scopus umbretta* AF

22 WHALE-HEADED STORK Balaenicipitidae

Whale-headed Stork *Balaeniceps rex* ● AF

23 STORKS Ciconiidae

American Wood Ibis *Mycteria americana* NA NT
Milky Stork *Mycteria cinerea* ● OR
Yellow-billed Stork *Mycteria ibis* AF
Painted Stork *Mycteria leucocephala* OR

Asian Open-bill Stork *Anastomus oscitans* OR
African Open-bill Stork *Anastomus lamelligerus* AF

Black Stork *Ciconia nigra* WP EP AF OR
Abdim's Stork *Ciconia abdimii* AF
Woolly-necked Stork *Ciconia episcopus* AF OR
Storm's Stork *Ciconia stormi* ● OR
Maguari Stork *Ciconia maguari* NT

White Stork *Ciconia ciconia* WP AF OR
Oriental White Stork *Ciconia boyciana* ● EP

Black-necked Stork *Ephippiorhynchus asiaticus* OR AU
Saddle-bill Stork *Ephippiorhynchus senegalensis* AF
Jabiru *Ephippiorhynchus mycteria* NT

Lesser Adjutant Stork *Leptoptilos javanicus* ● OR
Greater Adjutant Stork *Leptoptilos dubius* ● OR
Marabou Stork *Leptoptilos crumeniferus* AF

24 IBISES AND SPOONBILLS Threskiornithidae

Threskiornithinae

Sacred Ibis *Threskiornis aethiopicus* AF
Oriental Ibis *Threskiornis melanocephalus* EP OR
Australian White Ibis *Threskiornis molucca* AU
Straw-necked Ibis *Threskiornis spinicollis* AU

Black Ibis *Pseudibis papillosa* OR

Giant Ibis *Thaumatibis gigantea* ● OR

Hermit (Waldrapp) Ibis *Geronticus eremita* ● WP AF
Bald Ibis *Geronticus calvus* ● AF

Japanese Crested Ibis *Nipponia nippon* ● EP

Olive Ibis *Lampribis olivacea* AF
Spot-breasted Ibis *Lampribis rara* AF

Hadada Ibis *Hagedashia hegedash* AF

Wattled Ibis *Bostrychia carunculata* AF

Plumbeous Ibis *Harpiprion caerulescens* NT

Buff-necked Ibis *Theristicus caudatus* NT
Black-faced Ibis *Theristicus melanopis* NT

Sharp-tailed Ibis *Cercibis oxycerca* NT

Green Ibis *Mesembrinibis cayennensis* NT

Bare-faced Ibis *Phimosus infuscatus* NT

White Ibis *Eudocimus albus* NA NT
Scarlet Ibis *Eudocimus ruber* NT

Glossy Ibis *Plegadis falcinellus* WP EP AF OR AU NT
White-faced Ibis *Plegadis chihi* NT
Puna Ibis *Plegadis ridgwayi* NT

Crested Wood Ibis *Lophotibis cristata* ● AF

Plataleinae

Spoonbill *Platalea leucorodia* WP EP AF OR
Black-faced Spoonbill *Platalea minor* ● EP OR
African Spoonbill *Platalea alba* AF
Royal Spoonbill *Platalea regia* AU
Yellow-billed Spoonbill *Platalea flavipes* AU

Roseate Spoonbill *Ajaia ajaja* NA NT

25 FLAMINGOS Phoenicopteridae

Greater Flamingo *Phoenicopterus ruber* WP EP AF OR NT
Chilean Flamingo *Phoenicopterus chilensis* NT

Lesser Flamingo *Phoeniconaias minor* AF OR

Andean Flamingo *Phoenicoparrus andinus* ● NT
James' Flamingo *Phoenicoparrus jamesi* ● NT

WATERFOWL Anseriformes

26 SCREAMERS Anhimidae

Horned Screamer *Anhima cornuta* NT

Crested Screamer *Chauna torquata* NT
Northern Screamer *Chauna chavaria* ● NT

27 DUCKS, GEESE AND SWANS Anatidae

Anseranatinae

Magpie Goose *Anseranas semipalmata* AU

Anserinae

Spotted Whistling Duck *Dendrocygna guttata* OR AU
Plumed Whistling Duck *Dendrocygna eytoni* AU

Fulvous Whistling Duck *Dendrocygna bicolor* AF OR NA NT
Wandering Whistling Duck *Dendrocygna arcuata* OR AU
Indian Whistling Duck *Dendrocygna javanica* OR
White-faced Whistling Duck *Dendrocygna viduata* AF NT
Black-billed (W Indian) Whistling Duck *Dendrocygna arborea* ● NT
Red-billed (Black-bellied) Whistling Duck *Dendrocygna autumnalis* NA NT

Mute Swan *Cygnus olor* WP EP AF OR
Black Swan *Cygnus atratus* AU
Black-necked Swan *Cygnus melanocoryphus* NT
Whooper Swan *Cygnus cygnus* WP EP OR
Trumpeter Swan *Cygnus buccinator* NA
Bewick Swan *Cygnus columbianus* WP EP NA

Coscoroba Swan *Coscoroba coscoroba* NT

Swan Goose *Anser cygnoides* EP OR
Bean Goose *Anser fabalis* WP EP
Pink-footed Goose *Anser brachyrhynchus* WP
White-fronted Goose *Anser albifrons* WP EP OR NA
Lesser White-fronted Goose *Anser erythropus* ● WP EP OR
Greylag Goose *Anser anser* WP EP OR
Bar-headed Goose *Anser indicus* EP OR
Snow Goose *Anser caerulescens* NA
Ross's Goose *Anser rossi* NA
Emperor Goose *Anser canagicus* EP NA

Hawaiian Goose *Branta sandvicensis* ● PA
Canada Goose *Branta canadensis* WP NA
Barnacle Goose *Branta leucopsis* WP
Brent Goose *Branta bernicla* WP EP NA
Red-breasted Goose *Branta ruficollis* ● EP

Cereopsis Goose *Cereopsis novaehollandiae* AU

Freckled Duck *Stictonetta naevosa* ● AU

Anatinae

Blue-winged Goose *Cyanochen cyanopterus* AF

Andean Goose *Chloephaga melanoptera* NT
Magellan Goose *Chloephaga picta* NT
Kelp Goose *Chloephaga hybrida* NT
Ashy-headed Goose *Chloephaga poliocephala* NT
Ruddy-headed Goose *Chloephaga rubidiceps* NT

Orinoco Goose *Neochen jubatus* NT

Egyptian Goose *Alopochen aegyptiacus* AF

Ruddy Shelduck *Tadorna ferruginea* WP EP OR
South African Shelduck *Tadorna cana* AF
Paradise Shelduck *Tadorna variegata* AU
Australian Shelduck *Tadorna tadornoides* AU
Common Shelduck *Tadorna tadorna* WP EP OR
Radjah Shelduck *Tadorna radjah* AU
Crested Shelduck *Tadorna cristata* ● EP

Flying Steamer Duck *Tachyeres patachonicus* NT
Flightless Steamer Duck *Tachyeres pteneres* NT
Falkland Is Flightless Steamer Duck *Tachyeres brachypterus* NT
White-headed Flightless Steamer Duck *Tachyeres leucocephalus* NT

Spur-winged Goose *Plectropterus gambensis* AF

Muscovy Duck *Cairina moschata* NT
White-winged Wood Duck *Cairina scutulata* ● OR

Comb Duck *Sarkidiornis melanotos* AF OR NT

Hartlaub's Duck *Pteronetta hartlaubii* AF

Green Pygmy Goose *Nettapus pulchellus* AU
Cotton Teal *Nettapus coromandelianus* OR AU
African Pygmy Goose *Nettapus auritus* AF

Ringed Teal *Callonetta leucophrys* NT

Wood Duck *Aix sponsa* NA
Mandarin Duck *Aix galericulata* EP

Maned Goose *Chenonetta jubata* AU

Brazilian Teal *Amazonetta brasiliensis* NT

Mountain Duck *Hymnenolaimus malacorhynchos* AU

Torrent Duck *Merganetta armata* NT

Salvadori's Duck *Anas waigiuensis* AU
African Black Duck *Anas sparsa* AF
Wigeon *Anas penelope* WP EP AF OR
American Wigeon *Anas americana* NA NT
Chiloe Wigeon *Anas sibilatrix* NT
Falcated Teal *Anas falcata* EP OR
Gadwall *Anas strepera* WP EP AF OR NA
Baikal Teal *Anas formosa* ● EP
Teal *Anas crecca* WP EP AF OR NA NT
Chilean Teal *Anas flavirostris* NT
Cape Teal *Anas capensis* AF
Grey Teal *Anas gibberifrons* OR AU
Madagascar Teal *Anas bernieri* ● AF
Chestnut-breasted Teal *Anas castanea* AU
New Zealand Teal *Anas aucklandica* ● AU
Mallard *Anas platyrhynchos* WP EP OR PA NA NT
North American Black Duck *Anas rubripes* NA
Meller's Duck *Anas melleri* AF
African Yellow-bill *Anas undulata* AF
Spotbill Duck *Anas poecilorhyncha* EP OR AU
Pacific Black Duck *Anas superciliosa* OR AU PA
Philippine Duck *Anas luzonica* OR
Bronze-winged Duck *Anas specularis* NT
Crested Duck *Anas specularioides* NT
Pintail *Anas acuta* WP EP AF OR NA
Georgian Teal *Anas georgica* NT
Bahama Pintail *Anas bahamensis* NT
Red-billed Pintail *Anas erythrorhynchos* AF
Versicolor Teal *Anas versicolor* NT
Hottentot Teal *Anas punctata* AF
Garganey *Anas querquedula* WP EP AF OR
Blue-winged Teal *Anas discors* NA NT
Cinnamon Teal *Anas cyanoptera* NA NT
Argentine Shoveler *Anas platalea* NT
Cape Shoveler *Anas smithii* AF
Australian Shoveler *Anas rhynchotis* AU
Shoveler *Anas clypeata* WP EP AF OR NA

Pink-eared Duck *Malacorhynchus membranaceus* AU

Marbled Teal *Marmaronetta angustirostris* ● WP OR

Pink-headed Duck *Rhodonessa caryophyllacea* ● OR

Red-crested Pochard *Netta rufina* WP OR
Southern Pochard *Netta erythrophthalma* AF NT
Rosybill *Netta peposaca* NT

Canvasback *Aythya valisineria* NA
Pochard *Aythya ferina* WP EP
Redhead *Aythya americana* NA
Ring-necked Duck *Aythya collaris* NA NT
Australian White-eyed Duck *Aythya australis* AU
Baer's Pochard *Aythya baeri* ● EP OR
Ferruginous Duck *Aythya nyroca* WP OR
Madagascar Pochard *Aythya innotata* AF
New Zealand Scaup *Aythya novaeseelandiae* AU
Tufted Duck *Aythya fuligula* WP EP OR
Greater Scaup *Aythya marila* WP EP OR NA
Lesser Scaup *Aythya affinis* NA NT

Eider *Somateria mollissima* WP EP NA
King Eider *Somateria spectabilis* WP EP NA
Spectacled Eider *Somateria fischeri* EP NA

Steller's Eider *Polysticta stelleri* EP NA

Harlequin Duck *Histrionicus histrionicus* WP EP NA

Long-tailed Duck *Clangula hyemalis* WP EP NA

Common (Black) Scoter *Melanitta nigra* WP EP NA
Surf Scoter *Melanitta perspicillata* EP NA
Velvet Scoter *Melanitta fusca* WP EP NA

Bufflehead *Bucephala albeola* NA
Barrow's Goldeneye *Bucephala islandica* WP NA
Goldeneye *Bucephala clangula* WP EP OR NA
Hooded Merganser *Mergus cucullatus* NA
Smew *Mergus albellus* WP EP OR
Brazilian Merganser *Mergus octosetaceus* ● NT
Red-breasted Merganser *Mergus serrator* WP EP NA
Chinese Merganser *Mergus squamatus* ● EP
Goosander *Mergus merganser* WP EP OR NA

Black-headed Duck *Heteronetta atricapilla* NT

Masked Duck *Oxyura dominica* NT
Ruddy Duck *Oxyura jamaicensis* NA NT
White-headed Duck *Oxyura leucocephala* ● WP EP OR
Maccoa Duck *Oxyura maccoa* AF

Argentine Lake Duck *Oxyura vittata* NT
Australian Blue-billed Duck *Oxyura australis* AU

Musk Duck *Biziura lobata* AU

White-backed Duck *Thalassornis leuconotos* AF

DIURNAL BIRDS OF PREY Falconiformes

28 NEW WORLD VULTURES Cathartidae

Turkey Vulture *Cathartes aura* NA NT
Lesser Yellow-headed Vulture *Cathartes burrovianus* NT
Greater Yellow-headed Vulture *Cathartes melambrotus* NT

American Black Vulture *Coragyps atratus* NA NT

King Vulture *Sarcorhamphus papa* NT

Californian Condor *Gymnogyps californianus* ● NA

Andean Condor *Vultur gryphus* NT

29 OSPREY Pandionidae

Osprey *Pandion haliaetus* WP EP AF OR AU NA NT

30 HAWKS AND EAGLES Accipitridae

African Cuckoo Hawk *Aviceda cuculoides* AF
Madagascar Cuckoo Hawk *Aviceda madagascariensis* AF
Jerdon's Baza *Aviceda jerdoni* OR AU
Crested Baza *Aviceda subcristata* OR AU
Black Baza *Aviceda leuphotes* OR

Grey-headed Kite *Leptodon cayanensis* NT

Hook-billed Kite *Chondrohierax uncinatus* NT

Long-tailed Honey Buzzard *Henicopernis longicauda* AU
Black Honey Buzzard *Henicopernis infuscata* ● AU

Honey Buzzard *Pernis apivorus* WP AF
Oriental Honey Buzzard *Pernis ptilorhynchus* EP OR
Barred Honey Buzzard *Pernis celebensis* OR AU

Swallow-tailed Kite *Elanoides forficatus* NA NT

Bat Hawk *Machaerhamphus alcinus* AF OR AU

Pearl Kite *Gampsonyx swainsonii* NT

White-tailed Kite *Elanus leucurus* NA NT
Black-shouldered Kite *Elanus caeruleus* AF OR AU
Australian Black-shouldered Kite *Elanus notatus* AU
Letter-winged Kite *Elanus scriptus* AU

Scissor-tailed Kite *Chelictinia riocourii* AF

Everglade Kite *Rostrhamus sociabilis* NA NT
Slender-billed Kite *Rostrhamus hamatus* NT

Double-toothed Kite *Harpagus bidentatus* NT
Rufous-thighed Kite *Harpagus diodon* NT

Plumbeous Kite *Ictinia plumbea* NT
Mississippi Kite *Ictinia mississippiensis* NA NT

Square-tailed Kite *Lophoictinia isura* AU

Black-breasted Buzzard Kite *Hamirostra melanosternon* AU

Black Kite *Milvus migrans* WP EP AF OR AU
Red Kite *Milvus milvus* ● WP

Whistling Kite *Haliastur sphenurus* AU
Brahminy Kite *Haliastur indus* OR AU

White-bellied Sea Eagle *Haliaeetus leucogaster* OR AU
Sanford's Sea Eagle *Haliaeetus sanfordi* ● PA
African Fish Eagle *Haliaeetus vocifer* AF
Madagascar Fish Eagle *Haliaeetus vociferoides* ● AF
Pallas' Sea Eagle *Haliaeetus leucoryphus* ● EP OR
American Bald Eagle *Haliaeetus leucocephalus* NA
White-tailed Sea Eagle *Haliaeetus albicilla* ● WP EP OR
Steller's Sea Eagle *Haliaeetus pelagicus* ● EP

Lesser Fishing Eagle *Ichthyophaga humilis* OR
Grey-headed Fishing Eagle *Ichthyophaga ichthyaetus* OR

Cinereous Vulture *Aegypius monachus* ● WP EP OR
Lappet-faced Vulture *Aegypius tracheliotus* WP AF
White-headed Vulture *Aegypius occipitalis* AF

Red-headed Vulture *Aegypius calvus* OR

Hooded Vulture *Necrosyrtes monachus* AF

Griffon Vulture *Gyps fulvus* WP OR
Long-billed Griffon *Gyps indicus* OR
Himalayan Griffon *Gyps himalayensis* EP OR
Ruppell's Griffon *Gyps rueppelli* AF
Cape Vulture *Gyps coprotheres* AF
Oriental White-backed Vulture *Gyps bengalensis* OR
African White-backed Vulture *Gyps africanus* AF

Egyptian Vulture *Neophron percnopterus* WP AF OR

Lammergeier *Gypaetus barbatus* WP EP AF

Palm-nut Vulture *Gypohierax angolensis* AF

Short-toed Eagle *Circaetus gallicus* WP EP AF OR
Brown Snake Eagle *Circaetus cinereus* AF
Southern Banded Snake Eagle *Circaetus fasciolatus* AF
Smaller Banded Snake Eagle *Circaetus cinerascens* AF

Bateleur *Terathopius ecaudatus* AF

Crested Serpent Eagle *Spilornis cheela* OR
Sulawesi Serpent Eagle *Spilornis rufipectus* OR
Kinabalu Serpent Eagle *Spilornis kinabaluensis* ● OR
Nicobar Serpent Eagle *Spilornis minimus* OR
Andaman Serpent Eagle *Spilornis elgini* ● OR

Congo Serpent Eagle *Dryotriorchis spectabilis* AF

Madagascar Serpent Eagle *Eutriorchis astur* ● AF

Dark Chanting-Goshawk *Melierax metabates* AF
Pale Chanting-Goshawk *Melierax canorus* AF
Somali Chanting-Goshawk *Melierax poliopterus* AF

Gabar Goshawk *Micronisus gabar* AF

African Harrier Hawk *Polyboroides typus* AF
Madagascar Harrier Hawk *Polyboroides radiatus* AF

Lizard Buzzard *Kaupifalco monogrammicus* AF

Grasshopper Buzzard-Eagle *Butastur rufipennis* AF
Rufous-winged Buzzard *Butastur liventer* OR
White-eyed Buzzard *Butastur teesa* OR
Grey-faced Buzzard Eagle *Butastur indicus* EP OR

Spotted Harrier *Circus assimilis* OR AU
Black Harrier *Circus maurus* AF
Hen Harrier (Marsh Hawk) *Circus cyaneus* WP EP NA NT
Cinereous Harrier *Circus cinereus* NT
Pallid Harrier *Circus macrourus* WP EP AF OR
Pied Harrier *Circus melanoleucos* EP OR
Montague's Harrier *Circus pygargus* WP EP AF OR
Marsh Harrier *Circus aeruginosus* WP EP
Eastern Marsh Harrier *Circus spilonotus* EP OR AU
Pacific Marsh Harrier *Circus approximans* AU PA
African Marsh Harrier *Circus ranivorus* AF
Malagasy Marsh Harrier *Circus maillardi* AF
Long-winged Harrier *Circus buffoni* NT

Grey-bellied Goshawk *Accipiter poliogaster* ● NT
Asian Crested Goshawk *Accipiter trivirgatus* OR
Sulawesi Crested Goshawk *Accipiter griseiceps* OR
African Goshawk *Accipiter tachiro* AF
Chestnut-bellied Sparrowhawk *Accipiter castanilus* AF
Levant Sparrowhawk *Accipiter brevipes* WP
Shikra *Accipiter badius* EP AF OR
Nicobar Shikra *Accipiter butleri* OR
Grey Frog Hawk *Accipiter soloensis* OR
Frances' Sparrowhawk *Accipiter francesii* AF
Spot-tailed Sparrowhawk *Accipiter trinotatus* OR
Brown (Australian) Goshawk *Accipiter fasciatus* OR AU
White Goshawk *Accipiter novaehollandiae* OR AU PA
Black-mantled Goshawk *Accipiter melanochlamys* AU
Pied Goshawk *Accipiter albogularis* PA
Fiji Goshawk *Accipiter rufitorques* PA
New Caledonia Sparrowhawk *Accipiter haplochrous* PA
Gray's Goshawk *Accipiter henicogrammus* OR
Blue & Grey Sparrowhawk *Accipiter luteoschistaceus* AU
Imitator Sparrowhawk *Accipiter imitator* ● PA
New Guinea Grey-headed Goshawk *Accipiter poliocephalus* AU

New Britain Grey-headed Goshawk *Accipiter princeps* AU
Tiny Sparrowhawk *Accipiter superciliosus* NT
Semi-collared Sparrowhawk *Accipiter collaris* ● NT
Red-thighed Sparrowhawk *Accipiter erythropus* AF
African Little Sparrowhawk *Accipiter minullus* AF
Japanese Sparrowhawk *Accipiter gularis* EP OR
Besra Sparrowhawk *Accipiter virgatus* OR
Dwarf Sparrowhawk *Accipiter nanus* ● OR
Australian Collared Sparrowhawk *Accipiter cirrhocephalus* AU
New Britain Collared Sparrowhawk *Accipiter brachyurus* ● AU
Grey Moluccan Collared Sparrowhawk *Accipiter erythrauchen* OR
Vinous-breasted Sparrowhawk *Accipiter rhodogaster* OR
Madagascar Sparrowhawk *Accipiter madagascariensis* AF
Ovampo Sparrowhawk *Accipiter ovampensis* AF
Sparrowhawk *Accipiter nisus* WP EP
Rufous-breasted Sparrowhawk *Accipiter rufiventris* AF
Sharp-shinned Hawk *Accipiter striatus* NA NT
Cooper's Hawk *Accipiter cooperii* NA NT
Gundlach's Hawk *Accipiter gundlachii* ● NT
Bicoloured Sparrowhawk *Accipiter bicolor* NT
Great Sparrowhawk *Accipiter melanoleucus* AF
Henst's Goshawk *Accipiter henstii* AF
Goshawk *Accipiter gentilis* WP EP NA
Meyer's Goshawk *Accipiter meyerianus* OR AU PA
Chestnut-shouldered Goshawk *Accipiter buergersi* AU
Red Goshawk *Accipiter radiatus* ● AU
Doria's Goshawk *Accipiter doriae* AU

African Long-tailed Hawk *Urotriorchis macrourus* AF

Crane Hawk *Geranospiza caerulescens* NT

Slate-coloured Hawk *Leucopternis schistacea* NT
Plumbeous Hawk *Leucopternis plumbea* ● NT
Barred Hawk *Leucopternis princeps* NT
Black-faced Hawk *Leucopternis melanops* NT
White-browed Hawk *Leucopternis kuhli* NT
White-necked Hawk *Leucopternis lacernulata* ● NT
Semi-plumbeous Hawk *Leucopternis semiplumbea* NT
White Hawk *Leucopternis albicollis* NT
Grey-backed Hawk *Leucopternis occidentalis* ● NT
Mantled Hawk *Leucopternis polionota* ● NT

Grey Hawk *Asturina nitida* NA NT

Common Black Hawk *Buteogallus anthracinus* NA NT
Mangrove Black Hawk *Buteogallus subtilis* NT
Rufous Crab Hawk *Buteogallus aequinoctialis* NT
Great Black Hawk *Buteogallus urubitinga* NT
Savanna Hawk *Buteogallus meridionalis* NT

Black Solitary Eagle *Harpyhaliaetus solitarius* ● NT
Crowned Solitary Eagle *Harpyhaliaetus coronatus* ● NT

Black-collared Hawk *Busarellus nigricollis* NT

Black-chested Buzzard Eagle *Geranoaetus melanoleucus* NT

Harris' Hawk *Parabuteo unicinctus* NA NT

Roadside Hawk *Buteo magnirostris* NT
White-rumped Hawk *Buteo leucorrhous* NT
Ridgway's Hawk *Buteo ridgwayi* ● NT
Red-shouldered Hawk *Buteo lineatus* NA
Broad-winged Hawk *Buteo platypterus* NA NT
Short-tailed Hawk *Buteo brachyurus* NA NT
White-throated Hawk *Buteo albigula* NT
Swainson's Hawk *Buteo swainsonii* NA NT
Galapagos Hawk *Buteo galapagoensis* ● NT
White-tailed Hawk *Buteo albicaudatus* NA NT
Red-backed Hawk *Buteo polyosoma* NT
Variable Hawk *Buteo poecilochrous* NT
Zone-tailed Hawk *Buteo albonotatus* NA NT
Hawaiian Hawk *Buteo solitarius* ● PA
Rufous-tailed Hawk *Buteo ventralis* ● NT
Red-tailed Hawk *Buteo jamaicensis* NA NT
Buzzard *Buteo buteo* WP EP AF
African Mountain Buzzard *Buteo oreophilus* AF

Madagascar Buzzard *Buteo brachypterus* AF
Rough-legged Buzzard *Buteo lagopus* WP EP NA
Long-legged Buzzard *Buteo rufinus* WP EP
Upland Buzzard *Buteo hemilasius* EP
Ferruginous Hawk *Buteo regalis* NA
African Red-tailed Buzzard *Buteo auguralis* AF
Augur Buzzard *Buteo augur* AF
Jackal Buzzard *Buteo rufofuscus* AF

Guiana Crested Eagle *Morphnus guianensis* ● NT

Harpy Eagle *Harpia harpyja* ● NT

New Guinea Harpy Eagle *Harpyopsis novaeguineae* ● AU

Philippine Eagle *Pithecophaga jefferyi* ● OR

Indian Black Eagle *Ictinaetus malayensis* OR

Lesser Spotted Eagle *Aquila pomarina* WP OR
Greater Spotted Eagle *Aquila clanga* WP EP OR
Tawny Eagle *Aquila rapax* WP EP AF OR
Imperial Eagle *Aquila heliaca* ● WP EP OR
Gurney's Eagle *Aquila gurneyi* AU
Golden Eagle *Aquila chrysaetos* WP EP NA
Wedge-tailed Eagle *Aquila audax* AU
Verreaux's Eagle *Aquila verreauxii* AF

Wahlberg's Eagle *Hieraaetus wahlbergi* AF
Bonelli's Eagle *Hieraaetus fasciatus* WP AF
African Hawk Eagle *Hieraaetus spilogaster* AF
Booted Eagle *Hieraaetus pennatus* WP EP
Little Eagle *Hieraaetus morphnoides* AU
Ayres' Hawk Eagle *Hieraaetus ayresii* AF
Chestnut-bellied Hawk Eagle *Hieraaetus kienerii* OR
Martial Eagle *Hieraaetus bellicosus* AF

Black & White Hawk Eagle *Spizastur melanoleucus* NT

Long-crested Eagle *Spizaetus occipitalis* AF
Cassin's Hawk Eagle *Spizaetus africanus* AF
Crested Hawk Eagle *Spizaetus cirrhatus* OR
Hodgson's Hawk Eagle *Spizaetus nipalensis* EP OR
Java Hawk Eagle *Spizaetus bartelsi* ● OR
Sulawesi Hawk Eagle *Spizaetus lanceolatus* OR
Philippine Hawk Eagle *Spizaetus philippensis* OR
Blyth's Hawk Eagle *Spizaetus alboniger* OR
Wallace's Hawk Eagle *Spizaetus nanus* ● OR
Black Hawk Eagle *Spizaetus tyrannus* NT
Ornate Hawk Eagle *Spizaetus ornatus* NT
Crowned Eagle *Spizaetus coronatus* AF
Black & Chestnut Eagle *Spizaetus isidori* NT

31 SECRETARY BIRD Sagittariidae

Secretary Bird *Sagittarius serpentarius* AF

32 FALCONS AND CARACARAS Falconidae

Yellow-throated Caracara *Daptrius ater* NT
Red-throated Caracara *Daptrius americanus* NT

Carunculated Caracara *Phalcoboenus carunculatus* NT
Mountain Caracara *Phalcoboenus megalopterus* NT
White-throated Caracara *Phalcoboenus albogularis* NT
Forster's Caracara *Phalcoboenus australis* NT

Common Caracara *Polyborus plancus* NA NT

Chimango *Milvago chimango* NT
Yellow-headed Caracara *Milvago chimachima* NT

Laughing Falcon *Herpetotheres cachinnans* NT

Barred Forest Falcon *Micrastur ruficollis* NT
Lined Forest Falcon *Micrastur gilvicollis* NT
Slaty-backed Forest Falcon *Micrastur mirandollei* NT
Collared Forest Falcon *Micrastur semitorquatus* NT
Traylor's Forest Falcon *Micrastur buckleyi* NT

Spot-winged Falconet *Spiziapteryx circumcinctus* NT

African Pygmy Falcon *Polihierax semitorquatus* AF
White-rumped Pygmy Falcon *Polihierax insignis* OR

Collared Falconet *Microhierax caerulescens* OR
Black-thighed Falconet *Microhierax fringillarius* OR
Bornean Falconet *Microhierax latifrons* OR
Philippine Falconet *Microhierax erythrogenys* OR
Pied Falconet *Microhierax melanoleucus* OR

Lesser Kestrel *Falco naumanni* WP EP AF
Greater Kestrel *Falco rupicoloides* AF

Fox Kestrel *Falco alopex* AF
American Kestrel *Falco sparverius* NA NT
Kestrel *Falco tinnunculus* WP EP AF OR
Madagascar Kestrel *Falco newtoni* AF
Mauritius Kestrel *Falco punctatus* ● AF
Seychelles Kestrel *Falco araea* ● OR
Moluccan Kestrel *Falco moluccensis* OR AU
Australian Kestrel *Falco cenchroides* AU
Grey Kestrel *Falco ardosiaceus* AF
Dickinson's Kestrel *Falco dickinsoni* AF
Madagascar Banded Kestrel *Falco zoniventris* AF
Western Red-footed Falcon *Falco vespertinus* WP EP AF
Eastern Red-footed Falcon *Falco amurensis* EP AF
Red-headed Falcon *Falco chicquera* AF OR
Merlin *Falco columbarius* WP EP AF NA NT
Brown Falcon *Falco berigora* AU
New Zealand Falcon *Falco novaezeelandiae* AU
Hobby *Falco subbuteo* WP EP AF OR
African Hobby *Falco cuvieri* AF
Oriental Hobby *Falco severus* OR AU
Australian Hobby *Falco longipennis* AU
Eleonora's Falcon *Falco eleonorae* WP AF
Sooty Falcon *Falco concolor* AF
Bat Falcon *Falco rufigularis* NT
Aplomado Falcon *Falco femoralis* NA NT
Grey Falcon *Falco hypoleucos* ● AU
Black Falcon *Falco subniger* AU
Lanner Falcon *Falco biarmicus* WP AF
Prairie Falcon *Falco mexicanus* NA
Laggar Falcon *Falco jugger* OR
Saker Falcon *Falco cherrug* WP EP OR
Gyr Falcon *Falco rusticolus* WP EP NA
Orange-breasted Falcon *Falco deiroleucus* ● NT
Taita Falcon *Falco fasciinucha* AF
Peregrine Falcon *Falco peregrinus* WP EP OR AU NA NT
Barbary Falcon *Falco pelegrinoides* WP

FOWL-LIKE BIRDS Galliformes

33 MEGAPODES Megapodiae

Nicobar Scrub Fowl *Megapodius nicobariensis* ● OR
Philippine Scrub Fowl *Megapodius cumingii* OR
Sula Scrub Fowl *Megapodius bernsteinii* ● AU
New Guinea Scrub Fowl *Megapodius affinis* AU
Orange-footed Scrub Fowl *Megapodius reinwardt* AU
Bismarck Scrub Fowl *Megapodius eremita* AU PA
Dusky Scrub Fowl *Megapodius freycinet* AU
Banks Is Scrub Fowl *Megapodius layardi* AU
Marianas Scrub Fowl *Megapodius laperouse* ● PA
Polynesian Scrub Fowl *Megapodius pritchardii* ● PA
Moluccas Scrub Fowl *Megapodius wallacei* ● AU

Mallee Fowl *Leipoa ocellata* ● AU

Brush Turkey *Alectura lathami* AU

Red-billed Brush Turkey *Talegalla cuvieri* AU
Black-billed Brush Turkey *Talegalla fuscirostris* AU
Brown-collared Brush Turkey *Talegalla jobiensis* AU

Wattled Brush Turkey *Aepypodius arfakianus* AU
Bruijn's Brush Turkey *Aepypodius bruijnii* ● AU

Maleo Fowl *Macrocephalon maleo* ● AU

34 CURASSOWS AND GUANS Cracidae

Plain Chachalaca *Ortalis vetula* NA NT
Grey-headed Chachalaca *Ortalis cinereiceps* NT
Chestnut-winged Chachalaca *Ortalis garrula* NT
Rufous-vented Chachalaca *Ortalis ruficauda* NT
Rufous-headed Chachalaca *Ortalis erythroptera* ● NT
West Mexican Chachalaca *Ortalis poliocephala* NT
Chaco Chachalaca *Ortalis canicollis* NT
White-bellied Chachalaca *Ortalis leucogastra* NT
Variable Chachalaca *Ortalis motmot* NT

Band-tailed Guan *Penelope argyrotis* NT
Bearded Guan *Penelope barbata* ● NT
Andean Guan *Penelope montagnii* NT
Baudo Guan *Penelope ortoni* NT
Marail Guan *Penelope marail* NT
Rusty-margined Guan *Penelope superciliaris* NT
Red-faced Guan *Penelope dabbenei* ● NT
Dusky-legged Guan *Penelope obscura* NT

Spix's Guan *Penelope jacquacu* NT
White-winged Guan *Penelope albipennis* ● NT
Cauca Guan *Penelope perspicax* ● NT
Crested Guan *Penelope purpurascens* NT
White-browed Guan *Penelope jacucaca* NT
Chestnut-bellied Guan *Penelope ochrogaster* ● NT
White-crested Guan *Penelope pileata* NT

Blue-throated Piping Guan *Aburria pipile* NT
Red-throated Piping Guan *Aburria cujubi* NT
Black-fronted Piping Guan *Aburria jacutinga* ● NT
Wattled Guan *Aburria aburri* NT

Black Guan *Chamaepetes unicolor* NT
Sickle-winged Guan *Chamaepetes goudotii* NT

Highland Guan *Penelopina nigra* ● NT

Horned Guan *Oreophasis derbianus* ● NT

Nocturnal Curassow *Nothocrax urumutum* NT

Lesser Razor-billed Curassow *Crax tomentosa* NT
Salvin's Curassow *Crax salvini* NT
Razor-billed Curassow *Crax mitu* NT
Helmeted Curassow *Crax pauxi* ● NT
Horned Curassow *Crax unicornis* NT
Great Curassow *Crax rubra* NT
Blue-billed Curassow *Crax alberti* ● NT
Yellow-knobbed Curassow *Crax daubentoni* NT
Black Curassow *Crax alector* NT
Wattled Curassow *Crax globulosa* ● NT
Bare-faced Curassow *Crax fasciolata* NT
Red-billed Curassow *Crax blumenbachii* ● NT

35 PHEASANTS AND GROUSE Phasianidae

Melagridinae

Common Turkey *Meleagris gallopavo* NA NT
Ocellated Turkey *Agriocharis ocellata* ● NT

Tetraoninae

Siberian Spruce Grouse *Dendragapus falcipennis* EP
Spruce Grouse *Dendragapus canadensis* NA
Blue Grouse *Dendragapus obscurus* NA

Willow/Red Grouse *Lagopus lagopus* WP EP NA
Ptarmigan *Lagopus mutus* WP EP NA
White-tailed Ptarmigan *Lagopus leucurus*

Caucasian Black Grouse *Tetrao mlokosiewiczi* WP
Black Grouse *Tetrao tetrix* WP EP
Black-billed Capercaillie *Tetrao parvirostris* EP
Capercaillie *Tetrao urogallus* WP EP
Severtzov's Hazel Grouse *Bonasa sewerzowi* EP
Hazel Grouse *Bonasa bonasia* WP EP
Ruffed Grouse *Bonasa umbellus* NA

Sage Grouse *Centrocercus urophasianus* NA

Sharp-tailed Grouse *Tympanuchus phasianellus* NA
Prairie Chicken *Tympanuchus cupido* NA
Lesser Prairie Chicken *Tympanuchus pallidicinctus* NA

Odontophorinae

Bearded Tree Quail *Dendrortyx barbatus* ● NT
Long-tailed Tree Quail *Dendrortyx macroura* NT
Buffy-crowned Tree Quail *Dendrortyx leucophrys* NT

Mountain Quail *Oreortyx picta* NA

Scaled Quail *Callipepla squamata* NA NT

California Quail *Lophortyx californica* NA
Gambel's Quail *Lophortyx gambelii* NA NT
Elegant Quail *Lophortyx douglasii* NT

Banded Quail *Philortyx fasciatus* NT

Northern Bobwhite *Colinus virginianus* NA NT
Black-throated Quail *Colinus nigrogularis* NT
Crested Bobwhite *Colinus cristatus* NT

Marbled Wood Quail *Odontophorus gujanensis* NT
Spot-winged Wood Quail *Odontophorus capueira* NT
Rufous-fronted Wood Quail *Odontophorus erythrops* NT
Black-fronted Wood Quail *Odontophorus atrifrons* NT

Black-backed Wood Quail *Odontophorus melanonotus* NT
Chestnut Wood Quail *Odontophorus hyperythrus* ● NT
Rufous-breasted Wood Quail *Odontophorus speciosus* NT
Gorgeted Wood Quail *Odontophorus strophium* ● NT
Tacarcuna Wood Quail *Odontophorus dialeucos* NT
Venezuela Wood Quail *Odontophorus colombianus* NT
White-throated Wood Quail *Odontophorus leucolaemus* NT
Stripe-faced Wood Quail *Odontophorus balliviani* NT
Starred Wood Quail *Odontophorus stellatus* NT
Spotted Wood Quail *Odontophorus guttatus* NT

Singing Quail *Dactylortyx thoracicus* NT

Montezuma's Quail *Cyrtonyx montezumae* NA NT
Salle's Quail *Cyrtonyx sallei* NT
Ocellated Quail *Cyrtonyx ocellatus* NT

Tawny-faced Quail *Rhynchortyx cinctus* NT

Phasianinae

Snow Partridge *Lerwa lerwa* EP OR

See See Partridge *Ammoperdix griseogularis* EP OR
Sand Partridge *Ammoperdix heyi* WP

Caucasian Snowcock *Tetraogallus caucasicus* WP
Caspian Snowcock *Tetraogallus caspius* WP
Tibetan Snowcock *Tetraogallus tibetanus* EP OR
Altai Snowcock *Tetraogallus altaicus* EP
Himalayan Snowcock *Tetraogallus himalayensis* EP

Verreaux's Monal Partridge *Tetraophasis obscurus* EP OR
Szechenyi's Monal Partridge *Tetraophasis szechenyii* EP OR

Rock Partridge *Alectoris graeca* WP EP OR
Chukar Partridge *Alectoris chukar* WP EP
Przewalski's Rock Partridge *Alectoris magna* EP
Philby's Rock Partridge *Alectoris philbyi* WP
Barbary Partridge *Alectoris barbara* WP
Red-legged Partridge *Alectoris rufa* WP
Arabian Chukar *Alectoris melanocephala* WP

Snow Mountain Quail *Anurophasis monorthonyx* AU

Black Partridge *Francolinus francolinus* WP EP OR
Painted Partridge *Francolinus pictus* OR
Chinese Francolin *Francolinus pintadeanus* OR
Red-necked Spurfowl *Francolinus afer* AF
Swainson's Francolin *Francolinus swainsonii* AF
Painted Francolin *Francolinus rufopictus* AF
Yellow-necked Francolin *Francolinus leucoscepus* AF
Erckel's Francolin *Francolinus erckelii* AF
Pale-bellied Francolin *Francolinus ochropectus* ● AF
Chestnut-naped Francolin *Francolinus castaneicollis* AF
Jackson's Francolin *Francolinus jacksoni* AF
Handsome Francolin *Francolinus nobilis* AF
Cameroun Mountain Francolin *Francolinus camerunensis* ● AF
Swierstra's Francolin *Francolinus swierstrai* AF
Ahanta Francolin *Francolinus ahantensis* ● AF
Scaly Francolin *Francolinus squamatus* AF
Grey-striped Francolin *Francolinus griseostriatus* AF
Double-spurred Francolin *Francolinus bicalcaratus* AF
Yellow-billed Francolin *Francolinus icterorhynchus* AF
Clapperton's Francolin *Francolinus clappertoni* AF
Hildebrandt's Francolin *Francolinus hildebrandti* AF
Natal Francolin *Francolinus natalensis* AF
Hartlaub's Francolin *Francolinus hartlaubi* AF
Harwood's Francolin *Francolinus harwoodi* AF
Red-billed Francolin *Francolinus adspersus* AF
Cape Francolin *Francolinus capensis* AF
Crested Francolin *Francolinus sephaena* AF
Ring-necked Francolin *Francolinus streptophorus* AF
Montane Francolin *Francolinus psilolaemus* AF
Kirk's Francolin *Francolinus rovuma* AF
Shelley's Francolin *Francolinus shelleyi* AF
Greywing Francolin *Francolinus africanus* AF
Archer's Greywing Francolin *Francolinus levalliantoides* AF

Red-winged Francolin *Francolinus levaillantii* AF
Finsch's Francolin *Francolinus finschi* AF
Coqui's Francolin *Francolinus coqui* AF
White-throated Francolin *Francolinus albogularis* AF
Schlegel's Francolin *Francolinus schlegelii* AF
Latham's Francolin *Francolinus lathami* AF
Nahan's Forest Francolin *Francolinus nahani* ● AF
Indian Grey Francolin *Francolinus pondicerianus* OR
Swamp Partridge *Francolinus gularis* ● OR

Grey Partridge *Perdix perdix* WP
Daurian Partridge *Perdix dauuricae* EP
Tibetan Partridge *Perdix hodgsoniae* EP OR

Long-billed Wood Partridge *Rhizothera longirostris* OR

Madagascar Partridge *Margaroperdix madagarensis* AF

Black Wood Partridge *Melanoperdix nigra* OR

Quail *Coturnix coturnix* WP EP AF OR
Japanese Quail *Coturnix japonica* EP OR
Rain Quail *Coturnix coromandelica* OR
Harlequin Quail *Coturnix delegorguei* AF
Pectoral Quail *Coturnix pectoralis* AU
Brown Quail *Coturnix australis* AU
Tasmanian Brown Quail *Coturnix ypsilophora* AU
Indian Blue Quail *Coturnix chinensis* AF OR AU

Jungle Bush Quail *Perdicula asiatica* OR
Rock Bush Quail *Perdicula argoondah* OR
Painted Bush Quail *Perdicula erythrorhyncha* OR
Manipur Bush Quail *Perdicula manipurensis* ● OR

Common Hill Partridge *Arborophila torqueola* EP OR
Rufous-throated Hill Partridge *Arborophila rufogularis* OR
White-cheeked Hill Partridge *Arborophila atrogularis* OR
White-throated Hill Partridge *Arborophila crudigularis* OR
Red-breasted Hill Partridge *Arborophila mandellii* OR
Brown-breasted Hill Partridge *Arborophila brunneopectus* OR
Boulton's Hill Partridge *Arborophila rufipectus* ● EP OR
Rickett's Hill Partridge *Arborophila gingica* ● OR
David's Tree Partridge *Arborophila davidi* ● OR
Chestnut-headed Tree Partridge *Arborophila cambodiana* OR
Sumatran Hill Partridge *Arborophila orientalis* OR
Chestnut-bellied Tree Partridge *Arborophila javanica* OR
Red-billed Tree Partridge *Arborophila rubrirostris* OR
Red-breasted Tree Partridge *Arborophila hyperythra* OR
Hainan Hill Partridge *Arborophila ardens* ● OR

Chestnut-breasted Tree Partridge *Tropicoperdix charltonii* ● OR
Green-legged Hill Partridge *Tropicoperdix chloropus* OR
Annamese Hill Partridge *Tropicoperdix merlini* OR

Ferruginous Wood Partridge *Caloperdix oculea* OR

Crimson-headed Wood Partridge *Haematortyx sanguiniceps* OR

Crested Wood Partridge *Rollulus roulroul* OR

Stone Partridge *Ptilopachus petrosus* AF

Bamboo Partridge *Bambusicola fytchii* OR
Chinese Bamboo Partridge *Bambusicola thoracica* OR

Red Spurfowl *Galloperdix spadicea* OR
Painted Spurfowl *Galloperdix lunulata* OR
Ceylon Spurfowl *Galloperdix bicalcarata* OR

Himalayan Mountain Quail *Ophrysia superciliosa* ● OR

Blood Pheasant *Ithaginis cruentus* EP OR

Western Tragopan *Tragopan melanocephalus* ● EP OR
Satyr Tragopan *Tragopan satyra* EP OR
Blyth's Tragopan *Tragopan blythii* ● OR

Temminck's Tragopan *Tragopan temminckii* EP OR
Cabot's Tragopan *Tragopan caboti* ● OR

Koklass Pheasant *Pucrasia macrolopha* EP OR

Himalayan Monal Pheasant *Lophophorus impeyanus* EP OR
Sclater's Monal Pheasant *Lophophorus sclateri* ● EP OR
Chinese Monal Pheasant *Lophophorus lhuysii* ● EP OR

Red Jungle-fowl *Gallus gallus* OR
Ceylon Jungle-fowl *Gallus lafayettii* OR
Grey Jungle-fowl *Gallus sonneratii* OR
Green Jungle-fowl *Gallus varius* OR

Kalij Pheasant *Lophura leucomelana* EP OR
Silver Pheasant *Lophura nycthemera* OR
Imperial Pheasant *Lophura imperialis* ● OR
Edwards' Pheasant *Lophura edwardsi* ● OR
Vo Quy's Pheasant *Lophura haitiensis* ● OR
Swinhoe's Pheasant *Lophura swinhoii* ● OR
Salvadori's Pheasant *Lophura inornata* ● OR
Crestless Fireback Pheasant *Lophura erythrophthalma* OR
Siamese Fireback Pheasant *Lophura diardi* ● OR
Bulwer's Pheasant *Lophura bulweri* ● OR

White Eared-Pheasant *Crossoptilon crossoptilon* EP OR
Brown Eared-Pheasant *Crossoptilon mantchuricum* ● EP
Blue Eared-Pheasant *Crossoptilon auritum* EP OR

Cheer Pheasant *Catreus wallichii* ● OR

Elliot's Pheasant *Syrmaticus ellioti* ● OR
Mrs Hume's Pheasant *Syrmaticus humiae* ● OR
Mikado Pheasant *Syrmaticus mikado* ● OR
Copper Pheasant *Syrmaticus soemmerringi* EP
Reeves' Pheasant *Syrmaticus reevesii* ● EP OR

Pheasant *Phasianus colchicus* WP EP OR
Green Pheasant *Phasianus versicolor* EP

Golden Pheasant *Chrysolophus pictus* EP OR
Lady Amherst's Pheasant *Chrysolophus amherstiae* EP OR

Sumatran Peacock-Pheasant *Polyplectron chalcurum* OR
Rothschild's Peacock-Pheasant *Polyplectron inopinatum* OR
Germain's Peacock-Pheasant *Polyplectron germaini* ● OR
Burmese Peacock-Pheasant *Polyplectron bicalcaratum* OR
Malay Peacock-Pheasant *Polyplectron malacense* ● OR
Palawan Peacock-Pheasant *Polyplectron emphanum* ● OR

Crested Argus *Rheinartia ocellata* ● OR

Great Argus Pheasant *Argusianus argus* OR

Common Peafowl *Pavo cristatus* OR
Green Peafowl *Pavo muticus* ● OR

Congo Peafowl *Afropavo congensis* ● AF

Numidinae

Black Guineafowl *Phasidus niger* AF

White-breasted Guineafowl *Agelastes meleagrides* ● AF

Helmeted Guineafowl *Numida meleagris* AF

Plumed Guineafowl *Guttera plumifera* AF
Crested Guineafowl *Guttera edouardi* AF

Vulturine Guineafowl *Acryllium vulturinum* AF

CRANES, RAILS AND ALLIES Gruiformes

36 MESITES Mesitornithidae

White-breasted Mesite *Mesitornis variegata* ● AF
Brown Mesite *Mesitornis unicolor* ● AF

Bensch's Monia *Monias benschi* ● AF

37 BUTTON-QUAILS Turnicidae

Little Button-quail *Turnix sylvatica* WP AF OR
Red-backed Button-quail *Turnix maculosa* OR AU
Sumba Button-quail *Turnix everetti* ● OR
Worcester's Button-quail *Turnix worcesteri* ● OR
Hottentot Button-quail *Turnix hottentotta* OR
Yellow-legged Button-quail *Turnix tanki* OR
Bustard Quail *Turnix suscitator* OR
Madagascar Button-quail *Turnix nigricollis* AF
Spotted Button-quail *Turnix ocellata* OR
Black-breasted Button-quail *Turnix melanogaster* ● AU

Painted Button-quail *Turnix varia* AU
Chestnut-backed Button-quail *Turnix castanota* AU
Red-chested Button-quail *Turnix pyrrhothorax* AU
Little Quail *Turnix velox* AU

Quail Plover *Ortyxelos meiffrenii* AF

38 PLAINS WANDERER Pedionomidae

Plains Wanderer *Pedionomus torquatus* ● AU

39 CRANES Gruidae

Gruinae

Crane *Grus grus* WP EP OR
Black-necked Crane *Grus nigricollis* ● EP OR
Hooded Crane *Grus monacha* ● EP
Sandhill Crane *Grus canadensis* NA NT
Manchurian Crane *Grus japonensis* ● EP
Whooping Crane *Grus americana* ● NA
Japanese White-naped Crane *Grus vipio* ● EP
Sarus Crane *Grus antigone* OR
Brolga *Grus rubicunda* AU
Great White Crane *Grus leucogeranus* ● EP OR

Wattled Crane *Bugeranus carunculatus* ● AF

Demoiselle Crane *Anthropoides virgo* WP EP
Blue Crane or Stanley Crane *Anthropoides paradisea* AF

Balearicinae

Crowned Crane *Balearica pavonina* AF
South African Crowned Crane *Balearica regulorum* AF

40 LIMPKIN Aramidae

Limpkin *Aramus guarauna* NA NT

41 TRUMPETERS Psophiidae

Common Trumpeter *Psophia crepitans* NT
White-winged Trumpeter *Psophia leucoptera* NT
Green-winged Trumpeter *Psophia viridis* NT

42 RAILS AND COOTS Rallidae

Nkulengu Rail *Himantornis haematopus* AF

Grey-throated Rail *Canirallus oculeus* AF
Madagascar Grey-throated Rail *Canirallus kioloides* AF

White-throated Rail *Canirallus cuvieri* AF

Chestnut-bellied Rail *Eulabeornis castaneoventris* AU
Bare-eyed Rail *Eulabeornis plumbeiventris* AU
Bald-faced Rail *Eulabeornis rosenbergii* ● OR
Red-winged Wood Rail *Eulabeornis calopterus* NT
Slaty-breasted Wood Rail *Eulabeornis saracura* NT
Giant Wood Rail *Eulabeornis ypecaha* NT
Brown Wood Rail *Eulabeornis wolfi* NT
Little Wood Rail *Eulabeornis mangle* NT
Grey-necked Wood Rail *Eulabeornis cajaneus* NT
Rufous-necked Wood Rail *Eulabeornis axillaris* NT
Uniform Crake *Eulabeornis concolor* NT

Platen's Rail *Rallus plateni* ● OR
Wallace's Rail *Rallus wallacii* ● OR
New Britain Rail *Rallus insignis* AU
New Caledonian Wood Rail *Rallus lafresnayanus* ● AU
Lord Howe Wood Rail *Rallus sylvestris* ● AU
Barred Wing Rail *Rallus poecilopterus* ● AU PA
Plumbeous Rail *Rallus sanguinolentus* NT
Blackish Rail *Rallus nigricans* NT
Spotted Rail *Rallus maculatus* NT
Buff-banded Rail *Rallus philippensis* OR AU PA

Blue-breasted Banded Rail *Rallus striatus* OR
Barred Rail *Rallus torquatus* OR AU
Guam Rail *Rallus owstoni* ● PA
Lewin's Rail *Rallus pectoralis* OR AU
Kaffir Rail *Rallus caerulescens* AF
Madagascar Rail *Rallus madagascariensis* AF
Water Rail *Rallus aquaticus* WP EP OR
Bogota Rail *Rallus semiplumbeous* ● NT
Clapper Rail *Rallus longirostris* NA NT
King Rail *Rallus elegans* NA
Plain-flanked Rail *Rallus wetmorei* ● NT
Virginia Rail *Rallus limicola* NA NT
Austral Rail *Rallus antarcticus* ● NT
Okinawa Rail *Rallus okinawae* ● EP

Inaccessible Island Rail *Atlantisia rogersi* ● AF

Weka Rail *Gallirallus australis* AU

Rouget's Rail *Rougetius rougetii* AF

Zapata Rail *Cyanolimnas cerverai* ● NT

New Guinea Chestnut Rail *Rallina rubra* AU
White-striped Chestnut Rail *Rallina leucospila* AU
Forbes' Chestnut Rail *Rallina forbesi* AU
Mayr's Chestnut Rail *Rallina mayri* AU
Chestnut-headed Crake *Rallina castaneiceps* NT
Red-necked Crake *Rallina tricolor* AU
Andaman Banded Crake *Rallina canningi* OR
Red-legged Crake *Rallina fasciata* OR
Banded Crake *Rallina eurizonoides* OR
Band-bellied Crake *Rallina paykullii* EP OR

Red-chested Crake *Coturnicops rufa* AF
White-spotted Crake *Coturnicops pulchra* AF
African Chestnut-headed Crake *Coturnicops lugens* AF
Streaky-breasted Crake *Coturnicops boehmi* AF
Buff-spotted Crake *Coturnicops elegans* AF
Chestnut-tailed Crake *Coturnicops affinis* AF
Madagascar Crake *Coturnicops insularis* AF
Waters' Crake *Coturnicops watersi* ● AF
White-winged Crake *Coturnicops ayresi* ● AF
Ocellated Crake *Coturnicops schomburgkii* NT
Darwin's Rail *Coturnicops notata* NT
Yellow Rail *Coturnicops noveboracensis* EP NA NT

Black-banded Crake *Laterallus fasciatus* NT
Rusty-flanked Crake *Laterallus levraudi* ● NT
Ruddy Crake *Laterallus ruber* NT
Russet-crowned Crake *Laterallus viridis* NT
Grey-breasted Crake *Laterallus exilis* NT
Galapagos Rail *Laterallus spilonotus* NT
Rufous-sided Crake *Laterallus melanophaius* ● NT
White-throated Crake *Laterallus albigularis* NT
Red & White Crake *Laterallus leucopyrrhus* NT
American Black Crake *Laterallus jamaicensis* NA NT
Rufous-faced Crake *Laterallus xenopterus* ● NT

Corncrake *Crex crex* ● WP EP AF

African Crake *Porzana egregia* AF
African Black Crake *Porzana flavirostra* AF
Olivier's Rail *Porzana olivieri* ● AF
Yellow-breasted Crake *Porzana flaviventer* NT
White-browed Rail *Porzana cinerea* OR AU PA
Dot-winged Crake *Porzana spiloptera* ● NT
White-throated Crake *Porzana albicollis* NT
Striped Crake *Porzana marginalis* AF
Paint-billed Crake *Porzana erythrops* NT
Colombian Crake *Porzana columbiana* NT
Sooty Crake *Porzana tabuensis* AU PA
Henderson Island Crake *Porzana atra* ● PA
Little Crake *Porzana parva* EP
Baillon's Crake *Porzana pusilla* WP EP AF OR AU
Australian Spotted Crake *Porzana fluminea* AU
Spotted Crake *Porzana porzana* WP EP OR
Sora Crake *Porzana carolina* NA NT
Ruddy-breasted Crake *Porzana fusca* EP OR

Rufous-tailed Moorhen *Amaurornis olivaceus* OR AU
Sulawesi Waterhen *Amaurornis isabellinus* OR
New Guinea Flightless Rail *Amaurornis ineptus* AU
Brown Crake *Amaurornis akool* OR
Elwes' Crake *Amaurornis bicolor* OR
White-breasted Waterhen *Amaurornis phoenicurus* OR

Watercock *Gallicrex cinerea* EP OR

Black-tailed Native Hen *Gallinula ventralis* AU
Tasmanian Native Hen *Gallinula mortierii* AU
San Cristobal Mountain Rail *Gallinula sylvestris* ● PA
Samoan Wood Rail *Gallinula pacifica* PA
Gough Is Coot *Gallinula nesiotis* AF
Dusky Moorhen *Gallinula tenebrosa* OR AU
Moorhen *Gallinula chloropus*
Lesser Moorhen *Gallinula angulata* AF
Spot-flanked Gallinule *Gallinula melanops* NT
Azure Gallinule *Gallinula flavirostris* NT
Allen's Gallinule *Gallinula alleni* AF
American Purple Gallinule *Gallinula martinica* NA NT

Purple Swamphen *Porphyrio porphyrio* WP AF OR AU PA
Takahe *Porphyrio mantelli* ● AU

Red-gartered Coot *Fulica armillata* NT
White-winged Coot *Fulica leucoptera* NT
Red-fronted Coot *Fulica rufifrons* NT
Giant Coot *Fulica gigantea* NT
Horned Coot *Fulica cornuta* ● NT
Caribbean Coot *Fulica caribaea* NT
American Coot *Fulica americana* PA NA NT
Coot *Fulica atra* WP OR AU
Red-knobbed Coot *Fulica cristata* WP AF

43 SUNGREBES or FINFOOTS Heliornithidae

Peters' Finfoot *Podica senegalensis* AF

Masked Finfoot *Heliopais personata* ● OR

American Finfoot (Sungrebe) *Heliornis fulica*

44 KAGU Rhynochetidae

Kagu *Rhynochetos jubatus* ● PA

45 SUN-BITTERNS Eurypygidae

Sun-bittern *Eurypyga helias* NT

46 SERIEMAS Cariamidae

Red-legged Seriema *Cariama cristata* NT

Black-legged Seriema *Chunga burmeisteri* NT

47 BUSTARDS Otididae

Little Bustard *Tetrax tetrax* ● WP OR

Denham's Bustard *Neotis denhami* AF
Ludwig's Bustard *Neotis ludwigii* AF
Nubian Bustard *Neotis nuba* AF
Heuglin's Bustard *Neotis heuglinii* AF

Houbara Bustard *Chlamydotis undulata* ● WP AF OR

Arabian Bustard *Ardeotis arabs* AF
Kori Bustard *Ardeotis kori* AF

Great Indian Bustard *Choriotis nigriceps* ● OR
Australian Bustard *Choriotis australis* AU

Great Bustard *Otis tarda* ● WP EP

Crested Bustard *Eupodotis ruficrista* AF
Black Bustard *Eupodotis afra* AF
Vigors' Bustard *Eupodotis vigorsii* AF
Ruppell's Bustard *Eupodotis rueppellii* AF
Little Brown Bustard *Eupodotis humilis* AF
Blue Bustard *Eupodotis caerulescens* AF
White-bellied Bustard *Eupodotis senegalensis* AF
Black-bellied Bustard *Eupodotis melanogaster* AF
Hartlaub's Bustard *Eupodotis hartlaubii* AF

Bengal Florican *Houbaropsis bengalensis* ● OR

Lesser Florican *Sypheotides indica* ● OR

SHOREBIRDS, GULLS, AUKS
Charadriiformes

48 JACANAS Jacanidae

Smaller Jacana *Microparra capensis* AF

African Jacana *Actophilornis africana* AF
Madagascar Jacana *Actophilornis albinucha* AF

Comb-crested Jacana *Irediparra gallinacea* OR AU

Pheasant-tailed Jacana *Hydrophasianus chirurgus* OR

Bronze-winged Jacana *Metopidius indicus* OR

Northern Jacana *Jacana spinosa* NT
Wattled Jacana *Jacana jacana* NT

49 PAINTED SNIPE Rostratulidae

Painted Snipe *Rostratula benghalensis* AF OR AU

South American Painted Snipe *Nycticryphes semicollaris* NT

50 CRAB PLOVER Dromadidae

Crab Plover *Dromas ardeola* OR NA

51 OYSTERCATCHERS Haematopodidae

Oystercatcher *Haematopus ostralegus* WP EP
Canary Is Oystercatcher *Haematopus meadewaldoi* ● WP
American Oystercatcher *Haematopus palliatus* NA NT
Pied Oystercatcher *Haematopus longirostris* AU
American Black Oystercatcher *Haematopus bachmani* NA
African Black Oystercatcher *Haematopus moquini* AF
Variable Oystercatcher *Haematopus unicolor* AU
Chatham Island Oystercatcher *Haematopus chathamensis* ● AU
Magellanic Oystercatcher *Haematopus leucopodus* NT
Blackish Oystercatcher *Haematopus ater* NT
Sooty Oystercatcher *Haematopus fuliginosus* AU

52 IBISBILL Ibidorhynchidae

Ibisbill *Ibidorhyncha struthersii* EP OR

53 AVOCETS AND STILTS Recurvirostridae

Black-winged Stilt *Himantopus himantopus* WP EP AF OR
Australian Stilt *Himantopus leucocephalus* OR AU
Black-tailed Stilt *Himantopus melanurus* NT
Black-necked Stilt *Himantopus mexicanus* NA NT
Srk Lanka Stilt *Himantopus ceylonensis* OR
Hawaiian Stilt *Himantopus knudseni* ● PA
New Zealand Stilt *Himantopus novaezealandiae* ● AU
South African Stilt *Himantopus meridionalis* AF

Banded Stilt *Cladorhynchus leucocephalus* AU

Avocet *Recurvirostra avosetta* WP EP AF OR
American Avocet *Recurvirostra americana* NA NT
Red-necked Avocet *Recurvirostra novaehollandiae* AU
Andean Avocet *Recurvirostra andina* NT

54 STONE-CURLEWS or THICK-KNEES Burhinidae

Stone-curlew *Burhinus oedicnemus* WP EP AF OR
Senegal Stone-curlew *Burhinus senegalensis* AF
Water Dikkop *Burhinus vermiculatus* AF
Cape Dikkop *Burhinus capensis* AF
Double-striped Stone-curlew *Burhinus bistriatus* NT
Peruvian Stone-curlew *Burhinus superciliaris* NT
Australian Stone-curlew *Burhinus magnirostris* AU

Great Stone Plover *Esacus recurvirostris* OR
Great Australian Stone Plover *Esacus magnirostris* OR AU

55 COURSERS AND PRATINCOLES Glareolidae

Cursoriinae

Egyptian Plover *Pluvianus aegyptius* AF

Cream-coloured Courser *Cursorius cursor* WP AF OR
Burchell's Courser *Cursorius rufus* AF
Indian Courser *Cursorius coromandelicus* OR
Temminck's Courser *Cursorius temminckii* AF

Two-banded Courser *Rhinoptilus africanus* AF
Heuglin's Courser *Rhinoptilus cinctus* AF
Bronze-winged Courser *Rhinoptilus chalcopterus* AF
Jerdon's Courser *Rhinoptilus bitorquatus* ● OR

Glareolinae

Australian Pratincole *Stiltia isabella* OR AU

Pratincole *Glareola pratincola* WP AF OR

Eastern Collared Pratincole *Glareola maldivarus* EP OR
Black-winged Pratincole *Glareola nordmanni* WP EP AF
Madagascar Pratincole *Glareola ocularis* AF
White-collared Pratincole *Glareola nuchalis* AF
Cream-coloured Pratincole *Glareola cinerea* AF
Little Pratincole *Glareola lactea* OR

56 PLOVERS Charadriidae

Lapwing *Vanellus vanellus* WP EP
Long-toed Lapwing *Vanellus crassirostris* AF
Spur-winged Plover *Vanellus spinosus* WP AF OR
River Lapwing *Vanellus duvaucelii* OR
Black-headed Plover *Vanellus tectus* AF
Yellow-wattled Lapwing *Vanellus malabaricus* OR
White-crowned Wattled Plover *Vanellus albiceps* AF
Senegal Plover *Vanellus lugubris* AF
Black-winged Plover *Vanellus melanopterus* AF
Crowned Plover *Vanellus coronatus* AF
Senegal Wattled Plover *Vanellus senegallus* AF
Spot-breasted Plover *Vanellus melanocephalus* AF
Brown-chested Wattled Plover *Vanellus superciliosus* AF
Sociable Plover *Vanellus gregarius* ● EP AF OR
White-tailed Plover *Vanellus leucurus* WP EP OR
Cayenne Plover *Vanellus cayanus* NT
Southern Lapwing *Vanellus chilensis* NT
Andean Lapwing *Vanellus resplendens* NT
Grey-headed Lapwing *Vanellus cinereus* EP OR
Red-wattled Lapwing *Vanellus indicus* WP OR
Javanese Wattled Lapwing *Vanellus macropterus* ● OR

Banded Plover *Vanellus tricolor* AU
Masked Plover *Vanellus miles* OR AU

Blacksmith Plover *Anitibyx armatus* AF

Golden Plover *Pluvialis apricaria* WP EP OR
American Golden Plover *Pluvialis dominica* NA NT
Pacific Golden Plover *Pluvialis fulva* EP OR AU NA
Grey Plover *Pluvialis squatarola* WP EP AF OR NA NT
New Zealand Dotterel *Pluvialis obscura* AU

Ringed Plover *Charadrius hiaticula* WP EP AF NA
Semi palmated Plover *Charadrius semipalmatus* NA NT
Long-billed Ringed Plover *Charadrius placidus* EP OR
Little Ringed Plover *Charadrius dubius* WP EP AF OR AU
Wilson's Plover *Charadrius wilsonia* NA NT
Killdeer Plover *Charadrius vociferus* NA NT
Piping Plover *Charadrius melodus* ● NA NT
Black-banded Sand Plover *Charadrius thoracicus* ● AF
Kittlitz's Sand Plover *Charadrius pecuarius* AF
St Helena Sand Plover *Charadrius sanctaehelenae* ● AF
Three-banded Plover *Charadrius tricollaris* AF
Kentish Plover *Charadrius alexandrinus* WP EP AF OR AU
White-fronted Sand Plover *Charadrius marginatus* AF
Snowy Plover *Charadrius occidentalis* NA NT
Red-capped Dotterel *Charadrius ruficapillus* AU
Malaysian Plover *Charadrius peronii* OR
Chestnut-banded Sand Plover *Charadrius venustus* AF
Collared Plover *Charadrius collaris* NT
Double-banded Plover *Charadrius bicinctus* AU
Two-banded Plover *Charadrius falklandicus* NT
Puna Plover *Charadrius alticola* NT
Lesser Sand Plover *Charadrius mongolus* EP AF OR AU
Great Sand Plover *Charadrius leschenaultii* EP AF OR AU
Caspian Plover *Charadrius asiaticus* EP AF OR
Eastern Sand Plover *Charadrius veredus* EP OR
Rufous-chested Dotterel *Charadrius modestus* NT
Mountain Plover *Charadrius montanus* NA NT
Black-fronted Plover *Charadrius melanops* AU
Red-kneed Dotterel *Charadrius cinctus* AU
Hooded Plover *Charadrius rubricollis* ● AU
Long-billed Plover *Charadrius novaeseelandiae* ● AU

Wrybill *Anarhynchus frontalis* AU
Mitchell's Plover *Phegornis mitchellii* NT
Australian Courser *Peltohyas australis* AU
Dotterel *Eudromias morinellus* WP EP
Tawny-throated Dotterel *Eudromias ruficollis* NT
Magellanic Plover *Pluvianellus socialis* NT

57 SANDPIPERS AND SNIPE Scolopacidae

Tringinae

Black-tailed Godwit *Limosa limosa* WP EP AF OR AU
Hudsonian Godwit *Limosa haemastica* NA NT
Bar-tailed Godwit *Limosa lapponica* WP EP AF OR AU PA NA
Marbled Godwit *Limosa fedoa* NA NT

Little Curlew *Numenius minutus* EP OR AU
Eskimo Curlew *Numenius borealis* ● NA NT
Whimbrel *Numenius phaeopus* WP EP AF OR AU PA NA NT
Bristle-thighed Curlew *Numenius tahitiensis* ● PA NA
Slender-billed Curlew *Numenius tenuirostris* ● WP EP
Curlew *Numenius arquata* WP EP AF OR
Far Eastern Curlew *Numenius madagascariensis* EP OR AU
Long-billed Curlew *Numenius americanus* NA NT

Upland Sandpiper *Bartramia longicauda* NA NT

Spotted Redshank *Tringa erythropus* WP EP AF OR
Redshank *Tringa totanus* WP EP AF OR
Marsh Sandpiper *Tringa stagnatilis* WP EP AF OR AU
Greenshank *Tringa nebularia* WP EP AF OR AU
Spotted Greenshank *Tringa guttifer* EP OR
Greater Yellowlegs *Tringa melanoleuca* NA NT
Lesser Yellowlegs *Tringa flavipes* NA NT
Green Sandpiper *Tringa ochropus* WP EP AF OR
Solitary Sandpiper *Tringa solitaria* NA NT
Wood Sandpiper *Tringa glareola* WP EP AF OR AU

Willet *Catoptrophorus semipalmatus* NA NT

Terek Sandpiper *Xenus cinereus* WP EP AF OR AU

Common Sandpiper *Actitis hypoleucos* WP EP AF OR AU
Spotted Sandpiper *Actitis macularia* NA NT

Grey-tailed Tattler *Heteroscelus brevipes* EP OR AU
Wandering Tattler *Heteroscelus incanus* PA NA

Tuamotu Sandpiper *Prosobonia cancellata* ● PA

Arenariinae

Turnstone *Arenaria interpres* WP EP AF OR AU NA NT
Black Turnstone *Arenaria melanocephala* NA

Phalaropodinae

Wilson's Phalarope *Phalaropus tricolor* NA NT
Red-necked Phalarope *Phalaropus lobatus* WP EP OR AU PA NA NT
Grey Phalarope *Phalaropus fulicarius* WP EP AF NA NT

Scolopacinae

Woodcock *Scolopax rusticola* WP EP OR
Amami Woodcock *Scolopax mira* EP
Rufous Woodcock *Scolopax saturata* OR AU
Sulawesi Woodcock *Scolopax celebensis* ● OR
Obi Woodcock *Scolopax rochussenii* ● EP
American Woodcock *Scolopax minor* NA

Gallinagoninae

Sub-Antarctic Snipe *Coenocorypha aucklandica* ● AU

Solitary Snipe *Gallinago solitaria* EP OR
Japanese Snipe *Gallinago hardwickii* EP AU

Wood Snipe *Gallinago nemoricola* ● EP OR
Pintail Snipe *Gallinago stenura* EP OR
Swinhoe's Snipe *Gallinago megala* EP OR AU
African Snipe *Gallinago nigripennis* AF
Madagascar Snipe *Gallinago macrodactyla* AF
Great Snipe *Gallinago media* WP EP AF
Snipe *Gallinago gallinago* WP EP AF OR NA NT
Magellan Snipe *Gallinago paraguaiae* NT
Noble Snipe *Gallinago nobilis* NT
Giant Snipe *Gallinago undulata* NT
Cordilleran Snipe *Gallinago stricklandii* NT
Andean Snipe *Gallinago jamesoni* NT
Imperial Snipe *Gallinago imperialis* NT

Jack Snipe *Lymnocryptes minima* WP EP AF OR

Short-billed Dowitcher *Limnodromus griseus* NA NT
Long-billed Dowitcher *Limnodromus scolopaceus* NA NT
Asiatic Dowitcher *Limnodromus semipalmatus* ● EP OR

Calidridinae

Surfbird *Aphriza virgata* NA NT

Knot *Calidris canutus* WP EP AF AU NA NT
Great Knot *Calidris tenuirostris* EP OR AU
Sanderling *Calidris alba* WP EP OR AU NA NT
Semipalmated Sandpiper *Calidris pusilla* NA NT
Western Sandpiper *Calidris mauri* NA NT
Rufous-necked Stint *Calidris ruficollis* EP OR AU NA
Little Stint *Calidris minuta* WP AF OR
Temminck's Stint *Calidris temminckii* WP EP AF OR
Long-toed Stint *Calidris subminuta* EP OR
Least Sandpiper *Calidris minutilla* NA NT
White-rumped Sandpiper *Calidris fuscicollis* NA NT
Baird's Sandpiper *Calidris bairdii* EP NA NT
Pectoral Sandpiper *Calidris melanotos* EP NA NT
Sharp-tailed Sandpiper *Calidris acuminata* EP AU PA
Cox's Sandpiper *Calidris paramelanotus* ● AU
Purple Sandpiper *Calidris maritima* WP NA
Rock Sandpiper *Calidris ptilocnemis* NA
Dunlin *Calidris alpina* WP EP AF OR
Curlew Sandpiper *Calidris ferruginea* WP EP AF OR AU

Spoon-billed Sandpiper *Eurynorhynchus pygmeus* ● EP OR

Broad-billed Sandpiper *Limicola falcinellus* WP EP OR AU

Stilt Sandpiper *Micropalama himantopus* NA NT

Buff-breasted Sandpiper *Tryngites subruficollis* NA NT

Ruff *Philomachus pugnax* WP EP AF OR

58 SEEDSNIPE Thinocoridae

Rufous-bellied Seedsnipe *Attagis gayi* NT
White-bellied Seedsnipe *Attagis malouinus* NT

Grey-breasted Seedsnipe *Thinocorus orbignyianus* NT
Least Seedsnipe *Thinocorus rumicivorus* NT

59 SHEATHBILLS Chionididae

Snowy Sheathbill *Chionis alba* NT AN
Black-faced Sheathbill *Chionis minor* AN

60 SKUAS or JAEGERS Stercorariidae

Great Skua *Catharacta skua* WP AF NA
Chilean Skua *Catharacta chilensis* NT
South Polar Skua *Catharacta maccormicki* PA NT AN
Antarctic Skua *Catharacta antarctica* AF NT AN
Pomarine Skua *Stercorarius pomarinus* WP EP AF OR AU NA NT
Arctic Skua *Stercorarius parasiticus* WP EP AF OR AU NA NT
Long-tailed Skua *Stercorarius longicaudus* WP EP OR NA NT

61 GULLS AND TERNS Laridae

Larinae

Pacific Gull *Gabianus pacificus* AU
Magellan Gull *Gabianus scoresbii* NT

Ivory Gull *Pagophila eburnea* WP EP NA

Dusky Gull *Larus fuliginosus* NT
Grey Gull *Larus modestus* NT
Heermann's Gull *Larus heermanni* NA NT
White-eyed Gull *Larus leucophthalmus* ● AF
Sooty Gull *Larus hemprichii* AF OR
Band-tailed Gull *Larus belcheri* NT
Japanese Gull *Larus crassirostris* EP
Audouin's Gull *Larus audouinii* WP
Ring-billed Gull *Larus delawarensis* NA NT
Common Gull *Larus canus* WP NA
Eastern Mew Gull *Larus kamtschatschensis* EP
Herring Gull *Larus argentatus* WP EP AF NA NT
Armenian Gull *Larus armenicus* WP
Thayer's Gull *Larus thayeri* NA
Lesser Black-backed Gull *Larus fuscus* WP AF
Californian Gull *Larus californicus* NA NT
Western Gull *Larus occidentalis* NA
Yellow-footed Gull *Larus livens* NA
Southern Black-backed Gull *Larus dominicanus* AF AU NT
Slaty-backed Gull *Larus schistisagus* EP NA
Greater Black-backed Gull *Larus marinus* WP AF
Glaucous-winged Gull *Larus glaucescens* EP OR NA
Glaucous Gull *Larus hyperboreus* WP EP NA
Iceland Gull *Larus glaucoides* WP EP NA
Great Black-headed Gull *Larus ichthyaetus* EP OR
Laughing Gull *Larus atricilla* NA NT
Indian Black-headed Gull *Larus brunnicephalus* EP OR
Grey-headed Gull *Larus cirrocephalus* AF NT
Andean Gull *Larus serranus* NT
Franklin's Gull *Larus pipixcan* NA NT
Silver Gull *Larus novaehollandiae* AF AU
Mediterranean Gull *Larus melanocephalus* WP EP
Relict Gull *Larus relictus* ● EP
Buller's Gull *Larus bulleri* AU
Brown-hooded Gull *Larus maculipennis* NT
Black-headed Gull *Larus ridibundus* WP EP OR
Slender-billed Gull *Larus genei* WP
Bonaparte's Gull *Larus philadelphia* NA
Little Gull *Larus minutus* WP EP
Saunders' Gull *Larus saundersi* ● EP

Ross's Gull *Rhodostethia rosea* EP NA

Kittiwake *Rissa tridactyla* WP EP AF NA
Red-legged Kittiwake *Rissa brevirostris* NA

Swallow-tailed Gull *Creagrus furcatus* NT

Sabine's Gull *Xema sabini* WP EP AF NA

Sterninae

Whiskered Tern *Chlidonias hybrida* WP AF OR AU
White-winged Black Tern *Chlidonias leucoptera* WP EP AF OR AU
Black Tern *Chlidonias niger* WP EP AF NA NT

Large-billed Tern *Phaetusa simplex* NT

Gull-billed Tern *Gelochelidon nilotica* WP EP AF OR AU NA NT

Caspian Tern *Hydroprogne caspia* WP EP AF OR AU NA NT

Indian River Tern *Sterna aurantia* OR
South American Tern *Sterna hirundinacea* NT
Common Tern *Sterna hirundo* WP EP AF OR AU PA NA NT
Arctic Tern *Sterna paradisaea* WP EP AF AU NA NT AN
Antarctic Tern *Sterna vittata* AF AU NT AN
Kerguelen Tern *Sterna virgata* ● AU AN
Forster's Tern *Sterna forsteri* NA
Trudeau's Tern *Sterna trudeaui* NT
Roseate Tern *Sterna dougallii* WP EP AF OR AU PA NA NT AN
White-fronted Tern *Sterna striata* AU
White-cheeked Tern *Sterna repressa* EP AF OR

Black-naped Tern *Sterna sumatrana* EP OR AU PA
Black-bellied Tern *Sterna melanogaster* OR
Aleutian Tern *Sterna aleutica* EP NA
Spectacled Tern *Sterna lunata* PA
Bridled Tern *Sterna anaethetus* WP EP AF OR AU PA NA NT
Sooty Tern *Sterna fuscata* AF OR AU PA NA NT
Fairy Tern *Sterna nereis* AU
Black-fronted Tern *Sterna albistriata* ● AU
Amazon Tern *Sterna superciliaris* NT
Damara Tern *Sterna balaenarum* ● AF
Chilean Tern *Sterna lorata* NT
Little Tern *Sterna albifrons* WP EP AF OR AU
Least Tern *Sterna antillarum* NA NT
Black-shafted Tern *Sterna saundersii* EP OR

Greater Crested Tern *Thalasseus bergii* WP AF OR AU PA
Royal Tern *Thalasseus maximus* AF NA NT
Lesser Crested Tern *Thalasseus bengalensis* WP AF OR AU PA
Chinese Crested Tern *Thalasseus bernsteini* ● EP OR
Elegant Tern *Thalasseus elegans* NA NT
Sandwich Tern *Thalasseus sandvicensis* WP EP AF OR NA NT

Inca Tern *Larosterna inca* NT

Blue-grey Noddy *Procelsterna cerulea* AU PA NA

Common Noddy *Anous stolidus* AF OR AU PA NA NT
Lesser Noddy *Anous tenuirostris* AF OR AU PA NA NT
White-capped Noddy *Anous minutus* AF AU PA NT

White Tern *Gygis alba* AF OR AU PA

62 SKIMMERS Rynchopidae

Black Skimmer *Rynchops niger* NA NT
African Skimmer *Rynchops flavirostris* WP AF
Indian Skimmer *Rynchops albicollis* OR

63 AUKS Alcidae

Little Auk *Alle alle* WP NA

Razorbill *Alca torda* WP NA

Brunnich's Guillemot *Uria lomvia* WP NA
Guillemot *Uria aalge*

Black Guillemot *Cepphus grylle* WP NA
Pigeon Guillemot *Cepphus columba* EP NA
Spectacled Guillemot *Cepphus carbo* EP

Marbled Murrelet *Brachyramphus marmoratus* EP NA
Kittlitz's Murrelet *Brachyramphus brevirostris* EP NA
Xantus' Murrelet *Brachyramphus hypoleucus* NA NT
Craveri's Murrelet *Brachyramphus craveri* NA

Ancient Murrelet *Synthliboramphus antiquus* EP NA
Crested Murrelet *Synthliboramphus wumizusume* ● EP

Cassin's Auklet *Ptychoramphus aleuticus* NA NT

Parakeet Auklet *Cyclorrhynchus psittacula* EP NA

Crested Auklet *Aethia cristatella* EP NA
Least Auklet *Aethia pusilla* EP NA
Whiskered Auklet *Aethia pygmaea* EP NA

Rhinoceros Auklet *Cerorhinca monocerata* EP NA

Puffin *Fratercula arctica* WP NA
Horned Puffin *Fratercula corniculata* EP NA

Tufted Puffin *Lunda cirrhata* EP NA

PIGEONS AND SANDGROUSE Columbiformes

64 SANDGROUSE Pteroclididae

Tibetan Sandgrouse *Syrrhaptes tibetanus* EP OR
Pallas' Sandgrouse *Syrrhaptes paradoxus* EP OR

Pin-tailed Sandgrouse *Pterocles alchata* WP EP OR
Namaqua Sandgrouse *Pterocles namaqua* AF
Chestnut-bellied Sandgrouse *Pterocles exustus* WP EP AF OR
Spotted Sandgrouse *Pterocles senegallus* WP EP AF OR

Black-bellied Sandgrouse *Pterocles orientalis* EP OR

Crowned Sandgrouse *Pterocles coronatus* WP EP AF

Yellow-throated Sandgrouse *Pterocles gutturalis* AF
Variegated Sandgrouse *Pterocles burchelli* AF
Madagascar Sandgrouse *Pterocles personatus* AF
Black-faced Sandgrouse *Pterocles decoratus* AF
Lichtenstein's Sandgrouse *Pterocles lichtensteinii* WP EP AF
Double-banded Sandgrouse *Pterocles bicinctus* AF
Painted Sandgrouse *Pterocles indicus* OR
Four-banded Sandgrouse *Pterocles quadricinctus* AF

65 DOVES AND PIGEONS Columbidae

Rock Dove *Columba livia* WP EP AF OR AU NA
Eastern Rock Pigeon *Columba rupestris* EP OR
Snow Pigeon *Columba leuconota* EP OR
Speckled Pigeon *Columba guinea* AF
White-collared Pigeon *Columba albitorques* AF
Stock Dove *Columba oenas* WP EP OR
Yellow eyed Pigeon *Columba eversmanni* EP OR
Somali Pigeon *Columba oliviae* ● AF
Woodpigeon *Columba palumbus* WP EP OR
Trocaz Pigeon *Columba trocaz* ● WP
Bolle's Pigeon *Columba bollii* ● WP
African Wood Pigeon *Columba unicincta* ● AF
Laurel Pigeon *Columba junoniae* ● WP
Cameroun Olive Pigeon *Columba sjostedi* AF
Olive Pigeon *Columba arquatrix* AF
Sao Thome Olive Pigeon *Columba thomensis* ● AF
Comoro Olive Pigeon *Columba pollenii* AF
Speckled Wood Pigeon *Columba hodgsonii* EP OR
White-naped Wood Pigeon *Columba albinucha* AF
Ashy Wood Pigeon *Columba pulchricollis* EP OR
Nilgiri Wood Pigeon *Columba elphinstonii* ● OR
Sri Lanka Wood Pigeon *Columba torringtoni* ● OR
Pale-capped Pigeon *Columba punicea* ● OR
Silver Pigeon *Columba argentina* ● OR
Andaman Wood Pigeon *Columba palumboides* OR
Black Wood Pigeon *Columba janthina* EP OE
White-throated Pigeon *Columba vitiensis* OR AU PA
White-headed Pigeon *Columba leucomela* AU
Silver-banded Black Pigeon *Columba jouyi* EP
Yellow-legged Pigeon *Columba pallidiceps* ● PA
White-crowned Pigeon *Columba leucocephala* NA NT
Red-necked Pigeon *Columba squamosa* NA NT
Scaled Pigeon *Columba speciosa* NT
Picazuro Pigeon *Columba picazuro* NT
Bare-eyed Pigeon *Columba corensis* NT
Spotted Pigeon *Columba maculosa* NT
Band-tailed Pigeon *Columba fasciata* NA NT
Chilean Pigeon *Columba araucana* NT
Jamaican Band-tailed Pigeon *Columba caribaea* ● NT
Rufous Pigeon *Columba cayennensis* NT
Red-billed Pigeon *Columba flavirostris* NA NT
Salvin's Pigeon *Columba oenops* NT
Plain Pigeon *Columba inornata* NT
Plumbeous Pigeon *Columba plumbea* NT
Ruddy Pigeon *Columba subvinacea* NT
Short-billed Pigeon *Columba nigrirostris* NT
Dusky Pigeon *Columba goodsoni* NT
Delegorgue's Pigeon *Columba delegorguei* AF
Bronze-naped Pigeon *Columba iriditorques* AF
Sao Thome Bronze-naped Pigeon *Columba malherbii* AF
Pink Pigeon *Columba mayeri* ● AF

Turtle Dove *Streptopelia turtur* WP EP AF OR
Dusky Turtle Dove *Streptopelia lugens* AF
Pink-bellied Turtle dove *Streptopelia hypopyrrha* AF
Eastern Turtle Dove *Streptopelia orientalis* EP OR
Javanese Collared Dove *Streptopelia bitorquata* OR PA
Collared Dove *Streptopelia decaocto* WP EP OR
African Collared Dove *Streptopelia roseogrisea* AF
White-winged Collared Dove *Streptopelia reichenowi* AF
Mourning Collared Dove *Streptopelia decipiens* AF
Red-eyed Dove *Streptopelia semitorquata* AF
Ring-necked Dove *Streptopelia capicola* AF
Vinaceous Dove *Streptopelia vinacea* AF
Red-collared Dove *Streptopelia tranquebarica* EP OR

Madagascar Turtle Dove *Streptopelia picturata* AF
Spotted-necked Dove *Streptopelia chinensis* EP OR
Laughing Dove *Streptopelia senegalensis* WP EP AF OR

Lemon Dove *Aplopelia larvata* AF

Bar-tailed Cuckoo Dove *Macropygia unchall* OR
Pink-breasted Cuckoo Dove *Macropygia amboinensis* AU
Large Brown Cuckoo Dove *Macropygia phasianella* OR
Enggano Cuckoo Dove *Macropygia cinnamomea* AU
Indonesian Cuckoo Dove *Macropygia emiliana* OR AU
Large Cuckoo Dove *Macropygia magna* OR AU
Andaman Cuckoo Dove *Macropygia rufipennis* OR
Lesser Bar-tailed Cuckoo Dove *Macropygia nigrirostris* AU
Mackinlay's Cuckoo Dove *Macropygia mackinlayi* AU PA
Little Cuckoo Dove *Macropygia ruficeps* OR

Reinwardt's Long-tailed Pigeon *Reinwardtoena reinwardtsi* AU
Brown's Long-tailed Pigeon *Reinwardtoena browni* AU
Crested Long-tailed Pigeon *Reinwardtoena crassirostris* PA

White-faced Pigeon *Turacoena manadensis* OR AU
Timor Black Pigeon *Turacoena modesta* AU

Emerald-spotted Wood Dove *Turtur chalcospilos* AF
Black-billed Wood Dove *Turtur abyssinicus* AF
Blue-spotted Wood Dove *Turtur afer* AF
Tambourine Dove *Turtur tympanistria* AF
Blue-headed Wood Dove *Turtur brehmeri* AF

Namaqua Dove *Oena capensis* AF

Emerald Dove *Chalcophaps indica* OR AU PA
Brown-backed Emerald Dove *Chalcophaps stephani* AU PA

Black Bronzewing *Henicophaps albifrons* AU
New Britain Bronzewing *Henicophaps foersteri* AU

Common Bronzewing *Phaps chalcoptera* AU
Brush Bronzewing *Phaps elegans* AU
Flock Pigeon *Phaps histrionica* AU

Crested Pigeon *Ocyphaps lophotes* AU

White-bellied Plumed Spinifex Pigeon *Petrophassa plumifera* AU
Red-plumed Pigeon *Petrophassa ferruginea* AU
Partridge Bronzewing *Petrophassa scripta* AU
Bare-eyed Partridge Bronzewing *Petrophassa smithii* AU
Chestnut-quilled Rock Pigeon *Petrophassa rufipennis* AU
White-quilled Rock Pigeon *Petrophassa albipennis* AU

Diamond Dove *Geopelia cuneata* AU
Zebra Dove *Geopelia striata* OR AU PA
Timor Zebra Dove *Geopelia maugei* AU
Gould's Zebra Dove *Geopelia placida* AU
Bar-shouldered Dove *Geopelia humeralis* AU

Wonga Pigeon *Leucosarcia melanoleuca* AU

Mourning Dove *Zenaida macroura* NA NT
Eared Dove *Zenaida auriculata* NA NT
Zenaida Dove *Zenaida aurita* NT
Galapagos Dove *Zenaida galapagoensis* NT
White-winged Dove *Zenaida asiatica* NA NT

Common Ground Dove *Columbina passerina* NA NT
Plain-breasted Ground Dove *Columbina minuta* NT
Ecuadorean Ground Dove *Columbina buckleyi* NT
Ruddy Ground Dove *Columbina talpacoti* NT
Picui Dove *Columbina picui* NT
Gold-billed Ground Dove *Columbina cruziana* NT
Blue-eyed Ground Dove *Columbina cyanopis* ● NT

Blue Ground Dove *Claravis pretiosa* NT
Purple-barred Ground Dove *Claravis godefrida* ● NT
Purple-breasted Ground Dove *Claravis mondetoura* NT

Bare-faced Ground Dove *Metriopelia ceciliae* NT

Bare-eyed Ground Dove *Metriopelia morenoi* NT
Black-winged Ground Dove *Metriopelia melanoptera* NT
Bronze-winged Ground Dove *Metriopelia aymara* NT

Inca Dove *Scardafella inca* NA NT
Scaly Dove *Scardafella squammata* NT

Long-tailed Ground Dove *Uropelia campestris* NT

White-fronted Dove *Leptotila verreauxi* NA NT
White-faced Dove *Leptotila megalura* NT
Grey-fronted Dove *Leptotila rufaxilla* NT
Grey-headed Dove *Leptotila plumbeiceps* NT
Pallid Dove *Leptotila pallida* NT
Grenada Dove *Leptotila wellsi* ● NT
White-bellied Dove *Leptotila jamaicensis* NT
Cassin's Dove *Leptotila cassini* NT
Buff-bellied Dove *Leptotila ochraceiventris* ● NT
Tolima Dove *Leptotila conoveri* ● NT

Lawrence's Quail Dove *Geotrygon lawrencii* NT
Costa Rican Quail Dove *Geotrygon costaricensis* NT
Russet-crowned Quail Dove *Geotrygon goldmani* NT
Sapphire Quail Dove *Geotrygon saphirina* NT
Grey-faced Quail Dove *Geotrygon caniceps* ● NT
Crested Quail Dove *Geotrygon versicolor* NT
Olive-backed Quail Dove *Geotrygon veraguensis* NT
Lined Quail Dove *Geotrygon linearis* NT
White-throated Quail Dove *Geotrygon frenata* NT
Key West Quail Dove *Geotrygon chrysia* NA NT
Bridled Quail Dove *Geotrygon mystacea* NT
Violaceous Quail Dove *Geotrygon violacea* NT
Ruddy Quail Dove *Geotrygon montana* NA NT

Blue-headed Quail Dove *Starnoenas cyanocephala* ● OR

Nicobar Pigeon *Caloenas nicobarica* ● OR AU PA

Luzon Bleeding Heart *Gallicolumba luzonica* OR
Bartlett's Bleeding Heart *Gallicolumba criniger* OR
Mindoro Bleeding Heart *Gallicolumba platenae* ● OR
Negros Bleeding Heart *Gallicolumba keayi* ● OR
Tawitawi Bleeding Heart *Gallicolumba menagei* ● OR
Golden Heart *Gallicolumba rufigula* AU
Sulawesi Quail Dove *Gallicolumba tristigmata* OR
White-breasted Ground Pigeon *Gallicolumba jobiensis* AU PA
Truk Is Ground Dove *Gallicolumba kubaryi* PA
Society Is Ground Dove *Gallicolumba erythroptera* PA
White-throated Dove *Gallicolumba xanthonura* PA
Friendly Quail Dove *Gallicolumba stairi* PA
Santa Cruz Ground Dove *Gallicolumba sanctaecrucis* ● PA
Thick-billed Ground Dove *Gallicolumba salamonis* ● PA
Marquesas Ground Dove *Gallicolumba rubescens* ● PA
Grey-breasted Quail Dove *Gallicolumba beccarii* AU PA
Palau Ground Dove *Gallicolumba canifrons* PA
Wetar Ground Dove *Gallicolumba hoedtii* ● AU

Thick-billed Ground Pigeon *Trugon terrestris* AU

Solomon Is Ground Pigeon *Microgoura meeki* ● AU

Pheasant Pigeon *Otidiphaps nobilis* AU

Blue Crowned Pigeon *Goura cristata* ● AU
Maroon-breasted Crowned Pigeon *Goura scheepmakeri* ● AU
Victoria Crowned Pigeon *Goura victoria* ● AU

Tooth-billed Pigeon *Didunculus strigirostris* ● PA

Lesser Brown Fruit Dove *Phapitreron leucotis* OR
Greater Brown Fruit Dove *Phapitreron amethystina* OR

Cinnamon-headed Green Pigeon *Treron fulvicollis* OR
Little Green Pigeon *Treron olax* OR
Pink-necked Green Pigeon *Treron vernans* OR
Orange-breasted Green Pigeon *Treron bicincta* OR
Pompadour Green Pigeon *Treron pompadora* OR
Thick-billed Green Pigeon *Treron curvirostra* OR
Grey-faced Thick-billed Green Pigeon *Treron griseicauda* OR AU

Sumba Is Green Pigeon *Treron teysmanni* ● AU
Flores Green Pigeon *Treron floris* AU
Timor Green Pigeon *Treron psittacea* ● AU
Large Green Pigeon *Treron capellei* ● OR
Yellow-legged Green Pigeon *Treron phoenicoptera* OR
Yellow-bellied Green Pigeon *Treron waalia* AF
Madagascar Green Pigeon *Treron australis* AF
African Green Pigeon *Treron calva* AF
Pemba Is Green Pigeon *Treron pembaensis* AF
Sao Thome Green Pigeon *Treron sanctithomae* AF
Pin-tailed Green Pigeon *Treron apicauda* OR
Yellow-bellied Pin-tailed Green Pigeon *Treron oxyura* OR
Yellow-vented Pin-tailed Green Pigeon *Treron seimundi* OR
Wedge-tailed Green Pigeon *Treron sphenura* OR
Japanese Green Pigeon *Treron sieboldii* EP OR
Formosan Green Pigeon *Treron formosae* EP OR

Black-backed Fruit Dove *Ptilinopus cincta* OR AU
Black-banded Pigeon *Ptilinopus alligator* AU
Red-naped Fruit Dove *Ptilinopus dohertyi* ● AU
Pink-necked Fruit Dove *Ptilinopus porphyrea* PA
Marche's Fruit Dove *Ptilinopus marchei* OR
Merrill's Fruit Dove *Ptilinopus merrilli* OR
Yellow-breasted Fruit Dove *Ptilinopus occipitalis* OR
Fischer's Fruit Dove *Ptilinopus fischeri* OR
Jambu Fruit Dove *Ptilinopus jambu* OR
Dark-chinned Fruit Dove *Ptilinopus subgularis* OR AU
Black-chinned Fruit Dove *Ptilinopus leclancheri* OR
Scarlet-breasted Fruit Dove *Ptilinopus bernsteinii* AU
Magnificent Fruit Dove *Ptilinopus magnificus* AU
Pink-spotted Fruit Dove *Ptilinopus perlatus* AU
Ornate Fruit Dove *Ptilinopus ornatus* AU
Silver-shouldered Fruit Dove *Ptilinopus tannensis* PA
Orange-fronted Fruit Dove *Ptilinopus aurantiifrons* AU
Wallace's Fruit Dove *Ptilinopus wallacii* AO
Superb Fruit Dove *Ptilinopus superbus* AU
Many-coloured Fruit Dove *Ptilinopus perousii* PA
Purple-capped Fruit Dove *Ptilinopus porphyraceus* PA
Palau Fruit Dove *Ptilinopus pelewensis* OR
Rarotongan Fruit Dove *Ptilinopus rarotongensis* PA
Marianas Fruit Dove *Ptilinopus roseicapilla* ● PA
Pink-capped Fruit Dove *Ptilinopus regina* AU
Silver-capped Fruit Dove *Ptilinopus richardsii* PA
Grey-green Fruit Dove *Ptilinopus purpuratus* PA
Grey's Fruit Dove *Ptilinopus greyii* PA
Rapa I Fruit Dove *Ptilinopus huttoni* ● OR
White-capped Fruit Dove *Ptilinopus dupetithouarsii* OR
Red-moustached Fruit Dove *Ptilinopus mercierii* ● AU
Henderson I Fruit Dove *Ptilinopus insularis* AF
Lilac-capped Fruit Dove *Ptilinopus coronulatus* AU
Crimson-capped Fruit Dove *Ptilinopus pulchellus* AU
Blue-capped Fruit Dove *Ptilinopus monacha* AU
White-bibbed Fruit Dove *Ptilinopus rivoli* AU
Yellow-bibbed Fruit Dove *Ptilinopus solomonensis* AU PA
Red-bibbed Fruit Dove *Ptilinopus viridis* AU
White-headed Fruit Dove *Ptilinopus eugeniae* AU
Orange-bellied Fruit Dove *Ptilinopus iozonus* AU
Knob-billed Fruit Dove *Ptilinopus insolitus* AU
Grey-headed Fruit Dove *Ptilinopus hyogaster* OR
Carunculated Fruit Dove *Ptilinopus granulifrons* AU
Black-naped Fruit Dove *Ptilinopus melanospila* OR
Dwarf Fruit Dove *Ptilinopus naina* AU
Ripley's Fruit Dove *Ptilinopus arcanus* OR
Orange Dove *Ptilinopus victor* PA
Golden Dove *Ptilinopus luteovirens* PA
Yellow-headed Dove *Ptilinopus layardi* PA

Cloven-feathered Dove *Drepanoptila holosericea* ● PA

Madagascar Blue Pigeon *Alectroenas madagascariensis* AF
Comoro Blue Pigeon *Alectroenas sganzini* AF
Seychelles Blue Pigeon *Alectroenas pulcherrima* AF

Philippine Zone-tailed Pigeon *Ducula poliocephala* OR
Green & White Zone-tailed Pigeon *Ducula forsteni* OR

Mindoro Zone-tailed Pigeon *Ducula mindorensis* ● OR
Grey-headed Zone-tailed Pigeon *Ducula radiata* OR
Grey-necked Fruit Pigeon *Ducula carola* OR
Green Imperial Pigeon *Ducula aenea* OR
White-eyed Imperial Pigeon *Ducula perspicillata* OR
Blue-tailed Imperial Pigeon *Ducula concinna* AU
Pacific Pigeon *Ducula pacifica* AU
Micronesian Pigeon *Ducula oceanica* PA
Society Is Pigeon *Ducula aurorae* ● PA
Marquesas Pigeon *Ducula galeata* ● PA
Red-knobbed Pigeon *Ducula rubricera* PA
Black-knobbed Pigeon *Ducula myristicivora* AU
Rufous-bellied Fruit Pigeon *Ducula rufigaster* AU
Moluccan Rufous-bellied Fruit Pigeon *Ducula basilica* AU
Finsch's Rufous-bellied Fruit Pigeon *Ducula finschii* AU
Mountain Rufous-bellied Fruit Pigeon *Ducula chalconota* AU
Island Imperial Pigeon *Ducula pistrinaria* AU PA
Pink-headed Imperial Pigeon *Ducula rosacea* AU
Christmas I Imperial Pigeon *Ducula whartoni* ● AF
Grey Imperial Pigeon *Ducula pickeringii* ● OR
Peale's Pigeon *Ducula latrans* PA
Chestnut-bellied Pigeon *Ducula brenchleyi* PA
Baker's Pigeon *Ducula bakeri* PA
New Caledonian Pigeon *Ducula goliath* ● PA
Pinon Imperial Pigeon *Ducula pinon* AU
Black Imperial Pigeon *Ducula melanochroa* AU
Black-collared Fruit Pigeon *Ducula mullerii* AU
Banded Imperial Pigeon *Ducula zoeae* AU
Mountain Imperial Pigeon *Ducula badia* OR
Dark-backed Imperial Pigeon *Ducula lacernulata* OR
Timor Imperial Pigeon *Ducula cineracea* AU
Pied Imperial Pigeon *Ducula bicolor* OR AU
Sulawesi Pied Imperial Pigeon *Ducula luctuosa* AU
Australian Pied Imperial Pigeon *Ducula spilorrhoa* AU

Top-knot Pigeon *Lopholaimus antarcticus* AU

New Zealand Pigeon *Hemiphaga novaeseelandiae* AU

Sulawesi Dusky Pigeon *Cryptophaps poecilorrhoa* AU

Bare-eyed Mountain Pigeon *Gymnophaps albertisii* AU
Long-tailed Mountain Pigeon *Gymnophaps mada* AU
Pale Mountain Pigeon *Gymnophaps solomonensis* PA OR

PARROTS AND RELATIVES
Psittaciformes

66 LORIES Loridae

Black Lory *Chalcopsitta atra* AU
Duyvenbode's Lory *Chalcopsitta duivenbodei* AU
Yellow-streaked Lory *Chalcopsitta sintillata* AU
Cardinal Lory *Chalcopsitta cardinalis* PA

Black-winged Lory *Eos cyanogenia* ● AU
Violet-necked Lory *Eos squamata* AU
Blue-streaked Lory *Eos reticulata* ● AU
Red & Blue Lory *Eos histrio* ● AU
Red Lory *Eos bornea* AU
Blue-eared Lory *Eos semilarvata* AU

Dusky Lory *Pseudeos fuscata* AU

Ornate Lory *Trichoglossus ornatus* OR
Rainbow Lory *Trichoglossus haematodus* OR AU PA
Ponape Lory *Trichoglossus rubiginosus* PA
Johnstone's Lorikeet *Trichoglossus johnstoniae* OR
Yellow & Green Lorikeet *Trichoglossus flavoviridis* OR
Scaly-breasted Lorikeet *Trichoglossus chlorolepidotus* AU
Perfect Lorikeet *Trichoglossus euteles* AU
Varied Lorikeet *Trichoglossus versicolor* AU
Iris Lorikeet *Trichoglossus iris* AU
Goldie's Lorikeet *Trichoglossus goldiei* AU

Purple-bellied Lory *Lorius hypoinochrous* AU
Black-capped Lory *Lorius lory* OR AU
White-naped Lory *Lorius albidinuchus* AU
Stresemann's Lory *Lorius amabilis* AU
Yellow-bibbed Lory *Lorius chlorocercus* AU
Purple-naped Lory *Lorius domicellus* ● AU

Blue-thighed Lory *Lorius tibialis* AU
Chattering Lory *Lorius garrulus* OR AU

Collared Lory *Phigys solitarius* PA

Blue-crowned Lory *Vini australis* PA
Kuhl's Lory *Vini kuhlii* ● PA
Stephen's Lory *Vini stepheni* ● PA
Tahitian Lory *Vini peruviana* ● PA
Ultramarine Lory *Vini ultramarina* ● PA

Musk Lorikeet *Glossopsitta concinna* AU
Little Lorikeet *Glossopsitta pusilla* AU
Purple-crowned Lorikeet *Glossopsitta porphyrocephala* AU

Palm Lorikeet *Charmosyna palmarum* PA
Red-chinned Lorikeet *Charmosyna rubrigularis* AU
Meek's Lorikeet *Charmosyna meeki* PA
Blue-fronted Lorikeet *Charmosyna toxopei* ● AU
Striated Lorikeet *Charmosyna multistriata* AU
Wilhelmina's Lorikeet *Charmosyna wilhelminae* AU
Red-spotted Lorikeet *Charmosyna rubronotata* AU
Red-flanked Lorikeet *Charmosyna placentis* AU PA
New Caledonian Lorikeet *Charmosyna diadema* ● PA
Red-throated Lorikeet *Charmosyna amabilis* PA
Duchess Lorikeet *Charmosyna margarethae* PA
Fairy Lorikeet *Charmosyna pulchella* AU
Josephine's Lory *Charmosyna josefinae* AU
Papuan Lory *Charmosyna papou* AU

Whiskered Lorikeet *Oreopsittacus arfaki* AU

Musschenbroek's Lorikeet *Neopsittacus musschenbroekii* AU
Emerald Lorikeet *Neopsittacus pullicauda* AU

67 COCKATOOS Cacatuidae

Cacatuinae

Palm Cockatoo *Probosciger aterrimus* AU

Black Cockatoo *Calyptorhynchus funereus* AU
Red-tailed Cockatoo *Calyptorhynchus magnificus* AU
Glossy Cockatoo *Calyptorhynchus lathami* AU

Gang-gang Cockatoo *Callocephalon fimbriatum* AU

Galah *Eolophus roseicapillus* AU

Major Mitchell's Cockatoo *Cacatua leadbeateri* AU
Lesser Sulphur-crested Cockatoo *Cacatua sulphurea* ● OR
Sulphur-crested Cockatoo *Cacatua galerita* AU
Blue-eyed Cockatoo *Cacatua ophthalmica* AU
Salmon-crested Cockatoo *Cacatua moluccensis* ● AU
White Cockatoo *Cacatua alba* ● OR AU
Red-vented Cockatoo *Cacatua haematuropygia* ● OR
Goffin's Cockatoo *Cacatua goffini* ● AU
Little Corella *Cacatua sanguinea* AU
Long-billed Corella *Cacatua tenuirostris* AU
Eastern Long-billed Corella *Cacatua pastinator* AU
Ducorp's Cockatoo *Cacatua ducorps* PA

Nymphicinae

Cockatiel *Nymphicus hollandicus* AU

68 PARROTS Psittacidae

Nestorinae

Kea *Nestor notabilis* AU
Kaka *Nestor meridionalis* AU

Micropsittinae

Buff-faced Pygmy Parrot *Micropsitta pusio* AU
Yellow-capped Pygmy Parrot *Micropsitta keiensis* AU
Geelvink Pygmy Parrot *Micropsitta geelvinkiana* AU
Meek's Pygmy Parrot *Micropsitta meeki* AU
Finsch's Pygmy Parrot *Micropsitta finschii* AU
Red-breasted Pygmy Parrot *Micropsitta bruijnii* PA

Psittacinae

Orange-breasted Fig Parrot *Opopsitta gulielmitertii* AU

Double-eyed Fig Parrot *Opopsitta diophthalma* AU

Desmarest's Fig Parrot *Psittaculirostris desmarestii* AU

Edwards' Fig Parrot *Psittaculirostris edwardsii* AU

Salvadori's Fig Parrot *Psittaculirostris salvadorii* E AU

Guaiabero *Bolbopsittacus lunulatus* OR

Blue-rumped Parrot *Psittinus cyanurus* OR

Brehm's Parrot *Psittacella brehmii* AU
Painted Parrot *Psittacella picta* AU
Modest Parrot *Psittacella modesta* AU
Maderasz's Parrot *Psittacella maderaszi* AU

Red-cheeked Parrot *Geoffroyus geoffroyi* AU
Blue-collared Parrot *Geoffroyus simplex* AU
Singing Parrot *Geoffroyus heteroclitus* AU

Green Racket-tailed Parrot *Prioniturus luconensis* ● OR

Blue-crowned Racket-tailed Parrot *Prioniturus discurus* OR

Palawan Racket-tailed Parrot *Prioniturus platenae* OR

Sulu Racket-tailed Parrot *Prioniturus verticalis* OR
Mindanao Racket-tailed Parrot *Prioniturus waterstradti* OR
Mountain Racket-tailed Parrot *Prioniturus montanus* OR

Red-spotted Racket-tailed Parrot *Prioniturus flavicans* OR
Golden-mantled Racket-tailed Parrot *Prioniturus platurus* OR
Buru Racket-tailed Parrot *Prioniturus mada* ● OR

Great-billed Parrot *Tanygnathus megalorhynchos* OR AU

Blue-naped Parrot *Tanygnathus lucionensis* OR
Muller's Parrot *Tanygnathus sumatranus* OR
Rufous-tailed Parrot *Tanygnathus heterurus* AU
Black-lored Parrot *Tanygnathus gramineus* AU

Eclectus Parrot *Eclectus roratus* AU PA

Pesquet's Parrot *Psittrichas fulgidus* AU

Red Shining Parrot *Prosopeia tabuensis* PA
Kandavu Shining Parrot *Prosopeia splendens* PA
Masked Shining Parrot *Prosopeia personata* OR PA

Australian King Parrot *Alisterus scapularis* AU
Green-winged King Parrot *Alisterus chloropterus* AU
Amboina King Parrot *Alisterus amboinensis* AU

Red-winged Parrot *Aprosmictus erythropterus* AU
Timor Red-winged Parrot *Aprosmictus jonquillaceus* AU

Superb Parrot *Polytelis swainsonii* AU
Regent Parrot *Polytelis anthopeplus* AU
Princess Parrot *Polytelis alexandrae* ● AU

Red-capped Parrot *Purpureicephalus spurius* AU

Mallee Ringneck Parrot *Barnardius barnardi* AU
Port Lincoln Parrot *Barnardius zonarius* AU

Green Rosella *Platycercus caledonicus* AU
Crimson Rosella *Platycercus elegans* AU
Yellow Rosella *Platycercus flaveolus* AU
Adelaide Rosella *Platycercus adelaidae* AU
Eastern Rosella *Platycercus eximius* AU
Pale-headed Rosella *Platycercus adscitus* AU
Northern Rosella *Platycercus venustus* AU
Western Rosella *Platycercus icterotis* AU

Red-rumped Parrot *Psephotus haematonotus* AU
Mulga Parrot *Psephotus varius* AU
Blue Bonnet *Psephotus haematogaster* AU
Golden-shouldered Parrot ● AU *Psephotus chrysopterygius*
Paradise Parrot *Psephotus pulcherrimus* ● AU

Antipodes Green Parakeet *Cyanoramphus unicolor* ●

Red-fronted Parakeet *Cyanoramphus novaezelandiae* AU PA
Yellow-fronted Parakeet *Cyanoramphus auriceps* AU
Horned Parakeet *Cyanoramphus cornutus* PA

Bourke's Parrot *Neophema bourkii* AU

Blue-winged Parrot *Neophema chrysostoma* AU
Elegant Parrot *Neophema elegans* AU
Rock Parrot *Neophema petrophila* AU
Orange-bellied Parrot *Neophema chrysogaster* ● AU
Turquoise Parrot *Neophema pulchella* ● AU
Scarlet-chested Parrot *Neophema splendida* ● AU

Swift Parrot *Lathamus discolor* AU

Budgerigar *Melopsittacus undulatus* AU

Ground Parrot *Pezoporus wallicus* ● AU

Night Parrot *Geopsittacus occidentalis* ● AU

Vasa Parrot *Coracopsis vasa* AF
Black Parrot *Coracopsis nigra* AF

Grey Parrot *Psittacus erithacus* AF

Brown-necked Parrot *Poicephalus robustus* AF
Red-fronted Parrot *Poicephalus gulielmi* AF
Brown-headed Parrot *Poicephalus cryptoxanthus* AF
Niam-Niam Parrot *Poicephalus crassus* AF
Senegal Parrot *Poicephalus senegalus* AF
Red-bellied Parrot *Poicephalus rufiventris* AF
Brown Parrot *Poicephalus meyeri* AF
Ruppell's Parrot *Poicephalus rueppellii* AF
Yellow-faced Parrot *Poicephalus flavifrons* AF

Grey-headed Lovebird *Agapornis cana* AF
Red-faced Lovebird *Agapornis pullaria* AF
Black-winged Lovebird *Agapornis taranta* AF
Black-collared Lovebird *Agapornis swinderniana* AF
Peach-faced Lovebird *Agapornis roseicollis* AF
Fischer's Lovebird *Agapornis fischeri* AF
Masked Lovebird *Agapornis personata* AF
Nyasa Lovebird *Agapornis lilianae* AF
Black-cheeked Lovebird *Agapornis nigrigenis* ● AF

Vernal Hanging Parrot *Loriculus vernalis* OR
Ceylon Hanging Parrot *Loriculus beryllinus* OR
Philippine Hanging Parrot *Loriculus philippensis* OR
Blue-crowned Hanging Parrot *Loriculus galgulus* OR
Sulawesi Hanging Parrot *Loriculus stigmatus* AU
Sangihe Hanging Parrot *Loriculus catamene* ● AU
Moluccan Hanging Parrot *Loriculus amabilis* AU
Green Hanging Parrot *Loriculus exilis* OR
Wallace's Hanging Parrot *Loriculus flosculus* OR
Yellow-throated Hanging Parrot *Loriculus pusillus* OR
Orange-fronted Hanging Parrot *Loriculus aurantiifrons* AU

Alexandrine Parakeet *Psittacula eupatria* OR
Rose-ringed Parakeet *Psittacula krameri* AF OR
Mauritius Parakeet *Psittacula echo* ● AF
Slaty-headed Parakeet *Psittacula himalayana* EP OR
Plum-headed Parakeet *Psittacula cyanocephala* OR
Blossom-headed Parakeet *Psittacula roseata* OR
Intermediate Parrot *Psittacula intermedia* ● OR
Malabar Parakeet *Psittacula columboides* OR
Emerald-collared Parakeet *Psittacula calthorpae* OR
Lord Derby's Parakeet *Psittacula derbiana* EP OR
Moustached Parakeet *Psittacula alexandri* OR
Blyth's Parakeet *Psittacula caniceps* OR
Long-tailed Parakeet *Psittacula longicauda* OR

Hyacinth Macaw *Anodorhynchus hyancinthus* ● NT
Glaucous Macaw *Anodorhynchus glaucus* ● NT
Indigo Macaw *Anodorhynchus leari* ● NT

Spix's Macaw *Cyanopsitta spixii* ● NT

Blue & Yellow Macaw *Ara ararauna* NT
Wagler's Macaw *Ara caninde* ● NT
Military Macaw *Ara militaris* NT
Buffon's Macaw *Ara ambigua* NT
Scarlet Macaw *Ara macao* NT
Green-winged Macaw *Ara chloroptera* NT
Red-fronted Macaw *Ara rubrogenys* ● NT
Yellow-collared Macaw *Ara auricollis* NT
Chestnut-fronted Macaw *Ara severa* NT
Red-bellied Macaw *Ara manilata* NT
Illiger's Macaw *Ara maracana* NT
Blue-headed Macaw *Ara couloni* NT
Red-shouldered Macaw *Ara nobilis* NT

Blue-crowned Conure *Aratinga acuticauda* NT
Golden Conure *Aratinga guarouba* ● NT
Green Conure *Aratinga holochlora* NT
Pacific Parakeet *Aratinga strenua* NT

Finsch's Conure *Aratinga finschi* NT
Red-fronted Conure *Aratinga wagleri* NT
Mitred Conure *Aratinga mitrata* NT
Red-masked Conure *Aratinga erythrogenys* NT
White-eyed Conure *Aratinga leucophthalmus* NT
Hispaniolan Conure *Aratinga chloroptera* NA
Cuban Conure *Aratinga euops* NT
Golden-capped Conure *Aratinga auricapilla* ● NT
Jandaya Conure *Aratinga jandaya* NT
Sun Conure *Aratinga solstitialis* NT
Dusky-headed Conure *Aratinga weddellii* NT
Olive-throated Conure *Aratinga nana* NT
Orange-fronted Conure *Aratinga canicularis* NT
Brown-throated Conure *Aratinga pertinax* NT
Cactus Conure *Aratinga cactorum* NT
Peach-fronted Conure *Aratinga aurea* NT

Nanday Conure *Nandayus nenday* NT

Golden-plumed Conure *Leptosittaca branickii* ● NT

Yellow-eared Conure *Ognorhynchus icterotis* ● NT

Thick-billed Parrot *Rhynchopsitta pachyrhyncha* ● NT

Maroon-fronted Parrot *Rhynchopsitta terrisi* ● NT

Patagonian Conure *Cyanoliseus patagonus* NT

Blue-throated Conure *Pyrrhura cruentata* ● NT
Blaze-winged Conure *Pyrrhura devillei* NT
Maroon-bellied Conure *Pyrrhura frontalis* NT
Pearly Conure *Pyrrhura perlata* ● NT
Crimson-bellied Conure *Pyrrhura rhodogaster* NT
Green-cheeked Conure *Pyrrhura molinae* NT
White-eared Conure *Pyrrhura leucotis* NT
Painted Conure *Pyrrhura picta* NT
Santa Marta Conure *Pyrrhura viridicata* NT
Fiery-shouldered Conure *Pyrrhura egregia* NT
Maroon-tailed Conure *Pyrrhura melanura* NT
El Oro Parakeet *Pyrrhura orcesi* NT
Black-capped Conure *Pyrrhura rupicola* NT
White-necked Conure *Pyrrhura albipectus* ● NT
Brown-breasted Conure *Pyrrhura calliptera* ● NT
Red-eared Conure *Pyrrhura hoematotis* NT
Rose-crowned Conure *Pyrrhura rhodocephala* NT
Hoffmann's Conure *Pyrrhura hoffmanni* NT

Austral Conure *Enicognathus ferrugineus* NT
Slender-billed Conure *Enicognathus leptorhynchus* NT

Monk Parakeet *Myiopsitta monachus* NT

Sierra Parakeet *Bolborhynchus aymara* NT
Mountain Parakeet *Bolborhynchus aurifrons* NT
Barred Parakeet *Bolborhynchus lineola* NT
Andean Parakeet *Bolborhynchus orbygnesius* NT
Rufous-fronted Parakeet *Bolborhynchus ferrugineifrons* NT

Mexican Parrotlet *Forpus cyanopygius* NT
Green-rumped Parrotlet *Forpus passerinus* NT
Blue-winged Parrotlet *Forpus xanthopterygius* NT
Spectacled Parrotlet *Forpus conspicillatus* NT
Dusky-billed Parrotlet *Forpus sclateri* NT
Pacific Parrotlet *Forpus coelestis* NT
Yellow-faced Parrotlet *Forpus xanthops* NT

Plain Parakeet *Brotogeris tirica* NT
Canary-winged Parakeet *Brotogeris versicolurus* NT
Grey-cheeked Parakeet *Brotogeris pyrrhopterus* ● NT
Orange-chinned Parakeet *Brotogeris jugularis* NT
Cobalt-winged Parakeet *Brotogeris cyanoptera* NT
Golden-winged Parakeet *Brotogeris chrysopterus* NT
Tui Parakeet *Brotogeris sanctithomae* NT

Tepui Parrotlet *Nannopsittaca panychlora* NT

Lilac-tailed Parrotlet *Touit batavica* NT
Scarlet-shouldered Parrotlet *Touit huetii* NT
Red-winged Parrotlet *Touit dilectissima* NT
Sapphire-rumped Parrotlet *Touit purpurata* NT
Black-eared Parrotlet *Touit melanonota* ● NT
Golden-tailed Parrotlet *Touit surda* ● NT
Spot-winged Parrotlet *Touit stictoptera* ● NT

Black-headed Caique *Pionites melanocephala* NT
White-bellied Caique *Pionites leucogaster* NT

Red-capped Parrot *Pionopsitta pileata* NT
Brown-hooded Parrot *Pionopsitta haematotis* NT

Rose-faced Parrot *Pionopsitta pulchra* NT
Orange-cheeked Parrot *Pionopsitta barrabandi* NT
Saffron-headed Parrot *Pionopsitta pyrilia* NT
Caica Parrot *Pionopsitta caica* NT

Vulturine Parrot *Gypopsitta vulturina* NT

Black-winged Parrot *Hapalopsittaca melanotis* NT
Rusty-faced Parrot *Hapalopsittaca amazonina* ● NT

Short-tailed Parrot *Graydidascalus brachyurus* NT

Blue-headed Parrot *Pionus menstruus* NT
Red-billed Parrot *Pionus sordidus* NT
Scaly-headed Parrot *Pionus maximiliani* NT
Plum-crowned Parrot *Pionus tumultuosus* NT
White-headed Parrot *Pionus seniloides* NT
White-capped Parrot *Pionus senilis* NT
Bronze-winged Parrot *Pionus chalcopterus* NT
Dusky Parrot *Pionus fuscus* NT

Yellow-billed Amazon *Amazona collaria* NT
Cuban Amazon *Amazona leucocephala* NT
Hispaniolan Amazon *Amazona ventralis* NT
White-fronted Amazon *Amazona albifrons* NT
Yellow-lored Amazon *Amazona xantholora* NT
Black-billed Amazon *Amazona agilis*
Puerto Rican Amazon *Amazona vittata* ● NT
Tucuman Amazon *Amazona tucumana* NT
Red-spectacled Amazon *Amazona pretrei* ● NT
Green-cheeked Amazon *Amazona viridigenalis* ● NT
Lilac-crowned Amazon *Amazona finschi* NT
Red-lored Amazon *Amazona autumnalis* NT
Red-tailed Amazon *Amazona brasiliensis* ● NT
Blue-cheeked Amazon *Amazona dufresniana* NT
Festive Amazon *Amazona festiva* NT
Yellow-faced Amazon *Amazona xanthops* ● NT
Yellow-shouldered Amazon *Amazona barbadensis* ● NT
Blue-fronted Amazon *Amazona aestiva* NT
Yellow-naped Amazon *Amazona auropalliata* NT
Yellow-crowned Amazon *Amazona ochrocephala* NT
Yellow-headed Amazon *Amazona oratrix* NT
Orange-winged Amazon *Amazona amazonica* NT
Scaly-naped Amazon *Amazona mercenaria* NT
Mealy Amazon *Amazona farinosa* NT
Vinaceous Amazon *Amazona vinacea* ● NT
St Lucia Amazon *Amazona versicolor* ● NT
Red-necked Amazon *Amazona arausiaca* ● NT
St Vincent Amazon *Amazona guildingii* ● NT
Imperial Amazon *Amazona imperialis* ● NT

Hawk-headed (Red-fan) Parrot *Deroptyus accipitrinus* NT

Purple-bellied Parrot *Triclaria malachitacea* ● NT

Strigopinae

Kakapo *Strigops habroptilus* ● AU AU

CUCKOOS AND ALLIES Cuculiformes

69 TURACOS Musophagidae

Great Blue Turaco *Corythaeola cristata* AF

Grey Plantain-eater *Crinifer piscator* AF
Eastern Grey Plantain-eater *Crinifer zonurus* AF

Go-away Bird *Corythaixoides concolor* AF
Bare-faced Go-away Bird *Corythaixoides personata* AF

White-bellied Go-away Bird *Criniferoides leucogaster* AF

Violet-crested Turaco *Musophaga porphyreolopha* AF
Ruwenzori Turaco *Musophaga johnstoni* AF
Violet Turaco *Musophaga violacea* AF
Lady Ross's Turaco *Musophaga rossae* AF

Green Turaco *Tauraco persa* AF
Black-billed Turaco *Tauraco schuetti* AF
Fischer's Turaco *Tauraco fischeri* AF
Red-crested Turaco *Tauraco erythrolophus* AF
Bannerman's Turaco *Tauraco bannermani* ● AF
Crested Turaco *Tauraco macrorhynchus* AF
Hartlaub's Turaco *Tauraco hartlaubi* AF
White-cheeked Turaco *Tauraco leucotis* AF
Prince Ruspoli's Turaco *Tauraco ruspolii* ● AF
White-crested Turaco *Tauraco leucolophus* AF

70 CUCKOOS Cuculidae

Cuculinae

Great Spotted Cuckoo *Clamator glandarius* WP AF
Red-winged Crested Cuckoo *Clamator coromandus* OR

Black & White Cuckoo *Oxylophus jacobinus* WP AF OR
Levaillant's Cuckoo *Oxylophus levaillantii* AF

Thick-billed Cuckoo *Pachycoccyx audeberti* AF

Sulawesi Hawk Cuckoo *Cuculus crassirostris* AU
Large Hawk Cuckoo *Cuculus sparverioides* EP OR
Common Hawk Cuckoo *Cuculus varius* OR
Small Hawk Cuckoo *Cuculus vagans* OR
Fugitive Hawk Cuckoo *Cuculus fugax* EP OR
Red-chested Cuckoo *Cuculus solitarius* AF
Black Cuckoo *Cuculus clamosus* AF
Short-winged Cuckoo *Cuculus micropterus* EP OR
Cuckoo *Cuculus canorus* WP EP AF OR AU
African Cuckoo *Cuculus gularis* AF
Oriental Cuckoo *Cuculus saturatus* EP OR AU
Small Cuckoo *Cuculus poliocephalus* EP OR
Madagascar Cuckoo *Cuculus rochii* AF
Pallid Cuckoo *Cuculus pallidus* AU

Dusky Long-tailed Cuckoo *Cercococcyx mechowi* AF
Olive Long-tailed Cuckoo *Cercococcyx olivinus* AF
Mountain Long-tailed Cuckoo *Cercococcyx montanus* AF

Banded Bay Cuckoo *Penthoceryx sonneratii* OR

Plaintive Cuckoo *Cacomantis merulinus* OR
Brush Cuckoo *Cacomantis variolosus* AU PA
Indonesian Cuckoo *Cacomantis sepulcralis* OR AU
Chestnut-breasted Cuckoo *Cacomantis castaneiventris* AU
Heinrich's Brush Cuckoo *Cacomantis heinrichi* AU
Fan-tailed Cuckoo *Cacomantis pyrrhophanus* AU PA

Little Long-billed Cuckoo *Rhamphomantis megarhynchus* AU

Black-eared Cuckoo *Misocalius osculans* AU

African Emerald Cuckoo *Chrysococcyx cupreus* AF
Yellow-throated Green Cuckoo *Chrysococcyx flavigularis* AF
Klaas' Cuckoo *Chrysococcyx klaas* AF
Didric Cuckoo *Chrysococcyx caprius* AF

Asian Emerald Cuckoo *Chalcites maculatus* EP OR
Violet Cuckoo *Chalcites xanthorhynchus* OR
Horsfield's Bronze Cuckoo *Chalcites basalis* AU
Golden-Bronze Cuckoo *Chalcites lucidus* AU PA
New Guinea Bronze Cuckoo *Chalcites poecilurus* AU
Gould's Bronze Cuckoo *Chalcites russatus* OR AU
Green-cheeked Bronze Cuckoo *Chalcites rufomerus* ● AU
Little Bronze Cuckoo *Chalcites minutillus* AU
Pied Bronze Cuckoo *Chalcites crassirostris* AU
Reddish-throated Bronze Cuckoo *Chalcites ruficollis* AU
Meyer's Bronze Cuckoo *Chalcites meyeri* AU

White-crowned Koel *Caliechthrus leucolophus* AU

Drongo-Cuckoo *Surniculus lugubris* OR AU

Black-capped Cuckoo *Microdynamis parva* AU

Koel *Eudynamys scolopacea* OR AU PA
Australian Koel *Eudynamys cyanocephela* AU

Long-tailed Koel *Urodynamis taitensis* AU PA

Channel-billed Cuckoo *Scythrops novaehollandiae* AU

Phaenicophaeinae

Dwarf Cuckoo *Coccyzus pumilus* NT
Ash-coloured Cuckoo *Coccyzus cinereus* NT
Black-billed Cuckoo *Coccyzus erythrophthalmus* NA NT
Yellow-billed Cuckoo *Coccyzus americanus* NA NT
Pearly-breasted Cuckoo *Coccyzus euleri* NT
Mangrove Cuckoo *Coccyzus minor* NT
Cocos Cuckoo *Coccyzus ferrugineus* ● NT
Dark-billed Cuckoo *Coccyzus melacoryphus* NT

Grey-capped Cuckoo *Coccyzus lansbergi* NT

Rufous-breasted Cuckoo *Hyetornis rufigularis* NT
Chestnut-bellied Cuckoo *Hyetornis pluvialis* NT

Squirrel Cuckoo *Piaya cayana* NT
Black-bellied Cuckoo *Piaya melanogaster* NT
Little Cuckoo *Piaya minuta* NT

Great Lizard Cuckoo *Saurothera merlini* NT
Jamaican Lizard Cuckoo *Saurothera vetula* NT
Hispaniolan Lizard Cuckoo *Saurothera longirostris* NT
Puerto Rican Lizard Cuckoo *Saurothera vieilloti* NT

Yellow-bill *Ceuthmochares aereus* AF

Lesser Green-billed Malcoha *Rhopodytes diardi* OR
Rufous-bellied Malcoha *Rhopodytes sumatranus* OR
Greater Green-billed Malcoha *Rhopodytes tristis* EP OR
Small Green-billed Malcoha *Rhopodytes viridirostris* OR

Sirkeer Cuckoo *Taccocua leschenaultii* OR

Raffles' Malcoha *Rhinortha chlorophaea* OR

Red-billed Malcoha *Zanclostomus javanicus* OR

Fiery-billed Malcoha *Rhamphococcyx calyorhynchus* AU

Chestnut-breasted Malcoha *Rhamphococcyx curvirostris* OR

Red-faced Malcoha *Phaenicophaeus pyrrhocephalus* ● OR

Rough-crested Cuckoo *Dasylophus superciliosus* OR

Scale-feathered Cuckoo *Lepidogrammus cumingi* OR

Crotophaginae

Greater Ani *Crotophaga major* NT
Smooth-billed Ani *Crotophaga ani* NA NT
Groove-billed Ani *Crotophaga sulcirostris* NA NT

Guira Cuckoo *Guira guira* NT

Neomorphinae

Striped Cuckoo *Tapera naevia* NT

Lesser Ground Cuckoo *Morococcyx erythropygus* NT

Pheasant Cuckoo *Dromococcyx phasianellus* NT
Pavonine Cuckoo *Dromococcyx pavoninus* NT

Road-runner *Geococcyx californianus* NA NT
Lesser Road-runner *Geococcyx velox* NT

Rufous-vented Ground Cuckoo *Neomorphus geoffroyi* NT
Scaled Ground Cuckoo *Neomorphus squamiger* NT
Banded Ground Cuckoo *Neomorphus radiolosus* ● NT
Rufous-winged Ground Cuckoo *Neomorphus rufipennis* NT
Red-billed Ground Cuckoo *Neomorphus pucheranii* NT

Ground Cuckoo *Carpococcyx radiceus* ● OR
Coral-billed Ground Cuckoo *Carpococcyx renauldi* OR

Couinae

Giant Madagascar Coucal *Coua gigas* AF
Coquerel's Madagascar Coucal *Coua coquereli* AF
Roufous-breasted Madagascar Coucal *Coua serriana* AF
Red-footed Madagascar Coucal *Coua reynaudii* AF
Running Coucal *Coua cursor* AF
Red-capped Madagascar Coucal *Coua ruficeps* AF
Crested Madagascar Coucal *Coua cristata* AF
Southern Crested Madagascar Coucal *Coua verreauxi* AF
Blue Madagascar Coucal *Coua caerulea* AF

Centropodinae

Buff-headed Coucal *Centropus milo* PA
Large Coucal *Centropus goliath* AU
Violet Coucal *Centropus violaceus* AU

Greater Coucal *Centropus menbecki* AU
New Britain Coucal *Centropus ateralbus* AU
Biak Island Coucal *Centropus chalybeus* AU
Pheasant Coucal *Centropus phasianinus* AU
Kai Coucal *Centropus spilopterus* AU
Bernstein's Coucal *Centropus bernsteini* AU
Ceylon Coucal *Centropus chlororhynchus* ● OR
Short-toed Coucal *Centropus rectunguis* ● OR
Steere's Coucal *Centropus steerii* ● OR
Common Crow-Pheasant *Centropus sinensis* OR
Sunda Coucal *Centropus nigrorufus* ● OR
Philippine Coucal *Centropus viridis* OR
Black Coucal *Centropus toulou* AF OR
Lesser Coucal *Centropus bengalensis* OR AU
Black-chested Coucal *Centropus grillii* AF
Rufous-bellied Coucal *Centropus epomidis* AF
Black-throated Coucal *Centropus leucogaster* AF
Smaller Black-throated Coucal *Centropus neumanni* AF
Gabon Coucal *Centropus anselli* AF
Blue-headed Coucal *Centropus monachus* AF
Coppery-tailed Coucal *Centropus cupreicaudus* AF
Senegal Coucal *Centropus senegalensis* WP AF
White-browed Coucal *Centropus superciliosus* AF
Black-faced Coucal *Centropus melanops* OR
Bay Coucal *Centropus celebensis* AU
Rufous Coucal *Centropus unirufus* OR

71 HOATZIN Opisthocomidae

Hoatzin *Opisthocomus hoazin* NT

OWLS Strigiformes

72 BARN OWLS Tytonidae

Tytoninae

Madagascar Masked Owl *Tyto soumagnei* ● AF
Barn Owl *Tyto alba* WP EP AF OR AU PA NA NT
Ashy-faced Barn Owl *Tyto glaucops* NT
Sulawesi Masked Owl *Tyto rosenbergii* AU
Sula Masked Owl *Tyto nigrobrunnea* ● AU
Minnahassa Masked Owl *Tyto inexspectata* ● AU
Lesser Masked Owl *Tyto sororcula* ● AU
Australian Masked Owl *Tyto novaehollandiae* AU
New Britain Barn Owl *Tyto aurantia* ● AU
Greater Sooty Owl *Tyto tenebricosa* AU
Lesser Sooty Owl *Tyto multipunctata* AU
Grass Owl *Tyto capensis* AF OR AU

Phodilinae

Bay Owl *Phodilus badius* OR
Tanzanian Bay Owl *Phodilus prigoginei* ● AF

73 OWLS Strigidae

Buboninae

White-fronted Scops Owl *Otus sagittatus* ● OR
Rufous Scops Owl *Otus rufescens* OR
Cinnamon Scops Owl *Otus icterorhynchus* AF
Sokoke Scops Owl *Otus ireneae* ● AF
Spotted Scops Owl *Otus spilocephalus* OR
Andaman Scops Owl *Otus balli* OR
Luzon Scops Owl *Otus longicornis* OR
Mindoro Scops Owl *Otus mindorensis* ● OR
Mindanao Scops Owl *Otus mirus* OR
Sulawesi Scops Owl *Otus manadensis* AU
Flores Scops Owl *Otus alfredi* ● AU
Javan Scops Owl *Otus angelinae* ● AU
Simalur Scops Owl *Otus umbra* OR
Sao Thome Scops Owl *Otus hartlaubi* ● AF
Pallid Scops Owl *Otus brucei* WP AF OR
Flammulated Owl *Otus flammeolus* NA NT
African Scops Owl *Otus senegalensis* AF
Eurasian Scops Owl *Otus scops* WP EP AF OR
Oriental Scops Owl *Otus sunia* OR
Riukiu Scops Owl *Otus elegans* EP
South Philippines Scops Owl *Otus mantananensis* OR
Moluccan Scops Owl *Otus magicus* AU
Madagascar Scops Owl *Otus rutilus* AF
Pemba Scops Owl *Otus pembaensis* AF
Rajah Scops Owl *Otus brookii* OR
Indian Scops Owl *Otus bakkamoena* OR

Collared Scops Owl *Otus lempiji* EP OR
Philippine Scops Owl *Otus megalotis* OR
Palawan Scops Owl *Otus fuliginosus* OR
Sipora Scops Owl *Otus mentawi* OR
Wallace's Scops Owl *Otus silvicola* AU
Biak Island Scops Owl *Otus beccarii* ● AU
Bare-legged Scops Owl *Otus insularis* ● OR
White-faced Scops Owl *Otus leucotis* WP AF
Eastern Screech Owl *Otus asio* NA NT
Western Screech Owl *Otus kennicotti* NA NT
Balsas Screech Owl *Otus seductus* NT
Pacific Screech Owl *Otus cooperii* NT
Whiskered Screech Owl *Otus trichopsis* NA NT
Tropical Screech Owl *Otus choliba* NT
Long-tufted Screech Owl *Otus sanctaecatarinae* NT
Peruvian Screech Owl *Otus roboratus* NT
Maria Koepcke's Screech Owl *Otus koepckeae* NT
Bare-shanked Screech Owl *Otus clarkii* NT
Bearded Screech Owl *Otus barbarus* NT
Cloud-forest Screech Owl *Otus marshalli* NT
Rufescent Screech Owl *Otus ingens* NT
Colombian Screech Owl *Otus colombianus* NT
Cinnamon Screech Owl *Otus petersoni* NT
Tawny-bellied Screech Owl *Otus watsonii* NT
Black-capped Screech Owl *Otus atricapillus* NT
Vermiculated Screech Owl *Otus guatemalae* NT
Puerto Rico Screech Owl *Otus nudipes* NT
Cuban Screech Owl *Otus lawrencii* NT
Palau Scops Owl *Otus podarginus* PA
White-throated Screech Owl *Otus albogularis* NT

Giant Scops Owl *Mimizuku gurneyi* OR

Maned Owl *Jubula lettii* AF

Crested Owl *Lophostrix cristata* NT

Spectacled Owl *Pulsatrix perspicillata* NT
Tawny-browed Owl *Pulsatrix koeniswaldiana* NT
Band-tailed Owl *Pulsatrix melanota* NT

Great Horned Owl *Bubo virginianus* NA NT
Northern Eagle Owl *Bubo bubo* WP EP AF OR
Desert Eagle Owl *Bubo ascalaphus* AF
Indian Eagle Owl *Bubo bengalensis* OR
Cape Eagle Owl *Bubo capensis* AF
Spotted Eagle Owl *Bubo africanus* WP AF
Fraser's Eagle Owl *Bubo poensis* AF
Forest Eagle Owl *Bubo nipalensis* OR
Malay Eagle Owl *Bubo sumatrana* OR
Shelley's Eagle Owl *Bubo shelleyi* AF
Verreaux's Eagle Owl *Bubo lacteus* AF
Dusky Eagle Owl *Bubo coromandus* OR
Akun Eagle Owl *Bubo leucostictus* AF
Philippine Eagle Owl *Bubo philippensis* OR
Blakiston's Fish Owl *Bubo blakistoni* ● WP EP
Brown Fish Owl *Bubo zeylonensis* EP OR
Tawny Fish Owl *Bubo flavipes* EP OR
Malay Fish Owl *Bubo ketupu* OR

Pel's Fishing Owl *Scotopelia peli* AF
Rufous Fishing Owl *Scotopelia ussheri* ● AF
Vermiculated Fishing Owl *Scotopelia bouvieri* AF

Snowy Owl *Nyctea scandiaca* WP EP NA

Mottled Owl *Ciccaba virgata* NT
Black & White Owl *Ciccaba nigrolineata* NT
Black-banded Owl *Ciccaba huhula* NT
Rufous-banded Owl *Ciccaba albitarsus* NT

African Wood Owl *Strix woodfordi* AF
Spotted Wood Owl *Strix seloputo* OR
Mottled Wood Owl *Strix ocellata* OR
Brown Wood Owl *Strix leptogrammica* EP OR
Tawny Owl *Strix aluco* WP EP OR
Hume's Owl *Strix butleri* EP
Barred Owl *Strix varia* NA NT
Fulvous Owl *Strix fulvescens* NT
Spotted Owl *Strix occidentalis* ● NA NT
Ural Owl *Strix uralensis* WP EP
Great Grey Owl *Strix nebulosa* WP EP NA
Rusty Barred Owl *Strix hylophila* NT
Rufous-legged Owl *Strix rufipes* NT

Hawk Owl *Surnia ulula* WP EP NA

Collared Owlet *Glaucidium brodiei* EP OR
Eurasian Pygmy Owl *Glaucidium passerinum* ● WP EP

Pearl-spotted Owlet *Glaucidium perlatum* AF
Northern Pygmy Owl *Glaucidium gnoma* NA NT
Least Pygmy Owl *Glaucidium minutissimum* NT
Ferruginous Pygmy Owl *Glaucidium brasilianum* NA NT
Austral Pygmy Owl *Glaucidium nanum* NT
Andean Pygmy Owl *Glaucidium jardinii* NT
Cuban Pygmy Owl *Glaucidium siju* NT
Red-chested Owlet *Glaucidium tephronotum* AF
Chestnut-backed Owlet *Glaucidium sjostedti* AF
Jungle Owlet *Glaucidium radiatum* OR
Cuckoo Owlet *Glaucidium cuculoides* EP OR
Chestnut-winged Owlet *Glaucidium castanopterum* OR
African Barred Owlet *Glaucidium capense* AF
Ituri Owlet *Glaucidium castaneum* AF
Prigogine's Owlet *Glaucidium albertinum* AF
Eastern Barred Owlet *Glaucidium scheffleri* AF

Long-whiskered Owlet *Xenoglaux loweryi* NT

Elf Owl *Micrathene whitneyi* NA NT

Little Owl *Athene noctua* WP EP AF OR AU
Spotted Owlet *Athene brama* EP OR
Forest Owlet *Athene blewitti* ● OR
Burrowing Owl *Athene cunicularia* NA NT

Tengmalm's Owl *Aegolius funereus* WP EP OR NA
Saw-whet Owl *Aegolius acadicus* NA NT
Unspotted Saw-whet Owl *Aegolius ridgwayi* NT
Buff-fronted Owl *Aegolius harrisii* NT

Papuan Hawk Owl *Uroglaux dimorpha* AU

Rufous Hawk Owl *Ninox rufa* AU
Powerful Owl *Ninox strenua* AU
Barking Owl *Ninox connivens* AU
Boobook Owl *Ninox novaeseelandiae* AU
Sumba Boobook Owl *Ninox rudolfi* AU
Brown Hawk Owl *Ninox scutulata* EP OR
Andaman Hawk Owl *Ninox affinis* OR
Madagascar Hawk Owl *Ninox superciliaris* AF
Philippine Hawk Owl *Ninox philippensis* OR
Spotted Hawk Owl *Ninox spilonota* OR
Tweedale's Hawk Owl *Ninox spilocephala* OR
Ochre-bellied Hawk Owl *Ninox ochracea* AU
Moluccan Hawk Owl *Ninox squamipila* OR AU
Jungle Hawk Owl *Ninox theomacha* AU
Admiralty Islands Hawk Owl *Ninox meeki* AU
Speckled Hawk Owl *Ninox punctulata* AU
Bismarck Hawk Owl *Ninox variegata* AU
New Britain Hawk Owl *Ninox odiosa* AU
Solomon Islands Hawk Owl *Ninox jacquinoti* PA

White-faced Owl *Sceloglaux albifacies* AU

Striped Owl *Asio clamator* NT
Long-eared Owl *Asio otus* WP EP AF OR NA NT
Stygian Owl *Asio stygius* NT
Abyssinian Long-eared Owl *Asio abyssinicus* AF
Madagascar Long-eared Owl *Asio madagascariensis* AF
Short-eared Owl *Asio flammeus* WP EP AF OR AU PA NA NT
African Marsh Owl *Asio capensis* WP AF
Jamaican Owl *Pseudoscops grammacus* NT
Fearful Owl *Nesasio solomonensis* PA

NIGHTJARS AND ALLIES
Caprimulgiformes

74 OILBIRD Steatornithidae

Oilbird *Steatornis caripensis* NT

75 FROGMOUTHS Podargidae

Tawny Frogmouth *Podargus strigoides* AU
Papuan Frogmouth *Podargus papuensis* AU
Marbled Frogmouth *Podargus ocellatus* AU PA

Large Frogmouth *Batrachostomus auritus* OR
Dulit Frogmouth *Batrachostomus harterti* ● OR
Philippine Frogmouth *Batrachostomus septimus* OR
Gould's Frogmouth *Batrachostomus stellatus* OR
Ceylon Frogmouth *Batrachostomus moniliger* OR
Hodgson's Frogmouth *Batrachostomus hodgsoni* OR
Pale-headed Frogmouth *Batrachostomus poliolophus* OR

Sharpe's Frogmouth *Batrachostomus mixtus* OR
Javan Frogmouth *Batrachostomus javensis* OR
Bornean Frogmouth *Batrachostomus cornutus* OR

76 POTOOS Nyctibiidae

Great Potoo *Nyctibius grandis* NT
Long-tailed Potoo *Nyctibius aethereus* ● NT
Common Potoo *Nyctibius griseus* NT
White-winged Potoo *Nyctibius leucopterus* ● NT
Rufous Potoo *Nyctibius bracteatus* ● NT

77 OWLET-NIGHTJARS Aegothelidae

Halmahera Owletnightjar *Aegotheles crinifrons* AU
Large Owletnightjar *Aegotheles insignis* AU
Owletnightjar *Aegotheles cristatus* AU
New Caledonian Owletnightjar *Aegotheles savesi* PA
Barred Owletnightjar *Aegotheles bennettii* AU
Wallace's Owletnightjar *Aegotheles wallacii* AU
Mountain Owletnightjar *Aegotheles albertisi* AU
Eastern Mountain Owletnightjar *Aegotheles archboldi* AU

78 NIGHTJARS Caprimulgidae

Chordeilinae

Semi-collared Nighthawk *Lurocalis semitorquatus* NT

Least Nighthawk *Chordeiles pusillus* NT
Sand-coloured Nighthawk *Chordeiles rupestris* NT
Lesser Nighthawk *Chordeiles acutipennis* NA NT
Common Nighthawk *Chordeiles minor* NA NT
Antillean Nighthawk *Chordeiles gundlachii* NT

Band-tailed Nighthawk *Nyctiprogne leucopyga* NT

Nacunda Nighthawk *Podager nacunda* NT

Caprimulginae

Spotted Nightjar *Eurostopodus argus* AU
White-throated Nightjar *Eurostopodus mystacalis* AU PA
Satanic Nightjar *Eurostopodus diabolicus* ● AU
Papuan Nightjar *Eurostopodus papuensis* AU
Archbold's Nightjar *Eurostopodus archboldi* AU
Malaysian Eared Nightjar *Eurostopodus temminckii* AU
Great Eared Nightjar *Eurostopodus macrotis* OR AU

Brown Nightjar *Veles binotatus* AF

Common Pauraque *Nyctidromus albicollis* NA NT

Common Poorwill *Phalaenoptilus nuttallii* NA NT

Least Pauraque *Siphonorhis brewsteri* NT

Eared Poorwill *Otophanes mcleodii* NT
Yucatan Poorwill *Otophanes yucatanicus* NT

Ocellated Poorwill *Nyctiphrynus ocellatus* NT

Chuck Will's Widow *Caprimulgus carolinensis* NA NT
Rufus Nightjar *Caprimulgus rufus* NT
St Lucian Nightjar *Caprimulgus otiosus* NT
Greater Antillean Nightjar *Caprimulgus cubanensis* NT
Silky-tailed Nightjar *Caprimulgus sericocaudatus* NT
Tawny-collared Nightjar *Caprimulgus salvini* NT
Yucatan Tawny-collared Nightjar *Caprimulgus badius* NT
Ridgway's Whippoorwill *Caprimulgus ridgwayi* NT
Whip-poor-will *Caprimulgus vociferus* NA NT
Puerto Rican Nightjar *Caprimulgus noctitherus* ● NT
Dusky Nightjar *Caprimulgus saturatus* NT
Band-winged Nightjar *Caprimulgus longirostris* NT
White-tailed Nightjar *Caprimulgus cayennensis* NT
White-winged Nightjar *Caprimulgus candicans* ● NT
Spot-tailed Nightjar *Caprimulgus maculicaudus* NT
Little Nightjar *Caprimulgus parvulus* NT
Cayenne Nightjar *Caprimulgus maculosus* NT
Blackish Nightjar *Caprimulgus nigrescens* NT
Roraiman Nightjar *Caprimulgus whitelyi* NT
Pygmy Nightjar *Caprimulgus hirundinaceus* NT
Red-necked Nightjar *Caprimulgus ruficollis* WP AF
Jungle Nightjar *Caprimulgus indicus* EP OR AU PA
Nightjar *Caprimulgus europaeus* WP EP AF OR

Sykes' Nightjar *Caprimulgus mahrattensis* EP OR
Vaurie's Nightjar *Caprimulgus centralasicus* ● EP
Nubian Nightjar *Caprimulgus nubicus* WP AF
Egyptian Nightjar *Caprimulgus aegyptius* WP AF
Golden Nightjar *Caprimulgus eximius* WP AF
Indian Long-tailed Nightjar *Caprimulgus atripennis* OR
Long-tailed Nightjar *Caprimulgus macrurus* OR AU
African Dusky Nightjar *Caprimulgus pectoralis* AF
Black-shouldered Nightjar *Caprimulgus nigriscapularis* AF
Rufous-cheeked Nightjar *Caprimulgus rufigena* AF
Sombre Nightjar *Caprimulgus fraenatus* AF
Donaldson Smith's Nightjar *Caprimulgus donaldsoni* AF
Abyssinian Nightjar *Caprimulgus poliocephalus* AF
Ruwenzori Nightjar *Caprimulgus ruwenzorii* AF
Indian Nightjar *Caprimulgus asiaticus* EP OR
African White-tailed Nightjar *Caprimulgus natalensis* AF
Plain Nightjar *Caprimulgus inornatus* WP OR
Star-spotted Nightjar *Caprimulgus stellatus* AF
Ludovic's Nightjar *Caprimulgus ludovicianus* AF
Franklin's Nightjar *Caprimulgus monticolus* EP OR
Allied Nightjar *Caprimulgus affinis* OR
Freckled Nightjar *Caprimulgus tristigma* AF
Bonaparte's Nightjar *Caprimulgus concretus* OR
Salvadori's Nightjar *Caprimulgus pulchellus* ● OR
Collared Nightjar *Caprimulgus enarratus* AF
Bates' Nightjar *Caprimulgus batesi* AF

Gabon Nightjar *Scotornis fossii* AF
Long-tailed Nightjar *Scotornis climacurus* AF
Slender-tailed Nightjar *Scotornis clarus* AF

Standard-winged Nightjar *Macrodipteryx longipennis* AF

Pennant-winged Nightjar *Macrodipteryx vexillarius* AF

Ladder-tailed Nightjar *Hydropsalis climacocerca* NT
Scissor-tailed Nightjar *Hydropsalis brasiliana* NT

Swallow-tailed Nightjar *Uropsalis segmentata* NT
Lyre-tailed Nightjar *Uropsalis lyra* NT

Long-trained Nightjar *Macropsalis creagra* ● NT

Sickle-winged Nightjar *Eleothreptus anomalus* ● NT

SWIFTS AND HUMMINGBIRDS
Apodiformes

79 SWIFTS Apodidae

Cypseloidinae

Sooty Swift *Cypseloides fumigatus* NT
Spot-fronted Swift *Cypseloides cherriei* NT
White-chinned Swift *Cypseloides cryptus* NT
White-chested Swift *Cypseloides lemosi* ● NT
Great Swift *Cypseloides major* NT
Tepui Swift *Cypseloides phelpsi* NT
Chestnut-collared Swift *Cypseloides rutilus* NT

Black Swift *Nephoecetes niger* NA NT

Great Dusky Swift *Aerornis senex* NT

White-collared Swift *Streptoprocne zonaris* NT
Biscutate Swift *Streptoprocne biscutatus* NT
White-naped Swift *Streptoprocne semicollaris* NT

Apodinae

Waterfall Swift *Hydrochous gigas* ● OR

Moluccan Swiftlet *Aerodramus infuscatus* AU
White-rumped Swiftlet *Aerodramus spodiopygius* AU PA
Mountain Swiftlet *Aerodramus hirundinaceus* AU
Edible-nest Swiftlet *Aerodramus fuciphaga* OR AU
Grey-rumped Swiftlet *Aerodramus francica* AF OR
Oustalet's Swiftlet *Aerodramus germani* OR
Philippine Swiftlet *Aerodramus amelis* OR
Uniform Swiftlet *Aerodramus vanikorensis* AU PA
Mossy Swiftlet *Aerodramus salangana* OR
Caroline Swiftlet *Aerodramus inquietus* PA
Mariana Swiftlet *Aerodramus bartschi* PA
Brown-rumped Swiftlet *Aerodramus mearnsi* OR
Guadalcanal Swiftlet *Aerodramus orientalis* PA

Idenburg River Swiftlet *Aerodramus papuensis* AU
Schrader Mountain Swiftlet *Aerodramus nuditarsus* AU
Whitehead's Swiftlet *Aerodramus whiteheadi* OR AU PA
Himalayan Swiftlet *Aerodramus brevirostris* EP OR
Lowe's Swiftlet *Aerodramus maximus* OR
Seychelles Cave Swiftlet *Aerodramus elaphra* ● AF
Indian Edible-nest Swiftlet *Aerodramus unicolor* OR
Tahitian Swiftlet *Aerodramus leucophaeus* ● PA
Marquesan Swiftlet *Aerodramus ocista* PA
Cook Islands Swiftlet *Aerodramus sawtelli* PA

White-bellied Swiftlet *Collocalia esculenta* OR AU PA
Philippine Swiftlet *Collocalia marginata* OR
Pygmy Swiftlet *Collocalia troglodytes* OR
Linchi Swiftlet *Collocalia linchi* OR AU

Scarce Swift *Schoutedenapus myioptilus* AF
Schouteden's Swift *Schoutedenapus schoutedeni* ● AF

Philippine Spinetailed Swift *Mearnsia picina* OR
New Guinea Spinetailed Swift *Mearnsia novaeguineae* AU

Madagascar Spinetailed Swift *Zoonavena grandidieri* AF
Sao Thome Spinetailed Swift *Zoonavena thomensis* AF
Indian White-rumped Spinetailed Swift *Zoonavena sylvatica* OR

Mottle-throated Spinetailed Swift *Telacanthura ussheri* AF
Ituri Mottle-throated Spinetailed Swift *Telacanthura melanopygia* AF

White-rumped Spinetailed Swift *Raphidura leucopygialis* OR
Sabine's Spinetailed Swift *Raphidura sabini* AF

Cassin's Spinetailed Swift *Neafrapus cassini* AF
Boehm's Spinetailed Swift *Neafrapus boehmi* AF

White-throated Spinetailed Swift *Hirundapus caudacuta* WP EP OR AU PA
White-vented Spinetailed Swift *Hirundapus cochinchinensis* OR
Brown Spinetailed Swift *Hirundapus gigantea* OR
Sulawesi Spinetailed Swift *Hirundapus celebensis* AU

Band-rumped Swift *Chaetura spinicauda* NT
Lesser Antillean Swift *Chaetura martinica* NT
Grey-rumped Swift *Chaetura cinereiventris* NT
Pale-rumped Swift *Chaetura egregia* NT
Chimney Swift *Chaetura pelagica* NA NT
Vaux's Swift *Chaetura vauxi* NA NT
Chapman's Swift *Chaetura chapmani* NT
Ashy-tailed Swift *Chaetura andrei* NT
Short-tailed Swift *Chaetura brachyura* NT

White-throated Swift *Aeronautes saxatilis* NA NT
White-tipped Swift *Aeronautes montivagus* NT
Andean Swift *Aeronautes andecolus* NT

Antillean Palm Swift *Tachornis phoenicobia* NT
Pygmy Swift *Tachornis furcata* NT
Fork-tailed Palm Swift *Tachornis squamata* NT

Great Swallow-tailed Swift *Panyptila sanctihieronymi* NT
Lesser Swallow-tailed Swift *Panyptila cayennensis* NT

Asian Palm Swift *Cypsiurus batasiensis* OR
African Palm Swift *Cypsiurus parvus* AF

Alpine Swift *Tachymarptis melba* WP EP AF OR
Mottled Swift *Tachymarptis aequatorialis* AF

Alexander's Swift *Apus alexandri* AF
African Black Swift *Apus barbatus* AF
Berlioz' Swift *Apus berliozi* AF
Bradfield's Swift *Apus bradfieldi* AF
Nyanza Swift *Apus niansae* AF
Pallid Swift *Apus pallidus* WP AF
Swift *Apus apus* WP EP AF OR
Plain Swift *Apus unicolor* WP
Dark-backed Swift *Apus acuticauda* OR
Fork-tailed Swift *Apus pacificus* EP OR AU PA NA
House Swift *Apus affinis* WP EP AF OR

Horus Swift *Apus horus* AF
White-rumped Swift *Apus caffer* WP AF
Bates' Black Swift *Apus batesi* AF

80 TREE SWIFTS Hemiprocnidae

Crested Tree Swift *Hemiprocne longipennis* OR
Whiskered Tree Swift *Hemiprocne mystacea* AU PA
Lesser Tree Swift *Hemiprocne comata* OR

81 HUMMINGBIRDS Trochilidae

Blue-fronted Lancebill *Doryfera johannae* NT
Green-fronted Lancebill *Doryfera ludoviciae* NT

Tooth-billed Hummingbird *Androdon aequatorialis* NT

Saw-billed Hermit *Ramphodon naevius* NT

Hook-billed Hermit *Glaucis dohrnii* ● NT
Bronzy Hermit *Glaucis aenea* NT
Rufous-breasted Hermit *Glaucis hirsuta* NT

Sooty Barbthroat *Threnetes niger* NT
Black Barbthroat *Threnetes grzimeki* ● NT
Pale-tailed Barbthroat *Threnetes leucurus* NT
Band-tailed Barbthroat *Threnetes ruckeri* NT

White-whiskered Hermit *Phaethornis yaruqui* NT
Green Hermit *Phaethornis guy* NT
Tawny-bellied Hermit *Phaethornis syrmatophorus* NT
Long-tailed Hermit *Phaethornis superciliosus* NT
Great-billed Hermit *Phaethornis malaris* NT
Klabin Farm Long-tailed Hermit *Phaethornis margarettae* ● NT
Scale-throated Hermit *Phaethornis eurynome* NT
Black-billed Hermit *Phaethornis nigrirostris* ● NT
White-bearded Hermit *Phaethornis hispidus* NT
Pale-bellied Hermit *Phaethornis anthophilus* NT
Koepcke's Hermit *Phaethornis koepckeae* NT
Straight-billed Hermit *Phaethornis bourcieri* NT
Needle-billed Hermit *Phaethornis philippii* NT
Dusky-throated Hermit *Phaethornis squalidus* NT
Sooty-capped Hermit *Phaethornis augusti* NT
Planalto Hermit *Phaethornis pretrei* NT
Buff-bellied Hermit *Phaethornis subochraceus* NT
Cinnamon-throated Hermit *Phaethornis nattereri* NT
Broad-tipped Hermit *Phaethornis gounellei* NT
Reddish Hermit *Phaethornis ruber* NT
White-browed Hermit *Phaethornis stuarti* NT
Grey-chinned Hermit *Phaethornis griseogularis* NT
Little Hermit *Phaethornis longuemareus* NT
Minute Hermit *Phaethornis idaliae* NT

White-tipped Sicklebill *Eutoxeres aquila* NT
Buff-tailed Sicklebill *Eutoxeres condamini* NT

Scaly-breasted Hummingbird *Phaeochroa cuvierii* NT

Wedge-tailed Sabrewing *Campylopterus curvipennis* NT
Long-tailed Sabrewing *Campylopterus excellens* NT
Grey-breasted Sabrewing *Campylopterus largipennis* NT
Rufous Sabrewing *Campylopterus rufus* NT
Rufous-breasted Sabrewing *Campylopterus hyperythrus* NT
Buff-breasted Sabrewing *Campylopterus duidae* NT
Violet Sabrewing *Campylopterus hemileucurus* NT
White-tailed Sabrewing *Campylopterus ensipennis* ●
Lazuline Sabrewing *Campylopterus falcatus* NT
Santa Marta Sabrewing *Campylopterus phainopeplus* NT
Napo Sabrewing *Campylopterus villaviscensio* NT

Swallow-tailed Hummingbird *Eupetomena macroura* NT

White-necked Jacobin *Florisuga mellivora* NT

Black Jacobin *Melanotrochilus fuscus* NT

Brown Violetear *Colibri delphinae* NT
Green Violetear *Colibri thalassinus* NT
Sparkling Violetear *Colibri coruscans* NT
White-vented Violetear *Colibri serrirostris* NT

Green-throated Mango *Anthrocothorax viridigula* NT
Green-breasted Mango *Anthracothorax prevostii* NT
Black-throated Mango *Anthracothorax nigricollis* NT

Veraguas Mango *Anthracothorax veraguensis* NT
Antillean Mango *Anthracothorax dominicus* NT
Green Mango *Anthracothorax viridis* NT
Jamaican Mango *Anthracothorax mango* NT

Fiery-tailed Awlbill *Avocettula recurvirostris* NT

Purple-throated Carib *Eulampis jugularis* NT

Green-throated Carib *Sericotes holosericeus* NT

Ruby-Topaz Hummingbird *Chrysolampis mosquitus* NT

Antillean Crested Hummingbird *Orthorhyncus cristatus* NT

Violet-headed Hummingbird *Klais guimeti* NT

Emerald-chinned Hummingbird *Abeillia abeillei* NT

Black-breasted Plovercrest *Stephanoxis lalandi* NT

Tufted Coquette *Lophornis ornata* NT
Dot-eared Coquette *Lophornis gouldii* NT
Frilled Coquette *Lophornis magnifica* NT
Rufous-crested Coquette *Lophornis delattrei* NT
Spangled Coquette *Lophornis stictolopha* NT
Festive Coquette *Lophornis chalybea* NT
Peacock Coquette *Lophornis pavonina* NT
Bearded Coquette *Lophornis insignibarbis* NT

Black-crested Coquette *Paphosia helenae* NT
White-crested Coquette *Paphosia adorabilis* NT

Wire-crested Thorntail *Popelairia popelairii* NT
Black-bellied Thorntail *Popelairia langsdorffi* NT
Coppery Thorntail *Popelairia letitiae* ● NT
Green Thorntail *Popelairia conversii* NT

Racquet-tailed Coquette *Discosura longicauda* NT

Blue-chinned Sapphire *Chlorestes notatus* NT

Blue-tailed Emerald *Chlorostilbon mellisugus* NT
Simon's Emerald *Chlorostilbon vitticeps* NT
Glittering-bellied Emerald *Chlorostilbon aureoventris* NT
Fork-tailed Emerald *Chlorostilbon canivetii* NT
Garden Emerald *Chlorostilbon assimilis* NT
Cuban Emerald *Chlorostilbon ricordii* NT
Hispaniolan Emerald *Chlorostilbon swainsonii* NT
Puerto Rican Emerald *Chlorostilbon maugaeus* NT
Red-billed Emerald *Chlorostilbon gibsoni* NT
Coppery Emerald *Chlorostilbon russatus* NT
Narrow-tailed Emerald *Chlorostilbon stenura* NT
Green-tailed Emerald *Chlorostilbon alice* NT
Short-tailed Emerald *Chlorostilbon poortmani* NT

Dusky Hummingbird *Cynanthus sordidus* NT
Broad-billed Hummingbird *Cynanthus latirostris* NA NT

Blue-headed Hummingbird *Cyanophaia bicolor* NT

Crowned Woodnymph *Thalurania colombica* NT
Fork-tailed Woodnymph *Thalurania furcata* NT
Long-tailed Woodnymph *Thalurania watertonii* NT
Violet-capped Woodnymph *Thalurania glaucopis* NT
Lerch's Woodnymph *Thalurania lerchi* NT

Blue-tailed Sylph *Neolesbia nehrkorni* NT

Fiery-throated Hummingbird *Panterpe insignis* NT

Violet-bellied Hummingbird *Damophila julie* NT

Sapphire-throated Hummingbird *Lepidopyga coeruleogularis* NT
Sapphire-bellied Hummingbird *Lepidopyga lilliae* ● NT
Shining Green Hummingbird *Lepidopyga goudoti* NT

Black-fronted Hummingbird *Hylocharis xantusii* NT
White-eared Hummingbird *Hylocharis leucotis* NA NT
Blue-throated Goldentail *Hylocharis eliciae* NT
Rufous-throated Sapphire *Hylocharis sapphirina* NT
White-chinned Sapphire *Hylocharis cyanus* NT
Flame-rumped Sapphire *Hylocharis pyropygia* NT
Gilded Hummingbird *Hylocharis chrysura* NT
Blue-headed Sapphire *Hylocharis grayi* NT

Golden-tailed Sapphire *Chrysuronia oenone* NT

Violet-capped Hummingbird *Goldmania violiceps* NT

Pirre Hummingbird *Goethalsia bella* NT

Streamertail *Trochilus polytmus* NT

White-throated Hummingbird *Leucochloris albicollis* NT

White-tailed Goldenthroat *Polytmus guainumbi* NT
Tepui Goldenthroat *Polytmus milleri* NT
Green-tailed Goldenthroat *Polytmus theresiae* NT

Buffy Hummingbird *Leucippus fallax* NT
Tumbes Hummingbird *Leucippus baeri* NT
Spot-throated Hummingbird *Leucippus taczanowskii* NT
Olive-spotted Hummingbird *Leucippus chlorocercus* NT

Many-spotted Hummingbird *Taphrospilus hypostictus* NT

White-bellied Hummingbird *Amazilia chionogaster* NT
Green & White Hummingbird *Amazilia viridicauda* NT
White-bellied Emerald *Amazilia candida* NT
White-chested Emerald *Amazilia chionopectus* NT
Versicoloured Emerald *Amazilia versicolor* NT
Honduras Emerald *Amazilia luciae* ● NT
Glittering-throated Emerald *Amazilia fimbriata* NT
Tachira Emerald *Amazilia distans* ● NT
Sapphire-spangled Emerald *Amazilia lactea* NT
Charming Hummingbird *Amazilia decora* NT
Blue-chested Hummingbird *Amazilia amabilis* NT
Purple-chested Hummingbird *Amazilia rosenbergi* NT
Mangrove Hummingbird *Amazilia boucardi* ● NT
Andean Emerald *Amazilia franciae* NT
Plain-bellied Emerald *Amazilia leucogaster* NT
Red-billed Azurecrown *Amazilia cyanocephala* NT
Small-billed Azurecrown *Amazilia microrhyncha* NT
Indigo-capped Hummingbird *Amazilia cyanifrons* NT
Berylline Hummingbird *Amazilia beryllina* NT
Blue-tailed Hummingbird *Amazilia cyanura* NT
Steely-vented Hummingbird *Amazilia saucerrottei* NT
Copper-rumped Hummingbird *Amazilia tobaci* NT
Green-bellied Hummingbird *Amazilia virdigaster* NT
Snowy-breasted Hummingbird *Amazilia edward* NT
Cinnamon Hummingbird *Amazilia rutila* NT
Buff-bellied Hummingbird *Amazilia yucatanensis* NA NT
Rufous-tailed Hummingbird *Amazilia tzacatl* NA NT
Chestnut-bellied Hummingbird *Amazilia castaneiventris* ● NT
Amazilia Hummingbird *Amazilia amazilia* NT
Green-fronted Hummingbird *Amazilia viridifrons* NT
Violet-crowned Hummingbird *Amazilia violiceps* NA NT

White-tailed Hummingbird *Eupherusa poliocerca* ●
Stripe-tailed Hummingbird *Eupherusa eximia* NT
Oaxaca Hummingbird *Eupherusa cyanophrys* ● NT
Black-bellied Hummingbird *Eupherusa nigriventris* NT

White-tailed Emerald *Elvira chionura* NT
Coppery-headed Emerald *Elvira cupreiceps* NT

Snowcap *Microchera albocoronata* NT

White-vented Plumeleteer *Chalybura buffonii* NT
Bronze-tailed Plumeleteer *Chalybura urochrysia* NT

Sombre Hummingbird *Aphantochroa cirrochloris* NT

Blue-throated Hummingbird *Lampornis clemenciae* NA NT
Amethyst-throated Hummingbird *Lampornis amethystinus* NT
Green-throated Mountain Gem *Lampornis viridipallens* NT
Green-breasted Mountain Gem *Lampornis sybillae* NT
White-bellied Mountain Gem *Lampornis hemileucus* NT
White-throated Mountain Gem *Lampornis castaneoventris* NT

Purple-throated Mountain Gem *Lampornis calolaema* NT
Grey-tailed Mountain Gem *Lampornis cinereicauda* NT

Garnet-throated Hummingbird *Lamprolaima rhami* NT

Speckled Hummingbird *Adelomyia melanogenys* NT

Blossomcrown *Anthocephala floriceps* NT

Whitetip *Urosticte benjamini* NT

Ecuadorean Piedtail *Phlogophilus hemileucurus* NT
Peruvian Piedtail *Phlogophilus harterti* NT

Brazilian Ruby *Clytolaema rubricauda* NT

Gould's Jewel-front *Polyplancta aurescens* NT

Fawn-breasted Brilliant *Heliodoxa rubinoides* NT
Violet-fronted Brilliant *Heliodoxa leadbeateri* NT
Green-crowned Brilliant *Heliodoxa jacula* NT
Velvet-browed Brilliant *Heliodoxa xanthogonys* NT
Black-throated Brilliant *Heliodoxa schreibersii* NT
Pink-throated Brilliant *Heliodoxa gularis* NT
Rufous-webbed Brilliant *Heliodoxa branickii* NT
Empress Brilliant *Heliodoxa imperatrix* NT

Rivoli's Hummingbird *Eugenes fulgens* NA NT

Scissor-tailed Hummingbird *Hylonympha macrocerca* ● NT

Violet-chested Hummingbird *Sternoclyta cyanopectus* NT

Crimson Topaz *Topaza pella* NT
Fiery Topaz *Topaza pyra* NT

Black-breasted Hillstar *Oreotrochilus melanogaster* NT
White-sided Hillstar *Oreotrochilus leucopleurus* NT
Wedgetailed Hillstar *Oreotrochilus adela* NT

White-tailed Hillstar *Urochroa bougueri* NT

Giant Hummingbird *Patagona gigas* NT

Shining Sunbeam *Aglaeactis cupripennis* NT
Purple-backed Sunbeam *Aglaeactis aliciae* ● NT
White-tufted Sunbeam *Aglaeactis castelnaudii* NT
Black-hooded Sunbeam *Aglaeactis pamela* NT

Mountain Velvetbreast *Lafresnaya lafresnayi* NT

Great Sapphirewing *Pterophanes cyanopterus* NT

Bronzy Inca *Coeligena coeligena* NT
Brown Inca *Coeligena wilsoni* NT
Black Inca *Coeligena prunellei* ● NT
Collared Inca *Coeligena torquata* NT
White-tailed Starfrontlet *Coeligena phalerata* NT
Golden-bellied Starfrontlet *Coeligena bonapartei* NT
Dusky Starfrontlet *Coeligena orina* NT
Blue-throated Starfrontlet *Coeligena helianthea* NT
Buff-winged Starfrontlet *Coeligena lutetiae* NT
Violet-throated Starfrontlet *Coeligena violifer* NT
Rainbow Starfrontlet *Coeligena iris* NT

Sword-billed Hummingbird *Ensifera ensifera* NT

Green-backed Firecrown *Sephanoides sephaniodes* NT
Fernandez Firecrown *Sephanoides fernandensis* ● NT

Buff-tailed Coronet *Boissonneaua flavescens* NT
Chestnut-breasted Coronet *Boissonneaua matthewsii* NT

Velvet-Purple Coronet *Boissonneaua jardini* NT

Orange-throated Sunangel *Heliangelus mavors* NT
Merida Sunangel *Heliangelus spencei* NT
Amethyst-throated Sunangel *Heliangelus amethysticollis* NT
Gorgeted Sunangel *Heliangelus strophianus* NT
Royal Sunangel *Heliangelus regalis* ● NT
Tourmaline Sunangel *Heliangelus exortis* NT
Purple-throated Sunangel *Heliangelus viola* NT
Little Sunangel *Heliangelus micraster* NT
Olive-throated Sunangel *Heliangelus squamigularis* NT

Black-breasted Puffleg *Eriocnemis nigrivestris* ● NT
Soderstrom's Puffleg *Eriocnemis soderstromi* NT
Glowing Puffleg *Eriocnemis vestitus* NT
Turquoise-throated Puffleg *Eriocnemis godini* ● NT

Coppery-bellied Puffleg *Eriocnemis cupreoventris* NT
Sapphire-vented Puffleg *Eriocnemis luciani* NT
Isaacson's Puffleg *Eriocnemis isaacsonii* NT
Golden-breasted Puffleg *Eriocnemis mosquera* NT
Blue-capped Puffleg *Eriocnemis glaucopoides* NT
Colourful Puffleg *Eriocnemis mirabilis* ● NT
Emerald-bellied Puffleg *Eriocnemis alinae* NT
Black-thighed Puffleg *Eriocnemis derbyi* ● NT

Greenish Puffleg *Haplophaedia aureliae* NT
Hoary Puffleg *Haplophaedia lugens* ● NT

Booted Racquet-tail *Ocreatus underwoodii* NT

Black-tailed Trainbearer *Lesbia victoriae* NT
Green-tailed Trainbearer *Lesbia nuna* NT

Red-tailed Comet *Sappho sparganura* NT

Bronze-tailed Comet *Polyonymus caroli* NT

Purple-tailed Comet *Zodalia glyceria* NT

Purple-backed Thornbill *Ramphomicron microrhynchum* NT
Black-backed Thornbill *Ramphomicron dorsale* NT

Black Metaltail *Metallura phoebe* NT
Coppery Metaltail *Metallura theresiae* NT
Purple-tailed Thornbill *Metallura purpureicauda* NT
Scaled Metaltail *Metallura aeneocauda* NT
Violet-throated Metaltail *Metallura baroni* ● NT
Fire-throated Metaltail *Metallura eupogon* NT
Neblina Metaltail *Metallura odomae* NT
Viridian Metaltail *Metallura williami* NT
Tyrian Metaltail *Metallura tyrianthina* NT
Perija Metaltail *Metallura iracunda* NT

Rufous-capped Thornbill *Chalcostigma ruficeps* NT
Olivaceous Thornbill *Chalcostigma olivaceum* NT
Blue-mantled Thornbill *Chalcostigma stanleyi* NT
Bronze-tailed Thornbill *Chalcostigma heteropogon* NT
Rainbow-bearded Thornbill *Chalcostigma herrani* NT

Bearded Helmetcrest *Oxypogon guerinii* NT

Mountain Avocetbill *Opisthoprora euryptera* NT

Grey-billed Comet *Taphrolesbia griseiventris* ● NT

Long-tailed Sylph *Aglaiocercus kingi* NT
Violet-tailed Sylph *Aglaiocercus coelestis* NT

Bearded Mountaineer *Oreonympha nobilis* NT

Hyacinth Visor-bearer *Augastes scutatus* ● NT
Hooded Visor-bearer *Augastes lumachellus* NT

Wedge-billed Hummingbird *Schistes geoffroyi* NT

Purple-crowned Fairy *Heliothryx barroti* NT
Black-eared Fairy *Heliothryx aurita* NT

Horned Sungem *Heliactin cornuta* NT

Marvellous Sungem *Loddigesia mirabilis* ● NT

Plain-capped Starthroat *Heliomaster constantii* NT
Long-billed Starthroat *Heliomaster longirostris* NT
Stripe-breasted Starthroat *Heliomaster squamosus* NT
Blue-tufted Starthroat *Heliomaster furcifer* NT

Oasis Hummingbird *Rhodopis vesper* NT

Peruvian Sheartail *Thaumastura cora* NT

Bahama Woodstar *Philodice evelynae* NT
Magenta-throated Woodstar *Philodice bryantae* NT
Purple-throated Woodstar *Philodice mitchellii* NT

Slender Sheartail *Doricha enicura* NT
Mexican Sheartail *Doricha eliza* NT

Sparkling-tailed Hummingbird *Tilmatura dupontii* NT

Slender-tailed Woodstar *Microstilbon burmeisteri* NT

Lucifer Hummingbird *Calothorax lucifer* NA NT
Beautiful Hummingbird *Calothorax pulcher* NT

Ruby-throated Hummingbird *Archilochus colubris* NA NT
Black-chinned Hummingbird *Archilochus alexandri* NA NT

Amethyst Woodstar *Calliphlox amethystina* NT

Vervain Hummingbird *Mellisuga minima* NT

Anna's Hummingbird *Calypte anna* NA NT
Costa's Hummingbird *Calypte costae* NA NT
Bee Hummingbird *Calypte helenae* ● NT

Calliope Hummingbird *Stellula calliope* NA NT

Bumblebee Hummingbird *Atthis heloisa* NA NT
Wine-throated Hummingbird *Atthis ellioti* NT

Purple-collared Woodstar *Myrtis fanny* NT

Chilean Woodstar *Eulidia yarrellii* ● NT

Short-tailed Woodstar *Myrmia micrura* NT

White-bellied Woodstar *Acestrura mulsant* NT
Little Woodstar *Acestrura bombus* ● NT
Gorgeted Woodstar *Acestrura heliodor* NT
Colombian Woodstar *Acestrura astreans* NT
Esmeralda's Woodstar *Acestrura berlepschi* ● NT
Hartert's Woodstar *Acestrura harterti* NT

Rufous-shafted Woodstar *Chaetocercus jourdanii* NT

Broad-tailed Hummingbird *Selasphorus platycercus* NA NT
Rufous Hummingbird *Selasphorus rufus* NA NT
Allen's Hummingbird *Selasphorus sasin* NA NT
Volcano Hummingbird *Selasphorus flammula* NT
Glow-throated Hummingbird *Selasphorus ardens* ● NT
Scintillant Hummingbird *Selasphorus scintilla* NT

COLIES OR MOUSEBIRDS Coliiformes

82 MOUSEBIRDS Coliidae

Speckled Mousebird *Colius striatus* AF
Red-backed Mousebird *Colius castanotus* AF
White-backed Mousebird *Colius colius* AF
White-headed Mousebird *Colius leucocephalus* AF

Red-faced Mousebird *Urocolius indicus* AF
Blue-naped Mousebird *Urocolius macrourus* AF

TROGONS Trogoniformes

83 TROGONS Trogonidae

Resplendent Quetzal *Pharomachrus mocinno* ● NT
Crested Quetzal *Pharomachrus antisianus* NT
White-tipped Quetzal *Pharomachrus fulgidus* NT
Golden-headed Quetzal *Pharomachrus auriceps* NT
Pavonine Quetzal *Pharomachrus pavoninus* NT

Eared Trogon *Euptilotis neoxenus* ● NT

Cuban Trogon *Priotelus temnurus* NT

Hispaniolan Trogon *Temnotrogon roseigaster* NT

Slaty-tailed Trogon *Trogon massena* NT
Lattice-tailed Trogon *Trogon clathratus* NT
Black-tailed Trogon *Trogon melanurus* NT
Blue-tailed Trogon *Trogon comptus* NT
Baird's Trogon *Trogon bairdii* ● NT
White-tailed Trogon *Trogon viridis* NT
Citreoline Trogon *Trogon citreolus* NT
Black-headed Trogon *Trogon melanocephalus* NT
Mexican Trogon *Trogon mexicanus* NT
Elegant Trogon *Trogon elegans* NA
Collared Trogon *Trogon collaris* NT
Orange-bellied Trogon *Trogon aurantiiventris* NT
Masked Trogon *Trogon personatus* NT
Black-throated Trogon *Trogon rufus* NT
Surucua Trogon *Trogon surrucura* NT
Blue-crowned Trogon *Trogon curucui* NT
Violaceous Trogon *Trogon violaceus* NT

Narina's Trogon *Apaloderma narina* AF
Bare-cheeked Trogon *Apaloderma aequatoriale* AF
Bar-tailed Trogon *Apaloderma vittatum* AF

Reinwardt's Blue-tailed Trogon *Harpactes reinwardtii* OR
Malabar Trogon *Harpactes fasciatus* OR
Red-naped Trogon *Harpactes kasumba* OR
Diard's Trogon *Harpactes diardii* OR
Philippine Trogon *Harpactes ardens* OR
Whitehead's Trogon *Harpactes whiteheadi* OR
Cinnamon-rumped Trogon *Harpactes orrhophaeus* OR
Scarlet-rumped Trogon *Harpactes duvaucelii* OR

Orange-breasted Trogon *Harpactes oreskios* OR
Red-headed Trogon *Harpactes erythrocephalus* OR
Ward's Trogon *Harpactes wardi* OR

KINGFISHERS AND ALLIES
Coraciiformes

84 KINGFISHERS Alcedinidae

Cerylinae

Greater Pied Kingfisher *Megaceryle lugubris* WP OR
Giant Kingfisher *Megaceryle maxima* AF
Ringed Kingfisher *Megaceryle torquata* NA NT
Belted Kingfisher *Megaceryle alcyon* NA NT

Lesser Pied Kingfisher *Ceryle rudis* WP AF OR

Amazon Kingfisher *Chloroceryle amazona* NT
Green Kingfisher *Chloroceryle americana* NA NT
Green & Rufous Kingfisher *Chloroceryle inda* NT
Pygmy Kingfisher *Chloroceryle aenea* NT

Alcedininae

Blyth's Kingfisher *Alcedo hercules* ● OR
Kingfisher *Alcedo atthis* WP EP OR AU PA
Half-collared Kingfisher *Alcedo semitorquata* AF
Blue-eared Kingfisher *Alcedo meninting* OR
Shining-blue Kingfisher *Alcedo quadribrachys* AF
Blue-banded Kingfisher *Alcedo euryzona* OR
Small Blue Kingfisher *Alcedo coerulescens* OR
Azure Kingfisher *Alcedo azureus* AU
Mangrove Kingfisher *Alcedo pusilla* AU PA
Malachite Kingfisher *Alcedo cristata* AF
Madagascar Malachite Kingfisher *Alcedo vintsioides* AF
White-bellied Kingfisher *Alcedo leucogaster* AF

African Dwarf Kingfisher *Ceyx lecontei* AF
African Pygmy Kingfisher *Ceyx picta* AF
Madagascar Pygmy Kingfisher *Ceyx madagascariensis* AF
Dwarf River Kingfisher *Ceyx cyanopectus* OR
Silvery Kingfisher *Ceyx argentatus* OR
Goodfellow's Kingfisher *Ceyx goodfellowi* OR
Dwarf Kingfisher *Ceyx lepidus* OR AU PA
Bismarck Pygmy Kingfisher *Ceyx websteri* AU
Three-toed Kingfisher *Ceyx erithacus* OR
Philippine Forest Kingfisher *Ceyx melanurus* OR
Sulawesi Pygmy Kingfisher *Ceyx fallax* AU

Banded Kingfisher *Lacedo pulchella* OR

Laughing Kookaburra *Dacelo novaeguineae* AU
Blue-winged Kookaburra *Dacelo leachii* AU
Aru Giant Kingfisher *Dacelo tyro* AU
Rufous-bellied Giant Kingfisher *Dacelo gaudichaud* AU

Shovel-billed Kingfisher *Clytoceyx rex* AU

Hook-billed Kingfisher *Melidora macrorrhina* AU

Sulawesi Blue-eared Kingfisher *Cittura cyanotis* AU

Brown-winged Kingfisher *Halcyon amauroptera* OR
Stork-billed Kingfisher *Halcyon capensis* OR AU
Great-billed Kingfisher *Halcyon melanorhyncha* AU
Ruddy Kingfisher *Halcyon coromanda* WP AU OR
Chocolate-backed Kingfisher *Halcyon badia* AF
White-breasted Kingfisher *Halcyon smyrnensis* OR
Black-capped Kingfisher *Halcyon pileata* EP OR AU
Java Kingfisher *Halcyon cyanoventris* OR
Grey-headed Kingfisher *Halcyon leucocephala* AF
Woodland Kingfisher *Halcyon senegalensis* AF
African Mangrove Kingfisher *Halcyon senegaloides* AF
Blue-breasted Kingfisher *Halcyon malimbica* AF
Brown-hooded Kingfisher *Halcyon albiventris* AF
Striped Kingfisher *Halcyon chelicuti* AF
Blue-black Kingfisher *Halcyon nigrocyanea* AU
Winchell's Kingfisher *Halcyon winchelli* OR
Moluccan Kingfisher *Halcyon diops* AU
South Moluccan Kingfisher *Halcyon lazuli* AU
Forest Kingfisher *Halcyon macleayii* AU
White-backed Kingfisher *Halcyon albonotata* AU
Ultramarine Kingfisher *Halcyon leucopygia* PA

Chestnut-bellied Kingfisher *Halcyon farquhari* PA
Red-backed Kingfisher *Halcyon pyrrhopygia* AU
Lesser Yellow-billed Kingfisher *Halcyon torotoro* AU
Mountain Yellow-billed Kingfisher *Halcyon megarhyncha* AU
Timor Kingfisher *Halcyon australasia* ● AU
Sacred Kingfisher *Halcyon sancta* OR AU PA
Micronesian Kingfisher *Halcyon cinnamomina* PA
Sombre Kingfisher *Halcyon funebris* AU
White-collared Kingfisher *Halcyon chloris* WP OR AU PA
Obscure Kingfisher *Halcyon enigma* AU
White-headed Kingfisher *Halcyon saurophaga* AU PA
Flat-billed Kingfisher *Halcyon recurvirostris* PA
Tahitian Kingfisher *Halcyon venerata* PA
Borabora Kingfisher *Halcyon tuta* PA
Mangaia Kingfisher *Halcyon ruficollaris* ● PA
Tuamotu Kingfisher *Halcyon gambieri* ● PA
Marquesas Kingfisher *Halcyon godeffroyi* ● PA

Glittering Kingfisher *Caridonax fulgida* AU

Moustached Kingfisher *Actenoides bougainvillei* ● PA
Chestnut-collared Kingfisher *Actenoides concreta* OR
Spotted Wood Kingfisher *Actenoides lindsayi* OR
Blue-headed Wood Kingfisher *Actenoides monacha* AU
Bar-headed Wood Kingfisher *Actenoides princeps* AU

Aru Paradise Kingfisher *Tanysiptera hydrocharis* AU
Common Paradise Kingfisher *Tanysiptera galatea* AU
Biak Paradise Kingfisher *Tanysiptera riedelii* ● AU
Numfor Paradise Kingfisher *Tanysiptera carolinae* AU
Kofiau Paradise Kingfisher *Tanysiptera ellioti* AU
Pink-breasted Paradise Kingfisher *Tanysiptera nympha* AU
Brown-backed Paradise Kingfisher *Tanysiptera danae* AU
White-tailed Kingfisher *Tanysiptera sylvia* AU

85 TODIES Todidae

Cuban Tody *Todus multicolor* NT
Narrow-billed Tody *Todus angustirostris* NT
Jamaican Tody *Todus todus* NT
Puerto Rican Tody *Todus mexicanus* NT
Broad-billed Tody *Todus subulatus* NT

86 MOTMOTS Momotidae

Tody-Motmot *Hylomanes momotula* NT

Blue-throated Motmot *Aspatha gularis* NT

Broad-billed Motmot *Electron platyrhynchum* NT
Keel-billed Motmot *Electron carinatum* ● NT

Turquoise-browed Motmot *Eumomota superciliosa* NT

Rufous-capped Motmot *Baryphthengus ruficapillus* NT

Rufous Motmot *Baryphthengus martii* NT

Russet-crowned Motmot *Momotus mexicanus* NT
Blue-crowned Motmot *Momotus momota* NT

87 BEE-EATERS Meropidae

Red-bearded Bee-eater *Nyctyornis amicta* OR
Blue-bearded Bee-eater *Nyctyornis athertoni* OR

Sulawesi Bearded Bee-eater *Meropogon forsteni* AU

Black Bee-eater *Merops gularis* AF
Blue-headed Bee-eater *Merops muelleri* AF
Red-throated Bee-eater *Merops bulocki* AF
White-fronted Bee-eater *Merops bullockoides* AF
Little Bee-eater *Merops pusillus* AF
Blue-breasted Bee-eater *Merops variegatus* AF
Cinnamon-chested Bee-eater *Merops oreobates* AF
Swallow-tailed Bee-eater *Merops hirundineus* AF
Black-headed Bee-eater *Merops breweri* AF
Somali Bee-eater *Merops revoilii* AF
White-throated Bee-eater *Merops albicollis* AF
Little Green Bee-eater *Merops orientalis* WP AF OR
Boehm's Bee-eater *Merops boehmi* AF
Chestnut-headed Bee-eater *Merops viridis* OR
Blue-cheeked Bee-eater *Merops superciliosus* WP AF OR AU

Blue-tailed Bee-eater *Merops philippinus* OR
Australian Bee-eater *Merops ornatus* AU PA
Bee-eater *Merops apiaster* WP EP AF OR
Bay-headed Bee-eater *Merops leschenaulti* OR
Rosy Bee-eater *Merops malimbicus* AF
Carmine Bee-eater *Merops nubicus* AF

88 ROLLERS Coraciidae

Roller *Coracias garrulus* WP EP AF OR
Abyssinian Roller *Coracias abyssinica* AF
Lilac-breasted Roller *Coracias caudata* AF
Racquet-tailed Roller *Coracias spatulata* AF
Rufous-crowned Roller *Coracias naevia* AF
Indian Roller *Coracias benghalensis* EP OR
Sulawesi Roller *Coracias temminckii* AU
Blue-bellied Roller *Coracias cyanogaster* AF

African Broad-billed Roller *Eurystomus glaucurus* AF
Blue-throated Roller *Eurystomus gularis* AF
Eastern Broad-billed Roller (Dollar Bird) *Eurystomus orientalis* EP OR AU PA
Azure Roller *Eurystomus azureus* AU

89 GROUND ROLLERS Brachypteraciidae

Short-legged Ground Roller *Brachypteracias leptosomus* ● AF
Scaly Ground Roller *Brachypteracias squamigera* ● AF

Pitta-like Ground Roller *Atelornis pittoides* ● AF
Crossley's Ground Roller *Atelornis crossleyi* ● AF

Long-tailed Ground Roller *Uratelornis chimaera* ● AF

90 COUROL Leptosomatidae

Courol *Leptosomus discolor* AF

91 HOOPOE Upupidae

Hoopoe *Upupa epops* WP EP AF OR

92 WOOD HOOPOES Phoeniculidae

Green Wood Hoopoe *Phoeniculus purpureus* AF
Black-billed Wood Hoopoe *Phoeniculus somaliensis* AF
Violet Wood Hoopoe *Phoeniculus damarensis* AF
White-headed Wood Hoopoe *Phoeniculus bollei* AF
Forest Wood Hoopoe *Phoeniculus castaneiceps* AF
Black Wood Hoopoe *Phoeniculus aterrimus* AF

Abyssinian Scimitarbill *Rhinopomastus minor* AF
Scimitarbill *Rhinopomastus cyanomelas* AF

93 HORNBILLS Bucerotidae

Indian Grey Hornbill *Tockus birostris* OR
Pied Hornbill *Tockus fasciatus* AF
Crowned Hornbill *Tockus alboterminatus* AF
Bradfield's Hornbill *Tockus bradfieldi* AF
Pale-billed Hornbill *Tockus pallidirostris* AF
African Grey Hornbill *Tockus nasutus* AF
Hemprich's Hornbill *Tockus hemprichii* AF
Monteiro's Hornbill *Tockus monteiri* AF
Malabar Grey Hornbill *Tockus griseus* OR
Black Dwarf Hornbill *Tockus hartlaubi* AF
Red-billed Dwarf Hornbill *Tockus camurus* AF
Red-billed Hornbill *Tockus erythrorhynchus* AF
Yellow-billed Hornbill *Tockus flavirostris* AF
Southern Yellow-billed Hornbill *Tockus leucomelas* AF

Von der Decken's Hornbill *Tockus deckeni* AF
Jackson's Hornbill *Tockus jacksoni* AF

Long-crested Hornbill *Berenicornis comatus* OR
African White-crested Hornbill *Berenicornis albocristatus* AF

Tickell's Hornbill *Ptilolaemus tickelli* OR

Bushy-crested Hornbill *Anorrhinus galeritus* OR

Rufous-tailed Hornbill *Penelopides panini* OR
Temminck's Hornbill *Penelopides exarhatus* OR

Rufous-necked Hornbill *Aceros nipalensis* ● OR
Wrinkled Hornbill *Aceros corrugatus* ● OR
White-headed Hornbill *Aceros leucocephalus* OR
Sulawesi Hornbill *Aceros cassidix* AU
Wreathed Hornbill *Aceros undulatus* OR
Blyth's Hornbill *Aceros plicatus* OR AU PA
Everett's Hornbill *Aceros everetti* ● AU

Narcondam Hornbill *Aceros narcondami* ● OR

Black Hornbill *Anthracoceros malayanus* OR
Indian Pied Hornbill *Anthracoceros malabaricus* EP OR
Malabar Pied Hornbill *Anthracoceros coronatus* OR
Sulu Hornbill *Anthracoceros montani* ● OR
Palawan Hornbill *Anthracoceros marchei* OR

Piping Hornbill *Ceratogymna fistulator* AF
Trumpeter Hornbill *Ceratogymna bucinator* AF
Brown-cheeked Hornbill *Ceratogymna cylindricus* AF
Black & White Casqued Hornbill *Ceratogymna subcylindricus* AF
Silvery-cheeked Hornbill *Ceratogymna brevis* AF
Black-casqued Hornbill *Ceratogymna atrata* AF
Yellow-casqued Hornbill *Ceratogymna elata* AF

Rhinoceros Hornbill *Buceros rhinoceros* OR
Great Indian Hornbill *Buceros bicornis* OR
Rufous Hornbill *Buceros hydrocorax* OR

Helmeted Hornbill *Rhinoplax vigil* ● OR

Abyssinian Ground Hornbill *Bucorvus abyssinicus* AF
Southern Ground Hornbill *Bucorvus leadbeateri* AF

WOODPECKERS AND ALLIES Piciformes

94 JACAMARS Galbulidae

Chestnut Jacamar *Galbalcyrhynchus leucotis* NT
Purus Jacamar *Galbalcyrhynchus purusianus* NT

White-throated Jacamar *Brachygalba albogularis* NT
Brown Jacamar *Brachygalba lugubris* NT
Pale-headed Jacamar *Brachygalba goeringi* NT
Dusky-backed Jacamar *Brachygalba salmoni* NT

Three-toed Jacamar *Jacamaralcyon tridactyla* ● NT

Yellow-billed Jacamar *Galbula albirostris* NT
Blue-necked Jacamar *Galbula cyanicollis* NT
Green-tailed Jacamar *Galbula galbula* NT
Rufous-tailed Jacamar *Galbula ruficauda* NT
White-chinned Jacamar *Galbula tombacea* NT
Bluish-fronted Jacamar *Galbula cyanescens* NT
Coppery-chested Jacamar *Galbula pastazae* NT
Bronzy Jacamar *Galbula leucogastra* NT
Paradise Jacamar *Galbula dea* NT

Great Jacamar *Jacamerops aurea* NT

95 PUFFBIRDS Bucconidae

White-necked Puffbird *Notharchus macrorhynchos* NT
Black-breasted Puffbird *Notharchus pectoralis* NT
Brown-banded Puffbird *Notharchus ordii* NT
Pied Puffbird *Notharchus tectus* NT

Chestnut-capped Puffbird *Bucco macrodactylus* NT
Spotted Puffbird *Bucco tamatia* NT
Sooty-capped Puffbird *Bucco noanamae* NT
Collared Puffbird *Bucco capensis* NT

Barred Puffbird *Nystalus radiatus* NT
White-eared Puffbird *Nystalus chacuru* NT
Striolated Puffbird *Nystalus striolatus* NT
Spot-backed Puffbird *Nystalus maculatus* NT

Russet-throated Puffbird *Hypnelus ruficollis* NT

Crescent-chested Puffbird *Malacoptila striata* NT
White-chested Puffbird *Malacoptila fusca* NT
Semicollared Puffbird *Malacoptila semicincta* NT
Black-streaked Puffbird *Malacoptila fulvogularis* NT
Rufous-necked Puffbird *Malacoptila rufa* NT
White-whiskered Puffbird *Malacoptila panamensis* NT
Moustached Puffbird *Malacoptila mystacalis* NT

Lanceolated Monklet *Micromonacha lanceolata* NT

Rusty-breasted Nunlet *Nonnula rubecula* NT
Fulvous-chinned Nunlet *Nonnula sclateri* NT
Brown Nunlet *Nonnula brunnea* NT
Grey-cheeked Nunlet *Nonnula ruficapilla* NT
Chestnut-headed Nunlet *Nonnula amaurocephala* NT

White-faced Nunbird *Hapaloptila castanea* NT

Black Nunbird *Monasa atra* NT
Black-fronted Nunbird *Monasa nigrifrons* NT

White-fronted Nunbird *Monasa morphoeus* NT
Yellow-billed Nunbird *Monasa flavirostris* NT

Swallow-wing Puffbird *Chelidoptera tenebrosa* NT

96 BARBETS Capitonidae

Scarlet-crowned Barbet *Capito aurovirens* NT
Spot-crowned Barbet *Capito maculicoronatus* NT
Orange-fronted Barbet *Capito squamatus* NT
White-mantled Barbet *Capito hypoleucus* ● NT
Black-girdled Barbet *Capito dayi* NT
Five-coloured Barbet *Capito quinticolor* NT
Black-spotted Barbet *Capito niger* NT

Lemon-throated Barbet *Eubucco richardsoni* NT
Red-headed Barbet *Eubucco bourcierii* NT
Scarlet-hooded Barbet *Eubucco tucinkae* NT
Versicoloured Barbet *Eubucco versicolor* NT

Prong-billed Barbet *Semnornis frantzii* NT
Toucan Barbet *Semnornis ramphastinus* ● NT

Fire-tufted Barbet *Psilopogon pyrolophus* OR

Great Barbet *Megalaima virens* OR
Red-vented Barbet *Megalaima lagrandieri* OR
Oriental Green Barbet *Megalaima zeylanica* OR
Lineated Barbet *Megalaima lineata* OR
Small Green Barbet *Megalaima viridis* OR
Green-eared Barbet *Megalaima faiostricta* OR
Brown-throated Barbet *Megalaima corvina* OR
Gold-whiskered Barbet *Megalaima chrysopogon* OR
Many-coloured Barbet *Megalaima rafflesii* OR
Gaudy Barbet *Megalaima mystacophanos* OR
Black-banded Barbet *Megalaima javensis* ● OR
Yellow-fronted Barbet *Megalaima flavifrons* OR
Golden-throated Barbet *Megalaima franklinii* OR
Muller's Barbet *Megalaima oorti* OR
Blue-throated Barbet *Megalaima asiatica* OR
Hume's Blue-throated Barbet *Megalaima incognita* OR
Yellow-crowned Barbet *Megalaima henricii* OR
Blue-crowned Barbet *Megalaima armillaris* OR
Golden-naped Barbet *Megalaima pulcherrima* OR
Blue-eared Barbet *Megalaima australis* OR
Black-throated Barbet *Megalaima eximia* OR
Crimson-throated Barbet *Megalaima rubricapilla* OR
Crimson-breasted Barbet *Megalaima haemacephala* OR

Brown Barbet *Calorhamphus fuliginosus* OR

Naked-faced Barbet *Gymnobucco calvus* AF
Bristle-nosed Barbet *Gymnobucco peli* AF
Sladen's Barbet *Gymnobucco sladeni* AF
Grey-throated Barbet *Gymnobucco bonapartei* AF

White-eared Barbet *Smilorhis leucotis* AF

Green Barbet *Cryptolybia olivacea* AF

Anchieta's Barbet *Stactolaema anchietae* AF
Whyte's Barbet *Stactolaema whytii* AF

Yellow-spotted Barbet *Buccanodon duchaillui* AF

Speckled Tinkerbird *Pogoniulus scolopaceus* AF
Moustached Green Tinkerbird *Pogoniulus leucomystax* AF
Green Tinkerbird *Pogoniulus simplex* AF
Western Green Tinkerbird *Pogoniulus coryphaeus* AF
Red-fronted Tinkerbird *Pogoniulus pusillus* AF
Yellow-fronted Tinkerbird *Pogoniulus chrysoconus* AF
Yellow-rumped Tinkerbird *Pogoniulus bilineatus* AF
White-chested Tinkerbird *Pogoniulus makawai* ● AF
Yellow-throated Tinkerbird *Pogoniulus subsulphureus* AF
Red-rumped Tinkerbird *Pogoniulus atroflavus* AF

Spotted-flanked Barbet *Tricholaema lacrymosa* AF
Pied Barbet *Tricholaema leucomelaina* AF
Red-fronted Barbet *Tricholaema diademata* AF
Miombo Pied Barbet *Tricholaema frontata* AF
African Black-throated Barbet *Tricholaema melanocephala* AF
Hairy-breasted Barbet *Tricholaema hirsuta* AF

Banded Barbet *Lybius undatus* AF
Vieillot's Barbet *Lybius vieilloti* AF
Black-collared Barbet *Lybius torquatus* AF

Black-billed Barbet *Lybius guifsobalito* AF
Red-faced Barbet *Lybius rubrifacies* AF
Chaplin's Barbet *Lybius chaplini* AF
White-headed Barbet *Lybius leucocephalus* AF
Black-backed Barbet *Lybius minor* AF
Brown-breasted Barbet *Lybius melanopterus* AF
Double-toothed Barbet *Lybius bidentatus* AF
Bearded Barbet *Lybius dubius* AF
Black-breasted Barbet *Lybius rolleti* AF

Yellow-billed Barbet *Trachyphonus purpuratus* AF
Levaillant's Barbet *Trachyphonus vaillantii* AF
Red & Yellow Barbet *Trachyphonus erythrocephalus* AF
d'Arnaud's Barbet *Trachyphonus darnaudii* AF
Usambiro Barbet *Trachyphonus usambiro* AF
Yellow-breasted Barbet *Trachyphonus margaritatus* AF

97 HONEYGUIDES Indicatoridae

Cassin's Honeybird *Prodotiscus insignis* AF
Green-backed Honeybird *Prodotiscus zambesiae* AF
Wahlberg's Honeybird *Prodotiscus regulus* AF

Zenker's Honeyguide *Melignomon zenkeri* AF
Yellow-footed Honeyguide *Melignomon eisentrauti* ● AF

Spotted Honeyguide *Indicator maculatus* AF
Scaly-throated Honeyguide *Indicator variegatus* AF
Black-throated Honeyguide *Indicator indicator* AF
Lesser Honeyguide *Indicator minor* AF
Thick-billed Honeyguide *Indicator conirostris* AF
Least Honeyguide *Indicator exilis* AF
Willcocks' Honeyguide *Indicator willcocksi* AF
Eastern Least Honeyguide *Indicator meliphilus* AF
Pygmy Honeyguide *Indicator pumilio* AF
Indian Honeyguide *Indicator xanthonotus* AF
Malay Honeyguide *Indicator archipelagus* OR

Lyre-tailed Honeyguide *Melichneutes robustus* AF

98 TOUCANS Ramphastidae

Emerald Toucanet *Aulacorhynchus prasinus* NT
Groove-billed Toucanet *Aulacorhynchus sulcatus* NT
Chestnut-tipped Toucanet *Aulacorhynchus derbianus* NT
Crimson-rumped Toucanet *Aulacorhynchus haematopygus* NT
Yellow-browed Toucanet *Aulacorhynchus huallagae* ● NT
Blue-banded Toucanet *Aulacorhynchus coeruleicinctis* NT

Green Araçari *Pteroglossus viridis* NT
Lettered Araçari *Pteroglossus inscriptus* NT
Red-necked Araçari *Pteroglossus bitorquatus* NT
Ivory-billed Araçari *Pteroglossus flavirostris* NT
Black-necked Araçari *Pteroglossus aracari* NT
Chestnut-eared Araçari *Pteroglossus castanotis* NT
Many-banded Araçari *Pteroglossus pluricinctus* NT
Collared Araçari *Pteroglossus torquatus* NT
Fiery-billed Araçari *Pteroglossus frantzii* NT
Curl-crested Araçari *Pteroglossus beauharnaesii* NT

Spot-billed Toucanet *Selenidera maculirostris* NT
Gould's Toucanet *Selenidera gouldii* NT
Golden-collared Toucanet *Selenidera reinwardtii* NT
Tawny-tufted Toucanet *Selenidera nattereri* NT
Guianan Toucanet *Selenidera culik* NT
Yellow-eared Toucanet *Selenidera spectabilis* NT

Saffron Toucanet *Baillonius bailloni* NT

Grey-breasted Mountain Toucan *Andigena hypoglauca* NT
Plate-billed Mountain Toucan *Andigena laminirostris* NT
Hooded Mountain Toucan *Andigena cucullata* NT
Black-billed Mountain Toucan *Andigena nigrirostris* NT

Red-breasted Toucan *Ramphastos dicolorus* NT
Channel-billed Toucan *Ramphastos vitellinus* NT
Choco Toucan *Ramphastos brevis* NT
Keel-billed Toucan *Ramphastos sulfuratus* NT
Toco Toucan *Ramphastos toco* NT
Red-billed Toucan *Ramphastos tucanus* NT
Chestnut-mandibled Toucan *Ramphastos swainsonii* NT Ariel (?)

Black-mandibled Toucan *Ramphastos ambiguus* NT

99 WOODPECKERS Picidae

Wryneck *Jynx torquilla* WP EP AF OR
Rufous-necked Wryneck *Jynx ruficollis* AF

Speckled Piculet *Picumnus innominatus* EP OR
Bar-breasted Piculet *Picumnus aurifrons* NT
Lafresnaye's Piculet *Picumnus lafresnayi* NT
Golden-spangled Piculet *Picumnus exilis* NT
Ecuadorean Piculet *Picumnus sclateri* NT
Scaled Piculet *Picumnus squamulatus* NT
White-bellied Piculet *Picumnus spilogaster* NT
Guianan Piculet *Picumnus minutissimus* NT
Spotted Piculet *Picumnus pygmaeus* NT
Speckle-chested Piculet *Picumnus steindachneri* NT
Varzea Piculet *Picumnus varzeae* NT
White-barred Piculet *Picumnus cirratus* NT
White-wedged Piculet *Picumnus albosquamatus* NT
Rusty-necked Piculet *Picumnus fuscus* NT
Rufous-breasted Piculet *Picumnus rufiventris* NT
Tawny Piculet *Picumnus fulvescens* NT
Ochraceous Piculet *Picumnus limae* NT
Mottled Piculet *Picumnus nebulosus* NT
Plain-breasted Piculet *Picumnus castelnau* NT
Fine-barred Piculet *Picumnus subtilis* NT
Olivaceous Piculet *Picumnus olivaceus* NT
Greyish Piculet *Picumnus granadensis* NT
Chestnut Piculet *Picumnus cinnamomeus* NT

African Piculet *Sasia africana* AF
Rufous Piculet *Sasia abnormis* OR
White-browed Piculet *Sasia ochracea* OR

Antillean Piculet *Nesoctites micromegas* NT

White Woodpecker *Melanerpes candidus* NT
Lewis' Woodpecker *Melanerpes lewis* NA NT
Guadeloupe Woodpecker *Melanerpes herminieri* NT
Puerto Rican Woodpecker *Melanerpes portoricensis* NT
Red-headed Woodpecker *Melanerpes erythrocephalus* NA
Acorn Woodpecker *Melanerpes formicivorus* NA NT
Red-fronted Woodpecker *Melanerpes cruentus* NT
Yellow-fronted Woodpecker *Melanerpes flavifrons* NT
Golden-naped Woodpecker *Melanerpes chrysauchen* NT
Black-cheeked Woodpecker *Melanerpes pucherani* NT
White-fronted Woodpecker *Melanerpes cactorum* NT
Hispaniolan Woodpecker *Melanerpes striatus* NT
Jamaican Woodpecker *Melanerpes radiolatus* NT
Golden-cheeked Woodpecker *Melanerpes chrysogenys* NT
Grey-breasted Woodpecker *Melanerpes hypopolius* NT
Yucatan Woodpecker *Melanerpes pygmaeus* NT
Red-crowned Woodpecker *Melanerpes rubricapillus* NT
Hoffmann's Woodpecker *Melanerpes hoffmanni* NT
Gila Woodpecker *Melanerpes uropygialis* NA NT
Golden-fronted Woodpecker *Melanerpes aurifrons* NA NT
Red-bellied Woodpecker *Melanerpes carolinus* NA
Great Red-bellied Woodpecker *Melanerpes superciliaris* NT

Yellow-bellied Sapsucker *Sphyrapicus varius* NA NT
Red-naped Sapsucker *Sphyrapicus nuchalis* NA
Red-breasted Sapsucker *Sphyrapicus ruber* NA
Williamson's Sapsucker *Sphyrapicus thyroideus* NA NT

Cuban Green Woodpecker *Xiphidiopicus percussus* NT

Fine-spotted Woodpecker *Campethera punctuligera* AF
Bennett's Woodpecker *Campethera bennettii* AF
Nubian Woodpecker *Campethera nubica* AF
Golden-tailed Woodpecker *Campethera abingoni* AF
Knysna Woodpecker *Campethera notata* AF
Green-backed Woodpecker *Campethera cailliautii* AF
Little Green Woodpecker *Campethera maculosa* AF
Tullberg's Woodpecker *Campethera tullbergi* AF

Buff-spotted Woodpecker *Campethera nivosa* AF
Brown-eared Woodpecker *Campethera caroli* AF
Ground Woodpecker *Geocolaptes olivaceus* AF
Little Grey Woodpecker *Dendropicos elachus* AF
Speckle-breasted Woodpecker *Dendropicos poecilolaemus* AF
Gold-mantled Woodpecker *Dendropicos abyssinicus* AF
Cardinal Woodpecker *Dendropicos fuscescens* AF
Gabon Woodpecker *Dendropicos gabonensis* AF
Stierling's Woodpecker *Dendropicos stierlingi* AF
Bearded Woodpecker *Dendropicos namaquus* AF
Yellow-crested Woodpecker *Dendropicos xantholophus* AF
Fire-bellied Woodpecker *Dendropicos pyrrhogaster* AF
Elliot's Woodpecker *Dendropicos elliotii* AF
Grey Woodpecker *Dendropicos goertae* AF
Olive Woodpecker *Dendropicos griseocephalus* AF

Temminck's Pygmy Woodpecker *Picoides temminckii* AU
Philippine Pygmy Woodpecker *Picoides maculatus* OR
Brown-capped Woodpecker *Picoides moluccensis* OR AU
Brown-backed Woodpecker *Picoides obsoletus* AF
Japanese Spotted Woodpecker *Picoides kizuki* EP
Grey-capped Woodpecker *Picoides canicapillus* EP OR
Lesser Spotted Woodpecker *Picoides minor* WP EP
Streak-bellied Woodpecker *Picoides macei* EP OR
Stripe-breasted Woodpecker *Picoides atratus* OR
Brown-fronted Woodpecker *Picoides auriceps* OR
Yellow-crowned Woodpecker *Picoides mahrattensis* OR
Arabian Woodpecker *Picoides dorae* WP
Rufous-bellied Woodpecker *Picoides hyperythrus* EP OR
Crimson-breasted Woodpecker *Picoides cathpharius* EP OR
Brown-throated Woodpecker *Picoides darjellensis* EP OR
Middle-spotted Woodpecker *Picoides medius* WP EP
White-backed Woodpecker *Picoides leucotos* WP EP
Himalayan Woodpecker *Picoides himalayensis* EP OR
Sind Woodpecker *Picoides assimilis* EP OR
Syrian Woodpecker *Picoides syriacus* WP
White-winged Woodpecker *Picoides leucopterus* EP
Great Spotted Woodpecker *Picoides major* WP EP OR
Checkered Woodpecker *Picoides mixtus* NT
Striped Woodpecker *Picoides lignarius* NT
Ladder-backed Woodpecker *Picoides scalaris* NA NT
Nuttall's Woodpecker *Picoides nuttallii* NA NT
Downy Woodpecker *Picoides pubescens* NA
Red-cockaded Woodpecker *Picoides borealis* ● NA
Strickland's Woodpecker *Picoides stricklandii* NA
Hairy Woodpecker *Picoides villosus* NA
White-headed Woodpecker *Picoides albolarvatus* NA
Three-toed Woodpecker *Picoides tridactylus* WP EP NA
Black-backed Woodpecker *Picoides arcticus* NA

Scarlet-backed Woodpecker *Veniliornis callonotus* NT
Yellow-vented Woodpecker *Veniliornis dignus* NT
Bar-bellied Woodpecker *Veniliornis nigriceps* NT
Smoky-brown Woodpecker *Veniliornis fumigatus* NT
Little Woodpecker *Veniliornis passerinus* NT
Dot-fronted Woodpecker *Veniliornis frontalis* NT
White-spotted Woodpecker *Veniliornis spilogaster* NT
Blood-coloured Woodpecker *Veniliornis sanguineus* NT
Yellow-eared Woodpecker *Veniliornis maculifrons* NT
Red-stained Woodpecker *Veniliornis affinis* NT
Golden-collared Woodpecker *Veniliornis cassini* NT

Red-rumped Woodpecker *Veniliornis kirkii* NT
White-throated Woodpecker *Piculus leucolaemus* NT
Yellow-throated Woodpecker *Piculus flavigula* NT
Golden-green Woodpecker *Piculus chrysochloros* NT
White-browed Woodpecker *Piculus aurulentus* NT
Golden-olive Woodpecker *Piculus rubiginosus* NT
Grey-crowned Woodpecker *Piculus auricularis* NT
Crimson-mantled Woodpecker *Piculus rivolii* NT

Black-necked Flicker *Colaptes atricollis* NT
Spot-breasted Flicker *Colaptes punctigula* NT
Green-barred Flicker *Colaptes melanochloros* NT
Common Flicker *Colaptes auratus* NA
Fernandina's Flicker *Colaptes fernandinae* ● NT
Chilean Flicker *Colaptes pitius* NT
Andean Flicker *Colaptes rupicola* NT
Campo Flicker *Colaptes campestris* NT
Field Flicker *Colaptes campestroides* NT

Rufous Woodpecker *Celeus brachyurus* NT
Cinnamon Woodpecker *Celeus loricatus* NT
Waved Woodpecker *Celeus undatus* NT
Scale-breasted Woodpecker *Celeus grammicus* NT
Chestnut-coloured Woodpecker *Celeus castaneus* NT
Chestnut Woodpecker *Celeus elegans* NT
Pale-crested Woodpecker *Celeus lugubris* NT
Blond-crested Woodpecker *Celeus flavescens* NT
Cream-coloured Woodpecker *Celeus flavus* NT
Rufous-headed Woodpecker *Celeus spectabilis* NT
Ringed Woodpecker *Celeus torquatus* NT

Helmeted Woodpecker *Dryocopus galeatus* ● NT
Black-bodied Woodpecker *Dryocopus schulzi* NT
Lineated Woodpecker *Dryocopus lineatus* NT
Pileated Woodpecker *Dryocopus pileatus* NT
White-bellied Woodpecker *Dryocopus javensis* EP OR
Black Woodpecker *Dryocopus martius* WP EP OR

Powerful Woodpecker *Campephilus pollens* NT
Crimson-bellied Woodpecker *Campephilus haematogaster* NT
Red-necked Woodpecker *Campephilus rubricollis* NT
Robust Woodpecker *Campephilus robustus* NT
Pale-billed Woodpecker *Campephilus guatemalensis* NT
Crimson-crested Woodpecker *Campephilus melanoleucos* NT
Guayaquil Woodpecker *Campephilus gayaquilensis* NT
Cream-backed Woodpecker *Campephilus leucopogon* NT
Magellanic Woodpecker *Campephilus magellanicus* NT
Ivory-billed Woodpecker *Campephilus principalis* ● NA NT
Imperial Woodpecker *Campephilus imperialis* ● NT

Banded Red Woodpecker *Picus miniaceus* OR
Crimson-winged Woodpecker *Picus puniceus* OR
Lesser Yellow-naped Woodpecker *Picus chlorolophus* OR
Checker-throated Woodpecker *Picus mentalis* OR
Greater Yellow-naped Woodpecker *Picus flavinucha* OR
Laced Woodpecker *Picus vittatus* OR
Streak-throated Woodpecker *Picus xanthopygaeus* EP OR
Scaly-bellied Woodpecker *Picus squamatus* EP OR
Wavy-bellied Woodpecker *Picus awokera* EP
Green Woodpecker *Picus viridis* WP EP
Red-collared Woodpecker *Picus rabieri* ● OR
Black-headed Woodpecker *Picus erythropygius* OR
Grey-faced Woodpecker *Picus canus* WP EP OR

Olive-backed Woodpecker *Dinopium rafflesii* OR
Himalayan Gold-backed Woodpecker *Dinopium shorii* OR
Common Gold-backed Woodpecker *Dinopium javanense* OR
Lesser Flame-backed Woodpecker *Dinopium benghalense* OR

Greater Flame-backed Woodpecker *Chrysocolaptes lucidus* OR
Black-rumped Woodpecker *Chrysocolaptes festivus* OR

Bamboo Woodpecker *Gecinulus grantia* OR
Okinawan Woodpecker *Sapheopipo noguchii* ● EP
Maroon Woodpecker *Blythipicus rubiginosus* OR
Bay Woodpecker *Blythipicus pyrrhotis* OR
Orange-backed Woodpecker *Reinwardtipicus validus* OR

Buff-rumped Woodpecker *Meiglyptes tristis* OR
Black & Buff Woodpecker *Meiglyptes jugularis* OR
Buff-necked Woodpecker *Meiglyptes tukki* OR

Grey and Buff Woodpecker *Hemicircus concretus* OR
Heart-spotted Woodpecker *Hemicircus canente* OR

Fulvous Woodpecker *Mulleripicus fulvus* AU
Sooty Woodpecker *Mulleripicus funebris* OR
Great Slaty Woodpecker *Mulleripicus pulverulentus* OR

PERCHING BIRDS Passeriformes

100 BROADBILLS Eurylaimidae

Eurylaiminae

African Broadbill *Smithornis capensis* AF
Red-sided Broadbill *Smithornis rufolateralis* AF
Grey-headed Broadbill *Smithornis sharpei* AF

Grauer's Broadbill *Pseudocalyptomena graueri* ● AF

Dusky Broadbill *Corydon sumatranus* OR

Black & Red Broadbill *Cymbirhynchus macrorhynchos* OR

Banded Broadbill *Eurylaimus javanicus* OR
Black & Yellow Broadbill *Eurylaimus ochromalus* OR
Wattled Broadbill *Eurylaimus steerii* OR

Silver-breasted Broadbill *Serilophus lunatus* OR

Long-tailed Broadbill *Psarisomus dalhousiae* OR

Calyptomeninae

Lesser Green Broadbill *Calyptomena viridis* OR
Magnificent Green Broadbill *Calyptomena hosii* OR
Black-throated Green Broadbill *Calyptomena whiteheadi* OR

101 WOODCREEPERS Dendrocolaptidae

Tyrannine Woodcreeper *Dendrocincla tyrannina* NT
Large Tyrannine Woodcreeper *Dendrocincla macrorhyncha* NT
Plain-brown Woodcreeper *Dendrocincla fuliginosa* NT
Tawny-winged Woodcreeper *Dendrocincla anabatina* NT
White-chinned Woodcreeper *Dendrocincla merula* NT
Ruddy Woodcreeper *Dendrocincla homochroa* NT

Long-tailed Woodcreeper *Deconychura longicauda* NT
Spot-throated Woodcreeper *Deconychura stictolaema* NT

Olivaceous Woodcreeper *Sittasomus griseicapillus* NT

Wedge-billed Woodcreeper *Glyphorhynchus spirurus* NT

Scimitar-billed Woodhewer *Drymornis bridgesii* NT

Long-billed Woodcreeper *Nasica longirostris* NT

Cinnamon-throated Woodcreeper *Dendrexetastes rufigula* NT

Red-billed Woodcreeper *Hylexetastes perrotii* NT
Bar-bellied Woodcreeper *Hylexetastes stresemanni* NT

Strong-billed Woodcreeper *Xiphocolaptes promeropirhynchus* NT
White-throated Woodcreeper *Xiphocolaptes albicollis* NT
Vila Nova Woodcreeper *Xiphocolaptes villanovae* NT

Moustached Woodcreeper *Xiphocolaptes falcirostris* ● NT
Snethlage's Woodcreeper *Xiphocolaptes franciscanus* ● NT
Great Rufous Woodcreeper *Xiphocolaptes major* NT

Barred Woodcreeper *Dendrocolaptes certhia* NT
Concolor Woodcreeper *Dendrocolaptes concolor* NT
Hoffmann's Woodcreeper *Dendrocolaptes hoffmannsi* NT
Black-banded Woodcreeper *Dendrocolaptes picumnus* NT
Planalto Woodcreeper *Dendrocolaptes platyrostris* NT

Straight-billed Woodcreeper *Xiphorhynchus picus* NT
Zimmer's Woodcreeper *Xiphorhynchus necopinus* NT
Striped Woodcreeper *Xiphorhynchus obsoletus* NT
Ocellated Woodcreeper *Xiphorhynchus ocellatus* NT
Spix's Woodcreeper *Xiphorhynchus spixii* NT
Elegant Woodcreeper *Xiphorhynchus elegans* NT
Chestnut-rumped Woodcreeper *Xiphorhynchus pardalotus* NT
Buff-throated Woodcreeper *Xiphorhynchus guttatus* NT
Dusky-billed Woodcreeper *Xiphorhynchus eytoni* NT
Ivory-billed Woodcreeper *Xiphorhynchus flavigaster* NT
Stripe-throated Woodcreeper *Xiphorhynchus striatigularis* NT
Black-striped Woodcreeper *Xiphorhynchus lachrymosus* NT
Spotted Woodcreeper *Xiphorhynchus erythropygius* NT
Olive-backed Woodcreeper *Xiphorhynchus triangularis* NT

White-striped Woodcreeper *Lepidocolaptes leucogaster* NT
Streak-headed Woodcreeper *Lepidocolaptes souleyetii* NT
Narrow-billed Woodcreeper *Lepidocolaptes angustirostris* NT
Spot-crowned Woodcreeper *Lepidocolaptes affinis* NT
Scaled Woodcreeper *Lepidocolaptes squamatus* NT
Lesser Woodcreeper *Lepidocolaptes fuscus* NT
Lineated Woodcreeper *Lepidocolaptes albolineatus* NT

Greater Scythebill *Campylorhamphus pucheranii* NT
Red-billed Scythebill *Campylorhamphus trochilirostris* NT
Black-billed Scythebill *Campylorhamphus falcularius* NT
Brown-billed Scythebill *Campylorhamphus pusillus* NT
Curve-billed Scythebill *Campylorhamphus procurvoides* NT

102 OVENBIRDS Furnariidae

Furnariinae

Campo Miner *Geositta poeciloptera* NT
Common Miner *Geositta cunicularia* NT
Greyish Miner *Geositta maritima* NT
Coastal Miner *Geositta peruviana* NT
Puna Miner *Geositta punensis* NT
Dark-winged Miner *Geositta saxicolina* NT
Creamy-rumped Miner *Geositta isabellina* NT
Short-billed Miner *Geositta antarctica* NT
Rufous-banded Miner *Geositta rufipennis* NT
Thick-billed Miner *Geositta crassirostris* NT
Slender-billed Miner *Geositta tenuirostris* NT

Chaco Earthcreeper *Upucerthia certhioides* NT
Bolivian Earthcreeper *Upucerthia harterti* NT
Straight-billed Earthcreeper *Upucerthia ruficauda* NT
Rock Earthcreeper *Upucerthia andaecola* NT
White-throated Earthcreeper *Upucerthia albigula* NT
Striated Earthcreeper *Upucerthia serrana* NT
Scale-throated Earthcreeper *Upucerthia dumetaria* NT
Buff-breasted Earthcreeper *Upucerthia validirostris* NT
Plain-breasted Earthcreeper *Upucerthia jelskii* NT

Bar-winged Cinclodes *Cinclodes fuscus* NT

Stout-billed Cinclodes *Cinclodes excelsior* NT
Olrog's Cinclodes *Cinclodes olrogi* NT
Comechingones Cinclodes *Cinclodes comechingonus* NT
Long-tailed Cinclodes *Cinclodes pabsti* NT
White-winged Cinclodes *Cinclodes atacamensis* NT
White-bellied Cinclodes *Cinclodes palliatus* ● NT
Grey-flanked Cinclodes *Cinclodes oustaleti* NT
Dark-bellied Cinclodes *Cinclodes patagonicus* NT
Seaside Cinclodes *Cinclodes nigrofumosus* NT
Taczanowski's Cinclodes *Cinclodes taczanowskii* NT
Blackish Cinclodes *Cinclodes antarcticus* NT

Crag Chilia *Chilia melanura* NT

Lesser Hornero *Furnarius minor* NT
White-banded Hornero *Furnarius figulus* NT
Tricolour Hornero *Furnarius tricolor* NT
Pale-legged Hornero *Furnarius leucopus* NT
Pale-billed Hornero *Furnarius torridus* NT
Rufous Hornero *Furnarius rufus* NT
Crested Hornero *Furnarius cristatus* NT

Synallaxinae

Des Murs' Wiretail *Sylviorthorhynchus desmursii* NT

Thorn-tailed Rayadito *Aphrastura spinicauda* NT
Masafuera Rayadito *Aphrastura masafuerae* ● NT

Brown-capped Tit-Spinetail *Leptasthenura fuliginiceps* NT
Tawny Tit-Spinetail *Leptasthenura yanacensis* NT
Tufted Tit-Spinetail *Leptasthenura platensis* NT
Plain-mantled Tit-Spinetail *Leptasthenura aegithaloides* NT
Araucaria Tit-Spinetail *Leptasthenura setaria* NT
Streaked Tit-Spinetail *Leptasthenura striata* NT
Striolated Tit-Spinetail *Leptasthenura striolata* NT
Rusty-crowned Tit-Spinetail *Leptasthenura pileata* NT
White-browed Tit-Spinetail *Leptasthenura xenothorax* ● NT
Andean Tit-Spinetail *Leptasthenura andicola* NT

White-chinned Spinetail *Schizoeaca fuliginosa* NT
Ochre-browed Thistletail *Schizoeaca coryi* NT
Mouse-coloured Thistletail *Schizoeaca griseomurina* NT
Eye-ringed Thistletail *Schizoeaca palpebralis* NT
Itatiaia Spinetail *Schizoeaca moreirae* NT

Chotoy Spinetail *Schoeniophylax phryganophila* NT

Rufous-capped Spinetail *Synallaxis ruficapilla* NT
Buff-browed Spinetail *Synallaxis superciliosa* NT
Sooty-fronted Spinetail *Synallaxis frontalis* NT
Azara's Spinetail *Synallaxis azarae* NT
Elegant Spinetail *Synallaxis elegantior* NT
Dark-breasted Spinetail *Synallaxis albigularis* NT
Pale-breasted Spinetail *Synallaxis albescens* NT
Chicli Spinetail *Synallaxis spixi* NT
Cinereous-breasted Spinetail *Synallaxis hypospodia* NT
Plain Spinetail *Synallaxis infuscata* ● NT
Slaty Spinetail *Synallaxis brachyura* NT
Apurimac Spinetail *Synallaxis courseni* ● NT
Dusky Spinetail *Synallaxis moesta* NT
McConnell's Spinetail *Synallaxis macconnelli* NT
Silvery-throated Spinetail *Synallaxis subpudica* NT
Blackish-headed Spinetail *Synallaxis tithys* NT
Grey-bellied Spinetail *Synallaxis cinerascens* NT
Maranon Spinetail *Synallaxis maranonica* NT
White-bellied Spinetail *Synallaxis propinqua* NT
Reiser's Spinetail *Synallaxis hellmayri* NT
Plain-crowned Spinetail *Synallaxis gujanensis* NT
Ochre-breasted Spinetail *Synallaxis albilora* NT
Ruddy Spinetail *Synallaxis rutilans* NT
Chestnut-throated Spinetail *Synallaxis cherriei* ● NT
Rufous Spinetail *Synallaxis unirufa* NT
Black-throated Spinetail *Synallaxis castanea* NT
Santa Marta Spinetail *Synallaxis fuscorufa* NT
Russet-bellied Spinetail *Synallaxis zimmeri* ● NT
Rufous-breasted Spinetail *Synallaxis erythrothorax* NT
Stripe-breasted Spinetail *Synallaxis cinnamomea* NT
Necklaced Spinetail *Synallaxis stictothorax* NT
White-whiskered Spinetail *Synallaxis candei* NT
Hoary-throated Spinetail *Synallaxis kollari* NT
Ochre-cheeked Spinetail *Synallaxis scutatus* NT

Lafresnaye's White-browed Spinetail *Hellmayrea gularis* NT

Red-faced Spinetail *Certhiaxis erythrops* NT
Tepui Spinetail *Certhiaxis demissa* NT
Line-cheeked Spinetail *Certhiaxis antisiensis* NT
Pallid Spinetail *Certhiaxis pallida* NT
Ash-browed Spinetail *Certhiaxis curtata* NT
Olive Spinetail *Certhiaxis obsoleta* NT
Streak-capped Spinetail *Certhiaxis hellmayri* NT
Crested Spinetail *Certhiaxis subcristata* NT
Stripe-crowned Spinetail *Certhiaxis pyrrhophia* NT
Marcapata Spinetail *Certhiaxis marcapatae* NT
Light-crowned Spinetail *Certhiaxis albiceps* NT
Grey-headed Spinetail *Certhiaxis semicinerea* NT
Creamy-chested Spinetail *Certhiaxis albicapilla* NT
Rusty-backed Spinetail *Certhiaxis vulpina* NT
Scaled Spinetail *Certhiaxis muelleri* NT
Speckled Spinetail *Certhiaxis gutturata* NT
Sulphur-throated Spinetail *Certhiaxis sulphurifera* NT
Yellow-throated Spinetail *Certhiaxis cinnamomea* NT
Red & White Spinetail *Certhiaxis mustelina* NT

Lesser Canastero *Thripophaga pyrrholeuca* NT
Short-billed Canastero *Thripophaga baeri* NT
Canyon Canastero *Thripophaga pudibunda* NT
Rusty-fronted Canastero *Thripophaga ottonis* NT
Iquico Canastero *Thripophaga heterura* NT
Cordillera Canastero *Thripophaga modesta* NT
Cactus Canastero *Thripophaga cactorum* NT
Creamy-breasted Canastero *Thripophaga dorbignyi* NT
Berlepsch's Canastero *Thripophaga berlepschi* NT
Chestnut Canastero *Thripophaga steinbachi* NT
Dusky-tailed Canastero *Thripophaga humicola* NT
Patagonian Canastero *Thripophaga patagonica* NT
Streak-throated Canastero *Thripophaga humilis* NT
White-tailed Canastero *Thripophaga usheri* ● NT
Austral Canastero *Thripophaga anthoides* ● NT
Streak-backed Canastero *Thripophaga wyatti* NT
Cordoba Canastero *Thripophaga sclateri* NT
Line-fronted Canastero *Thripophaga urubambensis* ● NT
Junin Canastero *Thripophaga virgata* NT
Scribble-tailed Canastero *Thripophaga maculicauda* NT
Many-striped Canastero *Thripophaga flammulata* NT
Orinoco Softtail *Thripophaga cherriei* NT
Striated Softtail *Thripophaga macroura* ● NT
Hudson's Canastero *Thripophaga hudsoni* NT
Great Spinetail *Thripophaga hypochondriacus* NT

Rufous-fronted Thornbird *Phacellodomus rufifrons* NT
Little Thornbird *Phacellodomus sibilatrix* NT
Streak-fronted Thornbird *Phacellodomus striaticeps* NT
Red-eyed Thornbird *Phacellodomus erythrophthalmus* NT
Freckle-breasted Thornbird *Phacellodomus striaticollis* NT
Chestnut-backed Thornbird *Phacellodomus dorsalis* NT
Greater Thornbird *Phacellodomus ruber* NT
Plain Softtail *Phacellodomus fusciceps* NT
Russet-mantled Softtail *Phacellodomus berlepschi* NT
Canebrake Groundcreeper *Phacellodomus dendrocolaptoides* ● NT

Bay-capped Wren Spinetail *Spartonoica maluroides* NT

Wren-like Rushbird *Phleocryptes melanops* NT

Curve-billed Reedhaunter *Limnornis curvirostris* NT
Straight-billed Reedhaunter *Limnornis rectirostris* NT

Firewood Gatherer *Anumbius annumbi* NT

Lark-like Brushrunner *Coryphistera alaudina* NT

Band-tailed Earthcreeper *Eremobius phoenicurus* NT

Spectacled Prickletail *Siptornis striaticollis* NT

Orange-fronted Plushcrown *Metopothrix aurantiacus* NT

Double-banded Greytail *Xenerpestes minlosi* NT

Equatorial Greytail *Xenerpestes singularis* NT

Philydorinae

Roraima Barbtail *Margarornis adustus* NT
Rusty-winged Barbtail *Margarornis guttuligera* NT
Spotted Barbtail *Margarornis brunnescens* NT
White-throated Barbtail *Margarornis tatei* ● NT
Ruddy Treerunner *Margarornis rubiginosus* NT
Fulvous-dotted Treerunner *Margarornis stellatus* NT
Beautiful Treerunner *Margarornis bellulus* NT
Pearled Treerunner *Margarornis squamiger* NT

Sharp-tailed Streamcreeper *Lochmias nematura* NT

Rufous Cachalote *Pseudoseisura cristata* NT
Brown Cachalote *Pseudoseisura lophotes* NT
White-throated Cachalote *Pseudoseisura gutturalis* NT

Buffy Tuftedcheek *Pseudocolaptes lawrencii* NT
Streaked Tuftedcheek *Pseudocolaptes boissonneautii* NT

Point-tailed Palmcreeper *Berlepschia rikeri* NT

Chestnut-winged Hookbill *Philydor strigilatus* NT
Striped Woodhaunter *Philydor subulatus* NT
Guttulated Foliage-gleaner *Philydor guttulatus* NT
Lineated Foliage-gleaner *Philydor subalaris* NT
Buff-browed Foliage-gleaner *Philydor rufosuperciliatus* NT
Montane Foliage-gleaner *Philydor striaticollis* NT
White-browed Foliage-gleaner *Philydor amaurotis* NT
Scaly-throated Foliage-gleaner *Philydor variegaticeps* NT
Rufous-tailed Foliage-gleaner *Philydor ruficaudatus* NT
Rufous-rumped Foliage-gleaner *Philydor erythrocercus* NT
Chestnut-winged Foliage-gleaner *Philydor erythropterus* NT
Lichtenstein's Foliage-gleaner *Philydor lichtensteini* NT
Rufous-backed Foliage-gleaner *Philydor erythronotus* NT
Novaes' Foliage-gleaner *Philydor novaesi* ● NT
Black-capped Foliage-gleaner *Philydor atricapillus* NT
Buff-fronted Foliage-gleaner *Philydor rufus* NT
Cinnamon-rumped Foliage-gleaner *Philydor pyrrhodes* NT
Russet-mantled Foliage-gleaner *Philydor dimidiatus* NT
White-collared Foliage-gleaner *Philydor fuscus* NT
Peruvian Recurvebill *Philydor ucayalae* NT
Bolivian Recurvebill *Philydor striatus* NT

Pale-browed Treehunter *Cichlocolaptes leucophrus* NT

Uniform Treehunter *Thripadectes ignobilis* NT
Streak-breasted Treehunter *Thripadectes rufobrunneus* NT
Black-billed Treehunter *Thripadectes melanorhynchus* NT
Striped Treehunter *Thripadectes holostictus* NT
Streak-capped Treehunter *Thripadectes virgaticeps* NT
Buff-throated Treehunter *Thripadectes scrutator* NT
Flammulated Treehunter *Thripadectes flammulatus* NT

Rufous-necked Foliage-gleaner *Automolus ruficollis* ● NT
Buff-throated Foliage-gleaner *Automolus ochrolaemus* NT
Olive-backed Foliage-gleaner *Automolus infuscatus* NT
Crested Foliage-gleaner *Automolus dorsalis* NT
White-eyed Foliage-gleaner *Automolus leucophthalmus* NT
Brown-rumped Foliage-gleaner *Automolus melanopezus* NT
White-throated Foliage-gleaner *Automolus albigularis* NT
Ruddy Foliage-gleaner *Automolus rubiginosus* NT
Chestnut-crowned Foliage-gleaner *Automolus rufipileatus* NT

Chestnut-capped Foliage-gleaner *Automolus rectirostris* NT
Henna-hooded Foliage-gleaner *Automolus erythrocephalus* ● NT

Tawny-throated Leafscraper *Sclerurus mexicanus* NT
Short-billed Leafscraper *Sclerurus rufigularis* NT
Grey-throated Leafscraper *Sclerurus albigularis* NT
Black-tailed Leafscraper *Sclerurus caudacutus* NT
Rufous-breasted Leafscraper *Sclerurus scansor* NT
Scaly-throated Leafscraper *Sclerurus guatemalensis* NT

Sharp-billed Treehunter *Xenops contaminatus* NT
Rufous-tailed Xenops *Xenops milleri* NT
Slender-billed Xenops *Xenops tenuirostris* NT
Plain Xenops *Xenops minutus* NT
Streaked Xenops *Xenops rutilans* NT

Great Xenops *Megaxenops parnaguae* ● NT

White-throated Treerunner *Pygarrhichas albogularis* NT

103 ANTBIRDS Formicariidae

Fasciated Antshrike *Cymbilaimus lineatus* NT

Spot-backed Antshrike *Hypoedaleus guttatus* NT

Giant Antshrike *Batara cinerea* NT

Large-tailed Antshrike *Mackenziaena leachii* NT
Tufted Antshrike *Mackenziaena severa* NT

Black-throated Antshrike *Frederickena viridis* NT
Undulated Antshrike *Frederickena unduligera* NT

Great Antshrike *Taraba major* NT

Black-crested Antshrike *Sakesphorus canadensis* NT
Silvery-cheeked Antshrike *Sakesphorus cristatus* NT
Collared Antshrike *Sakesphorus bernardi* NT
Black-backed Antshrike *Sakesphorus melanonotus* NT
Band-tailed Antshrike *Sakesphorus melanothorax* NT
Glossy Antshrike *Sakesphorus luctuosus* NT

White-bearded Antshrike *Biatas nigropectus* ● NT

Barred Antshrike *Thamnophilus doliatus* NT
Bar-crested Antshrike *Thamnophilus multistriatus* NT
Lined Antshrike *Thamnophilus palliatus* NT
Black-hooded Antshrike *Thamnophilus bridgesi* NT
Black Antshrike *Thamnophilus nigriceps* NT
Cocha Antshrike *Thamnophilus praecox* ● NT
Blackish-grey Antshrike *Thamnophilus nigrocinereus* NT
Castelnau's Antshrike *Thamnophilus cryptoleucus* NT
White-shouldered Antshrike *Thamnophilus aethiops* NT
Uniform Antshrike *Thamnophilus unicolor* NT
Black-capped Antshrike *Thamnophilus schistaceus* NT
Mouse-coloured Antshrike *Thamnophilus murinus* NT
Upland Antshrike *Thamnophilus aroyae* NT
Slaty Antshrike *Thamnophilus punctatus* NT
Amazonian Antshrike *Thamnophilus amazonicus* NT
Streak-backed Antshrike *Thamnophilus insignis* NT
Variable Antshrike *Thamnophilus caerulescens* NT
Rufous-winged Antshrike *Thamnophilus torquatus* NT
Rufous-capped Antshrike *Thamnophilus ruficapillus* NT

Spot-winged Antshrike *Pygiptila stellaris* NT

Pearly Antshrike *Megastictus margaritatus* NT

Black Bushbird *Neoctantes niger* NT

Recurvebill Bushbird *Clytoctantes alixii* ● NT

Speckle-breasted Antshrike *Xenornis setifrons* ● NT

Russet Antshrike *Thamnistes anabatinus* NT

Spot-breasted Antvireo *Dysithamnus stictothorax* NT
Plain Antvireo *Dysithamnus mentalis* NT
Streak-crowned Antvireo *Dysithamnus striaticeps* NT
Spot-crowned Antvireo *Dysithamnus puncticeps* NT

Rufous-backed Antvireo *Dysithamnus xanthopterus* NT

Grey-throated Antvireo *Dysithamnus ardesiacus* NT

Saturnine Antshrike *Thamnomanes saturninus* NT
Western Antshrike *Thamnomanes occidentalis* NT
Plumbeous Antshrike *Thamnomanes plumbeus* ● NT
Cinereous Antshrike *Thamnomanes caesius* NT
Bluish-slate Antshrike *Thamnomanes schistogynus* NT

Pygmy Antwren *Myrmotherula brachyura* NT
Short-billed Antwren *Myrmotherula obscura* NT
Sclater's Antwren *Myrmotherula sclateri* NT
Klages' Antwren *Myrmotherula klagesi* NT
Streaked Antwren *Myrmotherula surinamensis* NT
Yellow-throated Antwren *Myrmotherula ambigua* NT
Cherrie's Antwren *Myrmotherula cherriei* NT
Rufous-bellied Antwren *Myrmotherula guttata* NT
Stripe-chested Antwren *Myrmotherula longicauda* NT
Plain-throated Antwren *Myrmotherula hauxwelli* NT
Star-throated Antwren *Myrmotherula gularis* NT
Brown-bellied Antwren *Myrmotherula gutturalis* NT
Checker-throated Antwren *Myrmotherula fulviventris* NT
White-eyed Antwren *Myrmotherula leucophthalma* NT
Stipple-throated Antwren *Myrmotherula haematonota* NT
Ornate Antwren *Myrmotherula ornata* NT
Rufous-tailed Antwren *Myrmotherula erythrura* NT
Black-hooded Antwren *Myrmotherula erythronotos* ● NT
White-flanked Antwren *Myrmotherula axillaris* NT
Slaty Antwren *Myrmotherula schisticolor* NT
Rio Suno Antwren *Myrmotherula sunensis* NT
Long-winged Antwren *Myrmotherula longipennis* NT
Salvadori's Antwren *Myrmotherula minor* ● NT
Ihering's Antwren *Myrmotherula iheringi* NT
Rio de Janeiro Antwren *Myrmotherula fluminensis* NT
Ashy Antwren *Myrmotherula grisea* NT
Unicoloured Antwren *Myrmotherula unicolor* NT
Plain-winged Antwren *Myrmotherula behni* NT
Band-tailed Antwren *Myrmotherula urosticta* NT
Grey Antwren *Myrmotherula menetriesii* NT
Leaden Antwren *Myrmotherula assimilis* NT

Banded Antcatcher *Dichrozona cincta* NT

Stripe-backed Antbird *Myrmorchilus strigilatus* NT

Black-capped Antwren *Herpsilochmus pileatus* ● NT
Spot-tailed Antwren *Herpsilochmus sticturus* NT
Todd's Antwren *Herpsilochmus stictocephalus* NT
Spot-backed Antwren *Herpsilochmus dorsimaculatus* NT
Roraiman Antwren *Herpsilochmus roraimae* NT
Pectoral Antwren *Herpsilochmus pectoralis* ● NT
Large-billed A NT *Herpsilochmus longirostris*
Herpsilochmus parkerintwren *Ash-throated Antwren*
Yellow-breasted Antwren *Herpsilochmus axillaris* NT
Rufous-winged Antwren *Herpsilochmus rufimarginatus* NT

Dot-winged Antwren *Microrhopias quixensis* NT

Narrow-billed Antwren *Formicivora iheringi* ● NT
White-fringed Antwren *Formicivora grisea* NT
Serra Antwren *Formicivora serrana* NT
Black-bellied Antwren *Formicivora melanogaster* NT
Rusty-backed Antwren *Formicivora rufa* NT

Ferruginous Antbird *Drymophila ferruginea* NT
Bertoni's Antwren *Drymophila rubricollis* NT
Rufous-tailed Antbird *Drymophila genei* NT
Ochre-rumped Antbird *Drymophila ochropyga* NT
Striated Antbird *Drymophila devillei* NT
Long-tailed Antbird *Drymophila caudata* NT
Dusky-tailed Antbird *Drymophila malura* NT
Scaled Antbird *Drymophila squamata* NT

Streak-capped Antwren *Terenura maculata* NT
Orange-bellied Antwren *Terenura sicki* ● NT
Rufous-rumped Antwren *Terenura callinota* NT
Chestnut-shouldered Antwren *Terenura humeralis* NT

Yellow-rumped Antwren *Terenura sharpei* ● NT
Ash-winged Antwren *Terenura spodioptila* NT

Grey Antbird *Cercomacra cinerascens* NT
Rio de Janeiro Antbird *Cercomacra brasiliana* ● NT
Dusky Antbird *Cercomacra tyrannina* NT
Blackish Antbird *Cercomacra nigrescens* NT
Black Antbird *Cercomacra serva* NT
Jet Antbird *Cercomacra nigricans* NT
Rio Branco Antbird *Cercomacra carbonaria* ● NT
Matto Grosso Antbird *Cercomacra melanaria* NT
Bananal Antbird *Cercomacra ferdinandi* NT

Stub-tailed Antbird *Sipia berlepschi* NT
Esmeralda's Antbird *Sipia rosenbergi* NT

White-backed Fire-eye *Pyriglena leuconota* NT
Swainson's Fire-eye *Pyriglena atra* ● NT
White-shouldered Fire-eye *Pyriglena leucoptera* NT

Slender Antbird *Rhopornis ardesiaca* ● NT

White-browed Antcreeper *Myrmoborus leucophrys* NT
Ash-breasted Antcreeper *Myrmoborus lugubris* NT
Black-faced Antcreeper *Myrmoborus myotherinus* NT
Black-tailed Antcreeper *Myrmoborus melanurus* NT

Warbling Antbird *Hypocnemis cantator* NT
Yellow-browed Antbird *Hypocnemis hypoxantha* NT

Black-chinned Antcreeper *Hypocnemoides melanopogon* NT
Band-tailed Antcreeper *Hypocnemoides maculicauda* NT

Black & White Antcatcher *Myrmochanes hemileucus* NT

Bare-crowned Antcatcher *Gymnocichla nudiceps* NT

Silvered Antcatcher *Sclateria naevia* NT

Black-headed Antbird *Percnostola rufifrons* NT
White-lined Antbird *Percnostola macrolopha* NT
Slate-coloured Antbird *Percnostola schistacea* NT
Spot-winged Antbird *Percnostola leucostigma* NT
Caura Antbird *Percnostola caurensis* NT
Rufous-crested Antbird *Percnostola lophotes* NT

Swainson's Antcatcher *Myrmeciza longipes* NT
Chestnut-backed Antbird *Myrmeciza exsul* NT
Ferruginous-backed Antbird *Myrmeciza ferruginea* NT
Scalloped Antbird *Myrmeciza ruficauda* ● NT
White-bibbed Antbird *Myrmeciza loricata* NT
Squamate Antbird *Myrmeciza squamosa* NT
Dull-mantled Antbird *Myrmeciza laemosticta* NT
Yapacana Antbird *Myrmeciza disjuncta* NT
Grey-bellied Antbird *Myrmeciza pelzelni* NT
Chestnut-tailed Antbird *Myrmeciza hemimelaena* NT
Plumbeous Antbird *Myrmeciza hyperythra* NT
Goeldi's Antbird *Myrmeciza goeldii* NT
White-shouldered Antbird *Myrmeciza melanoceps* NT
Sooty Antbird *Myrmeciza fortis* NT
Immaculate Antbird *Myrmeciza immaculata* NT
Grey-headed Antbird *Myrmeciza griseiceps* NT
Black-throated Antbird *Myrmeciza atrothorax* NT
Spot-breasted Antbird *Myrmeciza stictothorax* ● NT

White-faced Antcatcher *Pithys albifrons* NT
White-masked Antcatcher *Pithys castanea* NT

Rufous-throated Antcatcher *Gymnopithys rufigula* NT
White-throated Antcatcher *Gymnopithys salvini* NT
Lunulated Antcatcher *Gymnopithys lunulata* NT
Bicoloured Antcatcher *Gymnopithys leucaspis* NT

Bare-eyed Antcatcher *Rhegmatorhina gymnops* NT
Harlequin Antcatcher *Rhegmatorhina berlepschi* NT
Chestnut-crested Antcatcher *Rhegmatorhina cristata* NT
White-breasted Antcatcher *Rhegmatorhina hoffmannsi* ● NT
Hairy-crested Antcatcher *Rhegmatorhina melanosticta* NT

Spotted Antbird *Hylophylax naevioides* NT
Spot-backed Antbird *Hylophylax naevia* NT
Dot-backed Antbird *Hylophylax punctulata* NT
Scale-backed Antbird *Hylophylax poecilonota* NT

Black-spotted Bare-eye *Phlegopsis nigromaculata* NT
Argus Bare-eye *Phlegopsis barringeri* ● NT
Reddish-winged Bare-eye *Phlegopsis erythroptera* NT

Pale-faced Bare-eye *Skutchia borbae* NT

Ocellated Antbird *Phaenostictus mcleannani* NT

Rufous-capped Antthrush *Formicarius colma* NT
Black-faced Antthrush *Formicarius analis* NT
Rufous-fronted Antthrush *Formicarius rufifrons* ●
Black-headed Antthrush *Formicarius nigricapillus* NT
Rufous-breasted Antthrush *Formicarius rufipectus* NT

Short-tailed Antthrush *Chamaeza campanisona* NT
Striated Antthrush *Chamaeza nobilis* NT
Rufous-tailed Antthrush *Chamaeza ruficauda* NT
Barred Antthrush *Chamaeza mollissima* NT

Wing-banded Antthrush *Myrmornis torquata* NT

Black-crowned Antpitta *Pittasoma michleri* NT
Rufous-crowned Antpitta *Pittasoma rufopileatum* NT

Undulated Antpitta *Grallaria squamigera* NT
Giant Antpitta *Grallaria gigantea* ● NT
Great Antpitta *Grallaria excelsa* NT
Variegated Antpitta *Grallaria varia* NT
Moustached Antpitta *Grallaria alleni* ● NT
Pale-billed Antpitta *Grallaria carrikeri* NT
Scaled Antpitta *Grallaria guatimalensis* NT
Tachira Antpitta *Grallaria chthonia* ● NT
Plain-backed Antpitta *Grallaria haplonota* NT
Ochre-striped Antpitta *Grallaria dignissima* NT
Elusive Antpitta *Grallaria eludens* NT
Chestnut-crowned Antpitta *Grallaria ruficapilla* NT
Watkins' Antpitta *Grallaria watkinsi* NT
Santa Marta Antpitta *Grallaria bangsi* NT
Stripe-headed Antpitta *Grallaria andicola* NT
Puna Antpitta *Grallaria punensis* NT
Bicoloured Antpitta *Grallaria rufocinerea* ● NT
Chestnut-naped Antpitta *Grallaria nuchalis* NT
Chestnut Antpitta *Grallaria blakei* NT
White-throated Antpitta *Grallaria albigula* NT
Chestnut-brown Antpitta *Grallaria erythroleuca* NT
Bay-backed Antpitta *Grallaria hypoleuca* NT
Grey-naped Antpitta *Grallaria griseonucha* NT
Rufous Antpitta *Grallaria rufula* NT
Rufous-faced Antpitta *Grallaria erythrotis* NT
Tawny Antpitta *Grallaria quitensis* ● NT
Brown-banded Antpitta *Grallaria milleri* ● NT

Streak-chested Antpitta *Hylopezus perspicillatus* NT
Spotted Antpitta *Hylopezus macularius* NT
Fulvous-bellied Antpitta *Hylopezus fulviventris* NT
Amazonian Antpitta *Hylopezus berlepschi* NT
Speckle-breasted Antpitta *Hylopezus ochroleucus* NT

Thrush-like Antpitta *Myrmothera campanisona* NT
Brown-breasted Antpitta *Myrmothera simplex* NT

Ochre-breasted Antpitta *Grallaricula flavirostris* NT
Rusty-breasted Antpitta *Grallaricula ferrugineipectus* NT
Slate-crowned Antpitta *Grallaricula nana* NT
Scallop-breasted Antpitta *Grallaricula loricata* NT
Peruvian Antpitta *Grallaricula peruviana* NT
Crescent-faced Antpitta *Grallaricula lineifrons* ● NT
Ochre-fronted Antpitta *Grallaricula ochraceifrons* NT
Hooded Antpitta *Grallaricula cucullata* ● NT

104 GNATEATERS Conopophagidae

Rufous Gnateater *Conopophaga lineata* NT
Chestnut-belted Gnateater *Conopophaga aurita* NT
Hooded Gnateater *Conopophaga roberti* NT
Ash-throated Gnateater *Conopophaga peruviana* NT
Slaty Gnateater *Conopophaga ardesiaca* NT
Chestnut-crowned Gnateater *Conopophaga castaneiceps* NT
Black-cheeked Gnateater *Conopophaga melanops* NT
Black-bellied Gnateater *Conopophaga melanogaster* NT

105 TAPACULOS Rhinocryptidae

Chestnut-throated Huet-huet *Pteroptochos castaneus* NT

Black-throated Huet-huet *Pteroptochos tarnii* NT
Moustached Turka *Pteroptochos megapodius* NT

White-throated Tapaculo *Scelorchilus albicollis* NT
Chucao Tapaculo *Scelorchilus rubecula* NT

Crested Gallito *Rhinocrypta lanceolata* NT

Sandy Gallito *Teledromas fuscus* NT

Rusty-belted Tapaculo *Liosceles thoracicus* NT

Collared Crescentchest *Melanopareia torquata* NT
Olive-crowned Crescentchest *Melanopareia maximiliani* NT
Maranon Crescentchest *Melanopareia maranonica* NT
Elegant Crescentchest *Melanopareia elegans* NT

Spotted Bamboowren *Psilorhamphus guttatus* NT

Slaty Bristlefront *Merulaxis ater* NT
Stresemann's Bristlefront *Merulaxis stresemanni* NT

Ochre-flanked Tapaculo *Eugralla paradoxa* NT

Ash-coloured Tapaculo *Myornis senilis* NT

Unicoloured Tapaculo *Scytalopus unicolor* NT
Mouse-coloured Tapaculo *Scytalopus speluncae* NT
Large-footed Tapaculo *Scytalopus macropus* NT
Rufous-vented Tapaculo *Scytalopus femoralis* NT
Silvery-fronted Tapaculo *Scytalopus argentifrons* NT
Pale-throated Tapaculo *Scytalopus panamensis* NT
Narino Tapaculo *Scytalopus vicinior* NT
Brown-rumped Tapaculo *Scytalopus latebricola* NT
Brasilia Tapaculo *Scytalopus novacapitalis* ● NT
White-breasted Tapaculo *Scytalopus indigoticus* NT
Andean Tapaculo *Scytalopus magellanicus* NT
White-browed Tapaculo *Scytalopus superciliaris* NT

Ocellated Tapaculo *Acropternis orthonyx* NT

106 TYRANT FLYCATCHERS Tyrannidae

Planalto Tyrannulet *Phyllomyias fasciatus* NT
Rough-legged Tyrannulet *Phyllomyias burmeisteri* NT
Greenish Tyrannulet *Phyllomyias virescens* NT
Sclater's Tyrannulet *Phyllomyias sclateri* NT
Grey-capped Tyrannulet *Phyllomyias griseocapilla* NT
Sooty-headed Tyrannulet *Phyllomyias griseiceps* NT
Plumbeous-crowned Tyrannulet *Phyllomyias plumbeiceps* NT
Black-capped Tyrannulet *Phyllomyias nigrocapillus* NT
Ashy-headed Tyrannulet *Phyllomyias cinereiceps* NT
Tawny-rumped Tyrannulet *Phyllomyias uropygialis* NT

Paltry Tyrannulet *Zimmerius vilissimus* NT
Bolivian Tyrannulet *Zimmerius bolivianus* NT
Red-billed Tyrannulet *Zimmerius cinereicapillus* NT
Slender-footed Tyrannulet *Zimmerius gracilipes* NT
Golden-faced Tyrannulet *Zimmerius viridiflavus* NT

White-lored Tyrannulet *Ornithion inerme* NT
Yellow-bellied Tyrannulet *Ornithion semiflavum* NT
Brown-capped Tyrannulet *Ornithion brunneicapillum* NT

Southern Beardless Tyrannulet *Camptostoma obsoletum* NT
Northern Beardless Tyrannulet *Camptostoma imberbe* NA NT

Mouse-coloured Tyrannulet *Phaiomyias murina* NT

Scrub Flycatcher *Sublegatus modestus* NT
Dusky Flycatcher *Sublegatus obscurior* NT

Suiriri Flycatcher *Suiriri suiriri* NT

Yellow-crowned Tyrannulet *Tyrannulus elatus* NT

Forest Elaenia *Myiopagis gaimardii* NT
Grey Elaenia *Myiopagis caniceps* NT
Pacific Elaenia *Myiopagis subplacens* NT
Yellow-crowned Elaenia *Myiopagis flavivertex* NT
Greenish Elaenia *Myiopagis viridicata* NT
Yellow Elaenia *Myiopagis cotta* NT

Grey & White Tyrannulet *Pseudelaenia leucospodia* NT

Yellow-bellied Elaenia *Elaenia flavogaster* NT
Caribbean Elaenia *Elaenia martinica* NT

Large Elaenia *Elaenia spectabilis* NT
White-crested Elaenia *Elaenia albiceps* NT
Small-billed Elaenia *Elaenia parvirostris* NT
Olivaceous Elaenia *Elaenia mesoleuca* NT
Slaty Elaenia *Elaenia strepera* NT
Mottle-backed Elaenia *Elaenia gigas* NT
Brownish Elaenia *Elaenia pelzelni* NT
Plain-crested Elaenia *Elaenia cristata* NT
Lesser Elaenia *Elaenia chiriquensis* NT
Rufous-crowned Elaenia *Elaenia ruficeps* NT
Mountain Elaenia *Elaenia frantzii* NT
Highland Elaenia *Elaenia obscura* NT
Great Elaenia *Elaenia dayi* NT
Sierran Elaenia *Elaenia pallatangae* NT
Greater Antillean Elaenia *Elaenia fallax* NT

White-throated Tyrannulet *Mecocerculus leucophrys* NT
White-tailed Tyrannulet *Mecocerculus poecilocercus* NT
Buff-banded Tyrannulet *Mecocerculus hellmayri* NT
Rufous-winged Tyrannulet *Mecocerculus calopterus* NT
Sulphur-bellied Tyrannulet *Mecocerculus minor* NT
White-banded Tyrannulet *Mecocerculus stictopterus* NT

Torrent Tyrannulet *Serpophaga cinerea* NT
River Tyrannulet *Serpophaga hypoleuca* NT
Sooty Tyrannulet *Serpophaga nigricans* NT
Bananal Tyrannulet *Serpophaga araguayae* ● NT
White-crested Tyrannulet *Serpophaga subcristata* NT

Slender-billed Tyrannulet *Inezia tenuirostris* NT
Plain Tyrannulet *Inezia inornata* NT
Pale-tipped Tyrannulet *Inezia subflava* NT

Lesser Wagtail Tyrant *Stigmatura napensis* NT
Greater Wagtail Tyrant *Stigmatura budytoides* NT

Ash-breasted Tit Tyrant *Anairetes alpinus* ● NT
Unstreaked Tit Tyrant *Anairetes agraphia* NT
Agile Tit Tyrant *Anairetes agilis* NT
Pied-crested Tit Tyrant *Anairetes reguloides* NT
Yellow-billed Tit Tyrant *Anairetes flavirostris* NT
Tufted Tit Tyrant *Anairetes parulus* NT
Juan Fernandez Tit Tyrant *Anairetes fernandezianus* NT

Many-coloured Rush Tyrant *Tachuris rubrigastra* NT

Sharp-tailed Tyrant *Culicivora caudacuta* ● NT

Grey-backed Tachuri *Polystictus superciliaris* ● NT
Bearded Tachuri *Polystictus pectoralis* ● NT

Dinelli's Doradito *Pseudocolopteryx dinellianus* NT
Crested Doradito *Pseudocolopteryx sclateri* NT
Subtropical Doradito *Pseudocolopteryx acutipennis* NT
Warbling Doradito *Pseudocolopteryx flaviventris* NT

Tawny-crowned Pygmy Tyrant *Euscarthmus meloryphus* NT
Rufous-sided Pygmy Tyrant *Euscarthmus rufomarginatus* NT

Streak-necked Flycatcher *Mionectes striaticollis* NT
Olive-striped Flycatcher *Mionectes olivaceus* NT
Ochre-bellied Flycatcher *Mionectes oleagineus* NT
MacConnell's Flycatcher *Mionectes macconnelli* NT
Grey-hooded Flycatcher *Mionectes rufiventris* NT

Rufous-breasted Leptopogon *Leptopogon rufipectus* NT
Inca Leptopogon *Leptopogon taczanowskii* NT
Sepia-capped Leptopogon *Leptopogon amaurocephalus* NT
White-bellied Leptopogon *Leptopogon superciliaris* NT

Black-fronted Tyrannulet *Phylloscartes nigrifrons* NT
Variegated Bristle Tyrant *Phylloscartes poecilotis* NT
Chapman's Tyrannulet *Phylloscartes chapmani* NT
Marble-faced Bristle Tyrant *Phylloscartes ophthalmicus* NT
Southern Bristle Tyrant *Phylloscartes eximius* NT
Ecuadorean Bristle Tyrant *Phylloscartes gualaquizae* NT

Yellow-bellied Bristle Tyrant *Phylloscartes flaviventris* NT
Venezuelan Bristle Tyrant *Phylloscartes venezuelanus* NT
Spectacled Bristle Tyrant *Phylloscartes orbitalis* NT
Antioquia Bristle Tyrant *Phylloscartes lanyoni* NT
Yellow Tyrannulet *Phylloscartes flaveolus* NT
Minas Geraes Tyrannulet *Phylloscartes roquettei* ●
Mottle-cheeked Tyrannulet *Phylloscartes ventralis* NT
Sao Paulo Tyrannulet *Phylloscartes paulistus* ● NT
Oustalet's Tyrannulet *Phylloscartes oustaleti* NT
Ihering's Tyrannulet *Phylloscartes difficilis* NT
Long-tailed Tyrannulet *Phylloscartes ceciliae* ● NT
Yellow-green Tyrannulet *Phylloscartes flavovirens* NT
Olive-green Tyrannulet *Phylloscartes virescens* NT
Rufous-browed Tyrannulet *Phylloscartes superciliaris* NT
Bay-ringed Tyrannulet *Phylloscartes sylviolus* NT

Bronze-olive Pygmy Tyrant *Pseudotriccus pelzelni* NT
Hazel-fronted Pygmy Tyrant *Pseudotriccus simplex* NT
Rufous-headed Pygmy Tyrant *Pseudotriccus ruficeps* NT

Delalande's Antpipit *Corythopis delalandi* NT
Ringed Antpipit *Corythopis torquata* NT

Eared Pygmy Tyrant *Myiornis auricularis* NT
Short-tailed Pygmy Tyrant *Myiornis ecaudatus* NT
Black-capped Pygmy Tyrant *Myiornis atricapillus* NT
White-breasted Pygmy Tyrant *Myiornis albiventris* NT

Scale-crested Pygmy Tyrant *Lophotriccus pileatus* NT
Long-crested Pygmy Tyrant *Lophotriccus eulophotes* NT
Double-banded Pygmy Tyrant *Lophotriccus vitiosus* NT
Helmeted Pygmy Tyrant *Lophotriccus galeatus* NT

Pale-eyed Pygmy Tyrant *Atalotriccus pilaris* NT

Rufous-crowned Tody Tyrant *Poecilotriccus ruficeps* NT

Black & White Tody Flycatcher *Poecilotriccus capitale* NT
Tricoloured Tody Flycatcher *Poecilotriccus tricolor* NT

Black-chested Tyrant *Poecilotriccus andrei* NT

Southern Bentbill *Oncostoma olivaceum* NT
Northern Bentbill *Oncostoma cinereigulare* NT

Snethlage's Tody Tyrant *Hemitriccus minor* NT
Boat-billed Tody Tyrant *Hemitriccus josephinae* NT
Drab-breasted Pygmy Tyrant *Hemitriccus diops* NT
Brown-breasted Pygmy Tyrant *Hemitriccus obsoletus* NT
Flammulated Pygmy Tyrant *Hemitriccus flammulatus* NT
White-eyed Tody Tyrant *Hemitriccus zosterops* NT
Zimmer's Tody Tyrant *Hemitriccus aenigma* NT
Olivaceous Tody Tyrant *Hemitriccus orbitatus* NT
Johanne's Pygmy Tyrant *Hemitriccus iohannis* NT
Stripe-necked Tody Tyrant *Hemitriccus striaticollis* NT
Hangnest Tody Tyrant *Hemitriccus nidipendulus* NT
Yungas Tody Tyrant *Hemitriccus spodiops* NT
Pearly-vented Tody Tyrant *Hemitriccus margaritaceiventer* NT
Pelzeln's Tody Tyrant *Hemitriccus inornatus* NT
Black-throated Tody Tyrant *Hemitriccus granadensis* NT
Buff-breasted Tody Tyrant *Hemitriccus mirandae* NT
Kaempfer's Tody Tyrant *Hemitriccus kaempferi* ●
Buff-throated Tody Tyrant *Hemitriccus rufigularis* NT
Fork-tailed Pygmy Tyrant *Hemitriccus furcatus* ● NT

Plumbeous-crowned Tody Flycatcher *Todirostrum senex* NT

Ruddy Tody Flycatcher *Todirostrum russatum* NT
Ochre-faced Tody Flycatcher *Todirostrum plumbeiceps* NT
Smoky-fronted Tody Flycatcher *Todirostrum fumifrons* NT
Rusty-fronted Tody Flycatcher *Todirostrum latirostre* NT
Slate-headed Tody Flycatcher *Todirostrum sylvia* NT
Spotted Tody Flycatcher *Todirostrum maculatum* NT
Grey-headed Tody Flycatcher *Todirostrum poliocephalum* NT
Common Tody Flycatcher *Todirostrum cinereum* NT
Short-tailed Tody Flycatcher *Todirostrum viridanum* ● NT
Painted Tody Flycatcher *Todirostrum pictum* NT
Yellow-browed Tody Flycatcher *Todirostrum chrysocrotaphum* NT
Black-headed Tody Flycatcher *Todirostrum nigriceps* NT
Golden-winged Tody Flycatcher *Todirostrum calopterum* NT

Brownish Flycatcher *Cnipodectes subbrunneus* NT

Large-headed Flatbill *Ramphotrigon megacephala* NT
Rufous-tailed Flatbill *Ramphotrigon ruficauda* NT
Dusky-tailed Flatbill *Ramphotrigon fuscicauda* NT

Eye-ringed Flatbill *Rhynchocyclus brevirostris* NT
Olivaceous Flatbill *Rhynchocyclus olivaceus* NT
Fulvous-breasted Flatbill *Rhynchocyclus fulvipectus* NT

Yellow-olive Flycatcher *Tolmomyias sulphurescens* NT
Yellow-margined Flycatcher *Tolmomyias assimilis* NT
Grey-crowned Flycatcher *Tolmomyias poliocephalus* NT
Yellow-breasted Flycatcher *Tolmomyias flaviventris* NT

Cinnamon-crested Spadebill *Platyrinchus saturatus* NT
Stub-tailed Spadebill *Platyrinchus cancrominus* NT
White-throated Spadebill *Platyrinchus mystaceus* NT
Golden-crowned Spadebill *Platyrinchus coronatus* NT
Yellow-throated Spadebill *Platyrinchus flavigularis* NT
White-crested Spadebill *Platyrinchus platyrhynchos* NT
Russet-winged Spadebill *Platyrinchus leucoryphus* ● NT

Royal Flycatcher *Onychorhynchus coronatus* NT

Ornate Flycatcher *Myiotriccus ornatus* NT

Ruddy-tailed Flycatcher *Terenotriccus erythrurus* NT

Tawny-breasted Flycatcher *Myiobius villosus* NT
Sulphur-rumped Flycatcher *Myiobius barbatus* NT
Black-tailed Flycatcher *Myiobius atricaudus* NT

Flavescent Flycatcher *Myiophobus flavicans* NT
Orange-crested Flycatcher *Myiophobus phoenicomitra* NT
Unadorned Flycatcher *Myiophobus inornatus* NT
Roraiman Flycatcher *Myiophobus roraimae* NT
Orange-banded Flycatcher *Myiophobus lintoni* NT
Handsome Flycatcher *Myiophobus pulcher* NT
Olive-crested Flycatcher *Myiophobus cryptoxanthus* NT
Ochraceous-breasted Flycatcher *Myiophobus ochraceiventris* NT
Bran-coloured Flycatcher *Myiophobus fasciatus* NT

Euler's Flycatcher *Lathrotriccus euleri* NT

Tawny-chested Flycatcher *Aphanotriccus capitalis* NT

Black-billed Flycatcher *Aphanotriccus audax* NT

Belted Flycatcher *Xenotriccus callizonus* NT
Pileated Flycatcher *Xenotriccus mexicanus* NT

Cinnamon Flycatcher *Pyrrhomyias cinnamomea* NT

Tufted Flycatcher *Mitrephanes phaeocercus* NT
Olive Flycatcher *Mitrephanes olivaceus* NT

Olive-sided Flycatcher *Contopus borealis* NA NT

Smoke-coloured Pewee *Contopus fumigatus* NT
Ochraceous Pewee *Contopus ochraceus* NT
Western Wood Pewee *Contopus sordidulus* NA
Eastern Wood Pewee *Contopus virens* NA NT
Tropical Pewee *Contopus cinereus* NT
White-throated Pewee *Contopus albogularis* NT
Blackish Pewee *Contopus nigrescens* NT
Greater Antillean Pewee *Contopus caribaeus* NT
Lesser Antillean Pewee *Contopus latirostris* NT

Yellow-bellied Flycatcher *Empidonax flaviventris* NA NT
Acadian Flycatcher *Empidonax virescens* NA NT
Traill's Flycatcher *Empidonax traillii* NA NT
Alder Flycatcher *Empidonax alnorum* NA NT
Grey-breasted Flycatcher *Empidonax griseipectus* NT
White-throated Flycatcher *Empidonax albigularis* NT
Least Flycatcher *Empidonax minimus* NA NT
Hammond's Flycatcher *Empidonax hammondii* NA NT
Wright's Flycatcher *Empidonax oberholseri* NA NT
Grey Flycatcher *Empidonax wrightii* NA NT
Pine Flycatcher *Empidonax affinis* NT
Western Flycatcher *Empidonax difficilis* NA NT
Yellowish Flycatcher *Empidonax flavescens* NT
Buff-breasted Flycatcher *Empidonax fulvifrons* NA
Black-capped Flycatcher *Empidonax atriceps* NT

Cocos I Flycatcher *Nesotriccus ridgwayi* NT

Fuscous Flycatcher *Cnemotriccus fuscatus* NT

Eastern Phoebe *Sayornis phoebe* NA NT
Black Phoebe *Sayornis nigricans* NA NT
Say's Phoebe *Sayornis saya* NA NT

Vermilion Flycatcher *Pyrocephalus rubinus* NA NT

Slaty-backed Chat Tyrant *Ochthoeca cinnamomeiventris* NT
Yellow-bellied Chat Tyrant *Ochthoeca diadema* NT
Crowned Chat Tyrant *Ochthoeca frontalis* NT
Jelski's Chat Tyrant *Ochthoeca jelskii* NT
Golden-browed Chat Tyrant *Ochthoeca pulchella* NT
Rufous-breasted Chat Tyrant *Ochthoeca rufipectoralis* NT
Brown-backed Chat Tyrant *Ochthoeca fumicolor* NT
D'Orbigny's Chat Tyrant *Ochthoeca oenanthoides* NT
Patagonian Chat Tyrant *Ochthoeca parvirostris* NT
White-browed Chat Tyrant *Ochthoeca leucophrys* NT
Piura Chat Tyrant *Ochthoeca piurae* NT
Drab Water Tyrant *Ochthoeca littoralis* NT

Streak-throated Bush Tyrant *Myiotheretes striaticollis* NT
Red-rumped Bush-Tyrant *Myiotheretes erythropygius* NT
Rufous-webbed Bush-Tyrant *Myiotheretes rufipennis* NT
Santa Marta Bush-Tyrant *Myiotheretes pernix* NT
Smoky Bush-Tyrant *Myiotheretes fumigatus* NT
Rufous-bellied Bush-Tyrant *Myiotheretes fuscorufus* NT

Fire-eyed Diucon *Xolmis pyrope* NT
Grey Monjita *Xolmis cinerea* NT
Black-crowned Monjita *Xolmis coronata* NT
White-rumped Monjita *Xolmis velata* NT
Black & White Monjita *Xolmis dominicana* ● NT
White Monjita *Xolmis irupero* NT

Rusty-backed Monjita *Neoxolmis rubetra* NT
Chocolate-vented Tyrant *Neoxolmis rufiventris* NT

Black-billed Shrike Tyrant *Agriornis montana* NT
White-tailed Shrike Tyrant *Agriornis andicola* ● NT
Great Shrike Tyrant *Agriornis livida* NT
Grey-bellied Shrike Tyrant *Agriornis microptera* NT
Mouse-brown Shrike Tyrant *Agriornis murina* NT
Spot-billed Ground Tyrant *Muscisaxicola maculirostris* NT
Dark-faced Ground Tyrant *Muscisaxicola macloviana* NT
Little Ground Tyrant *Muscisaxicola fluviatilis* NT
Cinnamon-bellied Ground Tyrant *Muscisaxicola capistrata* NT
Rufous-naped Ground Tyrant *Muscisaxicola rufivertex* NT

White-browed Ground Tyrant *Muscisaxicola albilora* NT
Puna Ground Tyrant *Muscisaxicola juninensis* NT
Plain-capped Ground Tyrant *Muscisaxicola alpina* NT
Cinereous Ground Tyrant *Muscisaxicola cinerea* NT
White-fronted Ground Tyrant *Muscisaxicola albifrons* NT
Ochre-naped Ground Tyrant *Muscisaxicola flavinucha* NT
Black-fronted Ground Tyrant *Muscisaxicola frontalis* NT

Rufous-backed Negrito *Lessonia rufa* NT
Salvin's Negrito *Lessonia oreas* NT

Cinereous Tyrant *Knipolegus striaticeps* NT
Hudson's Black Tyrant *Knipolegus hudsoni* NT
Amazonian Black Tyrant *Knipolegus poecilocercus* NT
Jelski's Bush Tyrant *Knipolegus signatus* NT
Blue-billed Black Tyrant *Knipolegus cyanirostris* NT
Rufous-tailed Tyrant *Knipolegus poecilurus* NT
Riverside Tyrant *Knipolegus orenocensis* NT
White-winged Black Tyrant *Knipolegus aterrimus* NT
Crested Black Tyrant *Knipolegus lophotes* NT
Velvety Black Tyrant *Knipolegus nigerrimus* NT

Spectacled Tyrant *Hymenops perspicillata* NT

Pied Water Tyrant *Fluvicola pica* NT
Masked Water Tyrant *Fluvicola nengeta* NT
White-headed Marsh Tyrant *Fluvicola leucocephala* NT

Long-tailed Tyrant *Colonia colonus* NT

Cock-tailed Tyrant *Alectrurus tricolor* NT
Strange-tailed Tyrant *Alectrurus risora* ● NT

Streamer-tailed Tyrant *Gubernetes yetapa* NT

Yellow-browed Tyrant *Satrapa icterophrys* NT

Tumbes Tyrant *Tumbezia salvini* NT

Short-tailed Field Tyrant *Muscigralla brevicauda* NT

Cliff Flycatcher *Hirundinea ferruginea* NT

Cattle Tyrant *Machetornis rixosus* NT

Shear-tailed Grey Tyrant *Muscipipra vetula* NT

Rufous-tailed Attila *Attila phoenicurus* NT
Cinnamon Attila *Attila cinnamomeus* NT
Ochraceous Attila *Attila torridus* NT
Citron-bellied Attila *Attila citriniventris* NT
Dull-capped Attila *Attila bolivianus* NT
Grey-hooded Attila *Attila rufus* NT
Bright-rumped Attila *Attila spadiceus* NT

Rufous Casiornis *Casiornis rufa* NT
Ash-throated Casiornis *Casiornis fusca* NT

Greyish Mourner *Rhytipterna simplex* NT
Pale-bellied Mourner *Rhytipterna immunda* NT
Rufous Mourner *Rhytipterna holerythra* NT

Cinereous Mourner *Laniocera hypopyrra* NT
Speckled Mourner *Laniocera rufescens* NT

Sirystes *Sirystes sibilator* NT

Rufous Flycatcher *Myiarchus semirufus* NT
Yucatan Flycatcher *Myiarchus yucatanensis* NT
Sad Flycatcher *Myiarchus barbirostris* NT
Olivaceous Flycatcher *Myiarchus tuberculifer* NA NT
Swainson's Flycatcher *Myiarchus swainsoni* NT
Venezuelan Flycatcher *Myiarchus venezuelensis* NT
Panama Flycatcher *Myiarchus panamensis* NT
Short-crested Flycatcher *Myiarchus ferox* NT
Apical Flycatcher *Myiarchus apicalis* NT
Pale-edged Flycatcher *Myiarchus cephalotes* NT
Sooty-crowned Flycatcher *Myiarchus phaeocephalus* NT
Ash-throated Flycatcher *Myiarchus cinerascens* NT
Pale-throated Flycatcher *Myiarchus nuttingi* NT
Great Crested Flycatcher *Myiarchus crinitus* NT
Brown Crested Flycatcher *Myiarchus tyrannulus* NT
Galapagos Flycatcher *Myiarchus magnirostris* NT
Grenada Flycatcher *Myiarchus nugator* NT
Rufous-tailed Flycatcher *Myiarchus validus* NT
Puerto Rican Flycatcher *Myiarchus antillarum* NT
La Sagra's Flycatcher *Myiarchus sagrae* NT

Stolid Flycatcher *Myiarchus stolidus* NT
Wied's Crested Flycatcher *Myiarchus oberi* NT

Flammulated Flycatcher *Deltarhynchus flammulatus* NT

Great Kiskadee *Pitangus sulphuratus* NA NT

Lesser Kiskadee *Philohydor lictor* NT

Boat-billed Flycatcher *Megarhynchus pitangua* NT

Rusty-margined Flycatcher *Myiozetetes cayanensis* NT
Social Flycatcher *Myiozetetes similis* NT
Grey-capped Flycatcher *Myiozetetes granadensis* NT
Dusky-chested Flycatcher *Myiozetetes luteiventris* NT

White-bearded Flycatcher *Phelpsia inornata* NT

White-ringed Flycatcher *Conopias parva* NT
Three-striped Flycatcher *Conopias trivirgata* NT
Lemon-browed Flycatcher *Conopias cinchoneti* NT

Golden-bellied Flycatcher *Myiodynastes hemichrysus* NT
Streaked Flycatcher *Myiodynastes maculatus* NT
Sulphur-bellied Flycatcher *Myiodynastes luteiventris* NA NT
Baird's Flycatcher *Myiodynastes bairdi* NT
Golden-crowned Flycatcher *Myiodynastes chrysocephalus* NT

Piratic Flycatcher *Legatus leucophaius* NT

Variegated Flycatcher *Empidonomus varius* NT

Crowned Slaty Flycatcher *Griseotyrannus aurantioatrocristatus* NT

Sulphury Flycatcher *Tyrannopsis sulphurea* NT

Snowy-throated Kingbird *Tyrannus niveigularis* NT
White-throated Kingbird *Tyrannus albogularis* NT
Tropical Kingbird *Tyrannus melancholicus* NA NT
Couch's Kingbird *Tyrannus couchii* NA NT
Cassin's Kingbird *Tyrannus vociferans* NA NT
Thick-billed Kingbird *Tyrannus crassirostris* NA NT
Western Kingbird *Tyrannus verticalis* NA NT
Scissor-tailed Flycatcher *Tyrannus forficatus* NA NT
Fork-tailed Flycatcher *Tyrannus savana* NA NT
Eastern Kingbird *Tyrannus tyrannus* NA NT
Grey Kingbird *Tyrannus dominicensis* NA NT
Loggerhead Kingbird *Tyrannus caudifasciatus* NT
Giant Kingbird *Tyrannus cubensis* NT

White-naped Xenopsaris *Xenopsaris albinucha* NT

Green-backed Becard *Pachyramphus viridis* NT
Barred Becard *Pachyramphus versicolor* NT
Slaty Becard *Pachyramphus spodiurus* NT
Cinereous Becard *Pachyramphus rufus* NT
Chestnut-crowned Becard *Pachyramphus castaneus* NT
Cinnamon Becard *Pachyramphus cinnamomeus* NT
White-winged Becard *Pachyramphus polychopterus* NT
Black-capped Becard *Pachyramphus marginatus* NT
Black & White Becard *Pachyramphus albogriseus* NT
Grey-collared Becard *Pachyramphus major* NT
Glossy-backed Becard *Pachyramphus surinamus* NT
Rose-throated Becard *Pachyramphus aglaiae* NA NT
One-coloured Becard *Pachyramphus homochrous* NT
Pink-throated Becard *Pachyramphus minor* NT
Plain Becard *Pachyramphus validus* NT
Jamaican Becard *Pachyramphus niger* NT

Black-tailed Tityra *Tityra cayana* NT
Masked Tityra *Tityra semifasciata* NT
Black-crowned Tityra *Tityra inquisitor* NT

107 MANAKINS Pipridae

Greater Manakin *Schiffornis major* NT
Greenish Manakin *Schiffornis virescens* NT
Thrush-like Manakin *Schiffornis turdinus* NT

Broad-billed Manakin *Sapayoa aenigma* NT

Grey-hooded Manakin *Piprites griseiceps* NT
Wing-barred Manakin *Piprites chloris* NT
Black-capped Manakin *Piprites pileatus* ● NT

Cinnamon Manakin *Neopipo cinnamomea* NT

Yellow-headed Manakin *Chloropipo flavicapilla* NT
Green Manakin *Chloropipo holochlora* NT
Olive Manakin *Chloropipo uniformis* NT
Jet Manakin *Chloropipo unicolor* NT

Black Manakin *Xenopipo atronitens* NT

Helmeted Manakin *Antilophia galeata* NT

Dwarf Tyrant Manakin *Tyranneutes stolzmanni* NT
Tiny Tyrant Manakin *Tyranneutes virescens* NT

Pale-bellied Tyrant Manakin *Neopelma pallescens* NT
Saffron-crested Tyrant Manakin *Neopelma chrysocephalum* NT
Wied's Tyrant Manakin *Neopelma aurifrons* NT
Sulphur-bellied Tyrant Manakin *Neopelma sulphureiventer* NT

Orange-crowned Manakin *Heterocercus aurantiivertex* NT
Yellow-crowned Manakin *Heterocercus flavivertex* NT
Flame-crowned Manakin *Heterocercus linteatus* NT

Striped Manakin *Machaeropterus regulus* NT
Fiery-capped Manakin *Machaeropterus pyrocephalus* NT
Club-winged Manakin *Machaeropterus deliciosus* NT

White-bearded Manakin *Manacus manacus* NT
Golden-collared Manakin *Manacus vitellinus* NT

White-ruffed Manakin *Corapipo leucorrhoa* NT
White-throated Manakin *Corapipo gutturalis* NT

Pin-tailed Manakin *Ilicura militaris* NT

Golden-winged Manakin *Masius chrysopterus* NT

Long-tailed Manakin *Chiroxiphia linearis* NT
Lance-tailed Manakin *Chiroxiphia lanceolata* NT
Blue-backed Manakin *Chiroxiphia pareola* NT
Swallow-tailed Manakin *Chiroxiphia caudata* NT

White-crowned Manakin *Pipra pipra* NT
Blue-crowned Manakin *Pipra coronata* NT
Blue-rumped Manakin *Pipra isidorei* NT
Caerulean-capped Manakin *Pipra caeruleocapilla* NT
Snow-capped Manakin *Pipra nattereri* NT
Golden-crowned Manakin *Pipra vilasboasi* ● NT
Opal-crowned Manakin *Pipra iris* NT
White-fronted Manakin *Pipra serena* NT
Crimson-hooded Manakin *Pipra aureola* NT
Band-tailed Manakin *Pipra fasciicauda* NT
Wire-tailed Manakin *Pipra filicauda* NT
Red-capped Manakin *Pipra mentalis* NT
Golden-headed Manakin *Pipra erythrocephala* NT
Red-headed Manakin *Pipra rubrocapilla* NT
Round-tailed Manakin *Pipra chloromeros* NT
Scarlet-horned Manakin *Pipra cornuta* NT

108 COTINGAS Cotingidae

Guianian Red Cotinga *Phoenicercus carnifex* NT
Black-necked Red Cotinga *Phoenicercus nigricollis* NT

Shrike-like Cotinga *Laniisoma elegans* ● NT

Swallow-tailed Cotinga *Phibalura flavirostris* ● NT

Black & Gold Cotinga *Tijuca atra* NT
Grey-winged Cotinga *Tijuca condita* ● NT

Hooded Berryeater *Carpornis cucullatus* NT
Black-headed Berryeater *Carpornis melanocephalus* ● NT

Red-crested Cotinga *Ampelion rubrocristata* NT
Chestnut-crested Cotinga *Ampelion rufaxilla* NT
Bay-vented Cotinga *Ampelion sclateri* NT
White-cheeked Cotinga *Ampelion stresemanni* ● NT

Green & Black Fruiteater *Pipreola riefferii* NT
Band-tailed Fruiteater *Pipreola intermedia* NT
Barred Fruiteater *Pipreola arcuata* NT
Golden-breasted Fruiteater *Pipreola aureopectus* NT
Scarlet-breasted Fruiteater *Pipreola frontalis* NT
Fiery-throated Fruiteater *Pipreola chlorolepidota* NT
Handsome Fruiteater *Pipreola formosa* NT
Red-banded Fruiteater *Pipreola whitelyi* NT

Scaled Fruiteater *Ampelioides tschudii* NT

Buff-throated Purpletuft *Iodopleura pipra* ● NT
Dusky Purpletuft *Iodopleura fusca* NT
White-browed Purpletuft *Iodopleura isabellae* NT

Kinglet Calyptura *Calyptura cristata* ● NT

Grey-tailed Piha *Lipaugus subalaris* NT
Olivaceus Piha *Lipaugus cryptolophus* NT
Dusky Piha *Lipaugus fuscocinereus* NT
Screaming Piha *Lipaugus vociferans* NT
Rufous Piha *Lipaugus unirufus* NT
Cinnamon-vented Piha *Lipaugus lanioides* ● NT
Rose-coloured Piha *Lipaugus streptophorus* NT

Scimitar-winged Piha *Chirocylla uropygialis* NT

Purple-throated Cotinga *Porphyrolaema porphyrolaema* NT

Lovely Cotinga *Cotinga amabilis* NT
Ridgway's Cotinga *Cotinga ridgwayi* ● NT
Blue Cotinga *Cotinga nattererii* NT
Plum-throated Cotinga *Cotinga maynana* NT
Purple-breasted Cotinga *Cotinga cotinga* NT
Banded Cotinga *Cotinga maculata* ● NT
Spangled Cotinga *Cotinga cayana* NT

Pompadour Cotinga *Xipholena punicea* NT
White-tailed Cotinga *Xipholena lamellipennis* NT
White-winged Cotinga *Xipholena atropurpurea* ● NT

Black-tipped Cotinga *Carpodectes hopkei* NT
Snowy Cotinga *Carpodectes nitidus* NT
Yellow-billed Cotinga *Carpodectes antoniae* ● NT

Black-faced Cotinga *Conioptilon mcilhennyi* NT

Bare-necked Fruitcrow *Gymnoderus foetidus* NT

Crimson Fruitcrow *Haematoderus militaris* NT

Purple-throated Fruitcrow *Querula purpurata* NT

Red-ruffed Fruitcrow *Pyroderus scutatus* NT

Bare-necked Umbrellabird *Cephalopterus glabricollis* ● NT
Amazonian Umbrellabird *Cephalopterus ornatus* NT
Long-wattled Umbrellabird *Cephalopterus penduliger* ● NT

Capuchin Bird *Perissocephalus tricolor* NT

Three-wattled Bellbird *Procnias tricarunculata* NT
Bearded Bellbird *Procnias averano* NT
Bare-throated Bellbird *Procnias nudicollis* NT

Guianan Cock of the Rock *Rupicola rupicola* NT
Andean Cock of the Rock *Rupicola peruviana* NT

109 SHARPBILL Oxyruncidae

Sharpbill *Oxyruncus cristatus* NT

110 PLANTCUTTERS Phytotomidae

Chilean Plantcutter *Phytotoma rara* NT
Red-breasted Plantcutter *Phytotoma rutila* NT
Peruvian Plantcutter *Phytotoma raimondii* ● NT

111 PITTAS Pittidae

Phayre's Pitta *Pitta phayrei* OR
Blue-naped Pitta *Pitta nipalensis* EP OR
Blue-backed Pitta *Pitta soror* OR
Fulvous Pitta *Pitta oatesi* OR
Schneider's Pitta *Pitta schneideri* ● OR
Giant Pitta *Pitta caerulea* OR
Koch's Pitta *Pitta kochi* ● OR
Red-breasted Pitta *Pitta erythrogaster* OR AU
Blue-banded Pitta *Pitta arcuata* OR
Garnet Pitta *Pitta granatina* OR
Black-crowned Garnet Pitta *Pitta venusta* OR
Blue Pitta *Pitta cyanea* OR
Elliot's Pitta *Pitta elliotii* ● OR
Blue-tailed Pitta *Pitta guajana* OR
Gurney's Pitta *Pitta gurneyi* ● OR
Blue-headed Pitta *Pitta baudi* OR
Hooded Pitta *Pitta sordida* OR AU
Blue-winged Pitta *Pitta brachyura* OR
Fairy Pitta *Pitta nympha* ● OR
African Pitta *Pitta angolensis* AF
Green-breasted Pitta *Pitta reichenowi* AF
Superb Pitta *Pitta superba* ● AU
Great Pitta *Pitta maxima* AU
Steere's Pitta *Pitta steerei* ● OR

Moluccan Pitta *Pitta moluccensis* OR
Noisy Pitta *Pitta versicolor* AU
Elegant Pitta *Pitta elegans* AU
Rainbow Pitta *Pitta iris* AU
Black-faced Pitta *Pitta anerythra* ● PA

112 NEW ZEALAND WRENS Xenicidae

Rifleman *Acanthisitta chloris* AU

Bush Wren *Xenicus longipes* ● AU
Rock Wren *Xenicus gilviventris* AU

113 ASITIES Philepittidae

Velvet Asity *Philepitta castanea* AF
Schlegel's Asity *Philepitta schlegeli* AF

Wattled False Sunbird *Neodrepanis coruscans* AF
Small-billed False Sunbird *Neodrepanis hypoxantha* ● AF

114 LYREBIRDS Menuridae

Superb Lyrebird *Menura novaehollandiae* AU
Prince Albert's Lyrebird *Menura alberti* AU

115 SCRUBBIRDS Atrichornithidae

Western Scrubbird *Atrichornis clamosus* ● AU
Rufous Scrubbird *Atrichornis rufescens* ● AU

116 LARKS Alaudidae

Singing Bushlark *Mirafra javanica* WP AF OR AU
Hova Lark *Mirafra hova* AF
Kordofan Bushlark *Mirafra cordofanica* AF
Marsabit Lark *Mirafra williamsi* AF
Southern Singing Bushlark *Mirafra cheniana* AF
Northern White-tailed Bushlark *Mirafra albicauda* AF
Monotonous Lark *Mirafra passerina* AF
Nyiro Bushlark *Mirafra candida* AF
Friedmann's Bushlark *Mirafra pulpa* AF
Ash's Lark *Mirafra ashi* ● AF
Red-winged Bushlark *Mirafra hypermetra* AF
Somali Long-billed Lark *Mirafra somalica* AF
Rufous-naped Bushlark *Mirafra africana* AF
Angolan Lark *Mirafra angolensis* AF
Flappet Lark *Mirafra rufocinnamomea* AF
Clapper Lark *Mirafra apiata* AF
Fawn-coloured Bushlark *Mirafra africanoides* AF
Long-clawed Lark *Mirafra ruddi* ● AF
Collared Lark *Mirafra collaris* AF
Rufous-winged Bushlark *Mirafra assamica* OR
Rusty Bushlark *Mirafra rufa* AF
Sidamo Bushlark *Mirafra sidamoensis* ● AF
Gillett's Bushlark *Mirafra gilletti* AF
Pink-breasted Lark *Mirafra poecilosterna* AF
Sabota Lark *Mirafra sabota* AF
Red-winged Bushlark *Mirafra erythroptera* OR

Dusky Bushlark *Pinarocorys nigricans* AF
Red-tailed Bushlark *Pinarocorys erythropygia* AF

Long-billed Lark *Certhilauda curvirostris* AF
Short-clawed Lark *Certhilauda chuana* AF
Red Lark *Certhilauda erythrochlamys* AF
Karoo Lark *Certhilauda albescens* AF
Spike-heeled Lark *Certhilauda albofasciata* AF

Black-eared Finch Lark *Eremopterix australis* AF
Chestnut-backed Finch Lark *Eremopterix leucotis* AF
Chestnut-headed Finch Lark *Eremopterix signata* AF
Grey-headed Finch Lark *Eremopterix verticalis* AF
Black-crowned Finch Lark *Eremopterix nigriceps* WP AF OR
Ashy-crowned Finch Lark *Eremopterix grisea* OR
Fischer's Finch Lark *Eremopterix leucopareia* AF

Bar-tailed Desert Lark *Ammomanes cincturus* WP AF OR
Rufous-tailed Desert Lark *Ammomanes phoenicurus* OR
Desert Lark *Ammomanes deserti* WP AF OR
Dunn's Lark *Ammomanes dunni* AF
Gray's Lark *Ammomanes grayi* AF
Ferruginous Lark *Ammomanes burra* AF

Hoopoe Lark *Alaemon alaudipes* WP AF OR
Lesser Hoopoe Lark *Alaemon hamertoni* AF

Thick-billed Lark *Ramphocoris clotbey* WP

Calandra Lark *Melanocorypha calandra* WP EP
Bimaculated Lark *Melanocorypha bimaculata* WP EP OR
Mongolian Lark *Melanocorypha mongolica* EP
Long-billed Calandra Lark *Melanocorypha maxima* EP
White-winged Lark *Melanocorypha leucoptera* EP
Black Lark *Melanocorypha yeltoniensis* WP EP

Raza Island Lark *Calandrella razae* ● AF
Short-toed Lark *Calandrella cinerea* WP EP AF OR
Blanford's Lark *Calandrella blanfordi* AF
Hume's Short-toed Lark *Calandrella acutirostris* EP OR
Indian Sand Lark *Calandrella raytal* OR
Lesser Short-toed Lark *Calandrella rufescens* WP EP AF
Mongolian Short-toed Lark *Calandrella cheleensis* EP

Pink-billed Lark *Spizocorys conirostris* AF
Stark's Short-toed Lark *Spizocorys starki* AF
Sclater's Short-toed Lark *Spizocorys sclateri* AF
Obbia Lark *Spizocorys obbiensis* AF
Masked Lark *Spizocorys personata* AF

Botha's Lark *Botha fringillaris* ● AF

DuPont's Lark *Chersophilus duponti* WP AF

Short-tailed Lark *Pseudalaemon fremantlii* AF

Crested Lark *Galerida cristata* WP EP AF OR
Thekla Lark *Galerida theklae* WP AF
Malabar Crested Lark *Galerida malabarica* OR
Sykes' Crested Lark *Galerida deva* OR
Sun Lark *Galerida modesta* AF

Thick-billed Lark *Calendula magnirostris* AF

Woodlark *Lullula arborea* WP EP

Skylark *Alauda arvensis* WP EP OR
Oriental Skylark *Alauda gulgula* EP OR

Shore Lark *Eremophila alpestris* WP EP OR NA NT
Temminck's Horned Lark *Eremophila bilopha* WP

117 SWALLOWS AND MARTINS Hirundinidae

Pseudochelidoninae

African River Martin *Pseudochelidon eurystomina* AF
White-eyed River Martin *Pseudochelidon sirintarae* ● OR

Hirundininae

Tree Swallow *Tachycineta bicolor* NA NT
Mangrove Swallow *Tachycineta albilinea* NT
White-winged Swallow *Tachycineta albiventer* NT
White-rumped Swallow *Tachycineta leucorrhoa* NT
Chilean Swallow *Tachycineta leucopyga* NT
Violet-green Swallow *Tachycineta thalassina* NA NT
Bahama Swallow *Tachycineta cyaneoviridis* ● NT
Golden Swallow *Tachycineta euchrysea* NT

Brown-chested Martin *Phaeoprogne tapera* NT

Purple Martin *Progne subis* NA NT
Cuban Martin *Progne cryptoleuca* NT
Caribbean Martin *Progne dominicensis* NT
Sinaloa Martin *Progne sinaloae* NT
Grey-breasted Martin *Progne chalybea* NA NT
Southern Martin *Progne modesta* NT

Brown-bellied Swallow *Notiochelidon murina* NT
Blue & White Swallow *Notiochelidon cyanoleuca* NT
Pale-footed Swallow *Notiochelidon flavipes* NT
Black-capped Swallow *Notiochelidon pileata* NT

White-banded Swallow *Atticora fasciata* NT
Black-collared Swallow *Atticora melanoleuca* NT

White-thighed Swallow *Neochelidon tibialis* NT

Tawny-headed Swallow *Alopochelidon fucata* NT

Northern Rough-winged Swallow *Stelgidopteryx serripennis* NA NT
Southern Rough-winged Swallow *Stelgidopteryx ruficollis* NT

White-backed Swallow *Cheramoeca leucosternum* AU
Ariel ~?

African Sand Martin *Riparia paludicola* WP AF OR
Congo Sand Martin *Riparia congica* AF
Sand Martin *Riparia riparia* WP EP AF OR NA NT
Banded Sand Martin *Riparia cincta* AF

Mascarene Martin *Phedina borbonica* AF

Brazza's Martin *Phedinopsis brazzae* AF

Grey-rumped Swallow *Hirundo griseopyga* AF
Crag Martin *Hirundo rupestris* EP
Pale Crag Martin *Hirundo obsoleta* WP EP AF OR
African Rock Martin *Hirundo fuligula* AF
Dusky Crag Martin *Hirundo concolor* OR
Swallow *Hirundo rustica* WP EP AF OR AU NA NT
Red-chested Swallow *Hirundo lucida* AF
Angola Swallow *Hirundo angolensis* AF
Pacific Swallow *Hirundo tahitica* EP OR AU PA
White-throated Swallow *Hirundo albigularis* AF
Ethiopian Swallow *Hirundo aethiopica* AF
Wire-tailed Swallow *Hirundo smithii* AF OR
Blue Swallow *Hirundo atrocaerulea* AF
White-throated Swallow *Hirundo nigrita* AF
Pied-winged Swallow *Hirundo leucosoma* AF
White-tailed Swallow *Hirundo megaensis* ● AF
Black & Rufous Swallow *Hirundo nigrorufa* AF
Pearl-breasted Swallow *Hirundo dimidiata* AF
Greater Striped Swallow *Hirundo cucullata* AF
Lesser Striped Swallow *Hirundo abyssinica* AF
Red-breasted Swallow *Hirundo semirufa* AF
Mosque Swallow *Hirundo senegalensis* AF
Red-rumped Swallow *Hirundo daurica* WP EP AF OR
Greater Striated Swallow *Hirundo striolata* OR
Red-throated Cliff Swallow *Hirundo rufigula* AF
Preuss' Cliff Swallow *Hirundo preussi* AF
Andean Swallow *Hirundo andecola* AF
Tree Martin *Hirundo nigricans* AU PA
Red Sea Cliff Swallow *Hirundo perdita* ● AF
South African Cliff Swallow *Hirundo spilodera* AF
American Cliff Swallow *Hirundo pyrrhonota* NA NT
Cave Swallow *Hirundo fulva* NA NT
Indian Cliff Swallow *Hirundo fluvicola* EP OR
Fairy Martin *Hirundo ariel* AU
Dusky Cliff Swallow *Hirundo fuliginosa* AF
Welcome Swallow *Hirundo neoxena* AU

House Martin *Delichon urbica* WP AF PR
Asian House Martin *Delichon dasypus* EP OR
Nepal House Martin *Delichon nipalensis* EP OR

Square-tailed Saw-wing *Psalidoprocne nitens* AF
Cameroun Saw-wing *Psalidoprocne fuliginosa* AF
White-headed Saw-wing *Psalidoprocne albiceps* AF
Black Saw-wing *Psalidoprocne pristoptera* AF
Kaffa Saw-wing *Psalidoprocne oleaginea* AF
Brown Saw-wing *Psalidoprocne antinorii* AF
Petit's Saw-wing *Psalidoprocne petiti* AF
Eastern Saw-wing *Psalidoprocne orientalis* AF
Fantee Saw-wing *Psalidoprocne obscura* AF

118 WAGTAILS AND PIPITS Motacillidae

Forest Wagtail *Dendronanthus indicus* EP OR

Yellow Wagtail *Motacilla flava* WP EP AF PR
Citrine Wagtail *Motacilla citreola* EP OR
Grey Wagtail *Motacilla cinerea* WP EP AF OR
White Wagtail *Motacilla alba* WP EP AF OR
Japanese Pied Wagtail *Motacilla grandis* EP
Large Pied Wagtail *Motacilla maderaspatensis* OR
African Pied Wagtail *Motacilla aguimp* AF
Mountain Wagtail *Motacilla clara* AF
Cape Wagtail *Motacilla capensis* AF
Madagascar Wagtail *Motacilla flaviventris* AF

Golden Pipit *Tmetothylacus tenellus* AF

Cape Longclaw *Macronyx capensis* AF
Yellow-throated Longclaw *Macronyx croceus* AF
Fulleborn's Longclaw *Macronyx fuelleborni* AF
Abyssinian Longclaw *Macronyx flavicollis* AF
Pangani Longclaw *Macronyx aurantiigula* AF
Rosy-breasted Longclaw *Macronyx ameliae* AF
Grimwood's Longclaw *Macronyx grimwoodi* AF

Sharpe's Longclaw *Hemimacronyx sharpei* AF
Yellow-breasted Pipit *Hemimacronyx chloris* ● AF

Richard's Pipit *Anthus novaeseelandiae* EP AF OR AU

Blyth's Pipit *Anthus godlewskii* EP OR
Tawny Pipit *Anthus campestris* WP EP OR
Long-billed Pipit *Anthus similis* WP EP AF OR
Woodland Pipit *Anthus nyassae* AF
Sandy Plain-backed Pipit *Anthus vaalensis* AF
Dark Plain-backed Pipit *Anthus leucophrys* AF
Long-legged Pipit *Anthus pallidiventris* AF
Meadow Pipit *Anthus pratensis* WP EP
Tree Pipit *Anthus trivialis* WP EP AF OR
Indian Tree Pipit *Anthus hodgsoni* EP OR
Hodgson's Pipit *Anthus roseatus* EP OR
Red-throated Pipit *Anthus cervinus* WP EP AF OR
Petchora Pipit *Anthus gustavi* EP OR AU
Buff-bellied Pipit *Anthus rubescens* EP OR NA NT
Water Pipit *Anthus spinoletta* WP
Rock Pipit *Anthus petrosus* WP
Nilgiri Pipit *Anthus nilghiriensis* OR
Upland Pipit *Anthus sylvanus* EP
Canarian Pipit *Anthus berthelotii* WP
Striped Pipit *Anthus lineiventris* AF
Short-tailed Pipit *Anthus brachyurus* AF
Bushveld Pipit *Anthus caffer* AF
Jackson's Pipit *Anthus latistriatus* AF
Sokoke Pipit *Anthus sokokensis* ● AF
Malindi Pipit *Anthus melindae* AF
Large Yellow-tufted Pipit *Anthus crenatus* AF
New Guinea Pipit *Anthus gutturalis* AU
Sprague's Pipit *Anthus spragueii* NA NT
Short-billed Pipit *Anthus furcatus* NT
Yellowish Pipit *Anthus lutescens* NT
Chaco Pipit *Anthus chacoensis* ● NT
Correndera Pipit *Anthus correndera* NT
South Georgia Pipit *Anthus antarcticus* AN
Ochre-breasted Pipit *Anthus nattereri* ● NT
Hellmayr's Pipit *Anthus hellmayri* NT
Paramo Pipit *Anthus bogotensis* NT

119 CUCKOO SHRIKES Campephagidae

Ground Cuckoo Shrike *Pteropodocys maxima* AU

Black-faced Cuckoo Shrike *Coracina novaehollandiae* EP OR AU
Wallacean Cuckoo Shrike *Coracina personata* AU
Buru Is Cuckoo Shrike *Coracina fortis* AU
Moluccan Cuckoo Shrike *Coracina atriceps* AU
Slaty Cuckoo Shrike *Coracina schistacea* ● AU
Melanesian Greybird *Coracina caledonica* PA
Stout-billed Greybird *Coracina caeruleogrisea* AU
Caerulean Cuckoo Shrike *Coracina temminckii* AU
Black-faced Greybird *Coracina larvata* OR
Barred Cuckoo Shrike *Coracina striata* OR
Pied Cuckoo Shrike *Coracina bicolor* AU
Lineated Cuckoo Shrike *Coracina lineata* AU PA
White-lored Cuckoo Shrike *Coracina boyeri* AU
White-rumped Cuckoo Shrike *Coracina leucopygia* AU
Papuan Cuckoo Shrike *Coracina papuensis* AU PA
Little Cuckoo Shrike *Coracina robusta* AU
Black-hooded Greybird *Coracina longicauda* AU
Halmahera Cuckoo Shrike *Coracina parvula* AU
Pygmy Cuckoo Shrike *Coracina abbotti* AU
Caledonian Greybird *Coracina analis* PA
African Grey Cuckoo Shrike *Coracina caesia* AF
White-breasted Cuckoo Shrike *Coracina pectoralis* AF
Grauer's Cuckoo Shrike *Coracina graueri* AF
Madagascar Cuckoo Shrike *Coracina cinerea* AF
African Blue Cuckoo Shrike *Coracina azurea* AF
Mauritius Greybird *Coracina typica* ● AF
Reunion Greybird *Coracina newtoni* ● AF
Philippine Black Greybird *Coracina coerulescens* ● OR
Sumba Cicadabird *Coracina dohertyi* AU
Kei Cicadabird *Coracina dispar* AU
Common Cicadabird *Coracina tenuirostris* AU PA
Sula Cicadabird *Coracina sula* ● AU
Sulawesi Cicadabird *Coracina morio* AU
Pale Cicadabird *Coracina ceramensis* AU
Philippine Cicadabird *Coracina incerta* OR AU
White-winged Cuckoo Shrike *Coracina ostenta* ● OR
New Guinea Greybird *Coracina schisticeps* AU
Black Greybird *Coracina melaena* AU
Black-bellied Greybird *Coracina montana* AU
Black-bellied Cuckoo Shrike *Coracina holopolia* PA
Sharp-tailed Greybird *Coracina mcgregori* OR

Philippines Greybird *Coracina panayensis* OR
Indochinese Cuckoo Shrike *Coracina polioptera* OR
Dark-grey Cuckoo Shrike *Coracina melaschistos* EP OR
Lesser Cuckoo Shrike *Coracina fimbriata* OR
Black-headed Cuckoo Shrike *Coracina melanoptera* OR

Orange Cuckoo Shrike *Campochaera sloetii* AU

Black-breasted Triller *Chlamydochaera jefferyi* OR

Black & White Triller *Lalage melanoleuca* OR
Varied Triller *Lalage nigra* OR AU
White-winged Triller *Lalage sueurii* OR AU
Red-bellied Triller *Lalage aurea* AU
Black-browed Triller *Lalage atrovirens* AU
White-browed Triller *Lalage leucomela* AU
Spotted Triller *Lalage maculosa* PA
Samoan Triller *Lalage sharpei* PA
Long-tailed Triller *Lalage leucopyga* PA

African Black Cuckoo Shrike *Campephaga sulphurata* AF
Red-shouldered Cuckoo Shrike *Campephaga phoenicea* AF
Petit's Cuckoo Shrike *Campephaga petiti* AF
Purple-throated Cuckoo Shrike *Campephaga quiscalina* AF
Wattled Cuckoo Shrike *Campephaga lobata* ● AF

Rosy Minivet *Pericrocotus roseus* EP OR
Ashy Minivet *Pericrocotus divaricatus* EP OR
Small Minivet *Pericrocotus cinnamomeus* OR
Flores Minivet *Pericrocotus lansbergei* AU
Jerdon's Minivet *Pericrocotus erythropygius* OR
Yellow-throated Minivet *Pericrocotus solaris* EP OR
Long-tailed Minivet *Pericrocotus ethologus* EP OR
Short-billed Minivet *Pericrocotus brevirostris* EP OR
Sunda Minivet *Pericrocotus miniatus* OR
Scarlet Minivet *Pericrocotus flammeus* EP OR

Bar-winged Flycatcher Shrike *Hemipus picatus* EP OR
Black-winged Flycatcher Shrike *Hemipus hirundinaceus* OR

Brown-tailed Wood Shrike *Tephrodornis gularis* EP OR
Common Wood Shrike *Tephrodornis pondicerianus* OR

120 BULBULS Pycnonotidae

Crested Finchbill *Spizixos canifrons* OR
Collared Finchbill *Spizixos semitorques* OR

Straw-crowned Bulbul *Pycnonotus zeylanicus* OR
Striated Green Bulbul *Pycnonotus striatus* EP OR
Striated Bulbul *Pycnonotus leucogrammicus* OR
Olive-crowned Bulbul *Pycnonotus tympanistrigus* OR
Black & White Bulbul *Pycnonotus melanoleucos* OR
Grey-headed Bulbul *Pycnonotus priocephalus* OR
Black-headed Bulbul *Pycnonotus atriceps* OR
Black-crested Bulbul *Pycnonotus melanicterus* EP OR
Scaly-breasted Bulbul *Pycnonotus squamatus* OR
Grey-bellied Bulbul *Pycnonotus cyaniventris* OR
Red-whiskered Bulbul *Pycnonotus jocosus* EP OR
Anderson's Bulbul *Pycnonotus xanthorrhous* OR
Chinese Bulbul *Pycnonotus sinensis* OR
Formosan Bulbul *Pycnonotus taivanus* OR
White-cheeked Bulbul *Pycnonotus leucogenys* WP EP OR
Red-vented Bulbul *Pycnonotus cafer* EP OR
White-eared Bulbul *Pycnonotus aurigaster* OR AU
Black-capped Bulbul *Pycnonotus xanthopygos* WP
Red-eyed Bulbul *Pycnonotus nigricans* AF
Cape Bulbul *Pycnonotus capensis* AF
Garden Bulbul *Pycnonotus barbatus* AF
Puff-backed Bulbul *Pycnonotus eutilotus* OR
Blue-wattled Bulbul *Pycnonotus nieuwenhuisii* ● OR
Yellow-wattled Bulbul *Pycnonotus urostictus* OR
Orange-spotted Bulbul *Pycnonotus bimaculatus* OR
Stripe-throated Bulbul *Pycnonotus finlaysoni* OR
Yellow-throated Bulbul *Pycnonotus xantholaemus* OR
Yellow-tufted Bulbul *Pycnonotus penicillatus* OR
Flavescent Bulbul *Pycnonotus flavescens* OR
Yellow-vented Bulbul *Pycnonotus goiavier* OR AU
White-browed Bulbul *Pycnonotus luteolus* OR
Olive-brown Bulbul *Pycnonotus plumosus* OR

Blanford's Olive Bulbul *Pycnonotus blanfordi* OR
White-eyed Brown Bulbul *Pycnonotus simplex* OR
Red-eyed Brown Bulbul *Pycnonotus brunneus* OR
Lesser Brown Bulbul *Pycnonotus erythrophthalmus* OR
Shelley's Greenbul *Pycnonotus masukuensis* AF
Mountain Little Greenbul *Pycnonotus montanus* AF
Little Greenbul *Pycnonotus virens* AF
Hall's Greenbul *Pycnonotus hallae* AF
Little Grey Greenbul *Pycnonotus gracilis* AF
Ansorge's Greenbul *Pycnonotus ansorgei* AF
Cameroun Sombre Greenbul *Pycnonotus curvirostris* AF
Zanzibar Sombre Greenbul *Pycnonotus importunus* AF
Yellow-whiskered Greenbul *Pycnonotus latirostris* AF
Slender-billed Greenbul *Pycnonotus gracilirostris* AF
Olive-breasted Mountain Greenbul *Pycnonotus tephrolaemus* AF
Stripe-cheeked Greenbul *Pycnonotus milanjensis* AF

Serine Greenbul *Calyptocichla serina* AF

Honeyguide Greenbul *Baeopogon indicator* AF
Sjostedt's Honeyguide Greenbul *Baeopogon clamans* AF

Spotted Greenbul *Ixonotus guttatus* AF

Yellow-necked Greenbul *Chlorocichla falkensteini* AF
Simple Greenbul *Chlorocichla simplex* AF
Yellow-throated Leaf-Love *Chlorocichla flavicollis* AF
Yellow-bellied Greenbul *Chlorocichla flaviventris* AF
Joyful Greenbul *Chlorocichla laetissima* AF
Prigogine's Greenbul *Chlorocichla prigoginei* ● AF

Swamp Bulbul *Thescelocichla leucopleura* AF

Leaf-Love *Phyllastrephus scandens* AF
Bristle-necked Brownbul *Phyllastrephus terrestris* AF
Northern Brownbul *Phyllastrephus strepitans* AF
Grey Olive Greenbul *Phyllastrephus cerviniventris* AF
Pale Olive Greenbul *Phyllastrephus fulviventris* AF
Cameroun Olive Greenbul *Phyllastrephus poensis* AF
Toro Olive Greenbul *Phyllastrephus baumanni* AF
Grey-headed Greenbul *Phyllastrephus poliocephalus* AF
Yellow-streaked Greenbul *Phyllastrephus flavostriatus* AF
Slender Greenbul *Phyllastrephus debilis* AF
Sassi's Olive Greenbul *Phyllastrephus lorenzi* AF
White-throated Greenbul *Phyllastrephus albigularis* AF
Fischer's Greenbul *Phyllastrephus fischeri* AF
Cabanis' Greenbul *Phyllastrephus cabanisi* AF
Olive Mountain Greenbul *Phyllastrephus placidus* AF
Icterine Greenbul *Phyllastrephus icterinus* AF
Xavier's Greenbul *Phyllastrephus xavieri* AF
Spot-winged Greenbul *Phyllastrephus leucolepis* ● AF
Madagascar Tetraka *Phyllastrephus madagascariensis* AF
Short-billed Tetraka *Phyllastrephus zosterops* AF
Appert's Tetraka *Phyllastrephus apperti* ● AF
Dusky Tetraka *Phyllastrephus tenebrosus* ● AF
Grey-crowned Greenbul *Phyllastrephus cinereiceps* ● AF

Common Bristle-Bill *Bleda syndactyla* AF
Green-tailed Bristle-Bill *Bleda eximia* AF
Grey-headed Bristle-Bill *Bleda canicapilla* AF

Bearded Greenbul *Criniger barbatus* AF
Red-tailed Greenbul *Criniger calurus* AF
White-bearded Bulbul *Criniger ndussumensis* AF
Yellow-throated Olive Bulbul *Criniger olivaceus* ● AF
Finsch's Bearded Bulbul *Criniger finschii* OR
Ashy-fronted Bearded Bulbul *Criniger flaveolus* EP
Olivaceous Bearded Bulbul *Criniger pallidus* OR
Ochraceous Bearded Bulbul *Criniger ochraceus* OR
Grey-cheeked Bearded Bulbul *Criniger bres* OR
Grey-headed Bearded Bulbul *Criniger phaeocephalus* OR

Long-billed Bulbul *Setornis criniger* OR

Blyth's Olive Bulbul *Hypsipetes viridescens* OR
Grey-eyed Bulbul *Hypsipetes propinquus* OR
Crested Olive Bulbul *Hypsipetes charlottae* OR
Golden-eyed Bulbul *Hypsipetes palawanensis* OR
Hairy-backed Bulbul *Hypsipetes criniger* OR
Rufous-breasted Bulbul *Hypsipetes philippinus* OR
Slaty-crowned Bulbul *Hypsipetes siquijorensis* ● OR
Yellow-washed Bulbul *Hypsipetes everetti* OR
Golden Bulbul *Hypsipetes affinis* AU
Golden-browed Bulbul *Hypsipetes indicus* OR
Mountain Streaked Bulbul *Hypsipetes mcclellandii* OR
Green-backed Bulbul *Hypsipetes malaccensis* OR
Sumatran Bulbul *Hypsipetes virescens* OR
Ashy Bulbul *Hypsipetes flavalus* EP OR
Chestnut-eared Bulbul *Hypsipetes amaurotis* EP OR
Thick-billed Bulbul *Hypsipetes crassirostris* OR
Reunion Bulbul *Hypsipetes borbonicus* AF
Black Bulbul *Hypsipetes madagascariensis* EP AF OR
Nicobar Bulbul *Hypsipetes nicobariensis* OR
Bingham's Bulbul *Hypsipetes thompsoni* OR

Malia *Malia grata* AU

Black-collared Bulbul *Neolestes torquatus* AF

Kinkimavo *Tylas eduardi* AF

121 LEAFBIRDS AND IORAS Irenidae

Common Iora *Aegithina tiphia* OR
Marshall's Iora *Aegithina nigrolutea* OR
Green Iora *Aegithina viridissima* OR
Great Iora *Aegithina lafresnayei* OR

Yellow-quilled Leafbird *Chloropsis flavipennis* OR
Palawan Leafbird *Chloropsis palawanensis* OR
Greater Green Leafbird *Chloropsis sonnerati* OR
Lesser Green Leafbird *Chloropsis cyanopogon* OR
Blue-winged Leafbird *Chloropsis cochinchinensis* OR
Golden-fronted Leafbird *Chloropsis aurifrons* EP OR
Orange-bellied Leafbird *Chloropsis hardwickii* EP OR
Blue-mantled Leafbird *Chloropsis venusta* OR

Blue-backed Fairy Bluebird *Irena puella* OR
Black-mantled Fairy Bluebird *Irena cyanogaster* OR

122 SHRIKES Laniidae

Prionopinae

Ruppell's White-crowned Shrike *Eurocephalus ruppelli* AF
White-crowned Shrike *Eurocephalus anguitimens* AF

Long-crested Helmet Shrike *Prionops plumata* AF
Grey-crested Helmet Shrike *Prionops poliolopha* AF
Red-billed Shrike *Prionops caniceps* AF
Yellow-crested Helmet Shrike *Prionops alberti* AF
Retz's Red-billed Helmet Shrike *Prionops retzii* AF
Rand's Red-billed Helmet Shrike *Prionops gabela* ● AF
Chestnut-fronted Helmet Shrike *Prionops scopifrons* AF

Malaconotinae

Chat-Shrike *Lanioturdus torquatus* AF

Brubru Shrike *Nilaus afer* AF

Pringle's Puffback *Dryoscopus pringlii* AF
Puffback *Dryoscopus gambensis* AF
Black-backed Puffback *Dryoscopus cubla* AF
Red-eyed Puffback *Dryoscopus senegalensis* AF
Pink-footed Puffback *Dryoscopus angolensis* AF
Sabine's Puffback *Dryoscopus sabini* AF

Lesser Tchagra *Tchagra minuta* AF
Black-headed Tchagra *Tchagra senegala* WP AF
Levaillant's Tchagra *Tchagra tchagra* AF
Brown-headed Tchagra *Tchagra australis* AF
Three-streaked Tchagra *Tchagra jamesi* AF
Rosy-patched Shrike *Tchagra cruenta* AF

Red-crowned Bush Shrike *Laniarius ruficeps* AF
Luhder's Bush Shrike *Laniarius luhderi* AF
Turati's Boubou *Laniarius turatii* AF
Tropical Boubou *Laniarius aethiopicus* AF
Gabon Boubou *Laniarius bicolor* AF

Southern Boubou *Laniarius ferrugineus* AF
Common Gonolek *Laniarius barbarus* AF
Black-headed Gonolek *Laniarius erythrogaster* AF
Mufumbiri Shrike *Laniarius mufumbiri* AF
Burchell's Gonolek *Laniarius atrococcineus* AF
Yellow-breasted Boubou *Laniarius atroflavus* AF
Fulleborn's Black Boubou *Laniarius fuelleborni* AF
Slate-coloured Boubou *Laniarius funebris* AF
Sooty Boubou *Laniarius leucorhynchus* AF

Grey-green Bush Shrike *Telophorus bocagei* AF
Sulphur-breasted Bush Shrike *Telophorus sulfureopectus* AF
Olive Bush Shrike *Telophorus olivaceus* AF
Black-fronted Bush Shrike *Telophorus nigrifrons* AF
Many-coloured Bush Shrike *Telophorus multicolor* AF
Serle's Bush Shrike *Telophorus kupeensis* ● AF
Bokmakierie Shrike *Telophorus zeylonus* AF
Perrin's Bush Shrike *Telophorus viridis* AF
Four-coloured Bush Shrike *Telophorus quadricolor* AF
Doherty's Bush Shrike *Telophorus dohertyi* AF

Fiery-breasted Bush Shrike *Malaconotus cruentus* AF
Lagden's Bush Shrike *Malaconotus lagdeni* AF
Green-breasted Bush Shrike *Malaconotus gladiator* ● AF
Grey-headed Bush Shrike *Malaconotus blanchoti* AF
Monteiro's Bush Shrike *Malaconotus monteiri* ● AF
Black-cap Bush Shrike *Malaconotus alius* ● AF

Western Nicator *Nicator chloris* AF
Eastern Nicator *Nicator gularis* AF
Yellow-throated Nicator *Nicator vireo* AF

Laniinae

Yellow-billed Shrike *Corvinella corvina* AF
Magpie Shrike *Corvinella melanoleuca* AF

Tiger Shrike *Lanius tigrinus* EP OR AU
Souza's Shrike *Lanius souzae* AF
Bull-headed Shrike *Lanius bucephalus* EP OR
Brown Shrike *Lanius cristatus* EP OR AU
Red-backed Shrike *Lanius collurio* WP EP AF OR
Burmese Shrike *Lanius collurioides* OR
Emin's Shrike *Lanius gubernator* AF
Bay-backed Shrike *Lanius vittatus* EP OR
Black-headed Shrike *Lanius schach* EP OR AU
Strong-billed Shrike *Lanius validirostris* OR
Mackinnon's Shrike *Lanius mackinnoni* AF
Lesser Grey Shrike *Lanius minor* WP EP AF
Loggerhead Shrike *Lanius ludovicianus* NA NT
Great Grey Shrike *Lanius excubitor* WP EP AF OR NA
Grey-backed Fiscal *Lanius excubitoroides* AF
Chinese Great Grey Shrike *Lanius sphenocercus* EP
Long-tailed Fiscal *Lanius cabanisi* AF
Taita Fiscal *Lanius dorsalis* AF
Somali Fiscal *Lanius somalicus* AF
Fiscal Shrike *Lanius collaris* AF
Newton's Fiscal *Lanius newtoni* ● AF
Woodchat Shrike *Lanius senator* WP
Masked Shrike *Lanius nubicus* WP

Pityriasinae

Bornean Bristlehead *Pityriasis gymnocephala* OR

123 VANGA SHRIKES Vangidae

Red-tailed Vanga *Calicalicus madagascariensis* AF

Rufous Vanga *Schetba rufa* AF

Hook-billed Vanga *Vanga curvirostris* AF

Lafresnaye's Vanga *Xenopirostris xenopirostris* AF
Van Dam's Vanga *Xenopirostris damii* ● AF
Pollen's Vanga *Xenopirostris polleni* ● AF

Sicklebill *Falculea palliata* AF

White-headed Vanga *Artamella viridis* AF

Chabert Vanga *Leptopterus chabert* AF
Blue Vanga *Leptopterus madagascarinus* AF

Bernier's Vanga *Oriolia bernieri* ● AF

Helmet Bird *Euryceros prevostii* AF

Coral-billed Nuthatch *Hypositta corallirostris* AF

124 WAXWINGS AND ALLIES Bombycillidae

Bombycillinae

Waxwing *Bombycilla garrulus* WP EP NA
Japanese Waxwing *Bombycilla japonica* EP
Cedar Waxwing *Bombycilla cedrorum* NA NT

Ptilogonatinae

Grey Silky Flycatcher *Ptilogonys cinereus* NT
Long-tailed Silky Flycatcher *Ptilogonys caudatus* NT

Phainopepla *Phainopepla nitens* NA NT

Black & Yellow Silky Flycatcher *Phainoptila melanoxantha* NA NT

Hypocoliinae

Grey Hypocolius *Hypocolius ampelinus* WP

125 PALM CHAT Dulidae

Palm Chat *Dulus dominicus* NT

126 DIPPERS Cinclidae

Dipper *Cinclus cinclus* WP EP
Brown Dipper *Cinclus pallasii* EP OR
North American Dipper *Cinclus mexicanus* NA NT
White-capped Dipper *Cinclus leucocephalus* NT
Rufous-throated Dipper *Cinclus schulzi* ● NT NT

127 WRENS Troglodytidae

Black-capped Mockingthrush (Donacobius) *Donacobius atricapillus* NT

Boucard's Wren *Campylorhynchus jocosus* NT
Spotted Wren *Campylorhynchus gularis* NT
Yucatan Cactus-Wren *Campylorhynchus yucatanicus* NT
Cactus-Wren *Campylorhynchus brunneicapillus* NA NT
Giant Wren *Campylorhynchus chiapensis* NT
Bicoloured Wren *Campylorhynchus griseus* NT
Rufous-naped Wren *Campylorhynchus rufinucha* NT
Thrush-like Wren *Campylorhynchus turdinus* NT
White-headed Wren *Campylorhynchus albobrunneus* NT
Stripe-backed Wren *Campylorhynchus nuchalis* NT
Fasciated Wren *Campylorhynchus fasciatus* NT
Banded-backed Wren *Campylorhynchus zonatus* NT
Grey-barred Wren *Campylorhynchus megalopterus* NT

Tooth-billed Wren *Odontorchilus cinereus* NT
Grey-mantled Wren *Odontorchilus branickii* NT

Rock Wren *Salpinctes obsoletus* NA NT

Canyon Wren *Catherpes mexicanus* NA NT

Slender-billed Wren *Hylorchilus sumichrasti* ● NT

Rufous Wren *Cinnycerthia unirufa* NT
Sepia-brown Wren *Cinnycerthia peruana* NT

Sedge Wren *Cistothorus platensis* NA NT
Paramo Wren *Cistothorus meridae* NT
Apolinar's Wren *Cistothorus apolinari* ● NT
Marsh Wren *Cistothorus palustris* NA NT

Bewick's Wren *Thryomanes bewickii* NA NT
Revillagigedo Wren *Thryomanes sissonii* ● NT

Zapata Wren *Ferminia cerverai* ● NT

Black-throated Wren *Thryothorus atrogularis* NT
Sooty-headed Wren *Thryothorus spadix* NT
Black-bellied Wren *Thryothorus fasciatoventris* NT
Plain-tailed Wren *Thryothorus euophrys* NT
Inca Wren *Thryothorus eisenmanni* NT
Moustached Wren *Thryothorus genibarbis* NT
Coraya Wren *Thryothorus coraya* NT
Happy Wren *Thryothorus felix* NT
Spot-breasted Wren *Thryothorus maculipectus* NT
Rufous-breasted Wren *Thryothorus rutilus* NT
Riverside Wren *Thryothorus semibadius* NT
Bay Wren *Thryothorus nigricapillus* NT
Stripe-breasted Wren *Thryothorus thoracicus* NT
Stripe-throated Wren *Thryothorus leucopogon* NT

Banded Wren *Thryothorus pleurostictus* NT
Carolina Wren *Thryothorus ludovicianus* NA NT
Rufous & White Wren *Thryothorus rufalbus* NT
Niceforo's Wren *Thryothorus nicefori* ● NT
Bar-vented Wren *Thryothorus sinaloa* NT
Plain Wren *Thryothorus modestus* NT
Buff-breasted Wren *Thryothorus leucotis* NT
Superciliated Wren *Thryothorus superciliaris* NT
Fawn-breasted Wren *Thryothorus guarayanus* NT
Long-billed Wren *Thryothorus longirostris* NT
Grey Wren *Thryothorus griseus* NT

Wren *Troglodytes troglodytes* WP EP OR NA
Clarion Wren *Troglodytes tanneri* ● NT
House Wren *Troglodytes aedon* NA NT
Mountain Wren *Troglodytes solstitialis* NT
Tepui Wren *Troglodytes rufulus* NT

Timberline Wren *Thryorchilus browni* NT

White-bellied Wren *Uropsila leucogastra* NT

White-breasted Wood Wren *Henicorhina leucosticta* NT

Grey-breasted Wood Wren *Henicorhina leucophrys* NT

Bar-winged Wood Wren *Henicorhina leucoptera* NT
Nightingale Wren *Microcerculus philomela* NT
Scaly-breasted Wren *Microcerculus marginatus* NT
Flutist Wren *Microcerculus ustulatus* NT
Wing-banded Wren *Microcerculus bambla* NT

Chestnut-breasted Wren *Cyphorinus thoracicus* NT
Musician Wren *Cyphorinus aradus* NT

128 MOCKING BIRDS AND THRASHERS
Mimidae

Catbird *Dumetella carolinensis* NA NT

Black Catbird *Melanoptila glabrirostris* NT

Blue Mockingbird *Melanotis caerulescens* NT
Blue & White Mockingbird *Melanotis hypoleucus* NT

Northern Mockingbird *Mimus polyglottos* NA NT
Tropical Mockingbird *Mimus gilvus* NT
Bahama Mockingbird *Mimus gundlachii* NT
Chilean Mockingbird *Mimus thenca* NT
Long-tailed Mockingbird *Mimus longicaudatus* NT
Chalk-browed Mockingbird *Mimus saturninus* NT
Patagonian Mockingbird *Mimus patagonicus* NT
White-banded Mockingbird *Mimus triurus* NT
Brown-backed Mockingbird *Mimus dorsalis* NT

Galapagos Mockingbird *Nesomimus trifasciatus* PA

Socorro Thrasher *Mimodes graysoni* ● NT

Sage Thrasher *Oreoscoptes montanus* NA NT

Brown Thrasher *Toxostoma rufum* NA
Long-billed Thrasher *Toxostoma longirostre* NA NT
Cozumel Thrasher *Toxostoma guttatum* NT
Grey Thrasher *Toxostoma cinereum* NT
Bendire Thrasher *Toxostoma bendirei* NA NT
Ocellated Thrasher *Toxostoma ocellatum* NT
Curve-billed Thrasher *Toxostoma curvirostre* NA NT
Le Conte Thrasher *Toxostoma lecontei* NA NT
California Thrasher *Toxostoma redivivum* NA NT
Crissal Thrasher *Toxostoma dorsale* NA NT

Brown Trembler *Cinclocerthia ruficauda* NT

White-breasted Trembler *Ramphocinclus brachyurus* ● NT

Scaly-breasted Thrasher *Allenia fusca* NT

Pearly-eyed Thrasher *Margarops fuscatus* NT

129 ACCENTORS Prunellidae

Alpine Accentor *Prunella collaris* WP EP
Himalayan Accentor *Prunella himalayana* EP
Robin Accentor *Prunella rubeculoides* EP OR
Rufous-breasted Accentor *Prunella strophiata* EP OR
Mountain Accentor *Prunella montanella* EP
Brown Accentor *Prunella fulvescens* EP
Radde's Accentor *Prunella ocularis* WP EP
Black-throated Accentor *Prunella atrogularis* EP
Koslov's Accentor *Prunella koslowi* EP
Dunnock *Prunella modularis* WP
Japanese Hedge Sparrow *Prunella rubida* EP
Maroon-backed Accentor *Prunella immaculata* EP

130 THRUSHES AND CHATS
Turdidae

Gould's Shortwing *Brachypteryx stellata* OR
Rusty-bellied Shortwing *Brachypteryx hyperythra* ● EP OR
White-bellied Shortwing *Brachypteryx major* OR
Lesser Shortwing *Brachypteryx leucophrys* EP OR AU
Blue Shortwing *Brachypteryx montana* EP OR AU

Sulawesi Shortwing *Heinrichia calligyna* AU

Karroo Scrub Robin *Erythropygia coryphaea* AF
White-winged Scrub Robin *Erythropygia leucoptera* AF
White-browed Scrub Robin *Erythropygia leucophrys* AF
Red-backed Scrub Robin *Erythropygia zambesiana* AF
Brown-backed Scrub Robin *Erythropygia hartlaubi* AF
Rufous Scrub Robin *Erythropygia galactotes* WP AF OR
Kalahari Sandy Scrub Robin *Erythropygia paena* AF
Western Bearded Scrub Robin *Erythropygia leucosticta* AF
Eastern Bearded Scrub Robin *Erythropygia quadrivirgata* AF
Central Bearded Scrub Robin *Erythropygia barbata* AF
Brown Scrub Robin *Erythropygia signata* AF

Herero Chat *Namibornis herero* AF

Black Scrub Robin *Cercotrichas podobe* WP AF

Boulder Chat *Pinarornis plumosus* AF

Cape Rockjumper *Chaetops frenatus* AF
Orange-breasted Rockjumper *Chaetops aurantius* AF

Southern Scrub Robin *Drymodes brunneopygia* AU
Northern Scrub Robin *Drymodes superciliaris* AU

Starred Robin *Pogonocichla stellata* AF
Swynnerton's Bush Robin *Pogonocichla swynnertoni* ● AF

Gabela Akelat *Erithacus gabela* ● AF
Common Akelat *Erithacus cyornithopsis* AF
Jackson's Akelat *Erithacus aequatorialis* AF
Forest Robin *Erithacus erythrothorax* AF
Sharpe's Akelat *Erithacus sharpei* AF
East Coast Akelat *Erithacus gunningi* ● AF
Robin *Erithacus rubecula* WP EP
Japanese Robin *Erithacus akahige* EP
Riukiu Robin *Erithacus komadori* EP
Swinhoe's Robin *Erithacus sibilans* EP
Thrush Nightingale *Erithacus luscinia* WP EP AF
Nightingale *Erithacus megarhynchos* WP EP AF
Siberian Rubythroat *Erithacus calliope* EP OR
Bluethroat *Erithacus svecicus* WP EP OR
Himalayan Rubythroat *Erithacus pectoralis* EP OR
Rufous-headed Robin *Erithacus ruficeps* ● EP
Black-throated Blue Robin *Erithacus obscurus* ● EP
David's Rubythroat *Erithacus pectardens* EP
Indian Bluechat *Erithacus brunneus* EP
Siberian Blue Robin *Erithacus cyane* EP OR
White-browed Bush Robin *Erithacus indicus* EP OR
Rufous-breasted Bush Robin *Erithacus hyperythrus* EP
Collared Bush Robin *Erithacus johnstoniae* EP

Red-flanked Bluetail *Tarsiger cyanurus* EP OR
Golden Bush Robin *Tarsiger chrysaeus* OR

White-bellied Robin Chat *Cossypha roberti* AF
Red-capped Robin Chat *Cossypha natalensis* AF
Chorister Robin Chat *Cossypha dichroa* AF
Black-tailed Robin Chat *Cossypha semirufa* AF
White-browed Robin Chat *Cossypha heuglini* AF
Blue-shouldered Robin Chat *Cossypha cyanocampter* AF
Cape Robin Chat *Cossypha caffra* AF
White-throated Robin Chat *Cossypha humeralis* AF
Ansorge's Robin Chat *Cossypha ansorgei* AF
Snowy-headed Robin Chat *Cossypha niveicapilla* AF
Rand's Robin Chat *Cossypha heinrichi* ● AF
White-crowned Robin Chat *Cossypha albicapilla* AF

Olive-flanked Robin Chat *Dryocichloides anomala* AF

Grey-headed Robin Chat *Dryocichloides bocagei* AF
Rufous-cheeked Robin Chat *Dryocichloides insulana* AF
Grey-winged Robin Chat *Dryocichloides polioptera* AF
Archer's Robin Chat *Dryocichloides archeri* AF
Mountain Robin Chat *Dryocichloides isabellae* AF
Usambara Robin Chat *Dryocichloides montana* ● AF
Iringa Robin Chat *Dryocichloides lowei* ● AF

Spot Throat *Modulatrix stictigula* AF

Dappled Mountain Robin *Arcanator orostruthus* ● AF

Spotted Morning Warbler *Cichladusa guttata* AF
Collared Morning Warbler *Cichladusa arquata* AF
Red-tailed Morning Warbler *Cichladusa ruficauda* AF

White-tailed Alethe *Alethe diademata* AF
Fire-crested Alethe *Alethe castanea* AF
Red-throated Alethe *Alethe poliophrys* AF
White-chested Alethe *Alethe fuelleborni* AF
Brown-chested Alethe *Alethe poliocephala* AF
Cholo Mountain Alethe *Alethe choloensis* ● AF

Magpie Robin *Copsychus saularis* OR
Seychelles Magpie Robin *Copsychus sechellarum* ● OR
Madagascar Magpie Robin *Copsychus albospecularis* AF
White-rumped Shama *Copsychus malabaricus* OR
Strickland's Shama *Copsychus stricklandii* OR
White-browed Shama *Copsychus luzoniensis* OR
Black Shama *Copsychus niger* OR
Orange-tailed Shama *Copsychus pyrrhopygus* OR

White-throated Robin *Irania gutteralis* WP AF

Przewalski's Redstart *Phoenicurus alaschanicus* EP
Eversmann's Redstart *Phoenicurus erythronotus* EP OR
Blue-headed Redstart *Phoenicurus caeruleocephalus* EP
Black Redstart *Phoenicurus ochruros* WP EP OR
Redstart *Phoenicurus phoenicurus* WP EP AF
Hodgson's Redstart *Phoenicurus hodgsoni* EP OR
Blue-fronted Redstart *Phoenicurus frontalis* EP OR
Plumbeous Water Redstart *Phoenicurus fuliginosus* EP OR
White-capped Redstart *Phoenicurus leucocephalus* EP OR
White-throated Redstart *Phoenicurus schisticeps* EP OR
Daurian Redstart *Phoenicurus auroreus* EP OR
Moussier's Redstart *Phoenicurus moussieri* WP
Guldenstadt's Redstart *Phoenicurus erythrogaster* WP EP OR

Philippine Water Redstart *Rhyacornis bicolor* ● OR

White-bellied Redstart *Hodgsonius phaenicuroides* EP OR

White-tailed Blue Robin *Cinclidium leucurum* OR
Sunda Blue Robin *Cinclidium diana* OR
Blue-fronted Callene *Cinclidium frontale* OR

Hodgson's Grandala *Grandala coelicolor* OR

Eastern Bluebird *Sialia sialis* NA NT
Western Bluebird *Sialia mexicana* NA NT
Mountain Bluebird *Sialia currucoides* NA NT

Little Forktail *Enicurus scouleri* EP OR
Lesser Forktail *Enicurus velatus* OR
Chestnut-backed Forktail *Enicurus ruficapillus* OR
Black-backed Forktail *Enicurus immaculatus* EP OR
Slaty-backed Forktail *Enicurus schistaceus* EP OR
White-crowned Forktail *Enicurus leschenaulti* OR
Spotted Forktail *Enicurus maculatus* EP OR

Purple Cochoa *Cochoa purpurea* OR
Green Cochoa *Cochoa viridis* OR
Malaysian Cochoa *Cochoa azurea* ● OR
Sumatran Cochoa *Cochoa beccarii* ● OR

Hawaiian Thrush *Myadestes obscurus* PA
Large Kauai Thrush *Myadestes myadestinus* ● PA
Small Kauai Thrush *Myadestes palmeri* ● PA
Lanai Thrush *Myadestes lanaiensis* ● PA

Townsend's Solitaire *Myadestes townsendi* NA NT
Brown-backed Solitaire *Myadestes occidentalis* NT
Cuban Solitaire *Myadestes elisabeth* NT
Rufous-throated Solitaire *Myadestes genibarbis* NT
Black-faced Solitaire *Myadestes melanops* NT
Varied Solitaire *Myadestes coloratus* NT
Andean Solitaire *Myadestes ralloides* NT
Slate-coloured Solitaire *Myadestes unicolor* NT
Rufous-brown Solitaire *Myadestes leucogenys* NT

White-eared Solitaire *Entomodestes leucotis* NT
Black Solitaire *Entomodestes coracinus* NT

Rufous Broad-billed Ant-thrush *Stizorhina fraseri* AF
Finsch's Broad-billed Ant-thrush *Stizorhina finschii* AF

Red-tailed Ant-thrush *Neocossyphus rufus* AF
White-tailed Ant-thrush *Neocossyphus poensis* AF

Sicklewing Chat *Cercomela sinuata* AF
Familiar Chat *Cercomela familiaris* AF
Layard's Chat *Cercomela tractrac* AF
Karoo Chat *Cercomela schlegelii* AF
Brown Rockchat *Cercomela fusca* OR
Sombre Rockchat *Cercomela dubia* AF
Blackstart *Cercomela melanura* WP AF
Brown-tailed Rockchat *Cercomela scotocerca* AF
Hill Chat *Cercomela sordida* AF

Whinchat *Saxicola rubetra* WP EP AF
Stoliczka's Bushchat *Saxicola macrorhyncha* ● EP OR
Hodgson's Bushchat *Saxicola insignis* ● EP OR
Canary Islands Chat *Saxicola dacotiae* ● WP
Stonechat *Saxicola torquata* WP EP AF OR
White-tailed Stonechat *Saxicola leucura* OR
Pied Stonechat *Saxicola caprata* OR AU
Jerdon's Bushchat *Saxicola jerdoni* OR
Grey Bushchat *Saxicola ferrea* OR
White-bellied Bushchat *Saxicola gutturalis* ● AU

Congo Moorchat *Myrmecocichla tholloni* AF
Anteater Chat *Myrmecocichla aethiops* AF
Southern Anteater Chat *Myrmecocichla formicivora* AF
Sooty Chat *Myrmecocichla nigra* AF
White-headed Black-Chat *Myrmecocichla arnoti* AF
White-fronted Black Chat *Myrmecocichla albifrons* AF
Ruppell's Chat *Myrmecocichla melaena* AF

Cliffchat *Thamnolaea cinnamomeiventris* AF
White-crowned Cliffchat *Thamnolaea coronata* AF
White-winged Cliffchat *Thamnolaea semirufa* AF

Buff-streaked Chat *Oenanthe bifasciata* AF
Isabelline Wheatear *Oenanthe isabellina* EP OR
Red-breasted Wheatear *Oenanthe bottae* WP AF
Red-tailed Wheatear *Oenanthe xanthoprymna* WP EP OR
Wheatear *Oenanthe oenanthe* WP EP AF NA
Somali Wheatear *Oenanthe phillipsi* AF
Desert Wheatear *Oenanthe deserti* WP EP AF OR
Black-eared Wheatear *Oenanthe hispanica* WP AF
Finsch's Wheatear *Oenanthe finschii* WP EP OR
Variable Wheatear *Oenanthe picata* EP OR
Mourning Wheatear *Oenanthe lugens* WP AF OR
Hooded Wheatear *Oenanthe monacha* WP EP OR
Hume's Wheatear *Oenanthe alboniger* EP OR
Pied Wheatear *Oenanthe pleschanka* EP AF
Cyprus Wheatear *Oenanthe cypriaca* WP AF
White-tailed Black Wheatear *Oenanthe leucopyga* WP AF
Black Wheatear *Oenanthe leucura* WP
Mountain Chat *Oenanthe monticola* AF
Red-rumped Wheatear *Oenanthe moesta* WP
Capped Wheatear *Oenanthe pileata* AF

Black-backed Robin *Saxicoloides fulicata* OR

Madagascar Robin Chat *Pseudocossyphus imerinus* AF
Farkas'Robin Chat *Pseudocossyphus bensoni* ● AF

Cape Rock Thrush *Monticola rupestris* AF
Sentinel Rock Thrush *Monticola explorator* AF
Short-toed Rock Thrush *Monticola brevipes* AF
Little Rock Thrush *Monticola rufocinereus* AF
Miombo Rock Thrush *Monticola angolensis* AF

Rock Thrush *Monticola saxatilis* WP EP AF OR
Blue-capped Rock Thrush *Monticola cinclorhynchus* EP OR
White-throated Rock Thrush *Monticola gularis* EP OR
Chestnut-bellied Rock Thrush *Monticola rufiventris* OR
Blue Rock Thrush *Monticola solitarius* WP EP AF OR

Ceylon Whistling Thrush *Myiophoneus blighi* ● OR
Shiny Whistling Thrush *Myiophoneus melanurus* OR
Sunda Whistling Thrush *Myiophoneus glaucinus* OR
Malayan Whistling Thrush *Myiophoneus robinsoni* OR
Malabar Whistling Thrush *Myiophoneus horsfieldii* OR
Formosan Whistling Thrush *Myiophoneus insularis* OR
Himalayan (Blue) Whistling Thrush *Myiophoneus caeruleus* EP OR

Sulawesi Mountain Thrush *Geomalia heinrichi* AU

Slaty-backed Ground Thrush *Zoothera schistacea* ● AU
Moluccan Ground Thrush *Zoothera dumasi* AU
Chestnut-capped Ground Thrush *Zoothera interpres* OR AU
Red-backed Ground Thrush *Zoothera erythronota* AU
Chestnut-backed Ground Thrush *Zoothera dohertyi* AU
Pied Ground Thrush *Zoothera wardii* OR
Ashy Ground Thrush *Zoothera cinerea* OR
Orange-banded Ground Thrush *Zoothera peronii* ●
AU
Orange-headed Ground Thrush *Zoothera citrina* EP OR
Everett's Ground Thrush *Zoothera everetti* ● OR
Siberian Ground Thrush *Zoothera sibirica* EP OR
Abyssinian Ground Thrush *Zoothera piaggiae* AF
Western Orange Ground Thrush *Zoothera tanganjicae* AF
Forest Ground Thrush *Zoothera oberlaenderi* ● AF
Orange Ground Thrush *Zoothera gurneyi* AF
Prigogine's Ground Thrush *Zoothera kibalensis* ● AF
Black-eared Ground Thrush *Zoothera cameronensis* AF
Grey Ground Thrush *Zoothera princei* AF
Crossley's Ground Thrush *Zoothera crossleyi* AF
Spotted Ground Thrush *Zoothera guttata* AF
Spotted-winged Thrush *Zoothera spiloptera* AF
Sunda Ground Thrush *Zoothera andromedae* OR
Plain-backed Mountain Thrush *Zoothera mollissima* EP OR
Long-tailed Mountain Thrush *Zoothera dixoni* EP OR
White's Thrush *Zoothera dauma* EP OR AU
Fawn-breasted Thrush *Zoothera machiki* ● AU
Heine's Ground Thrush *Zoothera heinei* AU PA
Australian Ground Thrush *Zoothera lunulata* AU
Melanesian Ground Thrush *Zoothera talaseae* AU PA
Greater Long-billed Thrush *Zoothera monticola* EP OR
Lesser Long-billed Thrush *Zoothera marginata* OR

Varied Thrush *Ixoreus naevia* NA

Aztec Thrush *Ridgwayia pinicola* NT

Greater New Guinea Thrush *Amalocichla sclateriana* AU
Lesser New Guinea Thrush *Amalocichla incerta* AU

Cataponera Thrush *Cataponera turdoides* AU

Tristan Thrush *Nesocichla eremita* AF

Forest Thrush *Cichlherminia lherminieri* NT

Slender-billed Nightingale Thrush *Catharus gracilirostris* NT
Orange-billed Nightingale Thrush *Catharus aurantiirostris* NT
Slaty-backed Nightingale Thrush *Catharus fuscater* NT
Russet Nightingale Thrush *Catharus occidentalis* NT
Frantzius' Nightingale Thrush *Catharus frantzii* NT
Black-headed Nightingale Thrush *Catharus mexicanus* NT

Spotted Nightingale Thrush *Catharus dryas* NT
Veery *Catharus fuscescens* NA NT
Grey-cheeked Thrush *Catharus minimus* EP NA NT
Swainson's Thrush *Catharus ustulatus* NA NT
Hermit Thrush *Catharus guttatus* NA NT

Wood Thrush *Hylocichla mustelina* NA NT

Yellow-legged Thrush *Platycichla flavipes* NT
Pale-eyed Thrush *Platycichla leucops* NT

Comoro Thrush *Turdus bewsheri* AF
Sao Thome Thrush *Turdus olivaceofuscus* AF
Olive Thrush *Turdus olivaceus* AF
Mountain (Northern Olive) Thrush *Turdus abyssinicus* AF
Taita Olive Thrush *Turdus helleri* ● AF
Kurrichane Thrush *Turdus libonyanus* AF
African Bare-eyed Thrush *Turdus tephronotus* AF
Yemen Thrush *Turdus menachensis* ●WP
Somali Blackbird *Turdus ludoviciae* AF
Groundscraper Thrush *Turdus litsipsirupa* AF
Black-breasted Thrush *Turdus dissimilis* EP OR
Tickell's Thrush *Turdus unicolor* OR
Japanese Grey Thrush *Turdus cardis* EP OR
White-collared Blackbird *Turdus albocinctus* EP OR
Ring Ouzel *Turdus torquatus* WP EP AF
Grey-winged Blackbird *Turdus boulboul* EP OR
Blackbird *Turdus merula* WP EP OR
Island Thrush *Turdus poliocephalus* OR AU PA
Red-billed Thrush *Turdus chrysolaus* EP
Seven Islands Thrush *Turdus celaenops* EP
Grey-headed Thrush *Turdus rubrocanus* EP
Kessler's Thrush *Turdus kessleri* EP
Fea's Thrush *Turdus feae* ● EP
Pale Thrush *Turdus pallidus* EP
Eye-browed Thrush *Turdus obscurus* EP OR
Black-throated Thrush *Turdus ruficollis* EP OR
Dusky Thrush *Turdus naumanni* EP OR
Fieldfare *Turdus pilaris* WP EP
Redwing *Turdus iliacus* WP EP
Song Thrush *Turdus philomelos* WP EP
Mongolian Song Thrush *Turdus mupinensis* EP
Mistle Thrush *Turdus viscivorus* WP EP OR
White-chinned Thrush *Turdus aurantius* NT
Grand Cayman Thrush *Turdus ravidus* NT
Red-legged Thrush *Turdus plumbeus* NT
Chiguanco Thrush *Turdus chiguanco* NT
Sooty Robin *Turdus nigrescens* NT
Great Thrush *Turdus fuscater* NT
Black Robin *Turdus infuscatus* NT
Glossy-black Thrush *Turdus serranus* NT
Slaty Thrush *Turdus nigriceps* NT
Plumbeous-backed Thrush *Turdus reevei* NT
Black-hooded Thrush *Turdus olivater* NT
Maranon Thrush *Turdus maranonicus* NT
Chestnut-bellied Thrush *Turdus fulviventris* NT
Rufous-bellied Thrush *Turdus rufiventris* NT
Austral Thrush *Turdus falcklandii* NT
Pale-breasted Thrush *Turdus leucomelas* NT
Creamy-bellied Thrush *Turdus amaurochalinus* NT
Mountain Robin *Turdus plebejus* NT
Black-billed Thrush *Turdus ignobilis* NT
Lawrence's Thrush *Turdus lawrencii* NT
Cocoa Thrush *Turdus fumigatus* NT
Lesser Antillean Thrush *Turdus personus* NT
Pale-vented Thrush *Turdus obsoletus* NT
Unicoloured Thrush *Turdus haplochrous* NT
Clay-coloured Thrush *Turdus grayi* NT
Bare-eyed Thrush *Turdus nudigenis* NT
White-eyed Thrush *Turdus jamaicensis* NT
White-necked Thrush *Turdus albicollis* NT
Rufous-backed Robin *Turdus rufopalliatus* NT
La Selle Thrush *Turdus swalesi* NT
Rufous-collared Robin *Turdus rufitorques* NT
American Robin *Turdus migratorius* NA NT

131 LOGRUNNERS Orthonychidae

Spine-tailed Logrunner *Orthonyx temminckii* AU
Spalding's Logrunner *Orthonyx spaldingii* AU

Green-backed Babbler *Androphobus viridis* AU

Eastern Whipbird *Psophodes olivaceus* AU
Western Whipbird *Psophodes nigrogularis* ● AU

Wedgebill *Sphenostoma cristatum* AU

Spotted Quail Thrush *Cinclosoma punctatum* AU
Chestnut Quail Thrush *Cinclosoma castanotum* AU
Nullarbor Quail Thrush *Cinclosoma alisteri* AU
Cinnamon Quail Thrush *Cinclosoma cinnamomeum* AU
Chestnut-breasted Quail Thrush *Cinclosoma castaneothorax* AU
Ajax Quail Thrush *Cinclosoma ajax* AU

High Mountain Rail Babbler *Ptilorrhoa leucosticta* AU
Lowland Rail Babbler *Ptilorrhoa caerulescens* AU
Mid-mountain Rail Babbler *Ptilorrhoa castanonota* AU

Malay Rail Babbler *Eupetes macrocerus* OR

Lesser Melampitta *Melampitta lugubris* AU
Greater Melampitta *Melampitta gigantea* AU

Blue-capped Babbler *Ifrita kowaldi* AU

132 BABBLERS Timaliidae

White-chested Jungle Babbler *Trichastoma rostratum* OR
Sulawesi Jungle Babbler *Trichastoma celebense* AU
Ferruginous Jungle Babbler *Trichastoma bicolor* OR
Bagobo Babbler *Trichastoma woodi* ● OR

Abbott's Jungle Babbler *Malacocincla abbotti* OR
Horsfield's Jungle Babbler *Malacocincla sepiaria* OR
Black-browed Jungle Babbler *Malacocincla perspicillata* ● OR
Short-tailed Jungle Babbler *Malacocincla malaccensis* OR
Ashy-headed Jungle Babbler *Malacocincla cinereiceps* OR

Tickell's Jungle Babbler *Pellorneum tickelli* OR
Plain Brown Babbler *Pellorneum albiventre* OR
Marsh Spotted Babbler *Pellorneum palustre* ● OR
Spotted Babbler *Pellorneum ruficeps* OR
Brown-capped Jungle Babbler *Pellorneum fuscocapillum* OR
Temminck's Jungle Babbler *Pellorneum pyrrogenys* OR
Black-capped Babbler *Pellorneum capistratum* OR

Moustached Tree Babbler *Malacopteron magnirostre* OR
Sooty-capped Babbler *Malacopteron affine* OR
Scaly-crowned Babbler *Malacopteron cinereum* OR
Rufous-crowned Tree Babbler *Malacopteron magnum* OR
Palawan Tree Babbler *Malacopteron palawanense* OR
Grey-breasted Babbler *Malacopteron albogulare* OR

Blackcap Thrush Babbler *Illadopsis cleaveri* OR
Scaly-breasted Thrush Babbler *Illadopsis albipectus* OR
Rufous-winged Thrush Babbler *Illadopsis rufescens* OR
Puvel's Thrush Babbler *Illadopsis puveli* OR
Pale-breasted Thrush Babbler *Illadopsis rufipennis* OR
Brown Thrush Babbler *Illadopsis fulvescens* OR
Mountain Thrush Babbler *Illadopsis pyrrhopterus* OR
African Hill Babbler *Illadopsis abyssinicus* OR

Grey-chested Thrush Babbler *Kakamega poliothorax* AF

African Thrush Babbler *Ptyrticus turdinus* AF

Long-billed Scimitar Babbler *Pomatorhinus hypoleucos* OR
Rusty-cheeked Scimitar Babbler *Pomatorhinus erythrogenys* EP OR
Travancore Scimitar Babbler *Pomatorhinus horsfieldii* OR
Slaty-headed Scimitar Babbler *Pomatorhinus schisticeps* EP OR
Chestnut-backed Scimitar Babbler *Pomatorhinus montanus* OR
Streak-breasted Scimitar Babbler *Pomatorhinus ruficollis* OR
Red-billed Scimitar Babbler *Pomatorhinus ochraceiceps* OR
Coral-billed Scimitar Babbler *Pomatorhinus ferruginosus* OR

Isidor's Rufous Babbler *Garritornis isidorei* AU

Grey-crowned Babbler *Pomatostomus temporalis* AU
White-browed Babbler *Pomatostomus superciliosus* AU
Hall Babbler *Pomatostomus halli* AU
Chestnut-crowned Babbler *Pomatostomus ruficeps* AU

Slender-billed Scimitar Babbler *Xiphirhynchus superciliaris* OR

Danjou's Babbler *Jabouilleia danjoui* ● OR

Long-billed Wren Babbler *Rimator malacoptilus* OR

Bornean Wren Babbler *Ptilocichla leucogrammica* OR
Streaked Ground Babbler *Ptilocichla mindanensis* OR
Palawan Wren Babbler *Ptilocichla falcata* OR

Striped Wren Babbler *Kenopia striata* OR

Sumatran Wren Babbler *Napothera rufipectus* OR
Black-throated Wren Babbler *Napothera atrigularis* OR
Large Wren Babbler *Napothera macrodactyla* OR
Marbled Wren Babbler *Napothera marmorata* OR
Limestone Wren Babbler *Napothera crispifrons* OR
Streaked Wren Babbler *Napothera brevicaudata* OR
Mountain Wren Babbler *Napothera crassa* OR
Luzon Wren Babbler *Napothera rabori* ● OR
Lesser Wren Babbler *Napothera epilepidota* OR

Scaly-breasted Wren Babbler *Pnoepyga albiventer* OR
Pygmy Wren Babbler *Pnoepyga pusilla* EP OR AU

Short-tailed Wren Babbler *Spelaeornis caudatus* ● OR
Mishmi Wren Babbler *Spelaeornis badeigularis* ● OR
Bar-winged Wren Babbler *Spelaeornis troglodytoides* EP OR
Spotted Wren Babbler *Spelaeornis formosus* OR
Godwin-Austin's Wren Babbler *Spelaeornis chocolatinus* OR
Long-tailed Wren Babbler *Spelaeornis longicaudatus* ● OR

Wedge-billed Wren Babbler *Sphenocicla humei* ● OR

Northern Jery *Neomixis tenella* AF
Southern Green Jery *Neomixis viridis* AF
Stripe-throated Jery *Neomixis striatigula* AF
Wedge-tailed Jery *Neomixis flavoviridis* AF

Deignan's Babbler *Stachyris rodolphei* ● OR
Red-fronted Tree Babbler *Stachyris rufifrons* OR
Buff-chested Babbler *Stachyris ambigua* OR
Red-headed Tree Babbler *Stachyris ruficeps* OR
Red-billed Tree Babbler *Stachyris pyrrhops* OR
Golden-headed Tree Babbler *Stachyris chrysaea* OR
Pygmy Tree Babbler *Stachyris plateni* OR
Rufous-crowned Tree Babbler *Stachyris capitalis* OR
Rough-templed Tree Babbler *Stachyris speciosa* ● OR
Whitehead's Tree Babbler *Stachyris whiteheadi* OR
Striped Tree Babbler *Stachyris striata* OR
Negros Tree Babbler *Stachyris nigrorum* ● OR
Palawan Tree Babbler *Stachyris hypogrammica* OR
White-breasted Tree Babbler *Stachyris grammiceps* ● OR
Sooty Tree Babbler *Stachyris herberti* ● OR
Black-throated Tree Babbler *Stachyris nigriceps* OR
Grey-headed Tree Babbler *Stachyris poliocephala* OR
Spot-necked Tree Babbler *Stachyris striolata* OR
Austen's Spotted Tree Babbler *Stachyris oglei* ● OR
Chestnut-rumped Tree Babbler *Stachyris maculata* OR
White-necked Tree Babbler *Stachyris leucotis* OR
Black-throated Tree Babbler *Stachyris nigricollis* OR
White-collared Tree Babbler *Stachyris thoracica* OR
Chestnut-winged Tree Babbler *Stachyris erythroptera* OR
Pearl-cheeked Tree Babbler *Stachyris melanothorax* OR
White-bellied Tree Babbler *Stachyris zantholeuca* OR

Rufous-bellied Babbler *Dumetia hyperythra* OR

Black-headed Babbler *Rhopocichla atriceps* OR

Striped Tit-Babbler *Macronous gularis* OR
Grey-faced Tit-Babbler *Macronous kelleyi* OR
Brown Tit-Babbler *Macronous striaticeps* OR
Fluffy-backed Tit-Babbler *Macronous ptilosus* OR

Leyte Tit-Babbler *Micromacronus leytensis* OR

Chestnut-capped Babbler *Timalia pileata* OR

Oriental Yellow-eyed Babbler *Chrysomma sinense* OR

Jerdon's Babbler *Moupinia altirostris* ● OR
Rufous-crowned Babbler *Moupinia poecilotis* EP OR

Wren-Tit *Chamaea fasciata* NA

Spiny Babbler *Turdoides nipalensis* OR
Iraq Babbler *Turdoides altirostris* OR
Common Babbler *Turdoides caudatus* WP OR
Striated Babbler *Turdoides earlei* OR
White-throated Babbler *Turdoides gularis* OR
Slender-billed Babbler *Turdoides longirostris* OR
Large Grey Babbler *Turdoides malcolmi* OR
Arabian Babbler *Turdoides squamiceps* WP AF
Fulvous Babbler *Turdoides fulvus* WP AF
Scaly Chatterer *Turdoides aylmeri* AF
Rufous Chatterer *Turdoides rubiginosus* AF
Rufous Babbler *Turdoides subrufus* OR
White-headed Jungle Babbler *Turdoides striatus* OR
Ceylon Jungle Babbler *Turdoides rufescens* OR
White-headed Babbler *Turdoides affinis* OR
Black-lored Babbler *Turdoides melanops* AF
Dusky Babbler *Turdoides tenebrosus* AF
Blackcap Babbler *Turdoides reinwardtii* AF
Brown Babbler *Turdoides plebejus* AF
Arrow-marked Babbler *Turdoides jardineii* AF
Scaly Babbler *Turdoides squamulatus* AF
White-rumped Babbler *Turdoides leucopygius* AF
Hinde's Pied Babbler *Turdoides hindei* ● AF
Northern Pied Babbler *Turdoides hypoleucus* AF
Pied Babbler *Turdoides bicolor* AF
Bare-cheeked Babbler *Turdoides gymnogenys* AF

Chinese Babax *Babax lanceolatus* EP OR
Giant Babax *Babax waddelli* EP
Koslow's Babax *Babax koslowi* EP

Ashy-headed Laughing Thrush *Garrulax cinereifrons* ● OR
Grey & Brown Laughing Thrush *Garrulax palliatus* OR
Red-fronted Laughing Thrush *Garrulax rufifrons* OR
Spectacled Laughing Thrush *Garrulax perspicillatus* EP OR
White-throated Laughing Thrush *Garrulax albogularis* EP OR
White-crested Laughing Thrush *Garrulax leucolophus* OR
Lesser Necklaced Laughing Thrush *Garrulax monileger* OR
Greater Necklaced Laughing Thrush *Garrulax pectoralis* OR
Black Laughing Thrush *Garrulax lugubris* OR
Striated Laughing Thrush *Garrulax striatus* EP OR
Tickell's Laughing Thrush *Garrulax strepitans* OR
Black-hooded Laughing Thrush *Garrulax milleti* ● OR
Maes' Laughing Thrush *Garrulax maesi* EP OR
Chestnut-backed Laughing Thrush *Garrulax nuchalis* OR
Black-throated Laughing Thrush *Garrulax chinensis* OR
White-cheeked Laughing Thrush *Garrulax vassali* OR
Austen's Laughing Thrush *Garrulax galbanus* OR
Rufous-vented Laughing Thrush *Garrulax delesserti* OR
Variegated Laughing Thrush *Garrulax variegatus* EP OR
David's Laughing Thrush *Garrulax davidi* EP
Black-fronted Laughing Thrush *Garrulax sukatschewi* ● EP
Ashy Laughing Thrush *Garrulax cineraceus* EP OR
Rufous-chinned Laughing Thrush *Garrulax rufogularis* OR
Bar-backed Laughing Thrush *Garrulax lunulatus* EP
Biet's Laughing Thrush *Garrulax bieti* ● EP
Giant Laughing Thrush *Garrulax maximus* EP
White-spotted Laughing Thrush *Garrulax ocellatus* EP

Grey-sided Laughing Thrush *Garrulax caerulatus* EP
OR
Rufous Laughing Thrush *Garrulax poecilorhynchus*
OR
Chestnut-capped Laughing Thrush *Garrulax mitratus* OR
Rufous-necked Laughing Thrush *Garrulax ruficollis*
OR
Spot-breasted Laughing Thrush *Garrulax merulinus*
OR
Melodious Laughing Thrush (Hwamei) *Garrulax canorus* EP OR
White-browed Laughing Thrush *Garrulax sannio* EP
OR
Nilgiri White-breasted Laughing Thrush *Garrulax cachinnans* OR
White-breasted Laughing Thrush *Garrulax jerdoni*
OR
Himalayan Streaked Laughing Thrush *Garrulax lineatus* EP
Striped Laughing Thrush *Garrulax virgatus* OR
Brown-capped Laughing Thrush *Garrulax austeni*
OR
Blue-winged Laughing Thrush *Garrulax squamatus*
OR
Plain-coloured Laughing Thrush *Garrulax subunicolor* OR
Elliot's Laughing Thrush *Garrulax elliotii* EP
Prince Henry's Laughing Thrush *Garrulax henrici*
EP
Black-faced Laughing Thrush *Garrulax affinis* EP
OR
Red-headed Laughing Thrush *Garrulax erythrocephalus* EP OR
Yersin's Laughing Thrush *Garrulax yersini* ● OR
Crimson-winged Laughing Thrush *Garrulax formosus* EP
Red-tailed Laughing Thrush *Garrulax milnei* OR

Red-faced Liocichla *Liocichla phoenicea* OR
Mount Omei Liocichla *Liocichla omeiensis* ● OR
Steere's Liocichla *Liocichla steerii* OR

Silver-eared Mesia *Leiothrix argentauris* OR
Pekin Robin *Leiothrix lutea* EP OR

Nepal Cutia *Cutia nipalensis* OR

Rufous-bellied Shrike Babbler *Pteruthius rufiventer*
OR
Red-winged Shrike Babbler *Pteruthius flaviscapis*
EP OR
Green Shrike Babbler *Pteruthius xanthochlorus* EP
OR
Black-eared Shrike Babbler *Pteruthius melanotis* OR
Chestnut-fronted Shrike Babbler *Pteruthius aenobarbus* OR

White-headed Shrike Babbler *Gampsorhynchus rufulus* OR

Rusty-fronted Barwing *Actinodura egertoni* OR
Spectacled Barwing *Actinodura ramsayi*
ORActinodura nipalensis
Austen's Barwing *Actinodura waldeni* OR
Streaked Barwing *Actinodura souliei* OR
Formosan Barwing *Actinodura morrisoniana* OR

Blue-winged Minla *Minla cyanouroptera* OR
Chestnut-tailed Minla *Minla strigula* EP OR
Red-tailed Minla *Minla ignotincta* OR
Chestnut-backed Minla *Minla annectens* OR

Golden-breasted Fulvetta *Alcippe chrysotis* EP OR
Variegated Fulvetta *Alcippe variegaticeps* ● EP
Yellow-throated Fulvetta *Alcippe cinerea* OR
Chestnut-headed Fulvetta *Alcippe castaneceps* OR
White-browed Fulvetta *Alcippe vinipectus* EP OR
Chinese Mountain Fulvetta *Alcippe striaticollis* EP
Spectacled Fulvetta *Alcippe ruficapilla* EP OR
Streak-throated Fulvetta *Alcippe cinereiceps* EP OR
Rufous-throated Fulvetta *Alcippe rufogularis* OR
Gould's Fulvetta *Alcippe brunnea* OR
Brown Fulvetta *Alcippe brunneicauda* OR
Brown-cheeked Fulvetta *Alcippe poioicephala* OR
Javanese Fulvetta *Alcippe pyrrhoptera* OR
Mountain Fulvetta *Alcippe peracensis* OR
Grey-cheeked Fulvetta *Alcippe morrisonia* EP OR
Nepal Fulvetta *Alcippe nipalensis* EP OR

Bush Blackcap *Lioptilus nigricapillus* AF
White-throated Mountain Babbler *Kupeornis gilberti*
● AF
Red-collared Flycatcher Babbler *Kupeornis rufocinctus* AF
Chapin's Flycatcher Babbler *Kupeornis chapini* AF
Abyssinian Catbird *Parophasma galinieri* AF
Capuchin Babbler *Phyllanthus atripennis* AF
Mt Langbian Sibia *Crocias langbianis* ● OR
Spotted Sibia *Crocias albonotatus* OR

Black-capped Sibia *Heterophasia capistrata* EP
Grey Sibia *Heterophasia gracilis* OR
Black-headed Sibia *Heterophasia melanoleuca* OR
White-eared Sibia *Heterophasia auricularis* OR
Beautiful Sibia *Heterophasia pulchella* EP OR
Long-tailed Sibia *Heterophasia picaoides* OR

Striated Yuhina *Yuhina castaniceps* OR
White-naped Yuhina *Yuhina bakeri* OR
Whiskered Yuhina *Yuhina flavicollis* EP OR
Striped-throated Yuhina *Yuhina gularis* EP OR
White-collared Yuhina *Yuhina diademata* EP OR
Rufous-vented Yuhina *Yuhina occipitalis* EP OR
Formosan Yuhina *Yuhina brunneiceps* OR
Black-chinned Yuhina *Yuhina nigrimenta* EP OR

Fire-tailed Myzornis *Myzornis pyrrhoura* EP OR

Dohrn's Thrush-Babbler *Horizorhinus dohrni* AF

Grey-crowned Oxylabes *Oxylabes cinereiceps* AF
White-throated Oxylabes *Oxylabes madagascariensis*
AF
Yellow-browed Oxylabes *Oxylabes xanthophrys* ●
AF

Crossley's Babbler *Mystacornis crossleyi* AF

133 PARROTBILLS Panuridae

Bearded Reedling *Panurus biarmicus* WP EP

Great Parrotbill *Conostoma oemodium* EP OR

Three-toed Parrotbill *Paradoxornis paradoxus* EP
Brown Parrotbill *Paradoxornis unicolor* EP OR
Gould's Parrotbill *Paradoxornis flavirostris* ● OR
Spot-breasted Parrotbill *Paradoxornis guttaticollis*
OR
Spectacled Parrotbill *Paradoxornis conspicillatus* EP
OR
Yunnan Parrotbill *Paradoxornis ricketti* OR
Vinous-throated Parrotbill *Paradoxornis webbianus*
EP OR
Ashy-throated Parrotbill *Paradoxornis alphonsianus*
EP OR
Zappey's Parrotbill *Paradoxornis zappeyi* ● EP
Grey-crowned Parrotbill *Paradoxornis przewalskii*
● EP
Fulvous-fronted Parrotbill *Paradoxornis fulvifrons*
EP OR
Blyth's Parrotbill *Paradoxornis nipalensis* EP OR
David's Parrotbill *Paradoxornis davidianus* ● OR
Lesser Red-headed Parrotbill *Paradoxornis atrosuperciliaris* OR
Greater Red-headed Parrotbill *Paradoxornis ruficeps*
● EP OR
Grey-headed Parrotbill *Paradoxornis gularis* EP OR
Heude's Parrotbill *Paradoxornis heudei* EP

134 BALD CROWS Picathartidae

White-necked Bald Crow (Rockfowl) *Picathartes gymnocephalus* ● AF
Grey-necked Bald Crow (Rockfowl) *Picathartes oreas* ● AF

135 GNATCATCHERS Polioptilidae

Collared Gnatwren *Microbates collaris* NT
Half-collared Gnatwren *Microbates cinereiventris* NT

Long-billed Gnatwren *Ramphocaenus melanurus* NT

Blue-grey Gnatcatcher *Polioptila caerulea* NT
Black-tailed Gnatcatcher *Polioptila melanura* NT
Cuban Gnatcatcher *Polioptila lembeyei* NT
White-lored Gnatcatcher *Polioptila albiloris* NT
Black-capped Gnatcatcher *Polioptila nigriceps* NT
Tropical Gnatcatcher *Polioptila plumbea* NT

Cream-bellied Gnatcatcher *Polioptila lactea* NT
Guianan Gnatcatcher *Polioptila guianensis* NT
Slate-throated Gnatcatcher *Polioptila schistaceigula*
NT
Masked Gnatcatcher *Polioptila dumicola* NT

136 OLD WORLD WARBLERS Sylviidae

Chestnut-headed Ground Warbler *Oligura castaneocoronata* EP OR

Java Ground Warbler *Tesia superciliaris* OR
Slaty-bellied Ground Warbler *Tesia olivea* OR
Grey-bellied Ground Warbler *Tesia cyaniventer* EP
OR
Russet-capped Stubtail *Tesia everetti* AU

Timor Stubtail *Urosphena subulata* AU
Bornean Stubtail *Urosphena whiteheadi* OR
Scaly-headed Stubtail *Urosphena squameiceps* EP OR

Pale-footed Stubtail *Cettia pallidipes* EP OR
Japanese Bush Warbler *Cettia diphone* EP OR
Palau Bush Warbler *Cettia annae* PA
Shade Warbler *Cettia parens* PA
Fiji Warbler *Cettia ruficapilla* PA
Strong-footed Bush Warbler *Cettia fortipes* EP
Muller's Bush Warbler *Cettia vulcania* OR AU
Large Bush Warbler *Cettia major* OR
Yamdena Warbler *Cettia carolinae* AU
Aberrant Bush Warbler *Cettia flavolivacea* OR
Swinhoe's Bush Warbler *Cettia robustipes* EP OR
Rufous-capped Bush Warbler *Cettia brunnifrons* EP
OR
Cetti's Warbler *Cettia cetti* WP EP OR

African Sedge Warbler *Bradypterus baboecalus* AF
Grauer's Warbler *Bradypterus graueri* ● AF
Ja River Warbler *Bradypterus grandis* ● AF
White-winged Warbler *Bradypterus carpalis* AF
Bamboo Warbler *Bradypterus alfredi* AF
Knysna Scrub Warbler *Bradypterus sylvaticus* AF
Lopez' Warbler *Bradypterus lopezi* AF
Evergreen Forest Warbler *Bradypterus mariae* AF
Scrub Warbler *Bradypterus barratti* AF
Victorin's Scrub Warbler *Bradypterus victorini* AF
Cinnamon Bracken Warbler *Bradypterus cinnamomeus* AF
Brown Feather-tailed Warbler *Bradypterus brunneus*
AF
Spotted Bush Warbler *Bradypterus thoracicus* EP OR
Large-billed Bush Warbler *Bradypterus major* ● EP
Chinese Bush Warbler *Bradypterus tacsanowskius*
EP OR
Brown Bush Warbler *Bradypterus luteoventris* OR
Palliser's Warbler *Bradypterus palliseri* OR
Mountain Scrub Warbler *Bradypterus seebohmi* OR
AU
Long-tailed Ground Warbler *Bradypterus caudatus*
OR
Kinabalu Scrub Warbler *Bradypterus accentor* OR
Chestnut-backed Bush Warbler *Bradypterus castaneus* AU

Black-capped Rufous Warbler *Bathmocercus cerviniventris* AF
Black-faced Rufous Warbler *Bathmocercus rufus* AF
Mrs Moreau's Warbler *Bathmocercus winifredae* ●
AF

Seebohm's Feather-tailed Warbler *Amphilais seebohmi* AF

Tsikirity Warbler *Nesillas typica* AF
Comoro Warbler *Nesillas mariae* AF
Aldabra Warbler *Nesillas aldabranus* ● AF

Kiritika Warbler *Thamnornis chloropetoides* AF

Moustached Grass Warbler *Melocichla mentalis* AF

Damaraland Rock Jumper *Achaetops pycnopygius* AF

Cape Grassbird *Sphenoeacus afer* AF

Japanese Marsh Warbler *Megalurus pryeri* ● EP
Tawny Marshbird *Megalurus timoriensis* OR AU
Striated Canegrass Warbler *Megalurus palustris* OR
Fly River Grass Warbler *Megalurus albolimbatus* ●
AU
Little Marshbird *Megalurus gramineus* AU
Fernbird *Megalurus punctatus* AU

Rufous Songlark *Cinclorhamphus mathewsi* AU
Brown Songlark *Cinclorhamphus cruralis* AU

Spinifex Bird *Eremiornis carteri* AU

Buff-banded Bushbird *Megalurulus bivittatus* AU
New Caledonian Grass Warbler *Megalurulus mariei* PA

Thicket Warbler *Cichlornis whitneyi* PA
Whiteman Mountains Warbler *Cichlornis grosvenori* AU

Bougainville Thicket Warbler *Cichlornis llaneae* PA

Rufous-faced Thicket Warbler *Ortygocichla rubiginosa* AU
Long-legged Warbler *Ortygocichla rufa* ● PA

Bristled Grass Warbler *Chaetornis striatus* ● OR

Large Grass Warbler *Graminicola bengalensis* EP OR

Broad-tailed Warbler *Schoenicola platyura* AF

Lanceolated Warbler *Locustella lanceolata* EP OR
Grasshopper Warbler *Locustella naevia* WP EP OR
Pallas's Grasshopper Warbler *Locustella certhiola* EP OR
Middendorff's Grasshopper Warbler *Locustella ochotensis* EP OR
Styan's Grasshopper Warbler *Locustella pleskei* EP OR
River Warbler *Locustella fluviatilis* WP EP OR
Savi's Warbler *Locustella luscinioides* WP EP
Gray's Grasshopper Warbler *Locustella fasciolata* EP OR AU
Stepanyan's Grasshopper Warbler *Locustella amnicola* EP

Moustached Warbler *Acrocephalus melanopogon* WP EP
Aquatic Warbler *Acrocephalus paludicola* ●WP EP
Sedge Warbler *Acrocephalus schoenobaenus* WP EP AF
Speckled Reed Warbler *Acrocephalus sorghophilus* ● EP OR
Schrenk's Reed Warbler *Acrocephalus bistrigiceps* EP OR
Paddyfield Warbler *Acrocephalus agricola* EP OR
Blunt-winged Paddyfield Warbler *Acrocephalus concinens* EP OR
Reed Warbler *Acrocephalus scirpaceus* WP EP AF
Reichenow's Reed Warbler *Acrocephalus cinnamomeus* AF
African Reed Warbler *Acrocephalus baeticatus* AF
Marsh Warbler *Acrocephalus palustris* WP EP AF
Blyth's Reed Warbler *Acrocephalus dumetorum* WP EP OR
Great Reed Warbler *Acrocephalus arundinaceus* WP EP AF
Clamorous Reed Warbler *Acrocephalus stentoreus* WP OR AU
Large-billed Reed Warbler *Acrocephalus orinus* OR
Oriental Great Reed Warbler *Acrocephalus orientalis* EP OR
Nightingale Reed Warbler *Acrocephalus luscinia* PA
Hawaiian Reed Warbler *Acrocephalus kingii* ● PA
Polynesian Reed Warbler *Acrocephalus aequinoctialis* PA
Long-billed Reed Warbler *Acrocephalus caffer* PA
Tuamotu Warbler *Acrocephalus atyphus* PA
Pitcairn Warbler *Acrocephalus vaughanii* PA
Rufous Swamp Warbler *Acrocephalus rufescens* AF
Cape Verde Swamp Warbler *Acrocephalus brevipennis* AF
Swamp Warbler *Acrocephalus gracilirostris* AF
Madagascar Swamp Warbler *Acrocephalus newtoni* AF

Thick-billed Warbler *Acrocephalus aedon* EP OR

Rodriguez Brush Warbler *Bebrornis rodericanus* ● AF
Seychelles Brush Warbler *Bebrornis sechellensis* ● AF

Booted Warbler *Hippolais caligata* EP OR
Olivaceous Warbler *Hippolais pallida* WP EP AF
Upcher's Warbler *Hippolais languida* WP EP AF
Olive-tree Warbler *Hippolais olivetorum* AF
Melodious Warbler *Hippolais polyglotta* WP
Icterine Warbler *Hippolais icterina* WP EP AF

Yellow Warbler *Chloropeta natalensis* AF
Mountain Yellow Warbler *Chloropeta similis* AF
Yellow Swamp Warbler *Chloropeta gracilirostris* ● AF

Red-faced Cisticola *Cisticola erythrops* AF
Angolan Cisticola *Cisticola lepe* AF
Singing Cisticola *Cisticola cantans* AF
Whistling Cisticola *Cisticola lateralis* AF
Trilling Cisticola *Cisticola woosnami* AF
Chattering Cisticola *Cisticola anonyma* AF
Bubbling Cisticola *Cisticola bulliens* AF
Chubb's Cisticola *Cisticola chubbi* AF
Hunter's Cisticola *Cisticola hunteri* AF
Black-browed Cisticola *Cisticola nigriloris* AF
Lazy Cisticola *Cisticola aberrans* AF
Boran Cisticola *Cisticola bodessa* AF
Rattling Cisticola *Cisticola chiniana* AF
Ashy Cisticola *Cisticola cinereola* AF
Red-pate Cisticola *Cisticola ruficeps* AF
Grey Cisticola *Cisticola rufilata* AF
Red-headed Cisticola *Cisticola subruficapilla* AF
Wailing Cisticola *Cisticola lais* AF
Tana River Cisticola *Cisticola restricta* ● AF
Churring Cisticola *Cisticola njombe* AF
Winding Cisticola *Cisticola galactotes* AF
Carruthers' Cisticola *Cisticola carruthersi* AF
Chirping Cisticola *Cisticola pipiens* AF
Levaillant's Cisticola *Cisticola tinniens* AF
Stout Cisticola *Cisticola robusta* AF
Aberdare Mountain Cisticola *Cisticola aberdare* AF
Croaking Cisticola *Cisticola natalensis* AF
Piping Cisticola *Cisticola fulvicapilla* AF
Tabora Cisticola *Cisticola angusticauda* AF
Slender-tailed Cisticola *Cisticola melanura* AF
Siffling Cisticola *Cisticola brachyptera* AF
Rufous Cisticola *Cisticola rufa* AF
Foxy Cisticola *Cisticola troglodytes* AF
Tiny Cisticola *Cisticola nana* AF
Incana Cisticola *Cisticola incana* AF
Zitting Cisticola (Fan-tailed Warbler) *Cisticola juncidis* WP EP AF OR AU
Socotra Cisticola *Cisticola haesitata* ● AF
Madagascar Cisticola *Cisticola cherina* AF
Desert Cisticola *Cisticola aridula* AF
Tink-tink Cisticola *Cisticola textrix* AF
Black-necked Cisticola *Cisticola eximia* AF
Cloud-scraping Cisticola *Cisticola dambo* AF
Pectoral-patch Cisticola *Cisticola brunnescens* AF
Wing-snapping Cisticola *Cisticola ayresii* AF
Gold-capped Cisticola *Cisticola exilis* OR AU PA

Streaked Scrub Warbler *Scotocerca inquieta* WP OR

White-browed Chinese Warbler *Rhopophilus pekinensis* EP

Long-tailed Prinia *Prinia burnesii* OR
Hill Prinia *Prinia criniger* OR
Brown Hill Prinia *Prinia polychroa* OR
White-browed Prinia *Prinia atrogularis* OR
Hodgson's Long-tailed Warbler *Prinia cinereocapilla* OR
Rufous-fronted Prinia *Prinia buchanani* OR
Lesser Brown Prinia *Prinia rufescens* OR
Franklin's Prinia *Prinia hodgsonii* OR
Graceful Warbler *Prinia gracilis* WP AF OR
Jungle Prinia *Prinia sylvatica* OR
Bar-winged Prinia *Prinia familiaris* OR
Yellow-bellied Prinia *Prinia flaviventris* OR
Ashy Prinia *Prinia socialis* OR
Tawny-flanked Prinia *Prinia subflava* EP AF OR
Pale Prinia *Prinia somalica* AF
Lake Chad Prinia *Prinia fluviatilis* ● AF
Karroo Prinia *Prinia maculosa* AF
Black-chested Prinia *Prinia flavicans* AF
White-breasted Prinia *Prinia substriata* AF
Sao Thome Prinia *Prinia molleri* AF
Roberts' Prinia *Prinia robertsi* AF
White-chinned Prinia *Prinia leucopogon* AF
Sierra Leone Prinia *Prinia leontica* AF
Banded Prinia *Prinia bairdii* AF
Red-winged Warbler *Prinia erythroptera* AF
Rufous-eared Prinia *Prinia pectoralis* AF

Red-winged Grey Warbler *Drymocichla incana* AF

Green Longtail *Urolais epichlora* AF

Cricket Warbler *Spiloptila clamans* AF

Bar-throated Apalis *Apalis thoracica* AF
Black-collared Apalis *Apalis pulchra* AF
Collared Apalis *Apalis ruwenzorii* AF
Black-capped Apalis *Apalis nigriceps* AF
Black-throated Apalis *Apalis jacksoni* AF
White-winged Apalis *Apalis chariessa* ● AF
Masked Apalis *Apalis binotata* AF
Mountain-masked Apalis *Apalis personata* AF
Yellow-chested Apalis *Apalis flavida* AF
Rudd's Apalis *Apalis ruddi* AF
Buff-throated Apalis *Apalis rufogularis* AF
Sharpe's Apalis *Apalis sharpii* AF
Gosling's Apalis *Apalis goslingi* AF
Bamenda Apalis *Apalis bamendae* AF
Chestnut-throated Apalis *Apalis porphyrolaema* AF
Black-headed Apalis *Apalis melanocephala* AF
Chirinda Apalis *Apalis chirindensis* AF
Grey Apalis *Apalis cinerea* AF
Brown-headed Apalis *Apalis alticola* AF
Karamoja Apalis *Apalis karamojae* ● AF
Red-faced Warbler *Apalis rufifrons* AF

Fairy Flycatcher *Stenostira scita* AF

Buff-bellied Warbler *Phyllolais pulchella* AF

Red-capped Forest Warbler *Orthotomus metopias* AF
Long-billed Forest Warbler *Orthotomus moreaui* ● AF
Mountain Tailor Bird *Orthotomus cucullatus* OR AU
Long-tailed Tailor Bird *Orthotomus sutorius* OR
Black-necked Tailor Bird *Orthotomus atrogularis* OR
Luzon Tailor Bird *Orthotomus derbianus* OR
Red-headed Tailor Bird *Orthotomus sericeus* OR
Grey Tailor Bird *Orthotomus ruficeps* OR
Ashy Tailor Bird *Orthotomus sepium* OR
White-eared Tailor Bird *Orthotomus cinereiceps* OR
Black-headed Tailor Bird *Orthotomus nigriceps* OR
Samar Tailor Bird *Orthotomus samarensis* OR

Green-backed Camaroptera *Camaroptera brachyura* AF
Grey-backed Camaroptera *Camaroptera brevicaudata* AF
Hartert's Cameroptera *Camaroptera harterti* AF
Yellow-browed Camaroptera *Camaroptera superciliaris* AF
Olive-green Camaroptera *Camaroptera chloronota* AF

Grey Wren Warbler *Calamonastes simplex* AF
Stierling's Wren Warbler *Calamonastes stierlingi* AF
Barred Camaroptera *Calamonastes fasciolata* AF

Kopje Warbler *Euryptila subcinnamomea* AF

White-tailed Warbler *Poliolais lopezi* AF

Grauer's Warbler *Graueria vittata* AF

Yellow-bellied Eremomela *Eremomela icteropygialis* AF
Yellow-vented Eremomela *Eremomela flavicrissalis* AF
Green-cap Eremomela *Eremomela scotops* AF
Smaller Green-backed Eremomela *Eremomela pusilla* AF
Green-backed Eremomela *Eremomela canescens* AF
Yellow-rumped Eremomela *Eremomela gregalis* AF
Brown-crowned Eremomela *Eremomela badiceps* AF
Turner's Eremomela *Eremomela turneri* ● AF
Black-necked Eremomela *Eremomela atricollis* AF
Burnt-neck Eremomela *Eremomela usticollis* AF

Marvantsetra Warbler *Randia pseudozosterops* AF

Tulear Newtonia *Newtonia amphichroa* AF
Common Newtonia *Newtonia brunneicauda* AF
Tabity Newtonia *Newtonia archboldi* AF
Fanovana Newtonia *Newtonia fanovanae* ● AF

Green Crombec *Sylvietta virens* AF
Lemon-bellied Crombec *Sylvietta denti* AF
White-browed Crombec *Sylvietta leucophrys* AF
Northern Crombec *Sylvietta brachyura* AF
Somali Short-billed Crombec *Sylvietta philippae* AF
Red-faced Crombec *Sylvietta whytii* AF
Red-capped Crombec *Sylvietta ruficapilla* AF
Long-billed Crombec *Sylvietta rufescens* AF
Somali Long-billed Crombec *Sylvietta isabellina* AF

Neumann's Short-tailed Warbler *Hemitesia neumanni* AF

Kemp's Longbill *Macrosphenus kempi* AF
Yellow Longbill *Macrosphenus flavicans* AF
Grey Longbill *Macrosphenus concolor* AF
Pulitzer's Longbill *Macrosphenus pulitzeri* ● AF
Kretschmer's Longbill *Macrosphenus kretschmeri* AF

Bocage's Longbill *Amaurocichla bocagei* ● AF

Oriole Warbler *Hypergerus atriceps* AF
Grey-capped Warbler *Hypergerus lepida* AF

Yellow-bellied Flycatcher *Hyliota flavigaster* AF
Southern Yellow-bellied Flycatcher *Hyliota australis* AF
Violet-backed Flycatcher *Hyliota violacea* AF

Green Hylia *Hylia prasina* AF

Yellow-throated Woodland Warbler *Phylloscopus ruficapilla* AF
Mrs Boulton's Woodland Warbler *Phylloscopus laurae* AF
Red-faced Woodland Warbler *Phylloscopus laetus* AF
Uganda Woodland Warbler *Phylloscopus budongoensis* AF
Black-capped Woodland Warbler *Phylloscopus herberti* AF
Brown Woodland Warbler *Phylloscopus umbrovirens* AF
Willow Warbler *Phylloscopus trochilus* WP EP AF
Chiffchaff *Phylloscopus collybita* WP AF OR
Eastern Chiffchaff *Phylloscopus sindianus* WP EP OR
Plain Willow Warbler *Phylloscopus neglectus* WP OR
Bonelli's Warbler *Phylloscopus bonelli* WP AF
Wood Warbler *Phylloscopus sibilatrix* WP AF
Dusky Warbler *Phylloscopus fuscatus* EP OR
Smoky Warbler *Phylloscopus fuligiventer* EP OR
Tickell's Willow Warbler *Phylloscopus affinis* EP OR
Sulphur-bellied Willow Warbler *Phylloscopus griseolus* EP OR
Milne-Edwards' Willow Warbler *Phylloscopus armandii* EP OR
Radde's Bush Warbler *Phylloscopus schwarzi* EP OR
Orange-barred Willow Warbler *Phylloscopus pulcher* EP OR
Grey-faced Willow Warbler *Phylloscopus maculipennis* EP OR
Pallas' Leaf Warbler *Phylloscopus proregulus* EP OR
Brooks's Willow Warbler *Phylloscopus subviridis* EP OR
Yellow-browed Warbler *Phylloscopus inornatus* OR
Arctic Warbler *Phylloscopus borealis* WP EP OR
Greenish Warbler *Phylloscopus trochiloides* WP EP OR
Green Willow Warbler *Phylloscopus nitidus* WP EP OR
Two-barred Greenish Warbler *Phylloscopus plumbeitarsus* EP OR
Japanese Pale-legged Willow Warbler *Phylloscopus borealoides* EP
Pale-legged Willow Warbler *Phylloscopus tenellipes* EP OR
Large-billed Willow Warber *Phylloscopus magnirostris* EP OR
Tytler's Willow Warbler *Phylloscopus tytleri* EP OR
Large Crowned Willow Warbler *Phylloscopus occipitalis* EP OR
Temminck's Crowned Willow Warbler *Phylloscopus coronatus* EP OR
Ijima's Willow Warbler *Phylloscopus ijimae* EP OR
Blyth's Crowned Willow Warbler *Phylloscopus reguloides* EP OR
White-tailed Willow Warbler *Phylloscopus davisoni* EP OR
Yellow-faced Leaf Warbler *Phylloscopus cantator* OR
Black-browed Leaf Warbler *Phylloscopus ricketti* OR
Philippine Leaf Warbler *Phylloscopus olivaceus* OR
Dubois' Leaf Warbler *Phylloscopus cebuensis* OR
Mountain Leaf Warbler *Phylloscopus trivirgatus* OR
Sulawesi Leaf Warbler *Phylloscopus sarasinorum* AU
Timor Leaf Warbler *Phylloscopus presbytes* AU
Island Leaf Warbler *Phylloscopus poliocephalus* AU PA
San Cristobal Leaf Warbler *Phylloscopus makirensis* PA

Kulambangra Warbler *Phylloscopus amoenus* ● PA
Yellow-eyed Flycatcher Warbler *Seicercus burkii* EP OR
Grey-headed Flycatcher Warbler *Seicercus xanthoschistos* OR
Allied Flycatcher Warbler *Seicercus affinis* EP OR
Grey-cheeked Flycatcher Warbler *Seicercus poliogenys* OR
Chestnut-headed Flycatcher Warbler *Seicercus castaniceps* OR
Yellow-breasted Flycatcher Warbler *Seicercus montis* OR AU
Sunda Flycatcher Warbler *Seicercus grammiceps* OR
Broad-billed Flycatcher Warbler *Tickellia hodgsoni* OR

White-throated Flycatcher Warbler *Abroscopus albogularis* OR
Black-faced Flycatcher Warbler *Abroscopus schisticeps* OR
Yellow-bellied Flycatcher Warbler *Abroscopus superciliaris* OR

Yemen Tit Warbler *Parisoma buryi* AF
Brown Tit Warbler *Parisoma lugens* AF
Banded Tit Warbler *Parisoma boehmi* AF
Layard's Tit Warbler *Parisoma layardi* AF
Southern Tit Warbler *Parisoma subcaeruleum* AF

Blackcap *Sylvia atricapilla* WP AF
Garden Warbler *Sylvia borin* WP EP AF
Whitethroat *Sylvia communis* WP EP AF
Lesser Whitethroat *Sylvia curruca* WP EP AF OR
Desert Whitethroat *Sylvia nana* WP EP AF OR
Barred Warbler *Sylvia nisoria* WP EP AF
Orphean Warbler *Sylvia hortensis* WP AF OR
Red Sea Warbler *Sylvia leucomelaena* WP AF
Rüppell's Warbler *Sylvia rueppelli* WP AF
Sardinian Warbler *Sylvia melanocephala* WP
Cyprus Warbler *Sylvia melanothorax* WU
Menetries' Warbler *Sylvia mystacea* WP AF
Subalpine Warbler *Sylvia cantillans* WP AF
Spectacled Warbler *Sylvia conspicillata* WP
Tristram's Warbler *Sylvia deserticola* WP
Dartford Warbler *Sylvia undata* WP
Marmora's Warbler *Sylvia sarda* WP

Firecrest *Regulus ignicapillus* WP
Taiwan Firecrest *Regulus goodfellowi* OR
Goldcrest *Regulus regulus* WP
Golden-crowned Kinglet *Regulus satrapa* NA NT
Ruby-crowned Kinglet *Regulus calendula* NA NT

Severtzov's Tit Warbler *Leptopoecile sophiae* EP OR
Crested Tit Warbler *Leptopoecile elegans* EP

137 OLD WORLD FLYCATCHERS Muscicapidae

Silverbird *Melaenornis semipartitus* AF
Pale Flycatcher *Melaenornis pallidus* AF
African Brown Flycatcher *Melaenornis infuscatus* AF
Mariqua Flycatcher *Melaenornis mariquensis* AF
Grey Flycatcher *Melaenornis microrhynchus* AF
Abyssinian Slaty Flycatcher *Melaenornis chocolatinus* AF
White-eyed Slaty Flycatcher *Melaenornis fischeri* AF
Angolan Flycatcher *Melaenornis brunneus* AF
Black Flycatcher *Melaenornis edolioides* AF
South African Black Flycatcher *Melaenornis pammelaina* AF
Berlioz' Black Flycatcher *Melaenornis ardesiaca* AF
Liberian Black Flycatcher *Melaenornis annamarulae* ● AF
Forest Flycatcher *Melaenornis ocreata* AF
White-browed Forest Flycatcher *Melaenornis cinerascens* AF
Fiscal Flycatcher *Melaenornis silens* AF

Buru Jungle Flycatcher *Rhinomyias addita* AU
Flores Jungle Flycatcher *Rhinomyias oscillans* AU
White-gorgetted Jungle Flycatcher *Rhinomyias brunneata* AU
Olive-backed Jungle Flycatcher *Rhinomyias olivacea* OR
White-throated Jungle Flycatcher *Rhinomyias umbratilis* OR
Rufous-tailed Jungle Flycatcher *Rhinomyias ruficauda* OR

Sula Jungle Flycatcher *Rhinomyias colonus* ● AU
White-browed Jungle Flycatcher *Rhinomyias gularis* OR
Lepanto Jungle Flycatcher *Rhinomyias insignis* OR
Mindanao Jungle Flycatcher *Rhinomyias goodfellowi* OR

Spotted Flycatcher *Muscicapa striata* WP EP AF
Gambaga Spotted Flycatcher *Muscicapa gambagae* AF
Grey-streaked Flycatcher *Muscicapa griseisticta* EP OR AU
Siberian Flycatcher *Muscicapa sibirica* EP OR
Brown Flycatcher *Muscicapa dauurica* EP OR
Brown-breasted Flycatcher *Muscicapa muttui* OR
Rufous-tailed Flycatcher *Muscicapa ruficauda* OR
Ferruginous Flycatcher *Muscicapa ferruginea* EP OR
Sri Lankan Dusky Blue Flycatcher *Muscicapa sordida* OR
Indian Verditer Flycatcher *Muscicapa thalassina* EP OR
Philippine Verditer Flycatcher *Muscicapa panayensis* OR AU
Nilgiri Verditer Flycatcher *Muscicapa albicaudata* OR
Indigo Flycatcher *Muscicapa indigo* OR
African Sooty Flycatcher *Muscicapa infuscata* AF
Ussher's Dusky Flycatcher *Muscicapa ussheri* AF
Bohm's Flycatcher *Muscicapa boehmi* AF
Swamp Flycatcher *Muscicapa aquatica* AF
Olivaceous Flycatcher *Muscicapa olivascens* AF
Chapin's Flycatcher *Muscicapa lendu* ● AF
Dusky Flycatcher *Muscicapa adusta* AF
Little Grey Flycatcher *Muscicapa epulata* AF
Yellow-footed Flycatcher *Muscicapa sethsmithi* AF
Dusky Blue Flycatcher *Muscicapa comitata* AF
Tessman's Flycatcher *Muscicapa tessmanni* AF
Cassin's Grey Flycatcher *Muscicapa cassini* AF
Ashy Flycatcher *Muscicapa coerulescens* AF
Grey-throated Flycatcher *Muscicapa griseigularis* AF

Grey Tit Flycatcher *Myioparus plumbeus* AF

Humblot's Flycatcher *Humblotia flavirostris* ● AF

Pied Flycatcher *Ficedula hypoleuca* WP EP AF
Collared Flycatcher *Ficedula albicollis* WP AF
Yellow-rumped Flycatcher *Ficedula zanthopygia* EP OR
Narcissus Flycatcher *Ficedula narcissina* EP OR
Mugimaki Flycatcher *Ficedula mugimaki* EP OR
Rusty-breasted Blue Flycatcher *Ficedula hodgsonii* OR
Orange-breasted Flycatcher *Ficedula dumetoria* OR AU
Orange-gorgetted Flycatcher *Ficedula strophiata* EP OR
Red-breasted Flycatcher *Ficedula parva* WP EP OR
Kashmir Flycatcher *Ficedula subrubra* OR
White-gorgetted Flycatcher *Ficedula monileger* OR
Rufous-browed Flycatcher *Ficedula solitaris* OR
Thicket Flycatcher *Ficedula hyperythra* EP OR AU
White-vented Flycatcher *Ficedula rufigula* AU
Little Slaty Flycatcher *Ficedula basilanica* OR
Cinnamon-chested Flycatcher *Ficedula buruensis* AU
Damar Flycatcher *Ficedula henrici* ● AU
Hartert's Flycatcher *Ficedula harterti* ● AU
Palawan Flycatcher *Ficedula platenae* OR
Vaurie's Flycatcher *Ficedula crypta* OR
Lompobattang Flycatcher *Ficedula bonthaina* ● AU
Little Pied Flycatcher *Ficedula westermanni* EP OR AU
White-browed Blue Flycatcher *Ficedula superciliaris* EP OR
Slaty Blue Flycatcher *Ficedula tricolor* EP OR
Sapphire-headed Flycatcher *Ficedula sapphira* EP OR
Black & Orange Flycatcher *Ficedula nigrorufa* OR
Black-banded Flycatcher *Ficedula timorensis* ● AU

Blue & White Flycatcher *Cyanoptila cyanomelaena* EP OR

Large Niltava *Niltava grandis* EP OR
Small Niltava *Niltava macgregoriae* EP OR
Fukien Niltava *Niltava davidi* OR
Rufous-bellied Niltava *Niltava sundara* EP OR
Sumatran Niltava *Niltava sumatrana* OR

Vivid Niltava *Niltava vivida* OR
Blue-backed Niltava *Niltava hyacinthina* AU
Blue-fronted Niltava *Niltava hoevelli* AU
Matinan Niltava *Niltava sanfordi* ● AU
White-tailed Niltava *Niltava concreta* OR
Rueck's Niltava *Niltava ruecki* ● OR
Blue-breasted Niltava *Niltava herioti* ● OR
Grant's Niltava *Niltava hainana* OR
White-bellied Niltava *Niltava pallipes* OR
Brooks' Niltava *Niltava poliogenys* EP OR
Pale Niltava *Niltava unicolor* EP OR
Blue-throated Niltava *Niltava rubeculoides* EP OR
Hill Blue Niltava *Niltava banyumas* EP OR
Bornean Niltava *Niltava superba* OR
Large-billed Niltava *Niltava caerulata* OR
Malaysian Niltava *Niltava turcosa* OR
Tickell's Niltava *Niltava tickelliae* OR
Mangrove Niltava *Niltava rufigastra* OR AU
Pygmy Blue Flycatcher *Niltava hodgsoni* EP OR

Grey-headed Canary Flycatcher *Culicicapa ceylonensis* OR AU
Citrine Canary Flycatcher *Culicicapa helianthea* OR AU

138 WATTLE-EYES AND PUFF-BACK FLYCATCHERS Platysteiridae

African Shrike Flycatcher *Bias flammulatus* AF
Black & White Flycatcher *Bias musicus* AF

Ward's Flycatcher *Pseudobias wardi* AF

Ruwenzori Puff-back Flycatcher *Batis diops* AF
Boulton's Puff-back Flycatcher *Batis margaritae* AF
Short-tailed Puff-back Flycatcher *Batis mixta* AF
Malawi Puff-back Flycatcher *Batis dimorpha* AF
Cape Puff-back Flycatcher *Batis capensis* AF
Zululand Puff-back Flycatcher *Batis fratrum* AF
Chin Spot Puff-back Flycatcher *Batis molitor* AF
Paler Chin Spot Puff-back Flycatcher *Batis soror* AF
Pirit Puff-back Flycatcher *Batis pirit* AF
Senegal Puff-back Flycatcher *Batis senegalensis* AF
Grey-headed Puff-back Flycatcher *Batis orientalis* AF
Black-headed Puff-back Flycatcher *Batis minor* AF
Pygmy Puff-back Flycatcher *Batis perkeo* AF
Angola Puff-back Flycatcher *Batis minulla* AF
Verreaux's Puff-back Flycatcher *Batis minima* ● AF
Chapin's Puff-back Flycatcher *Batis ituriensis* AF
Fernando Po Puff-back Flycatcher *Batis poensis* AF
Lawson's Puff-back Flycatcher *Batis occultus* AF

Brown-throated Wattle-eye *Platysteira cyanea* AF
White-fronted Wattle-eye *Platysteira albifrons* AF
Black-throated Wattle-eye *Platysteira peltata* AF
Bamenda Wattle-eye *Platysteira laticincta* ● AF
White-spotted Wattle-eye *Platysteira tonsa* AF
Chestnut Wattle-eye *Platysteira castanea* AF
Red-cheeked Wattle-eye *Platysteira blissetti* AF
Reichenow's Wattle-eye *Platysteira chalybea* AF
Jameson's Wattle-eye *Platysteira jamesoni* AF
Yellow-bellied Wattle-eye *Platysteira concreta* AF

139 AUSTRALASIAN WRENS Maluridae

Rufous Wren Warbler *Clytomias insignis* AU

Wallace's Wren Warbler *Malurus wallacii* AU
Broad-billed Wren Warbler *Malurus grayi* AU
Black & White Wren *Malurus alboscapulatus* AU
Red-backed Wren *Malurus melanocephalus* AU
White-winged Wren *Malurus leucopterus* AU
Blue Wren *Malurus cyaneus* AU
Banded Wren *Malurus splendens* AU
Variegated Wren *Malurus lamberti* AU
Blue-breasted Wren *Malurus pulcherrimus* AU
Red-winged Wren *Malurus elegans* AU
Purple-crowned Wren *Malurus coronatus* ● AU
Blue Wren Warbler *Malurus cyanocephalus* AU

Southern Emu Wren *Stipiturus malachurus* AU
Mallee Emu Wren *Stipiturus mallee* ● AU
Rufous-crowned Emu Wren *Stipiturus ruficeps* AU

Thick-billed Grass Wren *Amytornis textilis* ● AU
Dusky Grass Wren *Amytornis purnelli* AU
Black Grass Wren *Amytornis housei* AU
White-throated Grass Wren *Amytornis woodwardi* AU

Red-winged Grass Wren *Amytornis dorotheae* ● AU
Striped Grass Wren *Amytornis striatus* AU
Grey Grass Wren *Amytornis barbatus* ● AU
Eyrean Grass Wren *Amytornis goyderi* ● AU

140 THORNBILLS AND FLYEATERS Acanthizidae

Common Bristlebird *Dasyornis brachypterus* ● AU
Rufous Bristlebird *Dasyornis broadbenti*

Pilot Bird *Pycnoptilus floccosus* AU

Rock Warbler *Origma solitaria* AU

Fern Wren *Crateroscelis gutturalis* AU
Lowland Mouse Warbler *Crateroscelis murina* AU
Mid-mountain Mouse Warbler *Crateroscelis nigrorufa* AU
Mountain Mouse Warbler *Crateroscelis robusta* AU

Yellow-throated Sericornis *Sericornis citreogularis* AU
Spotted Scrub Wren *Sericornis maculatus* AU
Tasmanian Sericornis *Sericornis humilis* AU
White-browed Sericornis *Sericornis frontalis* AU
Little Sericornis *Sericornis beccarii* AU
Large Mountain Sericornis *Sericornis nouhuysi* AU
Large-billed Sericornis *Sericornis magnirostris* AU
Atherton Sericornis *Sericornis keri* AU
Pale-billed Sericornis *Sericornis spilodera* AU
Buff-faced Sericornis *Sericornis parspicillatus* AU
Arfak Buff-faced Sericornis *Sericornis rufescens* AU
Papuan Sericornis *Sericornis papuensis* AU
Grey-green Sericornis *Sericornis arfakianus* AU
Scrub Tit *Sericornis magnus* AU

Redthroat *Pyrrholaemus brunneus* AU

Little Field Wren *Chthonicola sagittata* AU

Striated Field Wren *Calamanthus fuliginosus* AU
Rufous Field Wren *Calamanthus campestris* AU

Chestnut-rumped Heath Wren *Hylacola pyrrhopygius* AU
Shy Heath Wren *Hylacola cautus* AU

De Vis' Tree Warbler *Acanthiza murina* AU
Western Thornbill *Acanthiza inornata* AU
Buff-tailed Thornbill *Acanthiza reguloides* AU
Samphire Thornbill *Acanthiza iredalei* AU
Mountain Thornbill *Acanthiza katharina* AU
Brown Thornbill *Acanthiza pusilla* AU
Broad-tailed Thornbill *Acanthiza apicalis* AU
Tasmanian Thornbill *Acanthiza ewingi* AU
Yellow-tailed Thornbill *Acanthiza chrysorrhoa* AU
Chestnut-tailed Thornbill *Acanthiza uropygialis* AU
Robust Thornbill *Acanthiza robustirostris* AU
Little Thornbill *Acanthiza nana* AU
Striated Thornbill *Acanthiza lineata* AU

Weebill *Smicrornis brevirostris* AU

Grey Flyeater *Gerygone cinerea* AU
Green-backed Flyeater *Gerygone chloronota* AU
Black-headed Flyeater *Gerygone palpebrosa* AU
White-throated Flyeater *Gerygone olivacea* AU
Rufous-sided Flyeater *Gerygone dorsalis* AU
Yellow-bellied Flyeater *Gerygone chrysogaster* AU
Rufous-tailed Flyeater *Gerygone ruficauda* AU
Large-billed Flyeater *Gerygone magnirostris* AU
Golden-bellied Flyeater *Gerygone sulphurea* OR AU
Plain Flyeater *Gerygone inornata* AU
Treefern Flyeater *Gerygone ruficollis* AU
White-tailed Flyeater *Gerygone fusca* AU
Dusky Flyeater *Gerygone tenebrosa* AU
Mangrove Flyeater *Gerygone laevigaster* AU
Fan-tailed Flyeater *Gerygone flavolateralis* AU
Brown Flyeater *Gerygone mouki* AU
Norfolk Island Flyeater *Gerygone modesta* AU
New Zealand Grey Flyeater *Gerygone igata* AU
Chatham Is Flyeater *Gerygone albofrontata* AU

Southern Whiteface *Aphelocephala leucopsis* AU
Chestnut-breasted Whiteface *Aphelocephala pectoralis*E AU
Banded Whiteface *Aphelocephala nigricincta* AU

Yellowhead *Mohoua ochrocephala* AU

New Zealand Creeper *Finschia novaeseelandiae* AU

White-faced Chat *Ephthianura albifrons* AU

Crimson Chat *Ephthianura tricolor* AU
Orange Chat *Ephthianura aurifrons* AU
Yellow Chat *Ephthianura crocea* AU

Desert Chat *Ashbyia lovensis* AU

141 MONARCHS AND FANTAILS Monarchidae

Little Yellow Flycatcher *Erythrocercus holochlorus* AF
Chestnut-capped Flycatcher *Erythrocercus mccallii* AF
Livingstone's Flycatcher *Erythrocercus livingstonei* AF

Blue Flycatcher *Elminia longicauda* AF
White-tailed Blue Flycatcher *Elminia albicauda* AF
Dusky Crested Flycatcher *Elminia nigromitrata* AF
White-bellied Crested Flycatcher *Elminia albiventris* AF
White-tailed Crested Flycatcher *Elminia albonotata* AF

Chestnut-winged Monarch *Philentoma pyrrhopterum* OR
Maroon-breasted Monarch *Philentoma velatum* OR
Black-naped Blue Monarch *Hypothymis azurea* OR AU
Short-crested Blue Monarch *Hypothymis helenae* ● OR
Celestial Blue Monarch *Hypothymis coelestis* ● OR

Rowley's Flycatcher *Eutrichomyias rowleyi* ● AU

Cape Crested Flycatcher *Terpsiphone cyanomelas* AF
Blue-headed Crested Flycatcher *Terpsiphone nitens* AF
Red-bellied Paradise Flycatcher *Terpsiphone rufiventer* AF
Bedford's Paradise Flycatcher *Terpsiphone bedfordi* AF
Rufous-vented Paradise Flycatcher *Terpsiphone rufocinerea* AF
African Paradise Flycatcher *Terpsiphone viridis* AF
Asiatic Paradise Flycatcher *Terpsiphone paradisi* WP EP OR AU
Black Paradise Flycatcher *Terpsiphone atrocaudata* EP OR
Blue Paradise Flycatcher *Terpsiphone cyanescens* OR
Rufous Paradise Flycatcher *Terpsiphone cinnamomea* OR
Sao Thome Paradise Flycatcher *Terpsiphone atrochalybea* AF
Madagascar Paradise Flycatcher *Terpsiphone mutata* AF
Seychelles Paradise Flycatcher *Terpsiphone corvina* ● AF
Mascarene Paradise Flycatcher *Terpsiphone bourbonnensis* AF

Elepaio *Chasiempis sandwichensis* PA

Rarotonga Flycatcher *Pomarea dimidiata* ● PA
Society Is Flycatcher *Pomarea nigra* ● PA
Marquesas Flycatcher *Pomarea mendozae* ● PA
Allied Flycatcher *Pomarea iphis* ● PA
Large Flycatcher *Pomarea whitneyi* ● PA

Small Slaty Flycatcher *Mayrornis schistaceus* PA
Ogea Flycatcher *Mayrornis versicolor* PA
Slaty Flycatcher *Mayrornis lessoni* PA

Buff-bellied Flycatcher *Neolalage banksiana* PA

Southern Shrikebill *Clytorhynchus pachycephaloides* PA
Fiji Shrikebill *Clytorhynchus vitiensis* PA
Black-throated Shrikebill *Clytorhynchus nigrogularis* PA
Rennell Shrikebill *Clytorhynchus hamlini* ● PA

Truk Monarch *Metabolus rugiensis* ● PA

Black Monarch *Monarcha axillaris* AU
Rufous Monarch *Monarcha rubiensis* AU
Island Grey-headed Monarch *Monarcha cinerascens* AU AU
Black-faced Monarch *Monarcha melanopsis* AU
Black-winged Monarch *Monarcha frater* AU
Bougainville Monarch *Monarcha erythrosticta* PA
Chestnut-bellied Monarch *Monarcha castaneiventris* PA

Richard's Monarch *Monarcha richardsii* PA
White-eared Monarch *Monarcha leucotis* AU
Spot-winged Monarch *Monarcha guttula* AU
Tanimbar Monarch *Monarcha mundus* AU
Mees' Monarch *Monarcha sacerdotum* AU
Black-chinned Monarch *Monarcha boanensis* ● AU
Spectacled Monarch *Monarcha trivirgatus* AU
Kei Monarch *Monarcha leucurus* AU
White-tipped Monarch *Monarcha everetti* ● AU
Black-tipped Monarch *Monarcha loricatus* AU
Kofiau Monarch *Monarcha julienae* AU
Biak Monarch *Monarcha brehmii* ● AU
Black & White Monarch *Monarcha manadensis* AU
Unhappy Monarch *Monarcha infelix* AU
St Matthias Monarch *Monarcha menckei* AU
Bismarck Monarch *Monarcha verticalis* AU
Pied Monarch *Monarcha barbatus* PA
Kulambangra Monarch *Monarcha browni* PA
San Cristobal Monarch *Monarcha viduus* PA
Yap Monarch *Monarcha godeffroyi* PA
Tinian Monarch *Monarcha takatsukasae* PA
Black & Yellow Monarch *Monarcha chrysomela* AU

Pied Flycatcher *Arses kaupi* AU
Frilled Flycatcher *Arses telescophthalmus* AU

Micronesian Myiagra Flycatcher *Myiagra oceanica* PA
Helmet Flycatcher *Myiagra galeata* AU
Black Myiagra Flycatcher *Myiagra atra* AU
Leaden Flycatcher *Myiagra rubecula* AU
Steel-blue Flycatcher *Myiagra ferrocyanea* PA
San Cristobal Myiagra Flycatcher *Myiagra cervinicauda* PA
New Caledonian Myiagra Flycatcher *Myiagra caledonica* PA
Red-bellied Flycatcher *Myiagra vanikorensis* PA
White-vented Flycatcher *Myiagra albiventris* PA
Blue-headed Flycatcher *Myiagra azureocapilla* PA
Broad-billed Flycatcher *Myiagra ruficollis* AU
Satin Flycatcher *Myiagra cyanoleuca* AU
Shining Flycatcher *Myiagra alecto* AU
Dull Flycatcher *Myiagra hebetior* AU
Restless Flycatcher *Myiagra inquieta* AU

Silktail *Lamprolia victoriae* PA

Yellow-breasted Flatbill Flycatcher *Machaerirhynchus flaviventer* AU
Black-breasted Flatbill Flycatcher *Machaerirhynchus nigripectus* AU

Lowland Peltops Flycatcher *Peltops blainvillii* AU
Mountain Peltops Flycatcher *Peltops montanus* AU

Yellow-bellied Fantail *Rhipidura hypoxantha* EP OR
Blue Fantail *Rhipidura superciliaris* AU
Blue-headed Fantail *Rhipidura cyaniceps* OR
Red-tailed Fantail *Rhipidura phoenicura* OR
Black & Cinnamon Fantail *Rhipidura nigrocinnamomea* OR
White-throated Fantail *Rhipidura albicollis* EP OR
White-bellied Fantail *Rhipidura euryura* OR
White-browed Fantail *Rhipidura aureola* OR
Pied Fantail *Rhipidura javanica* OR
Pearlated Fantail *Rhipidura perlata* OR
Black & White Fantail (Willy Wagtail) *Rhipidura leucophrys* AU PA
Northern Fantail *Rhipidura rufiventris* AU
Cockerell's Fantail *Rhipidura cockerelli* PA
Friendly Fantail *Rhipidura albolimbata* AU
Sooty Thicket Fantail *Rhipidura threnothorax* AU
Chestnut-bellied Fantail *Rhipidura hyperythra* AU
Black Thicket Fantail *Rhipidura maculipectus* AU
White-breasted Fantail *Rhipidura leucothorax* AU
Black Fantail *Rhipidura atra* AU
Collared Grey Fantail *Rhipidura fuliginosa* AU
Mountain Fantail *Rhipidura drownei* PA
Dusky Fantail *Rhipidura tenebrosa* PA
Rennell Fantail *Rhipidura rennelliana* PA
Spotted Fantail *Rhipidura spilodera* PA
Samoan Fantail *Rhipidura nebulosa* PA
Dimorphic Rufous Fantail *Rhipidura brachyrhyncha* AU
Kandavu Fantail *Rhipidura personata* PA
Seram Rufous Fantail *Rhipidura dedemi* AU
Moluccan Fantail *Rhipidura superflua* AU
Sulawesi Rufous Fantail *Rhipidura teysmanni* AU

Cinnamon-tailed Fantail *Rhipidura fuscorufa* AU
Brown-capped Fantail *Rhipidura diluta* AU
Long-tailed Fantail *Rhipidura opistherytha* AU
Palau Fantail *Rhipidura lepida* AU
Grey-breasted Rufous Fantail *Rhipidura rufidorsa* AU
Island Rufous Fantail *Rhipidura dahli* AU
St Matthias Rufous Fantail *Rhipidura matthiae* AU
Malaita Rufous Fantail *Rhipidura malaitae* ● PA
Rufous Fantail *Rhipidura rufifrons* AU PA

142 AUSTRALASIAN ROBINS Eopsaltridae

River Flycatcher *Monachella muelleriana* AU

Jacky Winter (Australian Brown Flycatcher) *Microeca leucophaea* AU
Lemon-breasted Flycatcher *Microeca flavigaster* AU
Tanimbar Microeca Flycatcher *Microeca hemixantha* AU
Yellow-footed Flycatcher *Microeca griseoceps* AU
Olive Microeca Flycatcher *Microeca flavovirescens* AU
Papuan Microeca Flycatcher *Microeca papuana* AU

Red-backed Warbler *Eugerygone rubra* AU

Forest Robin *Petroica bivittata* AU
Rock Robin *Petroica archboldi* AU
Scarlet Robin *Petroica multicolor* AU PA
Red-capped Robin *Petroica goodenovii* AU
Flame Robin *Petroica phoenicea* AU
Pink Robin *Petroica rodinogaster* AU
Rose Robin *Petroica rosea* AU
Hooded Robin *Petroica cucullata* AU
New Zealand Tit *Petroica macrocephala* AU
Dusky Robin *Petroica vittata* AU
New Zealand Robin *Petroica australis* AU
Chatham I Robin *Petroica traversi* ● AU

White-faced Robin *Tregellasia leucops* AU
Pale Yellow Robin *Tregellasia capito* AU

Yellow Robin *Eopsaltria australis* AU
White-breasted Robin *Eopsaltria georgiana* AU
Yellow-bellied Robin *Eopsaltria flaviventris* PA

Mangrove Robin *Peneoenanthe pulverulenta* AU

White-breasted Robin *Poecilodryas brachyura* AU
Black & White Robin *Poecilodryas hypoleuca* AU
Olive-yellow Robin *Poecilodryas placens* AU
Black-throated Robin *Poecilodryas albonotata* AU
White-browed Robin *Poecilodryas superciliosa* AU

White-winged Thicket Flycather *Peneothello sigillatus* AU
Grey Thicket Flycatcher *Peneothello cryptoleucus* AU
Slaty Thicket Flycatcher *Peneothello cyanus* AU
White-rumped Thicket Flycatcher *Peneothello bimaculatus* AU

Ground Thicket Robin *Heteromyias albispecularis* AU
Grey-headed Thicket Robin *Heteromyias cinereifrons* AU

Green Thicket Flycatcher *Pachycephalopsis hattamensis* AU
White-throated Thicket Flycatcher *Pachycephalopsis poliosoma* AU

143 WHISTLERS Pachycephalidae

Wattled Shrike Tit *Eulacestoma nigropectus* AU

Crested Shrike Tit *Falcunculus frontatus* AU

Crested Bellbird *Oreoica gutturalis* AU

Golden-faced Pachycare *Pachycare flavogrisea* AU

Mottled Whistler *Rhagologus leucostigma* AU

Buff-throated Thickhead *Hylocitrea bonensis* AU

Maroon-backed Whistler *Coracornis raveni* AU

Rufous-naped Whistler *Aleadryas rufinucha* AU

Sooty Whistler *Pachycephala tenebrosa* AU
Olive Whistler *Pachycephala olivacea* AU
Red-lored Whistler *Pachycephala rufogularis* ● AU
Gilbert Whistler *Pachycephala inornata* AU
Mangrove Whistler *Pachycephala grisola* OR
Palawan Whistler *Pachycephala plateni* OR

Island Whistler *Pachycephala phaionota* AU
Rufous-breasted Whistler *Pachycephala hyperythra* AU
Brown-backed Whistler *Pachycephala modesta* AU
Yellow-bellied Whistler *Pachycephala philippinensis* OR
Sulphur-vented Whistler *Pachycephala sulfuriventer* AU
Bornean Mountain Whistler *Pachycephala hypoxantha* OR
Vogelkop Whistler *Pachycephala meyeri* ● AU
Sclater's Whistler *Pachycephala soror* AU
Grey Whistler *Pachycephala simplex* AU
Fawn-breasted Whistler *Pachycephala orpheus* AU
Golden Whistler *Pachycephala pectoralis* OR AU PA
Tonga Whistler *Pachycephala melanops* PA
Mangrove Golden Whistler *Pachycephala melanura* AU
Yellow-fronted Whistler *Pachycephala flavifrons* PA
New Caledonian Whistler *Pachycephala caledonica* PA
Mountain Whistler *Pachycephala implicata* PA
Bare-throated Whistler *Pachycephala nudigula* AU
Lorentz's Whistler *Pachycephala lorentzi* AU
Schlegel's Whistler *Pachycephala schlegelii* AU
Yellow-backed Whistler *Pachycephala aurea* AU
Rufous Whistler *Pachycephala rufiventris* AU PA
Aru Whistler *Pachycephala monacha* AU
White-bellied Whistler *Pachycephala leucogastra* AU
Wallacean Whistler *Pachycephala arctitorquis* AU
Drab Whistler *Pachycephala griseonota* AU
White-breasted Whistler *Pachycephala lanioides* AU

Rufous Shrike-thrush *Colluricincla megarhyncha* AU
Stripe-breasted Shrike-thrush *Colluricincla boweri* AU
Grey Shrike-thrush *Colluricincla harmonica* AU
Sandstone Shrike-thrush *Colluricincla woodwardi* AU

Variable Pitohui *Pitohui kirhocephalus* AU
Black-headed Pitohui *Pitohui dichrous* AU
Mottle-breasted Pitohui *Pitohui incertus* AU
Rusty Pitohui *Pitohui ferrugineus* AU
Crested Pitohui *Pitohui cristatus* AU
Black Pitohui *Pitohui nigrescens* AU
Morning Bird *Pitohui tenebrosus* PA

New Zealand Thrush *Turnagra capensis* AU

144 LONG-TAILED TITS Aegithalidae

Long-tailed Tit *Aegithalos caudatus* WP EP
White-cheeked Tit *Aegithalos leucogenys* EP OR
Red-headed Tit *Aegithalos concinnus* EP OR
Blyth's Long-tailed Tit *Aegithalos iouschistos* EP OR
Sooty Long-tailed Tit *Aegithalos fuliginosus* EP

Pygmy Tit *Psaltria exilis* OR

Common Bushtit *Psaltriparus minimus* NA NT
Black-eared Bushtit *Psaltriparus melanotis* NA NT

145 PENDULINE TITS Remizidae

Penduline Tit *Remiz pendulinus* WP EP OR

Sennar Kapok Tit *Anthoscopus punctifrons* AF
Yellow Penduline Tit *Anthoscopus parvulus* AF
Mouse-coloured Tit *Anthoscopus musculus* AF
Yellow-fronted Tit *Anthoscopus flavifrons* AF
African Penduline Tit *Anthoscopus caroli* AF
Rungwe Penduline Tit *Anthoscopus sylviella* AF
Southern Kapok Tit *Anthoscopus minutus* AF

Verdin *Auriparus flaviceps* NT

Fire-capped Tit Warbler *Cephalopyrus flammiceps* EP OR

146 TITS AND CHICKADEES Paridae

Marsh Tit *Parus palustris* WP EP
Sombre Tit *Parus lugubris* WP
Iranian Sombre Tit *Parus hyrcanus* EP EP
Willow Tit *Parus montanus* WP
Black-capped Chickedee *Parus atricapillus* NA
Carolina Chickadee *Parus carolinensis* NA
Mexican Chickadee *Parus sclateri* NA NT
Mountain Chickadee *Parus gambeli* NA NT
White-browed Tit *Parus superciliosus* EP
Pere David's Tit *Parus davidi* EP

Siberian Tit *Parus cinctus* WP NA
Boreal Chickadee *Parus hudsonicus* NA
Chestnut-backed Chickadee *Parus rufescens* NA
Bridled Titmouse *Parus wollweberi* NA NT
Black-crested Tit *Parus rubidiventris* EP OR
Rufous-naped Tit *Parus rufonuchalis* EP OR
Vigors' Crested Tit *Parus melanolophus* EP
Coal Tit *Parus ater* WP EP
Yellow-bellied Tit *Parus venustulus* EP
Elegant Tit *Parus elegans* OR
Palawan Tit *Parus amabilis* OR
Crested Tit *Parus cristatus* WP EP
Brown Crested Tit *Parus dichrous* EP
Southern Grey Tit *Parus afer* AF
Acacia Grey Tit *Parus cinerascens* AF
Miombo Grey Tit *Parus griseiventris* AF
Southern Black Tit *Parus niger* AF
White-winged Black Tit *Parus leucomelas* AF
Carp's Tit *Parus carpi* AF
White-breasted Tit *Parus albiventris* AF
White-backed Black Tit *Parus leuconotus* AF
Dusky Tit *Parus funereus* AF
Stripe-breasted Tit *Parus fasciiventer* AF
Red-throated Tit *Parus fringillinus* AF
Rufous-bellied Tit *Parus rufiventris* AF
Great Tit *Parus major* WP EP OR
Turkestan Tit *Parus bokharensis* EP
Green-backed Tit *Parus monticolus* EP OR
White-naped Tit *Parus nuchalis* ● OR
Black-spotted Yellow Tit *Parus xanthogenys* OR
Chinese Yellow Tit *Parus spilonotus* EP OR
Formosan Yellow Tit *Parus holsti* ● EP
Blue Tit *Parus caeruleus* WP
Azure Tit *Parus cyanus* EP
Varied Tit *Parus varius* EP
White-fronted Tit *Parus semilarvatus* OR
Plain Titmouse *Parus inornatus* NA NT
Tufted Titmouse *Parus bicolor* NA NT

Sultan Tit *Melanochlora sultanea* EP OR

Yellow-browed Tit *Sylviparus modestus* EP OR

147 NUTHATCHES Siittidae

Sittinae

Nuthatch *Sitta europaea* WP EP OR
Chestnut-bellied Nuthatch *Sitta castanea* EP OR
White-tailed Nuthatch *Sitta himalayensis* EP OR
White-browed Nuthatch *Sitta victoriae* ● OR
Pygmy Nuthatch *Sitta pygmaea* NA NT
Brown-headed Nuthatch *Sitta pusilla* NA NT
Corsican Nuthatch *Sitta whiteheadi* WP
Yunnan Nuthatch *Sitta yunnanensis* ● EP
Red-breasted Nuthatch *Sitta canadensis* NA
Chinese Nuthatch *Sitta villosa* EP
White-cheeked Nuthatch *Sitta leucopsis* EP
White-breasted Nuthatch *Sitta carolinensis* NA NT
Kruper's Nuthatch *Sitta krueperi* WP
Kabylie Nuthatch *Sitta ledanti* ●WP
Rock Nuthatch *Sitta neumayer* WP
Eastern Rock Nuthatch *Sitta tephronota* EP
Velvet-fronted Nuthatch *Sitta frontalis* OR
Lilac Nuthatch *Sitta solangiae* ● OR
Azure Nuthatch *Sitta azurea* OR
Giant Nuthatch *Sitta magna* ● OR
Beautiful Nuthatch *Sitta formosa* ● EP OR

Tichodromadinae

Wallcreeper *Tichodroma muraria* WP EP

Daphoenosittinae

Varied Sitella *Neositta chrysoptera* AU
Papuan Sitella *Neositta papuensis* AU

Pink-faced Nuthatch *Daphoenositta miranda* AU

148 TREECREEPERS Certhiidae

Certhiinae

Treecreeper *Certhia familiaris* WP EP
American Treecreeper *Certhia americana* NA NT
Short-toed Treecreeper *Certhia brachydactyla* WP
Himalayan Treecreeper *Certhia himalayana* EP OR
Stoliczka's Treecreeper *Certhia nipalensis* EP OR

Brown-throated Treecreeper *Certhia discolor* EP OR

Salpornithinae

Spotted Grey Creeper *Salpornis spilonotus* AF

149 PHILIPPINE CREEPERS Rhabdornithidae

Stripe-headed Creeper *Rhabdornis mystacalis* OR
Plain-headed Creeper *Rhabdornis inornatus* OR

150 AUSTRALIAN CREEPERS Climacteridae

Red-browed Treecreeper *Climacteris erythrops* AU
White-browed Treecreeper *Climacteris affinis* AU
Brown Treecreeper *Climacteris picumnus* AU
Rufous Treecreeper *Climacteris rufa* AU
Black-tailed Treecreeper *Climacteris melanura* AU
White-throated Treecreeper *Climacteris leucophaea* AU
Papuan Treecreeper *Climacteris placens* AU

151 FLOWERPECKERS Dicaeidae

Obscure Berrypecker *Melanocharis arfakiana* AU
Black Berrypecker *Melanocharis nigra* AU
Mid-mountain Berrypecker *Melanocharis longicauda* AU
Fan-tailed Berrypecker *Melanocharis versteri* AU
Streaked Berrypecker *Melanocharis striativentris* AU

Spotted Berrypecker *Rhamphocharis crassirostris* AU

Olive-backed Flowerpecker *Prionochilus olivaceus* OR
Yellow-throated Flowerpecker *Prionochilus maculatus* OR
Crimson-breasted Flowerpecker *Prionochilus percussus* OR
Palawan Yellow-rumped Flowerpecker *Prionochilus plateni* OR
Borneo Yellow-rumped Flowerpecker *Prionochilus xanthopygius* OR
Scarlet-breasted Flowerpecker *Prionochilus thoracicus* OR

Sunda Flowerpecker *Dicaeum annae* AU
Thick-billed Flowerpecker *Dicaeum agile* OR AU
Brown-backed Flowerpecker *Dicaeum everetti* ● OR
Grey-breasted Flowerpecker *Dicaeum proprium* OR
Yellow-vented Flowerpecker *Dicaeum chrysorrheum* EP OR
Yellow-bellied Flowerpecker *Dicaeum melanoxanthum* EP
Legge's Flowerpecker *Dicaeum vincens* OR
Yellow-sided Flowerpecker *Dicaeum aureolimbatum* AU
Olive-capped Flowerpecker *Dicaeum nigrilore* OR
Yellow-crowned Flowerpecker *Dicaeum anthonyi* OR
Bicoloured Flowerpecker *Dicaeum bicolor* OR
Philippine Flowerpecker *Dicaeum australe* OR
Mindoro Flowerpecker *Dicaeum retrocinctum* OR
Orange-bellied Flowerpecker *Dicaeum trigonostigma* OR
White-bellied Flowerpecker *Dicaeum hypoleucum* OR
Tickell's Flowerpecker *Dicaeum erythrorhynchos* OR
Plain Flowerpecker *Dicaeum concolor* EP OR
Palawan Flowerpecker *Dicaeum pygmaeum* OR
Red-headed Flowerpecker *Dicaeum nehrkorni* AU
Ashy-fronted Flowerpecker *Dicaeum vulneratum* AU
White-throated Flowerpecker *Dicaeum erythrothorax* AU
Olive-crowned Flowerpecker *Dicaeum pectorale* AU
Red-capped Flowerpecker *Dicaeum geelvinkianum* AU
Louisiade Flowerpecker *Dicaeum nitidum* AU
New Ireland Flowerpecker *Dicaeum eximium* AU
Solomon Is Flowerpecker *Dicaeum aeneum* PA
San Cristobal Flowerpecker *Dicaeum tristrami* PA
Black-banded Flowerpecker *Dicaeum igniferum* AU
Blue-cheeked Flowerpecker *Dicaeum maugei* AU
Mistletoe Flowerpecker *Dicaeum hirundinaceum* AU
Black-sided Flowerpecker *Dicaeum celebicum* AU
Bornean Fire-breasted Flowerpecker *Dicaeum monticolum* OR
Green-backed Flowerpecker *Dicaeum ignipectus* EP OR
Blood-breasted Flowerpecker *Dicaeum sanguinolentum* OR AU

Scarlet-backed Flowerpecker *Dicaeum cruentatum* EP OR
Scarlet-headed Flowerpecker *Dicaeum trochileum* OR AU

Tit Berrypecker *Oreocharis arfaki* AU

Crested Berrypecker *Paramythia montium* AU

Forty-spotted Pardalote *Pardalotus quadragintus* ● AU
Spotted Pardalote *Pardalotus punctatus* AU
Yellow-tailed Pardalote *Pardalotus xanthopygus* AU
Red-browed Pardalote *Pardalotus rubricatus* AU
Yellow-tipped Pardalote *Pardalotus striatus* AU
Red-tipped Pardalote *Pardalotus ornatus* AU
Striated Pardalote *Pardalotus substriatus* AU
Black-headed Pardalote *Pardalotus melanocephalus* AU

152 SUNBIRDS Nectariniidae

Brown Sunbird *Anthreptes gabonicus* AF
Scarlet-tufted Sunbird *Anthreptes fraseri* AF
Plain-backed Sunbird *Anthreptes reichenowi* AF
Anchieta's Sunbird *Anthreptes anchietae* AF
Plain-coloured Sunbird *Anthreptes simplex* OR
Plain-throated Sunbird *Anthreptes malacensis* OR AU
Shelley's Sunbird *Anthreptes rhodolaema* OR
Ruby-cheeked Sunbird *Anthreptes singalensis* OR
Violet-backed Sunbird *Anthreptes longuemarei* AF
Kenya Violet-backed Sunbird *Anthreptes orientalis* AF
Uluguru Violet-backed Sunbird *Anthreptes neglectus* AF
Violet-tailed Sunbird *Anthreptes aurantium* AF
Amani Sunbird *Anthreptes pallidigaster* ● AF
Green Sunbird *Anthreptes rectirostris* AF
Banded Green Sunbird *Anthreptes rubritorques* ● AF
Collared Sunbird *Anthreptes collaris* AF
Pygmy Sunbird *Anthreptes platurus* AF

Blue-naped Sunbird *Hypogramma hypogrammicum* OR

Little Green Sunbird *Nectarinia seimundi* AF
Bates's Olive Sunbird *Nectarinia batesi* AF
Olive Sunbird *Nectarinia olivacea* AF
Fernando Po Sunbird *Nectarinia ursulae* AF
Mouse-coloured Sunbird *Nectarinia veroxii* AF
Socotra Sunbird *Nectarinia balfouri* AF
Reichenbach's Sunbird *Nectarinia reichenbachii* AF
Principe Sunbird *Nectarinia hartlaubii* AF
Newton's Yellow-breasted Sunbird *Nectarinia newtonii* AF
Sao Thome Giant Sunbird *Nectarinia thomensis* ● AF
Cameroun Blue-headed Sunbird *Nectarinia oritis* AF
Blue-headed Sunbird *Nectarinia alinae* AF
Bannerman's Sunbird *Nectarinia bannermani* AF
Green-headed Sunbird *Nectarinia verticalis* AF
Blue-throated Sunbird *Nectarinia cyanolaema* AF
Carmelite Sunbird *Nectarinia fuliginosa* AF
Green-throated Sunbird *Nectarinia rubescens* AF
Amethyst Sunbird *Nectarinia amethystina* AF
Scarlet-chested Sunbird *Nectarinia senegalensis* AF
Hunter's Sunbird *Nectarinia hunteri* AF
Buff-throated Sunbird *Nectarinia adelberti* AF
Purple-rumped Sunbird *Nectarinia zeylonica* OR
Small Sunbird *Nectarinia minima* OR
Van Hasselt's Sunbird *Nectarinia sperata* OR
Black Sunbird *Nectarinia sericea* AF
Macklot's Sunbird *Nectarinia calcostetha* OR
Seychelles Sunbird *Nectarinia dussumieri* AF
Loten's Sunbird *Nectarinia lotenia* OR
Olive-backed Sunbird *Nectarinia jugularis* OR AU PA
Sumba I Sunbird *Nectarinia buettikoferi* ● AU
Timor Sunbird *Nectarinia solaris* AU
Purple Sunbird *Nectarinia asiatica* WP EP OR
Souimanga Sunbird *Nectarinia souimanga* AF
Humblot's Sunbird *Nectarinia humbloti* AF
Anjouan Sunbird *Nectarinia comorensis* AF
Mayotte Sunbird *Nectarinia coquerellii* AF
Variable Sunbird *Nectarinia venusta* AF
Southern White-bellied Sunbird *Nectarinia talatala* AF

Oustalet's White-bellied Sunbird *Nectarinia oustaleti* AF
Dusky Sunbird *Nectarinia fusca* AF
Lesser Double-collared Sunbird *Nectarinia chalybea* AF
Miombo Double-collared Sunbird *Nectarinia manoensis* AF
Greater Double-collared Sunbird *Nectarinia afra* AF
Double-breasted Sunbird *Nectarinia prigoginei* ● AF
Northern Double-collared Sunbird *Nectarinia preussi* AF
Eastern Double-collared Sunbird *Nectarinia mediocris* AF
Neergaard's Sunbird *Nectarinia neergaardi* AF
Olive-bellied Sunbird *Nectarinia chloropygia* AF
Tiny Sunbird *Nectarinia minulla* AF
Regal Sunbird *Nectarinia regia* AF
Loveridge's Sunbird *Nectarinia loveridgei* AF
Moreau's Sunbird *Nectarinia moreaui* AF
Rockefeller's Sunbird *Nectarinia rockefelleri* ● AF
Orange-breasted Sunbird *Nectarinia violacea* AF
Shining Sunbird *Nectarinia habessinica* WP AF
Southern Orange-tufted Sunbird *Nectarinia bouvieri* AF
Northern Orange-tufted Sunbird *Nectarinia osea* WP AF
Coppery Sunbird *Nectarinia cuprea* AF
Tacazze Sunbird *Nectarinia tacazze* AF
Bocage's Sunbird *Nectarinia bocagii* AF
Purple-breasted Sunbird *Nectarinia purpureiventris* AF
Shelley's Sunbird *Nectarinia shelleyi* AF
Mariqua Sunbird *Nectarinia mariquensis* AF
Violet-breasted Sunbird *Nectarinia pembae* AF
Purple-banded Sunbird *Nectarinia bifasciata* AF
Splendid Sunbird *Nectarinia coccinigastra* AF
Red-chested Sunbird *Nectarinia erythrocerca* AF
Congo Black-bellied Sunbird *Nectarinia congensis* AF
Beautiful Sunbird *Nectarinia pulchella* AF
Smaller Black-bellied Sunbird *Nectarinia nectarinioides* AF
Yellow-tufted Malachite Sunbird *Nectarinia famosa* AF
Red-tufted Malachite Sunbird *Nectarinia johnstoni* AF
Noted Sunbird *Nectarinia notata* AF
Rufous-winged Sunbird *Nectarinia rufipennis* ● AF
Madame Verreaux's Sunbird *Nectarinia johannae* AF
Superb Sunbird *Nectarinia superba* AF
Bronze Sunbird *Nectarinia kilimensis* AF
Golden-winged Sunbird *Nectarinia reichenowi* AF

Hachisuka's Sunbird *Aethopyga primigenius* OR
Apo Sunbird *Aethopyga boltoni* OR
Flaming Sunbird *Aethopyga flagrans* OR
Mountain Sunbird *Aethopyga pulcherrima* OR
Sanghir Yellow-backed Sunbird *Aethopyga duyvenbodei* ● AU
Palawan Sunbird *Aethopyga shelleyi* OR
Mrs Gould's Sunbird *Aethopyga gouldiae* EP OR
Green-tailed Sunbird *Aethopyga nipalensis* EP OR
Kuhl's Sunbird *Aethopyga eximia* OR
Fork-tailed Sunbird *Aethopyga christinae* OR
Black-throated Sunbird *Aethopyga saturata* EP OR
Yellow-backed Sunbird *Aethopyga siparaja* OR AU
Scarlet Sunbird *Aethopyga mystacalis* OR
Fire-tailed Sunbird *Aethopyga ignicauda* EP OR
Little Spiderhunter *Arachnothera longirostra* OR
Thick-billed Spiderhunter *Arachnothera crassirostris* OR
Long-billed Spiderhunter *Arachnothera robusta* OR
Greater Yellow-eared Spiderhunter *Arachnothera flavigaster* OR
Lesser Yellow-eared Spiderhunter *Arachnothera chrysogenys* OR
Naked-faced Spiderhunter *Arachnothera clarae* OR
Grey-breasted Spiderhunter *Arachnothera affinis* OR
Streaked Spiderhunter *Arachnothera magna* OR
Everett's Spiderhunter *Arachnothera everetti* OR
Whitehead's Spiderhunter *Arachnothera juliae* OR

153 WHITE-EYES Zosteropidae

Chestnut-flanked White-eye *Zosterops erythropleura* EP
Japanese White-eye *Zosterops japonica* EP OR

Philippine White-eye *Zosterops meyeni* OR
Oriental White-eye *Zosterops palpebrosa* OR AU
Large Sri Lanka White-eye *Zosterops ceylonensis* OR
Bridled White-eye *Zosterops conspicillata* PA
Enggano White-eye *Zosterops salvadorii* AU
Black-capped White-eye *Zosterops atricapilla* OR
Everett's White-eye *Zosterops everetti* OR
Philippine Yellow White-eye *Zosterops nigrorum* OR
Mountain White-eye *Zosterops montanus* OR AU
Yellow-spectacled White-eye *Zosterops wallacei* AU
Javan White-eye *Zosterops flava* ● OR
Lemon-bellied White-eye *Zosterops chloris* AU
Ashy-bellied White-eye *Zosterops citrinella* AU
Pale-bellied White-eye *Zosterops consobrinorum* AU
Pearl-bellied White-eye *Zosterops grayi* AU
Golden-bellied White-eye *Zosterops uropygialis* ● AU
Lemon-throated White-eye *Zosterops anomala* ● AU
Creamy-throated White-eye *Zosterops atriceps* AU
Moluccan Black-fronted White-eye *Zosterops atrifrons* AU
New Guinea Black-fronted White-eye *Zosterops minor* AU
White-throated White-eye *Zosterops meeki* ● AU
Biak White-eye *Zosterops mysorensis* AU
Yellow-bellied Mountain White-eye *Zosterops fuscicapilla* AU
Buru I White-eye *Zosterops buruensis* AU
Ambon White-eye *Zosterops kuehni* ● AU
New Guinea Mountain White-eye *Zosterops novaeguineae* AU
Yellow-throated White-eye *Zosterops metcalfii* PA
Christmas Island White-eye *Zosterops natalis* OR
Yellow Silver-eye *Zosterops lutea* AU
Louisiades White-eye *Zosterops griseotincta* AU
Rennell I White-eye *Zosterops rennelliana* AU
Solomon Is White-eye *Zosterops rendovae* PA
Kulambangra Mountain White-eye *Zosterops murphyi* PA
Grey-throated White-eye *Zosterops ugiensis* PA
Malaita White-eye *Zosterops stresemanni* PA
Santa Cruz White-eye *Zosterops sanctaecrucis* ● PA
Savaii White-eye *Zosterops samoensis* PA
Layard's White-eye *Zosterops explorator* PA
Yellow-fronted White-eye *Zosterops flavifrons* PA
Small Lifu White-eye *Zosterops minuta* PA
New Caledonia White-eye *Zosterops xanthochroa* PA
Western Silver-eye *Zosterops gouldi* AU
Grey-backed White-eye *Zosterops lateralis* AU PA
Slender-billed White-eye *Zosterops tenuirostris* AU
White-chested White-eye *Zosterops albogularis* ● AU
Large Lifu White-eye *Zosterops inornata* PA
Grey-brown White-eye *Zosterops cinerea* PA
White-breasted White-eye *Zosterops abyssinica* AF
Cape White-eye *Zosterops pallida* AF
African Yellow White-eye *Zosterops senegalensis* AF
Bourbon White-eye *Zosterops borbonica* AF
Principe White-eye *Zosterops ficedulina* ● AF
Annobon White-eye *Zosterops griseovirescens* AF
Hova Grey-backed White-eye *Zosterops hovarum* AF
Madagascar White-eye *Zosterops maderaspatana* AF
Chestnut-sided White-eye *Zosterops mayottensis* AF
Seychelles Brown White-eye *Zosterops modesta* ● AF
Grand Comoro White-eye *Zosterops mouroniensis* ● AF
Olive White-eye *Zosterops olivacea* AF
Mauritius Olive White-eye *Zosterops chloronothos* ● AF
Pemba White-eye *Zosterops vaughani* AF

Woodford's White-eye *Woodfordia superciliosa* ● PA
Sanford's White-eye *Woodfordia lacertosa* PA

Palau White-eye *Rukia palauensis* PA
Yap White-eye *Rukia oleaginea* PA
Truk White-eye *Rukia ruki* ● PA
Ponape White-eye *Rukia longirostra* ● PA

Bicoloured White-eye *Tephrozosterops stalkeri* AU

Rufous-throated White-eye *Madanga ruficollis* ● AU

Grey-hooded White-eye *Lophozosterops pinaiae* AU
Goodfellow's White-eye *Lophozosterops goodfellowi* OR

Streaky-headed White-eye *Lophozosterops squamiceps* AU
Javan Grey-throated White-eye *Lophozosterops javanicus* OR
White-browed White-eye *Lophozosterops superciliaris* AU
Crested White-eye *Lophozosterops dohertyi* AU

Pygmy White-eye *Oculocincta squamifrons* OR

Timor White-eye *Heleia muelleri* ● AU
Stripe-headed White-eye *Heleia crassirostris* AU

Olive Black-eye *Chlorocharis emiliae* OR
Cinnamon White-eye *Hypocryptadius cinnamomeus*

Fernando Po Speirops *Speirops brunnea* AF
Prince's I Speirops *Speirops leucophaea* ● AF
Black-capped Speirops *Speirops lugubris* AF

154 HONEYEATERS Meliphagidae

Mountain Straight-billed Honeyeater *Timeliopsis fulvigula* AU
Lowland Straight-billed Honeyeater *Timeliopsis griseigula* AU

Long-billed Honeyeater *Melilestes megarhynchus* AU
Bougainville Honeyeater *Melilestes bougainvillei* PA

Yellow-bellied Longbill *Toxorhamphus novaeguineae* AU
Slaty-chinned Longbill *Toxorhamphus poliopterus* AU

Grey-bellied Honeyeater *Oedistoma iliolophum* AU
Pygmy Honeyeater *Oedistoma pygmaeum* AU

White-eyed Honeyeater *Glycichaera fallax* AU

Lombok Honeyeater *Lichmera lombokia* AU
Plain Olive Honeyeater *Lichmera argentauris* AU
Brown Honeyeater *Lichmera indistincta* OR AU
Silver-eared Honeyeater *Lichmera incana* PA
White-eared Honeyeater *Lichmera alboauricularis* AU
White-tufted Honeyeater *Lichmera squamata* AU
Buru Honeyeater *Lichmera deningeri* AU
Spectacled Honeyeater *Lichmera monticola* AU
Timor Honeyeater *Lichmera flavicans* AU
Black-chested Honeyeater *Lichmera notabilis* ● AU
White-streaked Honeyeater *Lichmera cockerelli* AU

Ambon Honeyeater *Myzomela blasii* AU
White-chinned Honeyeater *Myzomela albigula* AU
Bismarck Honeyeater *Myzomela cineracea* AU
Red-spot Honeyeater *Myzomela eques* AU
Dusky Honeyeater *Myzomela obscura* AU
Red Honeyeater *Myzomela cruentata* AU
Black Honeyeater *Myzomela nigrita* AU
New Ireland Honeyeater *Myzomela pulchella* AU
Crimson-hooded Honeyeater *Myzomela kuehni* ● AU
Mangrove Red-headed Honeyeater *Myzomela erythrocephala* AU
Mountain Red-headed Honeyeater *Myzomela adolphinae* AU
Scarlet Honeyeater *Myzomela sanguinolenta* AU PA
Cardinal Honeyeater *Myzomela cardinalis* PA
Scarlet-throated Honeyeater *Myzomela sclateri* AU
Small Bougainville Honeyeater *Myzomela lafargei* PA
Black-headed Honeyeater *Myzomela melanocephala* PA
Yellow-vented Honeyeater *Myzomela eichhorni* PA
Malaita Honeyeater *Myzomela malaitae* PA
Tristram's Honeyeater *Myzomela tristrami* PA
Orange-breasted Honeyeater *Myzomela jugularis* PA
Black-bellied Honeyeater *Myzomela erythromelas* AU
Sunda Honeyeater *Myzomela vulnerata* AU
Black & Red Honeyeater *Myzomela rosenbergii* AU

Black Honeyeater *Certhionyx niger* AU
Pied Honeyeater *Certhionyx variegatus* AU

Large Spot-breasted Honeyeater *Meliphaga mimikae* AU
White-eared Mountain Honeyeater *Meliphaga montana* AU
Small Spot-breasted Honeyeater *Meliphaga orientalis* AU

White-marked Honeyeater *Meliphaga albonotata* AU
Puff-backed Honeyeater *Meliphaga aruensis* AU
Mimic Meliphaga *Meliphaga analoga* AU
Louisiades Honeyeater *Meliphaga vicina* ● AU
Graceful Honeyeater *Meliphaga gracilis* AU
Lesser Lewin Honeyeater *Meliphaga notata* AU
Yellow-gaped Honeyeater *Meliphaga flavirictus* AU
Lewin Honeyeater *Meliphaga lewinii* AU
Yellow Honeyeater *Meliphaga flava* AU
White-striped Honeyeater *Meliphaga albilineata* AU
Singing Honeyeater *Meliphaga virescens* AU
Varied Honeyeater *Meliphaga versicolor* AU
Mangrove Honeyeater *Meliphaga fasciogularis* AU
Guadalcanal Honeyeater *Meliphaga inexpectata* PA
Fuscous Honeyeater *Meliphaga fusca* AU
Yellow-fronted Honeyeater *Meliphaga plumula* AU
Yellow-faced Honeyeater *Meliphaga chrysops* AU
Purple-gaped Honeyeater *Meliphaga cratitia* AU
Grey-headed Honeyeater *Meliphaga keartlandi* AU
White-plumed Honeyeater *Meliphaga penicillata* AU
Eungella Honeyeater *Meliphaga hindwoodi* AU
Mallee Honeyeater *Meliphaga ornata* AU
Reticulated Honeyeater *Meliphaga reticulata* AU
White-eared Honeyeater *Meliphaga leucotis* AU
Yellow-throated Honeyeater *Meliphaga flavicollis* AU
Yellow-tufted Honeyeater *Meliphaga melanops* AU
White-gaped Honeyeater *Meliphaga unicolor* AU
Tawny-breasted Honeyeater *Meliphaga flaviventer* AU
Spotted Honeyeater *Meliphaga polygramma* AU
Yellow-streaked Honeyeater *Meliphaga macleayana* AU
Bridled Honeyeater *Meliphaga frenata* AU
Black-throated Honeyeater *Meliphaga subfrenata* AU
Obscure Honeyeater *Meliphaga obscura* AU

Orange-cheeked Honeyeater *Oreornis chrysogenys* AU

Carunculated Honeyeater *Foulehaio carunculata* PA
Yellow-faced Honeyeater *Foulehaio provocator* PA

Golden Honeyeater *Cleptornis marchei* PA

Bonin Island Honeyeater *Apalopteron familiare* ● EP

Brown-headed Honeyeater *Melithreptus brevirostris* AU
White-naped Honeyeater *Melithreptus lunatus* AU
White-throated Honeyeater *Melithreptus albogularis* AU
Black-headed Honeyeater *Melithreptus affinis* AU
Black-chinned Honeyeater *Melithreptus gularis* AU
Golden-backed Honeyeater *Melithreptus laetior* AU
Strong-billed Honeyeater *Melithreptus validirostris* AU

Blue-faced Honeyeater *Entomyzon cyanotis* AU

Stitch-bird *Notiomystis cincta* ● AU

New Guinea Brown Honeyeater *Pycnopygius ixoides* AU
Grey-fronted Honeyeater *Pycnopygius cinereus* AU
Streak-capped Honeyeater *Pycnopygius stictocephalus* AU

Meyer's Friarbird *Philemon meyeri* AU
Brass's Friarbird *Philemon brassi* ● AU
Little Friarbird *Philemon citreogularis* AU
Plain Friarbird *Philemon inornatus* AU
Striated Friarbird *Philemon gilolensis* AU
Dusky Friarbird *Philemon fuscicapillus* ● AU
Grey-necked Friarbird *Philemon subcorniculatus* AU
Moluccas Friarbird *Philemon moluccensis* AU
Timor Helmeted Friarbird *Philemon buceroides* AU
Melville I Friarbird *Philemon gordoni* AU
New Guinea Friarbird *Philemon novaeguineae* AU
New Britain Friarbird *Philemon cockerelli* AU
New Ireland Friarbird *Philemon eichhorni* AU
White-naped Friarbird *Philemon albitorques* AU
Silver-crowned Friarbird *Philemon argenticeps* AU
Noisy Friarbird *Philemon corniculatus* AU
New Caledonian Friarbird *Philemon diemenensis* PA

Leaden Honeyeater *Ptiloprora plumbea* AU
Meek's Streaked Honeyeater *Ptiloprora meekiana* AU

Red-sided streaked Honeyeater *Ptiloprora erythropleura* AU
Red-backed Honeyeater *Ptiloprora guisei* AU
Mayr's Streaked Honeyeater *Ptiloprora mayri* AU
Black-backed Streaked Honeyeater *Ptiloprora perstriata* AU

Sooty Honeyeater *Melidectes fuscus* AU
Gilliard's Honeyeater *Melidectes whitemanensis* AU
Long-bearded Honeyeater *Melidectes princeps* ● AU
Short-bearded Honeyeater *Melidectes nouhuysi* AU
Mid-mountain Honeyeater *Melidectes ochromelas* AU
White-fronted Melidectes *Melidectes leucostephes* AU
Belford's Melidectes *Melidectes belfordi* AU
Reichenow's Melidectes *Melidectes rufocrissalis* AU
Foerster's Melidectes *Melidectes foersteri* AU
Cinnamon-breasted Wattle Bird *Melidectes torquatus* AU

Arfak Melipotes *Melipotes gymnops* AU
Huon Melipotes *Melipotes ater* AU
Common Melipotes *Melipotes fumigatus* AU

Brown Honeysucker *Myza celebensis* AU
Spot-headed Honeysucker *Myza sarasinorum* AU

San Cristobal Honeyeater *Meliarchus sclateri* PA

Green Honeyeater *Gymnomyza viridis* PA
Black-breasted Honeyeater *Gymnomyza samoensis* PA
Red-faced Honeyeater *Gymnomyza aubryana* PA

Kauai O-o *Moho braccatus* ● PA
Bishop's O-o *Moho bishopi* ● PA

Tawny-crowned Honeyeater *Glyciphilus melanops* AU

Crescent Honeyeater *Phylidonyris pyrrhoptera* AU
Yellow-winged Honeyeater *Phylidonyris novaehollandiae* AU
White-cheeked Honeyeater *Phylidonyris nigra* AU
White-fronted Honeyeater *Phylidonyris albifrons* AU
Barred Honeyeater *Phylidonyris undulata* AU
White-bellied Honeyeater *Phylidonyris notabilis* PA

Bar-breasted Honeyeater *Ramsayornis fasciatus* AU
Brown-backed Honeyeater *Ramsayornis modestus* AU

Striped Honeyeater *Plectorhyncha lanceolata* AU

Grey Honeyeater *Conopophila whitei* AU
Rufous-banded Honeyeater *Conopophila albogularis* AU
Red-throated Honeyeater *Conopophila rufogularis* AU
Painted Honeyeater *Conopophila picta* AU

Regent Honeyeater *Xanthomyza phrygia* ● AU

Banded Honeyeater *Cissomela pectoralis* AU

Eastern Spinebill *Acanthorhynchus tenuirostris* AU
Western Spinebill *Acanthorhynchus superciliosus* AU

Bell Miner *Manorina melanophrys* AU
Noisy Miner *Manorina melanocephala* AU
Dusky Miner *Manorina obscura* AU
Yellow-throated Miner *Manorina flavigula* AU
Black-eared Miner *Manorina melanotis* ● AU

New Zealand Bellbird *Anthornis melanura* AU

Spiny-cheeked Honeyeater *Acanthagenys rufogularis* AU

Little Wattle Bird *Anthochaera chrysoptera* AU
Red Wattle Bird *Anthochaera carunculata* AU
Yellow Wattle Bird *Anthochaera paradoxa* AU

Parson Bird (Tui) *Prosthemadura novaeseelandiae* AU

Cape Sugarbird *Promerops cafer* AF
Gurney's Sugarbird *Promerops gurneyi* AF

155 BUNTINGS, CARDINALS AND TANAGERS
Emberizidae

BUNTINGS Emberizinae

Crested Bunting *Melophus lathami* OR

Fokien Blue Bunting *Latoucheornis siemsseni* EP

Corn Bunting *Emberiza calandra* WP EP

Yellowhammer *Emberiza citrinella*
Pine Bunting *Emberiza leucocephala* EP OR
Rock Bunting *Emberiza cia* WP EP OR
Siberian Meadow Bunting *Emberiza cioides* EP
Jankowski's Bunting *Emberiza jankowskii* EP
Grey-necked Bunting *Emberiza buchanani* WP EP OR
White-capped Bunting *Emberiza stewarti* EP OR
Cinereous Bunting *Emberiza cineracea* WP AF
Ortolan Bunting *Emberiza hortulana* WP EP AF
Cretzschmar's Bunting *Emberiza caesia* WP AF
Cirl Bunting *Emberiza cirlus* WP
House Bunting *Emberiza striolata* WP AF OR
Larklike Bunting *Emberiza impetuani* AF
Cinnamon-breasted Rock Bunting *Emberiza tahapisi* AF
Socotra Mountain Bunting *Emberiza socotrana* AF
Cape Bunting *Emberiza capensis* AF
Japanese Reed Bunting *Emberiza yessoensis* EP
Tristram's Bunting *Emberiza tristrami* EP
Grey-hooded Bunting *Emberiza fucata* EP OR
Little Bunting *Emberiza pusilla* EP OR
Yellow-browed Bunting *Emberiza chrysophrys* EP OR
Rustic Bunting *Emberiza rustica* WP EP
Yellow-headed Bunting *Emberiza elegans* EP
Yellow-breasted Bunting *Emberiza aureola* EP OR
Somali Golden-breasted Bunting *Emberiza poliopleura* AF
Golden-breasted Bunting *Emberiza flaviventris* AF
Brown-rumped Bunting *Emberiza affinis* AF
Cabanis's Yellow Bunting *Emberiza cabanisi* AF
Chestnut Bunting *Emberiza rutila* EP OR
Koslow's Bunting *Emberiza koslowi* EP
Black-headed Bunting *Emberiza melanocephala* WP OR
Red-headed Bunting *Emberiza bruniceps* EP OR
Japanese Yellow Bunting *Emberiza sulphurata* ● EP OR
Black-faced Bunting *Emberiza spodocephala* EP OR
Japanese Grey Bunting *Emberiza variabilis* EP
Pallas' Reed Bunting *Emberiza pallasi* EP
Reed Bunting *Emberiza schoeniclus* WP EP OR

McCown's Longspur *Calcarius mccownii* NA NT
Lapland Bunting *Calcarius lapponicus* WP EP NA
Smith's Longspur *Calcarius pictus* NA
Chestnut-collared Longspur *Calcarius ornatus* NA NT

Snow Bunting *Plectrophenax nivalis* WP EP NA
McKay's Bunting *Plectrophenax hyperboreus* NA

Lark Bunting *Calamospiza melanocorys* NA NT

Fox Sparrow *Passerella iliaca* NA NT

Song Sparrow *Melospiza melodia* NA NT
Lincoln Sparrow *Melospiza lincolnii* NA NT
Swamp Sparrow *Melospiza georgiana* NA NT

Rufous-collared Sparrow *Zonotrichia capensis* NT
Harris' Sparrow *Zonotrichia querula* NA
White-crowned Sparrow *Zonotrichia leucophrys* NA NT
White-throated Sparrow *Zonotrichia albicollis* NA NT
Golden-crowned Sparrow *Zonotrichia atricapilla* NA NT

Volcano Junco *Junco vulcani* NT
Dark-eyed Junco *Junco hyemalis* NA NT
Grey-headed Junco *Junco caniceps* NA NT
Mexican Junco *Junco phaeonotus* NA NT

Savannah Sparrow *Passerculus sandwichensis* NA NT

Seaside Sparrow *Ammodramus maritimus* NA
Sharp-tailed Sparrow *Ammodramus caudacutus* NA
Le Conte's Sparrow *Ammodramus leconteii* NA
Baird's Sparrow *Ammodramus bairdii* NA NT
Henslow's Sparrow *Ammodramus henslowii* NA
Grasshopper Sparrow *Ammodramus savannarum* NA NT

Sierra Madre Sparrow *Xenospiza baileyi* ● NT

Grassland Sparrow *Myospiza humeralis* NT
Yellow-browed Sparrow *Myospiza aurifrons* NT

American Tree Sparrow *Spizella arborea* NA
Chipping Sparrow *Spizella passerina* NA NT

Field Sparrow *Spizella pusilla* NA NT
Worthen's Sparrow *Spizella wortheni* NT
Black-chinned Sparrow *Spizella atrogularis* NA NT
Clay-coloured Sparrow *Spizella pallida* NA NT
Brewer's Sparrow *Spizella breweri* NA NT

Vesper Sparrow *Pooecetes gramineus* NA NT

Lark Sparrow *Chondestes grammacus* NA NT

Black-throated Sparrow *Amphispiza bilineata* NA NT

Sage Sparrow *Amphispiza belli* NA NT

Bridled Sparrow *Aimophila mystacalis* NT
Black-chested Sparrow *Aimophila humeralis* NT
Stripe-headed Sparrow *Aimophila ruficauda* NT
Cinnamon-tailed Sparrow *Aimophila sumichrasti* NT
Stripe-capped Sparrow *Aimophila strigiceps* NT
Bachman's Sparrow *Aimophila aestivalis* NA
Botteri's Sparrow *Aimophila botterii* NA NT
Cassin's Sparrow *Aimophila cassinii* NA
Five-striped Sparrow *Aimophila quinquestriata* NT
Rufous-winged Sparrow *Aimophila carpalis* NA NT
Rufous-crowned Sparrow *Aimophila ruficeps* NA NT
Oaxaca Sparrow *Aimophila notosticta* NT
Rusty Sparrow *Aimophila rufescens* NT

Tumbes Sparrow *Rhynchospiza stolzmanni* NT

Zapata Sparrow *Torreornis inexpectata* ● NT

Striped Sparrow *Orituris superciliosus* NT

Black-hooded Sierra Finch *Phrygilus atriceps* NT
Grey-hooded Sierra Finch *Phrygilus gayi* NT
Patagonian Sierra Finch *Phrygilus patagonicus* NT
Mourning Sierra Finch *Phrygilus fruticeti* NT
Plumbeous Sierra Finch *Phrygilus unicolor* NT
Red-backed Sierra Finch *Phrygilus dorsalis* NT
White-throated Sierra Finch *Phrygilus erythronotus* NT
Ash-breasted Sierra Finch *Phrygilus plebejus* NT
Carbonated Sierra Finch *Phrygilus carbonarius* NT
Band-tailed Sierra Finch *Phrygilus alaudinus* NT

Black-throated Finch *Melanodera melanodera* NT
Yellow-bridled Finch *Melanodera xanthogramma* NT

Slaty Finch *Haplospiza rustica* NT
Uniform Finch *Haplospiza unicolor* NT

Peg-billed Sparrow *Acanthidops bairdii*

Black-crested Finch *Lophospingus pusillus* NT
Grey-crested Finch *Lophospingus griseocristatus* NT

Long-tailed Reed Finch *Donacospiza albifrons* NT

Gough I Finch *Rowettia goughensis* ● AF

Nightingale Finch *Nesospiza acunhae* ● AF
Wilkins's Finch *Nesospiza wilkinsi* ● AF

White-winged Finch *Diuca speculifera* NT
Common Diuca Finch *Diuca diuca* NT

Short-tailed Finch *Idiopsar brachyurus* NT

Cinereous Finch *Piezorhina cinerea* NT

Slender-billed Finch *Xenospingus concolor* ● NT

Great Inca Finch *Incaspiza pulchra* NT
Rufous-backed Inca Finch *Incaspiza personata* NT
Grey-winged Inca Finch *Incaspiza ortizi* ● NT
Buff-bridled Inca Finch *Incaspiza laeta* NT
Little Inca Finch *Incaspiza watkinsi* NT

Bay-chested Warbling-Finch *Poospiza thoracica* NT
Bolivian Warbling-Finch *Poospiza boliviana* NT
Plain-tailed Warbling-Finch *Poospiza alticola* ● NT
Rufous-sided Warbling-Finch *Poospiza hypochondria* NT
Rusty-browed Warbling-Finch *Poospiza erythrophrys* NT
Cinnamon Warbling-Finch *Poospiza ornata* NT
Black & Rufous Warbling-Finch *Poospiza nigrorufa* NT
Red-rumped Warbling-Finch *Poospiza lateralis* NT
Rufous-breasted Warbling-Finch *Poospiza rubecula* ● NT
Chestnut-breasted Mountain Finch *Poospiza caesar* NT
Collared Warbling-Finch *Poospiza hispaniolensis* NT
Ringed Warbling-Finch *Poospiza torquata* NT

Grey and White Warbling-Finch *Poospiza cinerea* NT

Stripe-tailed Yellow Finch *Sicalis citrina* NT
Puna Yellow Finch *Sicalis lutea* NT
Bright-rumped Yellow Finch *Sicalis uropygialis* NT
Citron-headed Yellow Finch *Sicalis luteocephala* NT
Greater Yellow Finch *Sicalis auriventris* NT
Greenish Yellow Finch *Sicalis olivascens* NT
Patagonian Yellow Finch *Sicalis lebruni* NT
Orange-fronted Yellow Finch *Sicalis colombiana* NT
Saffron Finch *Sicalis flaveola* NT
Grassland Yellow Finch *Sicalis luteola* NT
Raimondi's Yellow Finch *Sicalis raimondii* NT
Sulphur-breasted Finch *Sicalis taczanowskii* NT

Cochabamba Mountain Finch *Compospiza garleppi* ● NT
Tucuman Mountain Finch *Compospiza baeri* ● NT

Wedge-tailed Grass Finch *Emberizoides herbicola* NT
Lesser Grass Finch *Emberizoides ypiranganus* NT
Mt Duida Grass Finch *Emberizoides duidae* NT

Great Pampa Finch *Embernagra platensis* NT
Buff-throated Pampa Finch *Embernagra longicauda* NT

Blue-black Grassquit *Volatinia jacarina* NT

Buffy-throated Seedeater *Sporophila frontalis* ● NT
Temminck's Seedeater *Sporophila falcirostris* ● NT
Slate-coloured Seedeater *Sporophila schistacea* NT
Grey Seedeater *Sporophila intermedia* NT
Plumbeous Seedeater *Sporophila plumbea* NT
Variable Seedeater *Sporophila aurita* NT
Wing-barred Seedeater *Sporophila americana* NT
White-collared Seedeater *Sporophila torqueola* NA NT
Rusty-collared Seedeater *Sporophila collaris* NT
Lined Seedeater *Sporophila lineola* NT
Black & White Seedeater *Sporophila luctuosa* NT
Yellow-bellied Seedeater *Sporophila nigricollis* NT
Dubois' Seedeater *Sporophila ardesiaca* NT
Hooded Seedeater *Sporophila melanops* NT
Dull-coloured Seedeater *Sporophila obscura* NT
Double-collared Seedeater *Sporophila caerulescens* NT
White-throated Seedeater *Sporophila albogularis* NT
White-bellied Seedeater *Sporophila leucoptera* NT
Parrot-billed Seedeater *Sporophila peruviana* NT
Drab Seedeater *Sporophila simplex* NT
Black & Tawny Seedeater *Sporophila nigrorufa* NT
Capped Seedeater *Sporophila bouvreuil* NT
Tumaco Seedeater *Sporophila insulata* ● NT
Ruddy-breasted Seedeater *Sporophila minuta* NT
Tawny-bellied Seedeater *Sporophila hypoxantha* NT
Rufous-naped Seedeater *Sporophila hypochroma* ● NT
Dark-throated Seedeater *Sporophila ruficollis* NT
Marsh Seedeater *Sporophila palustris* ● NT
Chestnut-bellied Seedeater *Sporophila castaneiventris* NT
Chestnut Seedeater *Sporophila cinnamomea* ● NT
Black-bellied Seedeater *Sporophila melanogaster* NT
Chestnut-throated Seedeater *Sporophila telasco* NT

Nicaraguan Seed Finch *Oryzoborus nuttingi* NT
Large-billed Seed Finch *Oryzoborus crassirostris* NT
Great-billed Seed Finch *Oryzoborus maximiliani* NT
Black-billed Seed Finch *Oryzoborus atrirostris* NT
Lesser Seed Finch *Oryzoborus angolensis* NT

Blue Seedeater *Amaurospiza concolor* NT
Blackish-blue Seedeater *Amaurospiza moesta* ● NT

Cuban Bullfinch *Melopyrrha nigra* NT

White-naped Seedeater *Dolospingus fringilloides* NT

Band-tailed Seedeater *Catamenia analis* NT
Plain-coloured Seedeater *Catamenia inornata* NT
Paramo Seedeater *Catamenia homochroa* NT
Colombian Seedeater *Catamenia oreophila* NT

Cuban Grassquit *Tiaris canora* NT
Yellow-faced Grassquit *Tiaris olivacea* NT
Black-faced Grassquit *Tiaris bicolor* NT
Sooty Grassquit *Tiaris fuliginosa* NT

Yellow-shouldered Grassquit *Loxipasser anoxanthus* NT

Puerto Rican Bullfinch *Loxigilla portoricensis* NT
Greater Antillean Bullfinch *Loxigilla violacea* NT
Lesser Antillean Bullfinch *Loxigilla noctis* NT

St Lucia Black Finch *Melanospiza richardsoni* ● NT

Large Ground Finch *Geospiza magnirostris* PA
Medium Ground Finch *Geospiza fortis* PA
Small Ground Finch *Geospiza fuliginosa* PA
Sharp-beaked Ground Finch *Geospiza difficilis* PA
Cactus Ground Finch *Geospiza scandens* PA
Large Cactus Ground Finch *Geospiza conirostris* PA

Vegetarian Tree Finch *Camarhynchus crassirostris* PA
Large Insectivorous Tree Finch *Camarhynchus psittacula* PA
Charles Insectivorous Tree Finch *Camarhynchus pauper* PA
Small Insectivorous Tree Finch *Camarhynchus parvulus* PA
Woodpecker Finch *Camarhynchus pallidus* PA
Mangrove Finch *Camarhynchus heliobates* ● PA

Warbler Finch *Certhidea olivacea* PA

Cocos Finch *Pinaroloxias inornata* NT

Collared Towhee *Pipilo ocai* NT
Rufous-sided Towhee *Pipilo erythrophthalmus* NA

Brown Towhee *Pipilo fuscus* NA NT
Abert's Towhee *Pipilo aberti* NA NT
White-throated Towhee *Pipilo albicollis* NT

Green-tailed Towhee *Chlorurus chlorurus* NA NT

Rusty-crowned Ground Sparrow *Melozone kieneri* NT
Prevost's Ground Sparrow *Melozone biarcuatum* NT
White-eared Ground Sparrow *Melozone leucotis* NT

Pectoral Sparrow *Arremon taciturnus* NT
Saffron-billed Sparrow *Arremon flavirostris* NT
Orange-billed Sparrow *Arremon aurantiirostris* NT
Golden-winged Sparrow *Arremon schlegeli* NT
Black-capped Sparrow *Arremon abeillei* NT

Olive Sparrow *Arremonops rufivirgatus* NA NT
Tocuyo Sparrow *Arremonops tocuyensis* NT
Green-backed Sparrow *Arremonops chloronotus* NT
Black-striped Sparrow *Arremonops conirostris* NT

White-naped Brush Finch *Atlapetes albinucha* NT
Yellow-throated Brush Finch *Atlapetes gutturalis* NT
Pale-naped Brush Finch *Atlapetes pallidinucha* NT
Rufous-naped Brush Finch *Atlapetes rufinucha* NT
White-rimmed Brush Finch *Atlapetes leucopis* NT
Santa Marta Brush Finch *Atlapetes melanocephalus* NT
Rufous-capped Brush Finch *Atlapetes pileatus* NT
Olive-headed Brush Finch *Atlapetes flaviceps* ● NT
Dusky-headed Brush Finch *Atlapetes fuscoolivaceus* NT
Tricoloured Brush Finch *Atlapetes tricolor* NT
Moustached Brush Finch *Atlapetes albofrenatus* NT
Slaty Brush Finch *Atlapetes schistaceus* NT
Rusty-bellied Brush Finch *Atlapetes nationi* NT
White-winged Brush Finch *Atlapetes leucopterus* NT
White-headed Brush Finch *Atlapetes albiceps* NT
Pale-headed Brush Finch *Atlapetes pallidiceps* ● NT
Rufous-eared Brush Finch *Atlapetes rufigenis* NT
Ochre-breasted Brush Finch *Atlapetes semirufus* NT
Tepui Brush Finch *Atlapetes personatus* NT
Fulvous-headed Brush Finch *Atlapetes fulviceps* NT
Yellow-striped Brush Finch *Atlapetes citrinellus* NT
Plain-breasted Brush Finch *Atlapetes apertus* NT
Chestnut-capped Brush Finch *Atlapetes brunneinucha* NT
Green-striped Brush Finch *Atlapetes virenticeps* NT
Stripe-headed Brush Finch *Atlapetes torquatus* NT
Black-headed Brush Finch *Atlapetes atricapillus* NT

Big-footed Sparrow *Pezopetes capitalis* NT

Tanager Finch *Oreothraupis arremonops* ● NT

Yellow-thighed Sparrow *Pselliophorus tibialis* NT
Yellow-green Sparrow *Pselliophorus luteoviridis* NT

Olive Finch *Lysurus castaneiceps* NT
Sooty-faced Finch *Lysurus crassirostris* NT

Black-backed Bush Tanager *Urothraupis stolzmanni* NT

Coal-crested Finch *Charitospiza eucosma* NT

Many-coloured Chaco Finch *Saltatricula multicolor* NT

Black-masked Finch *Coryphaspiza melanotis* ● NT

Pileated Finch *Coryphospingus pileatus* NT
Red-crested Finch *Coryphospingus cucullatus* NT
Crimson Finch *Rhodospingus cruentus* NT

PLUSH-CAPPED FINCH Catamblyrhynchinae

Plush-capped Finch *Catamblyrhynchus diadema* NT

CARDINALS AND GROSBEAKS Cardinalinae

Yellow Cardinal *Gubernatrix cristata* ● NT

Red-crested Cardinal *Paroaria coronata* NT
Red-cowled Cardinal *Paroaria dominicana* NT
Red-capped Cardinal *Paroaria gularis* NT
Crimson-fronted Cardinal *Paroaria baeri* NT
Yellow-billed Cardinal *Paroaria capitata* NT

Dickcissel *Spiza americana* NA NT

Yellow Grosbeak *Pheucticus chrysopeplus* NT
Black-thighed Grosbeak *Pheucticus tibialis* NT
Yellow-bellied Grosbeak *Pheucticus chrysogaster* NT
Black-backed Grosbeak *Pheucticus aureoventris* NT
Rose-breasted Grosbeak *Pheucticus ludovicianus* NA NT
Black-headed Grosbeak *Pheucticus melanocephalus* NA NT

Common Cardinal *Cardinalis cardinalis* NA NT

Vermilion Cardinal *Pyrrhuloxia phoeniceus* NT
Pyrrhuloxia *Pyrrhuloxia sinuatus* NA NT

Yellow-green Grosbeak *Caryothraustes canadensis* NT

Black-faced Grosbeak *Caryothraustes poliogaster* NT
Yellow-shouldered Grosbeak *Caryothraustes humeralis* NT

Crimson-collared Grosbeak *Rhodothraupis celaeno* NT

Red and Black Grosbeak *Periporphyrus erythromelas* NT

Slate-coloured Grosbeak *Pitylus grossus* NT
Black-throated Grosbeak *Pitylus fuliginosus* NT

Black-headed Saltator *Saltator atriceps* NT
Buff-throated Saltator *Saltator maximus* NT
Black-winged Saltator *Saltator atripennis* NT
Green-winged Saltator *Saltator similis* NT
Greyish Saltator *Saltator coerulescens* NT
Orinocan Saltator *Saltator orenocensis* NT
Thick-billed Saltator *Saltator maxillosus* NT
Golden-billed Saltator *Saltator aurantiirostris* NT
Masked Saltator *Saltator cinctus* NT
Black-throated Saltator *Saltator atricollis* NT
Rufous-bellied Saltator *Saltator rufiventris* NT
Streaked Saltator *Saltator albicollis* NT

Indigo Grosbeak *Cyanoloxia glaucocaerulea* NT

Blue-black Grosbeak *Cyanocompsa cyanoides* NT
Ultramarine Grosbeak *Cyanocompsa brissonii* NT
Blue Bunting *Cyanocompsa parellina* NT

Blue Grosbeak *Guiraca caerulea* NA NT

Indigo Bunting *Passerina cyanea* NA NT
Lazuli Bunting *Passerina amoena* NA NT
Varied Bunting *Passerina versicolor* NA NT
Painted Bunting *Passerina ciris* NA NT
Rose-bellied Bunting *Passerina rositae* NT
Orange-breasted Bunting *Passerina leclancherii* NT

Blue Finch *Porphyrospiza caerulescens* NT

TANAGERS Thraupinae

Brown Tanager *Orchesticus abeillei* NT

Cinnamon Tanager *Schistoclamys ruficapillus* NT
Black-faced Tanager *Schistoclamys melanopis* NT

White-banded Tanager *Neothraupis fasciata* NT
White-rumped Tanager *Cypsnagra hirundinacea* NT

Black & White Tanager *Conothraupis speculigera* NT
Cone-billed Tanager *Conothraupis mesoleuca* ● NT

Red-billed Pied Tanager *Lamprospiza melanoleuca* NT

Magpie Tanager *Cissopis leveriana* NT

Grass-green Tanager *Chlorornis riefferii* NT

Scarlet-throated Tanager *Compsothraupis loricata* NT

White-capped Tanager *Sericossypha albocristata* NT

Puerto Rican Tanager *Nesospingus speculiferus* NT

Common Bush Tanager *Chlorospingus ophthalmicus* NT
Tacarcuna Bush Tanager *Chlorospingus tacarcunae* NT
Pirre Bush Tanager *Chlorospingus inornatus* NT
Dusky-bellied Bush Tanager *Chlorospingus semifuscus* NT
Pileated Bush Tanager *Chlorospingus pileatus* NT
Short-billed Bush Tanager *Chlorospingus parvirostris* NT
Yellow-throated Bush Tanager *Chlorospingus flavigularis* NT
Yellow-green Bush Tanager *Chlorospingus flavovirens* ● NT
Ash-throated Bush Tanager *Chlorospingus canigularis* NT

Grey-hooded Bush Tanager *Cnemoscopus rubrirostris* NT

Black-capped Hemispingus *Hemispingus atropileus* NT
Parodi's Tanager *Hemispingus parodii* NT
Superciliaried Hemispingus *Hemispingus superciliaris* NT
Grey-capped Hemispingus *Hemispingus reyi* NT
Oleaginous Hemispingus *Hemispingus frontalis* NT
Black-eared Hemispingus *Hemispingus melanotis* NT
Slaty-backed Hemispingus *Hemispingus goeringi* NT
Rufous-browed Hemispingus *Hemispingus rufosuperciliaris* NT
Black-headed Hemispingus *Hemispingus verticalis* NT
Drab Hemispingus *Hemispingus xanthophthalmus* NT
Three-striped Hemispingus *Hemispingus trifasciatus* NT

Chestnut-headed Tanager *Pyrrhocoma ruficeps* NT

Fulvous-headed Tanager *Thlypopsis fulviceps* NT
Rufous-chested Tanager *Thlypopsis ornata* NT
Brown-flanked Tanager *Thlypopsis pectoralis* NT
Orange-headed Tanager *Thlypopsis sordida* NT
Buff-bellied Tanager *Thlypopsis inornata* NT
Rust & Yellow Tanager *Thlypopsis ruficeps* NT

Guira Tanager *Hemithraupis guira* NT
Rufous-headed Tanager *Hemithraupis ruficapilla* NT
Yellow-backed Tanager *Hemithraupis flavicollis* NT

Black & Yellow Tanager *Chrysothlypis chrysomelas* NT
Scarlet & White Tanager *Chrysothlypis salmoni* NT

Hooded Tanager *Nemosia pileata* NT
Cherry-throated Tanager *Nemosia rourei* ● NT

Black-crowned Palm Tanager *Phaenicophilus palmarum* NT
Grey-crowned Palm Tanager *Phaenicophilus poliocephalus* NT

Chat-Tanager *Calyptophilus frugivorus* NT

Rose-breasted Thrush Tanager *Rhodinocichla rosea* NT

Dusky-faced Tanager *Mitrospingus cassinii* NT
Olive-backed Tanager *Mitrospingus oleagineus* NT

Carmiol's Tanager *Chlorothraupis carmioli* NT
Lemon-browed Tanager *Chlorothraupis olivacea* NT
Ochre-breasted Tanager *Chlorothraupis stolzmanni* NT

Olive-green Tanager *Orthogonys chloricterus* NT

Grey-headed Tanager *Eucometis penicillata* NT

Fulvous Shrike-Tanager *Lanio fulvus* NT
White-winged Shrike-Tanager *Lanio versicolor* NT
Black-throated Shrike-Tanager *Lanio aurantius* NT
White-throated Shrike-Tanager *Lanio leucothorax*

Rufous-crested Tanager *Creurgops verticalis* NT
Slaty Tanager *Creurgops dentata* NT

Scarlet-browed Tanager *Heterospingus xanthopygius* NT

Flame-crested Tanager *Tachyphonus cristatus* NT
Natterer's Tanager *Tachyphonus nattereri* NT
Yellow-crested Tanager *Tachyphonus rufiventer* NT
Fulvous-crested Tanager *Tachyphonus surinamus* NT
White-shouldered Tanager *Tachyphonus luctuosus* NT

Tawny-crested Tanager *Tachyphonus delatrii* NT
Ruby-crowned Tanager *Tachyphonus coronatus* NT
White-lined Tanager *Tachyphonus rufus* NT
Red-shouldered Tanager *Tachyphonus phoenicius* NT

Black-goggled Tanager *Trichothraupis melanops* NT

Red-crowned Ant-Tanager *Habia rubica* NT
Red-throated Ant-Tanager *Habia fuscicauda* NT
Black-cheeked Ant-Tanager *Habia atrimaxillaris* ● NT
Sooty Ant-Tanager *Habia gutturalis* ● NT
Crested Ant-Tanager *Habia cristata* NT

Flame-coloured Tanager *Piranga bidentata* NT
Hepatic Tanager *Piranga flava* NA NT
Summer Tanager *Piranga rubra* NA NT
Rose-throated Tanager *Piranga roseogularis* NT
Scarlet Tanager *Piranga olivacea* NA NT
Western Tanager *Piranga ludoviciana* NA NT
White-winged Tanager *Piranga leucoptera* NT
Red-headed Tanager *Piranga erythrocephala* NT
Red-hooded Tanager *Piranga rubriceps* NT

Vermilion Tanager *Calochaetes coccineus* NT

Crimson-collared Tanager *Ramphocelus sanguinolentus* NT
Masked Crimson Tanager *Ramphocelus nigrogularis* NT
Crimson-backed Tanager *Ramphocelus dimidiatus* NT
Black-bellied Tanager *Ramphocelus melanogaster* NT
Silver-beaked Tanager *Ramphocelus carbo* NT
Brazilian Tanager *Ramphocelus bresilius* NT
Scarlet-rumped Tanager *Ramphocelus passerinii* NT
Flame-rumped Tanager *Ramphocelus flammigerus* NT
Yellow-rumped Tanager *Ramphocelus icteronotus* NT

Stripe-headed Tanager *Spindalis zena* NT

Blue-grey Tanager *Thraupis episcopus* NT
Sayaca Tanager *Thraupis sayaca* NT
Glaucous Tanager *Thraupis glaucocolpa* NT
Azure-shouldered Tanager *Thraupis cyanoptera* NT
Golden-chevroned Tanager *Thraupis ornata* NT
Yellow-winged Tanager *Thraupis abbas* NT
Palm Tanager *Thraupis palmarum* NT
Blue-capped Tanager *Thraupis cyanocephala* NT
Blue & Yellow Tanager *Thraupis bonariensis* NT

Blue-backed Tanager *Cyanicterus cyanicterus* NT

Blue & Gold Tanager *Buthraupis arcaei* NT
Black & Gold Tanager *Buthraupis melanochlamys* ● NT
Golden-chested Tanager *Buthraupis rothschildi* NT
Moss-backed Tanager *Buthraupis edwardsi* NT
Gold-ringed Tanager *Buthraupis aureocincta* ● NT
Hooded Mountain Tanager *Buthraupis montana* NT
Black-chested Mountain Tanager *Buthraupis eximia* NT
Golden-backed Mountain Tanager *Buthraupis aureodorsalis* ● NT
Masked Mountain Tanager *Buthraupis wetmorei* NT

Orange-throated Tanager *Wetmorethraupis sterrhopteron* NT

Black-cheeked Mountain Tanager *Anisognathus melanogenys* NT

Lacrimose Mountain Tanager *Anisognathus lacrymosus* NT
Scarlet-bellied Mountain Tanager *Anisognathus igniventris* NT
Blue-winged Mountain Tanager *Anisognathus flavinuchus* NT
Black-chinned Mountain Tanager *Anisognathus notabilis* NT

Diademed Tanager *Stephanophorus diadematus* NT

Purplish-mantled Tanager *Iridosornis porphyrocephala* NT
Yellow-throated Tanager *Iridosornis analis* NT
Golden-collared Tanager *Iridosornis jelskii* NT
Golden-crowned Tanager *Iridosornis rufivertex* NT
Yellow-scarfed Tanager *Iridosornis reinwardti* NT

Buff-breasted Mountain Tanager *Dubusia taeniata* NT

Chestnut-bellied Mountain Tanager *Delothraupis castaneoventris* NT

Fawn-breasted Tanager *Pipraeidea melanonota* NT

Jamaican Euphonia *Euphonia jamaica* NT
Plumbeous Euphonia *Euphonia plumbea* NT
Scrub Euphonia *Euphonia affinis* NT
Yellow-crowned Euphonia *Euphonia luteicapilla* NT
Purple-throated Euphonia *Euphonia chlorotica* NT
Trinidad Euphonia *Euphonia trinitatis* NT
Velvet-fronted Euphonia *Euphonia concinna* NT
Orange-crowned Euphonia *Euphonia saturata* NT
Finsch's Euphonia *Euphonia finschi* NT
Violaceous Euphonia *Euphonia violacea* NT
Thick-billed Euphonia *Euphonia laniirostris* NT
Yellow-throated Euphonia *Euphonia hirundinacea* NT
Green-throated Euphonia *Euphonia chalybea* ● NT
Blue-hooded Euphonia *Euphonia elegantissima* NT
Golden-rumped Euphonia *Euphonia aureata* NT
Antillean Euphonia *Euphonia musica* NT
Fulvous-vented Euphonia *Euphonia fulvicrissa* NT
Spot-crowned Euphonia *Euphonia imitans* NT
Olive-backed Euphonia *Euphonia gouldi* NT
Golden-bellied Euphonia *Euphonia chrysopasta* NT
Bronze-green Euphonia *Euphonia mesochrysa* NT
White-vented Euphonia *Euphonia minuta* NT
Tawny-capped Euphonia *Euphonia anneae* NT
Orange-bellied Euphonia *Euphonia xanthogaster* NT
Rufous-bellied Euphonia *Euphonia rufiventris* NT
Chestnut-bellied Euphonia *Euphonia pectoralis* NT
Golden-sided Euphonia *Euphonia cayennensis* NT

Yellow-collared Chlorophonia *Chlorophonia flavirostris* NT
Blue-naped Chlorophonia *Chlorophonia cyanea* NT
Chestnut-breasted Chlorophonia *Chlorophonia pyrrhophrys* NT
Blue-crowned Chlorophonia *Chlorophonia occipitalis* NT
Golden-browed Chlorophonia *Chlorophonia callophrys* NT

Glistening-green Tanager *Chlorochrysa phoenicotis* NT

Orange-eared Tanager *Chlorochrysa calliparaea* NT
Multicoloured Tanager *Chlorochrysa nitidissima* ● NT

Sira Tanager *Tangara phillipsi* ● NT
Plain-coloured Tanager *Tangara inornata* NT
Azure-rumped Tanager *Tangara cabanisi* ● NT
Grey & Gold Tanager *Tangara palmeri* NT
Turquoise Tanager *Tangara mexicana* NT
Paradise Tanager *Tangara chilensis* NT
Seven-coloured Tanager *Tangara fastuosa* ● NT
Green-headed Tanager *Tangara seledon* NT
Red-necked Tanager *Tangara cyanocephala* NT
Brassy-breasted Tanager *Tangara desmaresti* NT
Gilt-edged Tanager *Tangara cyanoventris* NT
Blue-whiskered Tanager *Tangara johannae* NT
Green & Gold Tanager *Tangara schrankii* NT
Emerald Tanager *Tangara florida* NT
Golden Tanager *Tangara arthus* NT
Silver-throated Tanager *Tangara icterocephala* NT
Saffron-crowned Tanager *Tangara xanthocephala* NT
Golden-eared Tanager *Tangara chrysotis* NT

Flame-faced Tanager *Tangara parzudakii* NT
Yellow-bellied Tanager *Tangara xanthogastra* NT
Spotted Tanager *Tangara punctata* NT
Speckled Tanager *Tangara guttata* NT
Dotted Tanager *Tangara varia* NT
Rufous-throated Tanager *Tangara rufigula* NT
Bay-headed Tanager *Tangara gyrola* NT
Rufous-winged Tanager *Tangara lavinia* NT
Burnished-buff Tanager *Tangara cayana* NT
Lesser Antillean Tanager *Tangara cucullata* NT
Black-backed Tanager *Tangara peruviana* ● NT
Chestnut-backed Tanager *Tangara preciosa* NT
Scrub Tanager *Tangara vitriolina* NT
Green-capped Tanager *Tangara meyerdeschauenseei* ● NT
Rufous-cheeked Tanager *Tangara rufigenis* NT
Golden-naped Tanager *Tangara ruficervix* NT
Metallic-green Tanager *Tangara labradorides* NT
Blue-browed Tanager *Tangara cyanotis* NT
Blue-necked Tanager *Tangara cyanicollis* NT
Golden-masked Tanager *Tangara larvata* NT
Masked Tanager *Tangara nigrocincta* NT
Spangle-cheeked Tanager *Tangara dowii* NT
Green-naped Tanager *Tangara fucosa* NT
Beryl-spangled Tanager *Tangara nigroviridis* NT
Blue & Black Tanager *Tangara vassorii* NT
Black-capped Tanager *Tangara heinei* NT
Silvery Tanager *Tangara viridicollis* NT
Green-throated Tanager *Tangara argyrofenges* NT
Black-headed Tanager *Tangara cyanoptera* NT
Yellow-collared Tanager *Tangara pulcherrima* NT
Opal-rumped Tanager *Tangara velia* NT
Opal-crowned Tanager *Tangara callophrys* NT

White-bellied Dacnis *Dacnis albiventris* ● NT
Black-faced Dacnis *Dacnis lineata* NT
Yellow-bellied Dacnis *Dacnis flaviventer* NT
Turquoise Dacnis *Dacnis hartlaubi* ● NT
Black-legged Dacnis *Dacnis nigripes* ● NT
Scarlet-thighed Dacnis *Dacnis venusta* NT
Blue Dacnis *Dacnis cayana* NT
Viridian Dacnis *Dacnis viguieri* NT
Scarlet-breasted Dacnis *Dacnis berlepschi* ● NT

Green Honeycreeper *Chlorophanes spiza* NT

Short-billed Honeycreeper *Cyanerpes nitidus* NT
Shining Honeycreeper *Cyanerpes lucidus* NT
Purple Honeycreeper *Cyanerpes caeruleus* NT
Red-legged Honeycreeper *Cyanerpes cyaneus* NT

Tit-like Dacnis *Xenodacnis parina* NT

Giant Conebill *Oreomanes fraseri* NT

Slaty Flowerpiercer *Diglossa baritula* NT
Glossy Flowerpiercer *Diglossa lafresnayii* NT
Coal-black Flowerpiercer *Diglossa carbonaria* NT
Venezuelan Flowerpiercer *Diglossa venezuelensis* ● NT
White-sided Flowerpiercer *Diglossa albilatera* NT
Scaled Flowerpiercer *Diglossa duidae* NT
Greater Flowerpiercer *Diglossa major* NT
Indigo Flowerpiercer *Diglossa indigotica* NT

Deep-blue Flowerpiercer *Diglossopis glauca* NT
Bluish Flowerpiercer *Diglossopis caerulescens* NT
Masked Flowerpiercer *Diglossopis cyanea* NT

Orangequit *Euneornis campestris* NT

SWALLOW TANAGER Tersininae

Swallow Tanager *Tersina viridis* NT

156 BANANAQUIT Coerebidae

Bananaquit *Coereba flaveola* NT

157 NEW WORLD WARBLERS Parulidae

Black & White Warbler *Mniotilta varia* NA NT

Bachman's Warbler *Vermivora bachmanii* ● NA NT
Golden-winged Warbler *Vermivora chrysoptera* NA NT
Blue-winged Warbler *Vermivora pinus* NA NT
Tennessee Warbler *Vermivora peregrina* NA NT
Orange-crowned Warbler *Vermivora celata* NA NT
Nashville Warbler *Vermivora ruficapilla* NA NT
Virginia's Warbler *Vermivora virginiae* NA NT

Colima Warbler *Vermivora crissalis* NA NT
Lucy's Warbler *Vermivora luciae* NA NT
Flame-throated Warbler *Vermivora gutteralis* NT
Crescent-chested Warbler *Vermivora superciliosa* NT

Parula Warbler *Parula americana* NA NT
Olive-backed Warbler (Tropical Parula) *Parula pitiayumi* NT

Yellow Warbler *Dendroica petechia* NA NT
Chestnut-sided Warbler *Dendroica pensylvanica* NA NT
Cerulean Warbler *Dendroica cerulea* NA NT
Black-throated Blue Warbler *Dendroica caerulescens* NA NT
Plumbeous Warbler *Dendroica plumbea* NT
Arrow-headed Warbler *Dendroica pharetra* NT
Elfin Woods Warbler *Dendroica angelae* NT
Pine Warbler *Dendroica pinus* NA NT
Grace's Warbler *Dendroica graciae* NA NT
Adelaide's Warbler *Dendroica adelaidae* NT
Olive-capped Warbler *Dendroica pityophila* NT
Yellow-throated Warbler *Dendroica dominica* NA NT
Black-throated Grey Warbler *Dendroica nigrescens* NA NT
Townsend's Warbler *Dendroica townsendi* NA NT
Hermit Warbler *Dendroica occidentalis* NA NT
Golden-cheeked Warbler *Dendroica chrysoparcia* ● NA NT
Black-throated Green Warbler *Dendroica virens* NA NT
Prairie Warbler *Dendroica discolor* NA NT
Vitelline Warbler *Dendroica vitellina* ● NT
Cape May Warbler *Dendroica tigrina* NA NT
Blackburnian Warbler *Dendroica fusca* NA NT
Magnolia Warbler *Dendroica magnolia* NA NT
Yellow-rumped Warbler *Dendroica coronata* NA NT
Palm Warbler *Dendroica palmarum* NA NT
Kirtland's Warbler *Dendroica kirtlandii* ● NA NT
Blackpoll Warbler *Dendroica striata* NA NT
Bay-breasted Warbler *Dendroica castanea* NA NT

Whistling Warbler *Catharopeza bishopi* ● NT

American Redstart *Setophaga ruticilla* NA NT

Ovenbird *Seiurus aurocapillus* NA NT
Northern Waterthrush *Seiurus noveboracensis* NA NT
Louisiana Waterthrush *Seiurus motacilla* NA NT

Swainson's Warbler *Limnothlypis swainsonii* NA NT

Worm-eating Warbler *Helmitheros vermivorus* NA NT

Prothonotary Warbler *Protonotaria citrea* NA NT

Common Yellowthroat *Geothlypis trichas* NA NT
Peninsular Yellowthroat *Geothlypis beldingi* NT
Yellow-crowned Yellowthroat *Geothlypis flavovelata* ● NT
Bahama Yellowthroat *Geothlypis rostrata* NT
Olive-crowned Yellowthroat *Geothlypis semiflava* NT
Black-polled Yellowthroat *Geothlypis speciosa* ● NT
Hooded Yellowthroat *Geothlypis nelsoni* NT
Chiriqui Yellowthroat *Geothlypis chiriquensis* NT
Masked Yellowthroat *Geothlypis aequinoctialis* NT
Grey-crowned Yellowthroat *Geothlypis poliocephala* NT
Kentucky Warbler *Geothlypis formosa* NA NT
Connecticut Warbler *Geothlypis agilis* NA NT
Mourning Warbler *Geothlypis philadelphia* NA NT
MacGillivray's Warbler *Geothlypis tolmei* NA NT

Green-tailed Ground Warbler *Microligea palustris* NT

White-winged Ground Warbler *Xenoligea montana* ● NT

Yellow-headed Warbler *Teretistris fernandinae* NT
Oriente Warbler *Teretistris fornsi* NT

Semper's Warbler *Leucopeza semperi* ● NT

Hooded Warbler *Wilsonia citrina* NA NT
Wilson's Warbler *Wilsonia pusilla* NA NT
Canada Warbler *Wilsonia canadensis* NA NT

Red-faced Warbler *Cardellina rubifrons* NA NT

Red Warbler *Ergaticus ruber* NT
Pink-headed Warbler *Ergaticus versicolor* NT

Painted Redstart *Myioborus pictus* NA NT
Slate-throated Redstart *Myioborus miniatus* NT
Brown-capped Redstart *Myioborus brunniceps* NT
Yellow-faced Redstart *Myioborus pariae* ● NT
Saffron-breasted Redstart *Myioborus cardonai* NT
Collared Redstart *Myioborus torquatus* NT
Golden-fronted Redstart *Myioborus ornatus* NT
Spectacled Redstart *Myioborus melanocephalus* NT
White-fronted Redstart *Myioborus albifrons* NT
Yellow-crowned Redstart *Myioborus flavivertex* NT
White-faced Redstart *Myioborus albifacies* NT

Neotropic Fan-tailed Warbler *Euthlypis lachrymosa* NT

Grey & Gold Warbler *Basileuterus fraseri* NT
Two-banded Warbler *Basileuterus bivittatus* NT
Golden-bellied Warbler *Basileuterus chrysogaster* NT
Flavescent Warbler *Basileuterus flaveolus* NT
Citrine Warbler *Basileuterus luteoviridis* NT
Pale-legged Warbler *Basileuterus signatus* NT
Black-crested Warbler *Basileuterus nigrocristatus* NT
Grey-headed Warbler *Basileuterus griseiceps* ● NT
Santa Marta Warbler *Basileuterus basilicus* NT
Grey-throated Warbler *Basileuterus cinereicollis* ● NT

Russet-crowned Warbler *Basileuterus coronatus* NT
Golden-crowned Warbler *Basileuterus culicivorus* NT
Rufous-capped Warbler *Basileuterus rufifrons* NT
Golden-browed Warbler *Basileuterus belli* NT
Black-cheeked Warbler *Basileuterus melanogenys* NT
Pirre Warbler *Basileuterus ignotus* NT
Three-striped Warbler *Basileuterus tristriatus* NT
Three-banded Warbler *Basileuterus trifasciatus* NT
White-bellied Warbler *Basileuterus hypoleucus* NT
White-browed Warbler *Basileuterus leucoblepharus* NT
White-striped Warbler *Basileuterus leucophrys* NT
River Warbler *Basileuterus rivularis* NT

Pardusco *Nephelornis oneilli* NT

Wren Thrush *Zeledonia coronata* NT

Olive Warbler *Peucedramus taeniatus* NA NT

Red-breasted Chat *Granatellus venustus* NT
Grey-throated Chat *Granatellus sallaei* NT
Rose-breasted Chat *Granatellus pelzelni* NT

Yellow-breasted Chat *Icteria virens* NA NT

Chestnut-vented Conebill *Conirostrum speciosum* NT
White-eared Conebill *Conirostrum leucogenys* NT
Bicoloured Conebill *Conirostrum bicolor* NT
Pearly-breasted Conebill *Conirostrum margaritae* ●

Cinereous Conebill *Conirostrum cinereum* NT
Tamarugo Conebill *Conirostrum tamarugensis* ● NT
White-browed Conebill *Conirostrum ferrugineiventre* NT

Rufous-browed Conebill *Conirostrum rufum* NT
Blue-backed Conebill *Conirostrum sitticolor* NT
Capped Conebill *Conirostrum albifrons* NT

158 HAWAIIAN HONEYCREEPERS Drepanididae

Laysan Finch *Telespyza cantans* ● PA
Nihoa Finch *Telespyza ultima* ● PA

Ou *Psittirostra psittacea* ● PA

Palila *Loxioides bailleui* ● PA

Maui Parrotbill *Pseudonestor xanthophrys* ● PA

Amakihi *Hemignathus virens* PA
Lesser Amakihi *Hemignathus parvus* PA
Kauai Akialoa *Hemignathus procerus* ● PA
Nukupuu *Hemignathus lucidus* ● PA
Akiapolaau *Hemignathus munroi* ● PA

Kauai Creeper *Oreomystis bairdi* ● PA
Hawaii Creeper *Oreomystis mana* PA

Maui Creeper *Paroreomyza montana* PA
Molokai Creeper *Paroreomyza flammea* ● PA
Oahu Creeper *Paroreomyza maculata* ● PA

Akepa *Loxops coccineus* ● PA

Iiwi *Vestiaria coccinea* PA

Crested Honeycreeper *Palmeria dolei* ● PA

Apapane *Himatione sanguinea* PA

Po'ouli *Melamprosops phaeosoma* ● PA

159 VIREOS Vireonidae

Rufous-browed Pepper-shrike *Cyclarhis gujanensis* NT
Black-billed Pepper-shrike *Cyclarhis nigrirostris* NT

Vireolaniinae

Chestnut-sided Shrike Vireo *Vireolanius melitophrys* NT
Green Shrike Vireo *Vireolanius pulchellus* NT
Yellow-browed Shrike Vireo *Vireolanius eximius* NT
Slaty-capped Shrike Vireo *Vireolanius leucotis* NT

Vireoninae

Slaty Vireo *Vireo brevipennis* NT
Hutton's Vireo *Vireo huttoni* NA NT
Black-capped Vireo *Vireo atricapillus* ● NA NT
White-eyed Vireo *Vireo griseus* NA NT
Mangrove Vireo *Vireo pallens* NT
St Andrew Vireo *Vireo caribaeus* NT
Cozumel Vireo *Vireo bairdi* NT
Cuban Vireo *Vireo gundlachii* NT
Thick-billed Vireo *Vireo crassirostris* NT
Grey Vireo *Vireo vicinior* NA NT
Bell's Vireo *Vireo bellii* NA NT
Dwarf Vireo *Vireo nelsoni* NT
Golden Vireo *Vireo hypochryseus* NT
Jamaican White-eyed Vireo *Vireo modestus* NT
Flat-billed Vireo *Vireo nanus* NT
Puerto Rican Vireo *Vireo latimeri* NT
Blue Mountain Vireo *Vireo osburni* NT
Carmiol's Vireo *Vireo carmioli* NT
Solitary Vireo *Vireo solitarius* NA NT
Yellow-throated Vireo *Vireo flavifrons* NA NT
Philadelphia Vireo *Vireo philadelphicus* NA NT
Red-eyed Vireo *Vireo olivaceus* NA NT
Yellow-green Vireo *Vireo flavoviridis* NT
Yucatan Vireo *Vireo magister* NT
Black-whiskered Vireo *Vireo altiloquus* NA NT
Warbling Vireo *Vireo gilvus* NT
Brown-capped Vireo *Vireo leucophrys* NT

Rufous-crowned Greenlet *Hylophilus poicilotis* NT
Lemon-chested Greenlet *Hylophilus thoracicus* NT
Grey-chested Greenlet *Hylophilus semicinereus* NT
Ashy-headed Greenlet *Hylophilus pectoralis* NT
Tepui Greenlet *Hylophilus sclateri* NT
Buff-chested Greenlet *Hylophilus muscicapinus* NT
Brown-headed Greenlet *Hylophilus brunneiceps* NT
Rufous-naped Greenlet *Hylophilus semibrunneus* NT
Golden-fronted Greenlet *Hylophilus aurantiifrons* NT
Dusky-capped Greenlet *Hylophilus hypoxanthus* NT
Scrub Greenlet *Hylophilus flavipes* NT
Olivaceous Greenlet *Hylophilus olivaceus* NT
Tawny-crowned Greenlet *Hylophilus ochraceiceps* NT
Grey-headed Greenlet *Hylophilus decurtatus* NT

160 NEW WORLD BLACKBIRDS Icteridae

Icterinae

Casqued Oropendola *Psarocolius oseryi* NT
Band-tailed Oropendola *Psarocolius latirostris* NT
Crested Oropendola *Psarocolius decumanus* NT
Green Oropendola *Psarocolius viridis* NT
Dusky-green Oropendola *Psarocolius atrovirens* NT
Russet-backed Oropendola *Psarocolius angustifrons* NT
Chestnut-headed Oropendola *Psarocolius wagleri* NT
Montezuma Oropendola *Psarocolius montezuma* NT
Chestnut-mantled Oropendola *Psarocolius cassini* ● NT
Para Oropendola *Psarocolius bifasciatus* ● NT
Black Oropendola *Psarocolius guatimozinus* NT
Olive Oropendola *Psarocolius yuracares* NT

Yellow-rumped Cacique *Cacicus cela* NT
Red-rumped Cacique *Cacicus haemorrhous* NT

Scarlet-rumped Cacique *Cacicus uropygialis* NT
Golden-winged Cacique *Cacicus chrysopterus* NT
Selva Cacique *Cacicus koepckeae* ● NT
Mountain Cacique *Cacicus leucoramphus* NT
Ecuadorian Black Cacique *Cacicus sclateri* NT
Solitary Cacique *Cacicus solitarius* NT
Yellow-winged Cacique *Cacicus melanicterus* NT
Yellow-billed Cacique *Cacicus holosericeus* NT

Moriche Oriole *Icterus chrysocephalus* NT
Epaulet Oriole *Icterus cayanensis* NT
Yellow-backed Oriole *Icterus chrysater* NT
Yellow Oriole *Icterus nigrogularis* NT
Jamaican Oriole *Icterus leucopteryx* NT
Orange Oriole *Icterus auratus* NT
Yellow-tailed Oriole *Icterus mesomelas* NT
Orange-crowned Oriole *Icterus auricapillus* NT
White-edged Oriole *Icterus graceannae* NT
Yellow-throated Oriole *Icterus xantholaemus* NT
Spotted-breasted Oriole *Icterus pectoralis* NT
Lichtenstein's Oriole *Icterus gularis* NA NT
Streak-backed Oriole *Icterus pustulatus* NT
Hooded Oriole *Icterus cucullatus* NA NT
Troupial *Icterus icterus* NT
Northern Oriole *Icterus galbula* NA NT
Orchard Oriole *Icterus spurius* NA NT
Black-cowled Oriole *Icterus dominicensis* NT
Black-vented Oriole *Icterus wagleri* NT
St Lucia Oriole *Icterus laudabilis* NT
Martinique Oriole *Icterus bonana* ● NT
Montserrat Oriole *Icterus oberi* NT
Audubon's Oriole *Icterus graduacauda* NT
Bar-winged Oriole *Icterus maculialatus* NT
Scott's Oriole *Icterus parisorum* NA NT

Jamaican Blackbird *Nesopsar nigerrimus* NT
Saffron-cowled Blackbird *Xanthopsar flavus* ● NT

Oriole Blackbird *Gymnomystax mexicanus* NT

Yellow-headed Blackbird *Xanthocephalus xanthocephalus* NA NT

Yellow-eyed Blackbird *Agelaius xanthophthalmus* NT
Yellow-winged Blackbird *Agelaius thilius* NT
Red-winged Blackbird *Agelaius phoeniceus* NA NT
Tricoloured Blackbird *Agelaius tricolor* NA
Yellow-hooded Blackbird *Agelaius icterocephalus* NT
Tawny-shouldered Blackbird *Agelaius humeralis* NT
Yellow-shouldered Blackbird *Agelaius xanthomus* ● NT
Unicoloured Blackbird *Agelaius cyanopus* NT
Chestnut-capped Blackbird *Agelaius ruficapillus* NT

Bonaparte's Blackbird *Sturnella superciliaris* NT
Red-breasted Blackbird *Sturnella militaris* NT
Peruvian Red-breasted Meadowlark *Sturnella bellicosa* NT
Lesser Red-breasted Meadowlark *Sturnella defilippi* ● NT
Long-tailed Meadowlark *Sturnella loyca* NT
Eastern Meadowlark *Sturnella magna* NA NT
Western Meadowlark *Sturnella neglecta* NA NT

Yellow-rumped Marshbird *Psuedoleistes guirahuro* NT

Brown-yellow Marshbird *Pseudoleistes virescens* NT

Scarlet-headed Blackbird *Amblyramphus holosericeus* NT

Red-bellied Grackle *Hypopyrrhus pyrohypogaster* ● NT

Austral Blackbird *Curaeus curaeus* NT
Forbes's Blackbird *Curaeus forbesi* ● NT

Chopi Blackbird *Gnorimopsar chopi* NT

Bolivian Blackbird *Oreopsar bolivianus* NT

Velvet-fronted Grackle *Lampropsar tanagrinus* NT

Mountain Grackle *Macroagelaius subalaris* NT
Golden-tufted Grackle *Macroagelaius imthurni* NT

Cuban Blackbird *Dives atroviolacea* NT
Melodious Blackbird *Dives dives* NT
Scrub Blackbird *Dives warszewiczi* NT

Great-tailed Grackle *Quiscalus mexicanus* NA NT
Boat-tailed Grackle *Quiscalus major* NA

Nicaraguan Grackle *Quiscalus nicaraguensis* NT
Common Grackle *Quiscalus quiscula* NA
Antillean Grackle *Quiscalus niger* NT
Carib Grackle *Qusicalus lugubris* NT

Rusty Blackbird *Euphagus carolinus* NA
Brewer's Blackbird *Euphagus cyanocephalus* NA NT

Bay-winged Cowbird *Molothrus badius* NT
Screaming Cowbird *Molothrus rufoaxillaris* NT
Shiny Cowbird *Molothrus bonariensis* NT
Bronzed Cowbird *Molothrus aeneus* NA NT
Brown-headed Cowbird *Molothrus ater* NA NT

Giant Cowbird *Scaphidura oryzivora* NT

Dolichonychinae

Bobolink *Dolichonyx oryzivorus* NA NT

161 FINCHES Fringillidae

Fringillinae

Chaffinch *Fringilla coelebs* WP EP
Blue Chaffinch *Fringilla teydea* ● WP
Brambling *Fringilla montifringilla* WP EP OR

Carduelinae

Red-fronted Serin *Serinus pusillus* WP EP
Serin *Serinus serinus* WP
Syrian Serin *Serinus syriacus* WP
Island Canary *Serinus canaria* WP
Citril Finch *Serinus citrinella* WP
Tibetan Siskin *Serinus thibetanus* EP OR
Yellow-crowned Canary *Serinus canicollis* AF
Black-headed Siskin *Serinus nigriceps* AF
African Citril Finch *Serinus citrinelloides* AF
Black-faced Canary *Serinus capistratus* AF
Papyrus Canary *Serinus koliensis* AF
Forest Canary *Serinus scotops* AF
White-rumped Seedeater *Serinus leucopygius* AF
Yellow-rumped Seedeater *Serinus atrogularis* WP AF
Lemon-breasted Seedeater *Serinus citrinipectus* AF
Yellow-fronted Canary *Serinus mozambicus* AF
Grosbeak Canary *Serinus donaldsoni* AF
Yellow Canary *Serinus flaviventris* AF
White-bellied Canary *Serinus dorsostriatus* AF
Brimstone Canary *Serinus sulphuratus* AF
White-throated Seedeater *Serinus albogularis* AF
Stripe-breasted Seedeater *Serinus reichardi* AF
Streaky-headed Seedeater *Serinus gularis* AF
Black-eared Seedeater *Serinus mennelli* AF
Brown-rumped Seedeater *Serinus tristriatus* AF
Ankober Serin *Serinus ankoberensis* ● AF
Yemen Serin *Serinus menachensis* WP
Streaky Seedeater *Serinus striolatus* AF
Thick-billed Seedeater *Serinus burtoni* AF
Principe Seedeater *Serinus rufobrunneus* AF
White-winged Seedeater *Serinus leucopterus* AF
Cape Siskin *Serinus tottus* AF
Drakensburg Sisken *Serinus symonsi* AF
Black-headed Canary *Serinus alario* AF
Malay Goldfinch *Serinus estherae* OR AU

Grosbeak-Weaver *Neospiza concolor* AF

Oriole-Finch *Linurgus olivaceus* AF

Golden-winged Grosbeak *Rhynchostruthus socotranus* AF

Greenfinch *Carduelis chloris* WP
Oriental Greenfinch *Carduelis sinica* EP
Black-headed Greenfinch *Carduelis spinoides* OR
Yunnan Greenfinch *Carduelis ambigua* EP OR
Siskin *Carduelis spinus* WP EP
Pine Siskin *Carduelis pinus* NA NT
Black-capped Siskin *Carduelis atriceps* NT
Andean Siskin *Carduelis spinescens* NT
Yellow-faced Siskin *Carduelis yarrellii* ● NT
Red Siskin *Carduelis cucullata* NT
Thick-billed Siskin *Carduelis crassirostris* NT
Hooded Siskin *Carduelis magellanica* NT
Antillean Siskin *Carduelis dominicensis* NT
Saffron Siskin *Carduelis siemiradzkii* ● NT
Olivaceous Siskin *Carduelis olivacea* NT
Black-headed Siskin *Carduelis notata* NT

Yellow-bellied Siskin *Carduelis xanthogastra* NT
Black Siskin *Carduelis atrata* NT
Yellow-rumped Siskin *Carduelis uropygialis* NT
Black-chinned Siskin *Carduelis barbata* NT
American Goldfinch *Carduelis tristis* NA NT
Dark-backed Goldfinch *Carduelis psaltria* NA NT
Lawrence's Goldfinch *Carduelis lawrencei* NA NT
Goldfinch *Carduelis carduelis* WP EP OR

Redpoll *Acanthis flammea* WP EP NA
Twite *Acanthis flavirostris* WP EP OR
Linnet *Acanthis cannabina* WP OR
Yemeni Linnet *Acanthis yemensis* WP
Warsangli Linnet *Acanthis johannis* ● AF

Hodgson's Rosy Finch *Leucosticte nemoricola* EP OR
Brandt's Rosy Finch *Leucosticte brandti* EP
Rosy Finch *Leucosticte arctoa* EP NA

Red-browed Rosefinch *Callacanthis burtoni* EP

Crimson-winged Finch *Rhodopechys sanguinea* WP
Trumpeter Finch *Rhodopechys githaginea* WP AF OR
Mongolian Trumpeter Finch *Rhodopechys mongolica* EP OR
Lichtenstein's Desert Finch *Rhodopechys obsoleta* WP

Long-tailed Rosefinch *Uragus sibiricus* EP

Przewalski's Rosefinch *Urocynchramus pylzowi* EP

Blanford's Rosefinch *Carpodacus rubescens* EP
Dark Rosefinch *Carpodacus nipalensis* EP OR
Common Rosefinch *Carpodacus erythrinus* WP EP OR
Purple Finch *Carpodacus purpureus* NA NT
Cassin's Finch *Carpodacus cassinii* NA NT
House Finch *Carpodacus mexicanus* NA NT
Beautiful Rosefinch *Carpodacus pulcherrimus* EP
Stresemann's Rosefinch *Carpodacus eos* EP
Pink-browed Rosefinch *Carpodacus rhodochrous* EP
Vinaceous Rosefinch *Carpodacus vinaceus* EP
Large Rosefinch *Carpodacus edwardsii* EP OR
Sinai Rosefinch *Carpodacus synoicus* WP EP
Pallas's Rosefinch *Carpodacus roseus* EP
Three-banded Rosefinch *Carpodacus trifasciatus* EP
Spot-winged Rosefinch *Carpodacus rhodopeplus* EP OR
White-browed Rosefinch *Carpodacus thura* EP
Red-mantled Rosefinch *Carpodacus rhodochlamys* EP
Eastern Great Rosefinch *Carpodacus rubicilloides* EP
Caucasian Great Rosefinch *Carpodacus rubicilla* WP EP
Rose-breasted Rosefinch *Carpodacus puniceus* EP
Tibet Rosefinch *Carpodacus roborowskii* EP

Pine Grosbeak *Pinicola enucleator* WP EP NA
Red-headed Finch *Pinicola subhimachalus* EP

Scarlet Finch *Haematospiza sipahi* EP OR

Parrot Crossbill *Loxia pytyopsittacus* WP EP
Scottish Crossbill *Loxia scotica* WP
Crossbill *Loxia curvirostra* WP EP OR NA NT
White-winged Crossbill *Loxia leucoptera* WP EP NA NT

Brown Bullfinch *Pyrrhula nipalensis* EP OR
Philippine Bullfinch *Pyrrhula leucogenys* OR
Orange Bullfinch *Pyrrhula aurantiaca* EP
Red-headed Bullfinch *Pyrrhula erythrocephala* EP
Beavan's Bullfinch *Pyrrhula erythaca* EP
Bullfinch *Pyrrhula pyrrhula* WP EP

Hawfinch *Coccothraustes coccothraustes* WP EP
Black-tailed Hawfinch *Coccothraustes migratorius* EP
Masked Hawfinch *Coccothraustes personata* EP
Black & Yellow Grosbeak *Coccothraustes icterioides* EP OR
Allied Grosbeak *Coccothraustes affinis* EP OR
Spotted-wing Grosbeak *Coccothraustes melanozanthos* EP OR
White-winged Grosbeak *Coccothraustes carnipes* WP EP
Evening Grosbeak *Coccothraustes vespertinus* NA NT
Hooded Grosbeak *Coccothraustes abeillei* NT

Gold-headed Finch *Pyrrhoplectes epauletta* EP OR

162 WAXBILLS Estrildidae

Flowerpecker Weaver Finch *Parmoptila woodhousei* AF

Red-fronted Flowerpecker Weaver Finch *Parmoptila jamesoni* AF

White-breasted Negro Finch *Nigrita fusconota* AF
Chestnut-breasted Negro Finch *Nigrita bicolor* AF
Pale-fronted Negro Finch *Nigrita luteifrons* AF
Grey-crowned Negro Finch *Nigrita canicapilla* AF

Fernando Po Olive-back *Nesocharis shelleyi* AF
White-collared Olive-back *Nesocharis ansorgei* AF
Grey-headed Olive-back *Nesocharis capistrata* AF

Crimson-winged Pytilia (Aurora Finch) *Pytilia phoenicoptera* AF
Red-faced Pytilia *Pytilia hypogrammica* AF
Orange-winged Pytilia *Pytilia afra* AF
Green-winged Pytilia *Pytilia melba* AF

Green-backed Twin-spot *Mandingoa nitidula* AF

Red-faced Crimson-wing *Cryptospiza reichenovii* AF
Ethiopian Crimson-wing *Cryptospiza salvadorii* AF
Dusky Crimson-wing *Cryptospiza jacksoni* AF
Shelley's Crimson-wing *Cryptospiza shelleyi* AF

Crimson Seedcracker *Pyrenestes sanguineus* AF
Black-bellied Seedcracker *Pyrenestes ostrinus* AF
Lesser Seedcracker *Pyrenestes minor* AF

Grant's Bluebill *Spermophaga poliogenys* AF
Western Bluebill *Spermophaga haematina* AF
Red-headed Bluebill *Spermophaga ruficapilla* AF

Brown Twin-spot *Clytospiza monteiri* AF

Rosy Twin-spot *Hypargos margaritatus* AF
Peter's Twin-spot *Hypargos niveoguttatus* AF

Dybowski's Dusky Twin-spot *Euschistospiza dybowskii* AF
Dusky Twin-spot *Euschistospiza cinereovinacea* AF

Black-bellied Firefinch *Lagonosticta rara* AF
Bar-breasted Firefinch *Lagonosticta rufopicta* AF
Brown Firefinch *Lagonosticta nitidula* AF
Red-billed Firefinch *Lagonosticta senegala* AF
Kuli Koro Firefinch *Lagonosticta virata* AF
African Firefinch *Lagonosticta rubricata* AF
Pale-billed Firefinch *Lagonosticta landanae* AF
Jameson's Firefinch *Lagonosticta rhodopareia* AF
Masked Firefinch *Lagonosticta larvata* AF
Vinaceous Firefinch *Lagonosticta vinacea* AF

Cordon-bleu *Uraeginthus angolensis* AF
Red-cheeked Cordon-bleu *Uraeginthus bengalus* AF
Blue-capped Cordon-bleu *Uraeginthus cyanocephala* AF
Common Grenadier *Uraeginthus granatina* AF
Purple Grenadier *Uraeginthus ianthinogaster* AF

Lavender Waxbill *Estrilda caerulescens* AF
Black-tailed Waxbill *Estrilda perreini* AF
Cinderella Waxbill *Estrilda thomensis* AF
Swee Waxbill *Estrilda melanotis* AF
East African Swee Waxbill *Estrilda quartinia* AF
Anambra Waxbill *Estrilda poliopareia* ● AF
Fawn-breasted Waxbill *Estrilda paludicola* AF
Orange-cheeked Waxbill *Estrilda melpoda* AF
Crimson-rumped Waxbill *Estrilda rhodopyga* AF
Arabian Waxbill *Estrilda rufibarba* WP
Black-rumped Waxbill *Estrilda troglodytes* AF
Common Waxbill *Estrilda astrild* AF
Black-faced Waxbill *Estrilda nigriloris* ● AF
Black-crowned Waxbill *Estrilda nonnula* AF
Black-headed Waxbill *Estrilda atricapilla* AF
Black-cheeked Waxbill *Estrilda erythronotos* AF
Red-rumped Waxbill *Estrilda charmosyna* AF

Red Munia *Amandava amandava* OR AU
Green Munia *Amandava formosa* ● OR
Zebra Waxbill *Amandava subflava* AF

African Quailfinch *Ortygospiza atricollis* AF
Red-billed Quailfinch *Ortygospiza gabonensis* AF
Locust Finch *Ortygospiza locustella* AF

Red-browed Waxbill *Aegintha temporalis* AU

Painted Finch *Emblema picta* AU
Beautiful Firetail Finch *Emblema bella* AU
Red-eared Firetail Finch *Emblema oculata* AU
Diamond Firetail Finch *Emblema guttata* AU

Crimson-sided Mountain Finch *Oreostruthus fuliginosus* AU

Crimson Finch *Neochmia phaeton* AU
Star Finch *Neochmia ruficauda* AU

(Spotted-sided) Zebra-Finch *Poephila guttata* AU
Double-barred Finch *Poephila bichenovii* AU
Masked Finch *Poephila personata* AU
Long-tailed Finch *Poephila acuticauda* AU
Black-throated Finch *Poephila cincta* AU

Bamboo Parrot-Finch *Erythrura hyperythra* OR AU
Pin-tailed Parrot-Finch *Erythrura prasina* OR
Green-faced Parrot-Finch *Erythrura viridifacies* ● OR
Three-coloured Parrot-Finch *Erythrura tricolor* AU
Mount Katanglad Parrot-Finch *Erythrura coloria* ● OR
Blue-faced Parrot-Finch *Erythrura trichroa* AU PA
Papuan Parrot-Finch *Erythrura papuana* AU
Red-throated Parrot-Finch *Erythrura psittacea* PA
Fiji Parrot-Finch *Erythrura pealii* PA
Red-headed Parrot-Finch *Erythrura cyaneovirens* PA
Pink-billed Parrot-Finch *Erythrura kleinschmidti* ● PA

Gouldian Finch *Chloebia gouldiae* AU

Plum-headed Finch *Aidemosyne modesta* AU

Bib-Finch *Lepidopygia nana* AF

African Silverbill *Lonchura cantans* AF
Indian Silverbill *Lonchura malabarica* WP OR
Grey-headed Silverbill *Lonchura griseicapilla* AF
Bronze Mannikin *Lonchura cucullata* AF
Black & White Mannikin *Lonchura bicolor* AF
Magpie Mannikin *Lonchura fringilloides* AF
White-backed Munia *Lonchura striata* EP OR
Javanese Mannikin *Lonchura leucogastroides* OR AU
Dusky Mannikin *Lonchura fuscans* OR
Moluccan Mannikin *Lonchura molucca* AU
Nutmeg Mannikin *Lonchura punctulata* EP OR AU
Rufous-bellied Mannikin *Lonchura kelaarti* OR
White-headed Munia *Lonchura leucogastra* OR
Streak-headed Mannikin *Lonchura tristissima* AU
White-spotted Mannikin *Lonchura leucosticta* AU
Coloured Finch *Lonchura quinticolor* AU
Chestnut Mannikin *Lonchura malacca* OR AU
Pale-headed Mannikin *Lonchura maja* OR
Pallid Finch *Lonchura pallida* AU
Great-billed Mannikin *Lonchura grandis* AU
Arfak Mannikin *Lonchura vana* AU
Grey-headed Mannikin *Lonchura caniceps* AU
White-crowned Mannikin *Lonchura nevermanni* AU
New Britain Mannikin *Lonchura spectabilis* AU
New Ireland Finch *Lonchura forbesi* AU
White-headed Finch *Lonchura hunsteini* AU PA
Yellow-tailed Mannikin *Lonchura flaviprymna* AU
Chestnut-breasted Mannikin *Lonchura castaneothorax* AU
Black Mannikin *Lonchura stygia* AU
Grand Valley Mannikin *Lonchura teerinki* AU
Alpine Mannikin *Lonchura monticola* AU
Snow Mountain Mannikin *Lonchura montana* AU
New Britain Finch *Lonchura melaena* AU
Pictorella Finch *Lonchura pectoralis* AU

Timor Dusky Sparrow *Padda fuscata* AU
Java Sparrow *Padda oryzivora* OR AU

Paradise Sparrow *Amadina erythrocephala* AF
Cut-throat Weaver *Amadina fasciata* AF

Tit-Hylia *Pholidornis rushiae* AF

163 WEAVERS AND SPARROWS Ploceidae

Viduinae

Village Indigobird *Vidua chalybeata* AF
Dusky Indigobird *Vidua purpurascens* AF
Variable Indigobird *Vidua funerea* AF
Pale-winged Indigobird *Vidua wilsoni* AF
Steel-blue Whydah *Vidua hypocherina* AF
Fischer's Whydah *Vidua fischeri* AF
Shaft-tailed Whydah *Vidua regia* AF
Pin-tailed Whydah *Vidua macroura* AF
Paradise Whydah *Vidua paradisaea* AF

Broad-tailed Paradise Whydah *Vidua orientalis* AF

Bublaornithinae

White-billed Buffalo Weaver *Bubalornis albirostris* AF
Red-billed Buffalo Weaver *Bubalornis niger* AF
White-headed Buffalo Weaver *Dinemellia dinemelli* AF

Passerinae

White-browed Sparrow Weaver *Plocepasser mahali* AF
Chestnut-crowned Sparrow Weaver *Plocepasser superciliosus* AF
Donaldson-Smith's Sparrow Weaver *Plocepasser donaldsoni* AF
Chestnut-mantled Sparrow Weaver *Plocepasser rufoscapulatus* AF

Rufous-tailed Weaver *Histurgops ruficauda* AF

Grey-headed Social Weaver *Pseudonigrita arnaudi* AF
Black-capped Social Weaver *Pseudonigrita cabanisi* AF

Sociable Weaver *Philetairus socius* AF

Saxaul Sparrow *Passer ammodendri* EP
House Sparrow *Passer domesticus* WP EP AF OR AU NA NT
Spanish Sparrow *Passer hispaniolensis* WP EP OR
Sind Jungle Sparrow *Passer pyrrhonotus* WP OR
Somali Sparrow *Passer castanopterus* AF
Cinnamon Sparrow *Passer rutilans* EP OR
Pegu House Sparrow *Passer flaveolus* OR
Dead Sea Sparrow *Passer moabiticus* WP AF
Great Sparrow *Passer motitensis* WP AF
Cape Sparrow *Passer melanurus* AF
Socotra Sparrow *Passer insularis* AF
Kenya Rufous Sparrow *Passer rufocinctus* AF
Grey-headed Sparrow *Passer griseus* AF
Swainson's Sparrow *Passer swainsonii* AF
Parrot-billed Sparrow *Passer gongonensis* AF
Swahili Sparrow *Passer suahelicus* AF
Desert Sparrow *Passer simplex* WP AF
Tree Sparrow *Passer montanus* WP EP OR

Sudan Golden Sparrow *Auripasser luteus* AF
Arabian Golden Sparrow *Auripasser euchlorus* AF

Chestnut Sparrow *Sorella eminibey* AF
Pale Rock Sparrow *Petronia brachydactyla* WP
Yellow-spotted Rock Sparrow *Petronia xanthosterna* AF
Streaked Rock Sparrow *Petronia petronia* WP EP
South African Rock Sparrow *Petronia superciliaris* AF
Lesser Rock Sparrow *Petronia dentata* AF

White-winged Snow Finch *Montifringilla nivalis* WP EP
Adams' Snow Finch *Montifringilla adamsi* EP
Mandelli's Snow Finch *Montifringilla taczanowskii* EP
Pere David's Snow Finch *Montifringilla davidiana* EP
Red-necked Snow Finch *Montifringilla ruficollis* EP
Blanford's Snow Finch *Montifringilla blanfordi* EP
Meinertzhagen's Snow Finch *Montifringilla theresae* EP

Scaly Weaver *Sporopipes squamifrons* AF
Speckle-fronted Weaver *Sporopipes frontalis* AF

Ploceinae

Grosbeak Weaver *Amblyospiza albifrons* AF

Baglafecht Weaver *Ploceus baglafecht* AF
Bannerman's Weaver *Ploceus bannermani* ● AF
Bates's Weaver *Ploceus batesi* AF
Black-chinned Weaver *Ploceus nigrimentum* ● AF
Bertrand's Weaver *Ploceus bertrandi* AF
Slender-billed Weaver *Ploceus pelzelni* AF
Loanga Slender-billed Weaver *Ploceus subpersonatus* ● AF
Little Masked Weaver *Ploceus luteolus* AF
Spectacled Weaver *Ploceus ocularis* AF
Black-necked Weaver *Ploceus nigricollis* AF

Strange Weaver *Ploceus alienus* AF
Black-billed Weaver *Ploceus melanogaster* AF
Cape Weaver *Ploceus capensis* AF
Bocage's Weaver *Ploceus temporalis* AF
Golden Weaver *Ploceus subaureus* AF
Kilombero Weaver *Ploceus burnieri* AF
Holub's Golden Weaver *Ploceus xanthops* AF
Orange Weaver *Ploceus aurantius* AF
Heuglin's Masked Weaver *Ploceus heuglini* AF
Golden Palm Weaver *Ploceus bojeri* AF

Parasitic Weaver *Anomalospiza imberbis* AF

Taveta Golden Weaver *Ploceus castaneiceps* AF
Proncipe Golden Weaver *Ploceus princeps* AF
Brown-throated Golden Weaver *Ploceus xanthopterus* AF
Northern Brown-throated Weaver *Ploceus castanops* AF
Rüppell's Weaver *Ploceus galbula* AF
Lake Victoria Weaver *Ploceus victoriae* ● AF
Northern Masked Weaver *Ploceus taeniopterus* AF
Lesser Masked Weaver *Ploceus intermedius* AF
African Masked Weaver *Ploceus velatus* AF
Lufira Masked Weaver *Ploceus ruweti* AF
Katanga Masked Weaver *Ploceus katangae* AF
Tanzanian Masked Weaver *Ploceus reichardi* AF
Vitelline Masked Weaver *Ploceus vitellinus* AF
Speke's Weaver *Ploceus spekei* AF
Fox's Weaver *Ploceus spekeoides* AF
Village Weaver *Ploceus cucullatus* AF
Layard's Black-headed Weaver *Ploceus nigriceps* AF
Giant Weaver *Ploceus grandis* AF
Vieillot's Black Weaver *Ploceus nigerrimus* AF
Weyns's Weaver *Ploceus weynsi* AF
Clarke's Weaver *Ploceus golandi* ● AF
Salvadori's Weaver *Ploceus dicrocephalus* AF
Black-headed Weaver *Ploceus melanocephalus* AF
Golden-backed Weaver *Ploceus jacksoni* AF
Cinnamon Weaver *Ploceus badius* AF
Chestnut Weaver *Ploceus rubiginosus* AF
Gold-naped Weaver *Ploceus aureonucha* ● AF
Yellow-mantled Weaver *Ploceus tricolor* AF
Maxwell's Black Weaver *Ploceus albinucha* AF
Nelicourvi Weaver *Ploceus nelicourvi* AF
Asian Golden Weaver *Ploceus hypoxanthus* OR
Compact Weaver *Ploceus superciliosus* AF
Bengal Weaver *Ploceus benghalensis* OR
Streaked Weaver *Ploceus manyar* OR
Baya Weaver *Ploceus philippinus* OR
Finn's Weaver *Ploceus megarhynchus* ● OR
Forest Weaver *Ploceus bicolor* AF
Golden-backed Weaver *Ploceus preussi* AF
Yellow-capped Weaver *Ploceus dorsomaculatus* AF
Olive-headed Golden Weaver *Ploceus olivaceiceps* AF
Usambara Weaver *Ploceus nicolli* ● AF
Brown-capped Weaver *Ploceus insignis* AF
Bar-winged Weaver *Ploceus angolensis* AF
Sao Thome Weaver *Ploceus sanctaethomae* AF

Yellow-legged Malimbe *Malimbus flavipes* ● AF
Red-crowned Malimbe *Malimbus coronatus* AF
Black-throated Malimbe *Malimbus cassini* AF
Rachel's Malimbe *Malimbus racheliae* AF
Tai Malimbe *Malimbus ballmani* AF
Red-vented Malimbe *Malimbus scutatus* AF
Ibadan Malimbe *Malimbus ibadanensis* ● AF
Gray's Malimbe *Malimbus nitens* AF
Red-headed Malimbe *Malimbus rubricollis* AF
Red-bellied Malimbe *Malimbus erythrogaster* AF
Crested Malimbe *Malimbus malimbicus* AF

Red-headed Weaver *Anaplectes melanotis* AF

Cardinal Quelea *Quelea cardinalis* AF
Red-headed Quelea *Quelea erythrops* AF
Red-billed Quelea *Quelea quelea* AF

Madagascan Red Fody *Foudia madagascariensis* AF
Mascarene Fody *Foudia eminentissima* AF
Red Forest Fody *Foudia omissa* AF
Mauritius Fody *Foudia rubra* ● AF
Seychelles Fody *Foudia sechellarum* ● AF
Rodriguez Fody *Foudia flavicans* ● AF
Sakalava Fody *Foudia sakalava* AF

Bob-tailed Weaver *Brachycope anomala* AF

Golden Bishop *Euplectes afer* AF

Fire-fronted Bishop *Euplectes diademata* AF
Gierow's Bishop *Euplectes gierowii* AF
Zanzibar Red Bishop *Euplectes nigroventris* AF
Red-crowned Bishop *Euplectes hordeacea* AF
Red Bishop *Euplectes orix* AF
Golden-backed Bishop *Euplectes aurea* AF
Yellow-rumped Bishop *Euplectes capensis* AF
Fan-tailed Whydah *Euplectes axillaris* AF
Yellow-mantled Whydah *Euplectes macrourus* AF
Marsh Whydah *Euplectes hartlaubi* AF
Mountain Marsh Whydah *Euplectes psammocromius* AF
White-winged Whydah *Euplectes albonotatus* AF
Red-collared Whydah *Euplectes ardens* AF
Long-tailed Whydah *Euplectes progne* AF
Jackson's Whydah *Euplectes jacksoni* AF

164 STARLINGS Sturnidae

Sturninae

New Hebrides Starling *Aplonis zelandica* PA
Mountain Starling *Aplonis santovestris* ● PA
Ponape Starling *Aplonis pelzelni* ● PA
Samoan Starling *Aplonis atrifusca* PA
Rarotonga Starling *Aplonis cinerascens* ● PA
Striped Starling *Aplonis tabuensis* PA
Striated Starling *Aplonis striata* PA
Norfolk I Starling *Aplonis fusca* PA
Micronesian Starling *Aplonis opaca* PA
Singing Starling *Aplonis cantoroides* AU
Tanimbar Starling *Aplonis crassa* AU
Fead Is Starling *Aplonis feadensis* AU PA
Rennell I Starling *Aplonis insularis* PA
San Cristobal Starling *Aplonis dichroa* PA
Large Glossy Starling *Aplonis grandis* AU
Moluccan Starling *Aplonis mysolensis* AU
Long-tailed Starling *Aplonis magna* AU
Short-tailed Starling *Aplonis minor* OR AU
Philippine Glossy Starling *Aplonis panayensis* OR AU
Shining Starling *Aplonis metallica* AU
Grant's Starling *Aplonis mystacea* AU
White-eyed Starling *Aplonis brunneicapilla* PA

Kenrick's Starling *Poeoptera kenricki* AF
Stuhlmann's Starling *Poeoptera stuhlmanni* AF
Narrow-tailed Starling *Poeoptera lugubris* AF

White-collared Starling *Grafisia torquata* AF

Waller's Red-winged Starling *Onychognathus walleri* AF
Pale-winged Starling *Onychognathus nabouroup* AF
African Red-winged Starling *Onychognathus morio* AF
Somali Chestnut-winged Starling *Onychognathus blythii* AF
Socotra Chestnut-winged Starling *Onychognathus frater* AF
Tristram's Starling *Onychognathus tristramii* WP
Chestnut-winged Starling *Onychognathus fulgidus* AF
Slender-billed Red-winged Starling *Onychognathus tenuirostris* AF
White-billed Starling *Onychognathus albirostris* AF
Bristle-crowned Starling *Onychognathus salvadorii* AF

Iris Glossy Starling *Lamprotornis iris* AF
Copper-tailed Glossy Starling *Lamprotornis cupreocauda* AF
Purple-headed Glossy Starling *Lamprotornis purpureiceps* AF
Superb Starling *Lamprotornis superbus* AF
Black-bellied Glossy Starling *Lamprotornis corruscus* AF
Purple Glossy Starling *Lamprotornis purpureus* AF
Red-shouldered Glossy Starling *Lamprotornis nitens* AF
Bronze-tailed Glossy Starling *Lamprotornis chalcurus* AF
Greater Blue-eared Glossy Starling *Lamprotornis chalybaeus* AF
Lesser Blue-eared Glossy Starling *Lamprotornis chloropterus* AF
Sharp-tailed Glossy Starling *Lamprotornis acuticaudus* AF
Splendid Glossy Starling *Lamprotornis splendidus* AF

Principe Glossy Starling *Lamprotornis ornatus* AF
Burchell's Starling *Lamprotornis australis* AF
Long-tailed Purple Starling *Lamprotornis mevesii* AF
Ruppell's Long-tailed Glossy Starling *Lamprotornis purpuropterus* AF
Long-tailed Purple Starling *Lamprotornis caudatus* AF

Abbott's Starling *Cinnyricinclus femoralis* ● AF
Sharpe's Starling *Cinnyricinclus sharpii* AF
Violet Starling *Cinnyricinclus leucogaster* AF

Magpie Starling *Speculipastor bicolor* AF

White-winged Starling *Neocichla gutturalis* AF

Fischer's Starling *Spreo fischeri* AF
Pied Starling *Spreo bicolor* AF
White-crowned Starling *Spreo albicapillus* AF
Chestnut-bellied Starling *Spreo pulcher* AF
Hildebrandt's Starling *Spreo hildebrandti* AF
Shelley's Starling *Spreo shelleyi* AF

Golden-breasted Starling *Cosmopsarus regius* AF
Ashy Starling *Cosmopsarus unicolor* AF

Madagascar Starling *Saroglossa aurata* AF
Spot-winged Starling *Saroglossa spiloptera* EP OR

Wattled Starling *Creatophora cinerea* AF

Ceylon White-headed Starling *Sturnus senex* OR
Ashy-headed Starling *Sturnus malabaricus* OR
White-headed Starling *Sturnus erythropygius* OR
Black-headed Starling *Sturnus pagodarum* OR
Silky Starling *Sturnus sericeus* EP
Violet-backed Starling *Sturnus philippensis* EP OR AU
Daurian Starling *Sturnus sturninus* EP OR
Rose-coloured Starling *Sturnus roseus* WP EP OR
Starling *Sturnus vulgaris* WP EP OR NA
Spotless Starling *Sturnus unicolor* WP
Grey Starling *Sturnus cineraceus* EP
Asian Pied Starling *Sturnus contra* OR
Black-collared Starling *Sturnus nigricollis* EP OR
Jerdon's Starling *Sturnus burmannicus* OR
Black-winged Starling *Sturnus melanopterus* OR AU
Chinese Starling *Sturnus sinensis* EP OR

Rothschild's Mynah *Leucopsar rothschildi* ● OR

Common Mynah *Acridotheres tristis* OR
Bank Mynah *Acridotheres ginginianus* OR
Indian Jungle Mynah *Acridotheres fuscus* OR AU
Great Mynah *Acridotheres grandis* OR
White-collared Mynah *Acridotheres albocinctus* OR
Chinese Jungle Mynah *Acridotheres cristatellus* EP OR

Gold-crested Mynah *Ampeliceps coronatus* OR

Golden-breasted Mynah *Mino anais* AU
Yellow-faced Mynah *Mino dumontii* AU PA

Sulawesi King Starling *Basilornis celebensis* AU
Greater King Starling *Basilornis galeatus* ● AU
Seram King Starling *Basilornis corythaix* AU
Mount Apo King Starling *Basilornis miranda* OR

Sulawesi Magpie *Streptocitta albicollis* AU
Sula Magpie *Streptocitta albertinae* ● AU

Bald Starling *Sarcops calvus* OR

Ceylon Grackle *Gracula ptilogenys* OR
Southern Grackle *Gracula religiosa* OR

Fiery-browed Enodes Starling *Enodes erythrophris* AU

Grosbeak Starling *Scissirostrum dubium* AU

Buphaginae

Yellow-billed Oxpecker *Buphagus africanus* AF
Red-billed Oxpecker *Buphagus erythrorhynchus* AF

165 ORIOLES Oriolidae

Brown Oriole *Oriolus szalayi* AU
Dusky Brown Oriole *Oriolus phaeochromus* AU
Grey-collared Oriole *Oriolus forsteni* AU
Black-eared Oriole *Oriolus bouroensis* AU
Olive Brown Oriole *Oriolus melanotis* AU
Olive-backed Oriole *Oriolus sagittatus* AU

Australian Yellow Oriole *Oriolus flavocinctus* AU
Dark-throated Oriole *Oriolus xanthonotus* OR
White-lored Oriole *Oriolus albiloris* OR
Isabella Oriole *Oriolus isabellae* ● OR
Golden Oriole *Oriolus oriolus* WP EP AF OR
African Golden Oriole *Oriolus auratus* AF
Black-naped Oriole *Oriolus chinensis* OR AU
Green-headed Oriole *Oriolus chlorocephalus* AF
Sao Thome Oriole *Oriolus crassirostris* AF
Western Black-headed Oriole *Oriolus brachyrhynchus* AF
Dark-headed Oriole *Oriolus monacha* AF
Montane Oriole *Oriolus percivali* AF
African Black-headed Oriole *Oriolus larvatus* AF
Black-winged Oriole *Oriolus nigripennis* AF
Asian Black-headed Oriole *Oriolus xanthornus* OR
Black Oriole *Oriolus hosii* OR
Crimson-breasted Oriole *Oriolus cruentus* OR
Maroon Oriole *Oriolus traillii* OR
Stresemann's Maroon Oriole *Oriolus mellianus* ● EP

Figbird *Sphecotheres viridis* AU

166 DRONGOS Dicruridae

Papuan Mountain Drongo *Chaetorhynchus papuensis* AU

Square-tailed Drongo *Dicrurus ludwigii* AF
Shining Drongo *Dicrurus atripennis* AF
Fork-tailed Drongo *Dicrurus adsimilis* AF
Comoro Drongo *Dicrurus fuscipennis* ● AF
Aldabra Drongo *Dicrurus aldabranus* AF
Crested Drongo *Dicrurus forficatus* AF
Mayotte Drongo *Dicrurus waldenii* ● AF
Black Drongo *Dicrurus macrocercus* EP OR
Ashy Drongo *Dicrurus leucophaeus* EP OR AU
White-bellied Drongo *Dicrurus caerulescens* OR
Crow-billed Drongo *Dicrurus annectans* OR
Bronzed Drongo *Dicrurus aeneus* OR
Lesser Racquet-tailed Drongo *Dicrurus remifer* OR
Balicassio Drongo *Dicrurus balicassius* OR
Spangled Drongo *Dicrurus bracteatus* OR AU
New Ireland Drongo *Dicrurus megarhynchus* AU
Wallacean Drongo *Dicrurus densus* AU
Sumatran Drongo *Dicrurus sumatranus* OR
Hair-crested Drongo *Dicrurus hottentottus* EP OR AU
Andaman Drongo *Dicrurus andamanensis* OR
Greater Racquet-tailed Drongo *Dicrurus paradiseus* OR

167 WATTLEBIRDS Callaeidae

Kokako *Callaeas cinerea* ● AU

Saddleback *Creadion carunculatus* ● AU

168 MAGPIE LARKS Grallinidae

Grallininae

Magpie Lark *Grallina cyanoleuca* AU
Torrent Lark *Grallina bruijni* AU

Corcoracinae

White-winged Chough *Corcorax melanorhamphos* AU

Apostle Bird *Struthidea cinerea* AU

169 WOODSWALLOWS Artamidae

Ashy Woodswallow *Artamus fuscus* OR
White-breasted Woodswallow *Artamus leucorhynchus* OR AU PA
White-backed Woodswallow *Artamus monachus* AU
Papuan Woodswallow *Artamus maximus* AU
Bismarck Woodswallow *Artamus insignis* AU
Masked Woodswallow *Artamus personatus* AU
White-browed Woodswallow *Artamus superciliosus* AU
Black-faced Woodswallow *Artamus cinereus* AU
Dusky Woodswallow *Artamus cyanopterus* AU
Little Woodswallow *Artamus minor* AU

170 BUTCHER BIRDS Cracticidae

Black-backed Butcher Bird *Cracticus mentalis* AU
Grey Butcher Bird *Cracticus torquatus* AU
Black-throated Butcher Bird *Cracticus nigrogularis* AU

Black-headed Butcher Bird *Cracticus cassicus* AU
White-rumped Butcher Bird *Cracticus louisiadensis* AU
Black Butcher Bird *Cracticus quoyi* AU

Black-backed Magpie *Gymnorhina tibicen* AU

Pied Currawong *Strepera graculina* AU
Black Currawong *Strepera fuliginosa* AU
Grey Currawong *Strepera versicolor* AU

171 BOWERBIRDS Ptilonorhynchidae

White-eared Catbird *Ailuroedus buccoides* AU
Green Catbird *Ailuroedus crassirostris* AU
Spotted Catbird *Ailuroedus melanotis* AU

Tooth-billed Catbird *Scenopoeetes dentirostris* AU

Archbold's Bowerbird *Archboldia papuensis* AU

Vogelkop Gardener Bowerbird *Amblyornis inornatus* AU
Macgregor's Gardener Bowerbird *Amblyornis macgregoriae* AU
Striped Gardener Bowerbird *Amblyornis subularis* AU
Yellow-fronted Gardener Bowerbird *Amblyornis flavifrons* ● AU

Newton's Golden Bowerbird *Prionodura newtoniana* AU

Flamed Bowerbird *Sericulus aureus* AU
Adelbert Bowerbird *Sericulus bakeri* ● AU
Regent Bowerbird *Sericulus chrysocephalus* AU

Satin Bowerbird *Ptilonorhynchus violaceus* AU
Spotted Bowerbird *Chlamydera maculata* AU
Great Grey Bowerbird *Chlamydera nuchalis* AU
Lauterbach's Bowerbird *Chlamydera lauterbachi* AU
Fawn-breasted Bowerbird *Chlamydera cerviniventris* AU

172 BIRDS OF PARADISE Paradisaeidae

Cnemophilinae

Loria's Bird of Paradise *Loria loriae* AU

Wattle-billed Bird of Paradise *Loboparadisea sericea* AU

Sickle-crested Bird of Paradise *Cnemophilus macgregorii* AU

Paradisaeinae

Macgregor's Bird of Paradise *Macgregoria pulchra* AU

Paradise Crow *Lycocorax pyrrhopterus* AU

Glossy-mantled Manucode *Manucodia ater* AU
Jobi Manucode *Manucodia jobiensis* AU
Crinkle-collared Manucode *Manucodia chalybatus* AU
Curl-crested Manucode *Manucodia comrii* AU

Trumpet Bird *Phonygammus keraudrenii* AU

Paradise Riflebird *Ptiloris paradiseus* AU
Queen Victoria Riflebird *Ptiloris victoriae* AU
Magnificent Riflebird *Ptiloris magnificus* AU

Wallace's Standardwing *Semioptera wallacei* AU

Twelve-wired Bird of Paradise *Seleucidis melanoleuca* AU

Long-tailed Paradigalla *Paradigalla carunculata* ● AU
Short-tailed Paradigalla *Paradigalla brevicauda* AU

Black-billed Sicklebill *Drepanornis albertisii* ● AU
Pale-billed Sicklebill *Drepanornis bruijnii* AU

Black Sicklebill *Epimachus fastosus* AU
Brown Sicklebill *Epimachus meyeri* AU

Arfak Bird of Paradise *Astrapia nigra* AU
Splendid Bird of Paradise *Astrapia splendidissima* AU

Ribbon-tailed Bird of Paradise *Astrapia mayeri* ● AU

Princess Stephanie's Bird of Paradise *Astrapia stephaniae* AU

Huon Bird of Paradise *Astrapia rothschildi* AU
Superb Bird of Paradise *Lophorina superba* AU

Arfak Parotia *Parotia sefilata* AU
Queen Carola's Parotia *Parotia carolae* AU
Lawes' Parotia *Parotia lawesii* AU
Wahnes' Parotia *Parotia wahnesi* ● AU

King of Saxony Bird of Paradise *Pteridophora alberti* AU

King Bird of Paradise *Cicinnurus regius* AU

Magnificent Bird of Paradise *Diphyllodes magnificus* AU
Wilson's Bird of Paradise *Diphyllodes respublica* AU

Greater Bird of Paradise *Paradisaea apoda* AU
Raggiana Bird of Paradise *Paradisaea raggiana* AU
Lesser Bird of Paradise *Paradisaea minor* AU
Goldie's Bird of Paradise *Paradisaea decora* ● AU
Red Bird of Paradise *Paradisaea rubra* AU
Emperor of Germany Bird of Paradise *Paradisaea guilielmi* AU
Blue Bird of Paradise *Paradisaea rudolphi* AU

173 CROWS AND JAYS Corvidae

Crested Shrike-jay *Platylophus galericulatus* OR

White-winged Magpie *Platysmurus leucopterus* OR

Pinyon Jay *Gymnorhinus cyanocephala* NA NT

Blue Jay *Cyanocitta cristata* NA
Steller's Jay *Cyanocitta stelleri* NA NT

Scrub Jay *Aphelocoma coerulescens* NA NT
Mexican Jay *Aphelocoma ultramarina* NA NT
Unicoloured Jay *Aphelocoma unicolor* NT

White-collared Jay *Cyanolyca viridicyana* NT
Collared Jay *Cyanolyca armillata* NT
Turquoise Jay *Cyanolyca turcosa* NT
Beautiful Jay *Cyanolyca pulchra* ● NT
Azure-hooded Jay *Cyanolyca cucullata* NT
Black-throated Jay *Cyanolyca pumilo* NT
Dwarf Jay *Cyanolyca nana* ● NT
White-throated Jay *Cyanolyca mirabilis* ● NT
Silvery-throated Jay *Cyanolyca argentigula* NT

Bushy-crested Jay *Cissilopha melanocyanea* NT
San Blas Jay *Cissilopha sanblasiana* NT
Yucatan Jay *Cissilopha yucatanica* NT
Purplish-backed Jay *Cissilopha beecheii* NT

Azure Jay *Cyanocorax caeruleus* NT
Purplish Jay *Cyanocorax cyanomelas* NT
Violaceous Jay *Cyanocorax violaceus* NT
Curl-crested Jay *Cyanocorax cristatellus* NT
Azure-naped Jay *Cyanocorax heilprini* NT
Cayenne Jay *Cyanocorax cayanus* NT
Black-chested Jay *Cyanocorax affinis* NT
Plush-crested Jay *Cyanocorax chrysops* NT
White-naped Jay *Cyanocorax cyanopogon* NT
White-tailed Jay *Cyanocorax mystacalis* NT
Tufted Jay *Cyanocorax dickeyi* NT
Green Jay *Cyanocorax yncas* NA NT

Brown Jay *Psilorhinus morio* NT

White-throated Magpie-jay *Calocitta formosa* NT
Collie's Magpie-jay *Calocitta colliei* NT

Jay *Garrulus glandarius* WP EP OR
Lanceolated Jay *Garrulus lanceolatus* EP OR
Purple Jay *Garrulus lidthi* EP

Grey Jay *Perisoreus canadensis* NA
Siberian Jay *Perisoreus infaustus* WP EP
Szechwan Grey Jay *Perisoreus internigrans* ● EP

Ceylon Blue Magpie *Urocissa ornata* ● OR
Formosan Blue Magpie *Urocissa caerulea* OR
Yellow-billed Blue Magpie *Urocissa flavirostris* EP OR
Red-billed Blue Magpie *Urocissa erythrorhyncha* EP OR
White-winged Magpie *Urocissa whiteheadi* OR

Green Magpie *Cissa chinensis* EP OR
Eastern Green Magpie *Cissa hypoleuca* EP OR
Short-tailed Green Magpie *Cissa thalassina* OR

Azure-winged Magpie *Cyanopica cyana* WP EP

Indian Treepie *Dendrocitta vagabunda* OR
Malaysian Treepie *Dendrocitta occipitalis* OR
Himalayan Treepie *Dendrocitta formosae* OR
Southern Treepie *Dendrocitta leucogastra* OR
Black-browed Treepie *Dendrocitta frontalis* OR
Andaman Treepie *Dendrocitta bayleyi* OR

Black Racquet-tailed Treepie *Crypsirina temia* OR
Hooded Racquet-tailed Treepie *Crypsirina cucullata* ● OR

Notch-tailed Treepie *Temnurus temnurus* OR

Magpie *Pica pica* WP EP OR NA
Yellow-billed Magpie *Pica nuttalli* NA

Stresemann's Bush Crow *Zavattariornis stresemanni* ● AF

Henderson's Ground Jay *Podoces hendersoni* EP
Biddulph's Ground Jay *Podoces biddulphi* EP
Pander's Ground Jay *Podoces panderi* WP
Pleske's Ground Jay *Podoces pleskei* WP

Hume's Ground Chough *Pseudopodoces humilis* EP

Clark's Nutcracker *Nucifraga columbiana* NA
Spotted Nutcracker *Nucifraga caryocatactes* WP EP OR

Chough *Pyrrhocorax pyrrhocorax* WP EP OR
Alpine Chough *Pyrrhocorax graculus* WP EP

Piapiac *Ptilostomus afer* AF

Jackdaw *Corvus monedula* WP EP OR
Daurian Jackdaw *Corvus dauuricus* EP
House Crow *Corvus splendens* OR
New Caledonian Crow *Corvus moneduloides* PA
Banggai Crow *Corvus unicolor* ● AU
Slender-billed Crow *Corvus enca* OR AU
Piping Crow *Corvus typicus* AU
Flores Crow *Corvus florensis* ● AU
Marianas Crow *Corvus kubaryi* ● PA
Moluccan Crow *Corvus validus* AU
White-billed Crow *Corvus woodfordi* PA
Brown-headed Crow *Corvus fuscicapillus* AU
Grey Crow *Corvus tristis* AU
Black Crow *Corvus capensis* AF
Rook *Corvus frugilegus* WP EP OR
American Crow *Corvus brachyrhynchos* NA
Northwestern Crow *Corvus caurinus* NA
Tamaulipas Crow *Corvus imparatus* NT
Sinaloa Crow *Corvus sinaloae* NT
Fish Crow *Corvus ossifragus* NA
Palm Crow *Corvus palmarum* NT
Jamaican Crow *Corvus jamaicensis* NT
Cuban Crow *Corvus nasicus* NT
White-necked Crow *Corvus leucognaphalus* NT
Carrion Crow *Corvus corone* WP EP OR
Jungle Crow *Corvus macrorhynchos* EP OR
Australian Crow *Corvus orru* AU
Little Crow *Corvus bennetti* AU
Australian Raven *Corvus coronoides* AU
Forest Raven *Corvus tasmanicus* AU
Little Raven *Corvus mellori* AU
Collared Crow *Corvus torquatus* EP OR
Pied Crow *Corvus albus* AF
Hawaiian Crow *Corvus hawaiiensis* ● PA
White-necked Raven *Corvus cryptoleucus* NA NT
Brown-necked Raven *Corvus ruficollis* WP AF OR
Raven *Corvus corax* WP EP OR NA
Fan-tailed Raven *Corvus rhipidurus* WP
African White-necked Raven *Corvus albicollis* AF
Thick-billed Raven *Corvus crassirostris* AF

A Complete Checklist of Birds of the World (3rd edition)
© Richard Howard and Alick Moore 1980, 1984, 1990
is published by the Academic Press Limited.

GLOSSARY

Altricial chicks blind and naked at hatching.
Dimorphic sexes different.
Jizz the general impression of a bird in thc ficld.
Lek an assembly ground or gathering of birds for communal display.
Monomorphic sexes alike.
Monophyletic descended from a single common ancestral group.
Monotypic group containing a single species.
Pelagic found out at sea when not breeding.
Polyandrous females having more than 1 male breeding partner in the same season.
Polygynous males having more than 1 female breeding partner in the same season.
Polymorphic having more than one plumage form in the same species, independent of sex.
Polyphyletic descended from more than 1 ancestral group.
Precocial chicks highly developed at hatching.
Syndactyl toes joined.
Zygodactyl 2 toes pointing forward and 2 toes pointing backward.

ABOUT THE AUTHORS

Dr Christopher Perrins (Foreword) is Director of the Edward Grey Institute of Field Ornithology at Oxford University and the foremost bird scientist in the UK.

Dr Andrew Gosler (General Editor) is a research ornithologist at the Edward Grey Institute of Field Ornithology at Oxford University. This is his third book.

Dr Edward H. Burtt (Neartic) is Professor of Zoology at Ohio Wesleyan University and editor of the US *Journal of Field Ornithology*.

Dr Martin Kelsey (Neotropical) is the South American officer for the International Council for Bird Preservation.

Dr Malcolm Ogilvie (Palaeartic) is a well-known ornithologist and author who specializes in wild-fowl.

Adrian Lewis (Afrotropical) lived in East Africa for many years and is the author of the *Atlas of Kenyan Birds*.

Martin W. Woodcock (Oriental) is an illustrator and author who has contributed to a number of books including *The Birds of South East Asia*.

Dr Hugh A. Ford (Australasian) of the University of New England, New South Wales, is one of the most respected ornithologists in Australia.

Alick Moore (World Checklist of Birds) is joint author of *A Complete Checklist of the Birds of the World*.

The illustrations in Bird Families of the World are by the following artists:

Sean Milne: pages 18, 19, 24, 25, 30, 31, 36, 37, 42, 43; Robert Morton/Bernard Thornton Artists: pages 22, 23, 28, 29, 34, 35, 40, 41, 46, 47; Maurice Pledger/Bernard Thornton Artists: pages 20, 21, 26, 27, 33; Jan Wilczure: pages 32, 38, 39, 44, 45.

INDEX

ACKNOWLEDGEMENTS

Note; the numbers of the pages on which the photographs occur are followed by letters indicating the position of the photographs on the particular page, moving horizontally across from left to right.

Academy of Natural Sciences 99I/*P. Alden* 131G/ *A.M. Bada* 105I/*J.R. Blake* 117C, 121J/*Erik N. Breden* 73J/*P. Canevari* 105C/*B. Chudleigh* 295J/*H. Clarke* 77J, 83J/*H. Cruickshank* 61G/*P. Davey* 193I/ *R. Dierrebaard* 105G/*J. Dunning* 97B, 97D, 101A, 107C, 107E, 107F, 111D, 113I, 115C, 115G, 117I, 121B, 121K, 121L, 123A, 123C, 123F, 123K, 125C, 125E, 125I, 127A, 127C, 127D, 127E, 127G, 127H, 127I, 127J, 127K, 127L, 129A, 129C, 131F, 131K, 133A, 133B, 133C, 133E, 133F, 133H, 133I, 135A, 135C, 135J, 135K, 137A, 137B, 137H, 137K, 139H/*C.H. Greenewalt* 75B, 107L, 119B, 119D, 119E, 119F, 119I, 119J, 119L, 123G, 125J, 125K, 129I, 135D, 135F, 139B, 139C, 139E, 139G, 299C, 301F, 301G, 311B, 313B, 319B, 319D, 319E, 323D, 323E/*Steven Holt* 79E, 89H, 139I/*M.P. Kahl* 295H, 295I/*E.F. Knights* 245E/*S. Lafrance* 51F/*S.J. Lans* 77D/*S. Lipschutz* 131A, 131D/*A. Morris* 77I, 87G/*C. Munn* 105J/*J.P. Myers* 111E/*J. Olsen* 291E/*P. O' Neill* 129G/*W.S. Peckover* 269F, 291D, 303G, 309C, 315I, 317C, 319H, 323C, 323F, 323G/*D.S. Pettingill Jr.* 315G/*M.J. Rauzon* 69I, 283I/*G. Reynard* 313A/ *R. Ridgely* 105L, 121C, 125D, 133K, 245A/*D. Roby & K. Brink* 99E/*F.K. Schleicher* 63A/*B. Schorre* 83K, 85A, 85D, 87B, 87D, 87H, 87J/*T. Schulenberg* 129F/ *Johann Schumacher* 83B/*R. Shallenberber* 99C/ *N.B. Smith* 115I, 137G/*C. Speeble* 251D/*P.W. Sykes Jr.* 103G/*P. Tomkovich* 251F/*Doug Wechsler* 71B, 131B, 125A, 263G, 265K/*D.K. Wheeler* 105A/*D. and M. Zimmerman* 73G, 83C, 87C, 103B, 109E; P. Alden 73I, 117H; Aquila/*G. & G. Attwell* 229B, 231J/*G.G. & I.M. Bates* 181G/*J.B. Blossom* 57H, 103E, 199B, 241G, 289C, 289H, 289I, 289K, 293G/*J.J. Brooks* 79I, 101F, 137I, 137J, 201C, 203A, 215C, 217A, 217G, 219C, 219F, 223G, 225G, 225H, 225I, 231G, 231K, 233J, 275E, 283A, 287C, 287F, 287H, 289E, 289G, 291C, 291F, 293A, 293C, 299D, 299E, 299F, 301B, 301H, 303D, 305E, 311C, 315D, 315E, 319C, 321E, 321G/*Abraham Cardwell* 147C/*Kevin Carlson* 293E/*J.F. Carlyon* 105E, 191A, 193J, 195A, 209H, 213B, 217B, 217E, 221E, 225A/*J.F. Carlyon & M. Wilson* 195K/ *Anthony Cooper* 147E/*Dr J. Davies* 289F, 289J, 293D, 295C, 295F, 295G, 297A, 297F, 301C, 301E, 303B, 303C, 307B, 307C, 321A, 321C/*Paul Doherty* 165G, 189F/*Hanne & Jens Eriksen* 93K, 251C, 275A, 289B, 305H, 319I/*C. Haus Geblus* 175B/*K. Ghani* 251H/*R. Gill* 283G, 285F/*R. Glover* 149H, 291A, 297D/*Conrad Greaves* 69C, 99I, 111J, 203I, 203K, 249D, 251A, 253B, 257G, 267I, 291B, 311D/*Brian Hawkes* 59G, 67H, 69H, 111I/*S.F. Horton* 257H, 291G, 299B, 299G, 299H, 307F, 321B, 321D, 321F, 323A/*Edgar T. Jones* 55G, 65E, 71I, 85G, 87A/*Edgar T. Jones & Gary R. Jones* 63H/*Gary R. Jones* 59J/*Mike Lane* 55F, 169G/ *Wayne Lankinen* foreword, 51A, 51B, 51J, 55C, 55D, 55H, 55I, 55J, 57A, 57B, 57C, 57G, 57I, 59H, 59I, 61A, 61B, 61C, 61D, 61E, 61I, 63F, 63G, 63K, 65A, 65C, 65E, 65K, 67A, 67B, 67D, 67G, 67I, 67K, 71J, 71K, 73A, 73B, 75C, 75F, 75G, 75H, 75I, 75J, 75L, 77B, 77E, 79D, 79F, 79K, 81D, 81F, 83E, 83H, 85B, 85E, 85I, 85J, 89A, 89B, 89D, 89J, 89K, 89L, 91B, 91C, 91D, 91H, 91I, 93A, 93B, 93C, 93D, 93E, 93F, 93G, 93J, 93L/*Graham M. Lenton* 255C/*Ed Mackrill* 99J, 103J, 107D, 231E/*Geoff McIlleron* 191F, 205I, 207A, 209G, 215H, 225J, 229F/*J. Mills* 277F, 287G, 295D, 299A, 301A/*Richard T. Mills* 65G/*A.T. Moffett* 167A/*D.K. Richards* 197H, 199C, 201A, 207G/*J. Lawton Roberts* 169E/*Bryan L. Sage* 51D, 51E, 53D, 89E, 287B/*Anup Shah* 189J, 195D, 209A, 215B, 219H, 225E, 239D, 239G, 241F, 243C, 247E, 251B, 255D, 265G, 267E, 269B, 297I/ *Philip Shaw* 297G/*R. Stevens* 221K/*C.T. Stuart* 231C/ *Ray Tipper* 81B, 251E, 253D, 253I, 257B, 277C, 297E/ *Gary Weber* 287E, 295A, 305C, 305F, 305G, 309A, 309B, 309F, 309I, 311A, 311G, 313C, 313D, 313F, 313G, 315F, 317D, 319F, 319G, 319J, 321H/*P.J. White* 117D/

M.C. Wilkes 51I, 53G, 53H, 53I, 53K, 63B, 63J, 65I, 65L, 67J, 69F, 71F, 79C, 83G, 187D, 187F, 189G, 189H, 189I, 191J, 193B, 193C, 193G, 195B, 197G, 199A, 199E, 199F, 201E, 201F, 201I, 203H, 207J, 209I, 209K, 211H, 213F, 217H, 219E, 225F, 227B, 231I, 233E, 233I, 235D/ *G.D. Wilson* 297C; C. Bradshaw 261H, 269G; Jose Colon 121F; Simon Cook 101D, 109F, 123B; P. Cotton 125L; R.S. Daniell 105K, 121H, 205K, 207I, 209C, 209J, 213A, 213C, 215J, 219B, 219K, 221I, 225D, 227H, 227K, 231A, 231H, 233J, 235I; T.R. Dean 257F, 259G, 263C; Sue Earle 269C, 277A, 393E, 323H; Michael Fogden 101E, 105D, 107G, 107H, 109G, 109H, 111B, 117A, 117E, 117F, 117G, 119A, 121A, 129E, 131E, 135F; Michael & Patricia Fogden 129H; S.J.M. Gantlett 101C, 243G, 261E, 315B; John Holmes 247G, 259E, 269A, 273A, 273B, 273D, 275B; Eric Hosking 109C, 137F, 253E, 253E, 257E, 261C, 279G; Eric & David Hosking 51G, 57J, 99D, 115B, 115D, 115E, 123H, 139A, 239E, 245C, 247C, 257A, 257D, 283D, 291H, 291I, 305B, 315C; David Hosking 111F, 111H, 247B; M.G. Kelsey 103A, 107B, 113D, 113H, 117B, 125B, 125F, 125H, 129B, 133D, 135G, 137L; Frank Lane Picture Agency/*Ronald Austing* 59E, 59L, 71D, 73H, 81E, 81I, 87I/*Hans Dieter Brandl* 75D/*C. Carvalhd* 113E/*W.S. Clark* 63C, 103H, 107A, 255E, 257C, 271B/*Eichhorn & Zingel* 183D, 243F, 253G/*W. Elison* 101J/*Tom & Pam Gardner* 101G, 241D, 285G, 285H, 293F, 297B, 311E, 311H, 317E, 323B/*K. Ghani* 103E, 239A/*A.R. Hamblin* 71A, 79A/*Hannu Hautala* 75K/*R. Jones* 311F/*John Karmali* 235H/*Frank W. Lane* 103C, 105H, 107J, 107K, 121G, 123E, 137C, 139D, 211D, 239I, 243H, 245D, 249E, 259C, 259D, 265C, 267B, 267F, 271A, 271C, 273E/*Steve Maslowski* 71E, 73F, 77F, 77H, 79H, 81G, 83A, 85C, 87E, 89C, 89I/*S.S. Menon* 263B/*Geoff Moon* 283C, 287D, 293B, 307A, 313E/*Mark Newman* 101I/*R. Van Nostrand* 261A/ *Mandal Ranjit* 239H, 241I, 249B, 253H, 277G/*Len Robinson* 283B, 295E, 303A, 303E, 305A, 305D, 307D, 309E, 309F, 309H, 315A, 315H, 315J, 317A/*Leonard Lee Rue* 61H, 69G, 73D, 97I/*Len Rue Jr.* 59F/*Hanz Schrempp* 139F, 265B/*Silvestris* 177J/*Irene Vandermolen* 79J/*L. West* 83D, 85K/*Terry Whittaker* 245G/ *Roger Wilmshurst* 241E/*Dieter Zingel* 151B, 245B/ *Dr T.S.U. de Zylva* 255A, 255B, 261B, 265J, 267C, 271D, 279D; Tim Loseby 255F, 265E; C. Mulguiney 307G, 309D, 317B; Nature Photographers Ltd/*T. Andrewartha* 153C, 153J, 161E, 163L, 165E, 181J/ *S.C. Bisserot* 103I, 109A, 243A, 277E, 289D/*F.V. Blackburn* frontispiece, 151J, 153D, 153F, 161F, 167F, 169A, 169C, 171D, 175D, 175F, 179G/*Mark Bolton* 151L, 157C, 163K, 181H/*Derick Bonsall* 161B, 171A, 175A/*L.H. Brown* 12–13, 93H, 153H, 187I, 189D, 189E, 189L, 191E, 199L, 203J/*N.A. Callow* 241C/ *Kevin Carlson* 59D, 63E, 85H, 91A, 91F, 105F, 137E, 145B, 145C, 145E, 145G, 151D, 151I, 153A, 153B, 155B, 155E, 155F, 157E, 157H, 159G, 159J, 161C, 161I, 167I, 169H, 169J, 171H, 173H, 173L, 175H, 177E, 177H, 179H, 183F, 201H, 213G, 219J, 223E, 223H, 223I, 233B, 280, 287A, 319A/*Colin Carver* 147I, 159B, 159H, 165C, 171E, 173B, 173E, 173K, 175C, 175L, 177D, 177F, 179J, 181E, 183B, 183K/*R.J. Chandler* 163C, 181I, 203C/*Hugh Clark* 53B, 113F, 167D, 169I, 177J, 247D/ *Andrew Cleave* 53E, 61J, 143K, 163F/*R.S. Daniell* 101B, 115J, 121J, 139J, 184, 187D, 187G, 189B, 193A, 195I, 197C, 197I, 199D, 201B, 201G, 201H, 201K, 205A, 205J, 207B, 207D, 207F, 211B, 211E, 211F, 215D, 215I, 217F, 221B, 221D, 221F, 221H, 223D, 225B, 225C, 227A, 227C, 229G, 231F, 233F, 235A, 289A, 307E/*Peter Davey* 209B/*A.K. Davies* 177B/*Thomas Ennis* 183I/*Richard Fairbank* 109I, 133J, 249F, 251G, 263H/*Michael Gore* 97A, 97H, 99B, 115A, 121D, 121E, 140–141, 151K, 159A, 161K, 167K, 175K, 179K, 179L, 181B, 187A, 187C, 191C, 191G, 193K, 195C, 195G, 197D, 197E, 199G, 199H, 199K, 203D, 205C, 205D, 205G, 211A, 211I, 213D, 213I, 215F, 217D, 219D, 219I, 221J, 223A, 223F, 225K, 227E, 227F, 227G, 227I, 227J, 229A, 229C, 229D, 229E, 231B, 231D, 231L, 235E, 263F, 279A, 283E, 297H/*James A. Hancock* 48, 53F,

59C, 69A, 97E, 111A, 111C, 123J, 135B, 139K, 155H, 203L, 207E, 207H, 221A, 235F, 239F, 247C, 287J/*M.P. Harris* 8–9, 53C, 69J, 97F, 97J, 99E, 163G, 187B, 283F, 283H, 283J, 285A, 285B, 285C, 285D, 285E/*Dr M.R. Hill* 59N, 63D, 63L, 99I, 119G, 149G, 161J, 171D, 181C, 236, 239B, 239C, 241B, 243B, 261D, 263E, 265H, 279E, 279I, 287I/*David Hutton* 179F, 197F/*E.A. Janes* 55A, 55B, 69E, 91G, 147H, 147K, 151F, 165H, 165I, 165K, 181F, 183G, 187H, 201K, 209D, 209E, 235B/*John Karmali* 97G, 189A, 189C, 189K, 191I, 191K, 195F, 197J, 205B, 205E, 205F, 205H, 209F, 211K, 213H, 215A, 215E, 221C, 223C, 229H, 229I, 229J, 233A, 233C, 233D, 233G, 235C, 235J, 235K/*Paul Knight* 145J/ *Chris & Jo Knights* 151E, 155K/*E.C.G. Lemon* 151N, 285I/*B. Mearns* 157G, 157I, 161D, 183K/*R. Mearns* 65B/*Hugh Miles* 147D, 245F/*M. Muller & H. Wohlmuth* 151H, 255D/*C.K. Mylne* 143C, 279F/ *Philip J. Newman* 51C, 71H, 143B, 151G, 171I, 173A, 173G, 175E, 177G/*D. Osborn* 99K/*William S. Paton* 59B, 71G, 73C, 111G, 147A, 149D, 151A, 157A, 167G, 179A, 179E, 187J, 227D, 233H/*J.F. Reynolds* 157B, 159I, 159K, 195I, 195J, 199I, 211C, 213E, 219A, 221G, 223B, 233L, 235G/*Peter Roberts* 99A/*J. Russell* 147G, 155C, 167C, 183H/*Don Smith* 71C, 73E, 77A, 77G, 79B, 79G, 81J, 83F, 89G, 151C, 153E, 153I, 161A, 167B, 171C, 197B, 207C, 211G/*Paul Sterry* 53L, 55K, 57D, 57E, 57J, 59A, 63I, 65D, 65J, 67C, 67E, 67F, 69B, 77C, 81C, 81H, 81K, 85F, 89F, 91E, 93I, 95, 107I, 109B, 113C, 113G, 115F, 115H, 119H, 119K, 123D, 127D, 127K, 131C, 131H, 131I, 131J, 133G, 135I, 137D, 143F, 143G, 143H, 143I, 145F, 145H, 147F, 147L, 149A, 149C, 149F, 149J, 155D, 155G, 155I, 157J, 159E, 159L, 163A, 163B, 163D, 163J, 165A, 165D, 167H, 167J, 171F, 173F, 175G, 175J, 181D, 183J, 191B, 191D, 215G/*Roger Tidman* 11, 53A, 53J, 57F, 61F, 65H, 67L, 69D, 81A, 87F, 91J, 103D, 105B, 143E, 143J, 145I, 147B, 147J, 149B, 149E, 149I, 153C, 155A, 155J, 157D, 157F, 159C, 159D, 159F, 161G, 161H, 163H, 165B, 165J, 167E, 169D, 169F, 171B, 171G, 171J, 175I, 177C, 183A, 191H, 193D, 193E, 193F, 195E, 197A, 199J, 203B, 203E, 203F, 217C, 219G/*E.K. Thompson* 55E, 143A/*Maurice K. Walker* 143D, 169B, 171C, 173I, 173J, 177A, 177I, 179B, 179C, 179D, 181A, 183C/*Derek Washington* 163I/*Jonathan Wilson* 145A, 145D, 179I; NHPA/*K. Ghani* 245E/*Ralph & Daphne Keller* 267H/*Lacz Lemoine* 273B/*Mandal Ranjit* 241A/*E. Hanumantha Rao* 247F, 259A, 259B, 261G/*Morten Strange* 249A, 249C, 255I, 259E, 263A, 263D, 265A, 277D, 279C/*David Tomlinson* 265I; William S. Paton 163E; G. Seywald 253C, 265D, 267A, 267D, 269E, 271E, 273F, 273G, 275C, 277B, 279B, 279H; David Tipling 249H, 253A; Ray Tipper 249G, 261F, 265F, 269D, 275F; Alan Tye 113B, 123I, 125G, 127F, 129D, 135H; Hilary Tye 97C, 101H, 109D, 113A, 119C; Martin Williams 243D, 255G, 267G.

Jacket acknowledgements

Front (clockwise from top left): Aquila/M.C. Wilkes, Nature Photographers Ltd/Don Smith, Michael Fogden, Nature Photographers Ltd/Peter Roberts, Aquila/ A.T. Moffett, Michael Fogden, Frank Lane Picture Agency/K. Ghani, Michael & Patricia Fogden, Nature Photographers Ltd/E.A. Janes, Frank Lane Picture Agency/Ron Austing, Nature Photographers Ltd/Dr M.R. Hill, Jose Colon, Academy of Natural Sciences/J. Dunning, Eric & David Hosking, Ardea/M.D. England. *Back (clockwise from top left):* Nature Photographers Ltd/William S. Paton, Nature Photographers Ltd/ Roger Tidman, Nature Photographers Ltd/Don Smith, Aquila/Mike Wilkes, Nature Photographers Ltd/John Karmali, Nature Photographers Ltd/Mark Bolton, Nature Photographers Ltd/S.C. Bisserot, Nature Photographers Ltd/Roger Tidman, Nature Photographers Ltd/Andrew Cleave, Nature Photographers Ltd/L.H. Brown, Aquila/Mike Wilkes, Michael Fogden, Nature Photographers Ltd/R.S. Daniell, Nature Photographers Ltd/John Karmali.

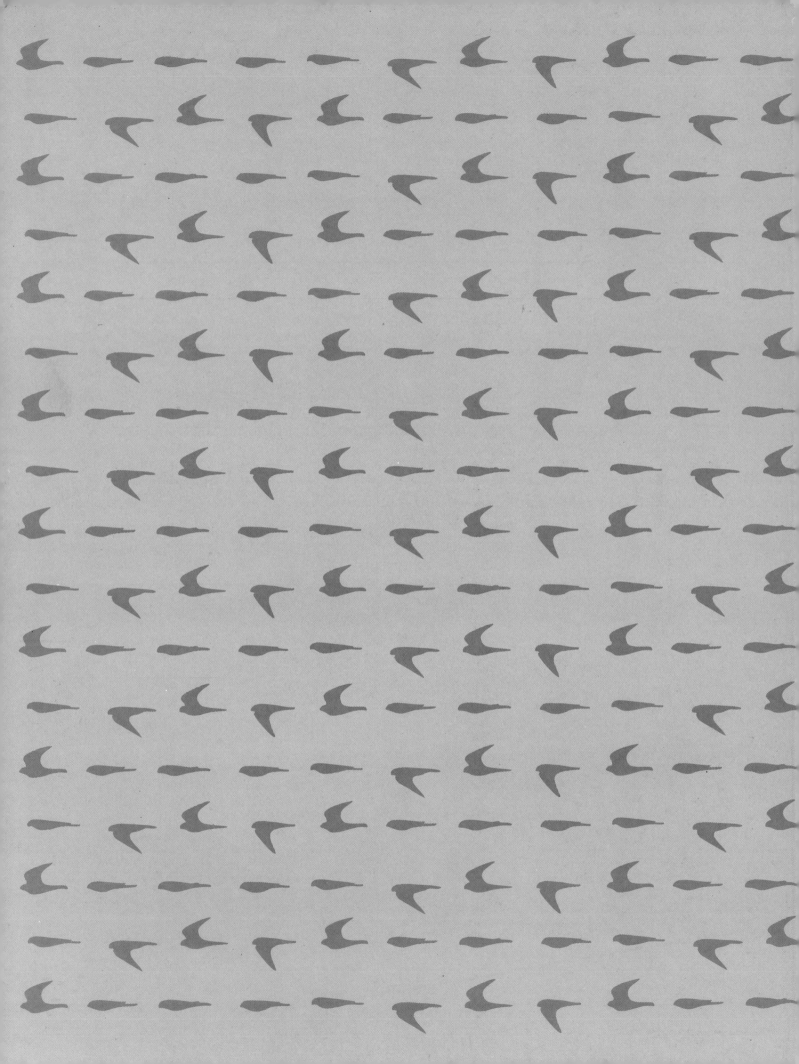